TOBY WILLIAMSON

FLUID POWER DESIGN HANDBOOK

FLUID POWER AND CONTROL

A Series of Textbooks and Reference Books

1. Hydraulic Pumps and Motors: Selection and Application for Hydraulic Power Control Systems, *by Raymond P. Lambeck*
2. Designing Pneumatic Control Circuits: Efficient Techniques for Practical Application, *by Bruce E. McCord*
3. Fluid Power Troubleshooting, *by Anton H. Hehn*
4. Hydraulic Valves and Controls: Selection and Application, *by John J. Pippenger*
5. Fluid Power Design Handbook, *by Frank Yeaple*

Other Volumes in Preparation

FLUID POWER DESIGN HANDBOOK

Frank Yeaple

Design News Magazine
Cahners Publishing Company
Boston, Massachusetts

MARCEL DEKKER, INC. New York and Basel

Library of Congress Cataloging in Publication Data

Yeaple, Franklin D.
Fluid power design handbook.

(Fluid power and control ; 5)
Includes index.
1. Fluid power technology. 2. Oil hydraulic machinery—Design and construction. I. Title. II. Series.
TJ843.Y43 1984 621.2 84-14270
ISBN 0-8247-7196-6

MARCEL DEKKER, INC.
270 Madison Avenue, New York, New York 10016

Current printing (last digit):
10 9 8 7 6 5 4 3 2 1

PRINTED IN THE UNITED STATES OF AMERICA

Preface

Hydraulic and pneumatic power and control are needed more than ever to solve force, torque, and speed problems in the fields of aerospace, transportation, agriculture, mining, construction, manufacturing, marine, processing, oil exploration, and power generation. All these industries, and more, are victims of the world's rush to increase efficiency and to pack more power into less space.

Hydraulic cylinders and motors have the unique capability of exerting—and precisely controlling—great force and torque in cramped quarters. Other kinds of systems are limited by bulk (gears) or by the need for high electrical currents (servomotors). Pneumatic systems are the answer where fast actuation and control—without electricity—are required.

Step-by-step techniques in this handbook, proved in practice, show how to incorporate all types of hydraulic and pneumatic components into systems, control them, and predict performance. Graphic shortcuts solve problems of flow, temperature, pressure, fluid shock, torque, speed, control, logic, and component application. Pumps, compressors, valves, microprocessors, servosystems, cylinders, actuators, motors, drives, accumulators, shock absorbers, filters, piping, fittings, and seals are critically analyzed. Practical tests for fatigue, noise, and overall performance are described in detail.

There is no comparable book. The material in this volume was developed to establish positive design procedures where none were previously available and to include enough theory to clearly explain the techniques.

All designers and users of equipment, vehicles, and systems that are powered or controlled with hydraulics or pneumatics will benefit from this book. Also, component manufacturers will find it useful because it explains how their products will be applied and what the performance and maintenance requirements must realistically be.

This handbook is based on a previous book by the author, *Hydraulic and Pneumatic Power and Control* (McGraw-Hill, New York), plus his articles published later in *Design Engineering* (Morgan Grampian, New York). The author thanks those publishers for their kind permission to adapt material for the handbook. Unless otherwise noted, all figures are reproduced courtesy of Morgan Grampian.

Thanks are due also to Jim Stone (New York City) for his skillful preparation of many of the original illustrations for the Morgan Grampian publications, and to the hundreds of design engineers who over the years have contributed data and concepts used in preparation of the original articles. Several engineers in particular helped in the earliest stages: H. M. Schiefer (Dow Corning), for fluids and lubricants; Louis Dodge (Engineering Consultant), for equations on pressure loss, heat transfer, fluid shock, accumulators, and labyrinth sealing; Dominic Lapera (Kemp Aero Products), for pneumatic flow equations; Dr. Warren E. Wilson (Harvey Mudd College), for pump equations; and Werner Holzbock (Sperry Vickers), for descriptions of hydraulic drives. Special thanks are due my wife and daughters for proofreading the entire manuscript.

What is the next step? Simply, to encourage realization worldwide that the role of fluid power is critical in solving tough performance problems that cannot be handled any other way.

Frank Yeaple

Contents

FLUID POWER
DESIGN HANDBOOK

1

Hydraulic Fluids and Lubes

There are fluid power and control fluids engineered for almost any purpose. But the properties are affected by how the fluids are used, and manufacturers' catalogs cannot give all the answers.

The parameters are numerous (see Table 1.1). The correct choice requires a team: fluids expert, equipment or vehicle design engineer, and the ultimate user of the equipment. Here are four basic steps to the selection:

- Determine acceptable limits of performance of system fluids: viscosity, density, vapor pressure, air solubility, bulk modulus, fire resistance, temperature range, thermal expansion, lubricity, compatibility, and toxicity.
- Seek spec sheets on as many fluids as possible that meet or nearly meet the objectives.
- Compromise as much as possible by altering system requirements.
- Get final advice from the fluid suppliers.

LANGUAGE OF PERFORMANCE

Let's get the technical terms straight first. See the accompanying drawings and graphs (Figs. 1.1 through 1.9) for average typical fluid characteristics and performance.

Viscosity and *density* together (Figs. 1.1 through 1.5) determine pressure loss for a given flow. Viscosity is frequently expressed in Saybolt Universal Seconds (SUS) or Saybolt Seconds Universal (SSU), which are two ways to say the same thing. It's a measure of the time it takes for 60 cc of oil to flow through an orifice 0.176 cm in diameter and 1.225 cm long. In Fig. 1.4 SUS is related to kinematic viscosity, centistokes (cSt). Values of viscosity as well as specific weight are given for several fluids. Kinematic viscosity, centistokes, equals absolute viscosity, centipoises (cp), divided by mass density.

Only absolute viscosity (centipoise and lb-sec/ft^2, for example) has physical units that can be visualized. Absolute viscosity also is called dynamic viscosity. Kinematic

Table 1.1 Parameters for Choosing Hydraulic Fluids and Lubes

Lubricating ability

Lubricity
Load-carrying
Traction coefficient
Preferential wetting of bearing surfaces

Viscosity

Viscosity index (change with temperature)
Shear loss (change with shearing rate)
Effect of pressure

Cold weather

Flowability at sub-zero temperatures

High Temperature

Thermal stability
Oxidation
Flashpoint
Volatility

Electrical properties

Impulse voltage strength
Dielectric strength
Conductivity
Polarization in magnetic fields

Compatibility

Compatibility with seals
" " " with other fluids
Acceptance of additives
Filterability
Water absorption or rejection
Liquid absorption or rejection (other fluids)
Air and gas absorption (equilibrium data)
Detergency (dispersancy)
Particle suspension
Reaction to metals in suspension or solution
Resistance to hydrolysis (chemical breakdown)

Ecology and safety

Fire resistance
Toxicity and odor
Noise damping ability
Ease of reclamation or disposal

Physical properties

Compressibility (bulk modulus)
Vapor pressure
Specific heat
Density
Thermal expansion
Latent heat of vaporization
Critical Reynold's number
Cavitation

viscosity is more or less a defined term. A simple way to visualize absolute viscosity—let's select lb-sec/ft^2 as the example—is to reproduce the definition pictorially (Fig. 1.1).

Imagine the sketched cube to be 1 ft on a side. Apply a sufficient force at the upper surface of the fluid to move that upper surface at a velocity of 1 ft/sec relative to the fixed lower surface. Absolute viscosity is a measure of the value of that shearing force, expressed as follows:

$$\mu = (F/A,\ \text{lb/ft}^2) \div (V/Y,\ \text{ft/sec per ft})$$
$$= FY/VA,\ \text{lb-sec/ft}^2$$

The slope as well as position of each viscosity-vs.-temperature curve is important. A flat slope indicates a large allowable temperature change before exceeding minimum and maximum values of viscosity.

The term viscosity index (VI) was invented to compare those slopes. It is based on two standard oils that have been given arbitrary values of zero and 100 VI, respectively. The zero VI is for the steeper slope (a naphthenic-petroleum oil, sensitive to

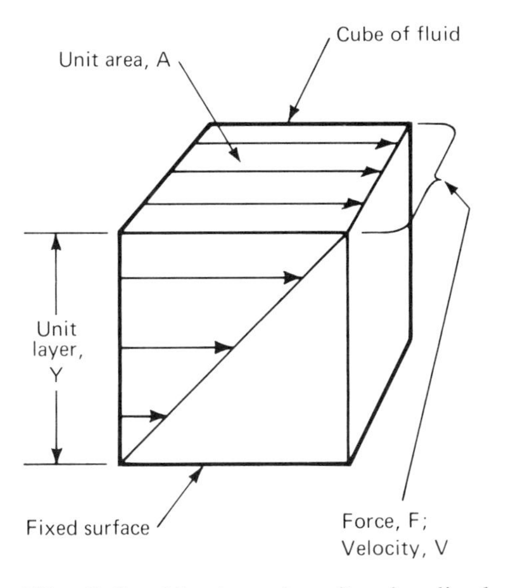

Fig. 1.1 Absolute viscosity visualized.

temperature changes); the 100 VI has less sensitivity to temperature (a paraffinic-base oil).

Although viscosity is usually considered constant for a given temperature and pressure regardless of the flow, it sometimes is greatly influenced by the rate of shear. In fluids that have large-polymer molecules, rapid shear rates temporarily lower the viscosity—suggesting that the very-long-polymer molecules align in the direction of flow through small clearances. This tends to break down large molecules.

Density not only influences flow rate and pressure drop but also affects the

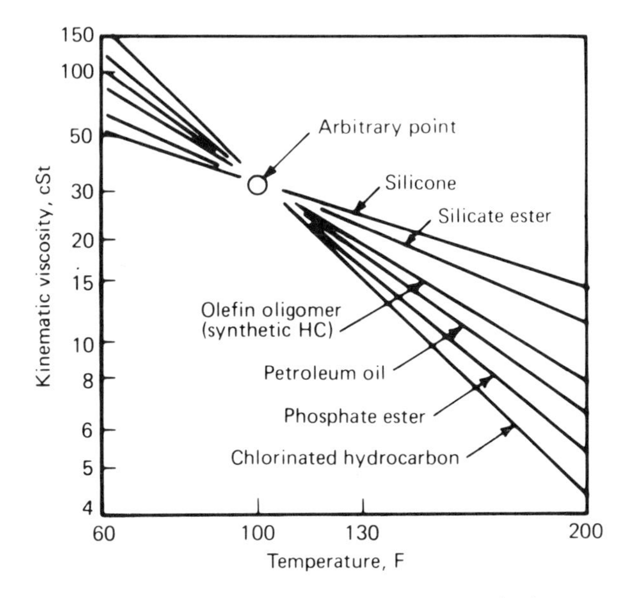

Fig. 1.2 Viscosity vs. temperature, typical.

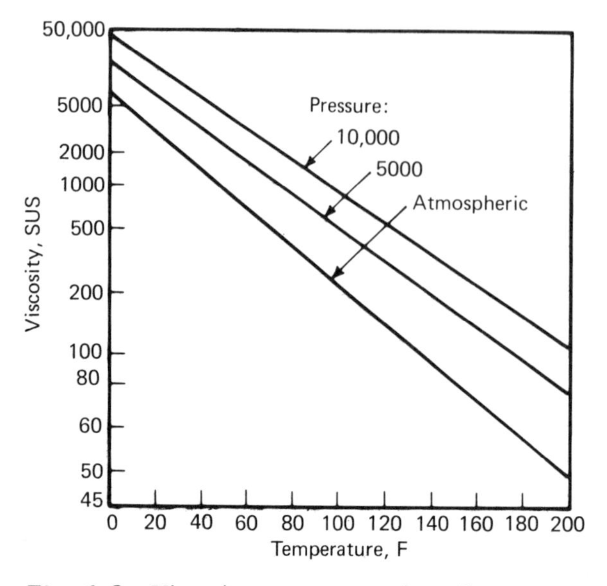

Fig. 1.3 Viscosity vs. pressure for oil.

natural frequency of a system. This frequency is proportional to the square root of the moving mass.

Air solubility in many liquids is directly proportional to absolute pressure of the system. High solubility is undesirable because the air (or other gas) dissolved at reservoir conditions may come out of solution at a lower pressure (pump inlet, for example). Pump cavitation can result. If the reservoir is pressurized, then the problem is compounded because more air can be dissolved, coming out of solution later. See Figs. 1.6 and 1.7.

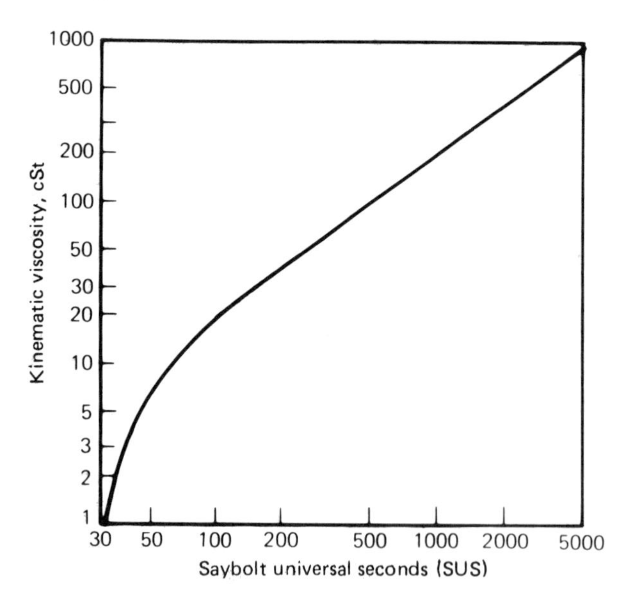

Fig. 1.4 Viscosity conversion, SUS to cSt.

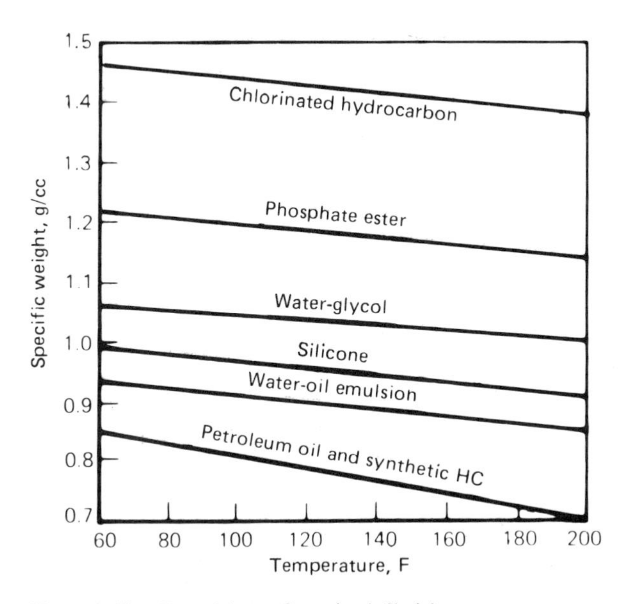

Fig. 1.5 Densities of typical fluids.

Vapor pressure of the fluid, if too high, will cause gas pockets to form in areas of low pressure—such as the inlet of a pump. This aggravates cavitation. See Fig. 1.8.

Bulk modulus—the reciprocal of compressibility—has units of psi. Bulk modulus is similar in concept to a mechanical spring rate and measures the pressure change needed to cause a given percent volume change. The higher this modulus, the stiffer the system and the higher the resonant frequency. A fast-responding servosystem requires a fluid with a high bulk modulus. See Fig. 1.9.

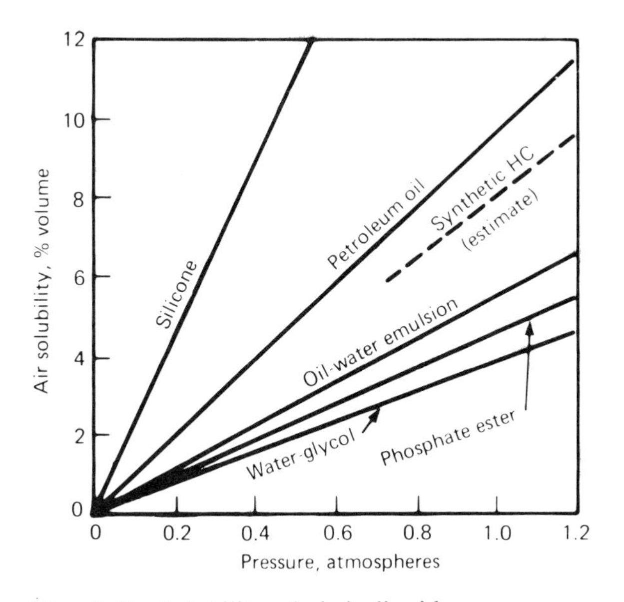

Fig. 1.6 Solubility of air in liquids.

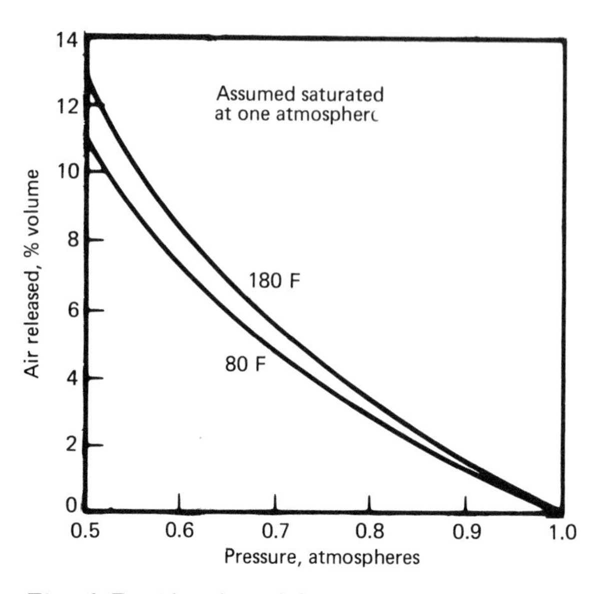

Fig. 1.7 Air released from petroleum oil.

Fire resistance is a tough requirement. Fire-resistant fluids include water, water-glycol, water-oil emulsions, phosphate esters, and chlorinated hydrocarbons. Water loss in the glycol type increases the viscosity, but in the oil-emulsion type, it decreases the viscosity. Change in viscosity can warn of a dangerous condition.

Water-based fluids usually are not recommended for high bearing loads found in ball bearings and axial-piston pumps or for fluid pressures over 1500 to 3000 psi.

Non-water-based types such as phosphate esters and chlorinated hydrocarbons are not volatile and can stand temperatures to 250 F and extremely high pressures.

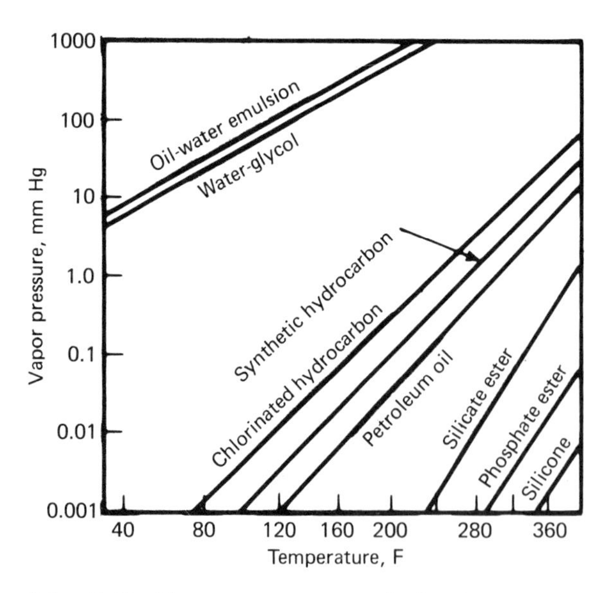

Fig. 1.8 Vapor pressures, typical values.

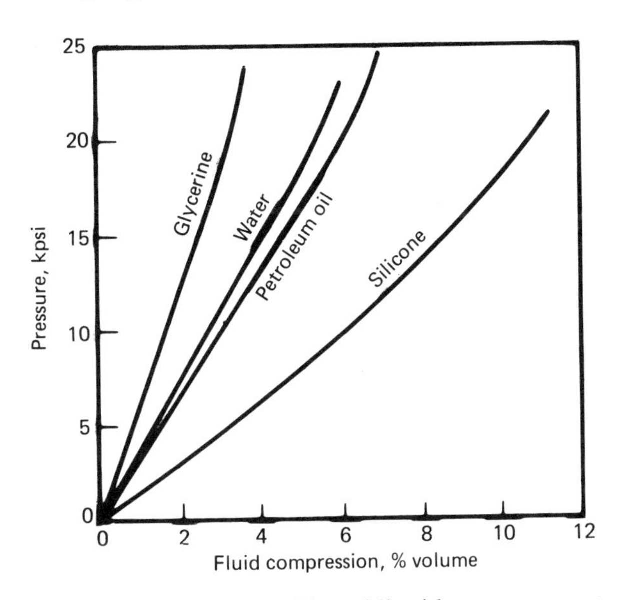

Fig. 1.9 Compressibility of liquids.

They are heavier than water-based fluids and petroleum and have the disadvantage of high viscosity at low temperatures. Auxiliary heating is often required for outdoor startups.

Compatibility of the fluid with the system must be assured. In no instance should a fire-resistant fluid be dumped into a system after merely draining the petroleum. Residual amounts of petroleum and sludge can cause emulsion fluids to "break" and water-based materials to emulsify.

Specify filters that can handle 4 times pump flow instead of the normal 1.5 times pump flow if water-based fluids are used. Choose only compatible seals and don't grease O-rings with noncompatible lube.

Fluid stability depends on the fluid's tolerance to temperature, pressure, atmospheric oxygen, water, radiation, and catalysts. Petroleum oils have good oxidation resistance, except at high temperature. Oxidation is speeded up by the presence of certain metals. In long-chain organic compounds, high temperatures will degrade the large molecules into smaller ones, reducing viscosity. There are fluids already designed to be stable in almost any atmosphere, but usually at a sacrifice in other properties.

Lubricity is a problem with many fluids that are formulated for special reasons, such as fire resistance or high temperature. Phosphate esters are better than other fire-resistant fluids, which helps justify their higher price. Water-glycol mixtures, unlike the fixed-composition esters, have varying proportions and ingredients, depending on the manufacturer. Therefore, the lubricity and other properties will vary.

Leakage of fluid is influenced by surface tension. Fluids that hold *dirt particles* in suspension sometimes can be more troublesome than those which let the particles drop harmlessly to the bottom of the tank.

Temperature often is the controlling parameter. Let's define high temperature as anything from 400 to 850 F. Above that, extreme temperature might be a more apt description. Moderate temperature goes from 250 to 400 F. Normal temperature is from −40 to 250 F. Low temperature starts at −40 and drops down to cryogenic temperature at −200 F.

No single fluid covers the whole range, although the silicones come as close as any. In fact, no fluid meets all the performance requirements in even a single range of temperature. The unvarnished truth is that you must sacrifice "important" characteristics to achieve "essential" characteristics. And remember that any special fluid is expensive compared with long-run refinery products.

Normal temperatures are no problem. Mineral oils usually are selected for temperatures to 300 F. Water-in-mineral oil emulsions are fire-resistant and suitable to 200 F. Water-glycols, also fire-resistant, are useful to 200 F. Between 250 and 400 F, many of the synthetics are applicable.

Superrefined oils, synthetic hydrocarbons, and alkyl silanes are developed for temperatures to 500 F. Viscosities are low at the higher temperatures, and oxidation will deteriorate the oil in open systems. However, closed systems and proper bearing and seal design help solve both problems.

Most problems are magnified as temperatures rise. Mild fluids become corrosive when hot, stable fluids break down into useless new compounds, and pumps lose suction when the liquid disappears into thin air.

REAL-LIFE APPLICATIONS

In hydraulic systems, viscosity ought to exceed 1.0 centistoke (e.g., water at zero C) yet not be more than 5000 cSt (maple syrup at room ambient) at any operating temperature. Lubricity should be at least equal to that of an oil-water emulsion, even though the fluid isn't used in an actual bearing. It still has to be pumped.

Sludging and varnishing are taboo, particularly in servosystems where clearances are small. Bulk modulus ought to be high, for fast response and good stiffness. Viscosity must not degrade even under shearing action. Hydrolytic stability (unaffected by water) must be good, and foaming must not be excessive. The fluid should be a good conductor of heat and have a low coefficient of thermal expansion.

In fluid couplings and shock absorbers, the fluid gets hot because of the shearing action. Viscosity ought to stay as constant as possible with temperature and be stable. Viscosity level depends on the application, and all values should be readily available from the fluid manufacturer. No deterioration is permitted, particularly in precision dashpots where timing is important.

In liquid springs, high compressibility (low bulk modulus) is the prime requirement. Viscosity should be as flat with temperature as possible.

In thermal control devices, a high thermal coefficient of expansion, linear with temperature, is required. Viscosity and vapor pressure should be low. All these characteristics must remain the same even after continued temperature cycling.

In all applications, remember the following:

- A continuous temperature slightly below the incipient thermal degradation level is more damaging than occasional hot-spot temperatures above the degradation level.
- Transient or hot-spot temperatures of 25 to 100 F over the incipient value will degrade only that portion of the fluid in contact. If the byproducts are

not objectionable, useful fluid life might not be reduced appreciably.

- For long-duration operation, the oxidation, thermal stability, and viscosity stability will determine the limiting temperature.
- Viscosities of certain fluids drop severely with an increase in temperature. For instance, aromatic phosphates usually will be of too low viscosity if operated at 500 F, even though they might withstand the temperature.
- Paints and elastomers might limit use of a fluid otherwise suitable for high temperatures. Aromatic phosphates and esters, for example, do damage to some paints and elastomers at 500 F.

Danger from Air Bubbles

One interesting theory is the adiabatic compression of air bubbles in high-pressure pumps. The peak temperatures are extremely high (see the table in Fig. 1.10). It's particularly serious in servosystems for numerically controlled machines, where unaccountable clogging of fine filters and failures of precision valves have puzzled users for years.

Essentially identical systems, each with the same maintenance programs, have differed dramatically in valve life. In each instance, the hydraulic fluid appeared clean and passed all conventional lab tests, yet certain of the systems had frequent failures that could be attributed only to degraded fluid.

It's claimed to be high-temperature cracking, oxidation, and nitration of the oil caused by tiny air bubbles that get compressed suddenly from atmospheric pressure to full system pressure, without time to be dissolved into the oil or to dissipate the heat (Fig. 1.10). The table lists typical theoretical temperatures created by adiabatic compression of air from 14.7 psia and 100 F to various pressure levels.

Only such high temperatures can explain the types of insoluble oxides and nitrogenous compounds discovered in the guilty hydraulic fluids. It's the first time outside of spark-ignition engines that the phenomenon of nitrogen fixation has been encountered. The nitrogen in the air apparently oxidizes to form highly reactive compounds, which then act on the oil and the additives.

To prove the theory, 96 servo-controlled machines in four factories were observed for about a year. Oil samples were taken every several months and analyzed with techniques developed by Mobil Oil. By extremely precise measurement of changes in concentration of these oil degradation products, it was proved possible to predict with considerable reliability when the oil would begin to damage the valves and filters in the system.

Parameters important to such a program include viscosity, wear particles, color, odor (burnt smell suggests "cracking"), oil history (including temperature), and chemical analysis.

The clue in any case is to take periodic samples and compare them with the original oil. Some systems are inherently clean enough to get by with annual tests; others require them monthly.

Where aeration seems to be the problem, getting rid of the air should be paramount. Studies indicate that oil rest time (reservoir capacity relative to flow-through) is critical. The reservoir ought to be big enough to supply full pump capacity for at

Air bubble temperature, F
(compressed from 14.7 psi and 100 F)

Pump press., psi	Bubble temp (adiabatic compr.)
1000	1410
2000	1820
3000	2100

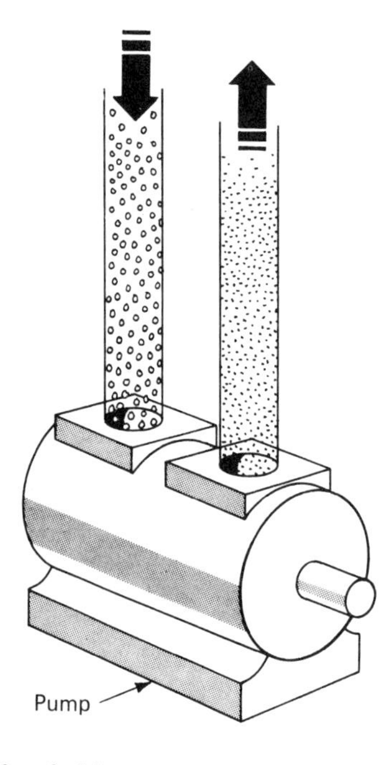

Fig. 1.10 Compression of air bubbles from 14.7 to 3000 psi creates high temperatures.

least 2.5 minutes. Some reservoirs were found to have 1-minute rest time, which is not enough.

FLUID TYPES AND CHEMISTRY

The commonest fluids are petroleum (napthenic, aromatic, paraffinic, inhibited, uninhibited), phosphate or silicate ester (an ester is formed by replacing the hydrogen of an acid with a hydrocarbon radical), various hydrocarbons (chemical compounds containing only carbon and hydrogen), and water-based fluids. See Figs. 1.11 and 1.12 for the performance of typical fluids.

Silicone fluids are not common in ordinary industrial hydraulic systems but find many applications in special circuits and devices, such as fluid springs, damping pistons, fluid couplings, braking systems, and some power transmission units.

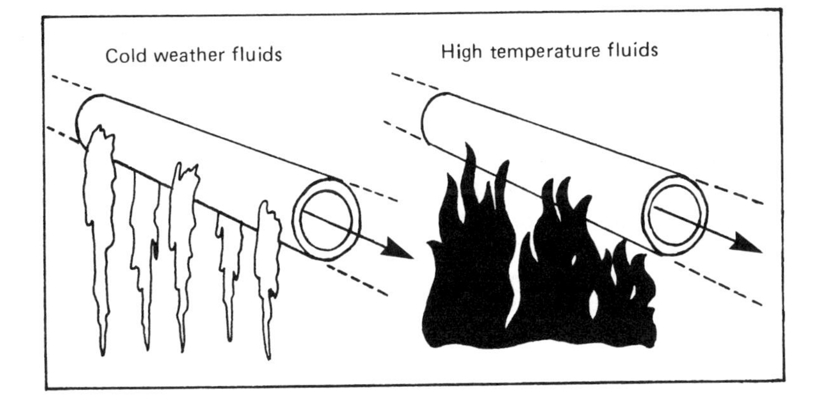

Fig. 1.11 Cold weather fluids and high temperature fluids.

Many variations of the above fluids are possible—chlorinated hydrocarbons, for example. Knowledge of chemistry is not essential, because the manufacturer will help select the exact type and grade to meet the need.

High-temperature fluids usually are synthetics. Moderate-temperature fluids can be superrefined mineral oils. Superrefined mineral oils are natural petroleum products consisting of complex mixtures of hydrocarbons (carbons and hydrogen in the form of paraffins, naphthenes, and aromatics). They differ from normal mineral oils in that additional refining steps are taken to eliminate weak spots in the molecules and to remove contaminants.

Synthetic hydrocarbons are similar to mineral oils in containing only carbon and hydrogen, except that they are built up to specific molecules by chemists. Silanes are similar to the synthetic hydrocarbons and are named silane because they contain one silicon atom per molecule.

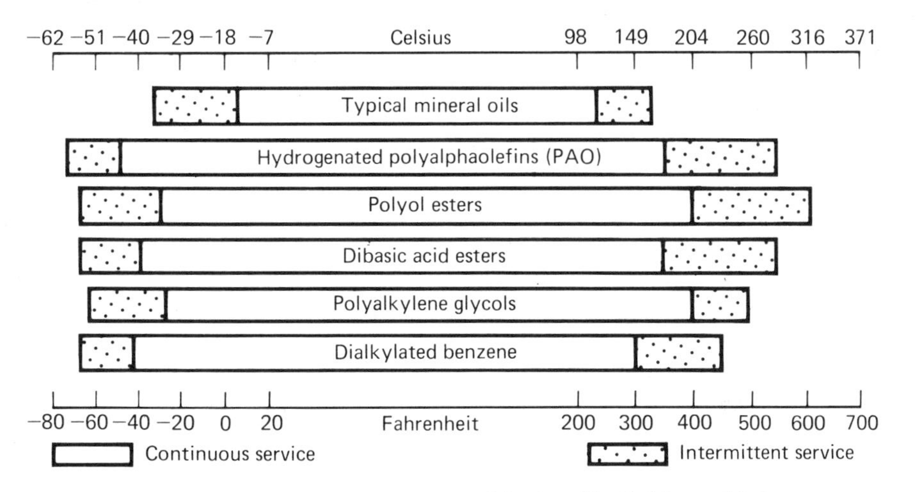

Fig. 1.12 Useful temperature range of hydraulic fluids and lubes.

Polyglycols (polyalkylene glycols, polymers of oxyalkylenes) contain a large portion of carbon and hydrogen in repeating alkylene groups (polymers) but also have an oxygen between each repeating group. The alkylene groups normally used are ethylene, propylene, butylene, or mixtures.

The phosphate ester structure is an organic alcohol (as in silicates) which is attached to a phosphorus atom. The phosphate differs from the silicate in that it contains only three carbon-oxygen groupings on phosphorus (four for silicates) and also has an additional oxygen attached.

The halogenated aromatic structures are exactly as the name implies—aromatic (phenyl) compounds that have hydrogen replaced with halogen groups, mainly chlorine (i.e., chlorinated polyphenyls).

Halogenated hydrocarbons are a bigger group classification which also includes the aliphatic (paraffins) compounds that have some or all of their hydrogen atoms replaced with chlorine, fluorine, or bromine atoms.

The silicones (organopolysiloxanes) have similarities to all hydrocarbon-type oils previously mentioned as well as to the structure of glass. They are polymers with the repeating unit being the stable silicon oxygen linkage (as present in glass) and containing side groups which are hydrocarbons (silicon carbon linkage). The hydrocarbon side groups vary: aliphatic (i.e., methyl), aromatic (i.e., phenyl), halogenated aromatic, and fluorocarbons.

Polyphenyl ethers contain repeating units of phenyl with an oxygen between them. Polyaromatics (polyphenyls) are similar but without the oxygen.

Now let's examine some key fluids in more detail.

Silicone Fluids

Silicone fluids have dramatic properties, particularly thermal stability. They resist physical and chemical change under severe heat, cold, shear, oxidation, and other operating conditions that break down organic fluids. Also, they are inert, noncorrosive, and nontoxic and have low volatility. Dow Corning (Midland, MI) and GE (Schenectady, NY) specialize in silicones.

Silicone fluids for hydraulic applications usually are dimethyl polysiloxanes. One feature often exploited is their high compressibility compared with petroleum oil or most synthetic fluids (Fig. 1.9). For example, they are found in liquid springs, automotive fan-drive couplings, and torsional vibration dampers.

Another important physical characteristic is a high viscosity index (VI), typically up to 400. It has the lowest change in viscosity with temperature of almost any hydraulic fluid or lube, which is another reason why silicone fluid is chosen for liquid springs and dampers (see Chap. 8).

As an example, the damping effect of silicone fluid decreases only 3 times from -40 to $+160$ F, compared with 2500 times for petroleum oil under those same extreme conditions. Also, silicone can be blended within an extremely large range of nominal viscosity: 1 cts to 2.5×10^6 cts.

Silicone brake fluids are already being used as original fill for sports cars, for some motorcycles, and in the aftermarket area for off-road equipment, racing cars, and fork-lift trucks. In comparison to glycols, they are a safer fluid since they do not pick up moisture, have better low-temperature properties, are noncorrosive, and do not blemish painted surfaces.

The U.S. government has chosen silicone brake fluids over traditional glycol for army vehicle brake systems. They are a one-time replacement for three different types of fluids used by the U.S. Army for over 30 years. Since silicone brake fluid is virtually boil-free under all operating conditions, it does not create vapor in brake systems in heavy use as may happen with glycol.

Dielectric coolants have become important. For instance, Dow corning's 561 silicone fluid is replacing the PCBs in transformers. The fluid is nontoxic, has good heat transfer and dielectric properties, and is safer than mineral oils. Another recent use is as a heat transfer fluid for solar energy systems.

Lubricity is sometimes a problem. It's not that silicone isn't a good lubricant—it is used successfully in many applications where absolute reliability for life of the instrument or machine is essential. For instance, it rates among the best lubricants for fiber and plastic gearing and bearings.

The problem is that standard silicone's boundary-lubrication ability at high loads is only fair for many combinations of metals, notably steel-on-steel. However some silicones, such as fluorosilicones, are as good as hydrocarbons. In fact, a new special silicone compressor fluid has replaced mineral oil in a rotary screw compressor with a 10-year warranty.

Synthetic Lubes for Extreme Cold

Logging companies in Maine and Canada have discovered that synthetic crankcase lube stays liquid in extreme cold, instead of gelling like most petroleum oils. So they now order synthetic engine oil for the winter logging months and actually save money in the process. How? Because vehicles start right up, auxiliary heaters are not necessary, ether bottles become far less important, fire insurance rates are lower, lubricity is better on the average, fuel economy is improved, and oil changes are less frequent. It works in any engine, including gas turbines and diesels.

Not only is performance better and overall operation more economical, but engine, gearing, and accessory wear is greatly reduced. For example, startups using sprayed ether as a booster shorten engine life and wear out starters rapidly. Also, petroleum oil light enough for easy cold-weather starts is not heavy enough to prevent wear when the engine is hot. The synthetics are much better, according to the many users who have tried both routes.

A typical fluid is Mobil's Delvac SHC, a synthetic hydrocarbon olefin oligomer blended with organic esters. The fluid is made from crude oil base stock reconstructed molecularly at the refinery to yield exactly the properties wanted, eliminating those not wanted. Gulf Oil Chemicals has a similar olefin oligomer called Synfluid.

The Mobil synthesized hydrocarbon fluids (SHF) are specifically designed to be completely compatible with regular petroleum oil, all conventional shaft seals, engine parts, and all hydraulic system components. Neither the oil nor SHF are fully compatible with phosphate esters (nor are the phosphate esters fully compatible with all conventional shaft seals). Normal additives can be incorporated in Delvac SHC, including detergent-dispersants, oxidation and corrosion inhibiters, and anti-wear agents.

The viscosity index (VI) of the base synthetic is between 135 and 140, compared with 95 to 100 for petroleum base stock. Viscosity change with temperature is low enough to eliminate the need for viscosity improvers. Pour point is −80 F (−62 C), which is far lower than that for petroleum oil.

Film strength of the Mobil synthetic hydrocarbon fluid is as great as that for a 140 weight gear oil. It's a great feature, particularly at startup. Engine wear and carbon formation are reported as practically nil, compared with previous experience on petroleum lubes. One test was run for 6000 hours at summer temperatures—the hardest kind of wear test.

Even colder applications are found in the Alaskan pipeline and the Arctic. Conoco's Polar Start DN-600 (an alkylated benzene), Emery's Frigid Go (a diester), and Chevron's Sub Zero (a polyalphaolefin), along with Mobil's Delvac SHC, are examples of fluids developed initially for extreme cold-start crankcase service. All will work in normal automotive applications too.

At the middle of the temperature spectrum are fork-lift trucks for outdoor use in normal climates. Here, conventional petroleum crankcase oils work well, and a more expensive lube shouldn't be needed. Yet Eaton's industrial truck division (Philadelphia, PA) has customers who specify organic diester-type synthetic crankcase oil for whole fleets of lift trucks, and Eaton itself has a leasing service for lift trucks that contain the same synthetic fluid.

Eaton sells the fluid to its dealers simply as Yale SL-1. It's completely compatible with mineral oil, seals, and engine parts and actually saves the user money even though cost per gallon is many times higher than that for petroleum lube. One saving is that the diester can be run for over 1000 hours without draining. Maintenance intervals thus are longer, saving labor costs as well.

A hidden benefit, not measurable in the field, is increased miles per gallon of fuel because of the better lubricity under typical operating conditions. Lab tests on automotive engines show fuel economy improvements of from 5 to 10% due solely to the use of synthetic crankcase lube.

Hot Applications for Synthetics

Rolling mills, die casting machines, extrusion presses, aircraft, and process conveyor lines use large quantities of phosphate esters, diesters, polyglycols, silicones, silicates, and water-based fluids, depending on the performance trade-offs available to the designer.

A recent hot application for diester-type synthetic lubes is in plant compressor lubrication. It's still somewhat controversial because diester fluids are compatible with most but not all seals and components. This is a problem if downstream equipment in the shop air system isn't supplied with the recommended compatible material. An example is the polycarbonate bowl in a lubricator (see Chap. 11). Some chemical crazing (and failure) can result. Also, certain seal materials will swell.

Performance of the diester has been excellent in the hot innards of the compressors, and major compressor manufacturers such as Ingersoll-Rand, Worthington, Joy, and Kellogg-American are supplying it in some cases as regular lubricant.

Ordinary hydraulic systems are not subject to the horrors of engine combustion, cold cranking starts, extreme heat, boiled-away fluids, hot-extreme-pressure sliding contact, and frequent dirt ingestion, so the conventional petroleum oils do quite well and cost far less than synthetics. It's one thing to fill a small crankcase and quite another to fill a huge hydraulic system.

However, there are applications where the same fluid is put into the hydraulic system as into the engine. Synthetics have proved satisfactory in both. Sometimes the anti-wear additive (such as zinc dithiophosphate) in ordinary crankcase lube can chemically react with copper-bearing metals in hydraulic piston pump valveplates, swashplate slippers, and system tubing, but the problem is easily avoided by adhering to component and equipment manufacturers' recommendations for additives. Abex Denison (Columbus, OH), for example, publishes a spec sheet describing specific mineral oils, synthetics, and water-based fluids that have proved suitable for the company's line of hydraulic pumps, motors, and controls.

Commercial aircraft hydraulic fluids and lubes are in a class by themselves. Fire resistance is mandatory in the hydraulic systems, and most planes use phosphate esters. The refinement of the base stock and the development of additives have required the best of the fluids art, because of the extremes in temperature and pressure experienced in flights in all climes and at all attitudes.

Sperry Vickers (Troy, MI) notes two problems to avoid by proper design: *deposits* and *erosion.*

The deposits stem from gradual thermal degradation of the fluid because of the high temperatures and can be kept under control fairly well.

The erosion is trickier. It is partly an electrochemical phenomenon caused by electrical charges generated by the streaming fluid. An ion layer forms next to the inside metal surface of the pipe, and the ion polarity is opposite to the metal polarity. Another layer of opposite charge is, in effect, swept downstream. Balancing currents flow out of the pipe wall into the fluid and cause the erosion.

Phosphate ester hydraulic fluids also are used on ships, in mines, and in industrial processes where fire is an extreme hazard. Typical brands include Skydrol and Pydraul (Monsanto), Hyjet (Chevron), Aero-Safe and Fyrquel (Stauffer), Houghto-Safe 1120 (Houghton), and Pyrogard 51 (Mobil).

Special-Purpose Fluids

There are certain properties of some fluids that serve in unique ways. Take the cycloaliphatic synthesized hydrocarbon fluids, for instance. They become stiff when subjected to extreme pressures and can transmit shear forces. One application is in rolling contact bearings where skid must be prevented; another is in traction (metal-to-metal rolling friction) drives where the torque is transmitted by fluid shear rather than gear teeth. A pioneer is Monsanto (St. Louis, MO) whose Santotrac fluid was developed for such drives. Figure 1.13 shows typical traction coefficients.

Slippery water is another: Union Carbide (New York) and other chemical companies have developed water additives such as polyethyleneoxide that decrease the viscosity of water, making it easier to pump. Ferrofluidics Corp. (Burlington, MA) offers fluids that congeal in an electric field, useful in flow control, sealing (see Chap. 21), and mechanical clutching. Petrolite Corp. (St. Louis, MO) has developed ways to combine immiscible nonreactive liquids—such as oil and water—into structured emulsions that behave as solids when at rest but as liquids when pumped.

WATER-BASED HYDRAULIC FLUIDS

Water-based fluids are of great interest today because of the need for lower-cost, safer, and environmentally purer products. Performance is not as good as that for conventional hydraulic fluids and lubricants, but with care in design of systems and components the results can be adequate.

The problems are the following: Water corrodes, evaporates, boils, freezes, cav-

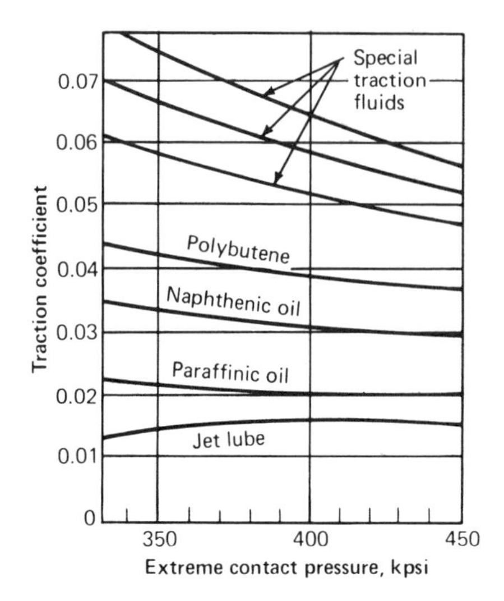

Fig. 1.13 Traction coefficients.

itates, encourages bacteria, is a poor lubricant, has low viscosity, and contains minerals (such as calcium). See Fig. 1.14.

There are several types available today: water-glycols, water-in-oil (invert) emulsions, oil-in-water emulsions (misnamed soluble oil), and synthetic solutions. Here are brief descriptions of each type (Fig. 1.15).

Water-glycol has been around for over 25 years and has found many applications in industry where petroleum fluids are not allowed (for fire safety or ecology reasons). The amount of water content ranges from 35 to 50%, and the rest is a mix of various glycols, polyglycols, and additives to provide freezing point compression, viscosity, shear stability, lubricity, vapor-phase corrosion protection, metal deactivation, and anti-wear properties. Most major chemical companies offer them. Water-glycol has the best lubricity and can take the highest pressures of the water-based hydraulic fluids but also is the most expensive. In the recent past it has cost about four times petroleum anti-wear oil.

An interesting property of water-glycol is that it can have inverse solubility in water. That is, high temperature (over 212 F) will make it less soluble, thus freeing some of the dissolved glycol polymer at points of high heat (as in a bearing), and actually increase effective load-carrying ability.

Water-in-oil emulsion (also called invert emulsion) is roughly 40% water, the rest oil, emulsifiers, and chemical additives. The water is in microscopic droplets surrounded by films of oil. Because the oil is the outside phase, lubricity is enhanced over the opposite oil-in-water-type emulsion. The fluid has many properties of petroleum oil yet has the advantage of considerable fire-snuffing ability because of the high water content. Cost is about 1.5 times the cost of conventional petroleum oil products.

High shear rates can halve the viscosity of water-in-oil emulsions. This can be an advantage because the fluid can be chosen with high viscosity (and therefore low leakage) yet will flow easily through high shear areas such as pumps and valves.

Oil-in-water emulsions are about 95% water and 5% soluble or emulsible oil and additives, including emulsifiers, anti-wear additives, rust and oxidation inhibitors, vapor-phase inhibitors, and bactericides. The oil is dispersed in fine droplets in the water, and each droplet is surrounded by water, the continuous phase.

Fire resistance is excellent, but lubricity suffers somewhat because of the small percentage of oil and because the outside phase is water. Just the same, hydraulic pumps and controls can function on this fluid if properly designed. A particular advantage is that it can leak harmlessly into machine tool coolants that have the same constituents. Cost in the past has been about 0.1 times the cost of conventional petroleum oil.

The size of the droplets in an oil-in-water emulsion changes the appearance. The fluid will be milky or opaque if the average droplet size is large (say about 50 micrometers) and translucent if small (about 2 micrometers). A micrometer is 0.0000394 in. Emulsifiers determine the degree of dispersion, the stability, and the size of the droplets.

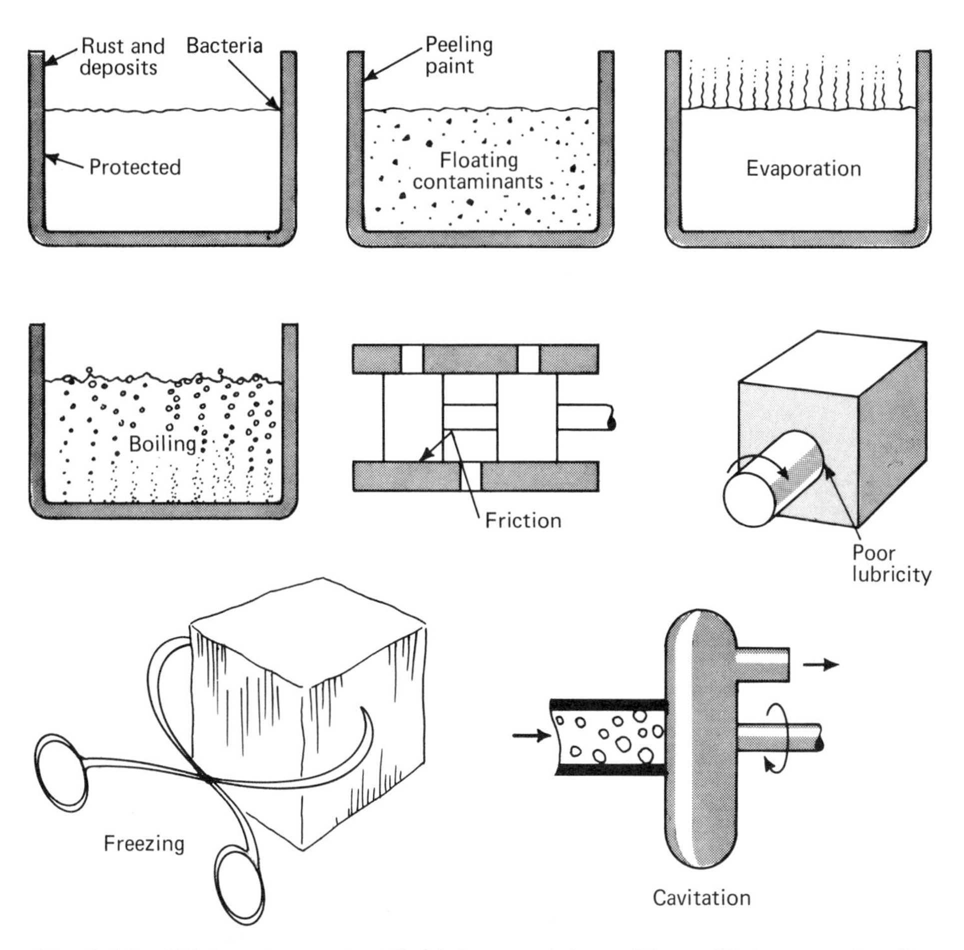

Fig. 1.14 High-water-content fluids have certain problems: Water corrodes, floats contaminants, evaporates, boils, has poor lubricity, freezes, and cavitates.

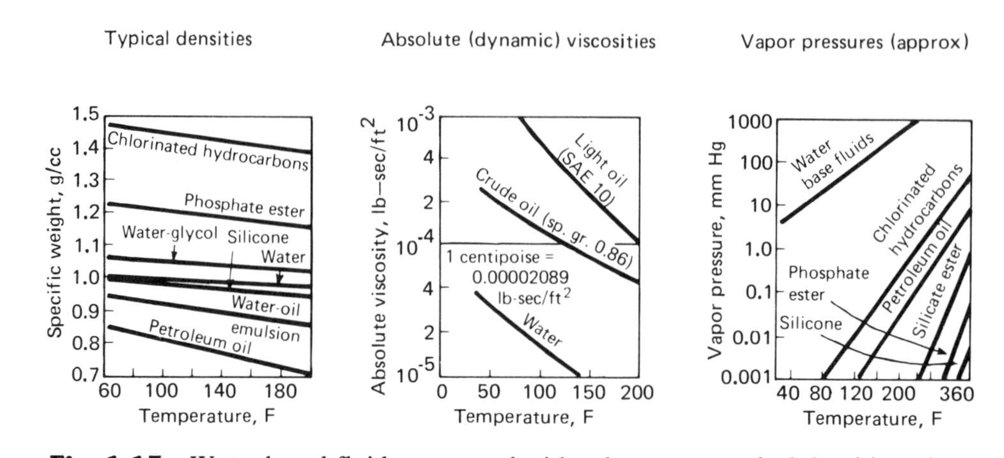

Fig. 1.15 Water-based fluids compared with other types: typical densities, absolute (dynamic) viscosities, vapor pressures.

A microemulsion has the smallest size droplets, almost a colloidal suspension, and tends to be more stable than ordinary emulsions. Many automotive manufacturers are applying microemulsions to production equipment with good results. A typical product is SUNSOL HWBF, developed jointly by Sun Petroleum and Lubrizol Corporation.

Synthetic solutions contain no oil but are about 95% water and 5% soluble salts and other chemicals, including anti-wear additives, rust and oxidation inhibitors, vapor-phase inhibitors, and bactericides. The fluids are clear, and dyes must be added to make them more visible.

There are continuing efforts to develop improved HWBFs. One problem with high water base hydraulic fluids (HWBF) has been low viscosity. Now chemists at BASF Wyandotte (Parsippany, NJ) have created a viscous HWBF with a nonoil additive. It has a thickener, based on polyoxyalkylene, that forms a matrix in the fluid.

The fluid, called Plurasafe, has viscosity equal to oil, resists breakdown even in high shear, remains stable throughout prolonged usage, and tests successfully in all types of pumps.

Two limitations are the following: It deteriorates if oil is added, and lubricity is much less than that for 95-5 oil-in-water HWBF.

High Water Base Fluid Properties

The discussions in the text chiefly apply to 95%–5% oil-in-water emulsions and 95%–5% synthetic solutions but not to water-glycol or 40% water-in-oil (invert) emulsions.

The point to remember is that we are dealing with a fluid that is 95% water, and it has many of the desirable and undesirable qualities of water.

Density of high water base fluid (also called high water content fluid—HWCF) is higher than that for oil (HWBF is about 1 g/cc; oil is about 0.8 g/cc). Bulk modulus of compression is over 330,000 psi for most operating conditions. Viscosity at ambient temperature is not much above 1 cSt or about 32 Saybolt Seconds (SUS), unless expensive viscosity improvers are added. Lubricity also is dependent on additives, because water is a poor lubricant.

HWBF boils at about 212 F and freezes at about 32 F, at atmospheric pressure. Even at normal running temperatures (say 120 F) evaporation can be a problem. A hidden advantage, however, is that water is a better coolant than oil, and a system will run cooler on HWBF.

Not every system can save money or even survive on high water base fluids. Some users say that component life at best will be half that for operation on oil. Only the automotive manufacturers seem to be extremely interested in converting from oil to 95% HWBF, and they usually limit the applications to machines that simply clamp, feed, or transfer. Most such machines operate at less than 1000 psig and 1200 rpm in a controlled plant environment. 10,000-hour life is one criterion, and it is being met.

Some engineers reported that success was achieved initially by reducing all the speed and force levels in the system, or if that didn't work, by increasing the cylinder sizes to develop the same force at less pressure.

Beyond 1000 psig, 1200 rpm, normal ambient, and simple hardware, it is a technical challenge to adapt to 95% HWBF. The fluids are rarely applied to high-speed hydraulic servosystems, outdoor mobile vehicles, or high-pressure equipment.

Where conditions are too severe for 95% HWBF, one of the other types might work. For example, the mining industry uses 40% water-in-oil emulsions, and the metals-producing industries use water-glycol. However, almost any machine could be designed to run on 95% fluid if necessary.

Pump Design

Piston pumps are now readily available for operation on high water base fluids up to 1500 rpm and 1000 psig. Vane and gear pumps are being developed to withstand the same loads.

Vickers (Troy, MI) offers special cartridges for vane pumps where HWBF service is specified. Rexnord (Racine, WI) reports success with special variable-volume vane pumps.

In any event, always get full recommendations from the pump manufacturer for HWBF applications. By the time you ask, the pressures and speeds may be higher.

Pump suction conditions are critical, and cavitation might result if the high vapor pressure of water is forgotten or ignored. Atmospheric inlet pressure is dissipated in several steps: inlet piping friction, strainer or filter loss, inlet porting and valving loss, and dynamic pressure loss as the incoming fluid increases its velocity to catch up with the moving elements of the pump. The pump cavity pressure might be only a few psia, and the water could flash to steam.

If you also subtract the losses in lifting the denser fluid to an above-tank pump suction and the losses caused by turbulence in the low-viscosity HWBF, there's no wonder that cavitation, noise, and wear sometimes result.

One way to prevent all of this is to pressurize the pump inlet. Overhead reservoirs are an answer, and many manufacturers now offer them. Supercharged pumps will work too, but they add more components and cost. A pressurized reservoir—say with a bladder over the fluid surface—is a method for pressurizing pump inlet and at the same time protecting the fluid and preventing evaporation (see Chap. 2).

Noise is no greater with HWBF than it is with oil, once the cavitation is gone. Also, pump wear or damage from collapsing air and vapor bubbles at the pump discharge will be halted.

Valve Design

Valve life is generally longer than pump life, and the chief problem is erosion and a wire-drawing phenomenon caused by throttling. Wire drawing is named after its appearance: a thin channel eroded along the line of flow. It's not generally a problem until pressures approach 1000 psig.

Cast iron does not resist wire drawing too well, and when valves are designed (or converted) for tough HWBF service, hardened stainless steel sleeves can be placed in the cast iron. Also, the annular clearances should be reduced below those that work for oil. The valves cost more to make because the tolerances are closer.

Solenoid spool valves have two potential problems: One, they are electrical and water is a danger, and two, the frictional breakaway force of the spool might be greater with HWBF than with oil. The solenoids can be protected against water by encapsulation or external drains. Friction doesn't have to be a problem if the spools are in hydraulic balance; then standard coils work.

Hard-chroming or Teflon coating of valve spools and sleeves will reduce friction. Grease fittings can reduce friction too, but the grease is likely to end up floating in the reservoir.

Flow control valves, including servovalves, have special problems too: HWBF has low viscosity, and the throttling areas must be much smaller than for the same flow with oil. Also, internal leakage is much greater, and this affects overall control at low flows. Fortunately, there are valves specifically designed for water service, such as the Sperry Vickers Salem valve. It has elastomer dynamic seals to eliminate internal leakage along the spool.

Teledyne Republic has a line of valves with highly polished lapped seals for extremely low leakage, even on HWBF. Operating pressure can be as high as 2000 psig.

Turbulent flow is 10 times more likely with HWBF than with oil because the Reynolds number is 10 times higher. Why? Because Re is a function of density over viscosity, and HWBF has somewhat higher density and much lower viscosity.

Overall System Design

The design problems are not basically different with HWBF than with oil, but the ranges and tolerances are different.

For example, line velocities should be kept near the lower limits of those recommended for oil—which are 2 to 4 ft/sec for inlet and 7 to 15 ft/sec for pressure lines—because HWBF is more likely than oil to go turbulent or cause cavitation.

Leaks are still leaks, but the priorities are reversed: Through a given clearance HWBF might leak 10 times as much because of its lower viscosity, but it evaporates and is less slippery than oil on the factory floor.

A more serious problem is internal leakage: The pumps and reservoirs might have to be bigger to make up the flow. It pays to specify better seals and tighter tolerances all around, just to keep the fluid where it belongs.

HWBF will foam more readily than oil, so good baffling should be designed into reservoirs. Also, atmospheric contaminants will deteriorate HWBF more rapidly, and protection is suggested. Furthermore, the fluid itself is a solvent and will loosen most coatings on reservoirs and elsewhere. Stainless steel reservoirs are one answer, or special coatings.

System temperature with HWBF will be 15 to 25 F lower than equivalent oil systems because water has better heat conductivity.

Filtration can be a major problem. A subtle hazard is the fact that HWBF is denser than oil, and lightweight particles of contamination aren't as apt to settle. Suspended contaminants might recirculate, clog passages, and load up filters. Also, inlet filtering is touchier with HWBF because every pressure loss in suction lines will increase the chances for cavitation.

Contamination carried to bearing surfaces can weaken the thinner lubricant film. Metal contaminant particles can be pulled out with tank-mounted magnets, and it's advised.

Filter design must include these parameters: hard surfaces to resist impingement, chemically compatible metals to avoid galvanic corrosion, anodizing of aluminum to resist corrosion (and to harden the surface), resin impregnation of absorbent fibers to prevent swelling and disintegration, epoxy rather than plastisol bonding to avoid washout in HWBF, choice of materials not harmed by HWBF, and the coarsest media that meets filtering requirements to keep pressure losses low.

Most experts recommend three filters at least: One before the pump (say 75 to 150 micrometers), one right after the pump (say 10 to 25 micrometers), and one in the tank return line (say 10 to 25 micrometers) should do it.

Compatibility involves more than corrosion and deterioration. A unique problem once encountered (and now solved) is the tendency of some HWBF types to form copper soaps in the presence of brass or copper metals in pumps. Strange chemistry will haunt some pioneers, but it is being solved.

Avoid materials that deteriorate or react with HWBF: cork, paper, asbestos, leather, paint, butyl, ethylene-propylene, lead, cadmium, magnesium, zinc, and some alloys of aluminum.

Make sure that: oil can be put into the system if the water base fluid fails to do the job; and the fluid is not being counted on for a secondary function that has requirements beyond what HWBF can perform. For example, if the return flow is expected to lubricate high-speed gears, forget it.

Installation and Maintenance

We'll take the simplest example: an oil system converted to run on high water base fluid. It's sometimes easy, and there are examples of machines that ran perfectly when the oil was drained and HWBF put in its place. However, the usual case is where great care and checking for problem areas must precede the changeover.

Wise recommendations are the following: Drain and flush the old oil with a preload of the new fluid, because small amounts of some oils will break an emulsion. Scrap the flushing fluid. Remove all traces of sludge from the system. Remove all paint from the inside surfaces that will be contacted by the new fluid.

Replace any filters, seals, or other components that are not recommended for HWBF application. Install larger inlet piping, slower pumps, and whatever else is needed to halt cavitation.

Shipping and storage of the HWBF, and maintenance of the system, are not turning out to be as big a problem as some had anticipated. Fact is that each plant manager becomes his or her own fluid blender, tester, maintenance supervisor, and chemist. The additives are shipped and stored cheaply, because they are only 5% of the final mix.

Minerals such as calcium in water will break emulsions and create other maintenance problems, so distilled or chemically deionized water is recommended.

The additive is mixed into the water, not vice versa. The pH (acidity) must be measured and controlled so that the fluid is basic enough to discourage bacteria and algae, which can clog systems.

Maintenance programs should be set up to check the condition of the high water base fluid, because of the sensitivity of the mix to a wide variety of disturbances. Special storage tanks, labeled HIGH WATER BASE FLUIDS and painted "alert orange," already are installed in some plants. (See Chap. 2.)

All in all, we have the makings of a new technology, and there will be plenty of opportunities for engineers.

2

Reservoirs and Cooling

A reservoir (Fig. 2.1) ideally should be large enough to supply the system pump for at least several minutes, neglecting any return flow.

That rule is fairly reasonable for industrial applications but is generally ignored in mobile applications where you do what you can within whatever space is available. More on that later under special-purpose reservoirs.

The reason for the several-minute figure is to ensure enough reserve fluid to fill the hydraulic system at startup without exposing filter and strainer (make sure of this), ensure a fairly stable oil level despite normal fluctuations in flow, pump enough fluid to sustain the system while the rotating parts coast to a stop during emergency shutdown if a return line breaks, have the thermal capacity to absorb unexpected heat for short periods or store heat during idle periods in a cold environment, and provide enough surface area (tank walls) for natural cooling.

If tank volume in gallons is less than once or twice pump flow in gallons per minute—that is, if the tank can be pumped dry in less than a minute or two—add a heat exchanger to avoid excessive temperature fluctuation. For any size tank, specify an oil-level indicator or sight glass.

Here are approximate guidelines for flows into and inside a reservoir: Typical velocity of return flow usually is over 10 ft/sec just before it reaches the tank; this should be reduced to less than 5 ft/sec by manifolding and finally to 1 ft/sec by diffusers or return-line filters. Bulk flow of oil across baffles (Fig. 2.2) and dams (Fig. 2.3) should be limited to 1 ft/sec. Pump inlet flow velocity should be less than 5 ft/sec.

IN-TANK DESIGN (FOR OIL)

The suction-line strainer (or filter) should be ½ to ¾ in. above the tank bottom. Strainer flow capacity should be two to four times pump capacity. A vacuum gage on the pump suction will indicate if the strainer is clogged. A permanent-magnet plug to catch ferrous particles can be mounted on the baffle plate or elsewhere in a region of concentrated return flow (Fig. 2.4).

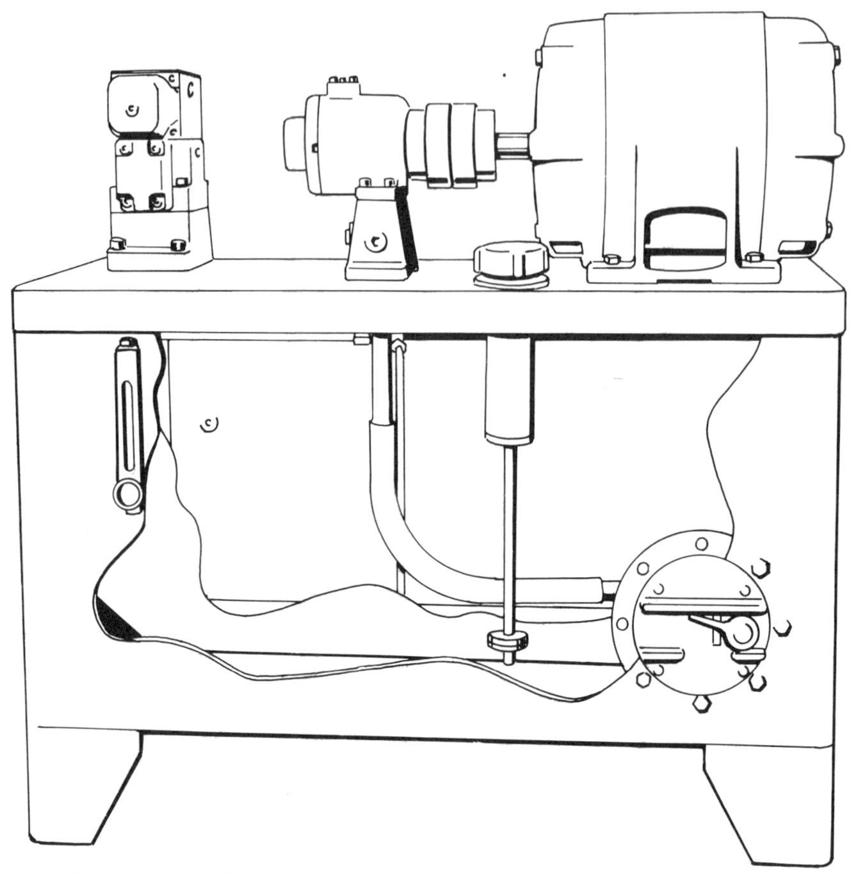

Fig. 2.1 Typical fluid reservoir, with pump affixed on top. (Courtesy of Rexnord, Inc., Milwaukee, WI.)

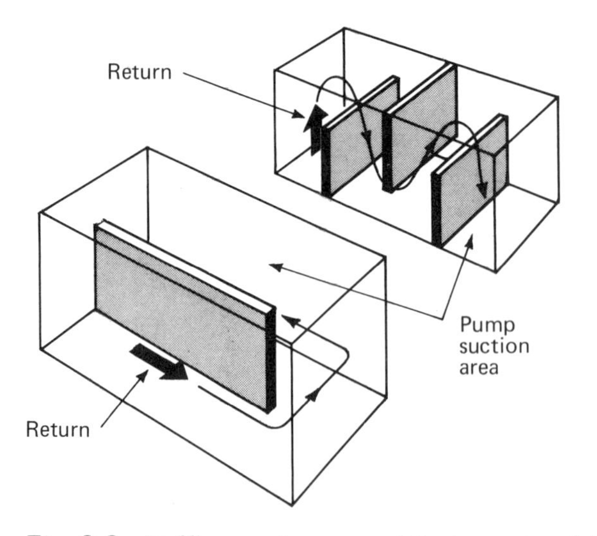

Fig. 2.2 Baffles may be arranged for longest path from return to suction, with good cooling.

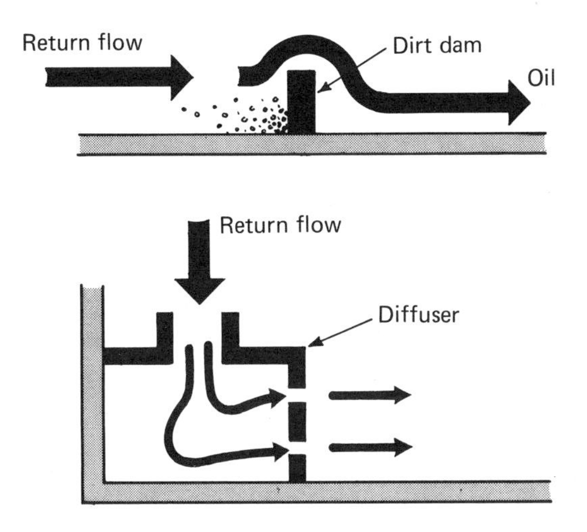

Fig. 2.3 Dirt dams help keep dirt away from suction, and diffusers help reduce turbulence of return.

The major return flow should discharge below oil level about 1 in. above the tank bottom; back pressure will be 5 to 10 psi and more. Atmospheric return lines, including seal leakage lines, are at zero pressure and should be discharged above oil level. (An exception is a pressurized tank; we'll discuss that later.)

If atmospheric lines have high flow and high content of air, they should be discharged above oil level into a chute sloping gradually (5 to 10 deg) into the fluid. The chute slows and fans out the flow, enabling the oil to free itself of air (Fig. 2.5). This is important, because oil saturated with air and operating at high pressure will run 25% hotter than air-free oil. This is caused partly by heat of compression of the air and partly by its low thermal conductivity.

Leakage lines are handled differently from main return lines. They must dis-

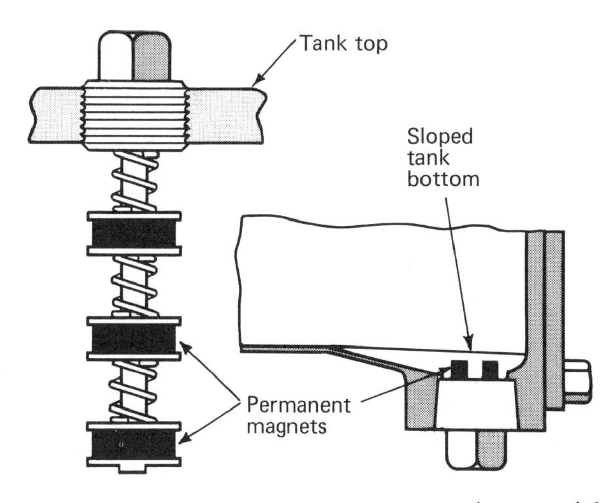

Fig. 2.4 Permanent magnets remove iron particles and can be mounted in various ways. (Courtesy of Lisle Corp., Clarinda, IO.)

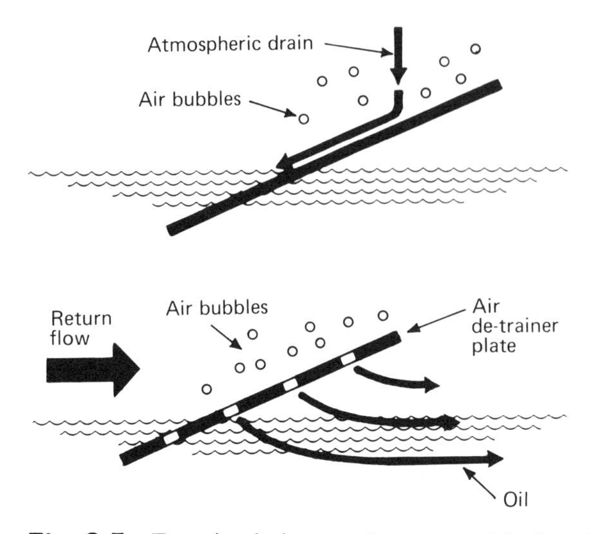

Fig. 2.5 Entrained air must be removed before it gets into the pump suction and spoils performance.

charge above oil level to release entrained or dissolved air and prevent airlock in these vital lines.

Internal baffles between the return pipe and pump inlet will slow down fluid circulation, help settle out dirt particles, give air a chance to escape, and allow dissipation of heat. The tops of the bottom baffles should be submerged about 30% below the surface of the fluid. Top-mounted baffles should have similar clearance at the bottom.

If dirt particles are expected to accumulate on a tank bottom, a wise addition is a low dam across the whole tank. The return oil can easily flow over it, but the dirt gets held safely until the next maintenance period.

A breather opening with an air filter can be installed on top of the tank. Its air capacity should be twice pump capacity to assure atmospheric pressure inside the tank even if the liquid level drops rapidly during peak demands of the hydraulic system. Remember that it is atmospheric pressure that forces fluid into the suction pipe of a pump in a simple reservoir.

A filler pipe for new oil should be on or near the top of the tank and easy to reach. Include a strainer to keep out foreign particles. An even better scheme is to have a totally separate smaller tank just for filling, with a hand pump for transfer to the main tank. Dirt is kept out with a fine filter (Fig. 2.6).

All top-mounted pipes and accessories should be protected against damage from dropped objects and workers' feet. Tank-top leakage can be kept from spilling on the floor if the tank cover is recessed ½ to 1 in. inside the tank walls.

Temperature and Viscosity

Keep oil temperature between 120 and 150 F—preferably the lower value for oil viscosities from 100 to 300 SSU (also called SUS; see Chap. 1) based on 100 F. Temperatures up to 160 F are permissible if viscosity is from 300 to 750 SSU based on 100 F. Higher operating temperatures require special design.

Without a heat exchanger, natural cooling (convection and radiation) must be depended on to dissipate generated heat. You should utilize all external and internal surface areas. Put baffles inside the tank to direct hot incoming oil against tank walls. Design legs under the tank to make the under-surface available for natural cooling and to help level the tank during assembly. Paint the tank exterior with a very thin (less than 0.001-in.) flat black lacquer for extra good radiation cooling.

Tank walls should be thin for good thermal conductivity. Make them approximately 1/16 in. thick for tank capacity up to 25 gal, ⅛ in. for capacity up to 100 gal, and ¼ in. for 100 gal or more. Use slightly heavier plate for the bottom. Give the top plate four times wall thickness to assure vibration-free operation and to hold alignment of pump and motor shafts.

To be on the safe side, calculate stresses in the tank walls, particularly if the reservoir is pressurized or has large areas unsupported. If the stresses are high, rearrange the baffles to add support.

In addition to whatever automatic temperature-limiting controls you provide, specify a thermometer to be mounted on the tank top where the operator can see it.

Special-Purpose Reservoirs

Not every reservoir can be roomy and accessible. Mobile vehicles and construction equipment have crowded accommodations, and some carry only enough fluid for a fraction of a minute operation unreplenished. If operation is intermittent, it works well.

Mobile designs must be exceptionally good to withstand the extremes of temperature, vibration, contaminated environment, and high-return-line velocities.

High water base fluids have a special problem and will be covered at the end of this chapter. Also see Chap. 1.

The first step in any system is to calculate total system flow and capacity at the extreme conditions. Don't forget that hydraulic cylinders have a greater contained volume when the rods are extended than when they are retracted. For the same rea-

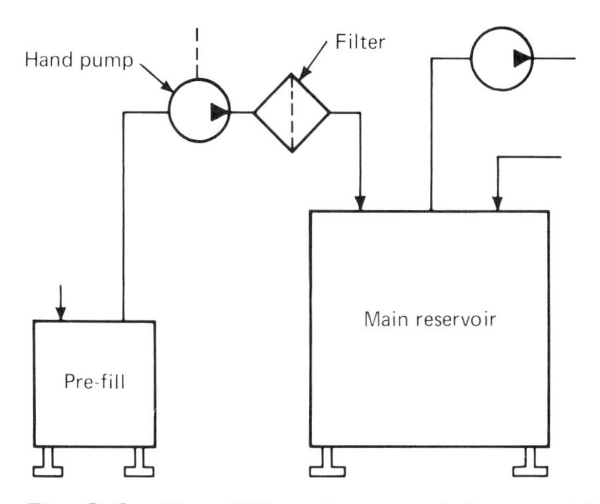

Fig. 2.6 Clean filling of a reservoir is assured if a separate prefill tank is included in design.

son, return flow from the blind end of a cylinder exceeds pumped flow to the rod end. Add to that the increased volume of oil as it heats up.

Air bubbles and cavitation can be a severe problem when the reservoir is cramped and when the pump is mounted on top. Return-line diffusers (perforated metal) help a lot (Fig. 2.3).

A better answer, where possible, is to pressurize the reservoir. It can be done with control air taken from elsewhere on the vehicle or machine, if available.

Another way is to replace the breather vent with a bladder accumulator that lets the reservoir breath but with a slight positive pressure. Greer Hydraulics (Los Angeles, CA) offers one that operates at atmospheric pressure, and the design is adaptable to positive pressures as well. The design also protects against dirty environments (Fig. 2.7).

A less expensive solution is to seal off the breather vent and let air-over-oil pressure rise as the oil heats up and expands against the trapped air.

Many manufacturers of hydraulic power units immerse the pump in the fluid to improve suction. For example, they mount the electric motor vertically on the cover to drive the immersed pump directly or drive it through a vertical wall—a horizontal configuration.

Some hydraulic power unit manufacturers immerse the pumps and also the controls in the fluid, then seal the unit and include full-immersed solenoid valves in the package. Advantages are protection from bad environments and compactness (Fig. 2.8).

Another scheme is to mount the entire hydraulic system inside the reservoir. The unit is surrounded by a diaphragm to maintain a slight positive pressure on the contained oil.

For higher inlet pressure, try overhead reservoirs. They are good for HWBF (high

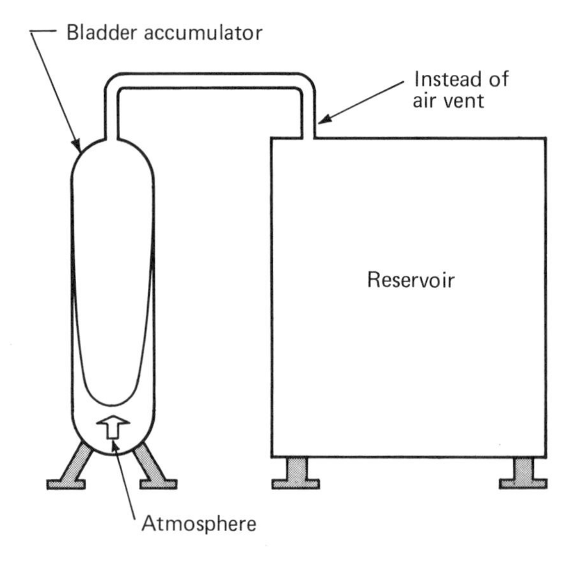

Fig. 2.7 Oil-level fluctuation in a reservoir can be absorbed with a built-in accumulator. (Courtesy of Greer Hydraulics, Inc., Los Angeles, CA.)

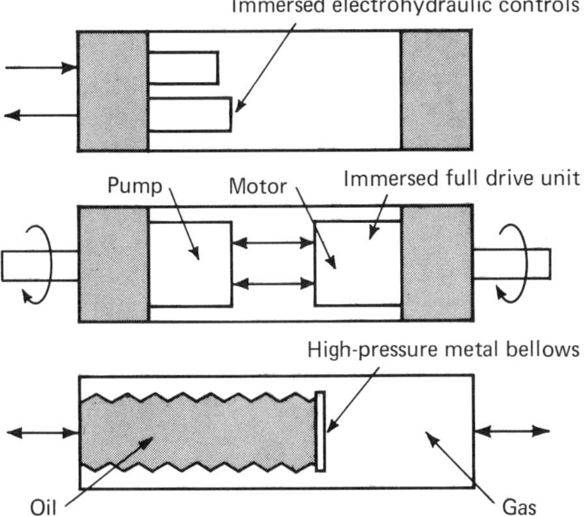

Fig. 2.8 Sealed reservoirs of special design protect hydraulic systems against bad environments.

water base fluid) systems. The increased head of pressure helps overcome a problem with HWBF: high vapor pressure and resulting cavitation (Fig. 2.9).

That isn't the only problem with HWBF: It freezes, evaporates, has low viscosity, is a poor lubricant, leaks more readily than oil, encourages bacterial growth, and corrodes unprotected steel. Just the same, HWBF is being successfully applied by knowledgeable designers. There are plenty of custom manufacturers who will build what you like.

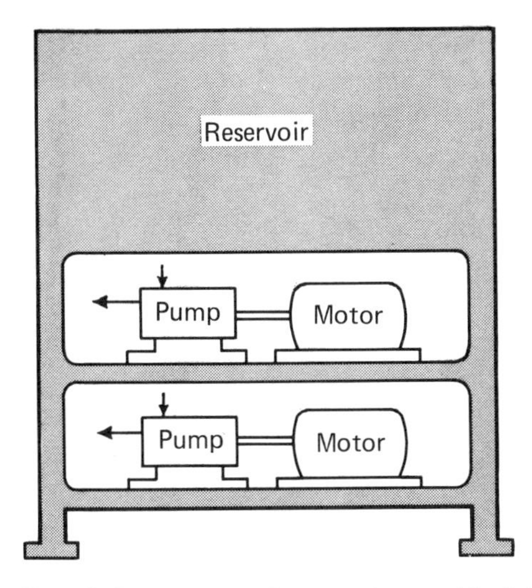

Fig. 2.9 Overhead reservoirs are ideal for any fluid but are essential for high water base systems.

Design for Maintenance

Integral tanks are the hardest to maintain. It is better to have a separate unit, accessible from all sides. Small tanks, in fact, are sometimes mounted on casters.

Equip the tank with cleanout doors and slope the bottom toward the doors. Provide a drain cock or discharge valve at the low point of the bottom and at other low points if this is necessary for complete drainage. Put a manhole cover on the tank top for removing filters and strainer.

Provide a connection for hooking up to a portable filtration unit. Look into the variety of filter types available that allow element change without flushing dirt into the system. Or even better, provide a prefill tank ahead of the main reservoir.

Write the assembly, painting, testing, and maintenance instructions before settling on final details of the design. Here are some important tips:

If the tank is made of cast iron, don't paint the interior surfaces, but be sure to remove all grit and core sand. Surfaces must be sandblasted.

If the tank is made of sheet steel, clean it thoroughly to remove all dirt, chips, and burrs. Use solvent and rags to remove any grease or oil and then dry with clean rags. If the tank is slightly rusted, clean by wire brushing or grinding. Heavily rusted and scaled surfaces should be shotblasted.

Painting technique depends on which hydraulic fluid will be used in your system. If the fluid has a petroleum base, coat all interior surfaces with an oil-resistant paint recommended by a reputable supplier. For fire-resistant fluids with a phosphate ester base, use no paint. Special paints resistant to phosphate esters are available, but the fluid itself is noncorrosive and provides the necessary protective coating.

If the fluid is an oil-water or water-glycol emulsion, don't apply paint to the surfaces because the emulsion will probably peel it off. But if you must paint, obtain detailed information from the manufacturer of the emulsion.

TAKING OUT THE HEAT

When the reservoir and piping cannot dissipate all the heat, specify a heat exchanger, but first estimate performance with this simple guide.

There are three ways to avoid overheating: Conserve energy by pumping only the amount of fluid needed to operate the system, provide plenty of cooling surface, or, if necessary, add a heat exchanger (Fig. 2.10).

Extra cooling in a hydraulic system probably is needed if oil temperature would otherwise exceed 120 F for long periods of time. There are much hotter systems that work well, but we'll base this method on 120 F.

The hardest calculation is the heat balance (details follow). The easiest task is the final selection of a heat exchanger because the manufacturers already have worked out the math for any combination of conditions. Given all the data from a heat balance—including inlet and outlet flows, pressures and temperatures, allowable pressure drops, water or air flows available for cooling, oil properties, duty cycle, and available space—the heat exchanger engineers practically can pick the right model from a chart.

The greatest problem encountered is lack of complete input/output conditions beforehand.

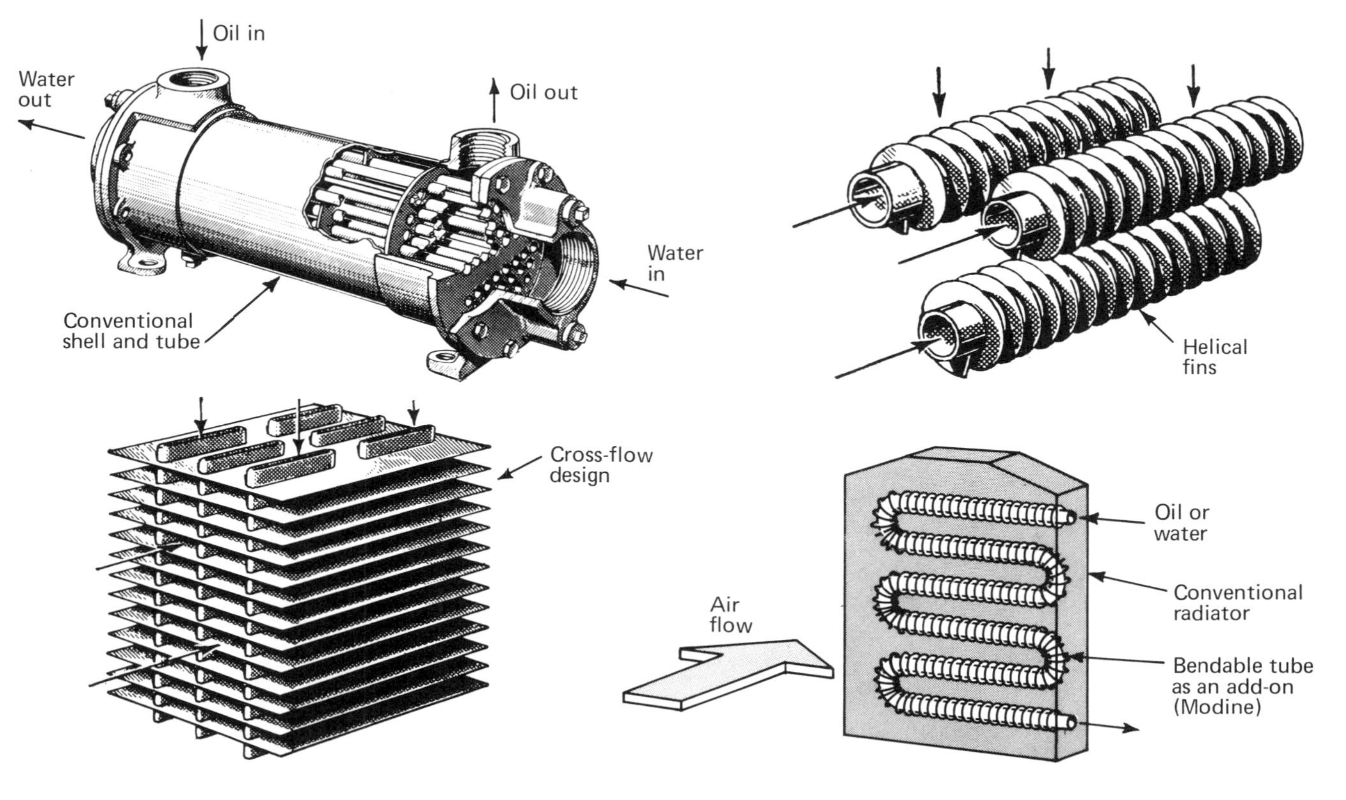

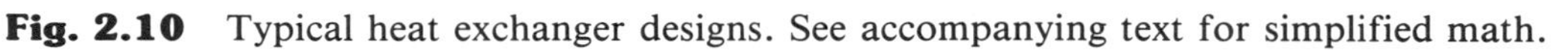

Fig. 2.10 Typical heat exchanger designs. See accompanying text for simplified math.

Choices in Basic Design

The first figures illustrate typical heat exchangers used in cooling the oil in engine and hydraulic systems. Most are intended for return-line service, where system pressures are lowest. Exchangers can be custom designed for any pressure level, but keeping them in the return line saves much bother and expense. Most exchangers are rated at a few hundred psi.

Water cooling is recommended for the most compact and efficient operation, but forced air is widely used as well. Air-to-oil exchangers are designed with generous surface areas to enhance heat transfer. In Fig. 2.11, note the turbulators, balls, and fins on the inside of some of the tubes. When viscous oil flows through these tubes, it is channeled against the heat transfer surfaces. Viscous oil otherwise would tend to develop parabolic flow distribution down the center of the tube. Low-viscosity fluids such as water or water-based fluids are less of a problem because they become turbulent more readily and make better contact with the tube walls.

Measuring Heat Loss

Heat is generated wherever fluids are throttled or otherwise restricted. Examples are pressure regulators, relief valves, undersize piping, dirty or undersize filters, internal leakage, pump and motor turbulence, and flow friction. Finally, add the gratuitous heat transferred from engines, motors, electrical controllers, and adjacent systems.

A basic measurement for overall heat loss is to subtract *useful energy* from *input energy:*

$$E_L = E_1 - E_2 \tag{2.1}$$

where E_L is energy lost, Btu/hr; E_1 is energy input to the pump, Btu/hr; and E_2 is useful energy output to drive cylinders, actuators, and hydraulic motors, Btu/hr. For symbols, see Table 2.1. The relationship also may be written as

$$E_L = E_1(1 - \mu) = 1.48QP(1 - \mu) \tag{2.2}$$

where μ is system efficiency; Q is pump flow, gpm; and P is pump discharge pressure, psig.

Energy loss of a given component—say a relief valve—can be calculated if the flow and pressure drop are known. A relief valve wastes all of its overflow, making the math simple:

$$\text{Btu loss} = 1.48Q\Delta P \tag{2.3a}$$

$$\text{hp loss} = 0.000582Q\Delta P \tag{2.3b}$$

where ΔP is pressure drop through the relief, psi.

For example, assume a 30-gpm fixed-displacement pump is supplying a 20-gpm motor. The excess flow (10 gpm) is discharged over a relief set at 1000 psi. The hp loss simply is $0.000582 \times 10 \times 1000 = 5.82$ hp, or 14,800 Btu/hr. That much hp dissipated for 1 hr at 20¢/kWh costs $5.82 \times 0.746 \times 0.20 = 87$¢. Or if the machine runs 5 hr/day for 250 days/yr, the cost is \$1087.

Actually, the cost is much higher because we've neglected electric motor and line losses. Add 20%, bringing the total cost of the energy loss to \$1305/yr.

In another case, if a 50-hp electric motor running at full load is necessary to operate a fluid power system for a final mechanical output of 30 hp to turn a drive

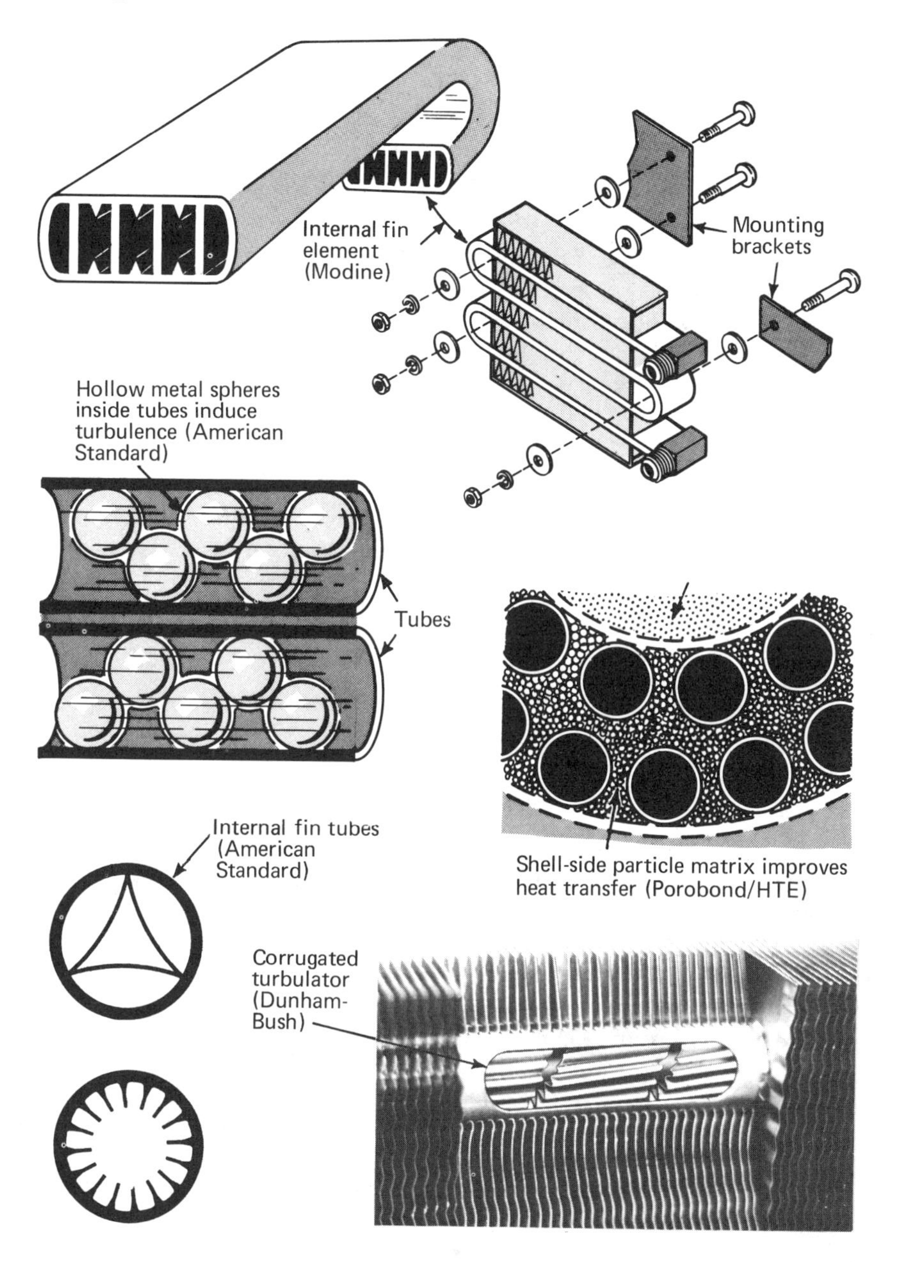

Fig. 2.11 Fins and turbulence inducers improve heat transfer.

shaft at 3000 rpm with 631-lb-in. torque, then the loss is 20 hp no matter how you figure it. That's $20 \times 0.746 \times 0.20 = \$2.98/\text{hr}$, or over \$3730/yr if the machine runs 5/hr day, 250 days/yr. If you can redesign the system to let a 40-hp electric motor carry the burden, the savings will be \$1865 in 10 years.

Here are some local losses to look out for: gross internal leakage somewhere in the circuits; extreme flow restrictions that increase pump pressure without benefiting the output; poorly sized components such as undersized filters or oversized

Table 2.1 Nomenclature for Heat Transfer Equations

Symbols

Heat loss and efficiency:

μ	=	System efficiency, E_2/E_1
E_1	=	Pump input power, Btu/hr
E_2	=	Energy used in system, Btu/hr
E_A	=	Heat absorbed by oil, tank, and components, Btu/hr
E_D	=	Heat dissipated to atmosphere or coolant, Btu/hr
E_{HE}	=	Heat exchanger load, Btu/hr
E_L	=	System heat loss, Btu/hr

Flow, pressure, temperature:

P	=	Pump pressure, psig
Q	=	Flow, gpm
t	=	Operating time, hr
T_D	=	Temperature-over-ambient for oil, F: $T_D = T_{OIL} - T_{AIR}$, mean values
ΔT	=	Heat exchanger only: $\Delta T_{WATER} = T_2 - T_1$; $\Delta T_{OIL} = T_1 - T_2$; ΔT_{MEAN} = log-mean ΔT, oil-to-water

Constants and operators:

A	=	Cooling surface area, ft^2
c	=	Specific heat, mean, Btu/ft^2-hr-F
e	=	Base, natural log, 2.718
k	=	Overall heat-transfer coefficient, Btu/ft^2-hr-F
W	=	Combined weight of oil and system components, lb
Σ	=	Summation. ΣcW is effective cW overall

Typical values for c and *k*:

c	=	0.40 for oil, 0.18 for aluminum, 0.11 for iron, and 0.09 for copper
k	=	2 to 5 for tank inside of machine, or just crowded
	=	5 to 10 for steel tank in normal air
	=	10 to 13 for tank with good guided air current
	=	25 to 60 for forced air, or oil-to-air exchanger
	=	80 to 100 for oil-to-water heat exchanger

pumps; inefficient, high-slip pumps and motors; fixed-flow pumps where adjustable flow pumps would prove more economical; excessive seal friction; and excessively viscous fluids.

An even larger saving is to idle or shut down all pumps that aren't needed during lulls in the load cycle. For instance, there may be no need to run a coolant pump if hot oil is not flowing through the exchanger.

Calculate Heat Balance

Figure 2.12 gives a clue to the problem. It shows performance of a fixed-displacement pump against a system with variable requirements. The area under each curve is energy, assuming flows are the same. The difference between the areas is energy lost.

A true heat balance is more complicated, being the mathematical summation of heat entering and leaving a system. What goes in must be accounted for either as a temperature increase or as a transfer of heat out of the system:

$$E_L = E_A + E_D \tag{2.4}$$

where E_L is internal heat loss generated by the pump and other components in the system; E_A is heat absorbed by the oil, tank, and components; and E_D is heat dissipated into the atmosphere or to a coolant.

The balance is usually a transient one, except where long-term steady-state operation is reached. The general relationship is

$$E_L dt = (\Sigma\, cW)dT_D + (\Sigma\, kA)T_D dt \tag{2.5}$$

where t is operation time, hr; c is mean specific heat, Btu/lb-F; W is combined weight, lb, of oil and system components; T_D is temperature difference, F, between oil system and surrounding air; k is overall heat transfer coefficient, Btu/ft^2-hr-F; and A is heat-dissipating area, ft^2. Typical values of c and k are given in Table 2.1.

The wanted quantity is temperature-over-ambient T_D. The differential equation first must be solved. Briefly,

$$dt = \frac{(\Sigma\, cW)dT_D}{E_L - (\Sigma\, kA)T_D} \tag{2.6}$$

Integrate the equation between limits of *initial* T_D and T_D and solve for the latter:

$$T_D = \frac{E_L}{\Sigma\, kA}(1 - e^{(-\Sigma\, kA/\Sigma\, cW)t}) + (\text{initial } T_D)e^{(-\Sigma\, kA/\Sigma\, cW)t} \tag{2.7}$$

where e is 2.718 and t is time, hr. Maximum T_D prevails when t approaches infinity; a steady-state temperature balance is reached, and oil temperature remains constant:

$$\max T_D = \Sigma\, E_L/kA \tag{2.8}$$

If the calculated maximum temperature difference exceeds the specified limit, and if adding surface area or increasing the oil capacity does not lower it sufficiently, then it's time to add a heat exchanger.

Water-cooled shell-and-tube exchangers, shown in the first sketch in Fig. 2.10, are typical, although air coolers are used where water is not readily available and where the air is at least 10 F cooler than the oil.

Water flows through the tubes, and oil flows across the tubes. This facilitates maintenance since water is more likely than oil to contaminate a metal surface and the inside of a tube is easier to clean than the outside. Baffles direct the oil across the tubes and also support them. Pressure drop of oil should not exceed 10 or 15 psi; typical flow velocity is about 3 ft/sec.

Single-pass and double-pass coolers are available. In single-pass coolers, the

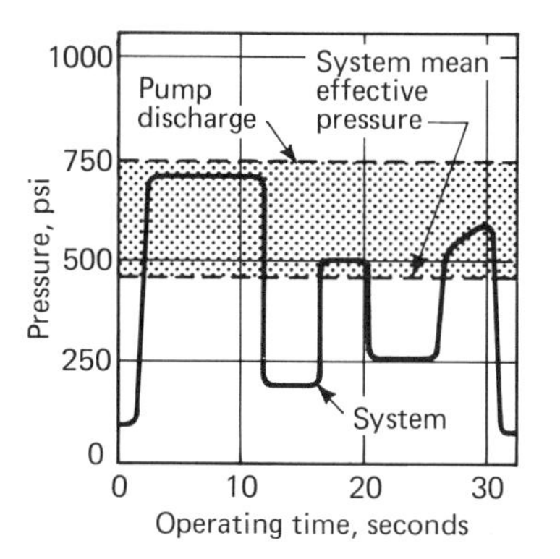

Fig. 2.12 Energy wasted in the shaded area converts to heat in the system.

water flows straight through once; the water in two-pass coolers makes one 180-deg turn. Choice depends on the calculated temperature difference between outlet water and outlet oil. If the outlet oil is considerably cooler than the outlet water, a single-pass cooler is sufficient. If the outlet oil temperature is likely to approach the outlet water temperature, multiple-pass exchangers can be justified, but you should increase the total calculated surface area 20% to compensate for lower efficiency of a multiple-pass exchanger.

Heat balance in an exchanger is

$$E_{HE} = \Delta T_{OIL} \times Q_{OIL} \times 210 = \Delta T_{WATER} \times Q_{WATER} \times 500 \tag{2.9}$$

where the specific heat of oil is assumed to be 0.4 Btu/lb-F, and ΔT_{OIL} and ΔT_{WATER} are input-to-output temperature changes. The balance depends on the overall heat transfer coefficient k between the water and the oil:

$$E_{HE} = kA\Delta T_{MEAN}, \text{ Btu/hr} \tag{2.10}$$

The value of k for most oil-to-water exchangers is 80 to 100 Btu/ft²-hr-F. Symbol A is the surface area, ft², and ΔT_{MEAN} is the logarithmic mean temperature difference as follows:

$$\Delta T_{MEAN} = \frac{\Delta T_{MAX} - \Delta T_{MIN}}{\ln(\Delta T_{MAX}/\Delta T_{MIN})} \tag{2.11}$$

where ΔT_{MAX} is the greatest and ΔT_{MIN} the least temperature difference of any measured between oil and water throughout the exchanger. Convenient values of ΔT_{MEAN} are plotted in Fig. 2.13. The method is valid for any parallel or counterflow exchanger using any materials and any fluids.

No Exchanger Required

Problem: Oil temperature must be predicted after a given period of time in a system without a heat exchanger.

Calculate the temperature of a standard 60-gal reservoir after 5 hr of operation. Pump discharge is 20 gpm at 750 ft². Cooling surface is 28 ft². An attached heat-dissipating working unit weighs 800 lb, and its effective surface area is 5.5 ft². Assuming an initial system temperature $T_{OIL} = 70$ F and an ambient temperature $T_{AIR} = 50$ F, the initial temperature-over-ambient initial $T_D = 20$ F. The estimated median value of $k = 4$ Btu/ft²-hr-F. Sixty gallons of oil weigh $7.4 \times 60 = 444$ lb.

$E_L = 9000$ Btu/hr, assumed

$\Sigma\, cW = 0.4 \times 444 + 0.1 \times 800 = 257$ Btu/F, based on Eq. (2.7)
$\Sigma\, kA$ for tank and working unit $= 4(28 + 5.5) = 134$ Btu/hr-F

The temperature above ambient T_D at $t = 5$ from the general heat-balance equation, Eq. (2.7):

$$T_D = \frac{9000}{134}(1 - e^{(-134/257)\times 5}) + 20e^{(-134/257)\times 5} = 67(1 - 0.0745) + 20 \times 0.0745 = 63.5 \text{ F}$$

The maximum operating temperature over ambient, from Eq. (2.8): max $T_D = 9000/134 = 67$ F. Oil temperature $= 50 + 67 = 117$ F.

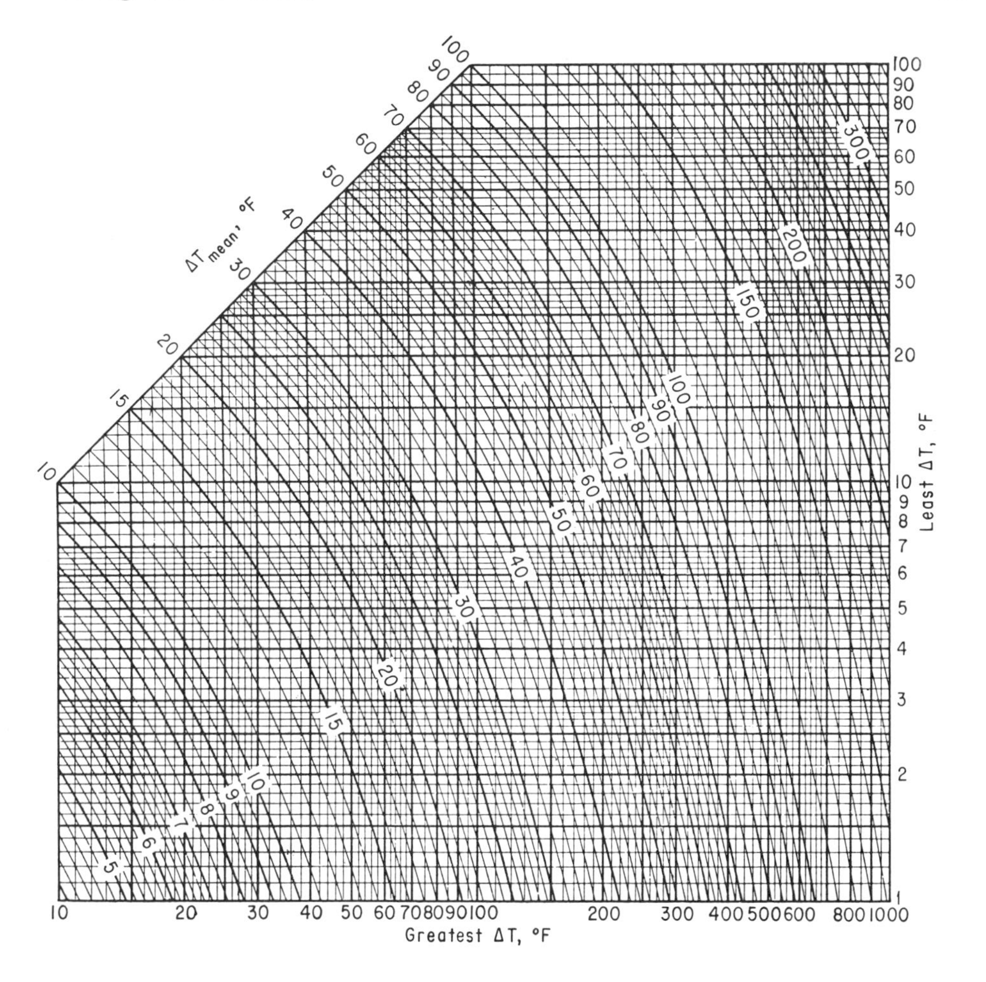

Fig. 2.13 Log mean temperature difference for any parallel or counterflow heat exchanger.

Conclusion: No heat exchanger is required.

Some Cooling Necessary

Problem: Determine if the reservoir can dissipate enough heat to keep the oil temperature below 120 F in a 70 F ambient (50 F difference T_D). The source of heat is a 20-gpm constant-delivery pump operating in a 60% efficient overall system. Assume first that the working unit, including piping, is not helping to dissipate heat. Tank size is small: The reservoir cooling surface is only 21 ft^2, based on a tank volume of twice pump flow, or 40 gal. System pressure is 750 psi; the overall heat transfer coefficient k is 5 Btu/ft^2-hr-F (conservative). Total generated heat is calculated with the heat-loss equation [Eq. (2.2)]:

$$E_L = 1.48 \times 20 \times 750 \times (1 - 0.60) = \text{about } 9000 \text{ Btu/hr}$$

Required surface area from maximum-temperature equation, Eq. (2.8):

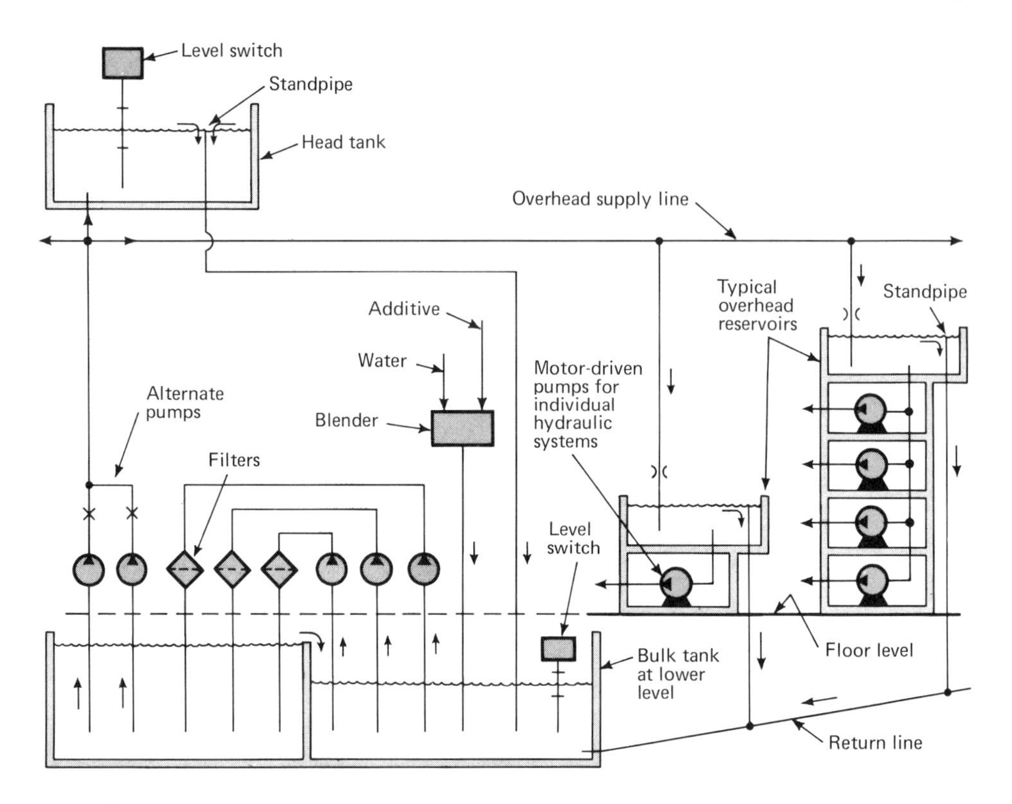

Fig. 2.14 One HWBF bulk tank, supplying dozens of remote tanks, is readily monitored and eliminates the need for individual chemical treatment and filtration for the remote tanks. (Courtesy of Buick Motor, Flint, MI.)

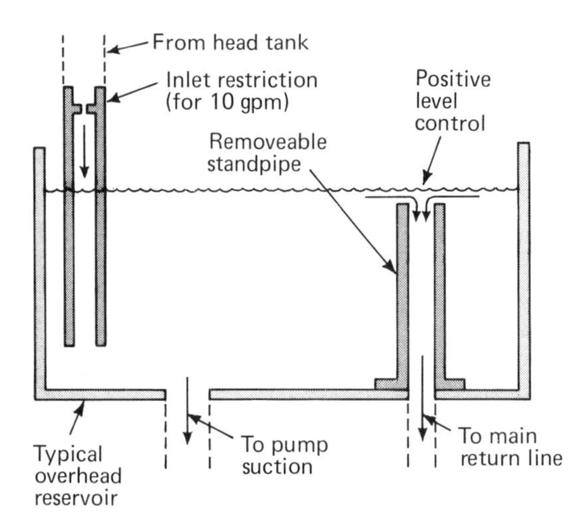

Fig. 2.15 Standpipe provides absolute level control; an orifice sets the desired fill flow rate.

$A = 9000/(50 \times 5) = 36\ ft^2$

Conclusion: The actual tank area is not great enough, and some means of cooling should be provided.

Choosing an Exchanger

Problem: Oil system return flow in a hypothetical system must be cooled continuously to125 F. The hottest drain temperature, uncooled, is 140 F. Flow is 12 gpm; water inlet = 65 F, and outlet = 85 F; k = 90 Btu/hr-ft^2-F. Assume a counterflow, single-pass exchanger.

Conclusion: The heat transfer area of the exchanger must be at least 7.32 ft^2 and water flow 3.78 gpm, calculated with Eqs. (2.9) and (2.10) as follows:

$$E_{HE} = (140 - 125) \times 12 \times 210 = 37{,}800 \text{ Btu/hr}$$
$$\Delta T_{MAX} = 125 - 65 = 60 \text{ F}$$
$$\Delta T_{MIN} = 140 - 85 = 55 \text{ F}$$
$$\Delta T_{MEAN} = 57.5 \text{ F (from chart, Fig. 2.13)}$$
$$A = 37{,}800/(90 \times 57.5) = 7.32 \text{ ft}^2$$
$$Q_{WATER} = 37{,}800/(500 \times 20) = 3.78 \text{ gpm}$$

CENTRALIZED SYSTEM FOR HIGH WATER BASE FLUIDS

A major drawback to conversion from petroleum hydraulic fluids to high water base fluids in a manufacturing plant is the problem of monitoring the relatively fragile HWBFs in remote and varied situations. Otherwise, there would be a much greater demand for machines and systems designed to run on them.

The Buick Division of GM (Flint, MI) has accomplished it (Fig. 2.14). Over 170 HWBF reservoirs are continuously drained and replenished with carefully monitored and controlled fluid from a huge central reservoir.

Each remote reservoir design has a hinged or removable top and a standpipe (Fig. 2.15). The standpipe automatically maintains a constant level, and computerized sensors keep constant watch. The individual reservoirs (typically 100 gal) will drain within 1 min when the standpipe is removed. While the tank is empty, it may be cleaned and flushed down the drain. It takes 10 min to refill a reservoir from the bulk tank after the standpipe is replaced. A calibrated orifice limits refill flow to 10 gpm. Faster filling is made possible by installing a bypass valve to take additional flow from the overhead piping network.

Fluid monitoring is done on a weekly basis by the fluid supplier, who checks pH (alkalinity), concentration ratios (additive to water), bacteria count and type, sediment content (with a centrifuge), and appearance (color and clarity). With each report comes a recommendation for correcting the chemistry.

3

Hydraulic Pumps

For pressurized hydraulic power and control systems, positive-displacement-type pumps (piston, gear, and vane, for example) are most often chosen. Centrifugal and other dynamic types are usually reserved for fluid handling rather than for pressure or power.

The majority of the text and illustrations in this chapter therefore will concentrate on positive-displacement pumps. However, some discussion of dynamic-type pumps is at the end of the chapter.

POSITIVE-DISPLACEMENT PUMPS

Positive-displacement pumps (Fig. 3.1) in general have good volumetric efficiency. They can develop high pressures, produce flow that is directly proportional to speed, and prime themselves, even at very low inlet pressure. *But* they must not be started or operated with the discharge valve closed; low-viscosity fluids increase the slip, and abrasive fluids cause wear. (A centrifugal pump is usually better than a positive-displacement pump at low discharge pressures.)

Of the positive-displacement pumps, the piston type has the highest volumetric efficiency because the leakage path between the piston and cylinder can be closely controlled in manufacture. Vane, screw, and lobe versions have the best suction characteristics and are least likely to cavitate. Many unusual types of positive-displacement pumps are available with special characteristics. (See Fig. 3.2 for 17 examples of pump types.)

Adjustable Displacement

Axial piston pumps used in hydrostatic drives (Chap. 6) have a wide range of controls and adjustments. Figures 3.3 through 3.6 are typical. The same or similar controls can be applied to other types of positive-displacement pumps as well—such as radial piston and vane pumps.

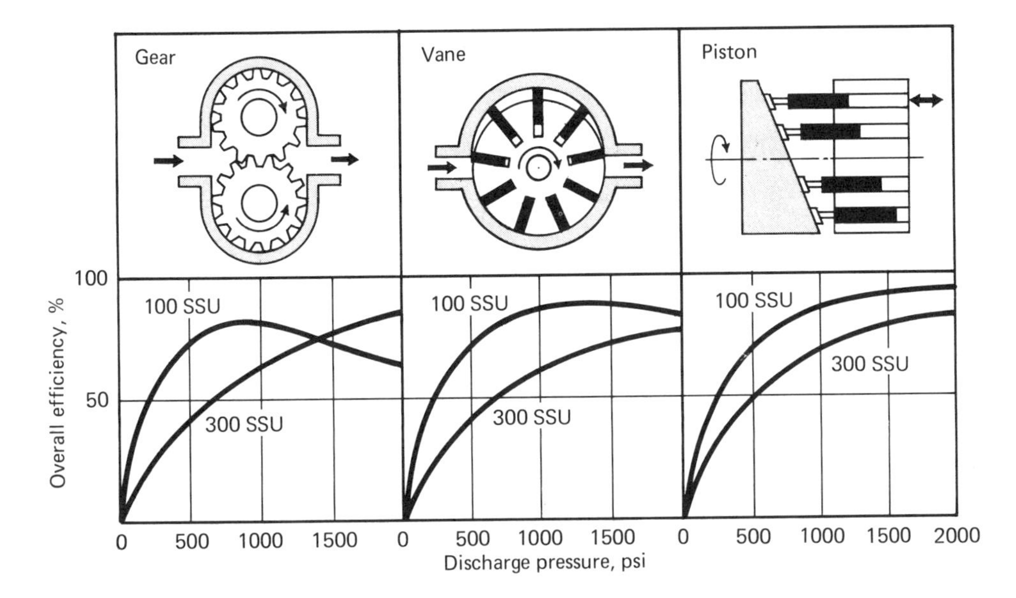

Fig. 3.1 Typical efficiencies of gear, vane, and piston pumps. SSU (or SUS) means Saybolt Seconds Universal; see Chap. 1.

The purpose of the controls is twofold: to pump exactly the pressure and flow needed by the hydraulic motors and actuators in the system; and to save money.

A positive-displacement piston pump is controlled by adjusting the piston displacement to match the desired flow and pressure—usually with a tiltable swashplate (Fig. 3.3). Pressure and flow sensors in the pump discharge line respond to variations from the set values and cause the servo piston to reposition the swashplate until the pump flow and pressure are corrected (Fig. 3.4). Details of a typical pressure control (called a pressure compensator) are given in Chap. 6.

There are special controls to compensate for any variable—even temperature. Figure 3.5 shows how a capillary in series with a sharp-edge orifice can reduce the control signal to a pump to limit flow during a cold start. The capillary creates a high restriction when the oil is cold, tending to increase the pressure in the second chamber relative to the first chamber, thus lowering the pressure differential signal.

The most sophisticated control for a pump is a servovalve (Fig. 3.6). A small electrical signal unbalances a pilot valve, and the resulting amplification of flow and pressure repositions the swashplate. Mechanical feedback rebalances the pilot valve. (See Chap. 7 for details on all kinds of servovalves.)

Theoretical Output Power

Figure 3.7 gives pump-output hp for any given flow at any given pressure. The actual power to drive the pump will, of course, be greater than the output power; it depends on the overall efficiency of the pump. The relationship is $hp_{out} = E \times hp_{in}$, where hp_{in} is the power needed to drive the pump, E is the overall efficiency of the pump, and hp_{out} is the theoretical pump-output hp based on flow and pressure. Symbols are explained in Table 3.1.

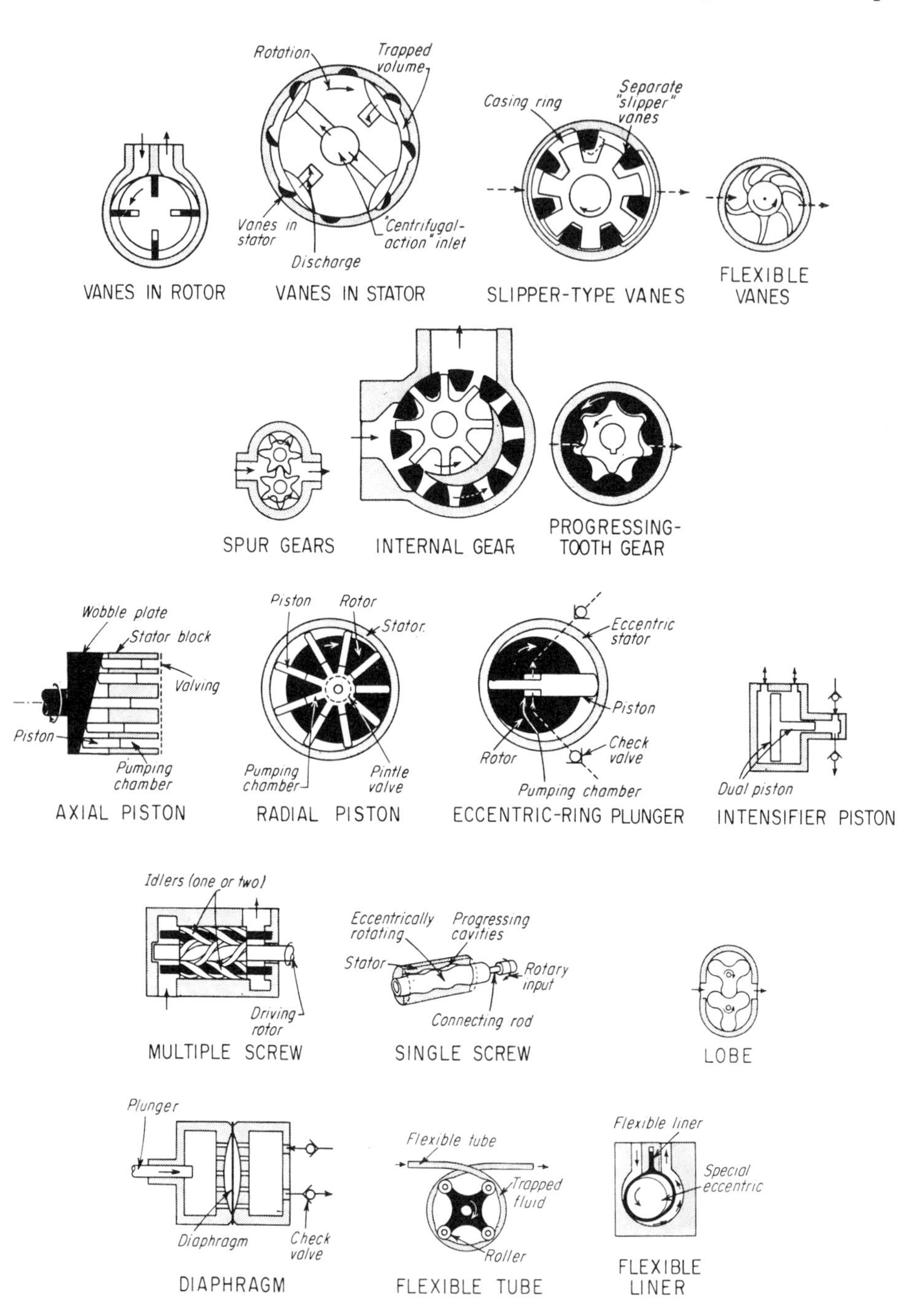

Fig. 3.2 All positive-displacement pumps—including the unusual types—have one thing in common: The liquid is forced out by reducing the volume of the pump enclosure. They differ in how the volume is reduced and in how the high-pressure liquid at the outlet is kept from leaking excessively to the inlet.

VANE PUMPS

VANES-IN-ROTOR version is most common; vanes move radially to follow pump contour. Leakage from outlet to inlet across top is kept small because vane is completely retracted, rotor-to-stator clearance is practically zero.

VANES-IN-STATOR PUMP has hollow-shaft inlet; centrifugal action reduces cavitation. Pumping occurs as rotor turns because trapped volume of liquid is squeezed out by vanes. Outlet is through rotor—flow leaves rotor via sliprings.

SLIPPER-VANE PUMP initially depends on centrifugal force to throw slippers against casing ring. But then liquid pressure, viscous drag, and oil-film cohesion hold vanes in position. Action is similar to that in Kingsbury thrust bearing.

FLEXIBLE-VANE PUMP depends on bending of vanes to decrease trapped volume, force out liquid.

GEAR PUMPS

SPUR-GEAR PUMP carries trapped liquid between pairs of outer teeth, forces it against existing outlet pressure. Outlet-to-inlet leakage is kept small by close-meshing teeth at center.

INTERNAL-GEAR PUMP is similar to vane pump in operating principle. Inlet liquid fills expanding pumping cavities, is carried around inner and outer path as shown. Outlet-to-inlet leakage is kept small by close-meshing spider teeth and vanes.

PROGRESSING-TOOTH GEAR PUMP has one less tooth in rotor than in stator, thereby moving trapped volume one tooth width each revolution. Performance is similar to vane pump.

PISTON PUMPS

Piston pumps have reciprocating action that allows inlet liquid to fill cavity on suction stroke, forces it out on pumping stroke.

AXIAL-PISTON PUMP depends on beveled "wobble" plate or equivalent off-center linkage system to reciprocate pistons. RADIAL-PISTON PUMP depends on eccentric stator (or rotor) to give reciprocating action.

ECCENTRIC-RING PLUNGER PUMP is single-piston version of radial-piston pump. It is shown in suction position; 90-deg rotation of rotor will close pumping chamber, forcing out liquid.

INTENSIFIER-PISTON PUMP transfers energy from one liquid system to another. Version shown converts high-volume low-pressure fluid (large chamber) to low-volume high-pressure fluid (small chamber).

SCREW PUMPS

MULTIPLE-SCREW PUMP drives through single rotor; the other meshing rotors (there can be one or more) help trap liquid in pumping cavity that moves axially as the rotor turns.

SINGLE-SCREW OR "PROGRESSING-CAVITY" PUMP has helical rotor rotated by connecting rod. However, rotor not only rotates about its own axis, but its own axis rotates relative to axis of input. This eccentric action keeps rotor in constant contact with stator, trapping liquid in axial-moving cavity.

LOBE PUMPS

LOBE PUMPS of simplest design are similar in operation to spur-gear pumps —liquid is carried in outside pockets.

DIAPHRAGM PUMPS

DIAPHRAGM PUMP shown transfers energy from plunger-driven fluid on left to pumped fluid on right. The amount pumped is equal to displacement of the plunger. Other versions can have cam-operated diaphragms.

SQUEEGEE PUMPS

FLEXIBLE-LINER PUMP traps liquid in "squeegee" action. Pumping and sealing stops as eccentric goes through top dead center, but momentum of fluid prevents backflow. Principle advantage of squeegee is isolation of moving element from pumped liquid.

FLEXIBLE-TUBE PUMP is also squeegee type, but multiple rollers keep liquid trapped at all times.

Fig. 3.2 (Continued)

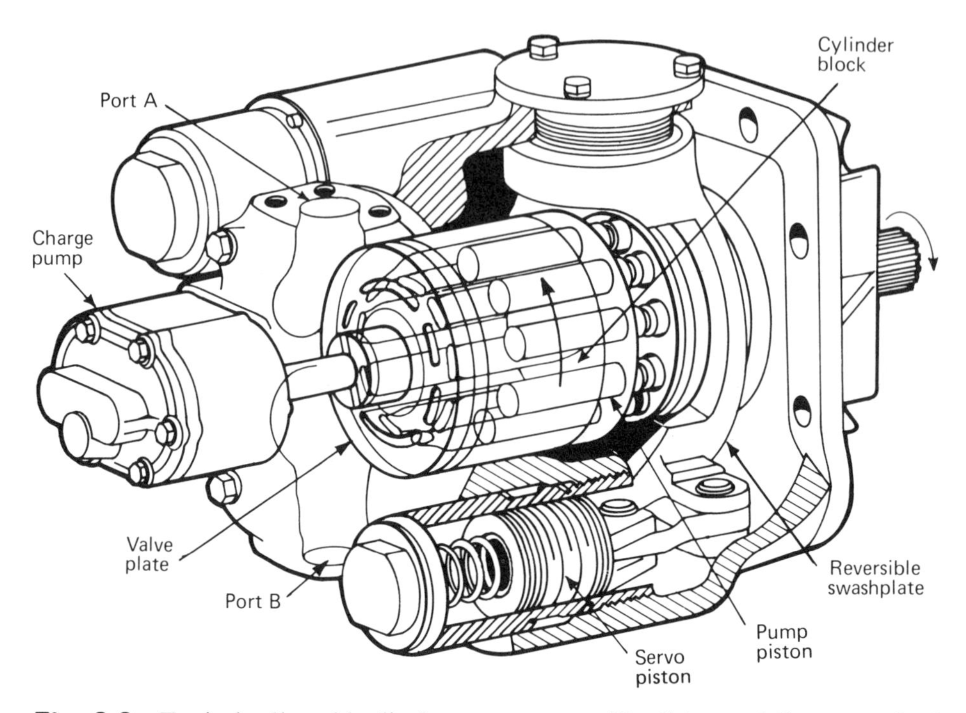

Fig. 3.3 Typical adjustable-displacement pump (Vardis) can deliver exactly the needed flow and therefore saves energy.

Pump efficiency depends on too many variables to allow easy calculation. Typical variables are internal backflow (slip), cavitation, and friction (dry and viscous). However, pumps of similar design tend to follow the same pattern of losses, and these can be estimated with "averaged" data (Fig. 3.8).

Volumetric efficiency—an important part of overall efficiency—is the ratio of delivered flow to ideal flow. Delivered flow $Q_d = E_V DN$, where E_V is volumetric efficiency, D is displacement, and N is pump speed. Volumetric efficiency depends on the shape and clearance of rotating and stationary elements, the suction characteristics of the pump, the viscosity of the liquid, and the speed and pressure of the pump.

If clearances are small and there is good dynamic sealing between the rotor and stator, internal backflow will be small and volumetric efficiency high. But this assumes the pump inlet is kept pressurized and full, so cavitation (see the next section) does not reduce delivery.

The viscosity of the liquid has a pronounced effect on volumetric and overall efficiency and also on friction load and cavitation. The effects are opposing: High viscosity helps reduce slip (improving the volumetric efficiency) but increases friction load and hinders suction. Low viscosity improves suction and decreases friction load but increases slip (lowers volumetric efficiency).

Slip (internal backflow) in gpm tends to conform to this formula for viscous flow between two flat plates: $Q_S = K \times (\Delta P/\text{viscosity}) \times (\text{width}/\text{length}) \times (\text{clearance})^3$, where K is the proportionality constant. Note that both the width and length of the clearance path can be increased without changing the absolute slip flow—if the width-to-length ratio and clearance are held constant. (See Chap. 9 under "Leakage Through Annular Clearances" for more details.)

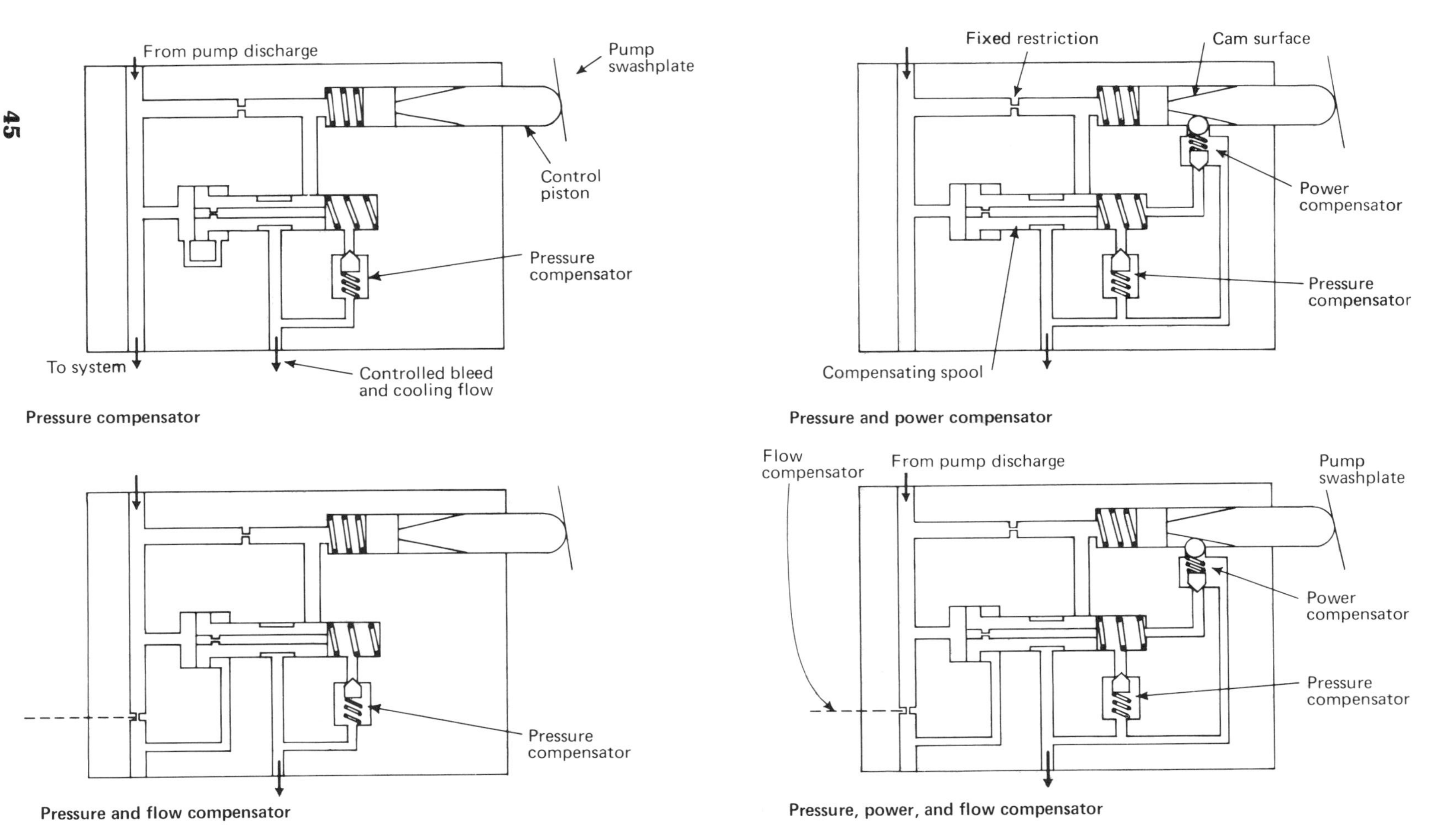

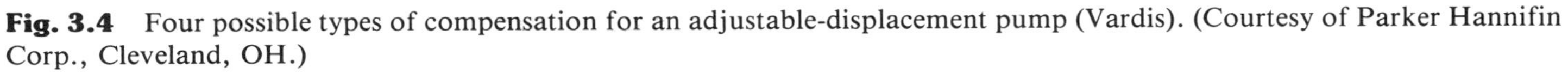

Fig. 3.4 Four possible types of compensation for an adjustable-displacement pump (Vardis). (Courtesy of Parker Hannifin Corp., Cleveland, OH.)

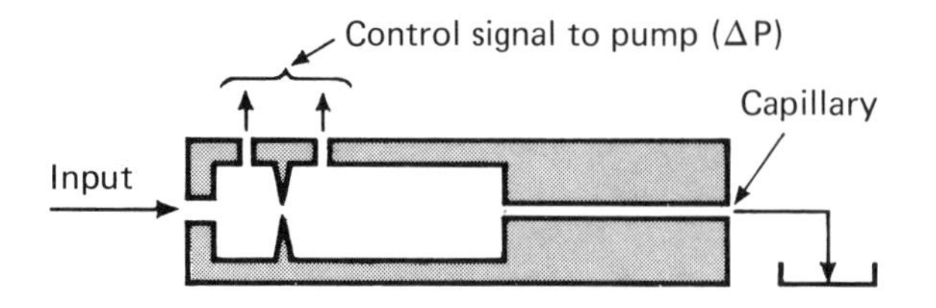

Fig. 3.5 Control signal for cold start of pump is automatically reduced by the capillary.

Actual pumps, however, do not have predictable leakage-path dimensions, and each pump must be tested separately to determine slip and volumetric efficiency. A convenient pump-slip equation, based on flat-plate flow, is $Q_S = (C_S D) \times (\Delta P/\mu)$, where C_S and D depend on the effective width-length-clearance relationship and pump size, and μ is fluid viscosity.

Note that slip is a function of pressure, not speed. Doubling speed does not double slip; slip flow stays about the same, so long as discharge pressure does not change.

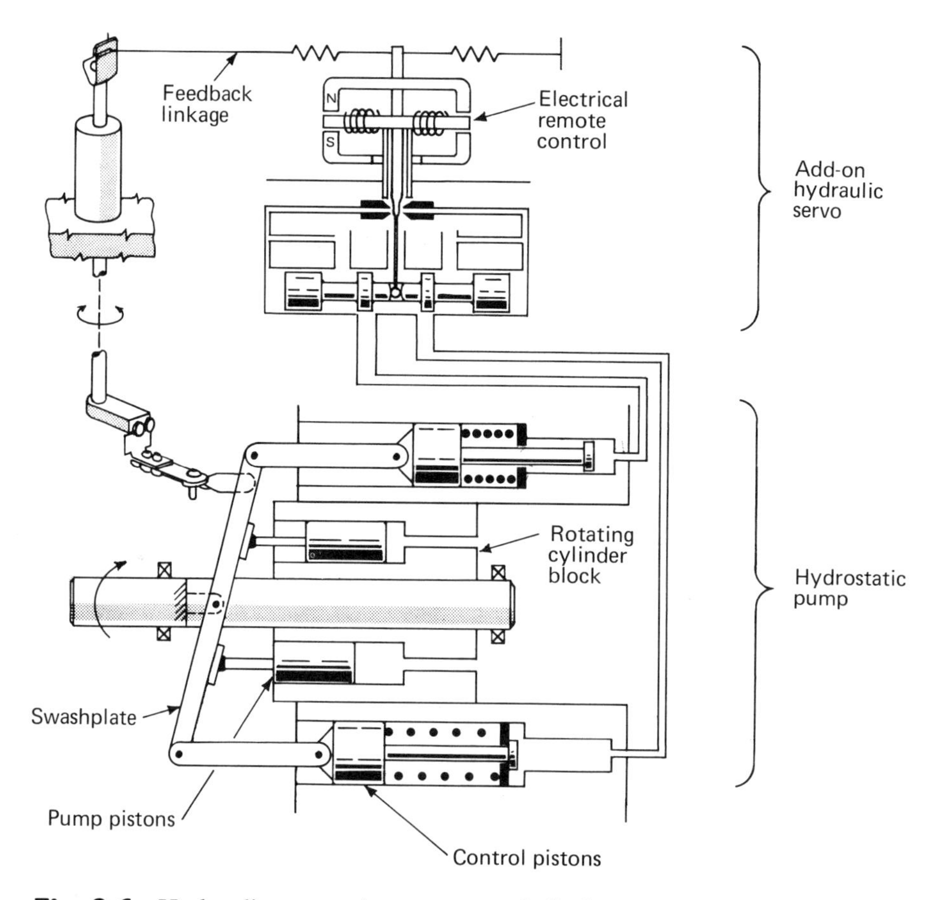

Fig. 3.6 Hydraulic servovalve can control displacement of pump.

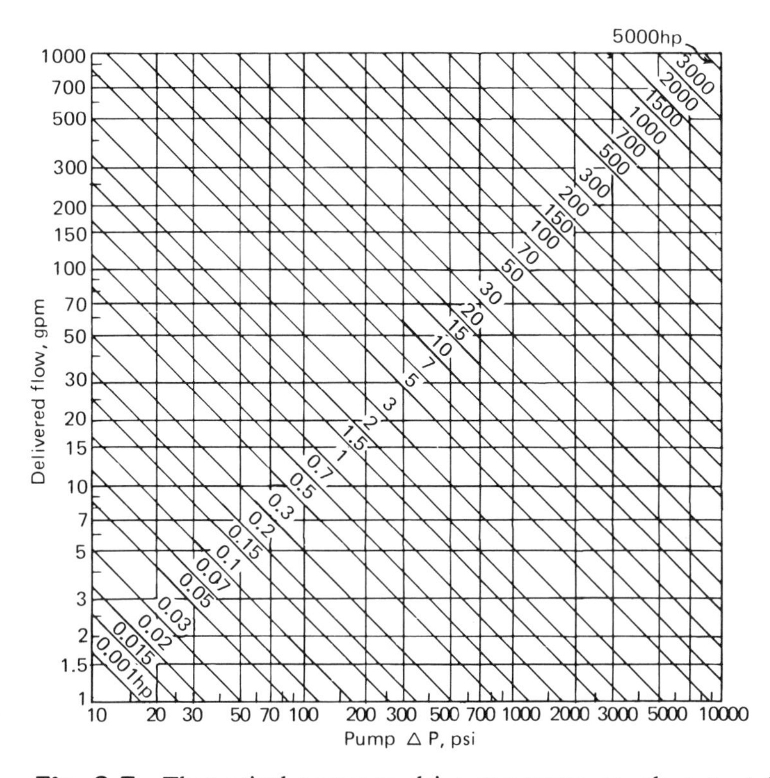

Fig. 3.7 Theoretical power to drive any pump equals output fluid horsepower, $5.82 \times 10^{-4} Q_d \, \Delta P$, plotted here. Actual power required to drive the pump is chart hp divided by overall efficiency.

Percent slip—slip per gpm delivered—actually becomes one-half the original value, because delivery flow doubles with speed. Also note that doubling viscosity halves slip.

Slip limits the maximum pressure of a positive-displacement pump. Assuming no cavitation, delivery flow $Q_d = DN - Q_S = DN - C_S D \, \Delta P/\mu$. Rearranging terms, pressure rise $\Delta P = \mu(DN - Q_d)/(C_S D)$. The maximum pressure (developed when $Q_d =$ 0) is $\mu N/C_S$.

All these effects—slip, friction, viscosity, speed, and pressure—have been taken into account in the averaged efficiency charts (Fig. 3.8). But individual pumps vary in design and workmanship. Hypothetical examples, using the graphs, are shown in Fig. 3.9.

Suction-Head Chart (NPSH)

For each condition of flow and speed there is a minimum value of inlet total pressure (static plus velocity head) that will prevent cavitation (Fig. 3.10). This minimum inlet head, based on P_v as the datum, is known as required net positive suction head (NPSH) and depends on the viscosity of the liquid, the square of the velocity (V^2) of the pumping element, the vapor pressure of the liquid, the design of the inlet flow path, the smoothness of the pumping cycle, and the percentage of dissolved air. Experience shows these combined effects tend to vary mostly with V^2. The simplified suc-

Table 3.1 Nomenclature and Equations for Pump Performance

Important pump equations

Pressure	$\Delta P = P_{out} - P_{in}$
Flow	$Q_d = Q_i - Q_S - Q_C = E_v Q_i$ $Q_i = ND$ $Q_S = C_S D \Delta P / \mu$
Torque	$T_m = T_i + T_F$ $T_i = 36.8\ D\Delta P$
Power	$hp_{in} = 1.585 \times 10^{-5}\ TN$ $hp_{out} = 5.82 \times 10^{-4}\ Q_d \Delta P$
Efficiency	$E = hp_{out}/hp_{in} = E_v\ E_T$ (varies with $\mu N/\Delta P$) $E_v = Q_d/Q_i$ $E_T = T_i/T_m$

Symbols for equations

Pressure and head		Torque	
P	Pressure, psi (P_A = atmos; P_v = vapor pressure)	T	Torque, lb-in. (subscripts: m = driving; i = ideal; F = friction, dry and viscous)
H	Head, ft of liquid		
Z	Elevation of inlet above pump centerline, ft		
NPSH	Net positive suction head, ft		
Flow and displacement		**Power and efficiency**	
Q	Flow, gpm (subscripts: d = delivered; i = ideal; S = slip; C = cavitation)	hp_{in}	Power to drive pump, hp
		hp_{out}	Power output (liquid), hp
N	Speed, rpm	E	Overall efficiency, % (subscripts: v = volumetric; T = torque)
V	Velocity, ft/sec (subscripts: 1 = inlet; 2 = pumping chamber)		
D	Displacement, gal/rev		
Coefficients and constants			
C_S	Coefficient of slip	SSU	Kinematic viscosity, Saybolt seconds, universal
g	Gravity constant, 32.2 ft/sec^2		
μ	Absolute viscosity, slugs/(ft × sec) or $lb_{force} \times sec/ft^2$	ρ	Mass density, slugs/cu ft
		R	Outlet restriction (pump load), psi per gpm

tion-pressure chart is based on V^2 and is fairly accurate for all positive-displacement pumps handling moderate liquid viscosities (up to 200 SSU). More detailed equations are given in the section on cavitation following.

To avoid cavitation, available NPSH must be greater than required NPSH. For example, if the liquid reaches the pump inlet with zero velocity and is at a pressure equal to or less than its vapor pressure, it cannot be forced into the pump. The same is true if its total inlet pressure (velocity head plus static head) is less than that of fluid adjacent to the moving pump element (Fig. 3.11).

Liquid at point 1 must enter the pumping chamber at point 2. Between 1 and 2 there may be gear teeth, vanes, rotor, piston, or other pump elements. At point 1 the liquid has velocity V_1 and pressure P_1. Energy (total head) per unit mass available to move liquid into the pump is the sum of its kinetic energy $V^2/2g$ and pressure (or potential) energy $P_1/\rho g + Z$, minus its vapor-pressure head $P_V/2\rho g$. Thus available $NPSH = V_1^2/2g + (P_1 - P_V)/\rho g + Z$.

If there were no losses of energy between points 1 and 2 and velocity V_1 was exactly equal to the required velocity V_2 at point 2, liquid would flow readily into the pumping chamber even when $P_1 = P_V$. In practice, many pumps have considerable loss in energy between 1 and 2—while others actually add energy to liquid in the inlet region by shearing action and other means. The required $NPSH = V_2^2/2g + H_L - H_G + P_V/\rho g$, where H_L is head loss and H_G is gain.

Available NPSH for piston pumps varies during each suction stroke, because flow pulses somewhat, depending on design. At the start of the stroke, fluid is at rest (or has a relatively low velocity), and available NPSH is reduced by the head needed to accelerate the fluid. At the end of the stroke, no acceleration is required, and available head depends on friction losses in inlet piping and passages. If inlet piping has an unusual number of fittings, the end-of-stroke (high-friction) condition is important in determining the NPSH available. The suction-head chart, however, is based on V^2 of the pumping element—the more usual criterion.

Dimensionless Cavitation Equations

There is another method related to NPSH. Thorough tests on vane and gear pumps have indicated a consistent relationship between type of pump and the amount of acceleration required by the fluid inside the pumping cavity.

It is this relation that leads to the simplified equation

$$\left(\frac{Q_L}{Q_T}\right) = 1 - \left(\frac{P_A}{\rho C V_p^2}\right)\left(\frac{P_i - P_V}{P_A}\right)$$

The terms (see Table 3.2) are kept dimensionless; Q_L/Q_T indicates loss in flow caused by cavitation, $(P_i - P_V)/P_A$ is a measure of static pressure available at the inlet to accelerate fluid up to the velocity of the pumping element, and $P_A/\rho C V_P^2$ is a coefficient that includes effects of density and acceleration. This coefficient will usually remain fairly constant at fixed speed for low values of accelerating pressure. The equation is the basis for Figs. 3.12 and 3.13. The "probability of cavitation" curves, for example, were plotted by letting $Q_L/Q_T = 0$ (zero cavitation) and solving for $(P_i - P_V)_0$.

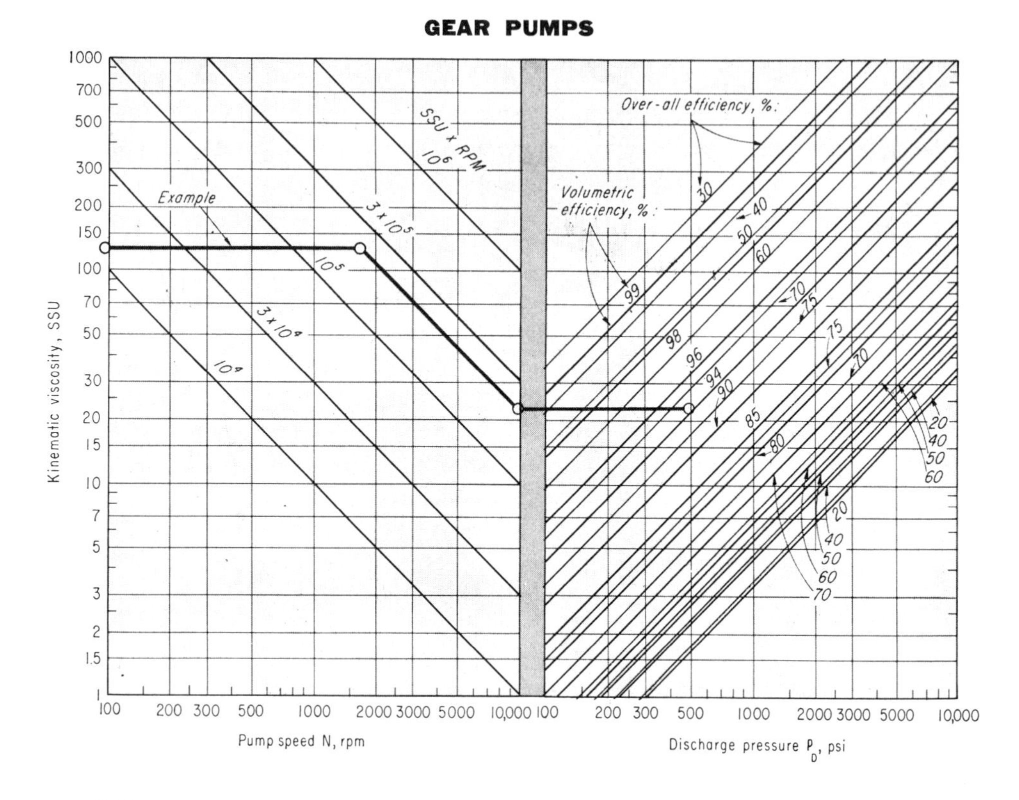

OVER-ALL EFFICIENCY and volumetric efficiency of gear, vane and piston pumps can be quickly estimated with these three nomographs—based on actual data from many pump tests. Plotting method assumes slip coefficient C_s is constant for each type pump, and $Q_s = C_s D \Delta P/\mu$. Dry friction is neglected because a well-designed pump has very little. Viscous drag is assumed proportional to speed and viscosity. Value for discharge pressure assumes inlet at atmospheric pressure. Oil used has specific gravity of 0.8, and is at normal operating temperatures.

Accuracy of these charts is probably little better than ±20% for most pumps because manufacturing and design variations are great. Better accuracy is possible if one or more performance points are known—effects of speed, viscosity and pressure can be easily predicted using charts.

Here is an example: oil at 130 SSU is pumped to 500 psi with 1800-rpm gear pump. What is probable over-all and volumetric efficiency? Answer: 94% volumetric, 72% over-all. Note: raising viscosity to 500 SSU will increase volumetric efficiency to over 98%, lower over-all to 45%. Best efficiency point can be found by trial and error.

Another example: pick gear pump to pump 1000-SSU liquid to 600 psi discharge. The best over-all efficiency is 76%—coinciding with horizontal line through SSU × rpm = 1.35×10^5. Following this line to left and upward to intersection with viscosity = 1000 SSU, best pump speed is seen to be 135 rpm. If 1000 rpm, 600 psi, and 1000 SSU are investigated, corresponding over-all efficiency is 47%, volumetric 98%. This is a decrease in over-all efficiency of 29% due to increase in speed and consequent increase in viscous drag. Cost of installing pump large enough to give required delivery at 135 rpm might offset the gain in efficiency. Compromise is intermediate speed of 600 rpm with 60% over-all efficiency.

Fig. 3.8 Quick calculation for performance of gear, vane, and piston pumps. SSU (or SUS) means Saybolt Seconds Universal; see Chap. 1.

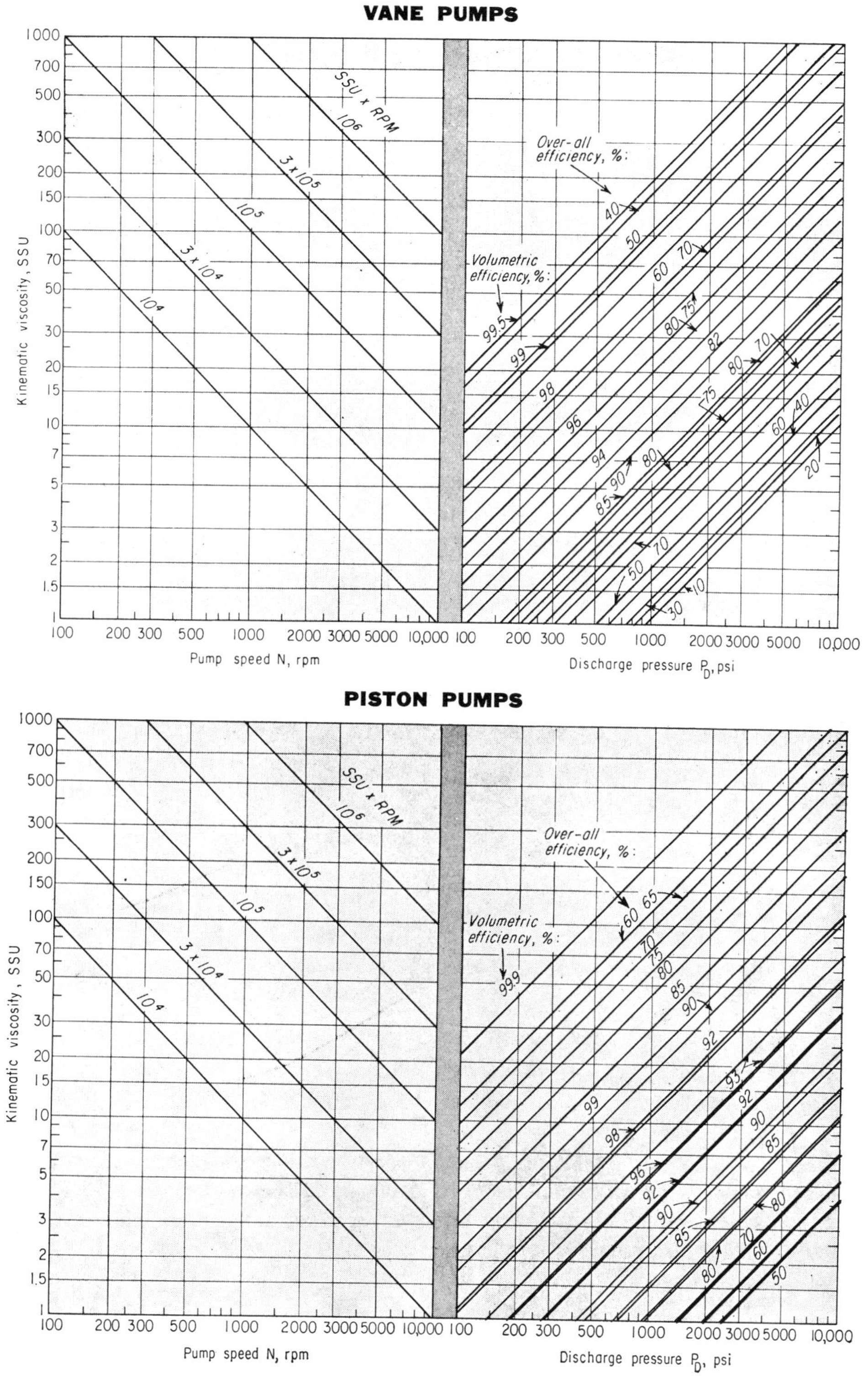

Fig. 3.8 (Continued)

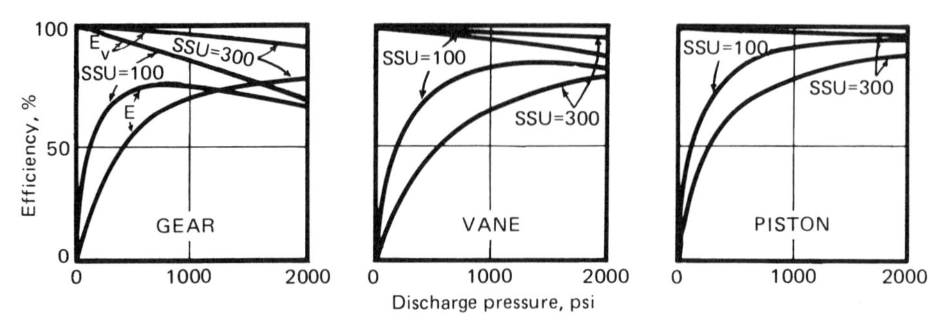

USING THE EFFICIENCY CHARTS to plot comparative performance of gear, vane, and piston pumps is easy—just follow lines on chart and read answers directly. These examples are for 1800-rpm pumps, but other speeds are evaluated just as readily. Over-all efficiency will reach peak value at a different pressure for each speed.

Fig. 3.9 Examples using the quick calculation method from Fig. 3.8.

Deriving the Equation. A rigorous analysis of cavitation in an actual pump is not possible, but the idealized sketch of pumping elements shows important effects and provides groundwork for a workable equation (Fig. 3.14).

Cavitation results if the liquid cannot keep up with the moving element and occurs between positions 1 and 2. Fluid pressure P_1 provides the only energy available to accelerate the fluid from velocity V_1 to V_2. The value of pressure necessary is directly proportional to the pump speed V_P. The fluid velocities inside the cavity are not readily measurable, but this general equation, based on $F = Ma$, is valid:

$$(P_1 - P_2)A_P = \rho Q_D(V_2 - V_1)$$

Tests have shown that each pump has a characteristic value for the ratio $C = (V_2 - V_1)/V_P$. Substituting CV_P for $V_2 - V_1$ gives

$$(P_1 - P_2)A_P = \rho Q_D C V_P$$

Let $P_2 = P_V$ (nearly correct) and $P_1 = P_i$ (valid for severe cavitation because flow is low). The equation can now be expressed as

$$Q_D = \frac{(P_i - P_V)A_P}{\rho C V_P}$$

From the accompanying key to nomenclature (Table 3.2), $Q_D = Q_T - Q_L$ and $Q_T = A_P V_P$. Introducing these factors and P_A gives the cavitation equation, which is a more useful tool.

Choosing Inlet Configuration. It is not enough to specify pump type. For any pump, suction performance will be improved by guiding and accelerating the fluid before it reaches the pumping chamber. It is much more efficient to control the fluid in a smooth inlet channel than in the pumping cavity.

Some positive-displacement rotary pumps prerotate the fluid by having a hollow rotor shaft serve as the inlet. Radial flow through special passages guides the fluid and pressurizes the pumping cavity.

Another type of rotary pump takes leakage flow from near the discharge, controls the pressure, and reduces cavitation by feeding it back to the inlet when necessary.

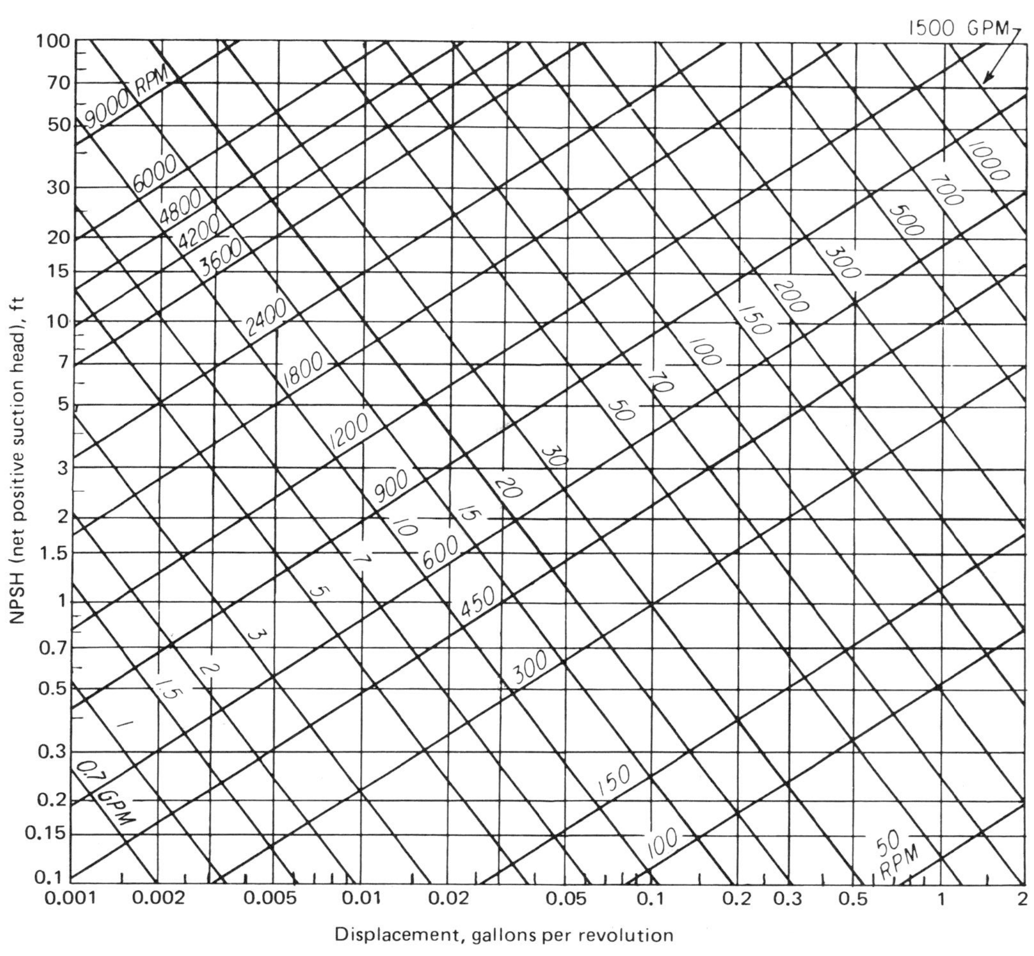

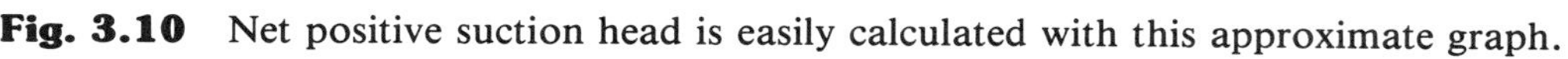

NET POSITIVE SUCTION HEAD here is based on square of velocity of pumping element, although other factors strongly influence suction requirements (see text). An example: a given pump has 10-gpm flow, runs at 1800 rpm (it can be gear, vane or piston type—the basic rules still apply.) What is NPSH? Answer: 5.2 ft of liquid pumped, absolute pressure.

Fig. 3.10 Net positive suction head is easily calculated with this approximate graph.

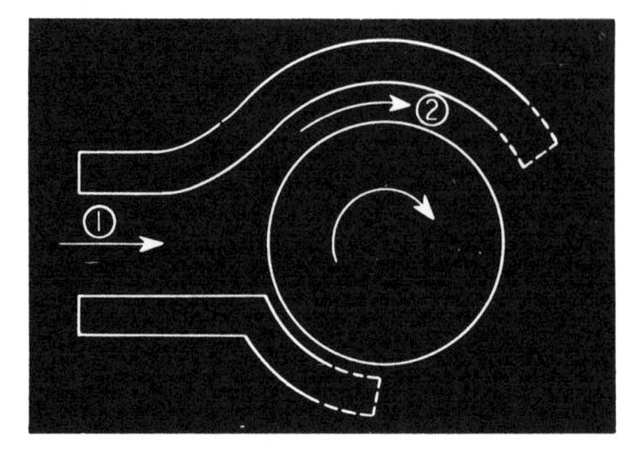

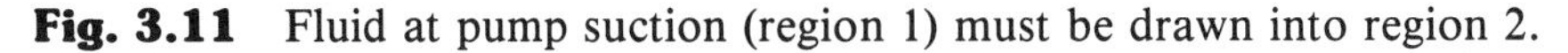
Fig. 3.11 Fluid at pump suction (region 1) must be drawn into region 2.

Limitations to the Method. Fluid that is cavitating may not flash to vapor as completely as theorized. In such cases, the pump gives better performance than predicted because back pressure is less. Conversely, fluids with vapor pressures higher than the inlet pressure can prevent pumping entirely—and pressurization or cooling of the inlet may be required.

Air or gas entrainment will increase cavitation, particularly if the inlet pressure drop is great, because trapped air expands and occupies pumping space. The simplified equation assumes air has not been entrained or dissolved and that the theoretical vapor pressure of liquid is reached.

Table 3.2 Nomenclature for Pump Cavitation Equations

- Volume flow, cu ft/sec
  - Q_T theoretical pump discharge, $A_P V_P$
  - Q_L loss due to cavitation
  - Q_D actual pump discharge, $Q_T - Q_L$
- Pressure, psf (1 in. hg = 70.7 psf)
  - P_i pump inlet
  - $P_{1,2}$ pumping cavity
  - P_A atmosphere
  - P_v vapor pressure of fluid
- Velocity, ft/sec
  - V_P piston, vane or tooth
  - V_i pump inlet
  - $V_{1,2}$ fluid in pumping cavity
- Other key units
  - A area, sq ft
  - ρ mass density, slugs/cu ft
  - g 32.2 ft/sec^2
  - C cavitation factor $(V_2 - V_1)/V_P$

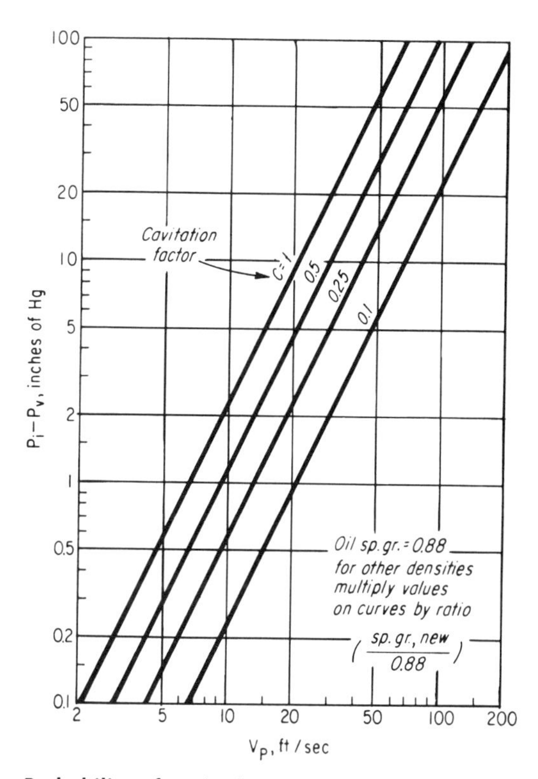

Probability of cavitation . . .
when oil is the fluid can be quickly predicted with this graph that relates effective inlet pressure to pump velocity. Cavitation factor C is a measure of the acceleration required by fluid to reach velocity of the moving element. High values of C mean fluid is initially at standstill—requires max acceleration. Low values indicate fluid has initial velocity that helps get oil into pumping cavity.

Value of C depends on inlet design as well as pump type. Vanes-in-stator pump shown in Fig. 3.2 gives pre-rotation to fluid because of centerline inlet—can have C as low as 0.25. Screw pumps may have value of 0.3, but spur gear and vanes-in-rotor pumps do no better than 0.4. Piston pumps will be poorer (0.7). Any pump can approach worst value (1.0) if inlet design is not correct.

EXAMPLE 1—hydraulic system has only 4 in. hg available at pump inlet; vapor pressure is 1 in. hg. Therefore, $P_1 - P_v = 3$ in. hg. Plot for $C = 1$ shows any pump with impeller velocity less than 11 ft/sec will operate without severe cavitation. A pump that has $C = 0.5$ can be run with V_P as high as 16 ft/sec.

EXAMPLE 2—if pump type, speed and size are known, a velocity reading on horizontal scale can give min permissible value of $P_1 - P_v$ for system on vertical scale.

Fig. 3.12 Predicting cavitation graphically.

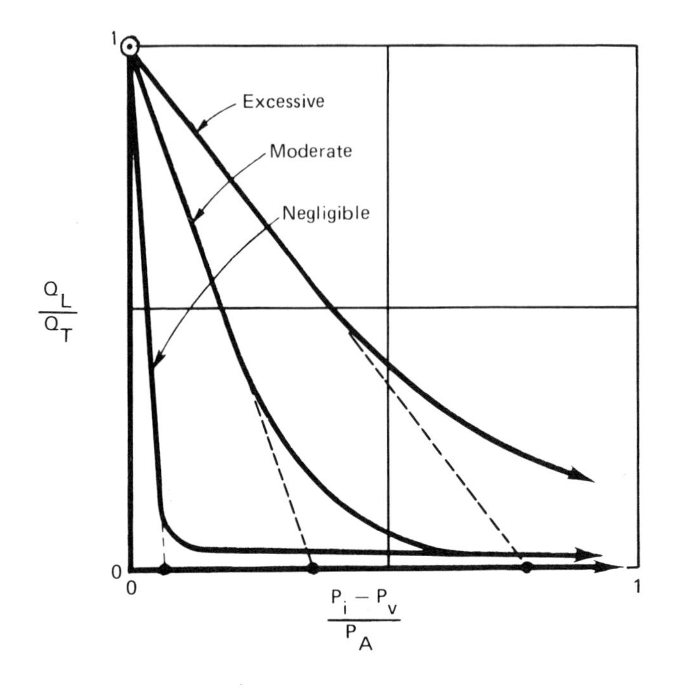

Nature of cavitation . . .
is seen in these curves plotted with dimensionless parameters. Q_L/Q_T, a measure of cavitation, varies almost linearly with available positive suction pressure in region of severest cavitation. A hypothetical pump with zero cavitation corresponds to curve on extreme left, where any pressure is sufficient to give full flow, $Q_LQ_T = 0$. Actual pumps will perform in area indicated by the three other curves. Curves are for pure cavitation (effects of entrained air and slip are not included).

Fig. 3.13 Dimensionless parameters show nature of cavitation.

Though temperature variations will affect vapor pressure, the slope of the curve Q_L/Q_T vs. $(P_i - P_V)/P_A$ remains the same. Any fluid system at any temperature can be examined by this method as long as there is a positive suction head.

Viscosities up to 2000 SSU do not affect performance, but above that, cavitation increases. Over 20,000 SSU, it may become necessary to pressurize or heat the fluid.

Any one or all of many separate effects can produce the family of Q_LQ_T vs. $(P_i - P_V)/P_A$ curves shown. For example, if speed is increased but other factors re-

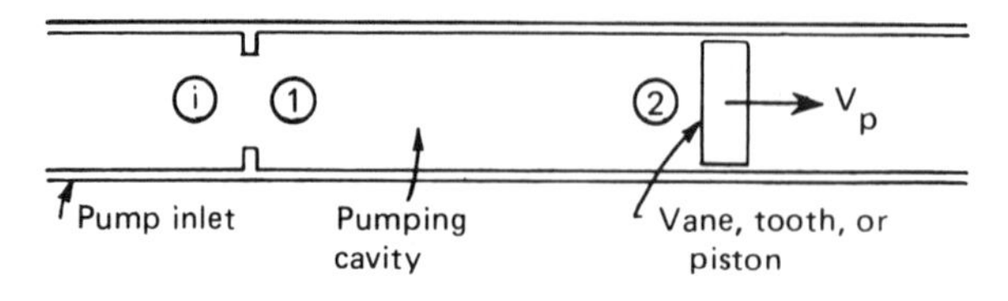

Fig. 3.14 Schematic for derivation of cavitation equation.

main the same, cavitation may become "excessive" instead of "moderate." But a severe inlet restriction or entrained air will give similar results. The method therefore is a guide only—careful analysis and testing will still be necessary.

What about NPSH? If desired, the new cavitation equation can be related to the older concept which defines net positive suction head. This value (NPSH) is total inlet pressure less the vapor pressure of the fluid:

$$\mathrm{NPSH} = \frac{P_i}{\rho g} + \frac{V_i^2}{2g} - \frac{P_V}{\rho g}$$

The relationship can be seen in the plot Q_L/Q_T vs. $(P_i - P_V)/P_A$ at the intersection of the straight portion of the curve extended to the horizontal axis. Here Q_L/Q_T is 0, and

$$\left(\frac{P_i - P_V}{P_A}\right)_0 = \left(\mathrm{NPSH} - \frac{V_i^2}{\rho g}\right)\frac{g\rho}{P_A}$$

Pump Pulsation

Freedom from pulsation—often a primary requirement—depends on geometric (shape), kinematic (velocity and flow path), and dynamic (pressure and force) considerations.

Little can be predicted about the pulsation problem in selecting a pump. Where the problem might be serious, choose a pump with very small slip (very high volumetric efficiency.) This is a good indication of fine workmanship and uniformity of clearances.

Reciprocating pumps inherently have periodic impulses. However, multipiston pumps reduce the magnitude of these variations. Gear, vane, and other rotary positive-displacement pumps have somewhat less pulsation. Theoretically, pumps can be built which eliminate the kinematic and geometric factors contributing to pulsation, but basic design and construction problems prevent elimination of pulsation completely.

Pulsation, although appearing as a fluctuation in pressure, actually arises as a fluctuation in flow rate because pressure is a dependent quantity. Here is why:

Discharge Q_d of a positive-displacement pump $= DN - C_S\Delta PD/\mu$. The system into which liquid is pumped will have some resistance R, where $R = \Delta P/Q_d$ or $Q_d = \Delta P/R$. (This is analogous to Ohm's electrical law, $I = E/R$.)

Equating, $Q_d = \Delta P/R = DN - C_S\Delta PD/\mu$. Noting that $\Delta P = P_{out} - P_{in}$, it can be shown that this relationship between pressure, flow, and discharge resistance applies:

$$P_{out} = (DN)/(1/R + C_S D/\mu) + P_{in}$$

If N, R, C_S, and P_{in} remain constant, the discharge pressure will be constant. But C_S depends on the cube of the pump internal clearances and can vary during each revolution, thus introducing large flow fluctuations. Flow and pressure fluctuations are directly related: $\Delta(P_{out})/P_{out} = \Delta(Q_S)/Q_i$, where Δ indicates a small change.

A byproduct of pulsation is noise (Fig. 3.15). Pump speed has the greatest effect; pressure and swept volume have lesser effects.

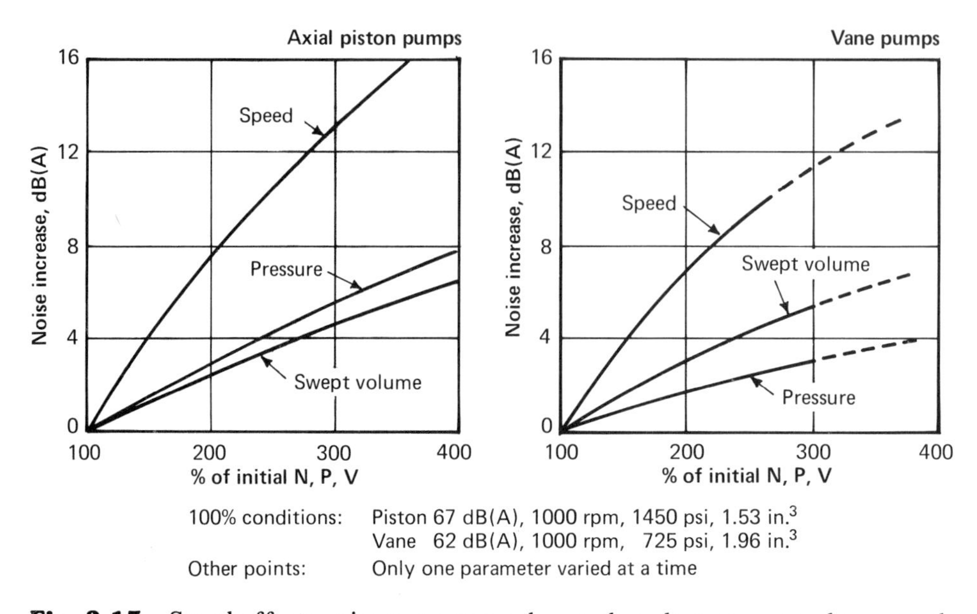

Fig. 3.15 Speed affects noise to a greater degree than do pressure and swept volume. Some designers pick 1500 rpm as the top practical speed for quiet pumps. (Courtesy of Rexroth Corp., Bethlehem, PA.)

EXAMPLES OF PUMP SELECTION

Liquid Transfer

Required: to pump 50 gpm of 100-SSU oil (70 F) to pressure of 250 psi. The motor drive will be at 1800 rpm. Inlet conditions are good, and the available pressure is at least 13 psi.

Solution: A large number of suitable pumps are available; however, piston pumps are somewhat expensive, and the low-pressure service does not justify them. A lower-cost turbine pump can also perform in this range but has low efficiency—less than 50%. Vane, gear, or lobe pumps seem indicated.

A power-output nomograph shows that the output hp is 7.2. Assuming an overall efficiency of 70%, which is reasonable for either a vane, gear, or lobe pump, the motor should have at least 10 hp. (A 50%-efficient turbine pump would require about 15 hp.) A gear pump would perform well and be quite economical. Its life would be long because the liquid being pumped has lubricating qualities, and service is not severe. A vane pump would be somewhat more expensive, have slightly higher efficiency, and perform well. If efficiency is not the primary need, choose a gear pump.

High Lift

Required: transfer of viscous liquid from a submerged tank 18 ft deep (the lowest depth is 18 ft, but the surface is essentially at ground level). The pump is to be placed at ground level. The vapor pressure of liquid at ordinary temperatures is 2 psia. Flow is 150 gpm; discharge, 550 psi; and viscosity, 1000 SSU at 70 F. Motors with speeds of 1800 or 2700 rpm are readily available; slower speeds will cost more.

Solution: Gear, vane, and piston pumps are available, but a suction-head nomograph indicates that the NPSH required is 32 ft of water. With an actual viscosity of 1000 SSU and an available NPSH of something in the order of 10 ft of liquid with a pump operating at 1800 or 2700 rpm, no pump is adequate. The manufacturer should be consulted, because the cavitation problem is serious. There will be little competition on an economic basis—only a few pumps are available, and none operate at speeds of 1800 or 2700 rpm. However, a pump with a displacement of 0.25 gal/revolution running at 600 rpm could handle the situation.

Hydraulic Control Pump

Required: 1500 psi, 25 gpm, viscosity 150 SSU at 100 F. Motor to run at 1800 rpm. There are no problems regarding inlet pressure.

Solution: Three types—gear, vane, and piston—are available. Therefore selection must be based on some special characteristic or on costs. If pulsation is a serious problem, a piston pump with its periodic-flow characteristic might not be suitable, but a well-designed vane or gear pump will have smooth flow free from pulsation.

High-Pressure Flow

Required: flow of 50 gpm at 4500 psi.

Solution: Only piston and gear pumps are available—and very few manufacturers are producing pumps for this service. However, inlet conditions are good, the viscosity of oil is 150 SSU at 100 F, and the motor runs at 1800 rpm. The gear pump is cheaper and is less likely to cavitate. But efficiency is higher for the piston type. Either pump meets the requirements.

High Altitude

Required: to pump fuel at 3600 rpm, with a flow of 100 gpm at 750 psi and viscosity at 30 SSU. The pressure available at the inlet is 3.5 in. of mercury absolute, and the vapor pressure is 2 in. Many pumps will deliver 100 gpm at 750 psi, but no standard pump will deliver satisfactorily under the inlet conditions prescribed.

Solution: A specially designed pump, free from cavitation even though the energy available at the inlet is extremely small. Any significant drop in pressure between the inlet and the pumping elements is intolerable because vapor pockets would form and cause serious cavitation.

SQUEEGEE PUMPS

How do you predict the performance of a pump that does part of its work pumping and the rest squeezing and unsqueezing itself? And why use such a pump?

Three common pump types—flexible vane, flexible liner, and flexible tube—fit that description. We'll call them squeegee pumps (Fig. 3.16) because the name is often used and seems to fit.

Hundreds of tests were run to find a common denominator. The result is the final graph (Fig. 3.17) for predicting the performance of any squeegee pump. All you need are two test points, usually provided by the manufacturer, and you can quickly estimate flow and efficiency under any normal conditions of speed, pressure, and viscosity.

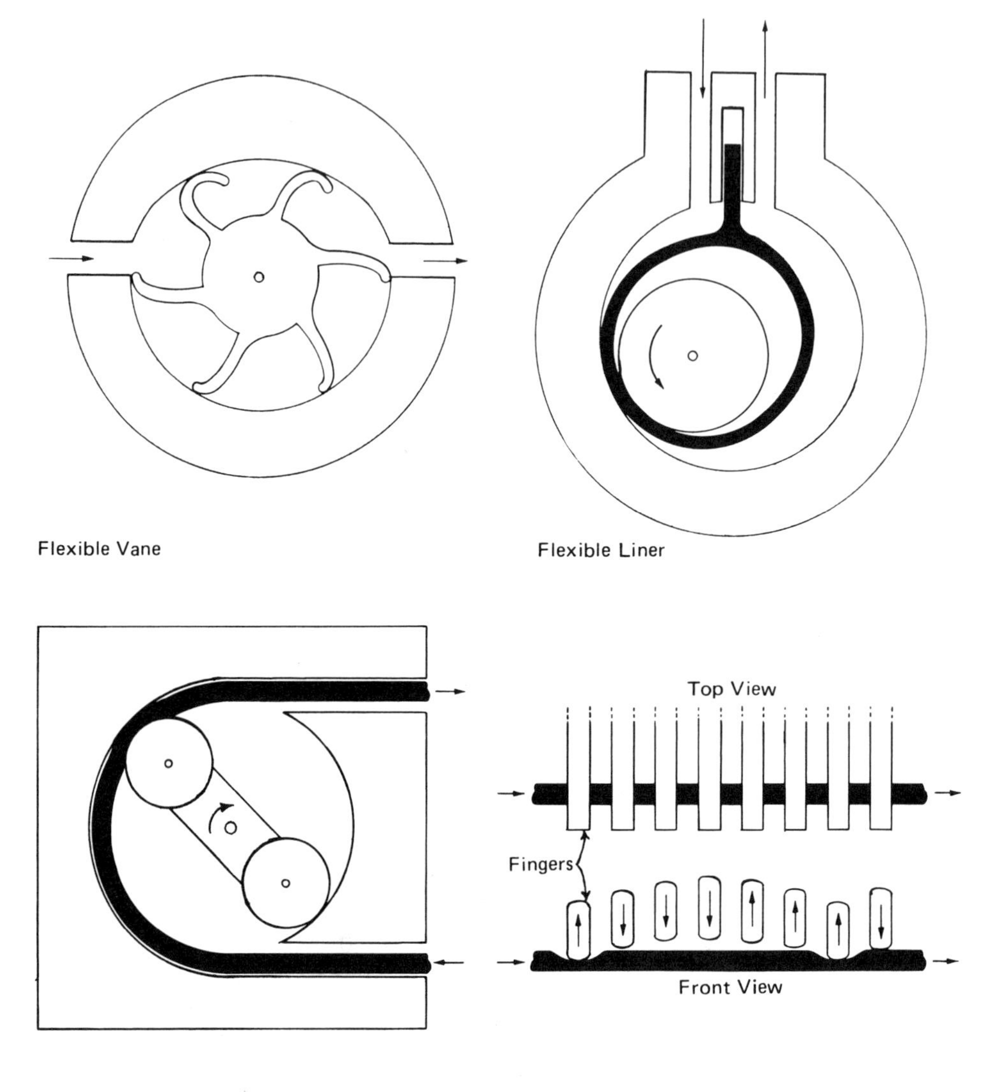

Fig. 3.16 Typical squeegee pumps.

The second question, why use a squeegee pump, is easy too. The elasticity gives each pump special advantages, even though flow calculations are less certain than those for conventional positive-displacement pumps.

For example, flexible vanes conform to the housing contour, pumping without the complexity of fitted vanes. Flexible liners eliminate vanes entirely, moving the fluid and at the same time sealing it from the mechanism. Flexible tubes are the ultimate in packless pumps: The fluid is squeezed like toothpaste and touches nothing but the inner wall. And all three types have low-cost, easily replaceable flexible elements which can pump dirty and corrosive fluids, even containing solids, without jamming the pump.

The parameters for calculating the overall performance of squeegee pumps are the same as those for positive-displacement pumps. Only the relative limits of performance differ. For example, pressure and speed will be lower because elastic parts are not as strong as rigid parts. And efficiencies are much less, but this is not usually a problem for such small pumps.

Test Results

The tests were run at room temperature and sea-level pressure with water and with glycerin (also spelled glycerine) as the fluids. At 70 F, water has an absolute viscosity of 1 cp and a kinematic viscosity of 0.000011 ft^2/sec. Glycerin at 70 F has an absolute viscosity of 1400 cp and a kinematic viscosity of 0.0115 ft^2/sec. Water is close to the "ideal" fluid, and glycerin represents an extremely viscous fluid.

Experimental data from all tests reveal clearly that with water, slip flow increases rapidly with pressure. With glycerin it increases much less rapidly. On the other hand, loss in delivery due to inlet restriction is much greater with glycerin and other high-viscosity liquids than it is with water. Torque requirements are significantly greater with glycerin than they are with water, resulting in lower overall efficiency. However, the increase in torque is not at all proportional to the increase in viscosity of the liquid itself. If it were, it would be essentially impossible to pump viscous oils with a positive-displacement pump.

Viscosities up to 2000 cp present no apparent difficulties. Between 2000 and 4000 cp, delivery is reduced in pumps with flexible liners and hoses. Above 4000 cp, special consideration must be given to the system design.

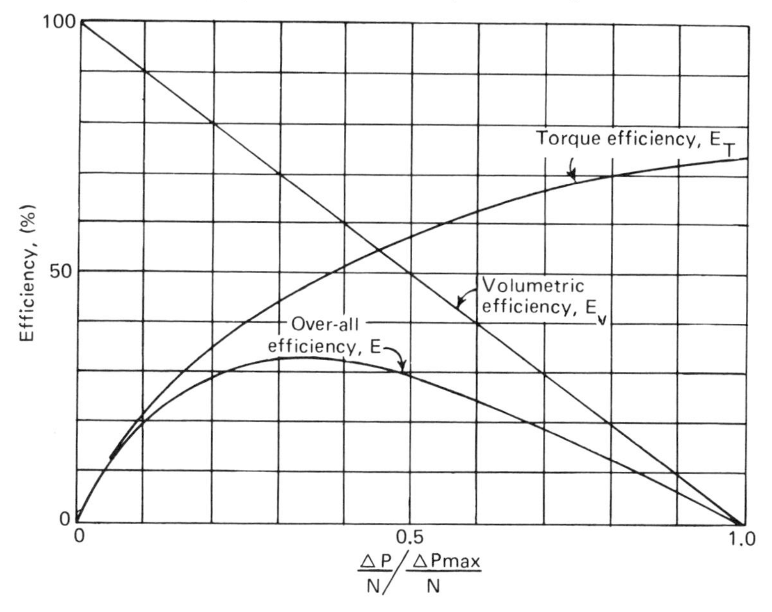

Fig. 3.17 Composite graphs help predict the performance of squeegee pumps.

Composite Performance Graph

Examination of all the data discloses a similarity in the shape of the performance curves. The maximum overall efficiencies are between 20 and 40% for each pump and seem to occur at a pressure nearly halfway between zero (open discharge) and shutoff (closed discharge) for a given rpm.

These observations and some known characteristics of positive-displacement pumps make possible a very crude but handy method of predicting performance with but the sparsest of published performance data. Figure 3.17, the composite performance graph, is the heart of the method. Here are the assumptions:

At any given operating speed, the pump discharge pressure will vary from zero (open discharge) to shutoff pressure (blocked discharge). The overall efficiency is highest, 33% for the composite, at rated delivery and pressure, which occurs at slightly less than half the shutoff pressure.

Volumetric efficiency is 100% when the discharge pressure is zero if there is no leakage or cavitation. Volumetric efficiency decreases linearly with pressure, reaching zero at shutoff pressure.

Torque efficiency, defined as the ratio of ideal driving torque to actual driving torque, is zero when pressure is zero and maximum when discharge pressure is maximum. The plotted curve is typical, ending at 75% shutoff efficiency.

Most important to the method is the ratio $(\Delta P/N)/(\Delta P_{max}/N)$, where ΔP is based on actual discharge pressure and ΔP_{max} is the shutoff pressure differential.

For example, if a pump is known to develop a maximum pressure rise of 45 psi at 450 rpm with blocked discharge, then the quotient $\Delta P_{max}/N = 45/450 = 0.10$ represents the maximum value of the abscissa in the composite graph. To determine its capabilities at 20 psi and 400 rpm, where $\Delta P/N = 20/400 = 0.05$ and $(\Delta P/N)/(\Delta P_{max}/N) = 0.5$, read the graph as follows: Overall efficiency = 30%, volumetric efficiency = 50%, and torque efficiency = 58%. This is very nearly the peak performance. Peak performance would be at 33% efficiency, or $(\Delta P/N)/(\Delta P_{max}/N) = 0.35$. Here, torque efficiency = 50%, volumetric efficiency = 65%, and $(\Delta P/N) = 0.035$. Note that efficiency stays relatively high (over 25%) from ratios 0.15 through 0.60.

Inlet conditions are of extreme importance in predicting performance. If the inlet pressure is less than 10 psia (5 psi below ambient), delivery will decrease. If the inlet pressure is 5 psia or less, the pump probably will not function satisfactorily. Between 5 psia and 10 psia, you can arbitrarily assume a linear variation of the delivery with absolute pressure. In other words, at 5 psia, delivery is zero, and at 10 psia, delivery is full, all at zero discharge pressure.

Speed in excess of some reasonable value, determined experimentally, affects delivery in much the same manner as reduction in inlet pressure: Higher delivery rates cause lower inlet pressures for the same given upstream conditions. One rule is to keep the inlet velocity less than 20 ft/sec.

Viscosity in excess of 500 cp at room temperature has an effect similar to that of a reduction in inlet pressure. A first approximation in determining the effect of viscosity is to assume a 25% reduction in volumetric efficiency for each 1000-cp increase in viscosity. Although this would indicate that volumetric efficiency is zero if viscosity is 4000 cp, it would not happen in most cases. But if viscosity is in that range, special design is in order.

Example of Predicting Performance

Suppose you want to pump 1.5 gpm of water against a 25-psi discharge with one of the squeegee pumps. Some information on the pump you selected indicates that the shutoff pressure is 40 psi at 300 rpm and that delivery is 1 gpm at zero discharge pressure and 200 rpm. Can the pump meet your requirements of 1.5 gpm and 25 psi? And at what speed should you run it?

Use the composite performance chart. The known value for $\Delta P_{max}/N = 40/300 = 0.133$. The best value for $(\Delta P/N)/(\Delta P_{max}/N) = 0.35$ based on peak overall efficiency. Thus $\Delta P/N$ ought to be about $0.35 \times 0.133 = 0.046$, or $N = \Delta P/0.046 = 25/0.046 = 540$ rpm. Torque efficiency is about 50%; volumetric efficiency, 65%; and overall efficiency, 33%.

Pump delivery is computed by converting known flow and speed to displacement, then applying the speed and efficiency factors. Known: Flow $Q = 1$ gpm at 200 rpm with open discharge. $D = Q/N = 1/200 = 0.005$ gal/rev. Delivery at 540 rpm and 25 psi discharge = ND × volumetric efficiency = $540 \times 0.005 \times 65\% = 1.76$ gpm, which is more than the requirements. Speed can be reduced slightly if you want, at some sacrifice in efficiency.

Torque to drive the pump is computed as follows:

$$T = \frac{\Delta PD}{2\pi} \times \frac{1}{E_T} = \frac{25}{2\pi} \times 0.005 \times \frac{231}{0.5} = 9.2 \text{ lb-in.}$$

Correspondingly, the input power is

$$Hp_{in} = 2\pi \times 540 \times \frac{9.2}{12(33{,}000)} = 0.079 \text{ hp}$$

The power output is

$$HP_{out} = 1.76 \times 231 \times \frac{25}{12(33{,}000)} = 0.026$$

yielding an overall efficiency of 33%, which is as assumed.

DYNAMIC-TYPE PUMPS

Centrifugal pumps (at least those that are most common, which are single-stage) rarely exceed 100-gpm flow or 100-psi discharge pressure. Larger-size single- and multistage types for pipelines and other applications are available but often are custom-made.

The output pressure of centrifugals depends on pump geometry and speed of rotation. It's best explained by so-called "specific speed."

The specific speed N_S of a centrifugal pump is the speed at which a geometrically similar pump would have to run to deliver 1 gpm against a head of 1 ft. Its formula is

$$N_S = \frac{N\sqrt{Q}}{H^{3/4}}$$

where Q = gpm pumped
N = pump speed, rpm

H = total head, ft

To visualize this, picture a pump reduced in size (Fig. 3.18) until it delivers a flow of 1 gpm against a total head of 1 ft.

Remember the fundamental rules of centrifugal-pump hydraulics: Capacity change is directly proportional to speed change, and head change is directly proportional to the square of speed change.

Now let's take a pump and shrink it down to get its specific speed. Assume a pump rated at 24 gpm, 300-ft total head, and 12,000 rpm. First we slow it down until it produces only 1-ft total head. We must reduce speed by the square root of the head, which means a cut to 17.31/300 = 0.0578 times 12,000, or 693 rpm. Capacity, too, will go down, to 0.0578 × 24 or 1.386 gpm.

But we have not yet changed the size of the impeller, merely slowed it down. Now shrink the impeller but maintain its proportions. To reduce capacity to 1 gpm and still pump against the 1-ft total head, keep all velocities the same. So reduce all areas down to 1/1.386 and all dimensions down to $1/\sqrt{1.386} = 0.85$. However, if we reduce the impeller diameter to 0.85 times its former size but still maintain the same peripheral velocity, to get the 1-ft total head, we must run it 1/0.85 times as fast.

We now have 1 gpm at 1-ft total head and 816 rpm. This 816 is the specific speed of the original pump.

As a check, take the original pump-design conditions, substitute them in the centrifugal-pump specific-speed formula, and see if we get the same answer:

$$N_S = \frac{12{,}000\sqrt{24}}{300^{3/4}} = 816$$

Figure 3.19 gives examples of specific speeds for various impeller designs from centrifugal to axial (propeller) and for a full range of flows.

Centrifugal pumps have a limited range of shaft speeds and cannot tolerate as much cavitation as other types. They perform best with low fluid viscosities and are better suited for transporting fluid than building up high pressures. Abrasive or dirty fluids can be easily handled by a properly designed centrifugal pump. However, in basic hydraulic power-and-control systems, pressure is the essence of operation, and positive-displacement pumps are usually recommended.

Propeller pumps (axial flow) and *mixed-flow* pumps (partly centrifugal, partly axial) also have output pressure dependent on geometry and speed of rotation and have the same limitations as centrifugals.

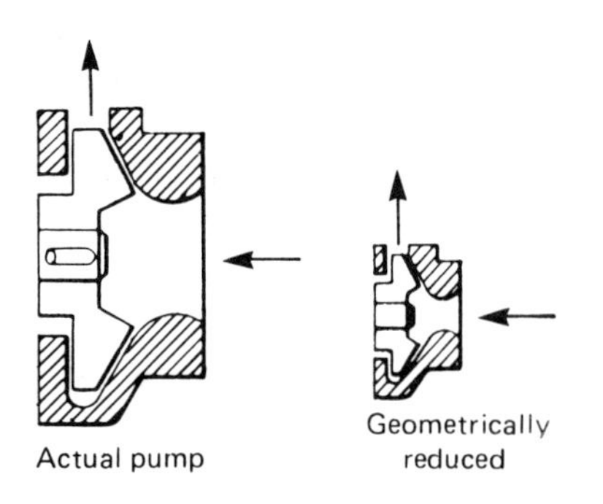

Fig. 3.18 Simple drawing helps explain specific speed concept.

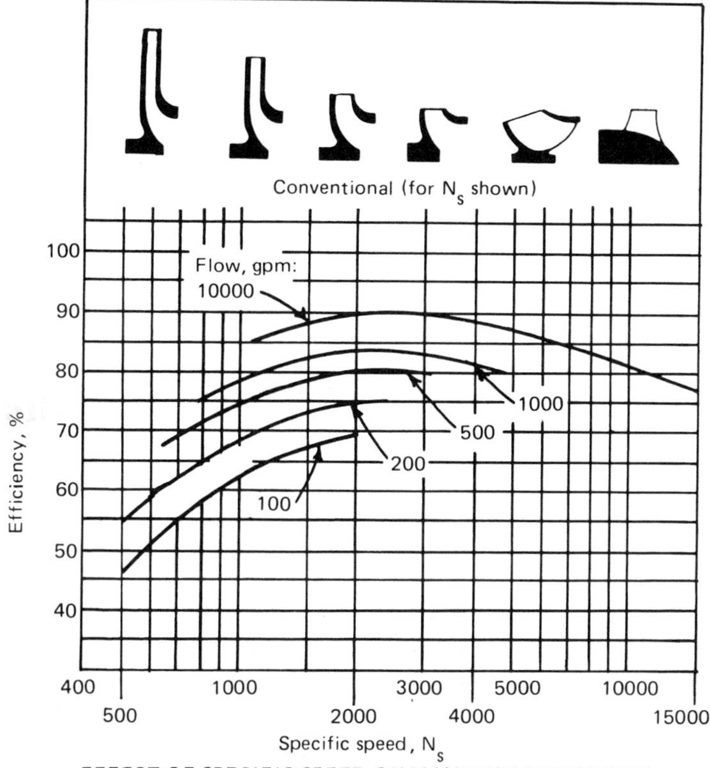

Fig. 3.19 Specific speeds, impeller shapes, flow, and efficiency of centrifugal (left), mixed-flow (center), and propeller (right) pumps have approximately the relationships shown here.

Turbine pumps, also known as regenerative, vortex, or periphery pumps, have a large number of radial vanes on the periphery and build up fairly high pressures by a succession of impulses in each revolution. They are similar in appearance and performance to the regenerative blowers described in Chap. 12.

The operating principle seems to be some form of momentum transfer of turbulent liquid. But performance is well known—discharge pressure responds somewhat linearly with speed, as it does in a positive-displacement pump. Discharge pressure is usually less than 300 psi in a single stage, but this is higher than that of a single-stage centrifugal. Low-viscosity fluids, such as water, gasoline, and alcohol—even at low suction pressures—can be easily handled.

Turbine pumps are not too efficient—about 40 to 50% overall—but their design simplicity overcomes that objection when efficiency is not a primary requirement.

Jet pumps, *pitot-tube* pumps, and other basic types are used for certain specialized applications. The jet type—common in well-pumping applications—entrains liquid from a low-pressure area into a jet of liquid that has high velocity head and low static head. A pitot-tube pump rotates a body of liquid in a cylinder and taps off total pressure via a stationary pitot tube inserted in the outer area where dynamic pressures are highest.

4

Hydraulic Valves and Circuits

Hydraulic valves for the control of systems, vehicles, or equipment usually are one or a combination of these three types: directional, pressure control, or flow control.

Physically, any valve is a controllable restriction, whether the working element be a flapper, ball, diaphragm, globe-disc, swing-disc, gate, needle, plug, poppet, shear-ball, shear-plug, shear-plate, butterfly, pinch tube, or spool.

Beyond those simple facts, valves become as complex as necessary to perform in a system. The other chapters in this book are filled with examples of valves applied to systems and equipment, and specific references will be made as necessary.

STANDARD DEFINITIONS

The American National Standards Institute (ANSI) glossary, developed under the sponsorship of the National Fluid Power Association (NFPA), defines all valves. The following explanations of valves and functions have been distilled from that glossary and apply to oil or air.

A *directional* valve directs or prevents flow through selected passages. (Descriptive schematics appear in Chap. 13.)

A *check* valve is a directional valve that allows flow in one direction only. A *straight-way* (also called a two-way) has two ports: one in and one out. A *selector* (diversion) valve interconnects two or more ports.

A *three-way* is a directional valve that pressurizes and exhausts one port. The "three" stands for the supply port, the exhaust port, and the outlet or working port that directs flow to the working device such as a cylinder, motor, or power valve.

A *four-way* directional valve pressurizes and exhausts two ports. It will have at least a supply port, an exhaust port, and two outlet ports. Many have an extra exhaust port for convenience and are called four-way, five-port directional valves. The ports in some versions can be reassigned, creating dual supply ports and other combinations.

Pressure control valves may control either upstream or downstream pressure.

For example, a *counterbalance* valve (Fig. 4.1) maintains back pressure in a system to prevent a load from drifting or falling. A *pressure reducer* controls downstream pressure. A *relief* valve (Fig. 4.2) limits upstream pressure. An *unloading* valve bypasses pump flow to reduce pressure to reduce load.

A *flow control* valve meters flow. A simple example is a needle valve; a complex example is a pressure-temperature compensated valve that controls the rate of flow independently of system pressure and temperature. A *deceleration* valve gradually reduces flow to slow down a cylinder or motor. A *flow divider* directs the flow from a single source into two or more branches.

Special-function valves abound. A *priority* valve directs flow to one operating circuit to satisfy the flow demand and directs excess flow to another operating circuit. As an example, a priority valve in a steering system makes sure that full steering control is available before allowing oil to flow into a lower-priority circuit such as a backhoe system.

Sequence valves (Fig. 4.3) direct flow to each circuit in a predetermined sequence. A *shuttle* valve directs flow to the outlet from whichever inlet is pressurized. A *surge-*

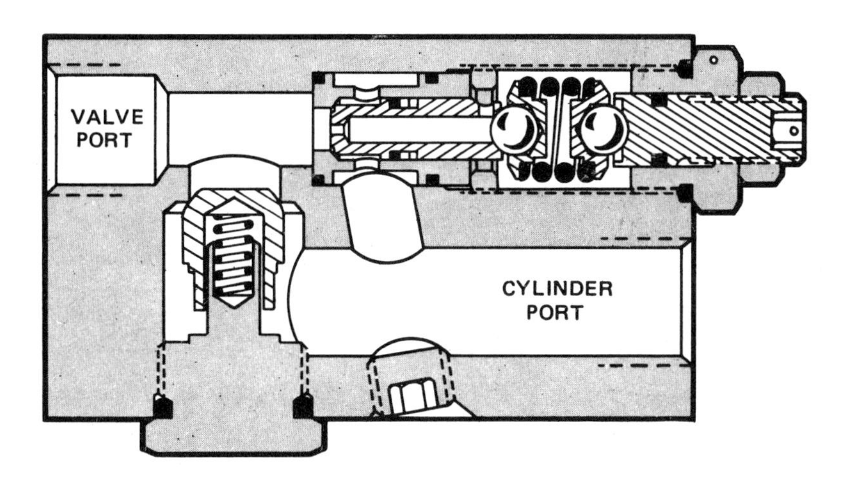

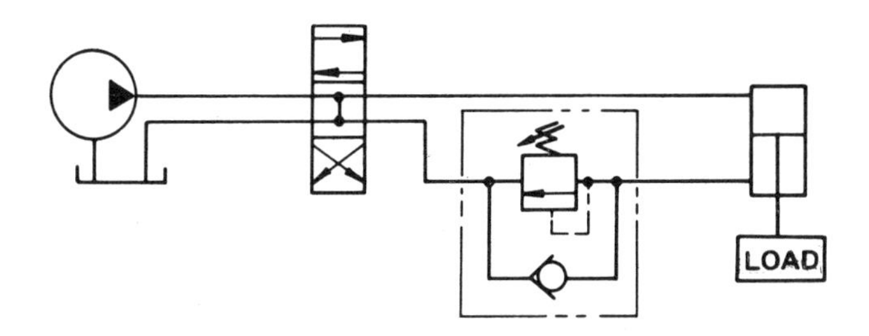

Fig. 4.1 Counterbalance valve can lock and position heavy loads in the up position and prevent drifting. (Courtesy of Fluid Controls, Inc., Mentor, OH.)

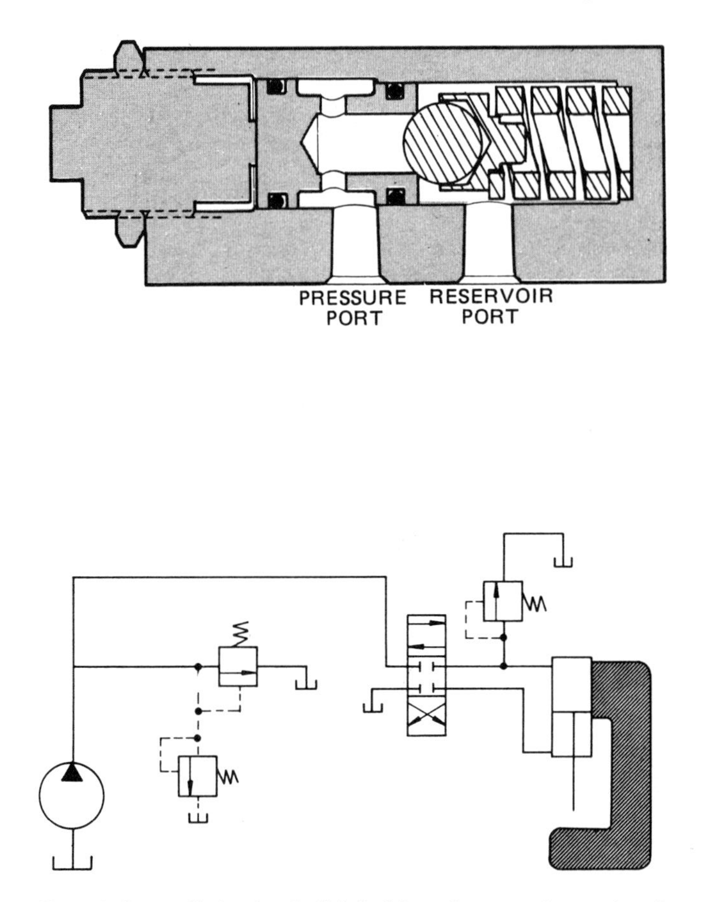

Fig. 4.2 Relief valve ball is held against seat by spring force. Schematic shows three locations for relief valve. (Courtesy of Fluid Controls, Inc., Mentor, OH.)

damping valve reduces shock by limiting the rate of acceleration of fluid flow. A *time-delay* valve prevents flow change until the desired time interval has elapsed.

Servovalves, solenoid valves, and microprocessor controlled valves are adaptations of the preceding but are covered in Chap. 7.

Typical Performance

Examples of many of the defined valves are scattered throughout the book. We'll limit the discussion here to one very common device: a direct spring-loaded relief valve. It is very simple in construction but needs to be understood to avoid problems.

The range of adjustment in pressure is narrow. If an orifice is plugged, the valve may fail to open at the expected pressure, or not at all. There is a tendency, unless highly damped, for the valve to chatter or hammer in response to rapid changes in circuit requirements. Sometimes this is the result of harmonics set up in the spring.

Any restriction in the line to the reservoir will increase the effective spring values so that the operating pressure will be higher. In certain applications it is common to connect returns from several different devices to a common return header. Pressure in this header will increase with large tank flows which, when reflected back to the relief valve, cause an increase in the entire circuit pressure. Therefore, individual and independent lines to the reservoir are desirable where economically feasible. Independent tank lines are also convenient in tracing down circuit losses.

Spring-loaded relief valves are inherently variable orifice devices. The spring requires a definite increment of pressure between barely cracked and full open. An 18-gpm pump might direct its full volume through the valve in certain parts of a machine cycle. At say 1000 psi the relief valve has established a maximum orifice. However, if work resistance during some other part of the cycle creates a pressure of only 900 psi, a portion of the flow may still be bypassed to the tank through a partly

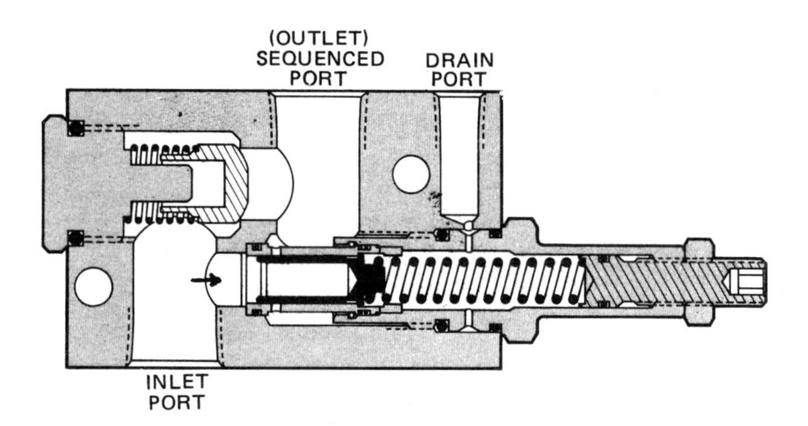

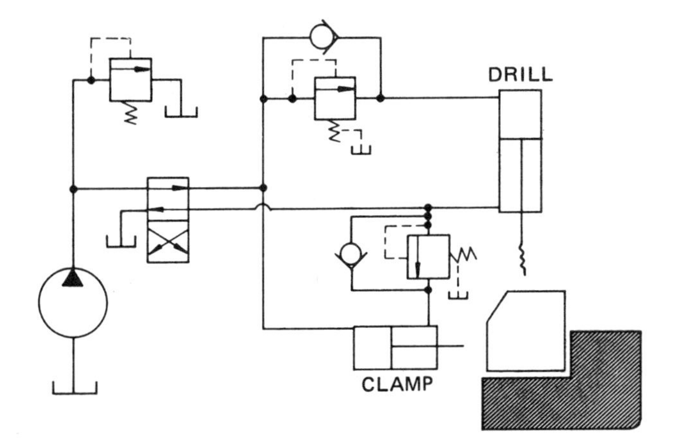

Fig. 4.3 Sequence valves. Two are shown in the circuit. Flow is not allowed into the DRILL cylinder until CLAMP pressure reaches minimum value at the upper sequence valve. Lower sequence valve will not release CLAMP until the DRILL cylinder is pressurized to return. (Courtesy of Fluid Controls, Inc., Mentor, OH.)

open orifice. Some valves do not entirely close until pressure drops to 50% of the relief valve setting.

Often this loss is unimportant. Some machines use the effect to intentionally slow operation as work resistance increases. When pressure reaches the set valve, the valve is full open, and the machine stops entirely. However, a penalty in efficiency is paid for this type of bleed-off speed control.

The piloted poppet relief valve solves many of the preceding problems. It probably is the most accurate and efficient relief valve for all-around use in hydraulic applications. In some cases it is higher priced on initial applications but usually saves in operating costs.

MODULAR VALVING

Valve functions can be consolidated, performance enhanced, and leaks eliminated by judicious choice of stacked, manifolded, and cartridge-type valves.

The basic techniques are stacked valves, stacked subplates, drilled plates, labyrinths, valve blocks, and assorted options such as cartridge inserts.

Stacked Valves

Stacked valves (Figs. 4.4 and 4.5) have mated ports with O-ring or gasket seals to eliminate interconnecting piping. Usually there is a common supply passing through all the valves and a common drain. Space is saved because each separate valve would need a minimum of five line connections: one pressure line, one or two exhaust lines, two service lines, and one drain line. For example, four individual valves have about 20 line connections; four stacked valves have only 11.

Stacked Subplates

Stacked subplates (Fig. 4.6) have a few advantages over stacked valves: Valves can be replaced without disturbing the piping, interconnections can be modified in the subplates without altering the valves, different valve sizes and porting can be mixed, and distortion in the subplate is not as likely to distort the valve.

Drilled Plates

Drilled plates (Fig. 4.7) put a wall between the valves and the piping. Appearance is better because the panel conceals the piping. Also, the valves are mounted on a flat surface and thus avoid strains from piping connections. Maintenance is simple because the front of the panel is mechanically independent of the back. Finally, the valves can be simply arranged and labeled.

Plates can be drilled straight through and the piping connections made directly to the back. The hole sometimes is used for clearance only, and the piping connection is made right to the valve.

Typically, however, the plates are drilled and cross-drilled to make interconnecting passages in one or more planes within the plate. Piping connections are made on the edges or back, and the valves are mounted on the plate and ported directly to it. A special advantage is that cartridge-type needle and check valves often can be inserted inside the plate, thereby eliminating extra lines and valve bodies.

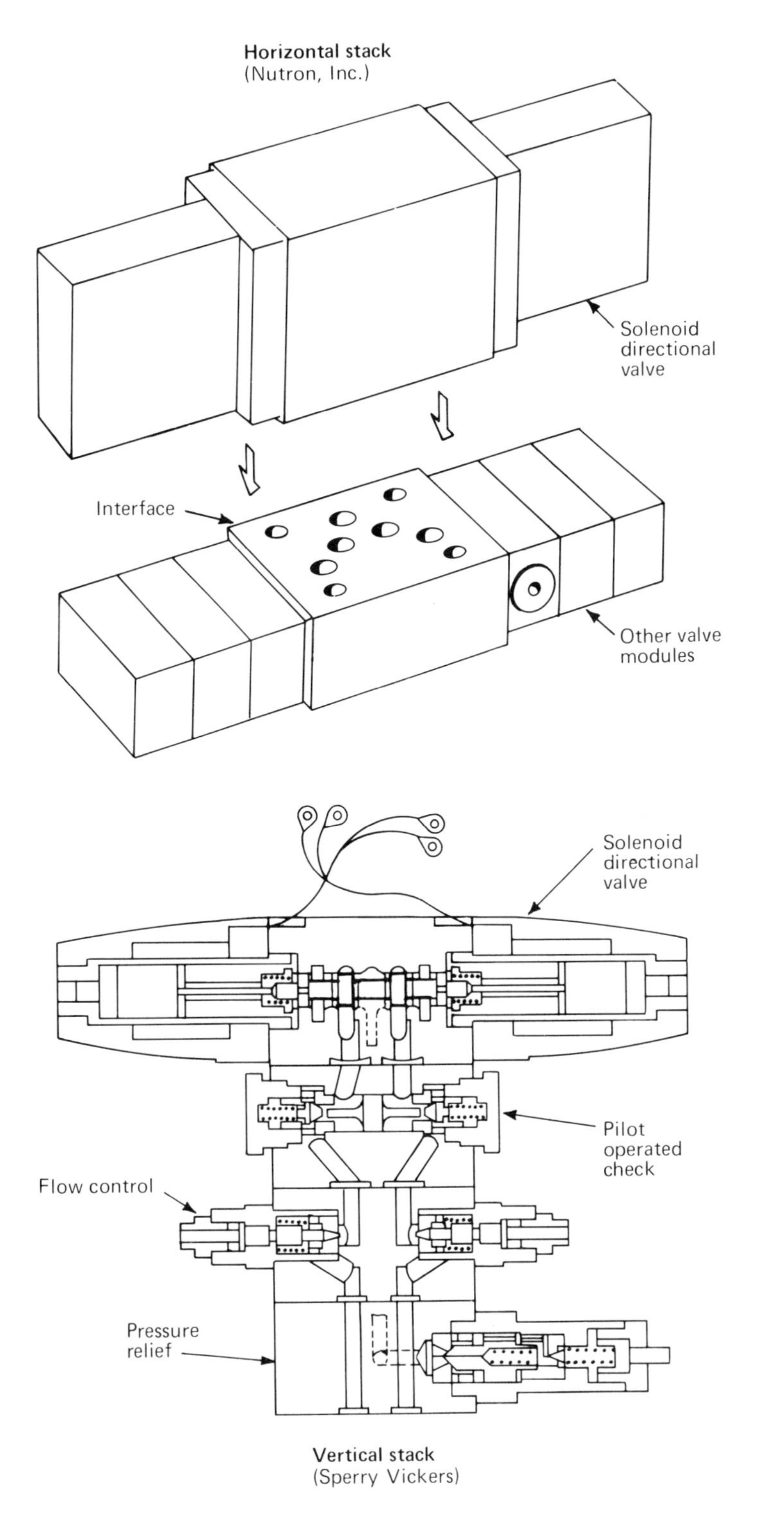

Fig. 4.4 Stacked valves eliminate tubing and increase speed and reliability.

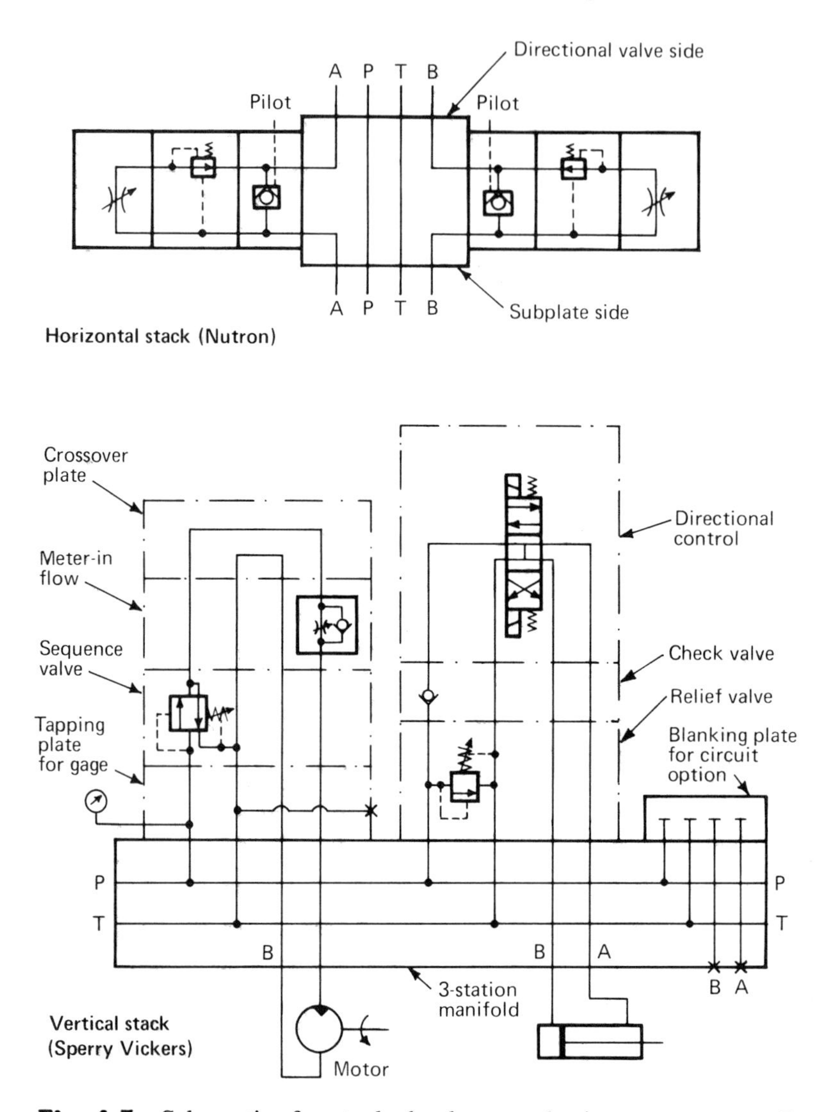

Fig. 4.5 Schematics for stacked valves emphasize compactness. Bottom example ensures that cylinder operation is complete before hydraulic motor turns.

Plates normally are made of steel or cast iron. Thicknesses up to 2 in. usually are made of cold-rolled steel, strain-relieved before machining. Plates up to 5 in. thick can be made of hot-rolled steel. When plates are made of cast iron, it must be fine-grained to avoid leakage paths along large flakes of graphite, particularly in the thin sections. All mounting surfaces for the valves must be ground.

Leave enough metal between adjacent passages: ⅛ in. for pipe tap sizes up to ½ in. in diameter, ¼ in. for taps up to 1 in. in diameter and 5/16 in. for taps above 1 in. in diameter. Be sure that runout in deep holes does not diminish the margin. Reduce runout by drilling the holes from both edges of the plate. The maximum practical drill depth is about 1 ft. Facilitate chip removal by specifying the tap drill for

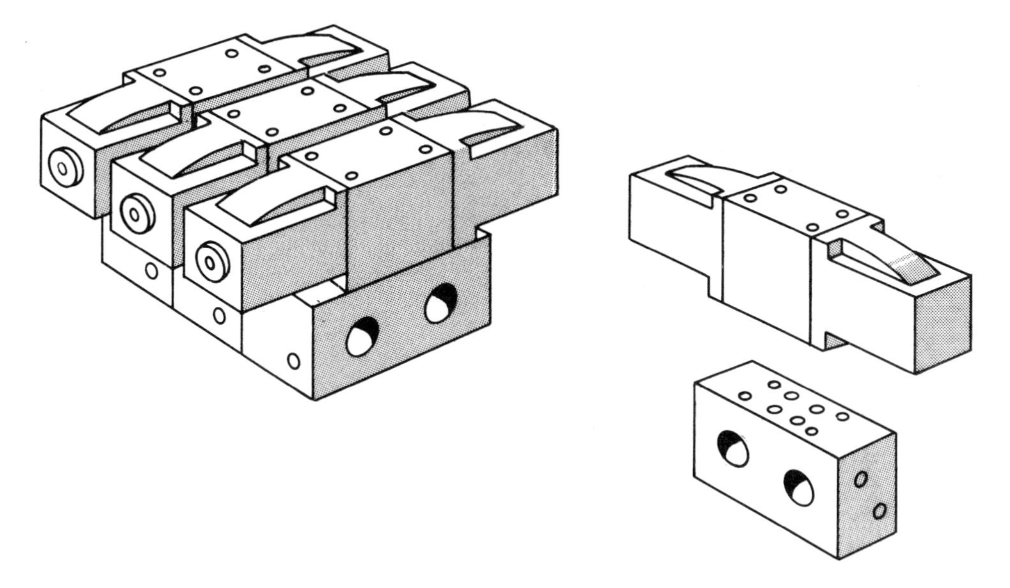

Fig. 4.6 Stacked subplates allow a greater variety of valve options.

the first 4 in., then reducing the drill size by 1/64 in. for each successive 4-in. depth. However, use the tap drill for the entire hole where possible.

Intersection holes are a special problem: The drill point should penetrate 1/16 in. beyond the far wall into solid metal. The centerlines of the two holes need not intersect exactly, but they should not be offset more than the radius of the smaller hole. In any case, make sure the breakthrough is complete. Also, taper-ream before tapping. Relieve the upper threads of the tapped hole about 1/16 in. below the finished surface.

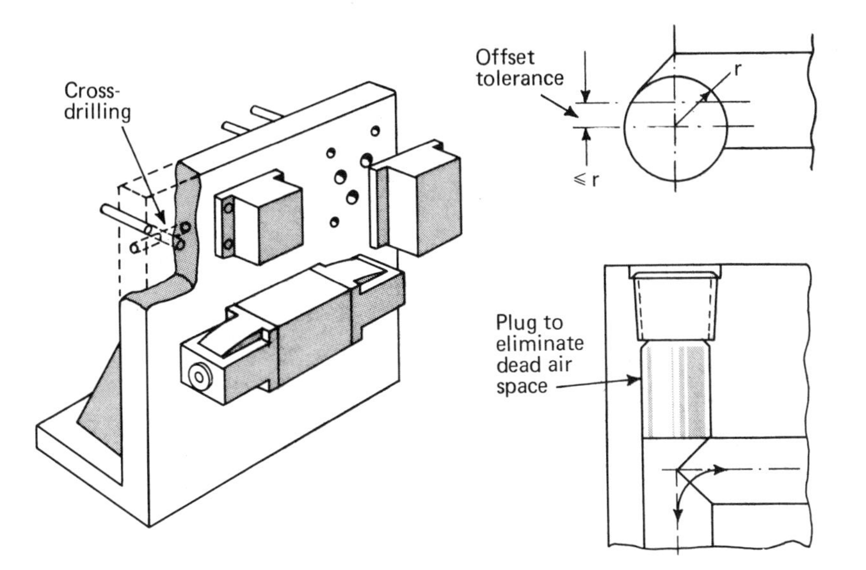

Fig. 4.7 Drilled plates serve the dual function of panel and interface.

Junction Blocks

Junction blocks on the back of a drilled plate help avoid unnecessary interdrilling and aid in combining similar functions. The supply ports can be combined, and so can the drain lines and pilot lines. It will be cheaper than using standard fittings. Exhaust lines usually can be combined too, but if a valve is sensitive to back pressure, provide a separate exhaust for that valve. Tubing can be enclosed in a sheet-metal box in the back of the main plate.

Maintain flatness on all mating surfaces. Avoid setting up stresses in the valve body and mounting surface during clamping for grinding. The surface will not stay as flat when the holding clamps are released.

Gaskets and O-rings will seal against moderate waviness. The O-rings can be placed in a simple counterbore or in a circular groove. The counterbore is less expensive and usually adequate. A groove is suggested for pressure higher than 1000 psi. The OD of the counterbore of the groove should be equal to the nominal OD of the O-ring. This prevents a radial pumping action during pressure rise and fall.

Labyrinths

Labyrinths (Fig. 4.8) also eliminate interconnecting piping. The difference is that the labyrinth type has milled or cast passages that can be made in any convenient shape, and more valves can be mounted in a given space. Materials for the plates can be steel, fine-grained cast iron, or hard aluminum.

Aluminum saves weight but has lower strength, is more difficult to braze, and

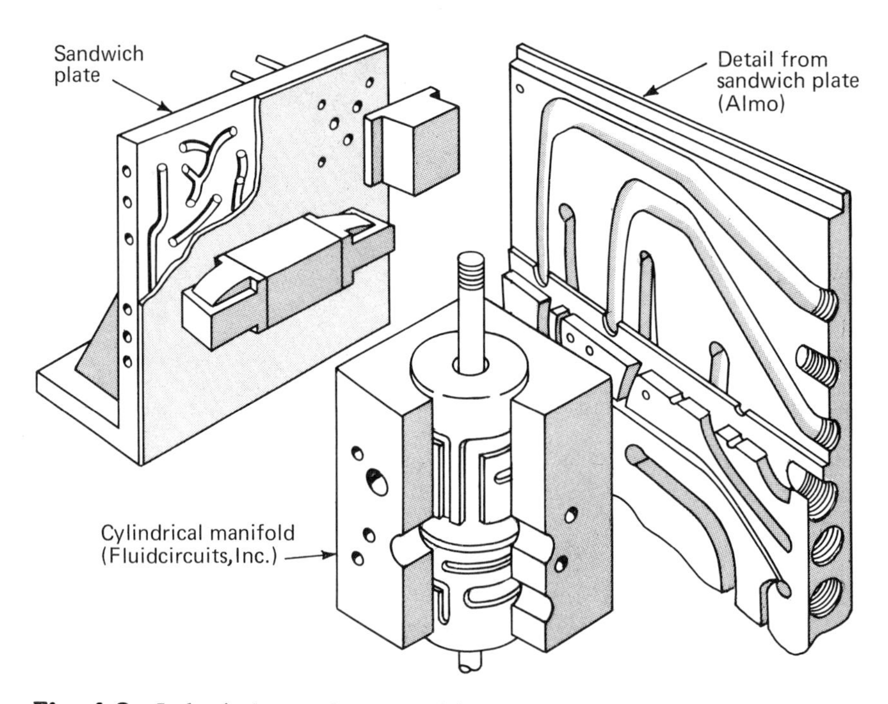

Fig. 4.8 Labyrinths can be created in many patterns and shapes.

is more easily scratched. Also, it has a much higher coefficient of expansion than steel and iron—a problem if the mounted valves are steel or iron.

If steel is the plate material, the passages usually are milled. If cast iron is chosen, the passages can be cast and need no further processing other than sandblasting to clean them. Casting costs less too, if the manifolds are made in quantity. Plates can be brazed, bolted, or bonded together to form one or more layers of grooved passages.

For example, sandwich plates can be copper-brazed under pressure and heated in an electric furnace. Capillary action causes the copper to flow to all contact surfaces, and the same action forces out the liquid flux. The excess copper will flow out with the flux through the open ports. A good joint must be flux-free, because trapped flux will block copper flow during brazing and create leakage paths.

Gaskets or O-rings work at low to moderate pressures. Or, if slight leakage is permissible, the plates can be lapped together and used without a gasket, provided that bolts are spaced close enough. To eliminate leakage, apply a thin coat of sealing compound before bolting. Other methods include epoxy resins, adhesives, and metallurgical bonding. The cylindrical example can be an interference fit.

Valve Blocks

Valve blocks (Fig. 4.9) have several valve-spool or valve-poppet assemblies contained within a single block. The valve assemblies can be self-contained—such as cartridge valves—or can be miscellaneous parts that perform the valve function when assembled into the block.

When the valve is in parts, and the block serves as the housing, the passage spacing can be closer than for any other combination of valves. Pressure drops are lower because passages are shorter; speed of response is improved.

The parts and machining steps required in valve-block assembly are fewer than those for separate valves stacked together, because there are no interconnection seals or joints. Fine-grained cast iron normally is selected as the block material to reduce porosity and provide mechanical strength.

Tips: Keep valve bores parallel where possible. Square the block, finish the valve bores, then interdrill. To prolong wear life and reduce maintenance, use press-fitted replaceable valve sleeves in the bores. Harden the sleeves to 32-35 Rockwell C; harden the spools to 58-62 Rockwell C.

If sleeves are not used, machine the valve bores and lap them in the block. Do the internal grooving and drilling after you've lapped the bores. Stone the edges of

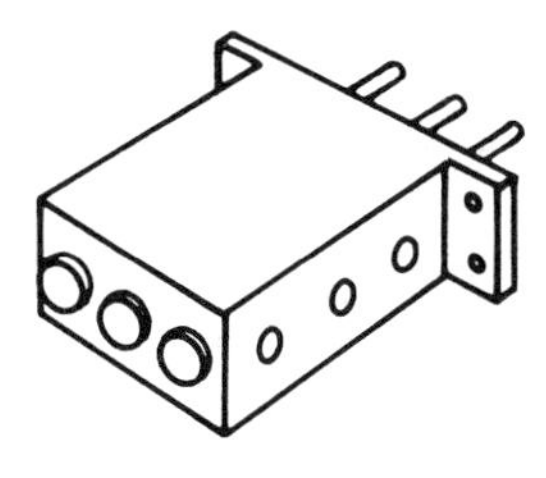

Fig. 4.9 Valve blocks contain several valve-spool or valve-poppet assemblies.

the grooves and clean them thoroughly. Afterwards, relap the bore. Without sleeves, the overall assembly can be smaller, valve stroke reduced, and response time decreased.

Don't forget that the tightening of cover screws can distort the block and the valve bores, interfering with valve operation. The problem is lessened if the ends of the valve bores are relieved 0.020 to 0.030 in. Also, pipe connections can leak if joints are distorted when the cover screws or fittings are tightened. Don't locate threaded ports close to a valve bore or to the block edge where mechanical strength is lower.

Cartridge Valves

Cartridge valves (Fig. 4.10) are available in variety and enhance the benefits of valve blocks, plates, and modular arrangements in general. The examples in the figure are direct-acting solenoid directional valves with conventional mounting threads and O-ring seals.

An alternative is an insert cartridge (like those by Kepner Products) that is slipped into a reamed bore, sealed with O-rings, and retained with a plate or plug. Another method (Lee Company) is to insert a nonthreaded but circumferentially grooved valve cartridge into the block, then swage it internally with a tapered pin to expand the metal into permanent contact with the bore.

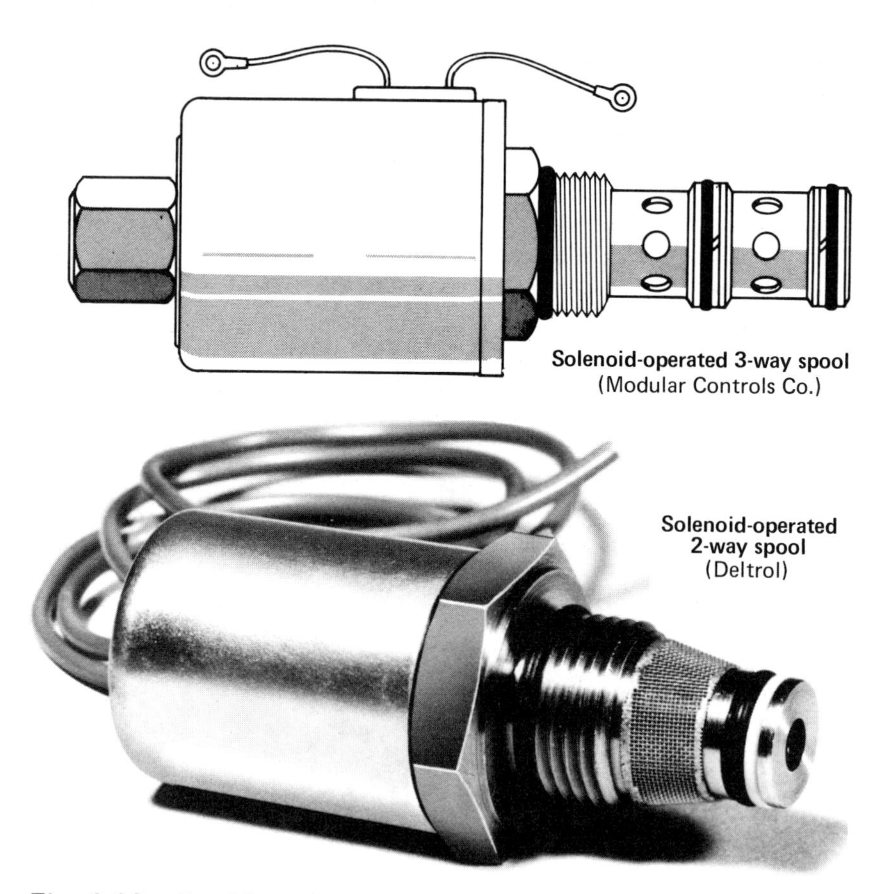

Fig. 4.10 Cartridge valves are an option with most of the other methods.

Many cartridges are full-function valves. At least several manufacturers (Sun Hydraulics, Modular Controls, and Oilgear Co. are examples) offer many of the following types: relief, counterbalance, regenerative-circuit, sequence, reducer, reducer-relief, check (including pilot and relief), needle, restrictor, pressure-compensated flow regulator, and shuttle.

5

Hydraulic Cylinders and Actuators

Hydraulic cylinders and actuators look simple but involve high forces and are frequently misapplied. A typical comment from hydraulic cylinder and actuator specialists is the following: Most mistakes in application are rudimentary—never mind about fancy computer programs for optimum performance. There seems to be a need for simple step-by-step design procedures for each kind of application.

Textbooks exist, but most are oriented either to classical hydrostatics and hydrodynamics (basic laws) or to simple application and troubleshooting of hardware (available types of pumps, motors, cylinders, and accessories). Only a few address the difficult design and application problems step by step.

THE KEY PROBLEMS

Heading the list of key problems seem to be off-axis loading, improper mounting, rod buckling, rod or cylinder sag, miscalculated acceleration and deceleration, extreme shock, wrong speeds and sequencing, unintended regeneration and intensification, and internal and external leakage. The addition of high water base fluids to the mix only makes matters worse.

Note: Also see Chap. 15 for additional illustrations of loading, mounting, sagging, and sequencing. (There are two chapters on cylinders because the application problems are different, though the physical appearances are often the same.) For more on built-in shock absorbers, see Chap. 8.

The Hit List

Off-axis loading (Fig. 5.1) is a problem with any linear or rotary actuator but perhaps most acute with hydraulic cylinders because of the extreme forces involved. It too often is forgotten that cylinders and actuators are not load guides or stops and need external bearings to carry anything but moderate off-axis forces.

Mounting is critical (Figs. 5.2, 5.3, and 5.4). It is not enough that every manu-

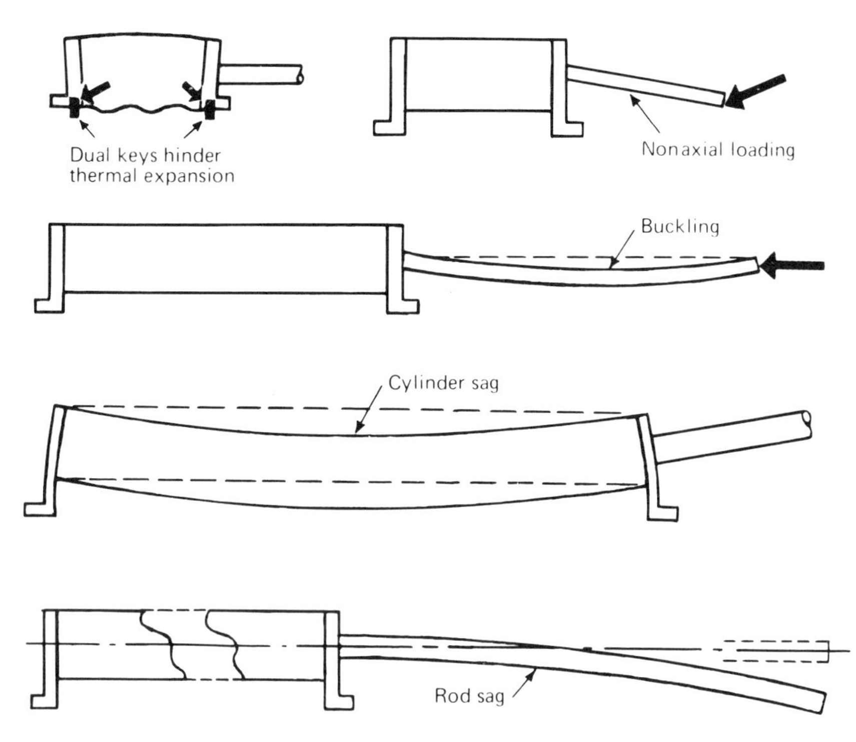

Fig. 5.1 Typical problems of misapplication experienced by users of hydraulic cylinders.

facturer does offer hinge mountings, universal rod eyes, and trunnions to help keep the rod or shaft in line. It also is essential that all force vectors be calculated and understood before assuming that the cylinder or actuator can survive. It becomes an interesting problem in engineering mechanics when a linear rod drives a rotary linkage.

Here are some tips for cylinders: Make sure that the clevis on the cap end is in the same motion plane as the clevis on the rod end. Avoid self-aligning mounts for trunnions, because trunnions are designed for shear and not bending. Prevent bending of threaded piston-to-rod or load-to-rod joints by providing shoulders to tighten them against.

A cylinder itself can provide some self-alignment if the rod head is mounted to float. It can be done by drilling the mounting holes in the cylinder lugs oversize relative to the positioning dowels in the baseplate.

The most troublesome strains are internal. For example, a fully extended rod creates high moments at the piston for relatively moderate sideloads at the rod end. Even the weight of the rod can cause strain. Specify a stop tube (sleeve) adjacent to the piston to prevent full extension if necessary (Fig. 5.5). Most major cylinder manufacturers publish handy graphs and tables to help determine how long the stop-tube should be. If the standard recommended cylinder rod is extremely long and heavy, perhaps the manufacturer can make it hollow to reduce weight.

A retracted rod induces another kind of strain. Suppose that the cylinder rod axis is slightly offset from (but parallel with) the load axis. There's no problem at ex-

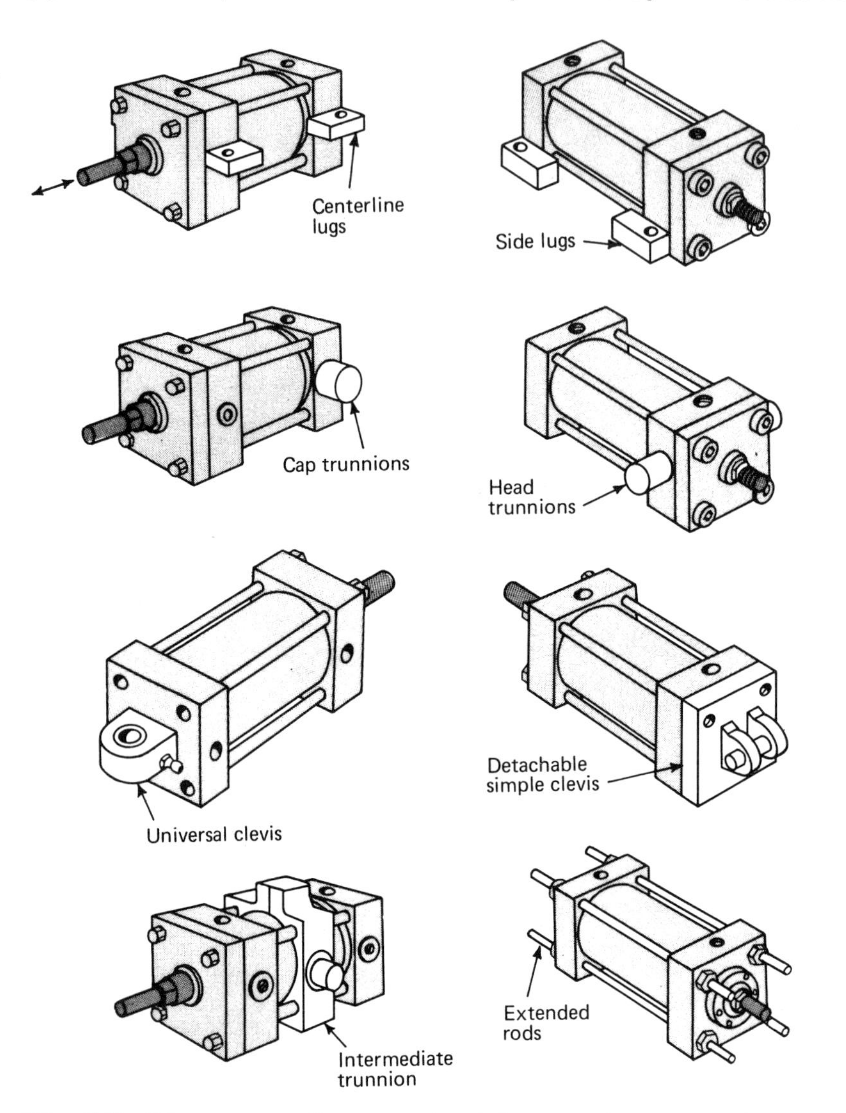

Fig. 5.2 These mounts for square-head tie-rod cylinders are widely available. See Fig. 5.3 for details on flange type.

tension if the rod is flexible. But the side load gets more severe as the rod is retracted, where there is less and less rod length to flex.

Another possible problem is noncenterline mounting—for example, cylinders attached at one edge of the head. Unless a cylinder is mounted along the axis of the load (centerline mounting), a bending moment results, equal to (load force) × (offset of axis from mounting surface). A short cylinder of course suffers most of all because the balancing moment arm (cylinder length) is short, and the bolt tension as a consequence is high.

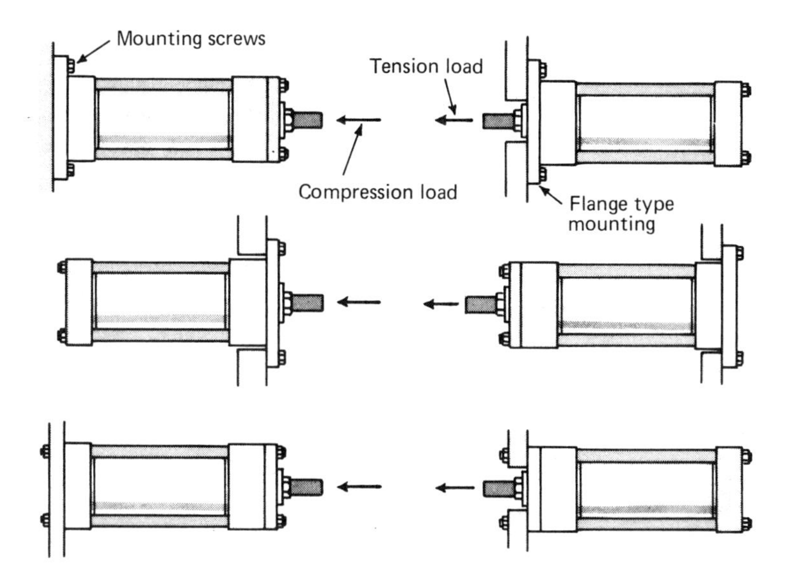

Fig. 5.3 Mounting screws for flange-type mounts are protected from tension stresses when arranged as shown. (Courtesy of Schrader-Bellows Div., Akron, OH.)

Rod Buckling

Rod Buckling should never be a problem but often is. Every manufacturer offers clear, conservative guidelines in its catalogs or manuals to help specify rod diameter for any stroke and pressure for push-type loads. Ignore them at your own risk.

The basic rule seems to be to allow compressive stress in the rod of up to about 10,000 to 20,000 psi as long as the effective rod length-to-diameter ratio (an abritrary value based on cylinder style) doesn't exceed about 6 : 1 at full extension. According to one catalog, a short stroke cylinder with a 1⅜ in. diameter rod (area = 1.485 in.^{2}) is allowed to push with 30,000-lb force. Rod compressive stress = 30,000/1.485 = 20,202 psi.

The picture changes when the length-to-diameter ratio exceeds 6 : 1. As an example, take a 25-in. stroke cylinder with a rod diameter of 1⅜ in., with the cylinder and rod end unsupported. According to the same catalog, the maximum push force should be less than 1000 lb. Then the compressive stress is less than 1000/1.485 = 673

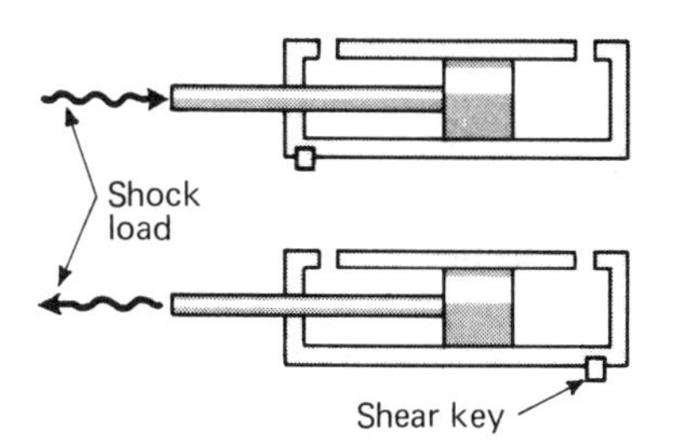

Fig. 5.4 Shear keys absorb shock best if they exploit cylinder elasticity as shown.

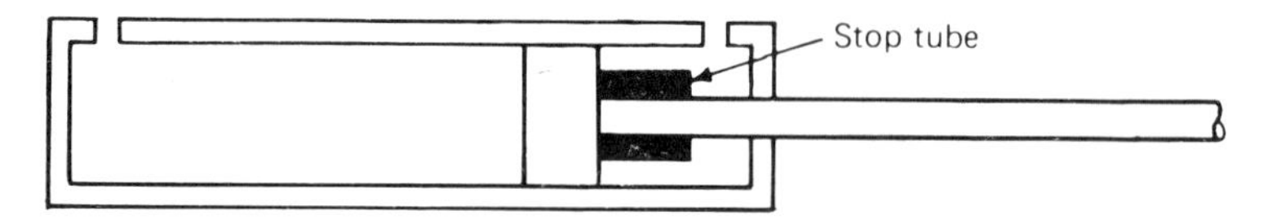

Fig. 5.5 Stop tube prevents overextension of piston rod.

psi. A more firmly guided rod can help prevent buckling and will allow at least four times as much extension.

If the manufacturer's recommendations are to be exceeded, then complex calculations involving all dimensions and materials of the cylinder and all force vectors of the load must be made. Such details as bearing stresses in the rod gland, piston moments, frequency of operation, local accelerations, and locations of pivot points become critical.

There is no readily available method to solve complex buckling problems, but some manufacturers and users of hydraulic cylinders have developed proprietary techniques that work well.

Design departments of Caterpillar, Bucyrus-Erie, J.I. Case, Deere, Harnischfeger, most aircraft companies, and other big users of hydraulics have great expertise in meeting their own design needs but aren't about to share secrets. They'll discuss the final products (lift trucks, aircraft, etc.) but not the proprietary design and test techniques that make them possible.

Eaton's industrial truck division (Philadelphia, PA), for example, has developed a computerized method for designing telescoping cylinders. Parker Hannifin's cylinder division (Des Plaines, IL) has a computerized iteration procedure for designing long-stroke pivot-mounted cylinders subjected to buckling loads (Fig. 5.6) based on original work by Fred Hoblit of Lockheed (*Product Engineering*, July 1950, p. 108).

Most manufacturers publish handy charts and simplified equations for estimating piston rod diameters, stop-tube lengths, actuator torque, pressures and forces for acceleration or deceleration, force vectors of mechanical output linkages, and general cylinder specs. They are good as far as they go.

Sagging rods on horizontal cylinders are a problem if only because of the ap-

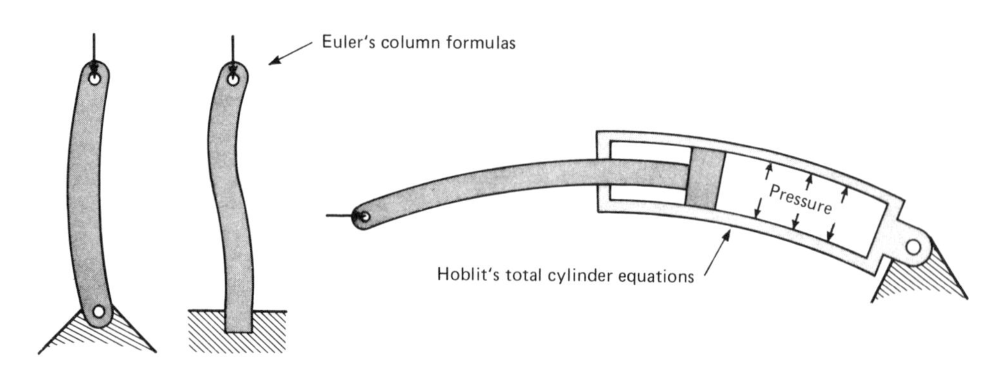

Fig. 5.6 Buckling force can be calculated by assuming one of the configurations shown, but the cylinder manufacturers have already done the calculations and offer good recommendations.

pearance. Special stiffer models can be specified. For example, a hollow rod will sag less than a solid rod. At least one manufacturer (Miller Fluid Power) offers to prestress the rod to hang straight out on extension. Just don't install it twisted 180 deg.

PERFORMANCE

Precise control of speed and position is possible with hydraulic cylinders and actuators, but oil under pressure is unforgiving. The final machine or vehicle is no place for trial and error.

The *acceleration* equations are simple and are summarized in Fig. 5.7 and Table 5.1. The conventional $F = Ma$ and $T = J\alpha$ equations are not too handy if the problem is finding the accelerating force or torque that will move the load a given distance with a given terminal velocity. It's better to use equations that have velocity and travel units instead of acceleration units.

Example 1: Linear Acceleration. Assume a horizontal cylinder, 10,000-lb load, 500-lb friction force, 1500 psig available at the cylinder port, zero initial piston velocity, and 100 ft/min terminal velocity reached after 3-in. travel at constant acceleration, rod extending. What net accelerating force is necessary, and what size cylinder will provide it? Use the equations in Table 5.1:

$$M = 10{,}000/32.17 = 310.85 \text{ slugs}$$
$$\Delta S = 3/12 = 0.25 \text{ ft}$$
$$\Delta V = 100/60 = 1.667 \text{ ft/sec}$$
$$F_A = 0.5 \times 310.85 \times (1.667)^2/0.25 = 1727.6 \text{ lb}$$

Add the friction force and compute the piston area and diameter:

$$\Sigma F = 1727.6 + 500 = 2227.6 \text{ lb}$$
$$A = F/P = 2227.6/1500 = 1.485 \text{ in.}^2$$
$$D = (4 \times 1.485/\pi)^{0.5} = 1.375 \text{ in.}$$

What maximum flow is needed?

$$Q = V \times A = (100 \times 12) \times (1.485)/231 = 7.7 \text{ gpm}$$

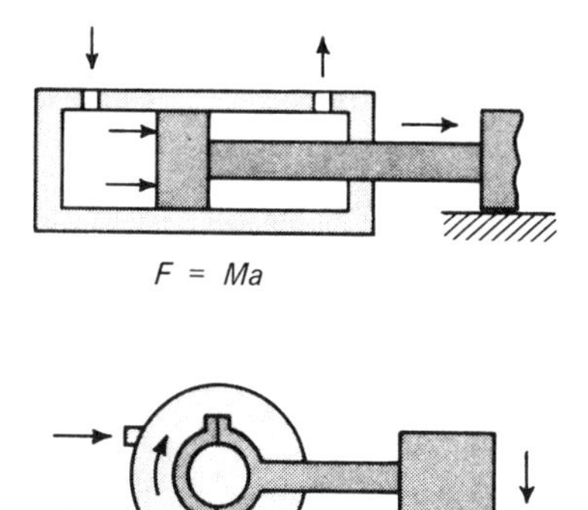

Fig. 5.7 Acceleration and deceleration of cylinders and actuators.

Table 5.1 Acceleration and Deceleration Equations

Linear:

F_A = $Ma = M\Delta V/\Delta t$

F_A = $0.5M(\Delta V)^2/\Delta S$, for special case where velocity = zero at start of accel or end of decel, so $\Delta t = \Delta S/(\text{average velocity}) = \Delta S/(0.5\Delta V)$

Rotary:

T_A = $J\alpha = MK^2 \times \text{rad/sec}^2 = 0.1047MK^2\Delta N/\Delta t = WK^2\Delta N/(307\Delta t)$

T_A = $0.0008725MK^2(\Delta N)^2/\Delta\text{revs}$, for special case where rpm = zero at start of accel or end of decel, so Δrevs = total revolutions = average rpm $\times \Delta t/60$ $= 0.5\Delta N\Delta t/60$. $\Delta t = 120 \times \Delta\text{revs}/\Delta N$

Symbols:

a	= linear acceleration, ft/sec^2 (assumed constant)
α	= angular acceleration, radians/sec^2
F_A	= net accelerating/decelerating force, lb
ΣF	= sum of forces acting on piston (pressure, friction, inertia, load)
J	= mass moment of inertia, slugs-ft^2 or lb-sec^2-ft
K	= radius of gyration, ft
M	= mass, slugs (or lb-sec^2/ft). 1 slug = 32.17 lb mass. $M = W/32.17$
ΔN	= rpm change during acceleration
Δrevs	= total revolutions turned during acceleration (1 rev = 2π radians)
ΔS	= linear piston travel during acceleration
Δt	= time to reach terminal velocity, sec
T_A	= net accelerating/decelerating torque, lb-ft
ΣT	= sum of torques on actuator (pressure, friction, inertia, load)
ΔV	= velocity change during acceleration, ft/sec
W	= mass in lbs

Example 2: Linear Deceleration. Assume a horizontal cylinder, rod extending, 500-lb_m load, 500-lb friction, 800-psig driving pressure at head end, and 80 ft/min initial velocity. If the rod diameter is 1 in. and the piston diameter is 1.5 in., what pressure at the rod end will stop the piston and load within 2 in. at constant deceleration?

$$\text{piston area} = \pi(1.5)^2/4 = 1.767 \text{ in.}^2$$
$$\text{rod area} = \pi(1.0)^2/4 = 0.7854 \text{ in.}^2$$
$$\text{differential area (piston} - \text{rod)} = 0.982 \text{ in.}^2$$

The driving force from pressure at the head end $= 800 \times 1.767 = 1413.6$ lb. The friction force $= 500$ lb; therefore the effective driving force $= 1413.6 - 500 = 913.6$ lb.

Calculate the decelerating forces:

$$M = 5000/32.17 = 155.4 \text{ slugs}$$
$$\Delta S = 2/12 = 0.1667 \text{ ft}$$
$$\Delta V = 80/60 = 1.333 \text{ ft/sec}$$
$$F_A = 0.5 \times 155.4 \times (1.333)^2/0.1667 = 828.2 \text{ lb}$$

Add the 913.6-lb driving force:

$$\Sigma F = 828.2 + 913.6 = 1741.8 \text{ lb}$$

Cushioning pressure in the annulus:

$$P = F/A = 1741.8/0.982 = 1773.7 \text{ psig}$$

A special problem is to quickly stop a cylinder piston and load at the end of stroke. Some cylinders have built-in cushioning systems. Others rely on meter-out circuits to create back pressure. One of the most direct methods of stopping a moving cylinder piston and load is with a separate decelerator or shock absorber (Chap. 8).

Even with careful design, some mechanical shock is bound to reach the cylinder. This can be absorbed in part by shear keys, pins, or plates in the cylinder mounts. Just be sure not to put the keys at both ends of a cylinder (they restrict thermal expansion) or at diagonally opposite corners (distortion under load will occur).

Put the keys or pins at the end which exploits cylinder elasticity the most (Fig. 5.4). If the shock load is likely to be in compression, put the key or keys at the rod end. If the shock load will be in tension, put the key or keys at the cap (blind) end. Then the body of the cylinder will help absorb the shock.

Pressure surges have another effect: the sudden mechanical expansion of the cylinder or actuator barrel. The stresses may be well within the design limits for ultimate strength, but the seal clearances might briefly open up enough to cause leakage or mechanical damage to the seals. If there is danger of this, pick a higher pressure cylinder or actuator (called derating).

Speed, Sequence, Regeneration

Wrong speeds and sequencing have little to do with the cylinders and actuators themselves but result from the complex movements the cylinders are asked to make. Yet there are plenty of directional, flow, and pressure control valves available, as well as meter-in, meter-out, bleed-off, feedback, dual-pressure, flow-divider, sequential, and electrohydraulic circuit schemes to exploit the valves.

Certain types of control and sequencing circuits are not as well understood as they should be. The regenerative circuit is one. Synchronizing circuits are another. Failsafe circuits are a third.

A *regenerative circuit* (Fig. 5.8) connects the rod end of a cylinder with the

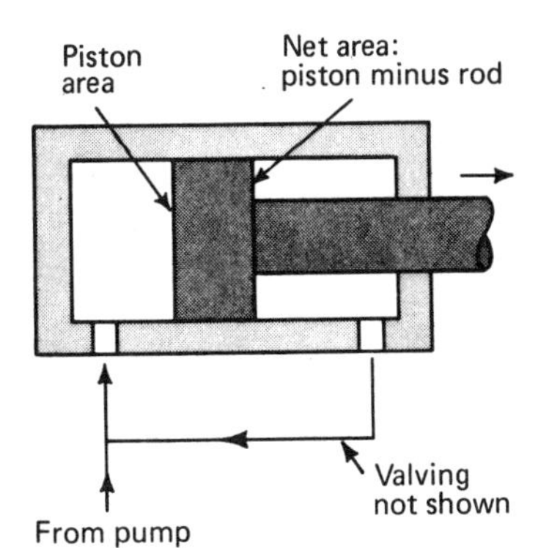

Fig. 5.8 Regeneration circuit for cylinders increases speed.

blind end. Both ends of the cylinder then are at approximately the same pressure, and the forward thrust (extension) is determined by the difference in areas between the rod and blind ends of the piston.

It seems strange to do that until the reason is understood. It's a tricky way to increase the speed of the piston without increasing the flow from the supply pump. The regenerated flow adds itself to the incoming pump flow and the piston moves proportionately faster. Speed = pump flow/rod area.

The rod-end oil readily forces itself into the inlet stream because of the normal intensification of pressure inherent at the rod end. The maximum pressure available is equal to the blind-end pressure times the ratio of the total piston area at the blind end to the effective piston area (piston minus rod) at the rod end. *Example:* Assume the piston diameter is 4 in. and the rod diameter is 3 in.; then the areas are 12.57 in.^{2} and 7.069 in.^{2} respectively. The effective piston area at the rod end = 12.57 − 7.069 = 5.501 in.^{2}

The ratio = 12.57/5.501 = 2.285, so if the blind-end pressure is 1000 psig, the rod-end pressure can be as high as 2.285 × 1000 = 2285 psig. However, the rod-end pressure never increases to the maximum value during normal regeneration but rises only enough to overcome flow resistance from the rod end to the blind end.

The rod output force is always diminished during regeneration, because of the back pressure at the rod end. Regeneration therefore is useful chiefly as a way to move a piston quickly into position against relatively light loads. When full force is needed again, the rod end must be ported to drain.

A problem arises in design when the intensification pressure level is forgotten or ignored and the system is built without the capability of withstanding that level of pressure, should it occur.

Another problem arises if the rapid speed is forgotten in design. If the rod diameter is small, most of the oil is regenerated, and the piston speed can be extremely high.

Intensification takes other forms, and they can be more severe than regeneration. For example, if a crane load is supported by oil pressure in a cylinder (Fig. 5.9), the oil can leak past the piston seals and fill the rod end of the cylinder. If that happens, and if the fluid is trapped by tight valving, the rod-end pressure will intensify as the load drifts down. The back pressure created will force the blind-end pressure to increase so that the load can still be supported. For example, if the initial pressure is 2000 psig and the piston area is four times the rod area, the final supporting pressure in the blind end can reach 8000 psig.

The sun can burst cylinders too. Oil expands with heat, and if a cylinder is shut down during the cold of the night and valved off too tightly (a tight closed-center valve, for instance), the heat of the day can burst it—and has. The answer is to provide relief valves or make sure the system can withstand any intensification of pressure that occurs.

Operation in Unison

Synchronized cylinders always are a design challenge because the rods will not travel at absolutely the same speed unless they are mechanically linked together, as with a bar, plate, or rack-and-pinion. The name of the game becomes "compromise."

Servo feedback probably is the most accurate synchronizing means short of

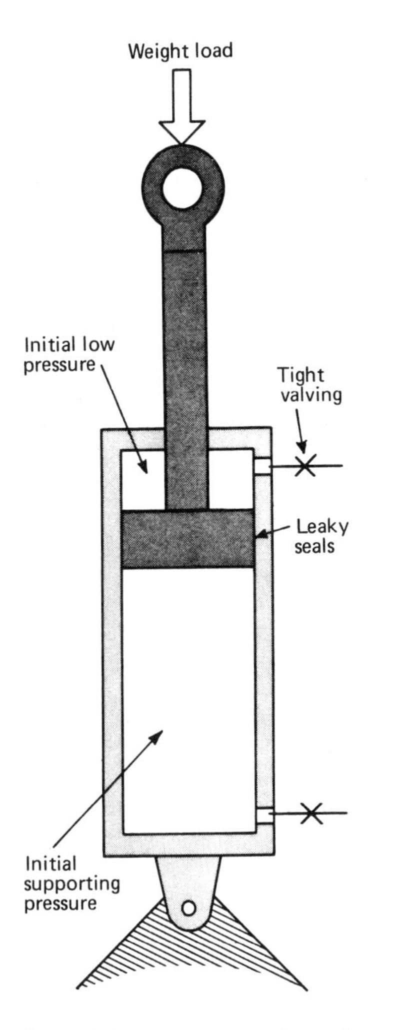

Fig. 5.9 Weight load can intensify pressure if seals leak.

mechanical linkage but is expensive and requires skill in design and application. Matched flow-control valves also work well, if the slight differences in performance of each cylinder caused by tolerance stackups and temperature effects are acceptable. Split-flow pumps, feeding equal flows to all cylinders, also are used.

Coupled hydrostatic motors (Fig. 5.10) work well if the internal leakages are matched and if the piston loads are equal. Otherwise one cylinder is bound to get ahead of the others and take most of the load.

One of the most interesting synchronizing methods is to drive the load with a single cylinder and constrain the movement with a separate group of synchronized cylinders that serve only to keep the load even. The synchronizing cylinders can be interlocked hydraulically by letting the effluent from one cylinder be the input to the next cylinder. Double-ended cylinders work best in theory because the effective areas (piston minus rod) are equal for all cylinders, and the flows are forced to be equal. Single-ended cylinders also will suffice, but the blind-end area of each following cylinder must equal the rod-end area of the preceding cylinder.

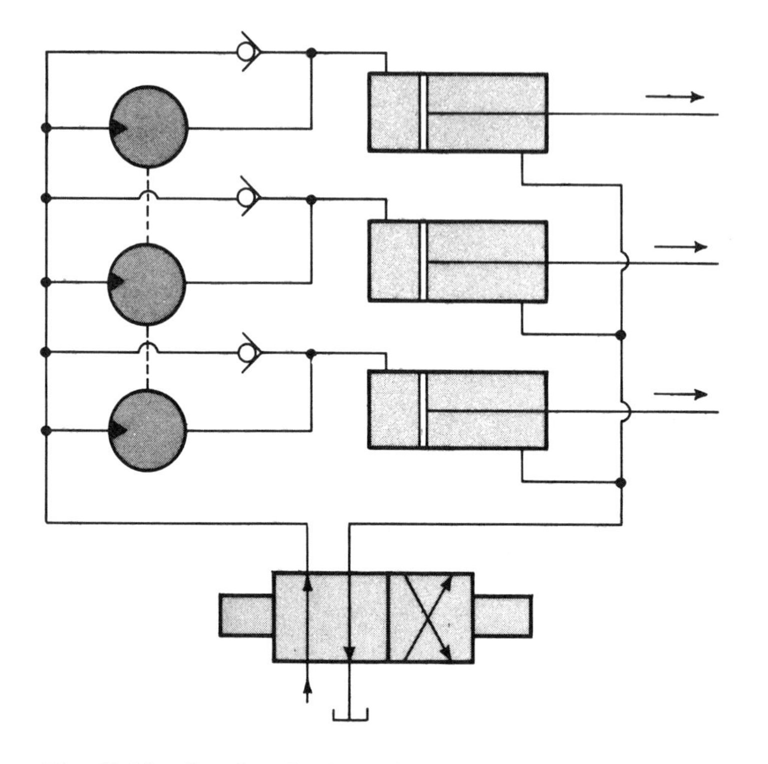

Fig. 5.10 Synchronization of three cylinders is possible with three freewheeling motors mechanically coupled as flow dividers.

The technique can be adapted to driving the load directly with the synchronized cylinders. However, unless the load (say, a vertically moving table) automatically loads each rod equally, one of the rods might pick up most of the load and let the other rods coast. Careful balancing of the flow restrictions will be required.

The biggest cause for loss of synchronism is leakage. It is nearly impossible to precisely match rod leakage, piston bypass leakage, and valve leakage for every combination of force, speed, pressure, and temperature. There are two choices: Learn to live with imperfect synchronism or design in automatic repositioning of each piston to a fixed point every cycle. For example, make sure all rods are completely retracted.

Failsafe cylinder circuits are another tricky problem. They are possible but take careful design. Previous paragraphs cover many of the ways to avoid danger. Also see the valve examples in Chap. 4.

Add these as well: interlocks such as two-hand anti-tiedown manual controls for hydraulic presses, so that nothing moves unless the operator has a hand on each valve lever; automatic safe retraction of all cylinders on loss of pressure or voltage, with mandatory manual reset; automatic bleed-off of accumulators on shutdown to avoid gratuitous pressure surges on the next startup; positive sequencing of cylinders so that cylinder B cannot move until cylinder A has finished its stroke; and limit valves or switches to shut down the system if a load drifts, shifts, or runs away. A smart idea is to challenge a co-worker to find flaws before a customer does.

Push-Pull-Twist Action

Triple-motion hybrid actuators by PHD (Ft. Wayne, IN) are a marriage of a pneumatic or hydraulic rotary actuator and a fixed-spline nonrotating piston rod. Figures 5.11 and 5.12 show the principles and also a hypothetical stackup of two of the actuators on top of a piston-controlled table.

The rack-and-pinion pistons are separately controlled with air or oil pressure supplied to the caps, providing force to rotate the pinion in both directions. The vertical splined rod is attached to the pinion and rotates with it. Thrust bearings at the base of the spline take the full thrust of the spline, and none of it is transferred to the pinion bearings. Optional features include adjustable hydraulic cushions or urethane shock pads at the end of any motion.

The splined piston moves freely in the axial direction under control of air or oil pressure from the ports of each end and also may be rotated at will with the splined rod. The resulting motion is any desired combination of thrust and twist, always under full control of the two pairs of pressurized ports, supplied from directional

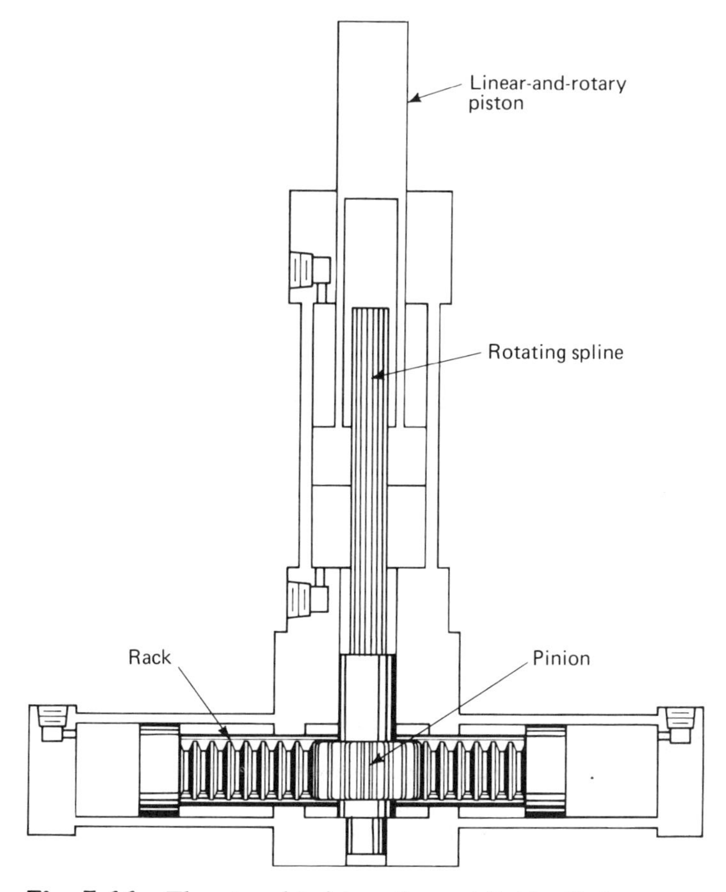

Fig. 5.11 Thrust-and-twist motions of fluid cylinder are separately controllable in each mode of a combination rotary actuator and linear piston. (Courtesy of PHD Co., Fort Wayne, IN.)

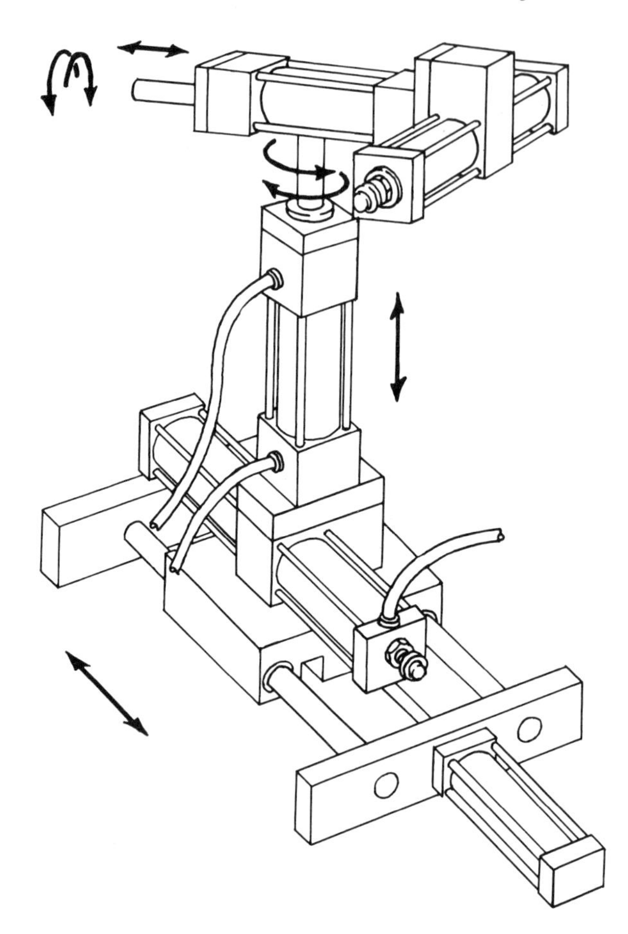

Fig. 5.12 Stacked thrust-and-twist actuators have several degrees of freedom and together serve as a simple fluid-mechanical robot. (Courtesy of PHD Co., Fort Wayne, IN.)

valves external to the unit. The action is much like that of a simple robot arm and can perform simple tasks at much less cost than a robot.

Other novel actuators are pictured in Figs. 5.13 through 5.17. The captions explain the functions.

High Water Base Fluids

High water base fluids (HWBF) are gaining popularity in industrial fluid power cylinder applications because of their much lower cost, greater safety, and biodegradability. That's despite the fact that cylinders operating on HWBF (90% to 95% water) are 10 times more likely to leak than those operating on oil. Also, HWBF fluids have poorer lubricity and do not protect metal from corrosion as well as oils do. See Chap. 1 for details.

Yet the cylinder manufacturers are willing and able to cope with the challenge. They've proved that cylinders will function well on HWBF if the cylinder specifications are tailored right. Anyway, they figure that pumps take a worse beating than

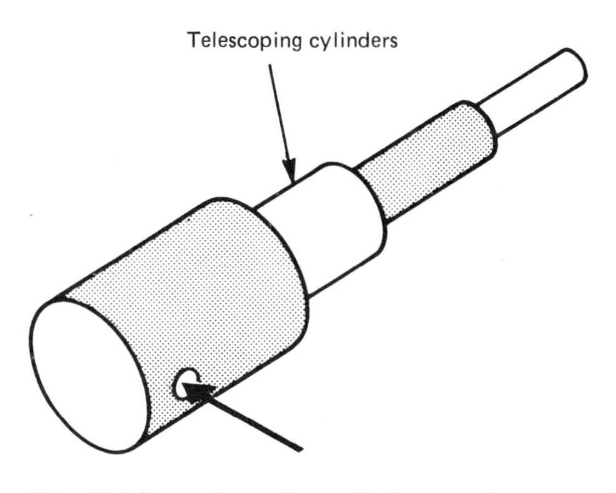

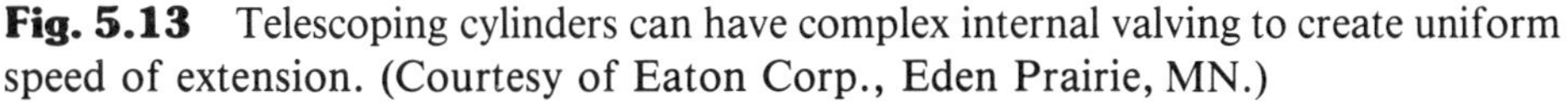

Fig. 5.13 Telescoping cylinders can have complex internal valving to create uniform speed of extension. (Courtesy of Eaton Corp., Eden Prairie, MN.)

cylinders and that if pumps can survive, so can cylinders. Some manufacturers offer cylinders to operate up to several thousand psig but most recommend a 1000-psig limit, same as for the pumps.

Here are some of the key features needed for operation on HWBF: nonmetal wear rings around the piston to protect the cylinder barrel from excessive wear; longer rod bushings to increase the bearing area; nonrusting materials; tougher, self-lubed, pressure-energized piston and rod seals to reduce leakage; and precision end-of-stroke cushions with closer clearances to maintain effective shock-absorbing action despite the low viscosity of the HWBF. Also, seals do not hydroplane on HWBF films as well as they do on oil films, and therefore they wear out faster.

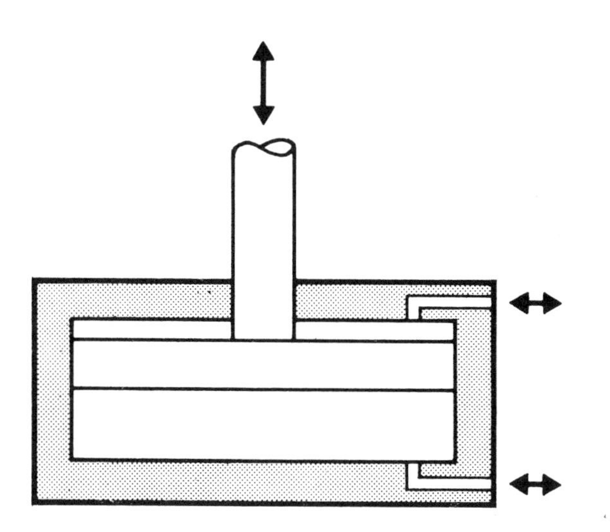

Fig. 5.14 Pancake cylinders generate high force-stroke ratios. (Courtesy of Mack Corp., Flagstaff, AZ.)

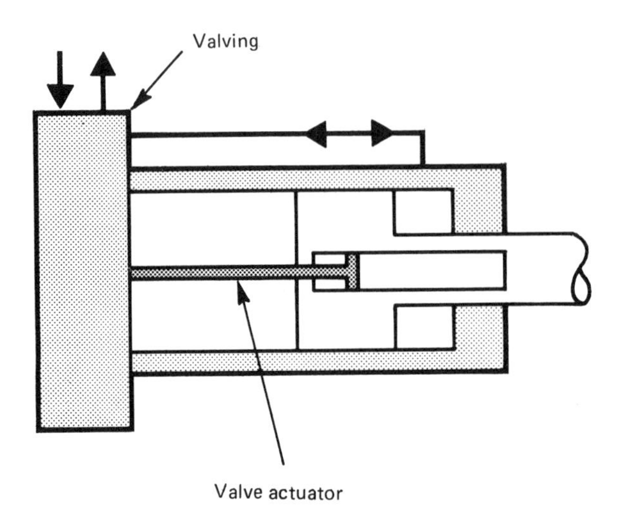

Fig. 5.15 This piston limits its own motion with a built-in stem. (Courtesy of Proteus Corp., Minneapolis, MN.)

TIE-ROD STRESS CALCULATIONS

John Berninger, manager of engineering at Parker Hannifin's Otsego, MI division, has mathematically related over eight variables that affect the maximum pressure rating of a tie-rod constructed hydraulic or pneumatic cylinder (Fig. 5.18) and has come up with a few simple equations that include them all. The key equation is

$$F_T = F_I + \eta P A_P \tag{5.1}$$

where F_T is tie-rod total tension force at any given pressure, F_I is tie-rod initial tension force (preload), P is internal cylinder pressure, and A_P is internal pressurized area (see Tables 5.2 and 5.3 for units). Constant η (Fig. 5.19) is called the pressure load coefficient and uniquely combines the complex effects of Hooke's law, Poisson's ratio, hoop and radial stress, cylinder relaxation, and balance of pressure forces. It can be calculated from known dimensions and characteristics for any given cylinder.

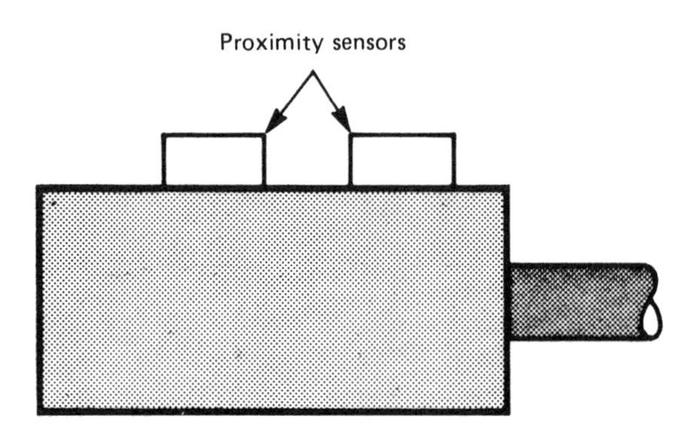

Fig. 5.16 Magnetic sensors detect piston position.

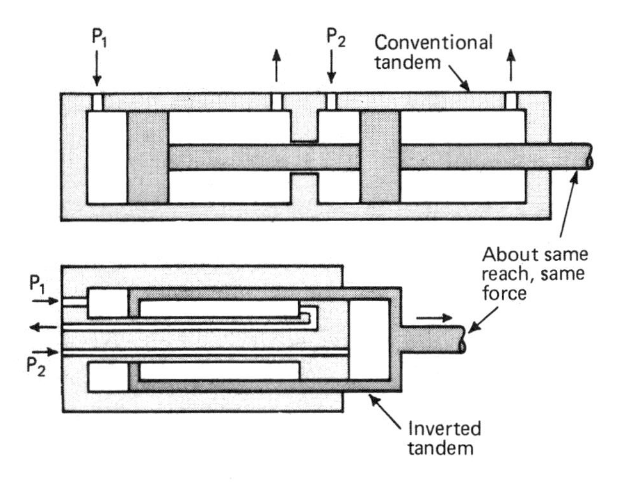

Fig. 5.17 Pressure-doubling cylinders add pressures of two pistons. Inverted tandem type is by B&K Co.

A useful derivation from Eq. (5.1) is one that relates the tie-rod initial tension (preload) to the internal cylinder pressure that causes the cylinder cap or head to separate from the cylinder body. That equation is

$$F_I = A_P(1 - \eta)P_S \tag{5.2}$$

where P_S is separation pressure. A plot of initial tie-rod stress vs. separation pressure is given in Fig. 5.20.

The accompanying additional equations and derivations explain the method and tie it in with more conventional bolt-torquing and fatigue formulas. Actual tests have verified the theory, and the company has used the calculations in specifications for special cylinders. The same procedure applies to any tie-rodded pressure container.

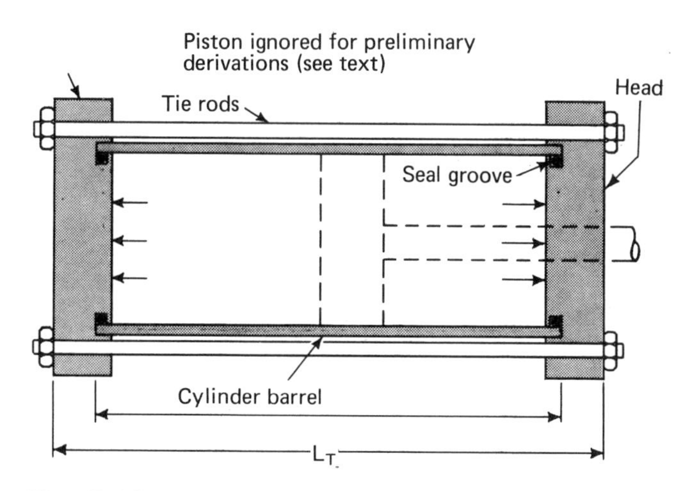

Fig. 5.18 Tie-rod constructed cylinder for accompanying derivations has been simplified as shown, but the method is valid for actual designs. (Courtesy of Parker Hannifin Corp., Cleveland, OH.)

Table 5.2 Nomenclature for Tie-Rod Equations

A	= cross-sect. area
A_C	= cyl. wall cross-sect. area
A_P	= internal pressurized area
A_T	= tie rod cross-sect. area
A_{Thd}	= tie rod stress area at thread
a	= cyl. inside radius
b	= cyl. outside radius
D_T	= nominal dia. of tie rod thread
E	= modulus of elasticity (stress/strain, where subscript C = cyl., T = tie rod)
F	= force
F_C	= cyl. total compression force
F_I	= tie rod initial tension force (preload)
F_T	= tie rod total tension force (pressurized)
K_C	= $A_C E_C / L_C$ (from Hooke's Law)
K_T	= $A_T E_T / L_T$
k	= friction coefficient (torque)
L	= length
L_C	= overall cyl. body length
L_T	= overall tie rod length
N	= number of tie rods
P	= internal cyl. pressure
P_S	= separation pressure
R	= radius at any point in cyl. wall
S	= stress (subscript C for cyl.)
T	= nut torque
Δ_C	= cyl. body re-expansion
Δ_T	= tie rod additional stretch
δ	= deformation (subscript C for cyl.)
ε_C	= unit axial strain in cyl.
η	= pressure load coeff.
μ	= Poisson's ratio
	Note: All units lb, in.

Included here are sample test values for the Parker-Hannifin 2H industrial hydraulic cylinder line.

Predicting Tie-Rod Stretch

The total method really is just a way to predict accurately the stretch and stress in cylinder tie rods under the interacting influences of preload torquing, piston position, cylinder body relaxation, maximum pressure, and pressure cycling. Beyond a certain pressure, depending on how the cylinder is applied, the cap or head will start to separate from the cylinder body. Slight separation is sometimes tolerable, depending on the seal type, but beyond some point leakage will occur. If the tie rods were not preloaded, the separation would start as soon as the cylinder was pressurized (Fig. 5.21). With preload, the cylinder barrel initially is precompressed axially by the force in the tie rods, and gradual pressurization of the cylinder stretches the tie rods and relaxes the barrel axially. No separation occurs until precompression has been removed by the combination of the tie rods stretching and the cylinder relaxing.

The most important equation for the derivation is

$$\Delta_T = (F_T - F_I)/K_T \tag{5.3}$$

where Δ_T is the tie-rod additional stretch resulting from pressurization of the cylinder, F_T is total tie-rod force, F_I is initial tie-rod force (preload), and K_T is $(A_T E_T)/L_T$, the constants from Hooke's law applications. (Hooke's law defines modulus of elasticity E as stress over strain, or $FL/A\delta$, where F is force, L is length, A is cross-sectional area, and δ is deformation.)

Cylinder Relaxation

This is not as simple as tie-rod stretch. It is a complex elastic function affected by three-dimensional stress-strain relationships involving not only Hooke's law but also radial and hoop stresses and Poisson's ratio:

$$\epsilon_C = \frac{Sc}{Ec} - \frac{\mu}{Ec}[S_{Hoop} + S_{Radial}] \tag{5.4}$$

where ϵ_C is unit axial strain of the cylinder barrel, S_C is cylinder stress, E_C is cylinder modulus of elasticity, and μ is Poisson's ratio. Poisson's ratio is the ratio of lateral deformation to longitudinal deformation, and a value of about 0.25 was selected for the steel used in typical cylinders.

Hoop and radial stresses, however, are not constant across the wall thickness of a cylinder body. Therefore, Berninger considered the stresses at an arbitrary point of radius R:

Table 5.3 Summary of Derived Tie-Rod Equations

Tie rod total tension

$$F_T = F_I + \eta P A_P \qquad \text{Eq (1)}$$

Tie rod initial tension

$$F_I = A_P(1 - \eta)P_S \qquad \text{Eq (2)}$$

Tie rod stretch

$$\Delta_T = (F_T - F_I)/K_T \qquad \text{Eq (3)}$$

Cylinder axial strain

$$\varepsilon_C = \frac{S_C}{E_C} - \frac{\mu}{E_C}[S_{Hoop} + S_{Radial}] \qquad \text{Eq (4)}$$

Cylinder re-expansion

$$\Delta_C = \frac{F_I - F_C - 2\mu(PAp)}{K_C} \qquad \text{Eq (5)}$$

Cylinder total compression

$$F_C = F_I - 2\mu(PAp) - \frac{K_C}{K_T}(F_T - F_I) \qquad \text{Eq (6)}$$

Final force balance

$$F_T = F_C + PA_P \qquad \text{Eq (7)}$$

Nut torque

$$T = k\left[(1 - \eta)\ \frac{D_T A p}{N}\right] P_S \qquad \text{Eq (8)}$$

Tie rod stress

$$S_T = S_I + \frac{A}{NA_{Thd}}\ (\eta)P \qquad \text{Eq (9)}$$

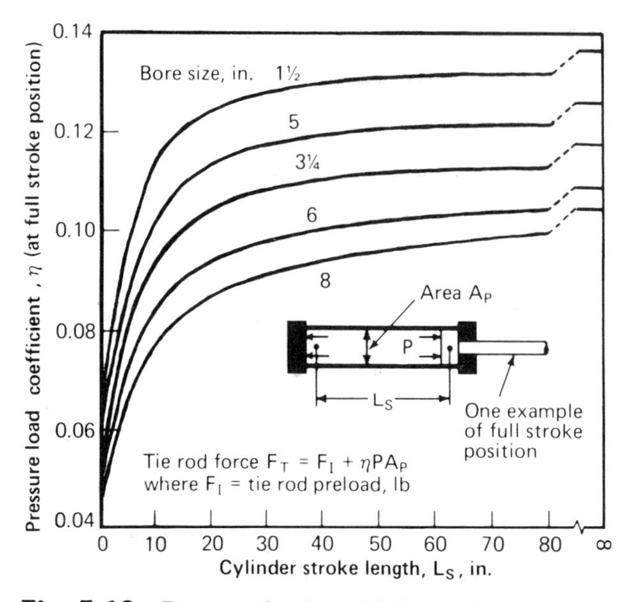

Fig. 5.19 Pressure load coefficient η for P-H series 2H hydraulic cylinders.

$$S_{\text{Hoop}} = \frac{a^2P}{b^2 - a^2}\left(1 + \frac{b^2}{R^2}\right)$$

$$S_{\text{Radial}} = \frac{a^2P}{b^2 - a^2}\left(1 - \frac{b^2}{R^2}\right)$$

where a is inside radius, b is outside radius, and R is the variable radius. Fortunately, the sum of these stresses becomes independent of radius R, thus simplifying the relationship to

$$S_{\text{Hoop}} + S_{\text{Radial}} = \frac{2a^2P}{b^2 - a^2}$$

which further reduces to

$$S_{\text{Hoop}} + S_{\text{Radial}} = 2P\left(\frac{A_P}{A_C}\right)$$

where the internal pressure area $A_P = \pi a^2$, and the cylinder wall cross-sectional area $A_C = \pi(b^2 - a^2)$. Substitute into Eq. (5.4):

$$\epsilon_C = \frac{S_C}{E_C} - \frac{\mu}{E_C}\left[2P\left(\frac{A_P}{A_C}\right)\right]$$

Convert to the same units as in Eq. (5.3) by recognizing that

$$\epsilon_C = (-\delta_C + \Delta_C)/L_C, \qquad S_C = -F_C/A_C$$

and

$$\delta_C = (-F_I L_C)/A_C E_C)$$

where Δ_C is cylinder body reexpansion during application of pressure and L_C is overall cylinder body length. The minus symbols distinguish compression from tension. The result, after manipulation, is

$$\Delta_C = \frac{F_I - F_C - 2\mu(PA_P)}{K_C} \tag{5.5}$$

where $K_C = (A_C E_C)/L_C$. Realizing that $\Delta_C = \Delta_T$, then Eq. (5.5) equals Eq. (5.3):

$$\frac{F_T + F_I}{K_T} = \frac{F_T - F_C - 2\mu(PA_P)}{K_C}$$

or

$$F_C = F_I - 2\mu(PA_P) - \frac{K_C}{K_T}(F_T - F_I) \tag{5.6}$$

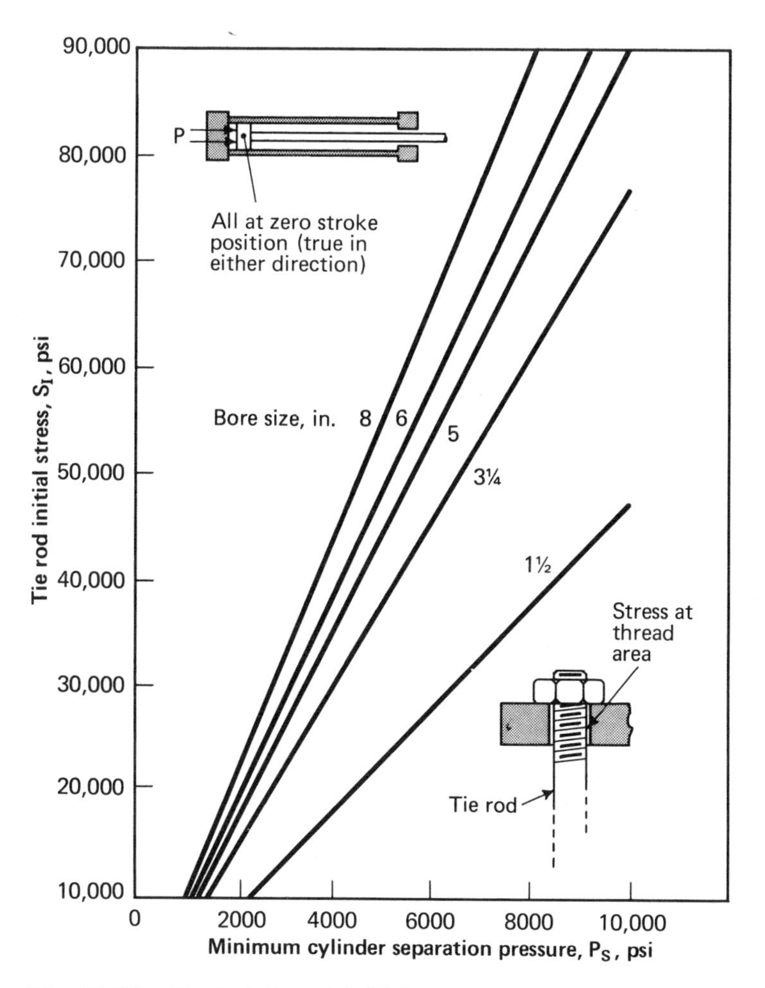

Fig. 5.20 Typical tie-rod initial stress.

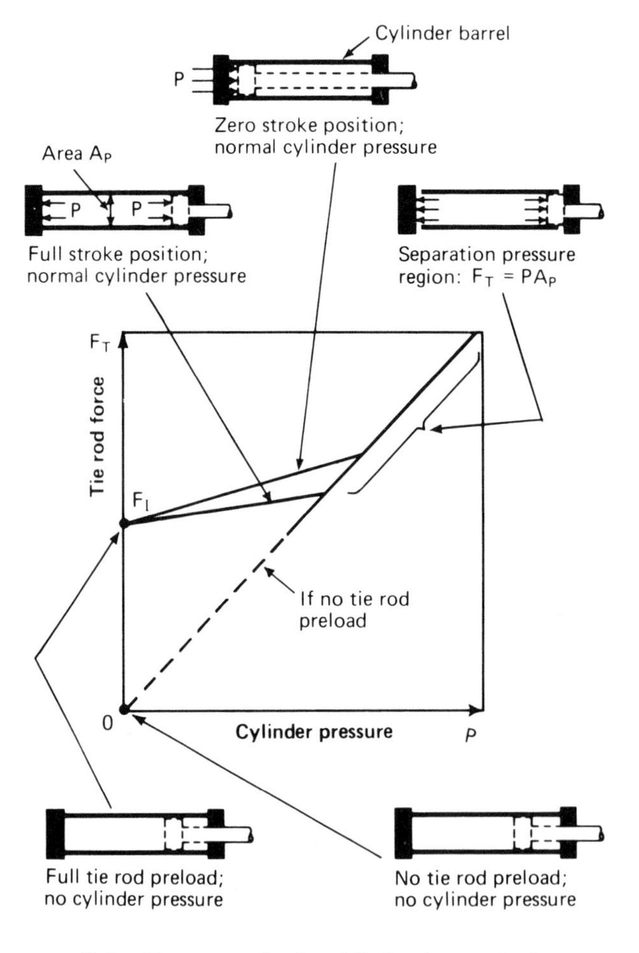

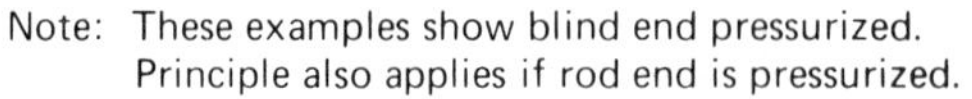
Note: These examples show blind end pressurized.
Principle also applies if rod end is pressurized.

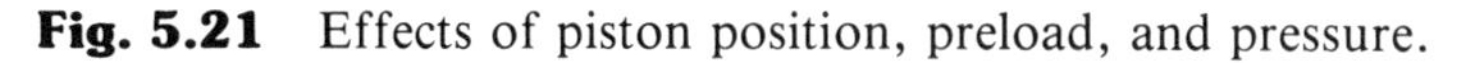
Fig. 5.21 Effects of piston position, preload, and pressure.

Final Balance of Forces

A simple axial balance of cylinder and pressure forces on either end head yields

$$F_T = F_C + PA_P \tag{5.7}$$

The original equation (5.1) is thus obtained by substituting Eq. (5.6) into Eq. (5.7):

$$F_T = \left[F_I - 2\mu(PA_P) - \frac{K_C}{K_T}(F_T - F_I)\right] + PA_P$$

which reduces to

$$F_T = F_I + \eta(PA_P) \tag{5.1}$$

where η is the pressure load coefficient:

$$\eta = \frac{1 - 2\mu}{1 + (K_C/K_T)} = \frac{1 - 2\mu}{1 + (A_C E_C L_T / A_T E_T L_C)}$$

Sample values of η are plotted in Fig. 5.19.

By the same token, Eq. (5.2) is obtained by combining Eq. (5.1) and Eq. (5.7) to get $F_I + \eta(PA_p) = F_C + PA_p$ and letting F_C = zero when $P = P_S$ because the head no longer loads against the cylinder. Thus,

$$F_I = A_P(1 - \eta)P_S$$

Effect of Piston Position

The preceding analysis was based on an empty cylinder barrel. It is valid only in the ideal instance of an exceptionally long cylinder with the piston in the full extended position and with the piston force taken fully by the head. However, the data can be readily adapted for actual cylinders by recognizing the effects of piston position and allowing for them (Fig. 5.21).

For instance, if the piston is in zero stroke position (with the cylinder pressurized only at the beginning of the stroke), then the entire cylinder length is not subjected to pressure. The effects of hoop and radial stress thus are nil, and the 2μ term drops out in the pressure load coefficient. This applies to the cylinder going in either direction. If $\mu = 0.25$ (a rough approximation for some cylinder barrel materials), then $1 - 2\mu = 0.5$, but without the 2μ term the value of η doubles. In other words, the pressure load coefficient for zero stroke position in either direction is roughly twice that for full stroke position.

In practical terms, it means that the rate of increase of F_T with pressure [Eq. (5.1)] is most rapid when the piston stroke is held at zero position. That's logical when you consider that most of the cylinder is not pressurized and therefore is not strained radially or circumferentially.

It also means that at zero stroke position separation will occur at higher pressure. That's because the cylinder relaxes less, and the tie rods have to stretch more before lifting the caps from the barrel.

Final Torquing of Nuts

The final part of the calculation is to determine the amount of tie-rod nut torque that will give the right preload for the head-separation pressure desired and to make sure that fatigue life will be essentially infinite within the maximum cylinder working pressure.

Several factors are involved: the nominal diameter of the threaded section of the tie rod, the coefficient of friction, the pressure load coefficient developed earlier, and the ultimate and fatigue stresses of the material. The basic equation is

$$T = kD_T \left(\frac{F_I}{N}\right)$$

where T is nut torque, k is friction coefficient, D_T is tie-rod nominal thread diameter, F_I is tie-rod preload force, and N is number of tie rods. The term F_I can be expressed as $A_P(1-\eta)P_S$ from Eq. (5.2), so the handy working equation becomes

$$T = k\left[(1-\eta)\frac{D_T A_P}{N}\right] P_S \tag{5.8}$$

Values of T should be determined at minimum values of η in order to keep parts from separating at the lowest conditions of separation pressure. This would be when the entire cylinder length is pressurized at full stroke position.

The stress S_T in the tie-rod thread area is F_T/NA_{Thd}, which when combined with Eq. (5.1) becomes

$$S_T = S_I + \frac{A}{NA_{Thd}}(\eta)P \tag{5.9}$$

The largest stress is at zero stroke position for reasons explained before. Also, long cylinders have higher values of stress than short cylinders at zero stroke position.

Fatigue life depends on excursion of load pressure and cycles of operation. The accompanying graphs of stress cycles and the modified Goodman diagram (Figs. 5.22 and 5.23) explain it sufficiently for this discussion. Here's how it works:

The tie rods are under initial stress (preload), which establishes the starting point (zero fluid pressure) for the pressure cycling. As the pressure increases from zero to maximum and back again, the stress level will cycle accordingly. Typical cycling tests are somewhat sinusoidal, and that is how the graphs are drawn. Average stress is midway between initial and maximum and range stress by definition is half the total excursion.

Fatigue life depends on the properties of the tie-rod material interpreted with the modified Goodman diagram. It is a straight-line plot joining two critical points: the yield point without cycling (right-hand end of the line) and the endurance limit when pressure is cycled from full tension to full compression (zero average stress). Goodman discovered that loadings below that arbitrary line generally result in infinite fatigue life and above it encourage failures. It is the intent of this tie-rod design method to avoid any loadings above the line. An example is shown in Fig. 5.23.

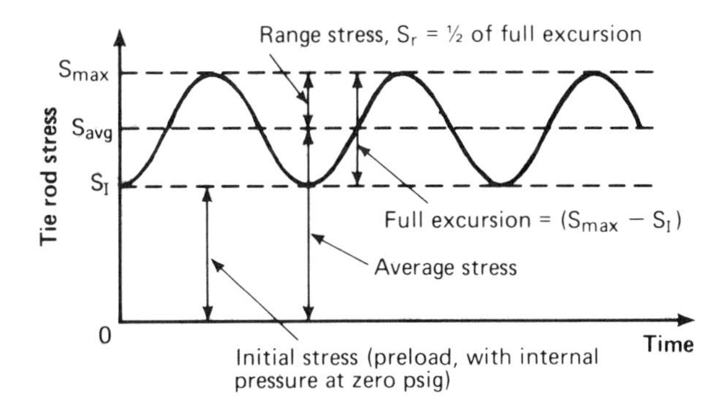

Fig. 5.22 Explanation of range stress cycling.

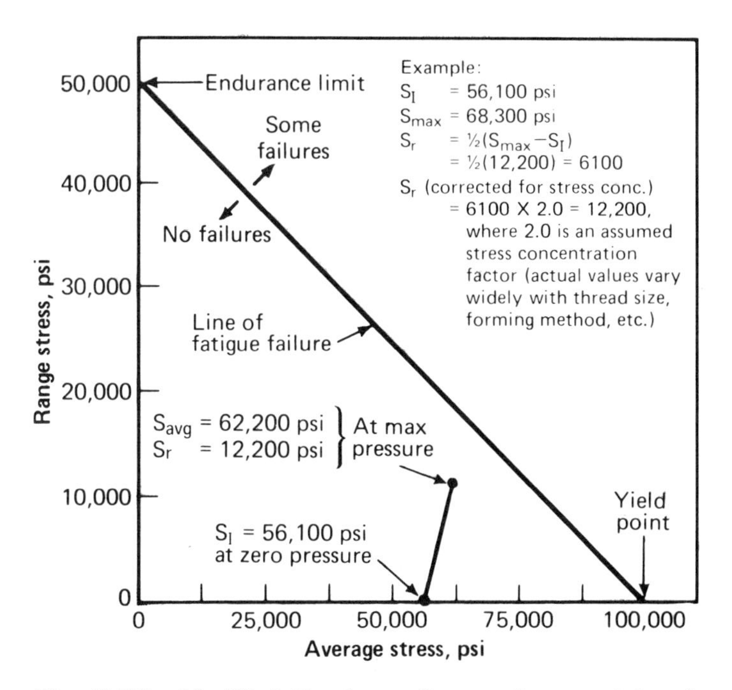

Fig. 5.23 Modified Goodman diagram for examining tie-rod stress.

Test Results Verification

Mr. Berninger calculated the stress values at different pressure levels for a test cylinder (5-in diameter) according to the equations and plotted them on a stress-pressure curve (Fig. 5.24). According to him, the calculated values very nearly match test values taken on an actual cylinder, instrumented to measure strain directly from preload to beyond final separation pressure.

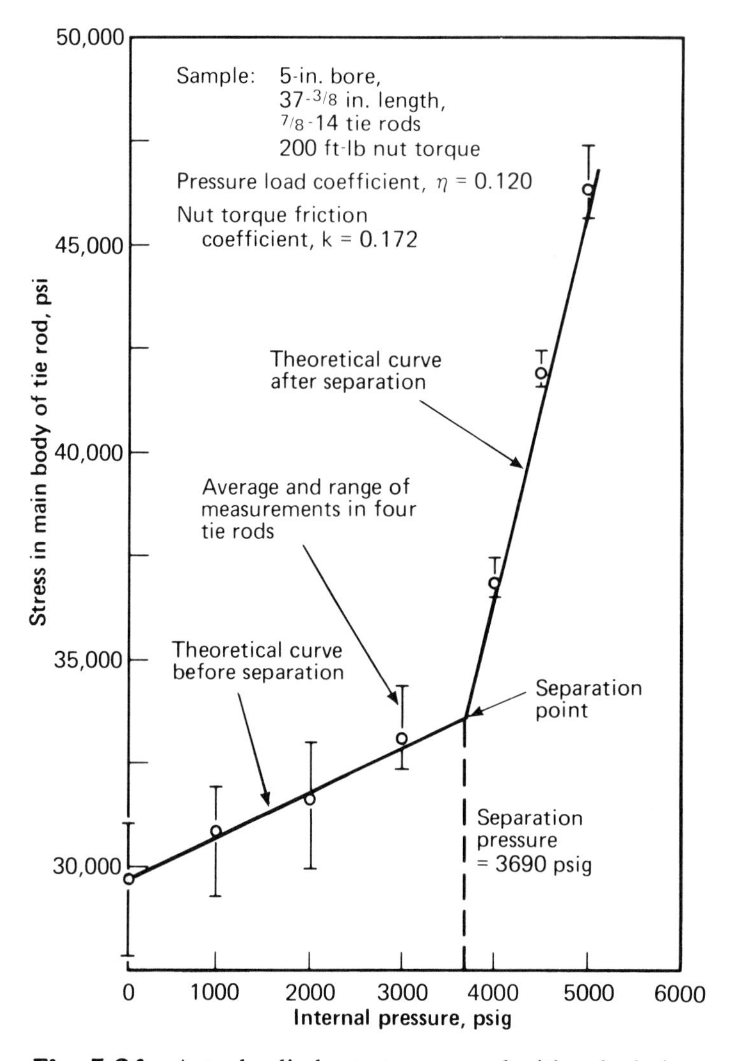

Fig. 5.24 Actual cylinder test compared with calculations.

6

Hydraulic Motors, Couplings, Drives

Four principles of operation cover the majority of hydraulic motors, couplings, and drives: (1) hydrodynamic, (2) hydroviscous, (3) hydromechanical, and (4) hydrostatic. Figure 6.1 briefly defines 13 common examples.

The bulk of applications of drives in the fluid power industry are hydrostatic—meaning a positive-displacement pump driving a positive-displacement motor—because they are compact, powerful, readily controlled, and widely available. Also, pressure, torque, and efficiency remain high even at partial loads and speeds.

Therefore, most of the discussions in this chapter pertain to positive-displacement hydrostatic motors and drives, including piston, gear, and vane types. However, some discussions of hydrodynamic, hydroviscous, and hydromechanical types are in order.

HYDRODYNAMIC COUPLINGS

The kinetic energy of a hydraulic fluid, put in motion by an input impeller, drives an output impeller. *Hydrodynamic* is synonymous with *hydrokinetic.*

The simplest adjustable-speed version is the single-stage scoop-type coupling (Fig. 6.2). When the impellers are at rest, the drive housing is partly filled with oil. As the input shaft rotates, the oil is put into motion by the input-shaft impeller, and kinetic energy drives the output impeller, called the runner. There must be a speed difference between the impeller and runner: At least 3% slip is required. Torque depends on oil volume and slip.

As shown in Fig. 6.2, outer casing A together with drive casing B and impeller C are driven by the input shaft. The output shaft carries runner D. As the impeller rotates, it loses oil through nozzle E, tending to empty the drive housing and reduce the coupling between the impeller and runner. However, scoop tube F prevents this by scooping up the oil and returning it to the drive casing.

The volume of oil in the drive casing and the speed of the runner are controlled, via the scoop tube, with lever G. Torque at 100% slip is very high, excellent for start-

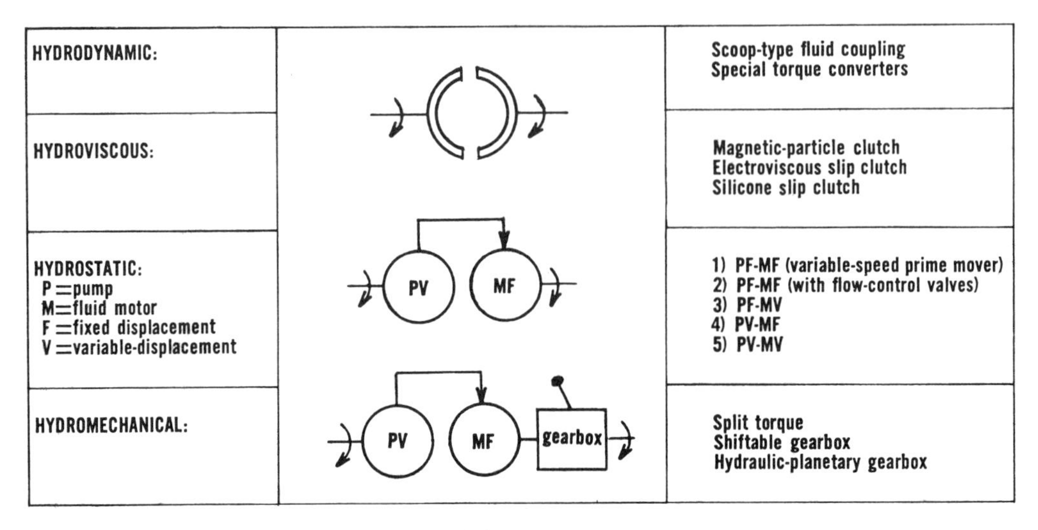

Category	Diagram	Examples
HYDRODYNAMIC:		Scoop-type fluid coupling Special torque converters
HYDROVISCOUS:		Magnetic-particle clutch Electroviscous slip clutch Silicone slip clutch
HYDROSTATIC: P =pump M=fluid motor F =fixed displacement V =variable-displacement	PV MF	1) PF-MF (variable-speed prime mover) 2) PF-MF (with flow-control valves) 3) PF-MV 4) PV-MF 5) PV-MV
HYDROMECHANICAL:	PV MF gearbox	Split torque Shiftable gearbox Hydraulic-planetary gearbox

Fig. 6.1 Categories of hydraulic drives.

ing up a still shaft. But speed droop is very great, not suitable for constant-speed operation with varying load.

The main advantage of a simple hydrodynamic drive is that it combines the characteristics of a fluid clutch with those of an adjustable-speed drive. A natural application is for frequent starting of heavy loads. Heavy shocks are smoothly absorbed with relative ease, and reliability is excellent. It is bulkier than positive-displacement drives such as the hydrostatic drives described later.

Torque converter: A hydrodynamic drive is also a torque converter if it has a third element, the stator, between the impeller and runner. Some torque converters have three stages (Fig. 6.3). The output torque at stall can be made many times the input. As output speed approaches input speed, output torque approaches zero.

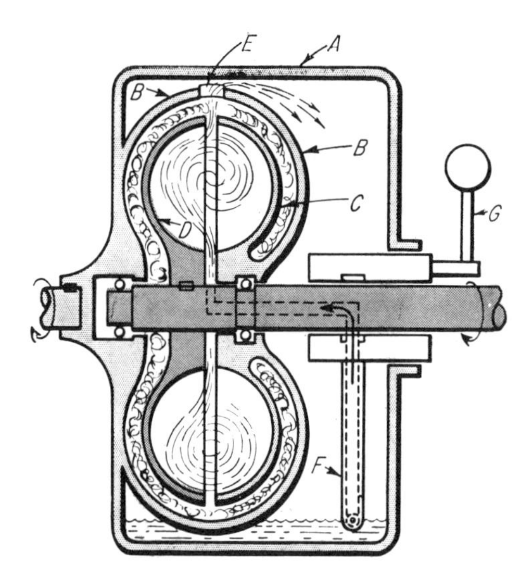

Fig. 6.2 Scoop-type hydraulic coupling.

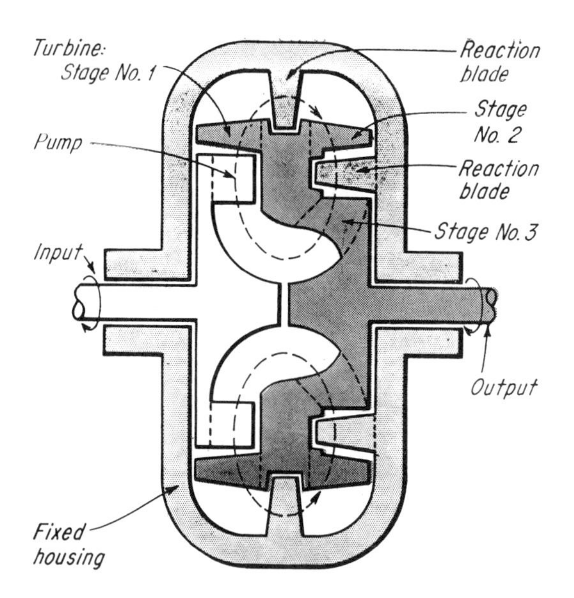

Fig. 6.3 Three-stage hydraulic torque converter.

In theory, speed can be changed by making the stator blades adjustable, but this normally is not done, and torque converters are not sold as adjustable-speed drives.

HYDROVISCOUS DRIVES

Hydroviscous is a coined word to describe drives that depend on viscosity change for torque and speed control. The most common is the magnetic-particle clutch (Fig. 6.4). The particles act as a fluid when unenergized. Lesser known, but also a true viscous drive, is the electroviscous hydraulic clutch (Fig. 6.5). A truer viscous drive is the liquid-slip clutch (Fig. 6.6). Although they are clutches, each in a limited way can control speed.

Magnetic-particle clutch: The space between the two rotating members in Fig. 6.4 is loosely filled with metallic particles, spherical in shape and screened to a uniform size. The electrical coil is stationary and produces a magnetic field which acts through an air gap; slip rings are not needed.

Unenergized, the magnetic particles lie at random and do not couple mechanically. When the coil is energized with dc, the resulting field aligns the magnetic particles in "chains," linking one rotating member with the other. The linking force can be increased or decreased by varying the current flow through the energized electrical coil.

One of the outstanding characteristics is that this force does not decrease with increasing speed as it does with friction clutches. Output torque is constant throughout the speed range.

Electroviscous hydraulic clutch: It is possible to change the viscosity of some liquids by applying an electrical field across them. If the space between the impeller and runner is filled with such a liquid, the amount of coupling between the input and output surfaces is a function of the electric field intensity. At a field potential of 75 V/mil of fluid thickness, measured between the plate surfaces, the viscosity is as much

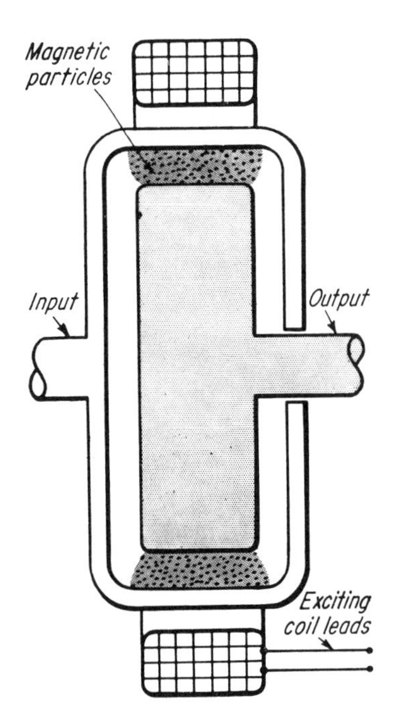

Fig. 6.4 Magnetic-particle clutch ("dry" fluid).

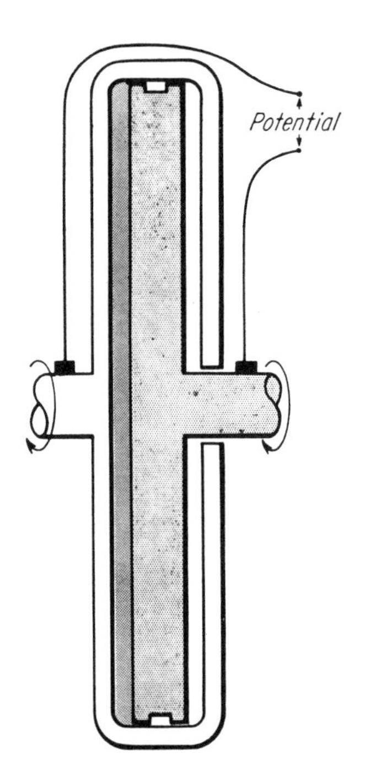

Fig. 6.5 Electroviscous clutch.

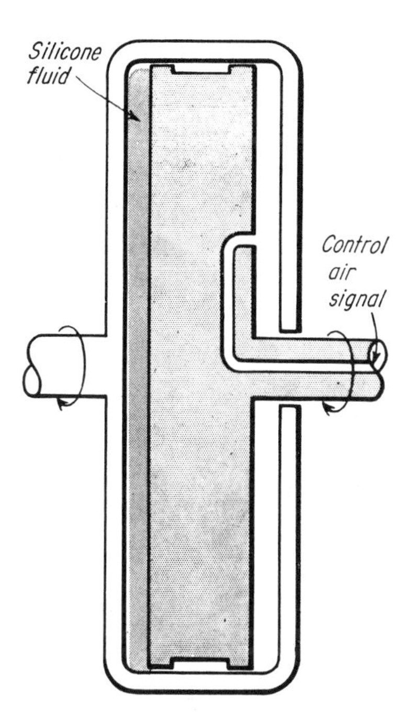

Fig. 6.6 Liquid-slip clutch.

as 200 times that of the nonenergized oil. This concept of electroviscosity is called the *Winslow effect.*

Liquid-slip clutch: Silicone oil, available in almost any desired viscosity and extremely stable at all normal temperatures, is the coupling medium in several clutch types. Transmitted torque and therefore speed can be varied by changing the fluid thickness or its shearing effect.

In one technique, the distance between the rotating surfaces is adjusted pneumatically, lending itself to feedback control. In another, centrifugally actuated valves permit some leakage of fluid, increasing slip and thus maintaining preset speed.

As in eddy-current clutches and simple hydrodynamic couplings, efficiency is highest when slip is lowest. Heat must be removed, usually by fins on the rotating parts or with auxiliary cooling fans.

HYDROMECHANICAL DRIVES

Really two drives in one, a hydromechanical drive is a hydrostatic or hydrodynamic drive combined with a gearbox. It can take one of three forms: (1) *split-torque* drive in which a variable-displacement motor adjusts the output speed of a mechanically driven differential planetary by varying the speed of one of the gears, (2) *mechanically shiftable* but separate gearbox driven by a hydrostatic or hydrodynamic drive, and (3) *hydraulic-planetary* gearbox in which the gears double as gear pumps and can be speed-controlled by varying the restriction at the pump outlets.

Split-Torque Drive

Figure 6.7 shows the most common arrangement, but many others will work. A highly compact design, originally intended for trucks, is shown in Fig. 6.8.

In any arrangement, the prime mover is connected to the load through two paths: directly to the differential planetary gearing and indirectly off the input shaft and through the hydrostatic drive. The planetary sums the two inputs to produce the output.

The speed relationship, based on Fig. 6.7, is

$$N_o = \frac{N_1 \pm 2N_C(1+a)}{1+2a}$$

where a is the ratio of radii of planet gears to sun gear and N_1, N_C, and N_o are input, planet-carrier, and output speeds, respectively.

Carrier speed N_C depends on the displacement of pump-motor A. If the displacement is zero, neither A nor B passes flow, and N_C is zero. Output speed N_o in this case equals $N_1(1+2a)$, called the base speed.

To run above base speed, adjust A to rotate the planet carrier in the same direction as N_o, shown in Fig. 6.7. To run below base speed, reverse N_C.

Pump motor A operates as a pump while N_o is above base speed but runs as a motor when N_o is below base speed. In either case, the pump is driven from the mechanical system, and the motor returns the energy to the same system. Theoretically, there is no energy loss.

But overall efficiency is highest if most of the power is transmitted through the mechanical gearing, because gears are more efficient than pumps or motors. Typical efficiency is considerably above 90% for a drive with an output speed ratio (max. N_o)/(min. N_o) of 2 : 1. A split-torque drive is not usually recommended unless the ratio is limited to 3 : 1. For top efficiency—over 95% is possible—select the narrowest possible speed range.

The size of the hydraulic portion of a split-torque drive is small compared with a straight hydrostatic drive because the gears do most of the work. The pump-motors only speed up or slow down the planet carrier as needed.

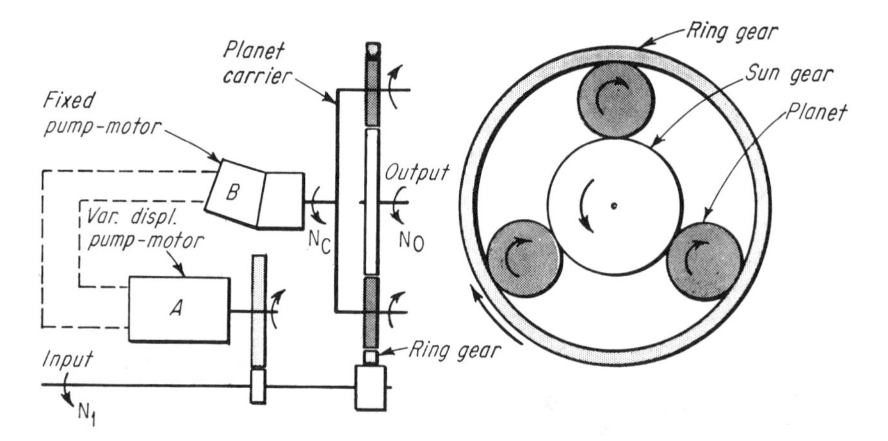

Fig. 6.7 Split-torque hydromechanical drive.

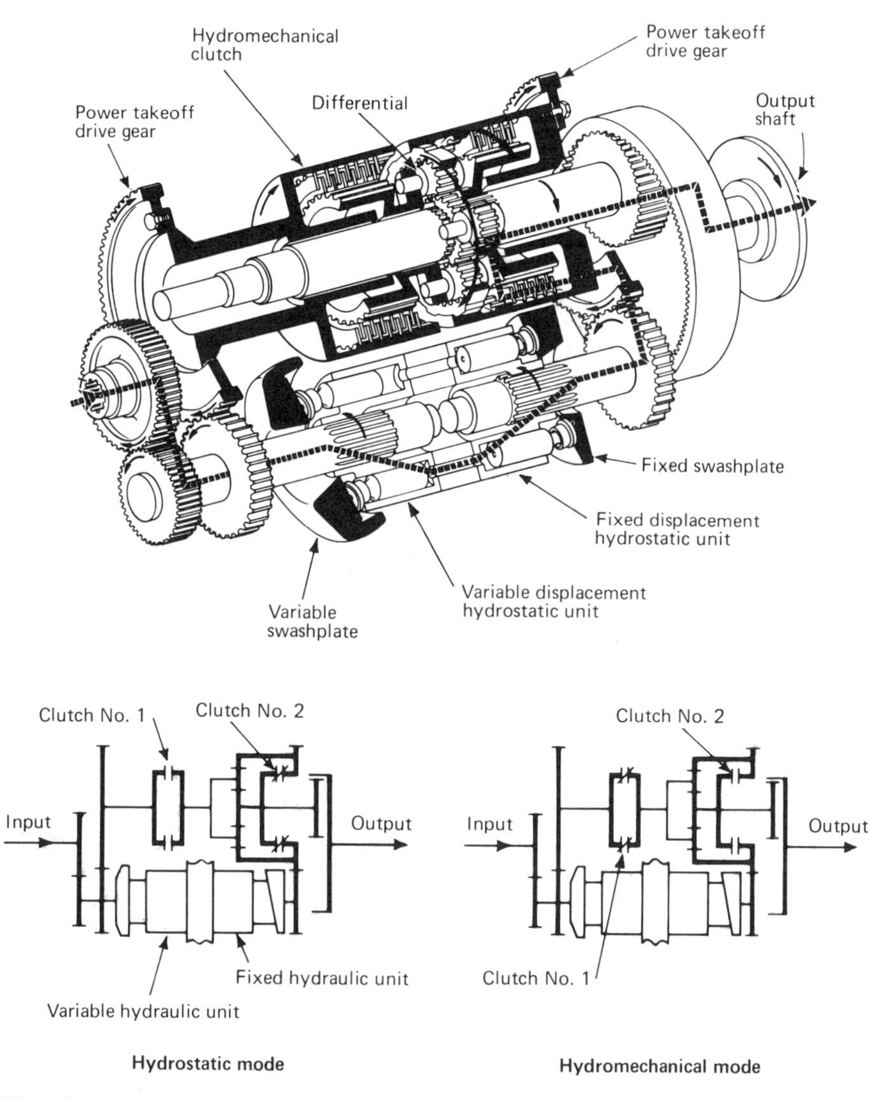

Fig. 6.8 Dual-path hydromechanical transmission exploits a small hydrostatic pump-motor drive for high torque at startup and at low speeds and then clutches in a planetary gear for sustained normal and high-speed operation. Sundstrand developed this design for trucks.

The initial cost of a split-torque drive is higher than that for a straight hydrostatic drive. So when simplicity or low first cost is more important than operating efficiency, as in low-horsepower applications, choose the straight drive. Where operating economy and small size are paramount, choose split-torque.

Shiftable Gearbox

The speed range of any hydrostatic or hydrodynamic drive can be extended by adding a gearbox with adjustable ratios. For example, a truck or car with a torque-converter hydrodynamic drive (Fig. 6.9) coupled to a five-speed gear transmission can utilize nearly full engine horsepower and full drive efficiency in an output-speed range of over 6 : 1. A few of the components are shown in Fig. 6.9.

Without the gearbox, output torque, power, and speed depend on the characteristic performance curve of the torque converter, and although the drive can be readily matched to almost any load, speed is not truly adjustable.

Hydraulic-Planetary Gearbox

If the sun gear and planets of a simple planetary—without ring gear—are designed to double as gear pumps, the rotation of the planets around the sun can be controlled by restricting the pumped flow (Fig. 6.10).

In the version shown, the input shaft rotates the planet cage, and the sun gear is keyed to the output shaft. With 100% restriction of pumped flow, the planets lock to the sun gear, and the output speed is identical to the input. With zero restriction, the planets rotate freely, and the output shaft can be held still. In between, any output/input speed ratio is possible.

Valves inside the drive housing restrict the pump flow, and their adjustment is external. The entire drive housing rotates at input speed; the centrifugal action maintains an annular bath of oil to supply the valves.

Performance is not unlike an eddy-current clutch in that a decrease of output speed requires slip and therefore lowers efficiency. Slip in this case is flow through a valve. However, by mixing air with the oil at the high-slip conditions, the pumping

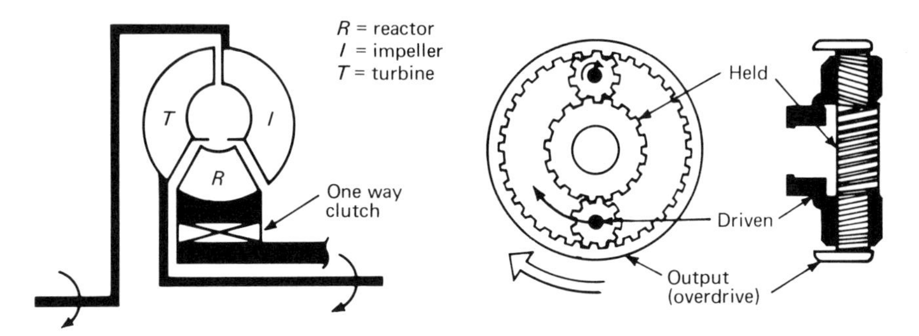

Fig. 6.9 Hydrodynamic drive, coupled to gearbox.

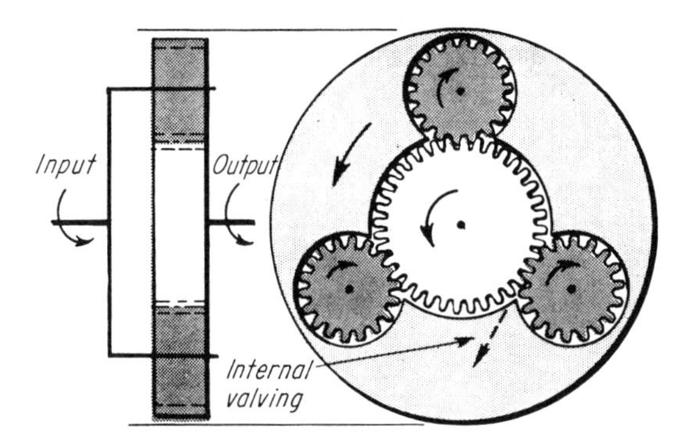

Fig. 6.10 Hydraulic-planetary drive uses planetary gears as hydrostatic motors.

losses are reduced. The gear-pump drive also has some of the advantages of an eddy-current clutch, which include compact design and good shock absorption characteristics.

HYDROSTATIC DRIVES

A hydrostatic drive consists of a positive-displacement pump driving a positive-displacement fluid motor, usually in one of the basic arrangements shown in Fig. 6.11. The pumps and motors are available in many forms, including external gear, internal gear, vane, radial piston, and axial piston. Each design (Fig. 6.12) with minor modifications can be either a pump or a motor.

Any pump or motor can be varied in speed; for example, with a valving system to vary the flow to the fluid motor or with a variable-speed prime mover to drive the pump.

However, only vane and piston-type pumps and motors can have variable displacement. Gear pumps have constant displacement and must be rotated at variable speed to adjust flow.

Adjustable radial-piston and vane pumps and motors have movable pressure rings that change the eccentricity of the vanes or pistons relative to the shaft, thus changing the displacement. Adjustable axial-piston pumps and motors usually have swash plates or wobble plates that determine the effective stroke of the pistons—or the valve block itself can be tilted. But the principles are the same.

Fluid motor speed is dependent on output flow of the pump divided by the displacement per revolution of the motor. Input torque to the pump (for a constant-speed electric motor) and output torque of the hydraulic motor for a given efficiency depend on pressure and displacement. In a positive-displacement pump-motor drive, pressure rises to whatever level is required to move the load at the set displacement

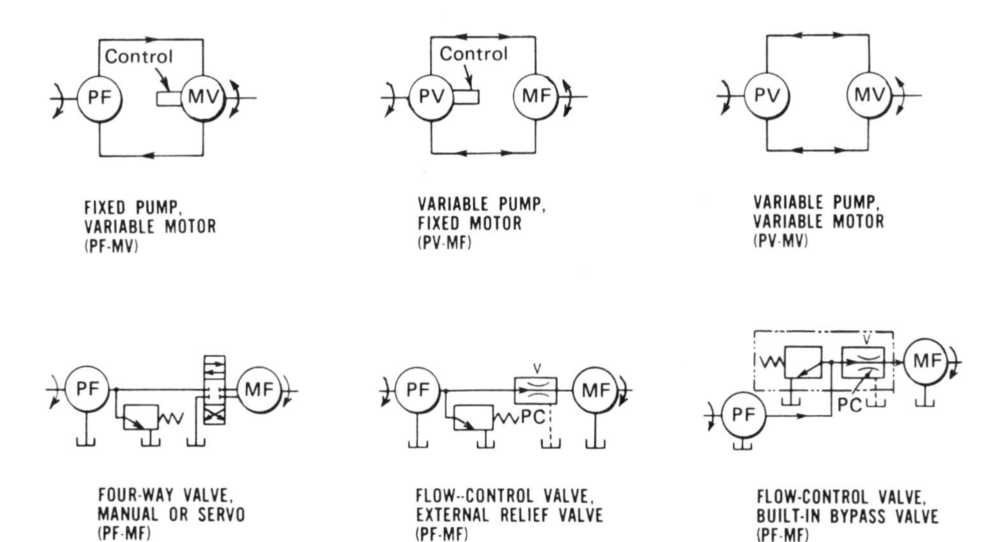

Fig. 6.11 Some typical arrangements for hydrostatic drive elements.

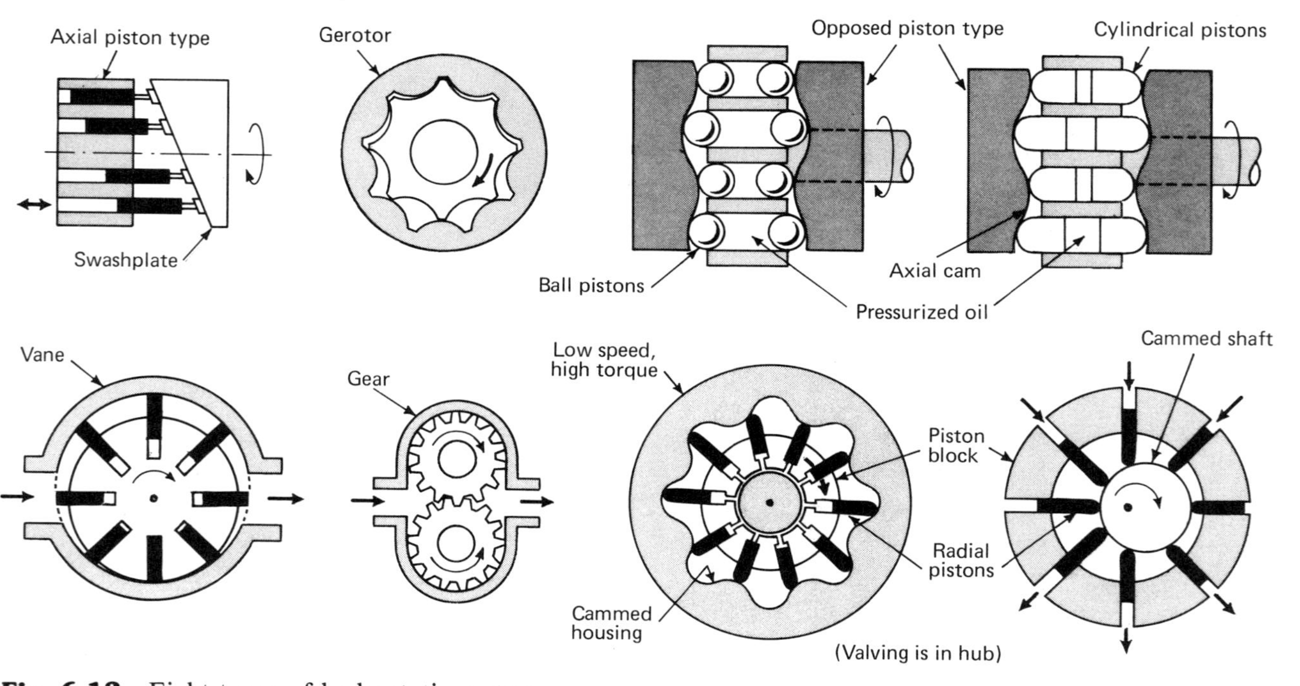

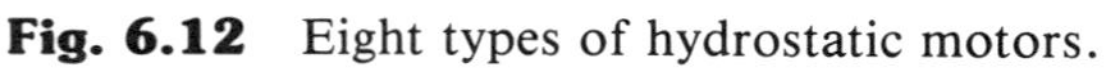
Fig. 6.12 Eight types of hydrostatic motors.

(speed) unless load torque exceeds the capacity of the fluid motor, the pump, or the prime mover. If the pressure exceeds the setting of any of the relief and safety devices, the drive slows down or stalls without damage because fluid is bypassed. As soon as the load returns to normal, the motor operates as before.

The most efficient units are the piston type; vanes are slightly less efficient.

Safe, Smooth Braking

Hydrostatic drives can be slowed down and stopped more smoothly and rapidly than most other drives. This is due in part to the overload capability of the fluid motor. There are two other important reasons: (1) The fluid motors are smaller and have considerably less inertia than equivalent electric motors, eddy-current clutches, or mechanical drives. (2) The motive fluid is easy to control or restrict at any of several flow points.

Two inherent braking systems are ideal for fluid drives: regenerative and dynamic.

With *regenerative* braking, the inertia of the driven machinery begins to drive the fluid motor as a pump, which in turn drives the pump as a motor. The kinetic energy is quickly absorbed, stopping the fluid motor and its connected equipment.

In *dynamic* braking, flow from the pump to the fluid motor is cut off by a valve. Load inertia drives the fluid motor as a pump, and the output is forced through a relief valve, creating a cushioned braking effect.

Conventional external braking with a friction clutch is often used, usually on the driven machinery. Inherent and external braking are combined in some applications. One example is a hydraulic-powered winch, where regenerative braking action is supplemented by a friction brake to hold the load.

Adjusting the Displacement

There are many conventional ways to change the position of the yoke or stroking ring of adjustable-speed pumps and motors, including hydraulic, pneumatic, electric, and mechanical. The pressure compensator (Fig. 6.13) is of particular interest because it is a feedback system in itself.

The *pressure compensator* refers to a control feedback device that senses pressure in the line to the fluid motor and uses it to adjust pump or motor displacement to correct or improve the existing operating conditions (compensate).

This pressure signal is useful for a variety of reasons. For a fixed-displacement fluid motor, you can determine torque by reading a pressure gage. And you can use the pressure as a feedback signal to reduce pump delivery, hence speed, if this has raised torque above a preset maximum.

In Fig. 6.13, the compensator is attached to the pump control, A. Hydraulic pressure is applied at the upper port. If pressure—representing torque—exceeds the setting of pilot valve screw F, the pump displacement is reduced until the torque returns to normal.

Initial pressure is below the minimum setting of the compensator. Spring B extends piston A for maximum pump displacement. Hydraulic pressure at R acts through orifice X and passage V to impose a thrust on spool H. Piston A remains extended when this pressure cannot overcome spring G to shift spool H.

When the pressure reaches the setting of pilot valve screw F, spool H shifts to

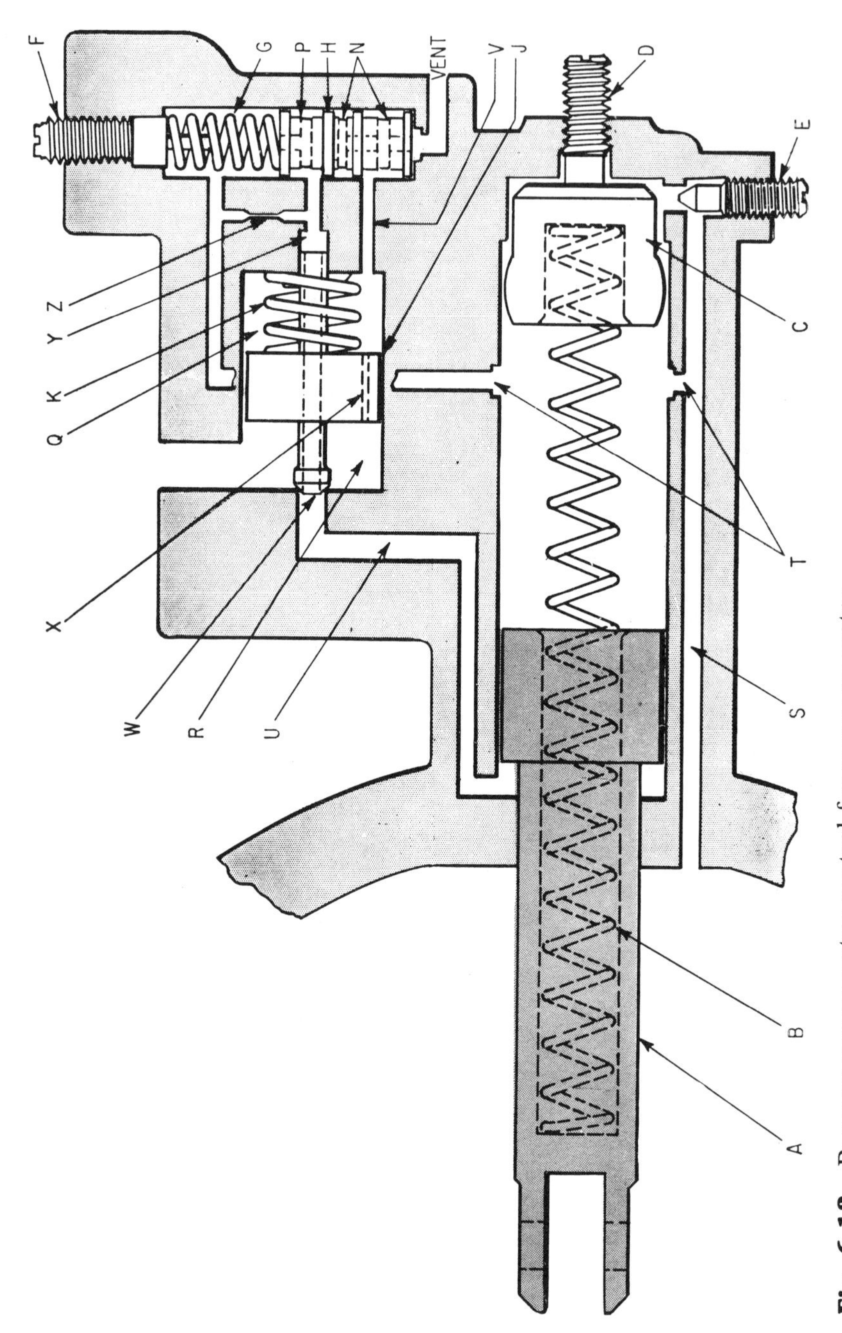

Fig. 6.13 Pressure-compensator control for pump or motor.

direct fluid from chamber Q and passage V through interconnected holes N of spool H to chamber Y. This chamber is connected to drain through orifice Z and to the annulus area of piston A through hole W of piston J and passage U. Flow through orifice X causes a pressure drop so that the pressure in chamber R becomes greater than in chamber Q. The pressure difference unseats piston J, and fluid is throttled from chamber R to passage U. Fluid in U moves piston A to decrease pump delivery until it equals system requirements.

The pressure compensator can also be mounted on a variable-displacement fluid motor. By adjusting motor displacement, you can change its speed to cope with a load change, holding the same hydraulic pressure as before. Thus speed is decreased as a function of load.

Hydrostatic-Drive Classifications

The drives shown in Fig. 6.11 can be further classified in the following five standard categories, all based on flow (Fig. 6.14).

PF-MF (variable-speed prime mover). Fixed-displacement pump, fixed-displacement motor; motor speed adjusted by varying the speed of the prime mover (electric motor or internal-combustion engine) to control pump output.

Prime mover characteristics and not pump-motor performance are involved here.

PF-MF (with flow-control valves). Fixed-displacement pump, fixed-displacement motor; motor speed adjusted with valving that controls the input or output flow of the motor.

In its simplest form the PF-MF (valved) is the lowest in cost of all hydrostatic drives. Efficiency is relatively low because control is by throttling, but there are many important uses for this basically simple drive. Several arrangements are described in detail under flow-control valves in a following section.

PF-MV. Fixed-displacement pump, variable-displacement motor; motor speed adjusted by changing the effective stroke of the pistons or vanes (gears not applicable).

This arrangement, and the PV-MF and PV-MV arrangements following, have well-defined characteristics.

In the PF-MV drive, torque decreases as speed increases, useful for driving a roll (Fig. 6.15) that winds paper or other web material. As the roll diameter increases, rotational speed must decrease to hold a constant linear velocity of the paper. At the same time, mass and resisting torque of the roll increases, matching the PF-MV performance characteristics.

Motor displacement is gradually increased by the control to reduce motor speed. The speed drops because the motor takes more fluid per revolution. Motor torque rises because the effective moment of the vanes or pistons is increased. The horsepower remains relatively constant over the entire speed range.

Using the constant torque drive described below would be uneconomical here because in such a drive the motor would have to be sized to withstand maximum torque at maximum speed—yet it would not normally run at that point.

For example, suppose the load has a torque of 40 lb-ft at 300 rpm and 10 lb-ft at 1200 rpm. A constant-torque drive would have to be sized to produce 40 lb-ft and

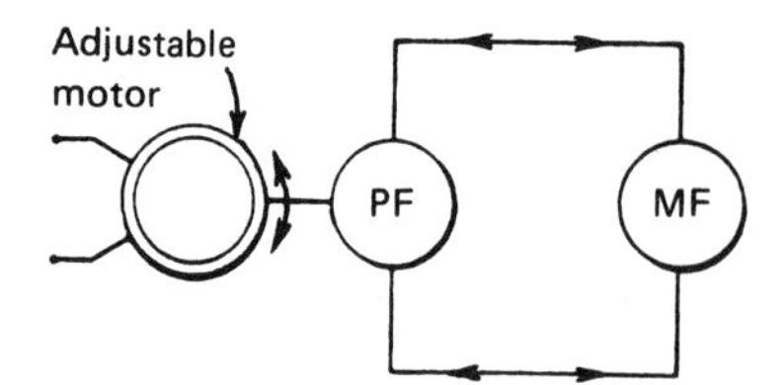

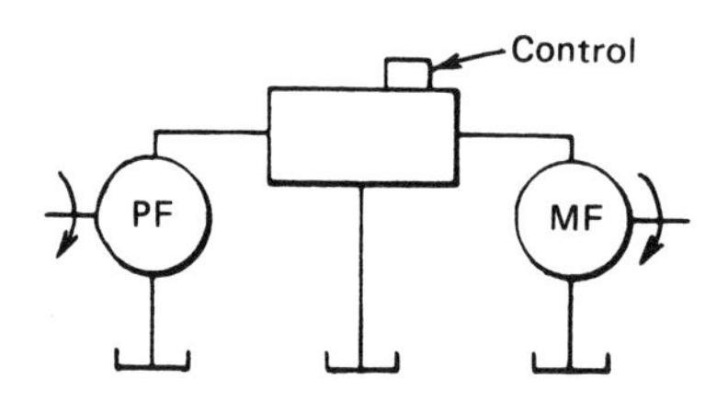

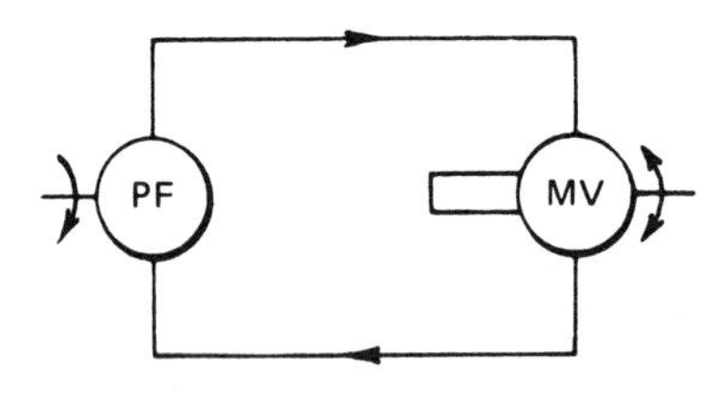

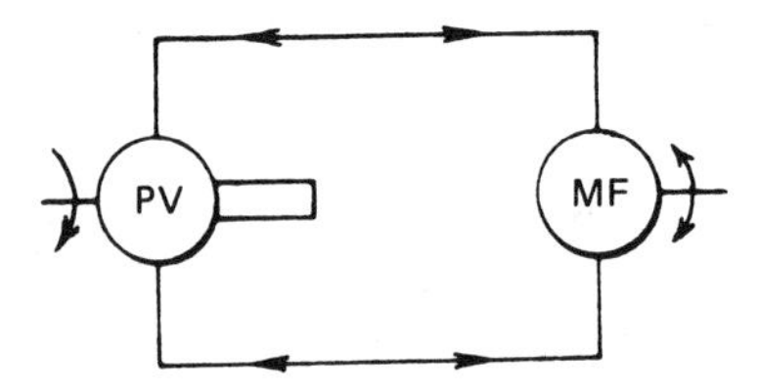

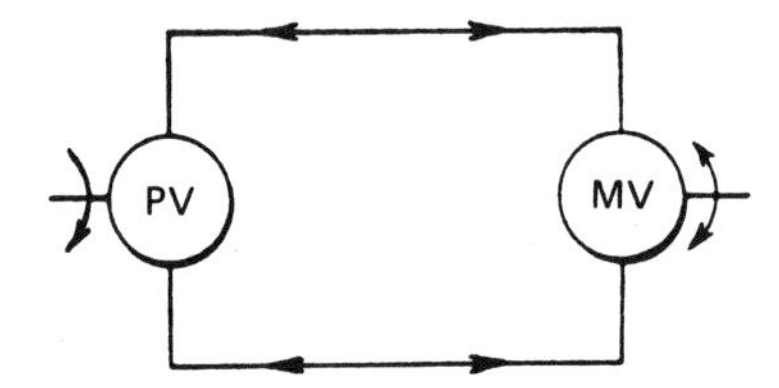

Fig. 6.14 Schematics to help classify hydrostatic drives.

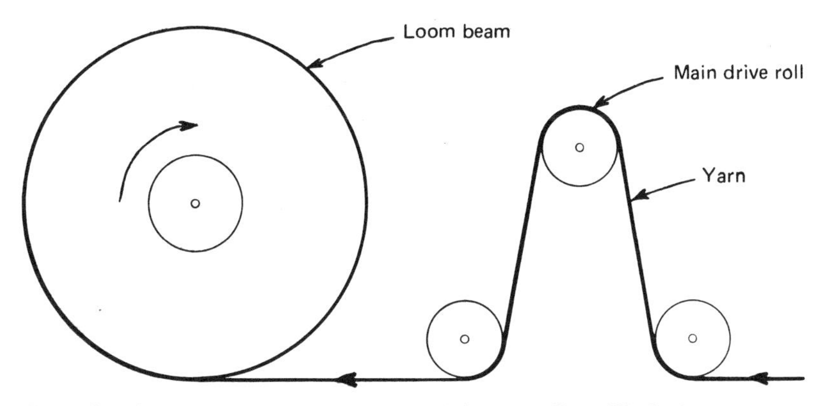

Fig. 6.15 Application of PV-MV drive: textile-mill slasher.

1200 rpm, or 9.6 hp, neglecting losses. A variable-torque drive would be sized for 10 lb-ft and 1200 rpm, or 2.4 hp, only one-fourth the size of the constant-torque drive.

PV-MF. Variable-displacement pump, fixed-displacement motor; motor speed adjusted by varying pump displacement to vary pump flow.

Torque remains constant over the entire speed range because motor displacement is fixed and the pump is able to maintain full pressure at all flows, that is, if you disregard changes in efficiency.

Motor speed is proportional to pump delivery rate, and pump power is proportional to motor speed. Pump speed is constant, even though delivery rate is adjustable.

PV-MV. Variable-displacement pump, variable-displacement motor; motor speed adjusted by a combination of adjustments of pump and motor displacements.

If neither the constant-torque nor the constant-horsepower characteristics of the PF-MV or the PV-MF drives is exactly suitable, then combining them will add versatility—and cost. Either the pump or the motor can be adjusted, or both can be adjusted together.

The speed range is greatly extended. For example, assume that the minimum speed of your driven equipment must be 50 rpm (a typical value) and maximum speed 3600 rpm.

Assume that the pump selected has a maximum displacement which is 75% of the full displacement of the fluid motor and that the pump speed is held constant at 1200 rpm. Thus the fluid motor will operate at $1200 \times 0.75 = 900$ rpm. If pump displacement is reduced to about 5% of its full stroke, the fluid motor speed will drop to about 50 rpm. With pump stroke at full displacement, and fluid-motor displacement at about 25% of full stroke (a recommended minimum), the output speed will increase to $900 \times 4 = 3600$ rpm. Total range $= 3600/50 = 72:1$, and much larger ranges are possible.

Where two machines or two separate parts of the same machine must be driven at different speeds and one of the speeds varies in a different manner from the other, the answer again is a PV-MV drive but with two fluid motors.

Speed Control with Valves

The second category of hydrostatic drives—PF-MF drive with control valves—depends on the valving for speed control and is discussed separately here for that reason.

One versatile pump-motor-valve arrangement has a so-called double or split-flow pump, each half having fixed displacement but of different magnitude. Valving can provide three basic flow rates to the fluid motor by utilizing one, the other, or both pump outputs (Fig. 6.16).

Another novel circuit is shown in Fig. 6.17. Flow control valves in a binary array give a choice of four parallel flow paths. The first path has a known fixed flow, and each successive path has double the one next to it. By opening them in every possible combination, 15 different flows are possible.

A more versatile speed control is pictured in Fig. 6.18. The directional valve in position #1 sends fluid at pump pressure through the motor and into drain. Torque and shock pressure are limited by a relief valve off the supply leg. Flow is set by subtracting from fixed pump flow with an adjustable flow control valve. Position #2 drains off inlet and discharge pressure so that the motor turns freely without doing work. Position #3 drains the input but forces the discharge to flow through the relief valve. The braking relief is an optional secondary adjustment for special characteristics and is not needed in the basic circuit.

Four Metering Configurations. Motor speed is controlled by the valving in one of four ways: bleed-off, meter-in, meter-out, or a combination of these (Fig. 6.19). *Bleed-off* control diverts a portion of pump delivery to the oil tank, and the remainder drives the fluid motor. *Meter-in* control adjusts input flow to the motor; *meter-out* control restricts output flow from the motor to vary its speed.

Combinations of bleed-off, meter-in, and meter-out circuits comprise the art

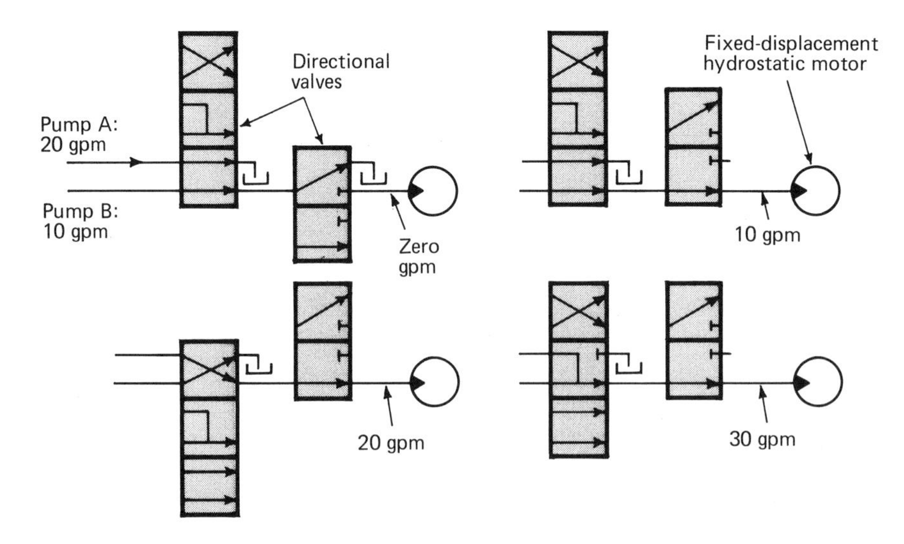

Fig. 6.16 Three flows to one motor can be created from two fixed-displacement pumps, using two valves. The pumps can be a single unit with split flow.

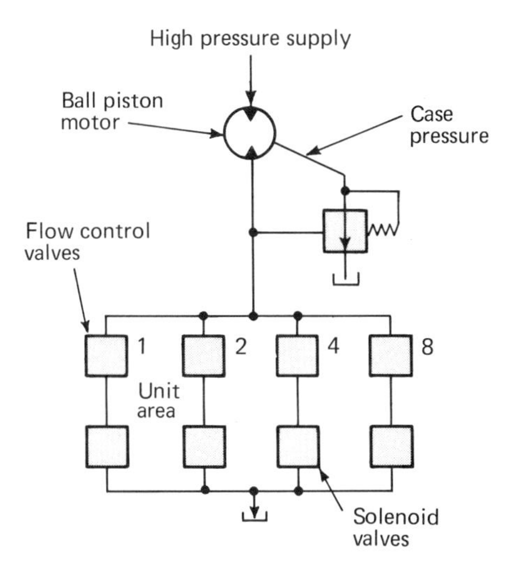

Fig. 6.17 Fifteen controlled flows to a single motor are possible with a binary array of valves at the motor outlet.

of valved hydraulic-transmission design. These combinations differ in almost every new circuit.

The choice of metering circuit is especially important. Drive performance is often based more on the circuit than on the pumps, motors, and valves you might choose to place in the circuit. Let's examine inherent advantages and limitations of each.

Bleed-off: Flow from the pump to the fluid motor is unrestricted. Control is with an adjustable bleed-off valve located in a branch of the pump discharge line. Assuming constant pump flow, whatever oil is bled to the tank never reaches the motor; thus the motor speed is inversely proportional to bled flow.

There are two advantages: The pump pressure adjusts to load, thus reducing pump effort, and the throttling losses are in the branch and not in the motor circuit. If an efficient flow-control valve is selected for the bleed-off, part-load efficiency is reasonably good.

Furthermore, if the fluid motor speed never goes to zero, the bleed-off valve never has to pass full pump flow and can be proportionately smaller. The other two configurations, meter-in or meter-out, pass full pump flow.

Meter-in: Flow from the pump to the fluid motor must pass through the control valve, which adjusts flow by automatic control of its valve port area. Pump excess flow is returned to the tank through a relief valve. Thus the pump must work against full pressure or higher at all times. This extra work is avoided in the integral bypass throttling valve.

Meter-out: All flow through the fluid motor must pass through the meter-out valve located downstream of the motor. The circuit is similar in performance to the meter-in circuit, in which the pump excess flow is discharged through a separate relief valve.

One advantage of meter-out control is its excellent shock-absorbing capability. The back pressure against the motor outlet damps sudden load changes.

Valve Types. There are three valve types used in most hydrostatic PF-MF (valved)

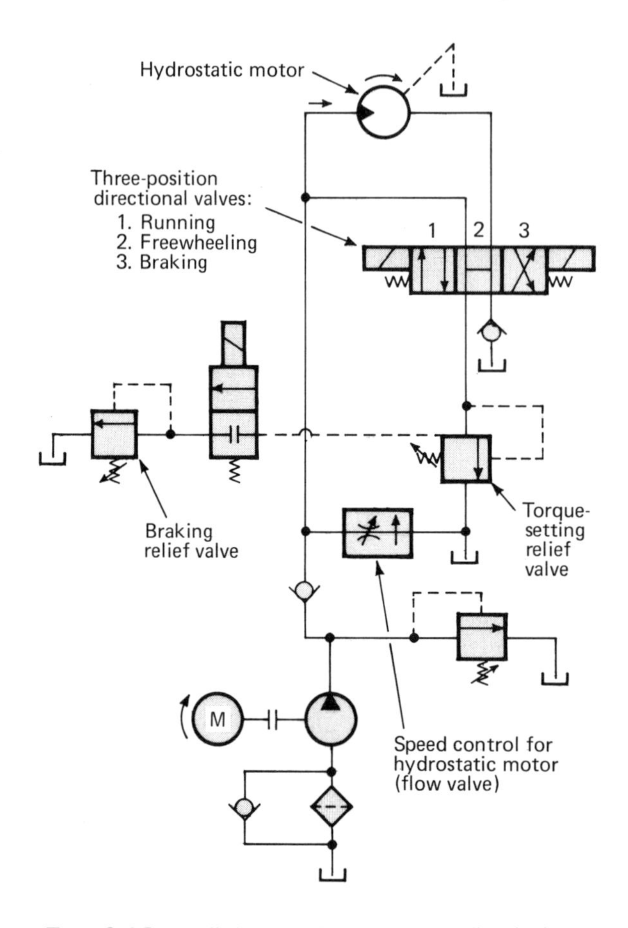

Fig. 6.18 Infinite speed control of a fixed-displacement hydrostatic motor plus control of torque, freewheeling, and braking are accomplished with special valving. (Courtesy of Parker Hannifin Corp., Cleveland, OH.)

drives: four-way throttle valve, compensated flow-control valve, and servovalve. Costs increase in the same order.

Four-way valve: A manually operated four-way valve is the simplest, cheapest, and least efficient of all the controls for hydrostatic drives. It operates basically as shown in the circuit of Fig. 6.20, which has a closed-center valve.

Move handle C to the left to drive the fluid motor in one direction at a speed determined by the throttled flow area of the valve. To reverse the motor direction, move the handle to the right.

Smooth stopping is assured by a bypass restriction placed between the two lines supplying the reversible fluid motor. This bleeds off the compression wave that is generated during quick stops.

The pump output is fully utilized only at full opening of the four-way valve; at partial flows the relief valve bleeds the excess flow to the oil tank. At the center position, fluid flow is blocked, stopping the motor.

An open-center four-way valve allows flow of the pump delivery back to the tank at greatly reduced pump-discharge pressure.

One disadvantage of the open-center valve is that the spool must be moved a considerable distance before the bypass closes off to allow flow to the motor. This ineffective motion is called *deadband.*

The biggest disadvantage of any four-way-valve control is that maximum pump power is needed whenever the valve is throttling the flow. This lost power is dissipated in heat which must be removed. For small drives, say ½ hp or so, the power consumption and temperature rise probably will be insignificant.

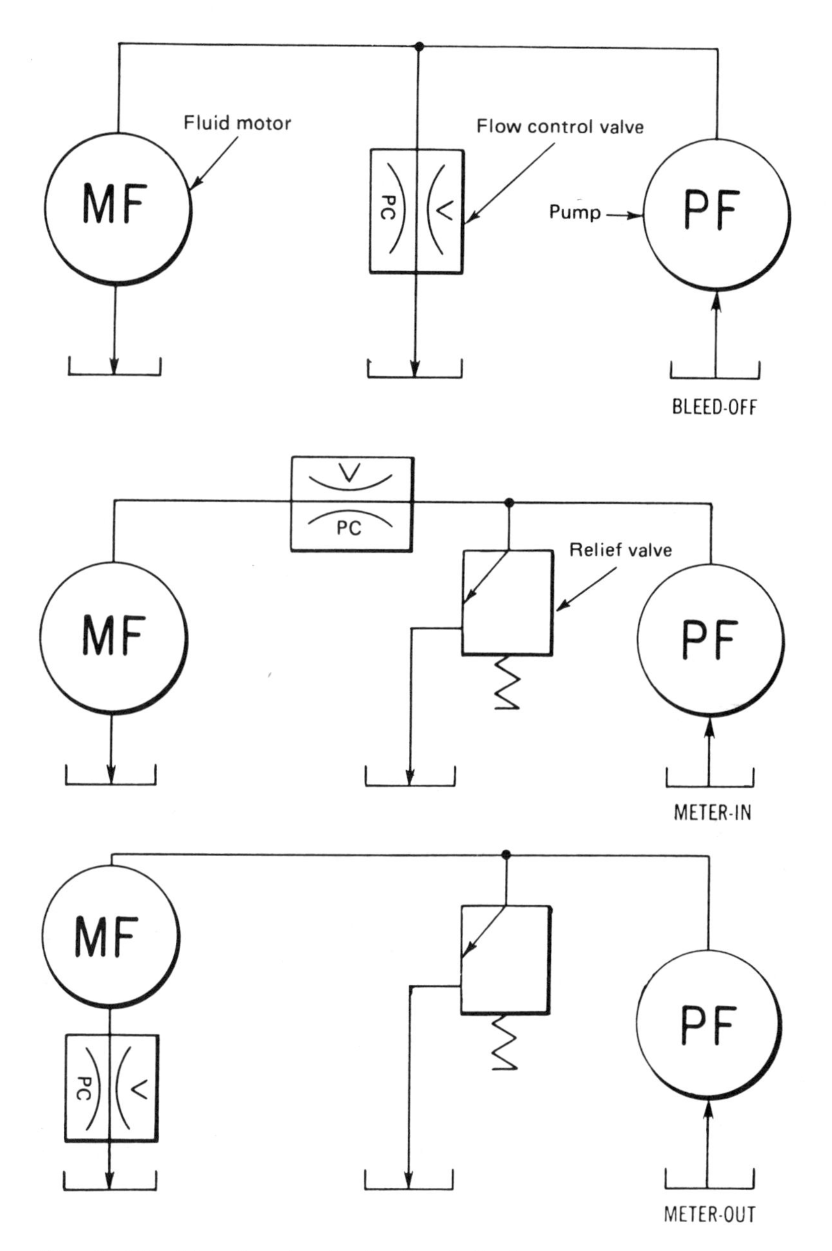

Fig. 6.19 Metering circuits for fixed-displacement motors usually are one or a combination of these basic arrangements.

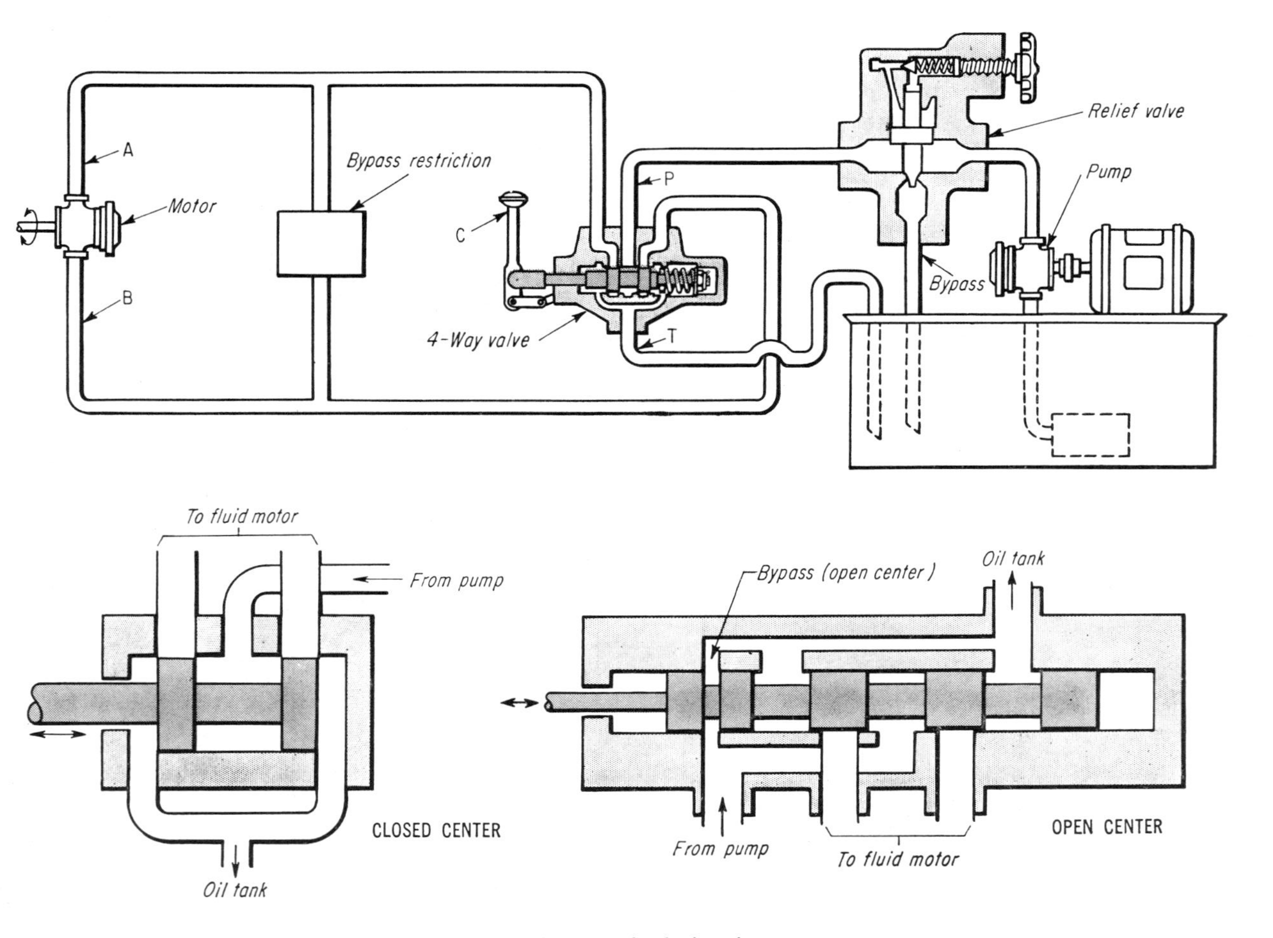

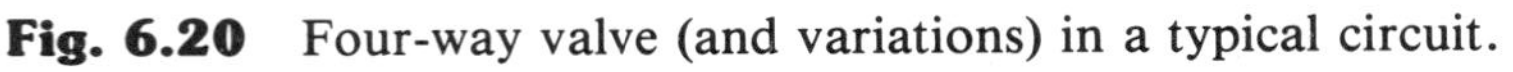

Fig. 6.20 Four-way valve (and variations) in a typical circuit.

A further limitation of the four-way control illustrated is speed droop, called *regulation*. It is the change in fluid-motor speed with load and can exceed 30% of normal speed. Here's an example:

Let the pump flow rate in the four-way valve control circuit (Fig. 6.20) be 30 gpm and the relief-valve setting be 1000 psi. With a valve opening that provides a flow area of 0.056 $in.^2$ from P to A and from B to T, respectively, the flow that passes at 1000 psi will just about equal the pump flow rate.

The basic flow equation is

$$Q = KA\sqrt{\Delta P}$$

where A is area, ΔP is pressure drop, and K is a constant. ΔP is 500 psi, not 1000 psi, because there are two pressure drops, P to A and B to T. This is true as long as there is no load on the fluid motor.

Suppose the motor is loaded and requires a pressure drop of 500 psi. For the same valve setting, the flow will now drop to 21.2 gpm because the available pressure drop at the valve has been halved. The excess is bled off by the relief valve through the bypass.

The speed of the fluid motor thus changes by approximately 30% from no load to full load. Hence the speed regulation of this circuit is 30%. Actually, it is greater if the speed regulation of the fluid motor is added.

Regulation can be reduced to zero if the valve opening is readjusted to increase the flow. For example, if the valve is opened from 0.056 $in.^2$ at no load to 0.079 $in.^2$ at full load, speed at no load will be the same as that at full load. A four-way valve can be designed for manual or automatic (servo) adjustment.

Flow-control valves: The valves at the top of Fig. 6.21 maintain steady flow despite changes in inlet or outlet pressures. It's done with the compensator piston C, which automatically adjusts its position to hold the pressure differential constant across the valve orifice. The equation is

$$Q = KA\sqrt{\Delta P}$$

and flow Q will stay constant if orifice area A and differential ΔP do not change.

With the spring, the compensator piston will move until the outlet pressure, together with the spring force, balances the inlet pressure, establishing the pressure differential. Flow is adjustable to any given constant value if the spring force or the orifice is adjustable.

In the first valve, the inlet pressure is not controlled, and a relief valve must be installed at the pump discharge to bleed off the excess (see meter-in circuit).

The second valve has a built-in pressure-relief valve H. When the pressure at the valve outlet exceeds the setting of the relief valve, its poppet is unseated. A small flow passes through the restriction, lowering the pressure at E. This further opens the compensator valve and increases the flow to the tank. Part-load efficiency is improved because pump pressure is reduced as bypass flow increases.

Servovalves: A servovalve (Fig. 6.21) can be a simple four-way valve or any other kind. But it is controlled electrically by a magnetic coil that positions the spool in response to a remote electrical signal, the feedback. It is called a *servovalve* because it is used in a servo system. (See Chap. 7.)

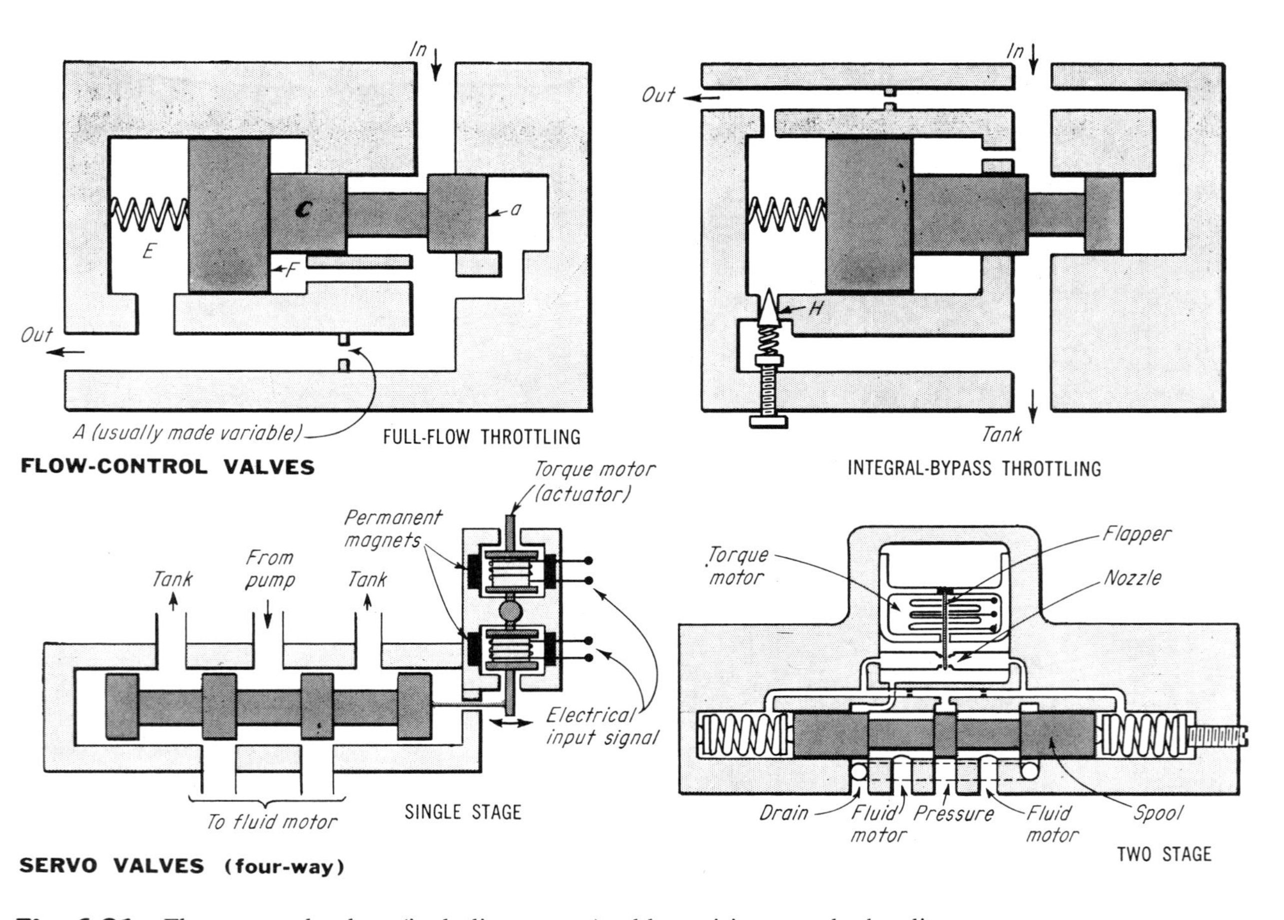

Fig. 6.21 Flow-control valves (including servos) add precision to a hydraulic system.

MATCHING MOTOR TO LOAD

The steps in matching the motor to the load are simple: (1) Determine load torque and speed, (2) pick a motor type; (3) find the displacement from catalogs (also see Table 6.1), and (4) calculate flow and pressure at each speed-torque point required, including stall, overspeed, braking, and creep.

Check out special conditions such as mechanical sideload or thrust, reverse rotation, and unusual fluids. Is the motor mounting flange standard? Will the motor fit in the space available? Will noise be a problem? Efficiency? Will the motor eventually have to run on high water base fluids (HWBF)?

It helps if you categorize the purposes of the application. It will be one or several of these: power, torque, velocity, and position.

Power-type hydrostatic drives use the motor as the output of a fairly complicated pump-motor system. The pump (and its pressure, flow, or load compensator) determines the basic characteristics of the drive. Power transmission efficiency and smoothness are the main concerns, along with noise, weight, and size per hp and, of course, cost.

Drives controlling *torque, velocity,* or *position* are generally much lower in power

Table 6.1 Fluid Power Equations and Nomenclature

Fluid power equations

Fluid hp	=	$5.83 \times 10^{-4} Q\Delta P = Q\Delta P/1715$
Mech. hp	=	$1.585 \times 10^{-5} TN = TN/63{,}025$
Torque, lb-in.	=	$36.8 \Delta P D_{gal}$
	=	$\Delta P D_{in.^3}/2\pi = 0.159 \Delta P D_{in.^3}$
Cost/hour	=	kW × ($/kWh)
	=	hp × 0.746 × ($/kWh)

Kinetic energy equations

Torque, lb-in. = $wK^2\Delta N/(3696\Delta t)$

Equiv. wK^2 of second rotary mass referred to main shaft = $(wK^2)_2 \times (N_2/N_M)^2$

Symbols

D_{gal}	=	displ., gal/rev	N	=	speed, rpm
$D_{in.^3}$	=	displ., in.^{3}/rev	ΔN	=	rpm change (const. accel)
Q	=	gpm	Δt	=	time interval, sec
ΔP	=	pressure drop, psi	w	=	weight, lb_m
T	=	torque, lb-in.	K	=	radius of gyration, in.

Conversions

1 hp	=	42.4 Btu/min	1 psi	=	6.875 kPa
	=	33,000 ft-lb/min	1 bar	=	14.5 psi
	=	0.746 kW		=	100 kPa
1 gpm	=	231 in.^{3}/min		=	10^5N/m^2

level—or at least in total energy consumption. If is often possible to ignore efficiency until all other performance factors have been examined.

The most important parameters involve the motor's ability to respond faithfully to pressure and flow commands. The pump simply provides constant pressure or flow to the valves that control the motor. If the application is positioning of a precision machine tool table, the pressure source should be smooth and constant and not the direct output of a pump. The exception is in applications with a very wide speed range. Constant pressure can become inefficient at the high-speed end where the load torque is often low.

Here it is advantageous to add a volume-limiting control to the pump (or use the entire pump output) and let the supply pressure fall to the actual load pressure in the full-speed condition.

A flow-control valve may be used for the critical lower portion of the speed range and be bypassed when it is not needed. Servovalves are excellent flow controls for precise control of motor motion and position. Just remember that servovalves are basically inefficient because they throttle.

Torquing applications are different from positioning applications. There are dynamic torque loads, such as web tensioning roll drives, and static torque loads, such as power torque wrenches (or "nut runners").

Motor torque variations can be masked by load inertia and compliance in most dynamic drives. However, the opposite occurs in static or low-speed drives where motor torque ripple, stiction, and leakage variations can combine to excite the load inertia and compliance into oscillatory movement and large inconsistencies in the stall torque. That's bad for precision drives.

Motor Performance

In fixed-displacement motors (Fig. 6.22 is one example: an axial piston type) the flow determines speed, pressure determines torque, and restriction in the discharge creates a braking action.

Adjustable-displacement motors are less common than adjustable-displacement pumps (see Chap. 3), and many of them are radial type with an adjustable center cam (Figs. 6.23 and 6.24). Some of the cams have only two positions; others are fully adjustable.

A fully adjustable, pressure-compensated motor has the most versatile performance (Fig. 6.25). It normally is at minimum displacement (maximum rpm). It remains at minimum displacement up to the pressure at which the compensator is set to open. Once the compensator opens, displacement increases with an increase in system pressure until maximum displacement (maximum torque, minimum speed) is reached. Torque varies inversely with speed at constant pressure and constant input flow.

Dual Fluid: Oil and Air

If the low-speed/high-torque and high-speed/low-torque requirements can be separated, then dramatic reductions can be made in the size, power, and cost of the system. One way is to use oil for low-speed/high-torque and switch to air for fast traverse. An example is shown in Fig. 6.26.

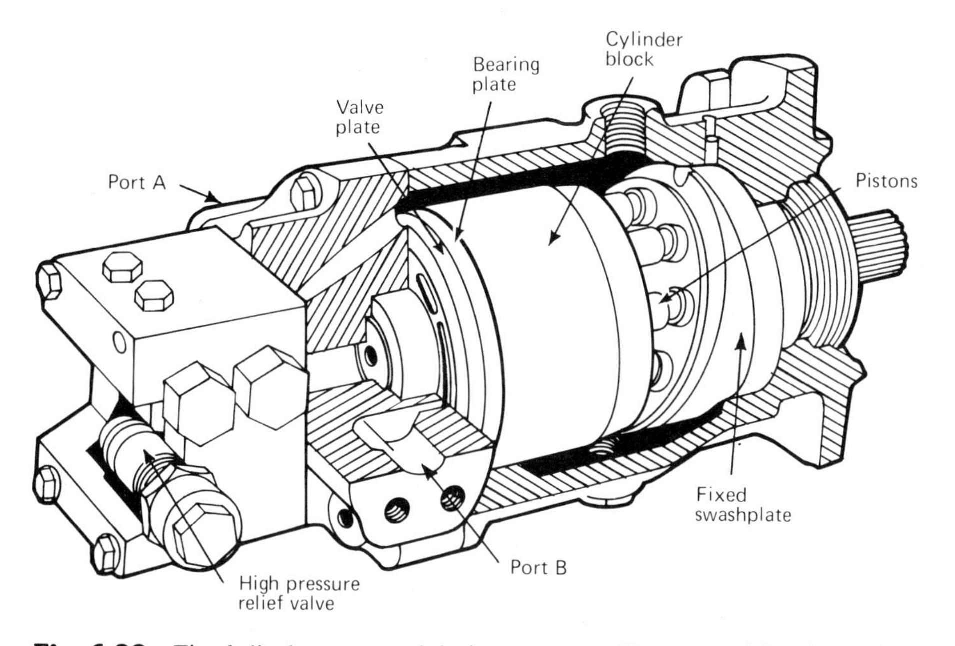

Fig. 6.22 Fixed-displacement axial-piston motor. (Courtesy of Sundstrand Corp., Rockford, IL.)

A feed drive operating at high thrust condition consumes very little power because of its low speed. High pressure at low flow can be furnished by a small hydraulic power unit of less than 1 hp, even for large machines. To achieve high speed (at low load) in rapid traverse, shop air pressure can be connected to the pump pressure line to augment the pump flow as the load pressure falls below the air supply pressure.

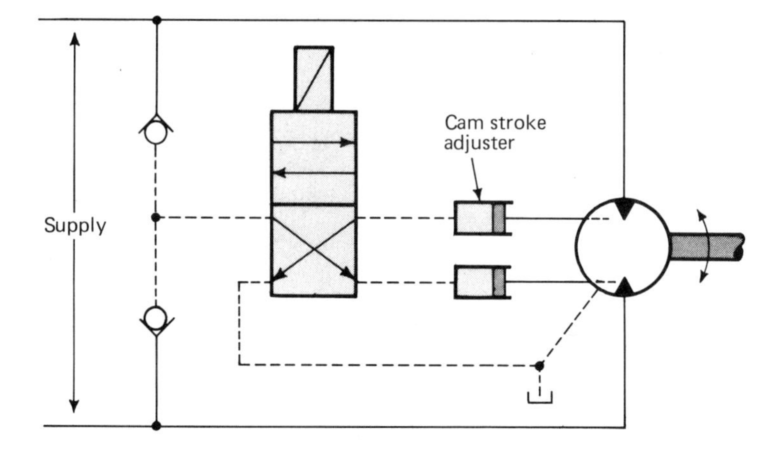

Fig. 6.23 Two-position piston motor has two displacements, set by a hydraulic cam adjuster. (Courtesy of Sperry Vickers, Troy MI.)

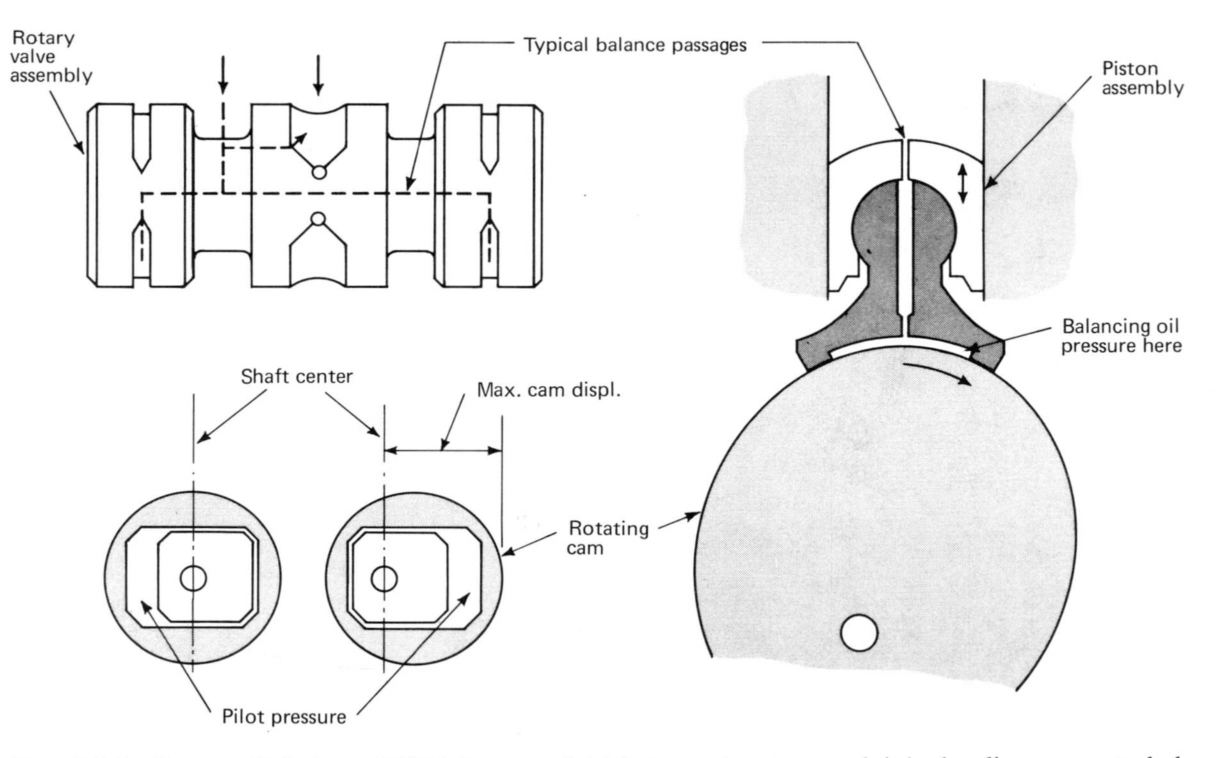

Fig. 6.24 Some radial-piston LSHT (low-speed, high-torque) motors exploit hydraulic pressure to balance the internal forces of rotary and oscillating parts and to adjust cam throw. (Courtesy of Double A Products Co., Inc., Manchester, MI.)

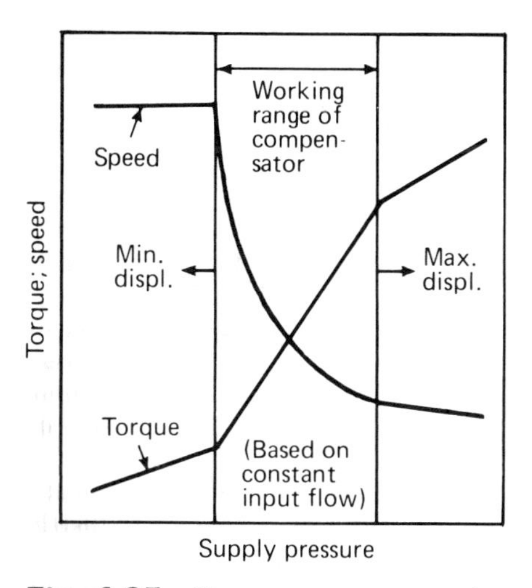

Fig. 6.25 Pressure-compensated motors slow down with pressure. (Courtesy of Sperry Vickers, Troy, MI.)

A low-pressure mixture of air and oil drives the motor; a check valve prevents backflow of hydraulic fluid into the air supply when the load pressure rises. The air bubbles that enter the reservoir through the return line are large and therefore rise and separate from the hydraulic fluid almost instantly and exhaust through the breather-vent. Fluid aeration is not a problem.

Leakage and Efficiency

Internal leakage (volumetric efficiency) usually is invariant with motor shaft speed. It is approximately a linear function, with pressure as the independent variable. If there is pressure on both ports of the motor, there will be internal losses from both pressures. Leakage flow rates usually do not vary significantly for the various displacements in a particular motor frame size. In general, smaller frame sizes will have lower leakage rates than larger frame sizes.

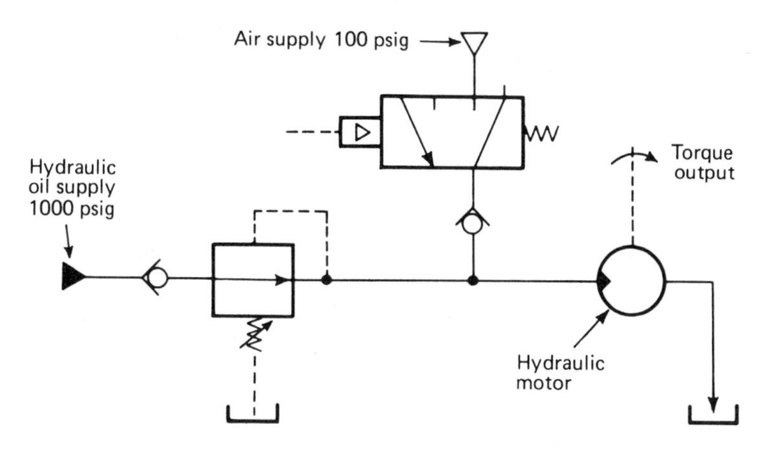

Fig. 6.26 Oil-plus-air system drives automatic rapid-rundown, slow-stallout nut runners. (Courtesy of Nutron Corp., Hingham, MA.)

Mechanical losses (also known as windage or tare losses) are caused by friction in bearings and rotary seals and between the moving parts of the motor. These losses are usually invariant with pressure and vary exponentially with speed. At high power levels, mechanical losses can be considered smaller in relationship to volumetric losses. Mechanical losses increase with frame size but do not vary with different displacements in a frame size.

Some tips to save energy: Since leakage flows are the same for different displacements in a frame size, using a larger displacement in that frame will lower system pressure and raise input flow without increasing internal leakage. Conversely, a similar displacement in a smaller frame size can increase overall efficiency because both volumetric and mechanical losses are smaller for smaller frames.

If noise is not a problem, increasing the motor speed increases flow while leakage remains constant. This increases mechanical losses, but the increase in volumetric efficiency will in most cases more than compensate.

Certain types of motors have typically higher leakage rates than others. Gear motors, whether of the spur, internal, or orbiting gear (gerotor) type, are examples. Piston motors, whether of the ball or cylindrical type, can be made with low internal leakage because of circular sealing.

Cross-port leakage—sometimes necessary for fluid slippage to reduce noise and to avoid damage from pressure peaks—spoils the motor's slow-speed controllability. Zero-lap valve ports, for minimum cross-port leakage, are best for low speed control, but care is needed to avoid sudden pressure surges, such as blocking of flow from the motor. High-leakage motors are better in that respect.

MOTOR EXAMPLES

The literature abounds with details of available hydrostatic motors and drives. We've already covered several typical types (Figs. 6.12 through 6.26). Now let's go through the details of design and control of two of the least understood types: gerotor gear motors and ripple-free axial piston motors.

Gerotor

The gerotor principle (Fig. 6.12), which uses an inner rotor with one less tooth than the outer stator, is as old as the hydraulic industry.

A full orbit of the inner rotor results in relative rotary displacement of only one tooth between the rotor and the stator. Thus, there is a mechanical advantage of 6 : 1 for a six-lobe/seven-lobe set.

In operation, oil pressure and flow against the lobes cause the rotor to orbit. A splined dogleg in the example pictured in Fig. 6.27 takes that rotation and delivers it as a very slow high-torque output. The dogleg also rotates internal valving to synchronize the oil flow.

There are thousands of applications in pumps, motors, and drives where this advantage is exploited. For example, they are widely used in steering mechanisms, self-propelled wheels, chain drives, rotary actuators, slow-speed drills, mixers, drum drives, and other applications calling for high torque and low speed.

Floating lobes: A novel way of reducing clearances between the meshing lobes is to substitute cylinders (roller elements) for the lobes in the stator (Fig. 6.27).

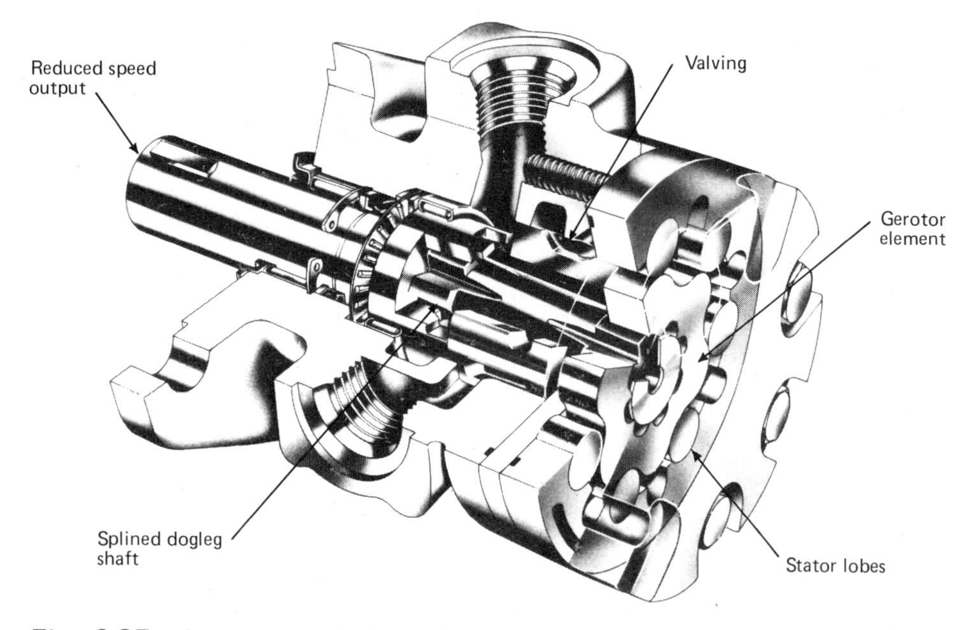

Fig. 6.27 Gerotor-type hydrostatic motor has speed reduction inherent in the nutating gerotor element. (Courtesy of Ross/TRW, Lafayette, IN.)

The cylindrical shape of the rollers is identical to the circular tooth form of conventional lobes, so the operation is the same. However, the cylindrical lobes are not mechanically attached to the stator shell. Instead they float on a film of pressurized oil.

Three advantages: The floating lobes are held in contact with the rotor by the pressure of the oil film, thus compensating for any wear that might occur. The lobes rotate freely and thus distribute wear over the surfaces. And the lobes are simple precision cylinders, which do not require shaping and grinding as do integral lobes.

Ripple-Free Piston Motors

Conventional swashplate-type axial-piston motors and sinusoidal cammed radial motors have a pressure ripple caused by the natural harmonic motion of the pistons. The flow into the casing at any given moment is not always exactly equal to the discharge flow, and therefore the contained volume of oil in the casing fluctuates. That's because ordinary harmonic cam shapes, such as sine waves, do not have a constant sum when used in multiple-piston arrays.

Special designs have been developed to eliminate this ripple; Fig. 6.28, the Nutron opposed-ball motor, designed by James and Beryl Denker, is an example. It has parabolic cam surfaces with slopes designed to assure that the sum of the velocities of the motoring balls and the sum of the velocities of the pumping balls are at all times equal and constant at any speed. Inlet flow is equal to outlet flow.

Four benefits result. The case may be operated at any pressure, or blocked off if desired, without creating trapped oil pockets that could crack the casing or creating voids that could cause cavitation in the rotating cylinders. No overlap of porting to cushion volume fluctuations is necessary, so the valve ports open and close with sharp cutoffs for minimum leakage.

Fig. 6.28 Opposed-ball hydrostatic motor by Nutron has parabolic cams that give balanced displacement of suction and discharge strokes.

The cam lobes even have dwell (flat spots) at the peaks and bases to further separate the motoring flow from the pumping flow. And finally, the balanced flows ensure smooth motor output without cogging.

No cross-porting is needed between inlet and discharge to prevent pressure traps that create noise and endanger the motor.

Four pressure regions: The Nutron motor provides for independent control of its supply, return, case pressure, and shaft drain. Any of those regions (Fig. 6.29) can withstand 5000 psi without damage, so there is wide flexibility in selection of differential pressures for driving the motor.

Only the shaft seals are vulnerable. If the drain is blocked off and all other ports are pressurized, the shaft seals will blow out because they are designed for 5 psid. Slight leakage (a string-sized stream at full pressure) occurs until the seals are replaced, but the motor continues to operate.

Various models operate at maximum speed from 500 to 2000 rpm and are reversible at full speed and load. Volumetric efficiency is claimed to be better than 99.5% at rated speed and over 90% at 10 rpm.

Operating principles: The maximum rating for each model is 1250 psid. Motor power is proportional to pressure differential (from supply to return) times flow. The differential is achieved either by controlling the supply pressure (metering-in) or by holding the supply pressure constant and building up back pressure in the return (metering-out).

Supply pressure enters the space between the balls in four of the nine rotating cylinders at a time. The oil forces the balls apart against the opposing cam surfaces to produce motoring torque.

Return oil is squeezed out of the space between the balls in four other chambers by action of the cam surfaces pushing the balls together. One cylinder chamber at any given instant is idle (on dwell) because of the flats provided on the cams. The total volume of oil involved in motoring and pumping is small, so the "cushion" effect (compliance) is small. Also, inertia and viscous drag are small.

The case oil in the annulus outside the piston chambers, being a separate region, is not part of the flow. The only communication between case and return is the slight internal leakage, except where both dump into the same drain.

The case oil serves multiple purposes. It maintains constant pressure against the rotating cylinder block to keep all balls in a firm bath of oil. It helps stabilize temperature shocks and damps noise. It enables the operator to force all ball pistons inward to the freewheeling condition simply by applying high pressure (above supply pressure) to the case control port.

Perhaps the most important use of steady case pressure is as a reliable source for internal pressure feedback in conjunction with integral valving. Fluctuations would make the valving unstable.

Function of integral valves: Some models have built-in integral valves to adjust certain differentials for special performance requirements. With the valves, the motor can hold speed droop to within less than 1 rpm, regardless of load fluctuations, without any external feedback control. It also can be run on leakage flow alone (return port blocked) by automatically balancing the leakage from the motoring cylinders. Extremely slow and stable performance results.

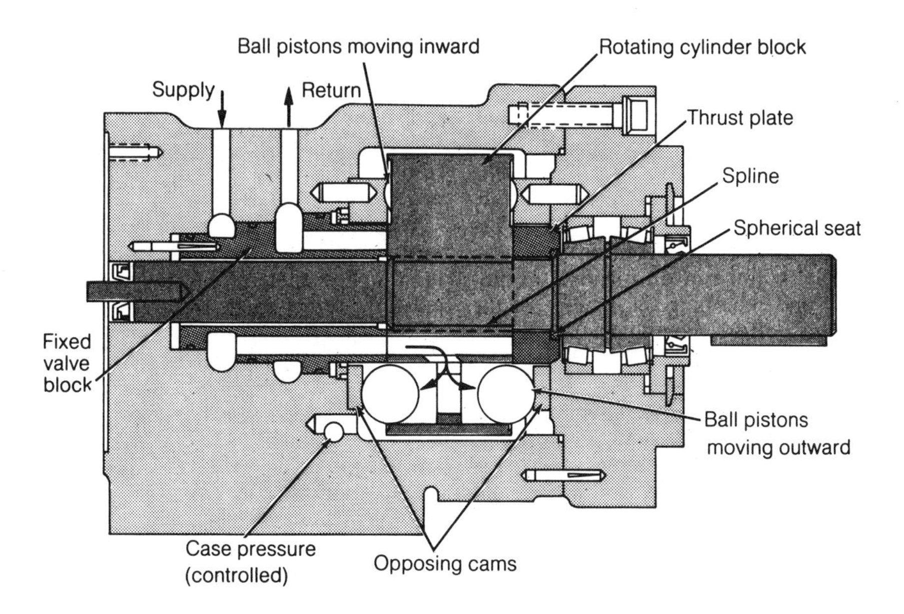

Fig. 6.29 Cylinder block of Nutron motor rotates against cam plates to produce torque. The splined output shaft could be removed even while the cylinder block is rotating under full pressure.

The integral valves are simple spring-biased, pressure-piloted relief valves (Fig. 6.30). The normal position is full open, held there with a spring bias of 50 psi. Only when the pilot pressure exceeds 50 psi will the valve close, blocking leakage flow from case to drain.

The valves come into play during leakage-flow motoring. For example, assume that the motor is turning slowly at low torque (back pressure in return line held slightly below supply pressure). If load is applied to the output shaft, the back pressure in the return line falls proportionately. This drop allows the integral valve to open and pass leakage to drain, thus increasing the motor torque.

The torque increases linearly with the drop in speed. The torque-to-rpm gain is 10 times that without the integral valves installed. In round numbers, it is 500 lb-ft/rpm.

Performance. The motors are designed for these common modes: metering-in flow, metering-out flow, series operation, freewheeling, and servo control (Fig. 6.31).

Metering-in operation. The motor speed is controlled by how much oil is permitted to enter the inlet. The return port goes directly to drain. The case pressure and vent pressure also go directly to drain.

Metering-in with freewheeling. By bringing out the case pressure through its own port, the pressure can be increased with oil from an external source. When the case pressure overcomes the supply pressure, or if the supply pressure is removed, the balls are forced inward and remain there. The rotor (rotating cylinder block) becomes freewheeling. This feature is useful where the load must be free-running or must be disconnected from the supply pressure. It has some features of a jaw clutch. The balls have to be either all the way in or in full engagement with the opposing cams. It would not be possible to use it as a slip clutch.

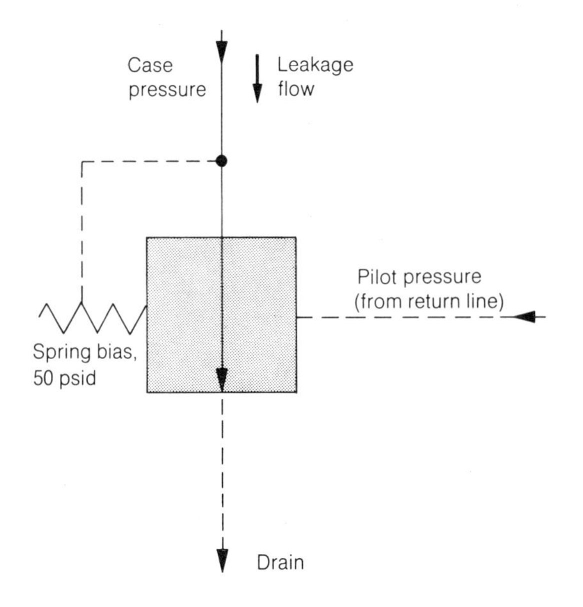

Fig. 6.30 Integral valve of Nutron motor establishes pressure differentials for speed and flow control without the need for external feedback.

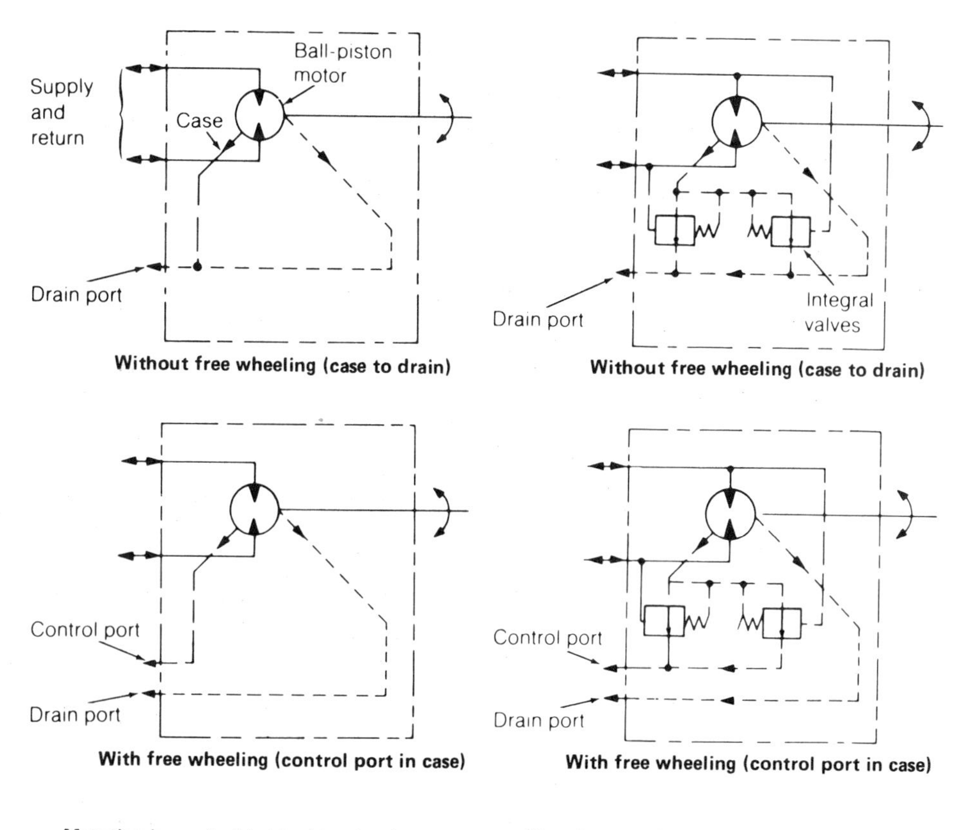

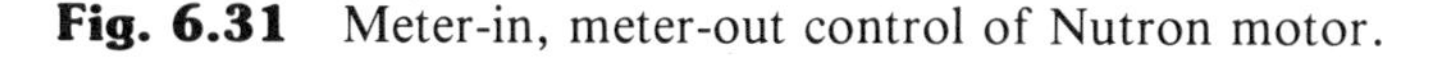
Fig. 6.31 Meter-in, meter-out control of Nutron motor.

Metering-out. When the motor speed is controlled by restricting the return, valves are needed to interconnect the case, the return, and the drain.

Series operation. This is where the output flow of one motor drives the input of the next. Combined with freewheeling controls, it permits two-speed or two-torque operation. The first motor can provide breakaway torque and low-speed operation, for instance. Then it can be made to freewheel, and the second motor can be actuated (unfreewheeled) to provide low torque at high speed. Both motors can be coupled or geared to the same final shaft, giving a simple two-speed hydrostatic drive.

Semiservo operation. Because of the low internal leakage, good speed control is possible without external feedback. For example, with the integral valving, the motor can be running at full speed (say, 1000 rpm) and no load and be suddenly loaded without losing more than 1 rpm. For adjustable speed, a flow-control valve is put in the return line.

The internal-pressure feedback permits velocity control with higher bandwidth than can be attained with expensive electronic means.

A Multimotor Example

With two or more pumps in series or parallel, the freewheeling makes it possible to create a multiple-speed drive (Fig. 6.32).

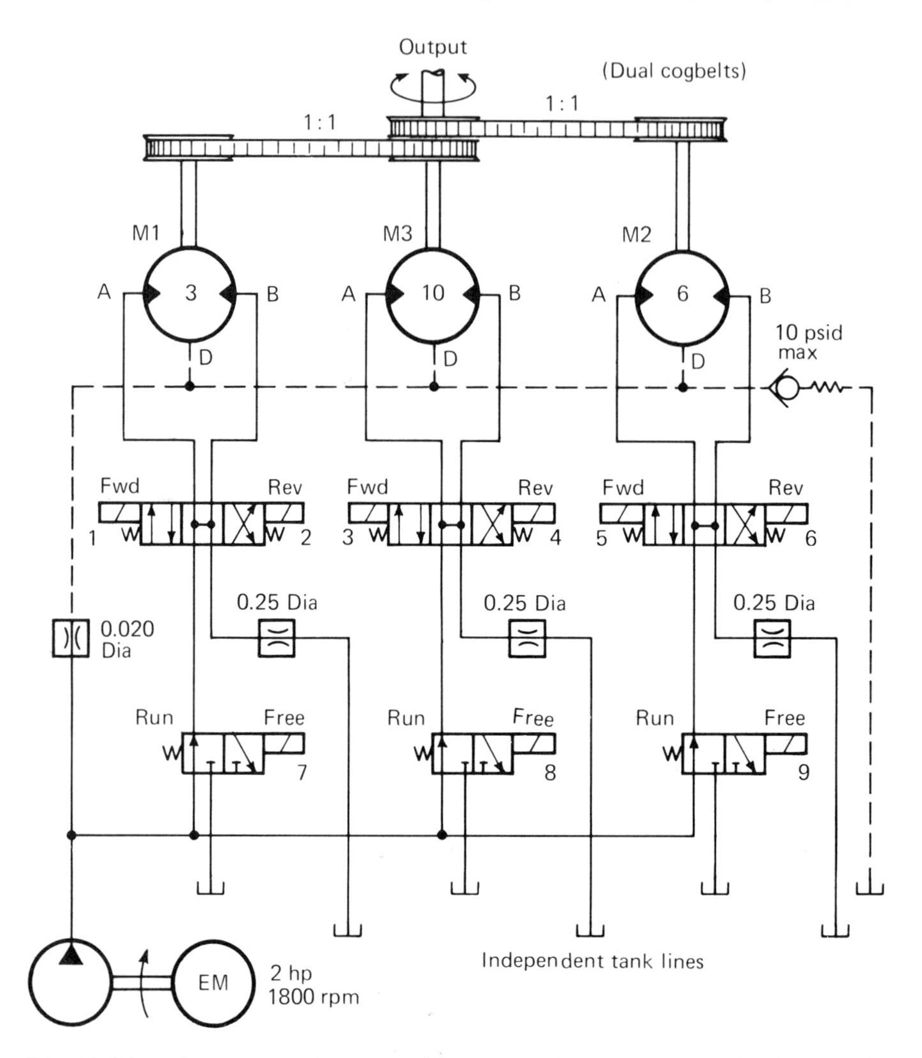

Fig. 6.32 Three coupled hydrostatic motors have seven speeds without the need for a servo. It's done with solenoid valves. (Courtesy of Nutron Corp., Hingham, MA.)

By freewheeling various combinations of motors M1, M2, and M3, the effective displacement at the combined output shaft can be any one of these: 3, 6, 9, 13, 16, or 19 in.^{3}/rev. The electric motor (EM) is shown driving a fixed-displacement 6-gpm pump, operating at 100 psig to 600 psig (relief set at 950 psig). If desired, the pump can be adjustable-displacement with pressure compensation or volume limiting. Then every one of the input combinations becomes full-variable speed.

7

Electrohydraulic Systems and Servovalves

The marriage of electronics with hydraulics can save energy and therefore money. The choice is among simple on-off solenoid directional valves, proportional pressure- and flow-control valves, servovalves, and digitally modulated valves. The last are the newest and are based on microprocessor-controlled stepping motors built into each valve.

The aerospace industry long ago swung over to electrohydraulics (EH) because it was impossible to control powerful aircraft with lengthy high-pressure plumbing or mechanical linkages from the cabin. In contrast, industrial users of hydraulics have fought off electrohydraulic systems, particularly servo systems, for decades. It always has been possible to control most machines and vehicles without the complexity and initial high expense of electronics by clever arrangements of purely hydromechanical elements.

Now designers are playing a different tune. Energy is too expensive to waste, and it's worth a buck or two to put more options at the fingertips of the operator or to replace him or her with a black box. An equally important reason for electrohydraulics is that the control can be remote.

FIVE LEVELS OF CONTROL

Simple *solenoids* are widely used to turn valves on and off, and this is the most basic electrohydraulic control. The flow and pressure are well defined by the size and nature of the circuit elements, but modulation is minimal.

A second level of sophistication is to modulate flow by rapidly opening and closing the solenoid valves. Some engineers call this *bang-bang* control; others call it pulse-width modulation. In either case, the variation in length of on-time vs. off-time establishes an average flow of any amount desired.

A typical system uses two valves, one for each direction of motion. If each valve is cycled in the open position for an equal time, the net motion of a cylinder or motor driven by these valves will be zero. Motion in one direction or the other is achieved by leaving the valve open for a longer period in that direction.

The bang-bang designation is fairly accurate, because the valves must be cycled fast and hard to effect any measure of control. A major problem at high power is noise and vibration.

A third level of electrical control is the *proportional valve.* It can be as simple as a linear-response solenoid valve, meaning only that spool position varies in direct proportion to the electrical current fed to the solenoid. Or it can be a complex EH valve that regulates either pressure or flow under close control of the input electrical signal. A sophisticated proportional valve is in many respects a servovalve except that it is not as likely to be used in a feedback system.

The fourth level of control is the true electrohydraulic *servovalve.* The accompanying tables, graphs, and illustrations help define servovalves and compare them with the very similar proportional valves.

The fifth level is the digital valve, wherein a microprocessor sends discrete signal pulses to a stepping motor, which in turn positions the pilot of a spool valve.

PROPORTIONAL VALVES

Let's start with the electrical solenoid. There are special breeds that adjust armature position with variable electrical input, usually against a spring load or hydraulic pressure. Europeans use more of them than Americans do, but a trend in the United States indicates growing acceptance.

A potential big application of *proportional solenoids* is the direct operation of spool valves. The coil current is high enough to move the spool reliably against friction and hydraulic forces.

HPI/Nichols (Sturtevand, WI) developed a two-solenoid pressure regulator pilot. It has a proportional solenoid at each end of a special spool (Fig. 7.1). Spool motion in one direction opens the supply port to pressurize the spool in the opposite direction, balancing the solenoid force. The controlled output pressure, up to 2000 psi, is used to position the spring-loaded spool of a separate flow-control valve.

Ledex (Vandalia, OH) developed a proportional solenoid (Fig. 7.2) whose plunger position is a function of the average current in the coil.

The current controller delivers a pulse-width-modulated signal that is full-on or full-off, and the average solenoid coil current is established by the percentage of on-vs.-off time. It's efficient because the transistors doing the controlling are full-on or full-off, avoiding the higher voltage drops that occur at partial loads. Solenoid power is sufficient to directly move the spool of a 3000-psi 150-gpm valve if the spool is balanced.

The solenoid plunger will position itself open loop (schematic), but better accuracy is possible with inner-loop or outer-loop feedback. Inner loop means to monitor the solenoid plunger position; outer loop means to monitor the final output such as the cylinder stroke or pump swashplate angle.

The loop feedback position signal is compared with the called-for position signal input; the error between the two tells the solenoid control to move the plunger in or out proportionally.

Midway between simple proportional valves and servovalves are torque motors or force motors used to operate valves directly.

For example, *torque motors* usually operate flappers or jet pipes in the pilot stage of two-stage and three-stage servovalves; force motors usually drive pilot spools. How-

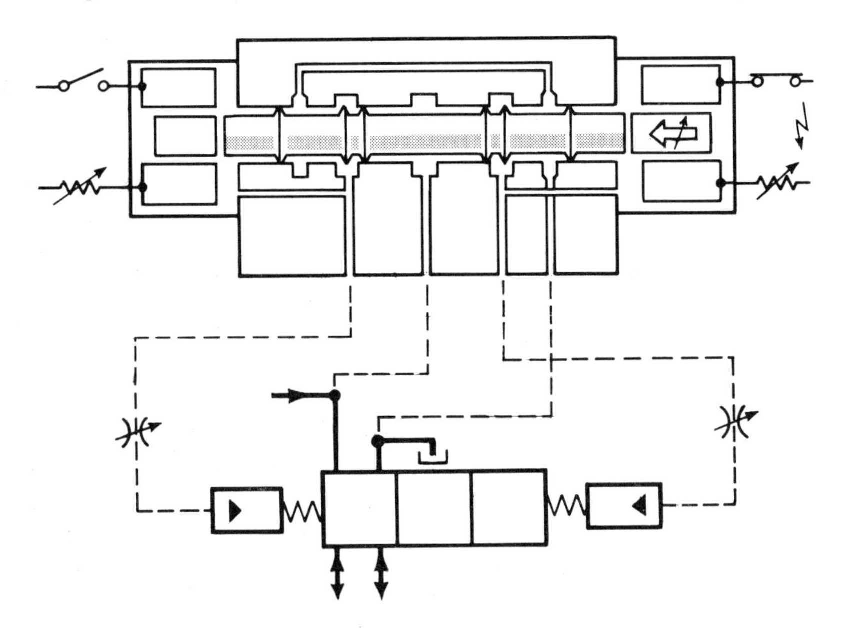

Fig. 7.1 Proportional solenoids (top picture) regulate pressure pilot of hydraulic flow-control valve. (Courtesy of HPI/Nichols, Sturtevand, WI.)

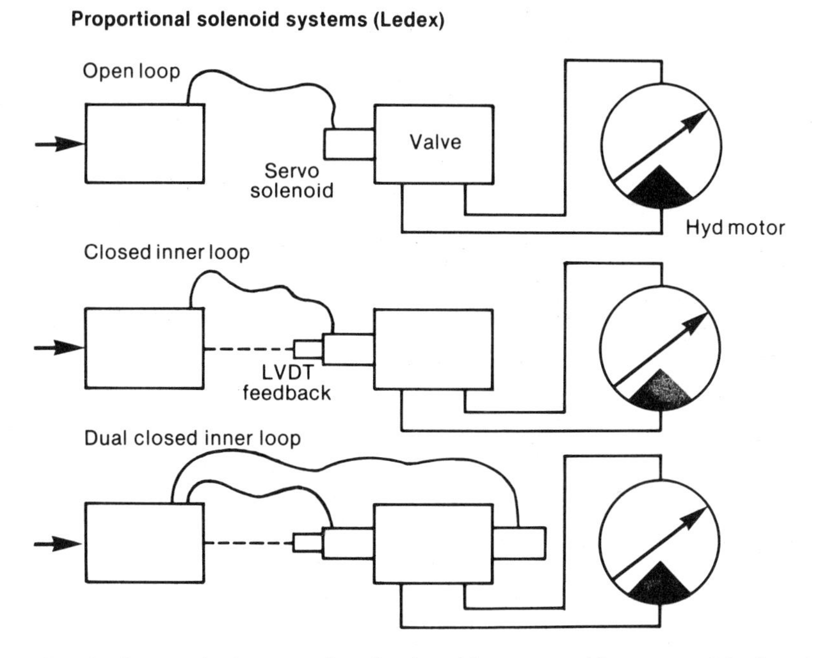

Fig. 7.2 Typical proportional solenoid systems. (Courtesy of Ledex, Inc., Vandalia, OH.)

ever, some designs are powerful enough to regulate flow or pressure in a single stage. One example is the torque-motor-flapper design made by Hydraulic Servocontrol Corp. (Buffalo, NY). Some users make it the first stage of a servovalve; others exploit it as a single-stage pressure controller.

Another example is the Honeywell pressure control pilot valve MCV-101A (Fig. 7.3). It has a pair of facing nozzles with the flapper between them. The net force of the torque motor is balanced when the flapper moves toward one of the jets, thus restricting the jet and increasing its back pressure. Differential pressure is proportional to command current. Honeywell also offers a single-stage flow valve (Fig. 7.4).

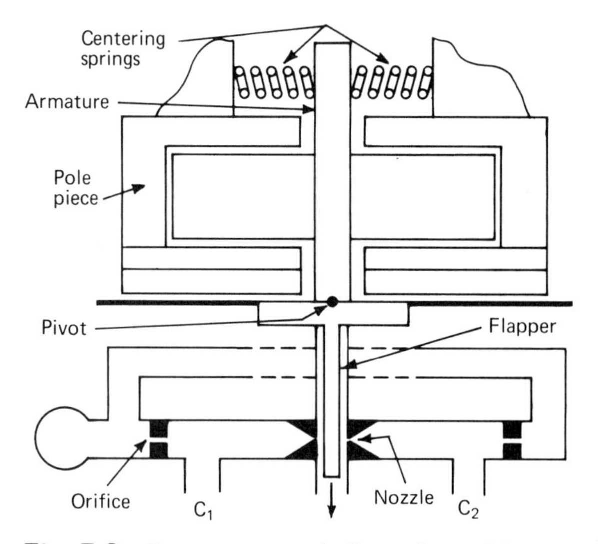

Fig. 7.3 Pressure-control pilot valve, with torque motor and flapper, regulates in one stage. (Courtesy of Honeywell, Bloomington, MN.)

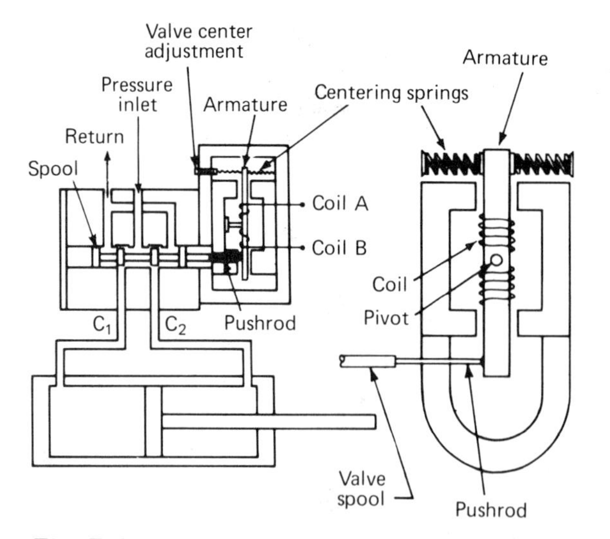

Fig. 7.4 One-stage flow-control servovalve is powered directly by the force of the torque motor. (Courtesy of Honeywell, Bloomington, MN.)

THE ART OF SERVOVALVES

Servo implies feedback control, and it's as much an art as a science. The chief difficulty is in keeping the system stable when the output is too much out of phase with the input or with the system disturbances. A 90-deg phase lag, based on an arbitrary sinusoidal input signal, is one measure of performance for individual components (see Fig. 7.5).

Manufacturers of servovalves and designers of servo systems have established simplified guidelines to encourage wider acceptance of their products. The details of proprietary designs are in company catalogs and manuals. We'll just point out areas of importance. Much of the discussion is based on comments by James Stegner of Moog Inc. (East Aurora, NY). Figures 7.6 through 7.10 are from many sources. Most of the comments apply to proportional valves as well as to servovalves.

Hydraulic amplifier stages: Servovalves generally are multistage devices of two or three stages. The largest share of the industrial/mobile market is served with two-stage servovalves.

The first stage is an electromechanical actuator and a hydraulic amplifier. Typical actuators are torque motors, force motors, or solenoids. The torque motor often is preferred because it has more usable output power per unit of input power. Also, it has low moving mass and therefore has high natural frequency and fast response. A low natural frequency limits the response of a servovalve.

Special design features in torque motors can enhance performance. For example, a flexure tube instead of a sliding bearing provides a limited-motion frictionless pivot for armature rotation. Symmetrical magnetic circuit design minimizes temperature-induced variations. Dual coils permit redundancy, choices of rated current, and choice of operation from two isolated command signal sources.

Two other frictionless amplifiers are the flapper-nozzle valve and the momentum jet valve. They are "open center," however, and consume a small hydraulic flow continuously.

The first-stage hydraulic amplifier also can be a sliding metering member such as a spool or plate, but dirt, friction, and wear can be a problem. Hydura/Oilgear (Milwaukee, WI) has solved the problem of friction and wear of sliding-plate valves by mounting the plate on flexure supports. The supports hold the plate next to the valve ports without quite touching. The torque motor and valve are immersed.

Sliding spools are popular for the second and subsequent stages. They are variable orifices and can be configured to provide linear or nonlinear flow in various circuit arrangements such as closed center and open center.

The first-stage output drives the second-stage spool. Two basic arrangements for controlling spool position are spring-centered spools and feedback of spool position to the first stage. (See the accompanying drawings for various feedback methods.)

The final position of a spring-centered spool is determined by the differential pressure output of the first stage, which corresponds to electrical current input. Limitations are as follows: Full spool driving force is available only with 100% of rated input current, spring loads on the spool tend to increase friction effects, and spool position gain (spool displacement per unit input current) depends on the pressure level supplying the first stage.

The feedback method has several advantages. For instance, feedback of the spool

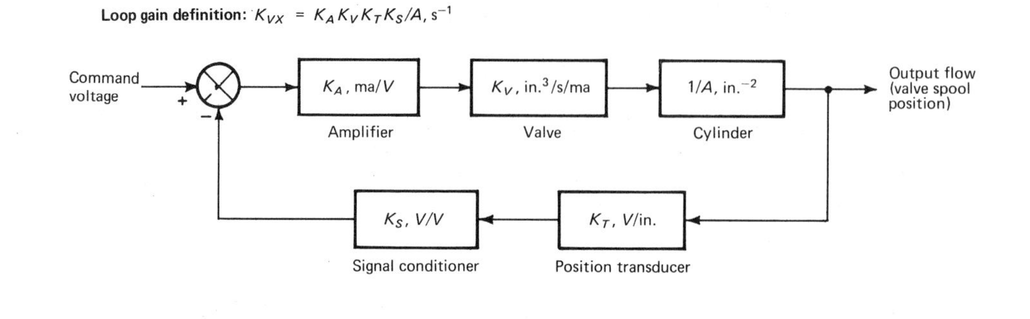

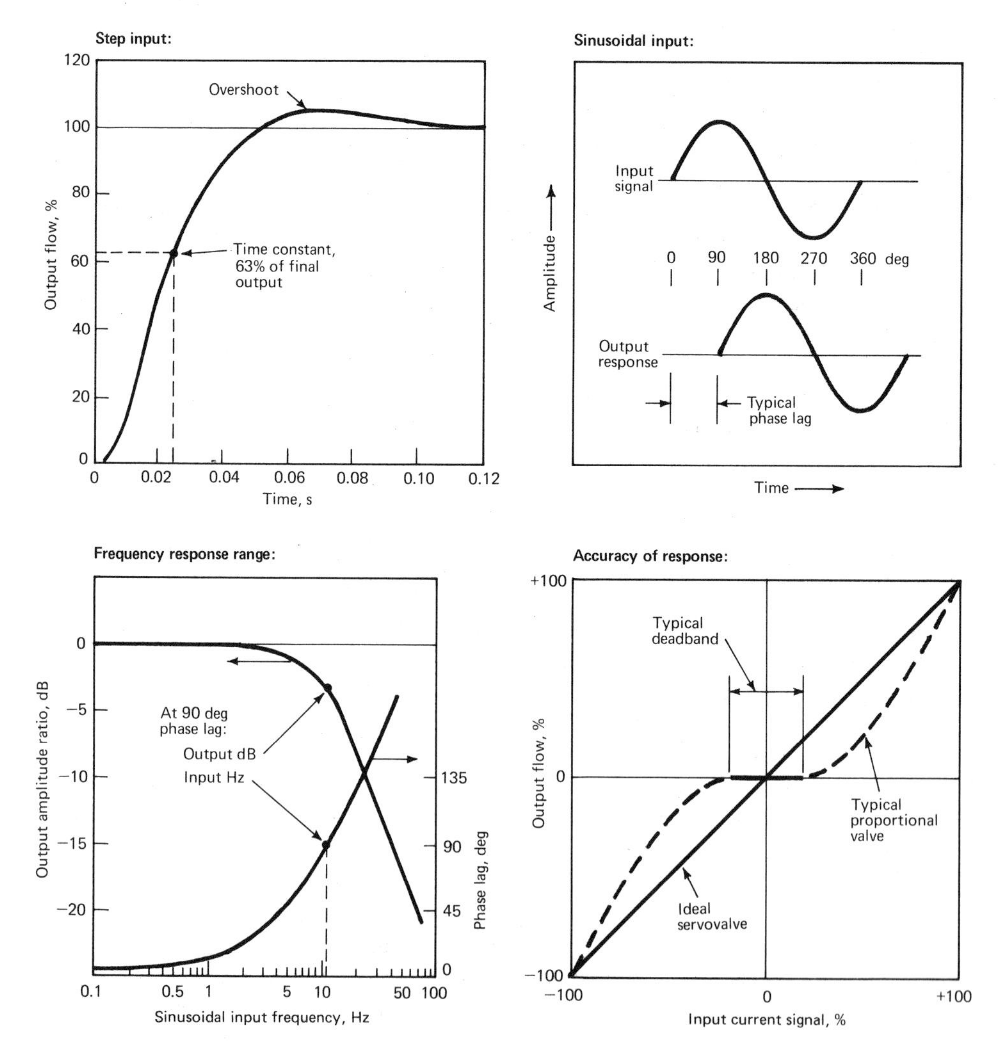

Fig. 7.5 Flow-control servovalve response theory. (Courtesy of Moog, Inc., East Aurora, NY.)

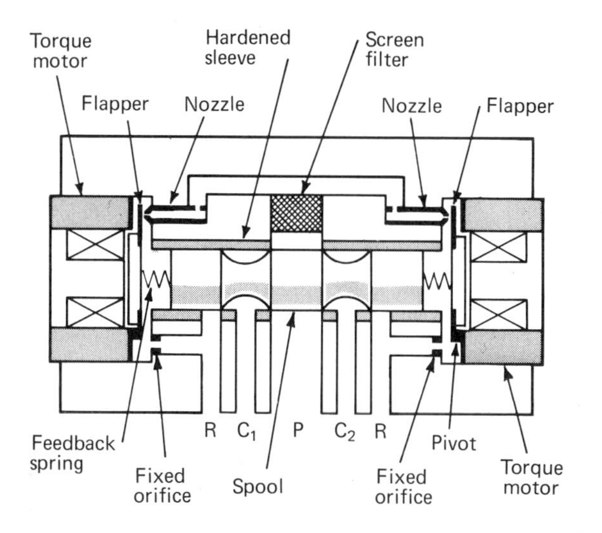

Fig. 7.6 Two-stage flow-control servovalve has dual nozzles and flappers to pressurize spool ends. (Courtesy of Olsen Controls, Bristol, CT.)

position to the first stage permits increased pilot stage gain as a function of spool position and eliminates the dependency of the final spool position on first-stage differential pressure output. Thus, the full spool driving force can be obtained at a small percentage of rated input current. The increased force is especially beneficial with open-center valves where spool flow force gradients can be steep.

In addition, full first-stage output differential pressure is available to work

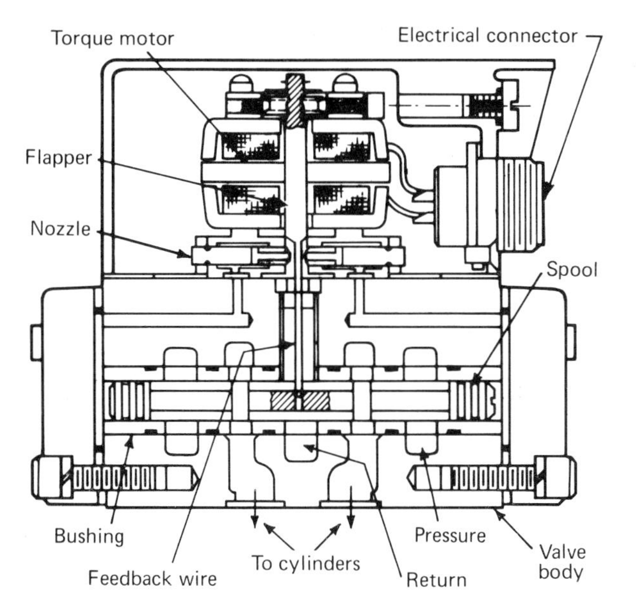

Fig. 7.7 Mechanical feedback wire is bent by spool motion to move flapper into null. (Courtesy of Moog, Inc., East Aurora, NY.)

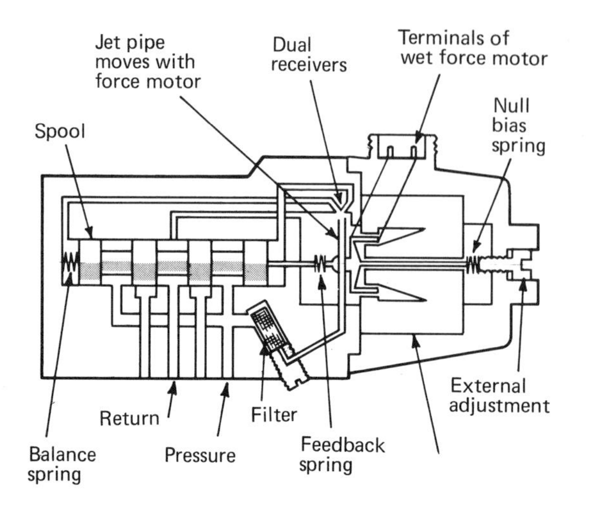

Fig. 7.8 Jet-pipe stage of flow-control servovalve pressurizes the spool ends. (Courtesy of Abex/Denison, Columbus, OH.)

against spool frictional forces. And spool position gain is independent of supply pressure.

Spool position feedback can be electrical or mechanical. Electrical feedback is the most flexible in gain selection and performs best. However, it needs a position transducer and a servoamplifier. Mechanical feedback is less flexible but costs less, and it has fewer parts.

Flow linearity is controlled by the shape of the metering orifices opened by the spool. Proportional valves usually have deadband at null, with increasing gain as spool displacement is increased. Servovalves provide linear flow output with spool displacement and have negligible gain variation at null (Fig. 7.5).

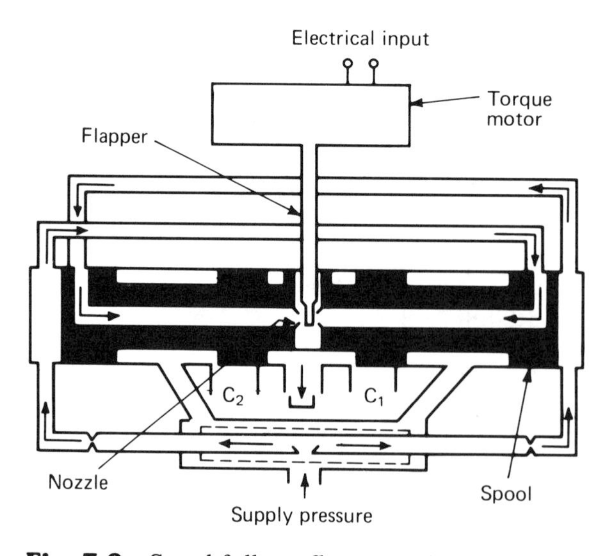

Fig. 7.9 Spool follows flapper to balance nozzle pressures in flow-control servovalve. (Courtesy of Pegasus, Troy, MI.)

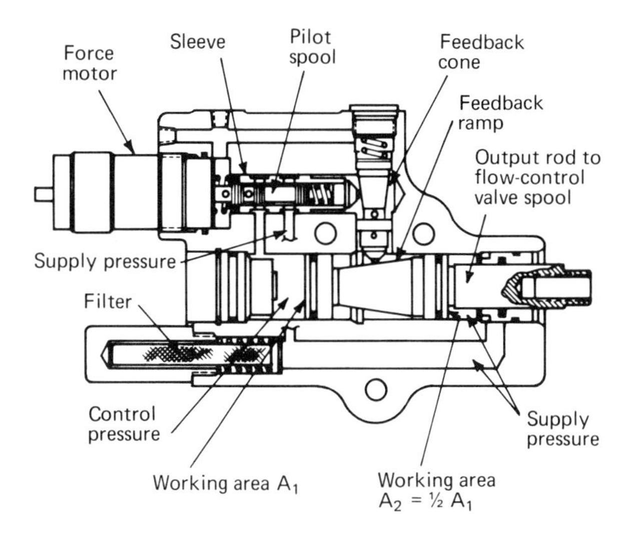

Fig. 7.10 Cone-type mechanical feedback from output rod moves sleeve, nulls pilot. (Courtesy of Dynex/Rivett, Pewaukee, WI.)

How first-stage amplifiers work: Take the jet pipe types, for instance (Fig. 7.8). They are mechanically deflected by the torque-motor, and they aim a small jet of hydraulic fluid at a set of receivers. The captured oil jet is converted into pressure in proportion to the flow received and is directed against the spool to position it. At null, the receiver pressures are equal, and the equal forces on the ends of the spool hold the spool positioned.

Other servovalves rely on torque-motor-operated flappers with mechanical feedback from the driven spool. A flapper moves against one orifice or the other and adjusts the pressure either downstream (Fig. 7.6) or upstream (Fig. 7.7). It is returned to null-balance by mechanical feedback of spool position to the torque motor.

Some flapper-type amplifiers allow the nozzles to follow the flapper to achieve null-balance. An example is the Pegasus/Koehring design in which the nozzles are part of the valve spool and move with it (Fig. 7.9).

Electrical Feedback

The simplest example of electrical feedback is a cylinder load under control of a flow-type servovalve. Flow continues into the cylinder until the cylinder rod reaches the called-for position. Then a position sensor, such as a linear variable differential transformer (LVDT), sends an electrical signal to null the input signal, and the cylinder stops and holds in that position.

Once electrical feedback control is specified, the options multiply. For example, it is then possible to program in special limits of performance that can be held without need of the operator's attention. Moog engineers cite several examples.

One is engine anti-stall control. The operator generally is too busy manipulating the machine hydraulic functions (for example, dozer blade elevation) to think much about engine power. A quickly dropping tachometer signal from the engine shaft, however, is a good indication of a stalling engine, and the signal can be tied directly to the electronic input of the hydrostatic pump swashplate control. The engine can be saved from stalling by reducing pump stroke quickly, automatically.

Another example is overload protection for an electric motor prime mover. The problem arises on winches, conveyors, machine tool spindles, and other industrial applications powered by motor-driven hydrostatic pumps. The operator might be unaware of an overheating motor, but a simple current-sensing circuit can automatically destroke the pump when overload occurs.

A third instance is crossover control of a pump-motor system that has variable stroke in both the pump and the motor. A conventional drive with stroke control just on the pump has a 4 : 1 to 5 : 1 speed range. This can be multiplied by making the hydrostatic motor adjustable as well. Electronics can phase the stroke control to hold full displacement of the motor until the pump has reached full flow and then destroke the motor to further increase the output speed.

A future step is to incorporate microcomputer technology. Computer-controlled industrial robots already are being given credit for a second industrial revolution. Many of these new robots are controlled by EH servovalves, and they have proved to be reliable and precise in hostile environments. This same technology can be applied to mobile and in-plant products.

Open-loop vs. closed-loop: Open-loop control systems (no feedback) cost least. They need a step response time constant of 0.3 sec or less. Closed-loop control systems are fastest. They are required for performance that is superior to human capability or for completely automatic or programmed control. Performance of a closed-loop system is generally specified in terms of position accuracy and *following error* (see theory and equations in Tables 7.1 and 7.2).

Following error is the difference in signals between the commanded position and the actual position of a cylinder or actuator when a constant velocity is commanded. The following error in a position loop is defined by Eq. 1 in Table 7.1. Stegner explains it as follows:

Loop gain K_{VX} is the product of the component gains around a closed system loop. The units are dependent on the number of integrations (or derivatives) around the loop. The most common position control system will have a single integration in the forward loop formed by the hydraulic actuator.

The output following error is inversely proportional to the loop gain. A high loop gain, therefore, is desirable. However, there are limits. Usually the most limiting condition is the resonance of the load mass on the controlling hydraulic actuator. The frequency of this resonance is given by Eq. 2. The net stiffness usually is dominated by compliance of the hydraulic fluid spring. Stiffness of the hydraulic fluid spring can be calculated from Eq. 3. If the mounting between the hydraulic cylinder piston and the load is not rigid, then its effect must be included in Eq. 4.

Having determined the load resonant frequency, the maximum achievable loop gain can be estimated. Generally, if the position loop gain (K_{VX}) is limited to less than 1.25 times f_v (see Eq. 5), a step response with negligible position overshoot will be achieved. In some cases a larger actuator area is desirable to increase the load resonant frequency.

To select an EH control valve that does not further limit the system loop gain, its 90-deg phase lag frequency should exceed the load resonant frequency by a factor of 3 or more.

Feedback transducers, signal conditioners, and valve drive amplifiers must have negligible dynamic effects (flat amplitude response and negligible phase lag beyond the frequency at which the servovalve exhibits 90 deg of phase lag). If a digital com-

Table 7.1 Servovalve Gain and Resonance Equations

1. Following error, in.	$X_F = \dfrac{dX/dt}{K_{VX}}$
2. Load resonant frequency, Hz	$f_n = \dfrac{1}{2\pi}\sqrt{\dfrac{K_n}{M}}$
3. Hydraulic stiffness, lb/in.	$K_O = \dfrac{4\beta A}{X}\eta$
4. Net stiffness, lb/in.	$K_n = \dfrac{K_O K_S}{K_O + K_S}$
5. Position loop gain, $\sec^{-1}$	$K_{VX} \leqslant 0.2\,(2\pi)\,f_n$

Where:

A = working area of double rod piston, in.^{2}

K_S = structural stiffness, lb/in.

M = moving mass, lb-sec^2/in.

V = total volume of hydraulic fluid in cylinder and porting to control valve spool, in.^{3}

X = total position stroke, in.

dX/dt = velocity, in./sec

X_F = following error, in.

β = fluid bulk modulus, lb/in.^{2} (usually $\approx$ 100,000)

η = volumetric coefficient = AX/V

Courtesy of Moog, Inc., East Aurora, NY.

puter is to be used within the loop, its update frequency must be at least three times faster than the servovalve frequency at 90 deg of phase lag.

Position accuracy is more difficult to predict. It is dependent on load breakout friction and other external forces; valve threshold, hysteresis, null shift, pressure, and flow gain; mechanical backlash; transducer resolution and linearity; and drive electronics resolution.

A well-designed system will have high electronic and transducer gains to help minimize the influence of valve inaccuracies. This is most easily done by careful sizing of the control valve. Using a valve with excessive flow capacity results in a more inaccurate system, according to Stegner.

Special Features

There is much custom work in servosystems, chiefly because of the many possibilities for greater performance. We'll describe a few.

Servovalve add-ons: A popular way to convert a hydrostatic pump to servo control is to add a package tailored to the pump (Fig. 7.11). Servovalve companies supply the valve and electronics, and the pump makers provide a mounting surface and interface for the servovalve.

Table 7.2 Servovalve Flow Equations

A simple example: cylinder load. Size the actuator area, A, in.^{2}, for the stall load force required, F_S, lb, at the supply pressure available, P_S, psi:

$$A = \frac{F_S}{P_S}$$

Next, calculate flow at the most critical load velocity, dX_L/dt, in./s; and actuator force required at this velocity, F_L, lb. Determine the servovalve loaded flow, Q_L, in.^{3}/s:

$$Q_L = \frac{AdX_L}{dt}$$

and the load pressure drop, ΔP_L, psi:

$$\Delta P_L = \frac{F_L}{A}$$

Then, the no-load flow, Q_{NL}, in.^{3}/s, is:

$$Q_{NL} = Q_L \sqrt{\frac{P_S}{P_S - \Delta P_L}}$$

Finally, determine the servovalve rated flow, Q_R, gpm, @ 1000 psi valve drop. Increase by 10% for margin:

$$Q_R = 1.1 \frac{Q_{NL}}{3.85} \sqrt{\frac{1000}{P_S}}$$

Courtesy of Moog, Inc., East Aurora, NY.

Servovalve-cylinder packages also are available. Moog, for example, offers its A085 line as a fully engineered integral package of cylinder plus valve for applications where a custom-designed system is not required. The thought is that although best economies and performance are realized when the servo system is designed to exactly match the need, an off-the-shelf packaged unit can save development time and cost.

Failsafe electrohydraulics: A hard-over failure of a flow-control servovalve can be avoided by redundancy. Several input signals in agreement will overwhelm any one signal that is false. It's called *majority voting,* and the space shuttle Columbia has them for critical flight-surface power actuators.

Another way to avoid failure is to select valves that inherently fail in the safe direction. One example is a pressure-control pilot, such as the Honeywell MCV 101A. It is designed to pressurize the swashplate control cylinders of variable-displacement pumps and motors and to pressurize open-end closed-center main control valves of up to 50-gpm ratings. In the event of electrical failure, the valve calls for zero differential pressure, and it automatically destrokes the pump or centers the valve spool.

Rotary-to-linear servos: Low-torque dc motors and stepping motors are good inputs to EH servo systems, but they require rotary-to-linear mechanisms for the conversion. The field is ripe because microcomputers have successfully invaded the controls market, particularly in steppers.

Olsen Controls (Bristol, CT) has a packaged rotary-to-linear EH servoactuator consisting of electrical stepper motor, rotary-mechanical valve, and cylinder with internal mechanical feedback. Digital pulses received by the stepper motor are translated by the valve into hydraulic pressure against a piston actuator. In one version (model LS-300), each pulse to the motor results in 0.001-in. output of the linear actuator. Force amplification is great because the cylinder actuator operates at pressures up to 3000 psi.

Commercial Shearing (Youngstown, OH) has a rotary-to-linear hydraulic servo that develops hundreds of pounds linear force with only in.-lb of input torque. It is driven by a permanent-magnet dc gearmotor, with a conductive plastic potentiometer mounted on the shaft of the gearbox. Any 12-V dc automotive battery will operate it.

A helical track within each device converts rotary motion to linear motion or vice versa. In the Olsen design, a ball screw within the piston rod converts linear motion of the piston into rotary feedback to reestablish null in the servovalve. In the Commercial Shearing design, the rotor servospool is machined with helixes that serve as pilot oil passages. Rotary position determines where the pressure is applied and in which linear direction the valve spool will move.

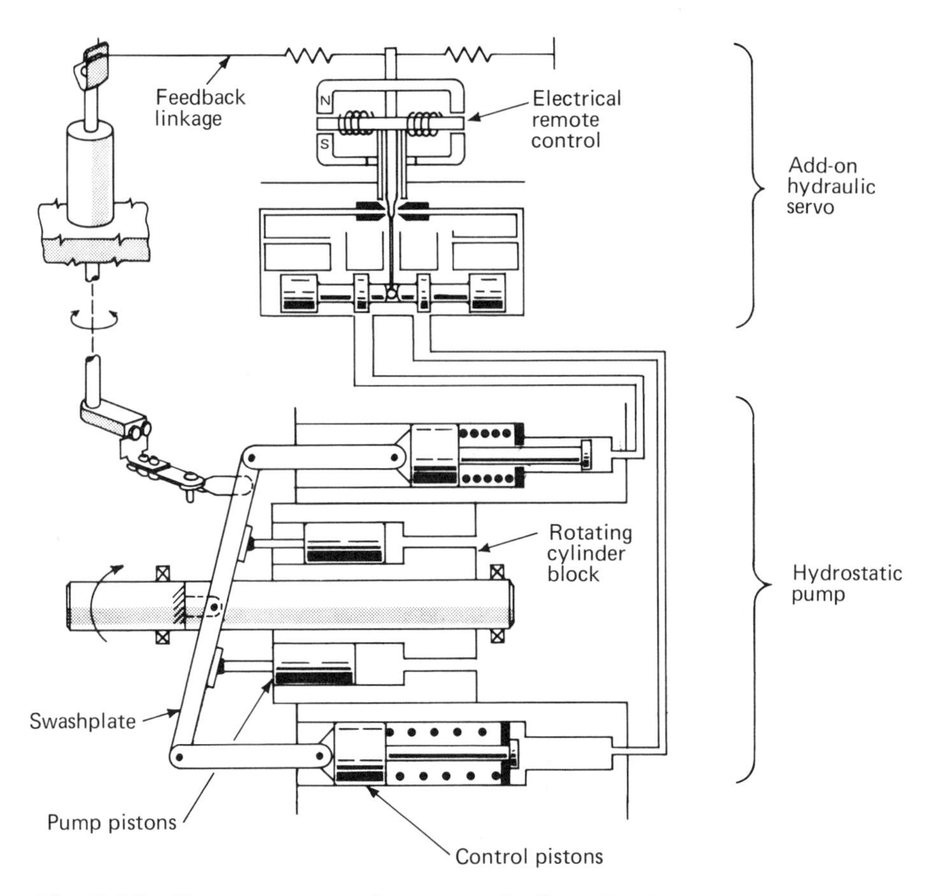

Fig. 7.11 Two-stage servovalve on top of adjustable-displacement pump determines the swashplate angle. (Courtesy of Moog, Inc., East Aurora, NY.)

MICROPROCESSOR CONTROL

A microprocessor (microcomputer) is to hydraulic power and control what a general is to the army. Signals and other information come to the general; he analyzes them and tells the army what to do. So it is with microprocessors.

One difference: The microprocessor (μP) is digital and deals only in discrete (single-value) signals. All of the analog input information first must be changed to digital pulses with an A/D (analog-to-digital) converter.

After the μP has analyzed the data (in milliseconds), it instantly generates instructions that the hydraulic valves and controls must follow. Those instructions are in digital code as they leave the μP and must be converted into whatever form is needed to do the following and more:

- Modulate servovalves
- Energize solenoid valves
- Call for new pump-stroke settings
- Reset flow and pressure regulators
- Display critical system conditions
- Annunciate warnings
- And, in the near future, *directly pulse digital valves*

That last item is the sleeper. Hardly anybody in the United States manufactures a line of microprocessor-controlled digital valves, but everybody says "it's coming and we're working on it."

Japan is another matter. Sperry Vickers (Troy, MI) in 1982 marketed in Japan a line of digital valves (Fig. 7.12) made by its affiliate, Tokyo Keiki (Tokyo, Japan). The valves seem to have all the elements described above.

In the meantime, at least several companies in the United States have developed special electrohydraulic actuator packages with μP digital input and rotary or linear actuator output. They contain the elements of digital valves, but they are not off-the-shelf valves.

Servovalves adapted to respond to microprocessor signals are another here-and-now technology, but they are not true digital valves. They require digital-to-analog (D/A) converters to create conventional analog input from digital signals. Just the same, there are many machines and vehicles with such controls.

Simple solenoid valves (on-off) can be readily triggered by microprocessor-generated pulses but are a limited exploitation of the true potential of microprocessors. See the following sections for work in that area.

Microhydraulic Marriage

The μP is the suitor; the hydraulic system engineer is the reluctant target. The world of the μP is foreign to most fluid power people, so computer specialists lead the way. It's also true that microprocessor people know very little about fluid power. We visited a few, and they had no idea how or why hydraulics would be a good application.

Despite that, pioneering groundwork has been laid in numerous special applications. Robots are the most dramatic example in that they not only are controlled with microprocessors but respond in a human way—including video eyesight. Hydraulic and pneumatic valves and cylinders abound in some types of robots.

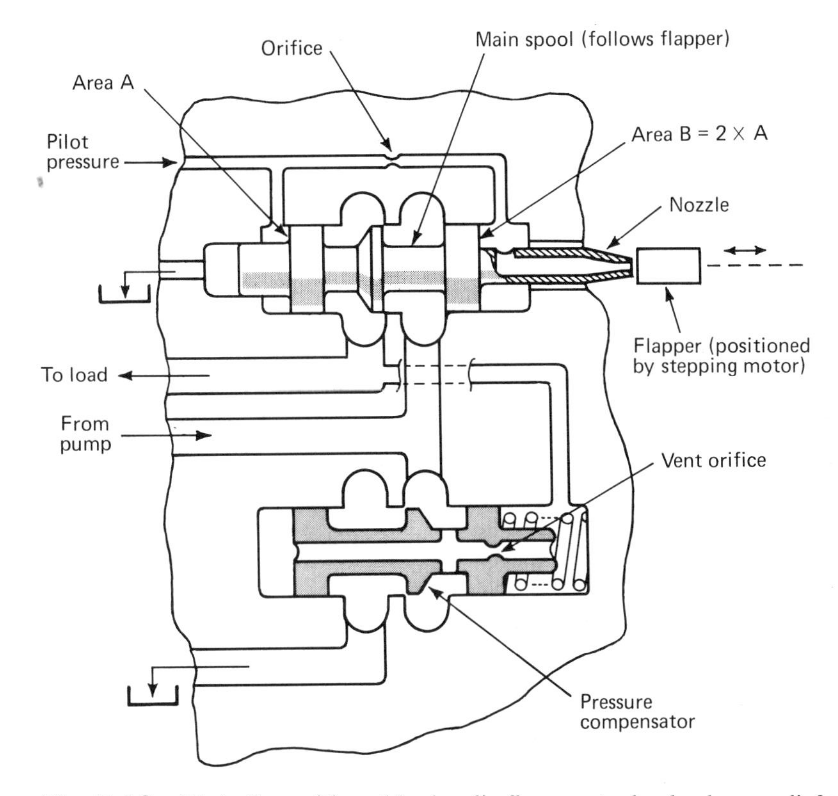

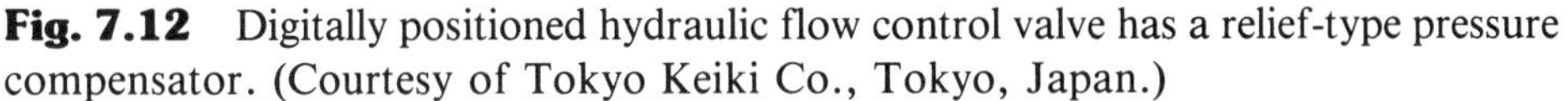

Fig. 7.12 Digitally positioned hydraulic flow control valve has a relief-type pressure compensator. (Courtesy of Tokyo Keiki Co., Tokyo, Japan.)

Other major microprocessor hydraulic applications are for vehicles and heavy equipment. For example, engineers at Motion Products Div. of MTS (Minneapolis, MN) and at GLI Div. of Baker International (Woodinville, WA) reported on successful designs and installations for earth-moving machines. Also, aerospace designers are well into the technology.

GLI Div. tells of μP controls in operation on power-shift controls and monitoring systems for agricultural tractors, mining trucks, oil fracturing equipment, construction vehicles, and other off-highway equipment; steering control systems for four-wheel-steer agricultural tractors, aircraft tow tractors, straddle cranes, and rough terrain cranes; automatic self-leveling controls for blasthole drills and agricultural equipment; and custom controls and indicator systems for off-highway vehicles.

Each design needs discrete electronic components for operation, and this is a natural application for the guiding hand of the μP. Others are remote and radio-remote control systems for underground mining machinery. The next step is to apply μPs to hydrostatic drives.

However, most of these operating systems are adaptations of existing hardware and include conventional servovalves, D/A and A/D converters, and all the other elements of two technologies: digital and analog. In those applications, it is well worth the customizing expense because the equipment itself is expensive. Garden tractors and ordinary machines are not candidates yet.

Another goal is digital valves and other hardware designed specifically for the new microprocessor hydraulics technology and widely available. According to Tokyo Keiki, there are several important benefits: A digital valve can operate open-loop from carefully counted pulses; the paraphernalia of feedback control are not necessary; valve tolerances don't have to be close, so contaminants are not a problem; and all of the control is digital, making it simpler to program total instructions into a computer.

There are exceptions, of course. Analog servovalves will continue to reign supreme where fast response and instant feedback are essential. No microprocessor chip has enough bits or speed to handle all pertinent control information for a high-response feedback servo. There, the μP must play the secondary role of making off-line decisions and setting up operating instructions.

Five Digital Elements

A digital valve system should contain a microprocessor control, a digital driver to amplify the pulse power, an electrical stepping motor integral with a digital valve, and a final actuator such as a cylinder or fluid motor. That's the "model" for this discussion (Fig. 7.13).

The *microprocessor* is a stored-program computer that processes input in predetermined ways and delivers coded pulses. The *driver* is an electronics package that

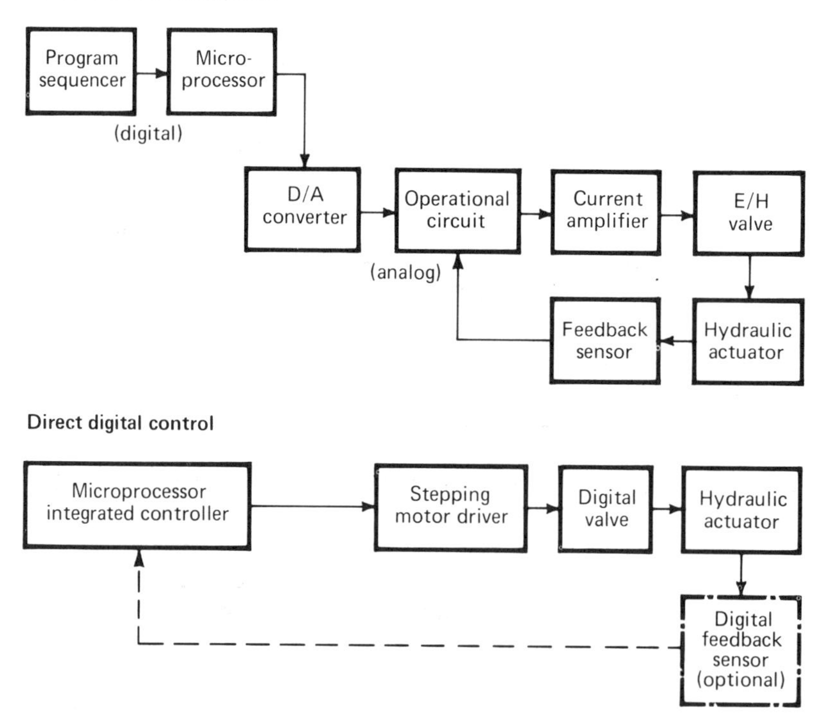

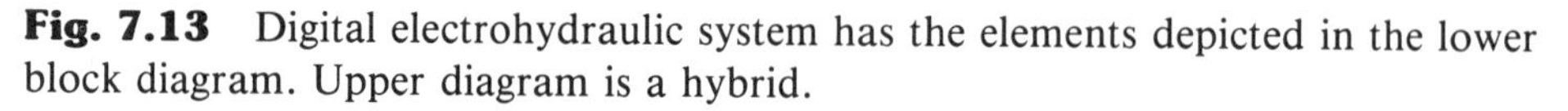

Fig. 7.13 Digital electrohydraulic system has the elements depicted in the lower block diagram. Upper diagram is a hybrid.

amplifies each pulse, without altering the pulse timing, to a level sufficient to operate the stepping motor against the resistance of friction and flow forces.

The electrical *stepping motor* (Fig. 7.14) rotates one step (1.8 deg in the Keiki design) for each pulse received. That's 200 steps per revolution. The motor is mounted integral with the digital valve (Fig. 7.12).

The pictured *digital valve* is a spool-type flow-control valve with a flapper pilot stage. The flapper is positioned by the integral stepping motor via a mechanical linkage; the spool follows the flapper and opens in proportion to the pulses received.

A pressure compensator spool also is included in the model pictured. It bypasses source flow to tank to maintain source pressure at a fixed differential above load pressure. The load pressure operates a cylinder or motor.

Practical Digital Systems

While we wait for off-the-shelf digital valves, three proven methods prevail: electrohydraulic stepper packages; μP-controlled solenoid valves; and modified servovalve systems.

A typical *EH stepper* (Fig. 7.15) has five matched elements: (1) the electrical stepping motor, (2) a helical-cam translator that converts rotary motion of the steping motor into linear motion of the valve spool, (3) the spool valve itself, (4) a sliding yoke pin at the opposite end of the spool valve to restrain the spool from rotary motion but to allow linear movement, and (5) a hydraulic motor (rolling-vane type shown). The amplification of torque and power is enormous. Relief valves are needed to limit pressure during certain conditions of loading.

Low-power pulses operate the stepping motor, just as in the preceding example. For each step of rotation, the valve spool is urged forward an amount that depends on the pitch of the helical groove. The hydraulic flow is proportional to the spool displacement and powers the hydraulic motor in a direction that rotates the valve spool back to the null position. Thus, for each rotary step of the electrical motor, there is a corresponding rotary step of the hydraulic motor, always ending with the valve spool at null.

The μP *solenoid system* operates the equipment with conventional solenoid valves, but the instructions to the solenoids can be extremely versatile. A good exam-

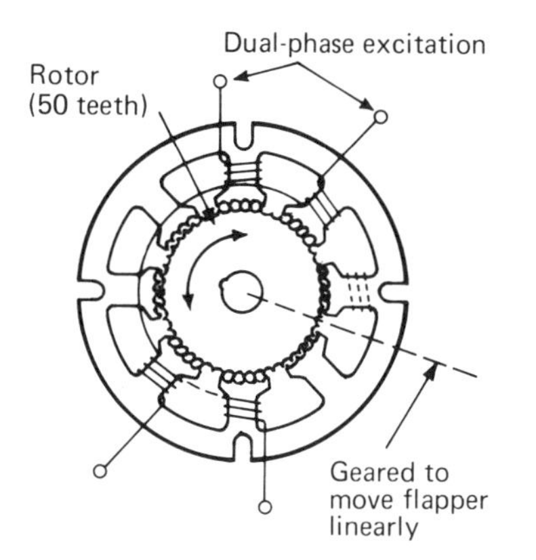

Fig. 7.14 This stepping motor is an integral part of the Tokyo Keiki digital valve.

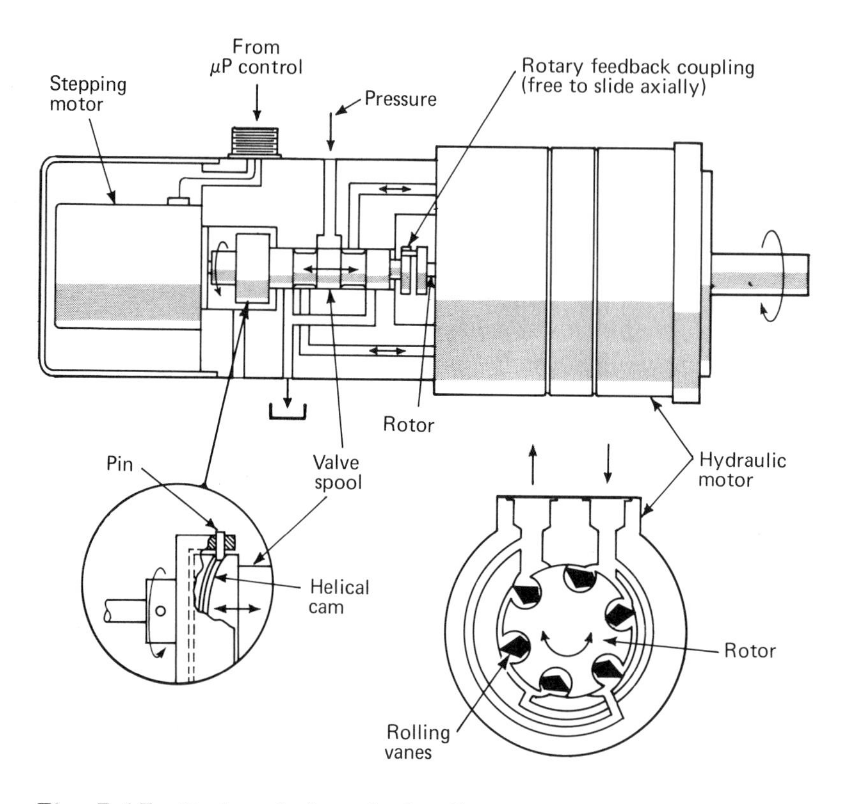

Fig. 7.15 Packaged electrohydraulic motor with stepper control operates with pulsed signals from a microprocessor. (Courtesy of Motion Products Div., MTS, Minneapolis, MN.)

ple is described by engineers at Grad-Line: a power-shift transmission for heavy trucks.

With microprocessors handling the logic, the shifting up and down can be controlled at a predetermined rate. The system can upshift or downshift automatically when engine speed changes. A speed-inhibit function may be added to prevent shifts to neutral (or reverse) above a programmed engine speed. Engine overspeed may actuate a retarding function.

The μP can handle far more than simple control. It can monitor as many conditions of temperature, stress, pressure, speed, vibration, position, energization, proximity, and environment as there are sensors. The major additional cost is in the sensors—not in the μP. It also can initiate and control displays of any function mentioned above, limited only by how much you want to spend for display readouts.

Performance Compared

The accompanying graph (Fig. 7.16) shows some inherent advantages of electrohydraulic stepping systems over analog systems. The advantages result from these basic facts about pulses: They can be counted with any speed and accuracy desired, they can be speeded up or slowed down at will, they can be generated in any code that can be imagined, and they will repeat with awesome reliability. There could be some problems of resonance at certain frequencies, but the biggest lack, compared with servovalves, is speed of response. A 500-pps stepper with 100 poles would take 0.2 sec/revolution.

Compared against on-off solenoid valves, the digital valve has the unique advantage of being able to control the initial response time at will to match changing load conditions. A solenoid valve has fixed opening characteristics, even though special machining, special coils, and special energizing voltages can be provided. Once the solenoid valve is installed, the response is known and fixed. With digital valves, response can be altered remotely with programming.

Care in system design is still essential, even though stepping is reasonably direct. The GIGO (garbage in, garbage out) principle applies, and the machine will do

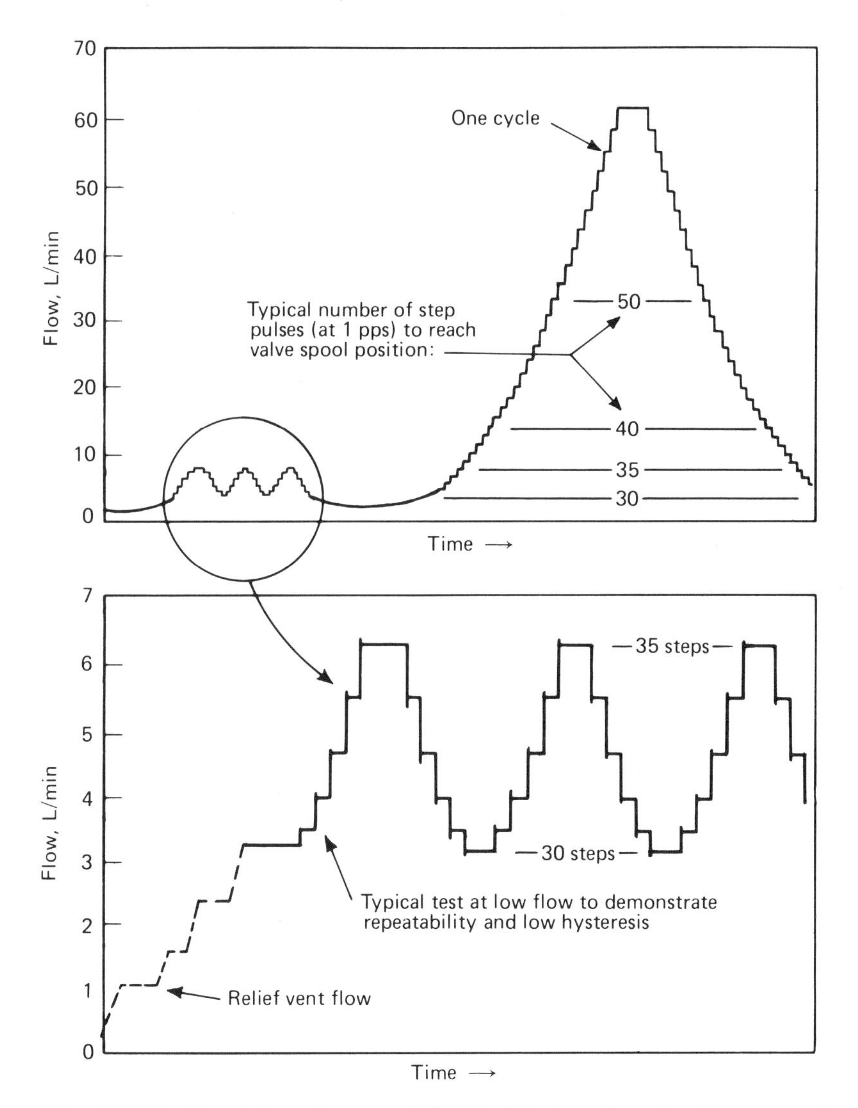

Fig. 7.16 Graph based on test of digital hydraulic valve shows good repeatability when input is stepped slowly. (Courtesy of Tokyo Keiki Co., Tokyo, Japan.)

precisely what it is told, even though wrong and possibly dangerous. Grad-Line includes failsafe operation as an important design feature. Some examples: What will happen if the machine or vehicle is running and a battery cable shorts out? What if a switch fails in the open position? What if the operator must override automatic features to gain control of the power-shift? Will transient spurious signals give false directions?

The answer of course is to program the microprocessor the same as you would a computer, which is what it is. Logic flow charts, printouts of hundreds of program steps, debugging, and all the other tools of the programmer's art must be skillfully applied. That's why only computer experts, guided by hydraulics experts, must do the system design when microprocessors are the control.

The need will be met initially by companies that specialize in tackling the interface between microprocessors and production hydraulic systems. Grad-Line, for example, designs μP control systems for major manufacturers of heavy vehicles and equipment and then produces the components to implement the designs. Eventually some of this expertise will move in-house, following the path that servo-system experts walked in the past.

8

Fluid Shock Absorbers and Accumulators

This chapter explains the theory of fluid shock absorbers and accumulators; other chapters are sprinkled with examples of the art. For instance, the cushioning of conventional cylinders is covered briefly in Chaps. 5 and 15, and some applications of accumulators are highlighted in Chap. 10.

Arbitrarily included under shock absorbers are damping devices to prevent fluid hammer (such as the water hammer heard in many home-piping systems) and a variety of special snubbers, cylinder cushions, liquid springs, and dynamic brakes.

Accumulators are not exactly related to shock absorbers, but they fit into this chapter better than any other.

SHOCK ABSORBERS

A shock absorber is a device that produces a dissipative output force over a given displacement to absorb energy and remove it from a system. It includes a reset or restoring means to reposition it after it has absorbed energy; the resetting energy is much smaller than the absorption capacity of the device.

That eliminates damped-spring systems, vibratory-isolation systems, pulse-absorbing accumulators, and other anti-vibration systems that are not what we know in the fluid power industry as shock absorbers.

It also discounts the common automotive shock absorber for two reasons: The shock absorber part does very little of the work (the spring does most of it), and the unit restores the energy and doesn't remove it from the system. The only absorption of energy is a damping action.

Arbitrarily included are built-in cushions for hydraulic cylinders, because the principles are the same.

This chapter will stick to fluid types (oil and air) for the most part (Figs. 8.1 and 8.2). Just the same, there are nonfluid types that handle similar jobs; let's discuss those first:

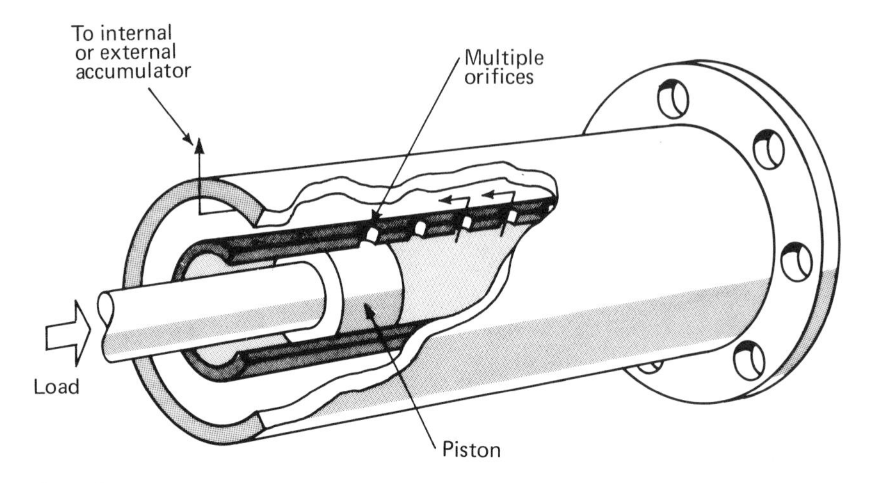

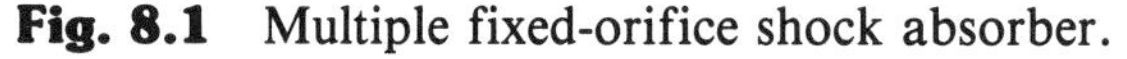
Fig. 8.1 Multiple fixed-orifice shock absorber.

- *Solid-elastomer types* for limited energy absorption (rubber-spring bumpers, stacked discs or doughnuts, and compressible-flowable.) See Fig. 8.3.
- *Mechanical types* for special modes of energy absorption (frangible metal and friction). Typical examples are sketched in Fig. 8.4.

The shearing of metal absorbs a lot of kinetic energy, but the action is one-time. Simple friction is a better way to absorb energy mechanically if repeat cycling is required. For example, Fig. 8.4, item B, shows the principle of operation of the friction ring springs made by Ringfeder (Westwood, NJ).

Engineers at the company say that travel is small, but the forces are very high and energy absorption can be from 1000 to 100,000 joules, depending on the number and size of the rings. The friction creates heat, which is dissipated rapidly by the metal. About two-thirds of the mechanical energy is dissipated, and one-third is retained as restoring force in the stretched and compressed rings. The smallest model offered has a designed impact force of 1 ton.

One characteristic of this friction-type shock is that the unit doesn't restore until two-thirds of the load is removed. If the load happens to be a gravity weight such as a truck, it is possible to hit a rut in the road hard enough to displace the shock but not have it restore after the truck has passed over the rut. Thus most applications are for heavy-duty buffers where quick rebound is not desired.

Fluid Absorber Facts

The ideal energy absorption curve (force vs. displacement) is rectangular (Fig. 8.5). That is, the output force is constant throughout the energy-absorbing stroke, and there are no peaks. Achieving it is difficult and in many applications is not worth the cost. A good compromise is to approximate a square wave, give or take a few peaks.

The area under the curve is the integral of output force over increments of displacement. Any shock absorber trace will have basically the same area under the curve as any other for a given load being stopped. The difference is in the wave shape. A simple orifice-type shock absorber will start with a high retarding force and a rapidly decreasing force vs. displacement as the piston slows down—in accordance with the

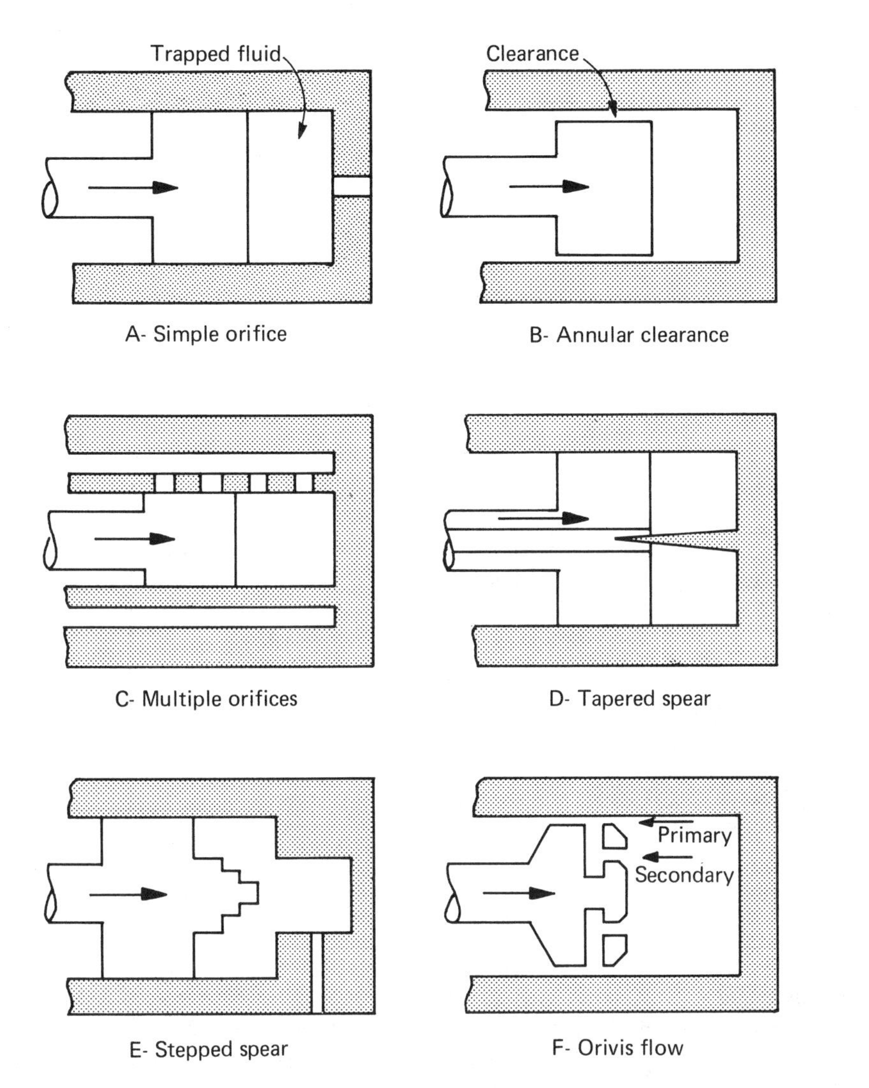

Fig. 8.2 Fluid shock-absorbing concepts.

square law of orifice flow. A perfectly designed variable-orifice or metering pin shock absorber will have a square trace. Multiple-orifice designs fall in between, depending on the number and placement of the orifices.

A useful mathematical relationship to compare traces is to calculate the hypothetical square-wave force: energy capacity divided by stroke. Then, divide that by the actual peak force to get what Taylor Devices Co. (Tonawanda, NY) calls "efficiency." Here's an example:

Take a 10,000-in.-lb capacity shock with a 4-in. stroke, cataloged for 4000-lb maximum force. The square-wave force is 10,000/4 or 2500 lb. Then efficiency is 2500/4000 or 62.5%. Efficiencies of 50% to 80% are typical for conventional industrial shocks and 80% to 90% for aerospace designs.The best reach 90% but only after costly effort.

That's a relatively unsophisticated way to rate a shock absorber, but a rigorous

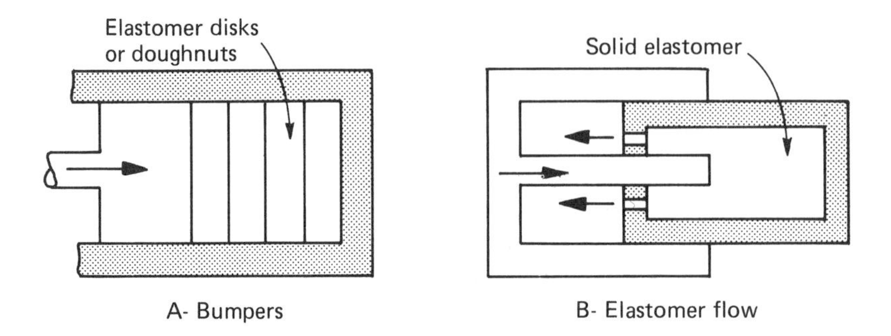

Fig. 8.3 Solid elastomers to absorb shock.

analysis is complicated. Navier-Stokes flow equations are correct enough and are easy to use if one ignores the higher-order terms. But the higher-order terms greatly affect performance.

The problem occurs when dynamic pressures reach 20,000 psi, and Taylor predicts 40,000 psi within a decade. Stroke speeds in some aerospace applications already are 30 ft/s to 40 ft/s, and flow velocity through the orifices is over 1200 ft/s. You can't solve flow problems like that out of a handbook, so testing always is an essential step.

Later in the chapter we'll tackle the math for some of the simpler shock absorbers, but that doesn't eliminate the need for tests.

Mistakes can be ruinous. The metal bottom of a shock absorber cylinder is no place to instantly dissipate the total kinetic energy of a heavy load, but that's what happens when the chosen unit is too small. Shock forces of a million lb are possible, and this will crush metal. Another mistake is to build in too much restoring force, either by bad design or a blocked orifice. The load ends up heading backwards at high velocity—not a good idea.

Fixed Adjustment Shocks. Fixed single-orifice designs are simplest but need at least one high-pressure seal to withstand the initial peak force. Also, fixed-orifice units cannot easily cope with a sustained propelling force that drives the piston at constant velocity smack into the end cap.

Multiple-orifice shock absorbers (Fig. 8.1) can be as efficient as desired, depending on the number and sizing of the orifices. One problem is that a propelling force

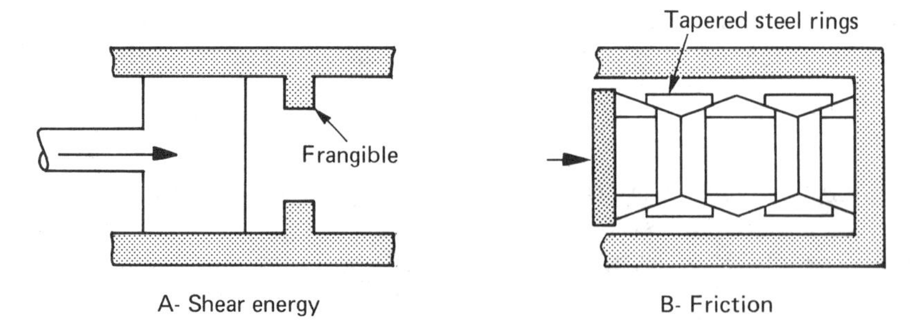

Fig. 8.4 Metal shear and friction to absorb shock.

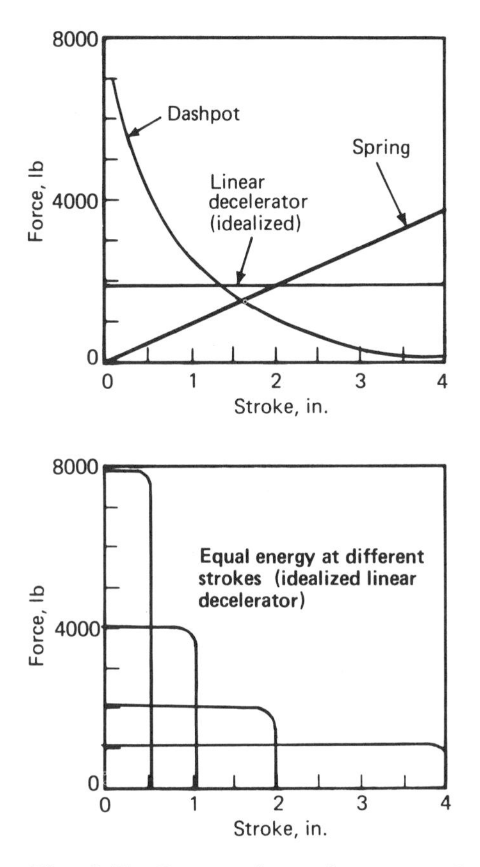

Fig. 8.5 Energy absorption curves for fluid shock absorbers.

will drive the piston into the end cap unless the final orifice is specially sized (very small) to match that force.

Metering pin shock absorbers of the parabolic type are so hard to build that designers settle for straight tapers or other variations, thus not improving much if any on multiple orifice types.

Another problem is the possible dieseling condition if a pressurized air chamber is integral with the moving piston, separated from the oil with a free piston. If oil leaks past the free piston, an explosive mixture can result. A heavy shock loading can compress the air adiabatically to ignition temperature. A safer way is to substitute a spring for the compressed air.

The entrained-flow orifice system patented by Taylor Devices (Fig. 8.2, item F) is designed for compressible fluids such as the silicones, which are 10% compressible at 20,000 psi. It works like this: The initial flow when the load impacts is through the primary passage. The velocity is high enough to create a low-pressure area at the junction with the secondary passage, where it entrains appreciable additional flow. The greater the speed of impact, the more oil is entrained. The result is a more square waveform because flow tends to be proportional to piston speed. This obeys fluidic principles, so the company calls it a fluidic control metering system and the product Fluidicshok.

Taylor says that the combination of the compressible fluid (essential to the entrainment) and the "fluidic" character of the flow yields high efficiency (75% to 95%). One disadvantage is that much trial and error is required for each new design to establish passage configuration. Once established, repeatability is good. Also, silicone fluids are not compatible with some common seal materials.

Adjustable Shocks. Adjustment devices are available for many of the shock absorber types shown. For example, the multiple-orifice design (Fig. 8.1) is built with a second orificed sleeve tightly fitted over the first. By manually adjusting the second sleeve in relation to the first, the effective orifice size can be varied (more details later).

The entrained-flow shock is not manually adjustable but can be supplied with a special inertia valve that senses the rise-time rate during initial impact and adjusts flow into the primary and secondary passages accordingly. This adjustment occurs automatically for each load condition, within a maximum piston speed range of from 10 in./s to 70 in./s.

Liquid-spring reset is possible with compressible-fluid shock absorbers. The stored energy of the compression is sufficient to return the rod. This does not affect the energy-absorbing action, which relies on throttling of flow, not storage of pressure. Rebound need not be a problem either if restrictive passages are included in the return direction.

Integral Cushioning. A lot of design work has been done on integral cushions for oil and air cylinders. One example, by Miller Fluid Power (Bensenville, IL), has a unique expandable cushion (Fig. 8.6). The figure is conceptual only.

The initial flow of hydraulic fluid at start of cushioning is through the annulus on the outside diameter of the cushion plunger, which is made of iron. This flow reaches high velocity as the plunger closes, and the pressure in that annulus decreases in accordance with Bernoulli's law. The pressure in the clearance at the inner diame-

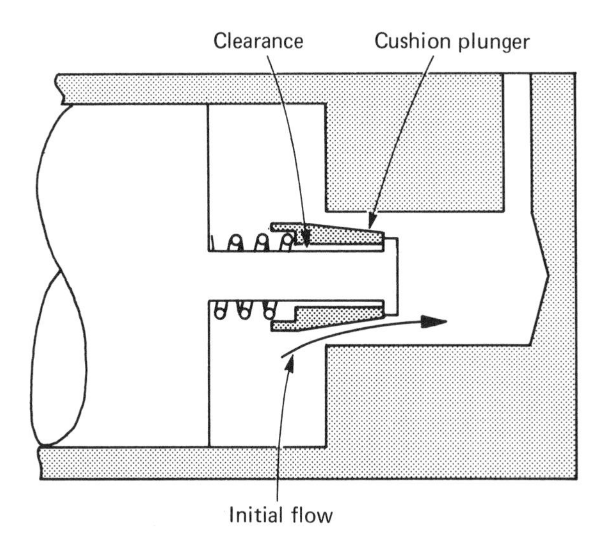

Fig. 8.6 Expandable-cushion plunger automatically varies clearances in response to shock pressures. (Courtesy of Miller Fluid Power, Bensenville, IL.)

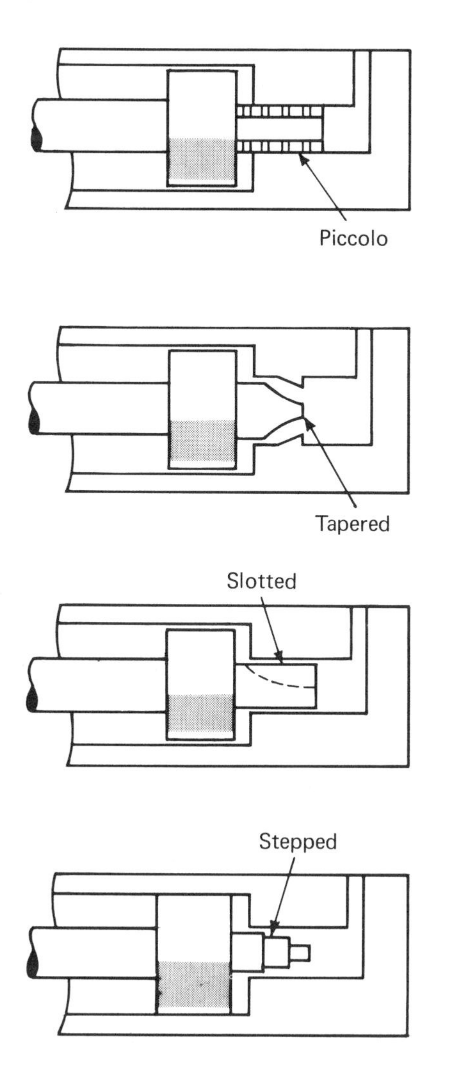

Fig. 8.7 Typical end-of-stroke cushions for hydraulic cylinders.

ter builds up during initial cushioning and stretches the cushion plunger radially outward to effectively seal off most of the annulus flow. The spring helps the chamber pressure keep the cushion pressed forward against the shoulder.

The result of this balance of forces is to trap pressure between the cylinder cap and the piston in proportion to the speed of the piston, automatically. Thus the retarding force is always adequate, whatever the mass and speed of the load. No cushion-adjusting screw is required.

On the return stroke the spring is compressed first, opening the clearance between the shoulder and the cushion plunger, allowing the full area of the piston to be pressurized immediately. Smooth, fast starts are achieved coming out of cushion, and no ball check is needed.

Parker Hannifin's Cylinder Div. (Des Plaines, IL) has further developed several other concepts for controlled-force cushions in hydraulic cylinders (Fig. 8.7).

The theory is that steady deceleration can be achieved if the orifice area is varied continuously to compensate for the piston velocity (Fig. 8.8).

The simplest is a stepped spear (Fig. 8.9): It has performance close to that of the ideal parabola and costs much less. (Not shown is the floating bushing that maintains the concentricity of the spear in the bore.)

The profile of an ideally tapered spear theoretically should be a concave parabola. The flow area is larger at the beginning of the cushion and rapidly reduces near the end. It is impractical, however, to produce a perfect parabola without a tape-controlled machine tool, so it is approximated by cutting a series of straight tapers along the spear.

The slotted spear is a variation of the tapered spear and is easier to make. A concave parabolic shape is required in the slot near one end but usually can be approximated with a milling cutter.

The multiple orifice design (dubbed "Piccolo") is similar to that of many industrial shock absorbers and is more expensive than the tapered or slotted spears. It has a series of orifice holes through its side and an exhaust hole through the center. When the spear first closes off the exhaust flow passage, all orifice holes are exposed to the cushion chamber, and maximum flow is obtained.

As the spear progresses into the cap, the orifice holes are blocked off to gradually reduce the cushion flow area. Parabolic spacing of constant-diameter holes will provide a stepwise approximation of the theoretical flow area. A large number of holes can closely approach the theoretical flow area curve.

For even better performance, thin-disc orifices with well-defined flow coefficients can be incorporated. Flow then remains independent of temperature-viscosity effects.

A needle valve is added to cushioned hydraulic cylinders. It is installed as a trimming device to more closely control the final contact velocity at the stroke end.

Applications

Constant-deceleration cylinders will cost more but can be justified wherever smooth stopping in minimum distance is essential, such as in handling sensitive sand molds or in moving pallets that carry unanchored parts. Conventional constant-area cushions still will suffice where pressure shocks will do no damage, such as in stops for clamping devices or where a quick initial slowdown followed by a slow, steady speed is wanted.

Be careful in writing application specifications, and don't neglect inertia and propelling forces of any moving part in the system—including rotary parts. See the equations in Table 8.1.

Another thing to watch out for is load orientation. A vertical load adds propelling force equal to its own weight, and this better not be forgotten. Horizontal loads add friction. Do your trigonometry.

Try to work within catalog ratings, but don't give up if you cannot find what you want. Bore and stroke actually can be any dimensions within reason in custom models, and the semistandard models range from a fraction of an inch stroke to over 5-ft stroke. Never hesitate to contact a shock absorber manufacturer: They all claim to welcome tough questions—even for small orders.

The biggest design challenge is to understand the difference between inertia and

How to calculate constant deceleration

The object of the calculation is to determine the plot of orifice area vs piston position that is needed for constant deceleration (constant slow-down force). Four basic equations are involved:

$F = Ma = \frac{W}{g}(\beta g)$	Newton's second law of motion
$\frac{P_t - P_e}{\gamma} = \frac{V_O^2}{2g}$	Bernoulli's equation
$A_2 V_P = C A_O V_O$	Law of continuity
$\text{Work} = E_k - E_p$	Conservation of energy

From those four basic equations, plus some general relationships, Berninger developed a composite equation that solves for area in the simplest case of supply and exhaust pressures held constant:

$$A_O = \frac{A_2}{C}\left(\frac{\beta\gamma A_2(l_3 - l)}{P_1 A_1 - W\cos\theta + W\beta - P_e A_2}\right)^{1/2}$$

Meanings of the symbols are given in the table. A typical plot of area vs piston position is shown in the graph. The equation is valid only for the special case of constant supply pressure and constant exhaust pressure.

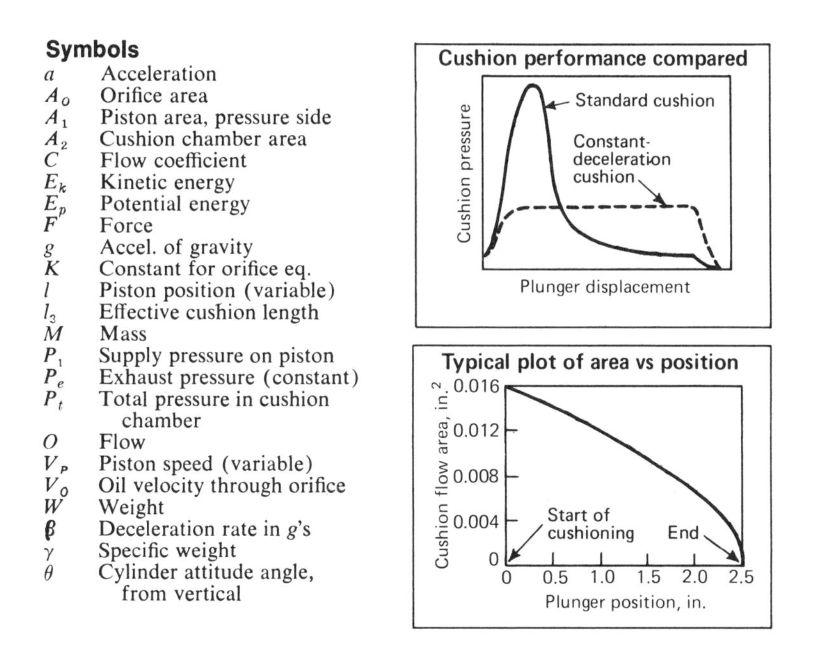

Symbols

a	Acceleration
A_O	Orifice area
A_1	Piston area, pressure side
A_2	Cushion chamber area
C	Flow coefficient
E_k	Kinetic energy
E_p	Potential energy
F	Force
g	Accel. of gravity
K	Constant for orifice eq.
l	Piston position (variable)
l_3	Effective cushion length
M	Mass
P_1	Supply pressure on piston
P_e	Exhaust pressure (constant)
P_t	Total pressure in cushion chamber
O	Flow
V_P	Piston speed (variable)
V_O	Oil velocity through orifice
W	Weight
β	Deceleration rate in g's
γ	Specific weight
θ	Cylinder attitude angle, from vertical

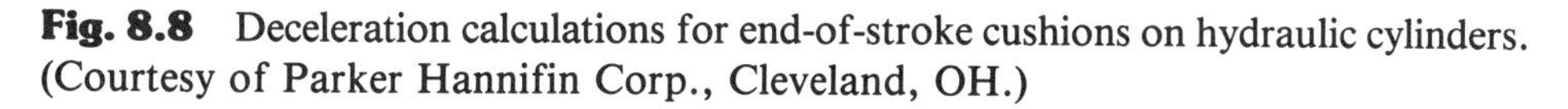

Fig. 8.8 Deceleration calculations for end-of-stroke cushions on hydraulic cylinders. (Courtesy of Parker Hannifin Corp., Cleveland, OH.)

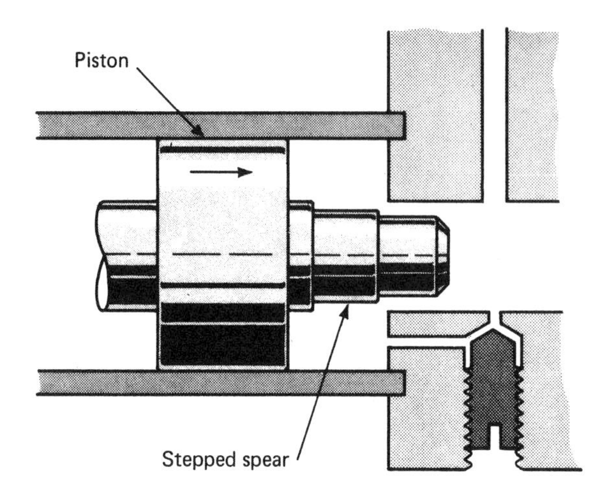

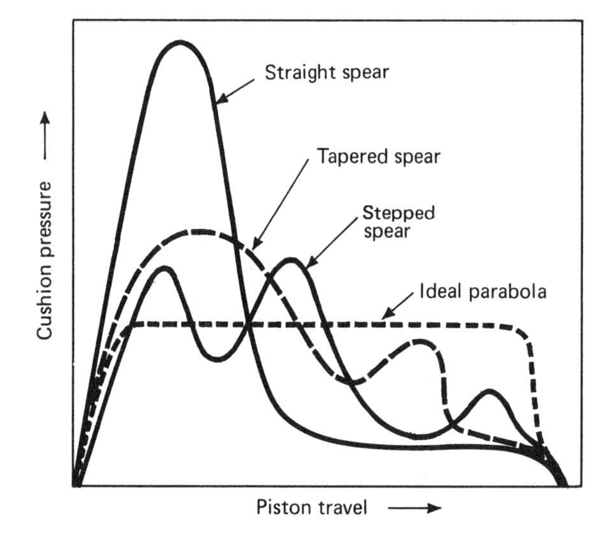

Fig. 8.9 Steps in spear are a good compromise to approximate the retarding pressure and travel of an ideal inverted parabola without the expense of machining the parabola. (Courtesy of Parker Hannifin Corp., Cleveland, OH.)

force and keep the two in perspective. Engineers at Ace Controls (Farmington, MI) strongly recommend that before plunging into the empirical world of shock absorbers that the designer ask himself or herself three questions: Is the velocity of the mass being stopped below 1 ft/s? Is it over 15 ft/s? Is the propelling force energy (force × stroke) more than three times the kinetic (inertial) energy? If the answer is yes to any of the three, take your problem to an expert.

Why? Because the specific parameters may not be in the range of most cataloged devices, even though the total energy absorbed is. For example, if the mass of the object is small relative to the absorber (a bullet impacting against the end of the rod), the rod might not even move before all of the energy is dissipated. At low velocities, a high proportion of the energy might be in the propelling force, instead of inertia.

Ace Controls recommends a simple "equivalent weight" equation for those applications where kinetic energy is combined with drive (propelling) energy. The company uses this weight W_E in estimating the proper size and capacity of shock absorbers. Here's the derivation:

$$E_T = E_K + E_D = 0.1865 W_E V^2$$

$$W_E = (E_K + E_D)/(0.1865 V^2)$$

where the units are explained in Table 8.1. The equivalent weight method can be used to choose sizes and ratings in catalogs from most manufacturers.

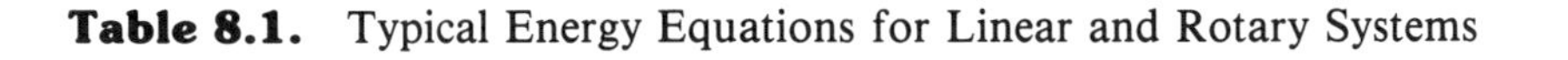

Table 8.1. Typical Energy Equations for Linear and Rotary Systems

Basic equations

Kinetic energy, in.-lb:
$E_K = 0.1865 WV^2$ for linear
$E_K = 6\ IV^2_R$ for rotary
$E_K = W(H + S)$ for vertical (free fall)

Drive (propelling) energy, in.-lb:
$E_D = F_D S$

Total energy, in.-lb:
$E_T = E_K + E_D$

Deceleration rate, linear:
$G\text{'s} = F/W = (0.1865 V^2)/(5\mu)$

Deceleration rate, rotary:
$G\text{'s} = (RV_R)^2/(772 S\mu)$

Deceleration time, s:
$t = S/6V\mu$

Example: A 2-in.-stroke shock, with 70% efficiency, is impacted at 2 ft/s. The deceleration time is:

$t = 2/(6 \times 2 \times 0.070) = 0.24$ s

This time is short enough to be of no consequence in the total cycle time of a typical industrial operation

Symbols:

W = weight in lb_M; V = velocity, ft/s; V_R = rad/s; I = lb-ft-sec²; H = height, in.; S = stroke, in.; F_D = drive force, lb; μ = shock absorber efficiency: (square wave peak force) divided by (actual peak force); R = radius of rotary mass that impacts shock, in.; G = acceleration of gravity; t = time, s; N = rpm; K = radius of gyration, ft; X as subscript = second shaft; M as subscript = main shaft (rotary); T = torque, lb-ft

Special rotary conversions

Average torque of rotary mass when slowed from N_1 to N_2 in time t:

$$T = WK^2\ (N_1 - N_2)/308t$$

Equivalent WK^2 of second rotary mass referred to main shaft:

$$(WK^2)_{\text{equiv}} = (WK^2)_X\ (N_X/N_M)^2$$

Conversion from rotary kinetic energy to linear kinetic energy:

$$\text{Linear } E_K = 0.1865 WV^2 = 7.37 WK^2N^2$$

In any case, list all the energy that can be input to the shock absorber, and use it to calculate both basic energy per cycle and energy per unit time that the shock must absorb.

Don't forget stall torque of electric motors and hydraulic motors. This typical source of energy often is overlooked in sizing shocks. When the shock absorbers do their job and bring a moving system to a stop, realize that the stall torque of any motor still energized will be several times its running torque. Ac electric motors have stall torques of about 2.5 times the running torque and dc motors about 3.5 times the running torque. Hydraulic motor torque also is much higher at stall.

For instance, take a 3000-lb weight moving at 2.0 ft/s, driven by a 1-hp ac electric motor, into a 4-in.-stroke shock absorber. Kinetic energy $E_K = 0.1856WV^2 = 0.1865 \times 3000 \times 4 = 2238$ in.-lb., where $W = lb_m$ and $V = ft/s$.

Nominal drive force resulting from rated motor torque is $550 \times (hp/V) = 550(1/2) = 275$ lb. Nominal drive energy $= 275 \times 4 = 1100$ in.-lb. Add that to the kinetic energy: $2238 + 1100 = 3338$ in.-lb total energy E_T to be absorbed by the shock, assuming only nominal torque.

But consider the fact that the motor actually is contributing 2.5 times its nominal running torque—at least at final stall. The drive force actually reaches $2.5 \times 275 = 688$ lb, and the maximum stalled-drive energy $= 688 \times 4 = 2752$ in.-lb. The shock must absorb $2238 + 2752 = 4990$, which is a 49% increase over the nominal assumptions.

Fluids for Shocks. Variations in normal fluids used in a shock absorber have little effect on output force in conventional applications. However, the fluid selected must be compatible with the seals, with the orificing methods, and with the environment.

For example, straight oils perform the same as silicone-base oils at room temperature, but viscosity and performance change drastically at temperatures below zero F or above 100 F. The viscosity of silicone fluid changes least. Silicone fluid also is better at the extreme temperatures in some foundry, mill, and aerospace applications, where its high flashpoint (over 600 F) is a safety feature.

Compressibility is needed for fluidic orifice shocks. This is one of the chief reasons for selecting silicones: compressibility is 10% at 20,000 psi, which is like squeezing 10 gal of oil into a 9-gal container.

What about "solid" fluids? Certain lightly cross-linked solid elastomers will behave as a fluid when squeezed through a restriction. Some shock absorbers are designed to exploit this phenomenon. Taylor says that it works, but because of the cross-linking of molecules, the material doesn't last too long when extruded like that. Also, the orificing generates heat that degrades the elastomer.

The usual application for solid fluids is to compress them rather than to totally extrude them. This yields an output similar to that of a high-force coil spring. About 80% of the energy input is returned to the system as a recoil; the balance is dissipated as heat generated by intermolecular friction. Typical applications are car bumpers, where recoil after impact can be tolerated.

Accumulators and Reservoirs. Accumulators are not shock absorbers in the strict sense, because they return the energy to the cycle. However, they are used in conjunction with shock absorbers as energy-storing devices and as reservoirs for excess flow. They also serve as dampers to smooth out flow pulsations. Some shocks have built-in accumulators. See details later.

FLUID SHOCK CALCULATIONS (TABLE 8.2)

By limiting the calculations to simple energy balances and by not getting into acceleration equations, it is possible to quickly approximate size and performance of ordinary dashpots, snubbers, buffers, shock absorbers, and decelerators. Nothing takes the place of actual tests, of course, especially when nonlinear variables such as friction, flow factors, and viscosity are involved.

Fancier calculations than these are being made by shock absorber experts, but

Table 8.2. Basic Equations for Simple Shock Absorbers

Laminar restriction (all fluids)

Stopping time, t_t, sec

Case I (see note)

$$t_t = \frac{1.18\mu\,(D_M L_B / C_{Rad})^3}{MV_i^2/2}$$

Case II

$$t_t = \frac{1.18\mu\,(D_M L_B / C_{Rad})^3}{(MV_i^2/2) - W_F}$$

Case III

$$t_t = \frac{1.18\mu\,(D_M L_B / C_{Rad})^3}{MV_i^2/2 + (F_1/2 + F_{Mech} - F_2)\,L_B - W_F}$$

Annular clearance, C_{Rad}, in.

Case I

$$C_{Rad} = D_M L_B \sqrt[3]{\frac{1.18\mu/t_t}{MV_i^2/2}}$$

Case II

$$C_{Rad} = D_M L_B \sqrt[3]{\frac{1.18\mu/t_t}{(MV_i^2/2) - W_F}}$$

Case III

$$C_{Rad} = D_M L_B \sqrt[3]{\frac{1.18\mu/t_t}{MV_i^2/2 + (F_1/2 + F_{Mech} - F_2)\,L_B - W_F}}$$

Bore length required, L_B, in. (trial and error)

Case I

$$L_B = \frac{C_{Rad}}{D_M}\sqrt[3]{\frac{MV_i^2 t_t}{2.36\mu}}$$

Case II

$$L_B = \frac{C_{Rad}}{D_M}\sqrt[3]{\frac{\left(\frac{MV_i^2}{2} - W_F\right) t_t}{1.18\mu}}$$

Case III

$$\frac{1}{L_B^3}\left[\frac{MV_i^2}{2} + (F_1/2 + F_{Mech} - F_2)\,L_B - W_F\right] = \frac{1.18\mu D_M^3}{t_t C_{Rad}^3}$$

Orifice restriction (Case IV; for oil)

Time, sec (to 0.1 V_i)

$$t\ @\ 0.1V_i = \frac{0.9MA_o^2}{V_i A_P^3 K\ 0.42\times 10^{-5}}$$

Orifice area, A_O (special)

$$A_o = \sqrt{\frac{(t\ @\ 0.1\,V_i)\,(V_i A_P^3 K\ 0.42\times 10^{-5})}{0.9M}}$$

Piston travel, x (general)

$$x = \frac{MA_o^2}{A_P^3 K\ 4.2\times 10^{-5}}\left[\ln\left(\frac{M}{V_i} + \frac{A_P^3}{A_o^2}K\ 4.2\times 10^{-5} t\right) - \ln\frac{M}{V_i}\right]$$

Symbols

Symbol	Meaning
A	= area, in.^{2}
C	= clearance, in.
C_F	= coefficient of friction
D	= diameter, in.
E_K	= kinetic energy, in.-lb
F	= force, lb
f	= friction factor for viscous flow, $64/R_E$
g	= acceleration of gravity = 386 in./sec^2
K	= K-factor for pressure loss
L	= length, in.
M	= mass, lb-sec^2/in.
P	= pressure, psi
ΔP	= differential pressure
Q	= flow, in.^{3}/sec
t	= time, sec
V	= velocity, in./sec
W	= work, in.-lb
x	= travel, in.
μ	= abs. viscos., lb-sec/in.^{2}
ρ	= density, lb/in.^{3}

Subscripts

B	= bore
E	= effective
F	= friction
i	= initial
L	= load
M	= mean, or average
O	= orifice
P	= piston
PL	= plunger
Rad	= radial
t	= total (to stopping)
1,2, 3	= pressure regions

Note: Case I = load inertia resisted by P_3; Case II = load inertia resisted by P_3 and friction; Case III = load inertia, P_1 and mechanical force resisted by P_2, P_3 and friction; Case IV = load inertia resisted by P_4 only.

most are proprietary and require computer programs. A few examples are discussed, with advice on how to take advantage of them.

In the meanwhile, examine the two most basic energy-absorbing flow restrictions: an annular restriction such as the clearance between a piston and cylinder and an orifice restriction.

An annular restriction develops a resisting pressure proportional to V if the flow is laminer and to V^2 if flow is turbulent. (See Chap. 9 for a detailed derivation of laminar flow through an annular clearance.)

An orifice restriction resists in proportion to V^2, thus responding nonlinearly with plunger speed. Orifice restrictions rapidly lose their damping effect as the plunger slows down, in contrast to laminar restrictions. The equations developed here cover both types.

Annular Restriction (Fig. 8.10, Item A)

Assumptions are as follows: Flow is linear with pressure, the plunger is centered (otherwise damping action would be only half as effective), the plunger travels the full length of the bore, the plunger velocity reduces linearly with travel, pressure in region 1 reduces linearly to zero (assuming the inlet valve is closed off at the same rate as the plunger moves), back pressure (region 2) is constant, friction is constant, and mechanical force is constant.

The examples following are derived from original work by Louis Dodge, hydraulic consultant (New Richmond, OH). Start with an energy balance:

$$E_K + W_{Mech} + W_1 = W_F + W_2 + W_3 \tag{8.1}$$

where

$$E_K = \tfrac{1}{2} M V_i^2 \tag{8.2}$$

$$W_{Mech} = (F_{Mech}) L_B \tag{8.3}$$

$$W_1 = \int_0^{L_B} P_1 A_E dx \tag{8.4}$$

$$W_F = F_F L_B \tag{8.5}$$

$$W_2 = P_2 A_E L_B \tag{8.6}$$

$$W_3 = W_{PL} = \int_0^{L_B} (\Delta P) A_{PL} dx \tag{8.7}$$

Now examine each of the six terms separately. Kinetic energy E_K is clear. Mechanical work W_{Mech} can also remain as is. Input work of source pressure can be handled easiest by carrying it as $F_1 L_B/2$, based on the assumption that pressure drops linearly to zero during stroke.

Friction work is impossible to express in simple terms because so much depends on the seal and slide design. A properly lubricated slide has a coefficient of friction of $C_F = 0.1$. A piston has negligible resistance in most instances. Rod-seal friction depends on contact pressure between the rod and seal and on fluid pressure in the cylinder. Seal $C_F = 0.06$ to 0.11 for lubricated leather and 0.08 to 0.1 for metal. Initial contact pressure will remain constant up to about 750-psi cylinder pressure in most designs. Rather than go into detail here, carry the friction work term as W_F to be computed for each design application.

Region 2 work W_2 is simple to compute because back pressure is assumed con-

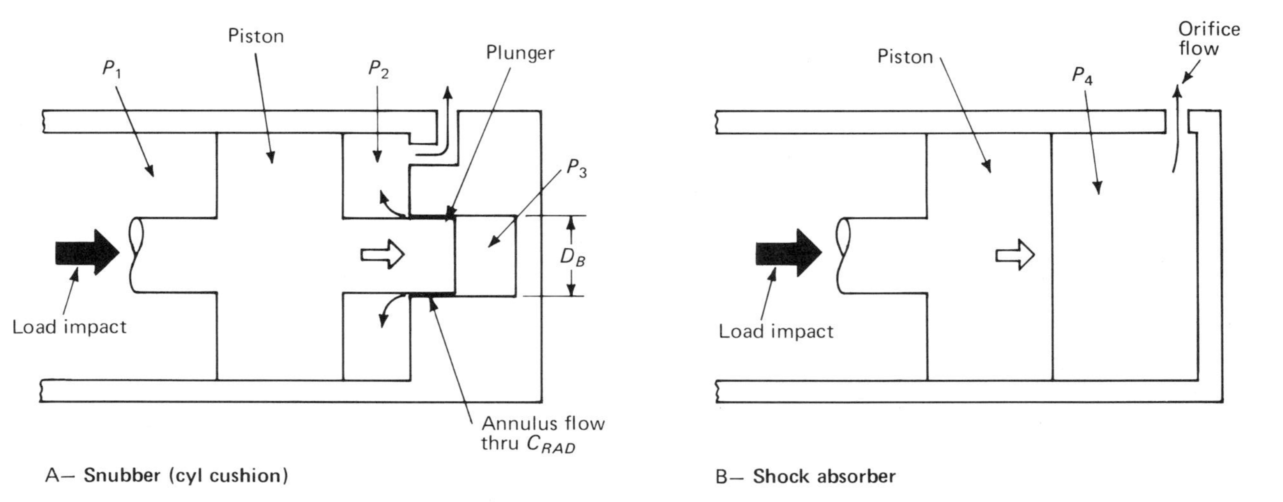

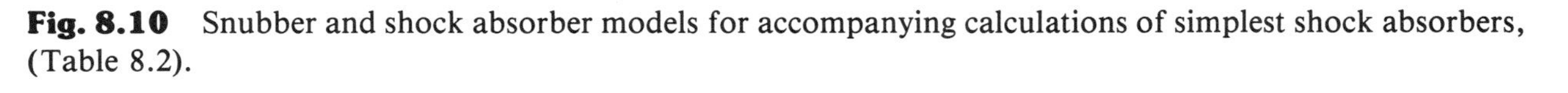

Fig. 8.10 Snubber and shock absorber models for accompanying calculations of simplest shock absorbers, (Table 8.2).

stant. Carry this term as F_2L_B to be consistent with W_1. Be careful with your computation of F_1 and F_2, because the effective areas involve several diameters. Remember that back pressure P_2 must be assumed to act over the entire piston area, including A_{PL}, if you also arbitrarily assume that $\Delta P = P_3 - P_2$.

Plunger work W_{PL} involves integral calculus and is derived as follows:

$$W_{PL} = \int_0^{L_B} (\Delta P)A_{PL}dx \tag{8.7}$$

where

$$\Delta P = P_3 - P_2 = \frac{12Q_M \mu L_M}{C_{Rad}^3 D_M \pi} \tag{8.8a}$$

Subscript M is mean, so that $L_M = L_B/2$ and $Q_M = V_i A_{PL}/2$. Thus ΔP is a mean value based on mean flow along the mean plunger length, as if the plunger were moving at a constant rate with plunger length fixed. Equation (8.8a) then becomes

$$\Delta P = \frac{12V_i A_{PL} \mu L_B/2}{2C_{Rad}^3 D_M \pi} \tag{8.8b}$$

Substitute $D_M^2/4$ for A_{PL} in Eqs. (8.7) and (8.8), and here is the result:

$$W_{PL} = \int_0^{L_B} \frac{0.59V_i \mu L_B D_M^3}{C_{Rad}^3} dx = \frac{0.59V_i \mu L_B^2 D_M^3}{C_{Rad}^3}$$

Get rid of V_i by recognizing that $V_i/2 = V_M$ and that $V_M = L_B/t_t$ by definition. Thus it follows that

$$W_{PL} = \frac{1.18\mu L_B^3 D_M^3}{C_{Rad}^3 t_t} \tag{8.9}$$

Equation (8.9) is the basic relationship used in the annular-restriction equations in the table. Going back to Eq. (8.1) and substituting the known terms,

$$\frac{1}{2}MV_i^2 + F_{Mech}L_B + \frac{F_1}{2}L_B = W_F + F_2L_B + \frac{1.18\mu L_B^3 D_M^3}{C_{Rad}^3 t_t}$$

Solve for C_{Rad}:

$$C_{Rad} = L_B D_M \times \sqrt[3]{\frac{1.18\mu/t_t}{0.5MV_i^2 + (F_{Mech} + 0.5F_1 - F_2)L_B - W_F}}$$

You can manipulate the individual forces to match your own situation. The mechanical force, for instance, can be negative (resistive), thus combining with the friction force. Also, back pressure in region 2 can be variable instead of fixed. There even can be a spring involved. And the flow restriction can be an orifice (text following) instead of an annulus.

Not included in the equations are terms to account for acceleration of the fluid as it is forced from the pressure region into the annulus. These forces can be considered negligible in all designs except those where the load and piston inertia are very small, as perhaps in an instrument dashpot with a small piston and wide clearances. The fluid accelerating force comes from the intensified fluid pressure, thus increasing the damping effect and lengthening the total stroke time.

Orifice Restriction (Fig. 8.10, Item B)

Any restriction can be equated with an orifice if flow depends on $(\Delta P)^{0.5}$. Thus a wide annular clearance qualifies, and a drilled hole. Just be careful to note the corrections and deviations.

A good example for analysis is a shock absorber with a drilled hole in the cylinder (Fig. 8.10, item B). Assume here that there is zero friction and no spring. Newton's law of motion is sufficient to equate force and acceleration:

$$F = M\frac{dV_P}{dt} \tag{8.10}$$

where $F = A_P\Delta P$ is the resisting force of the piston, M is load mass (the piston is assumed weightless), and V_P is the velocity of the load and piston after they collide and move together.

Head loss is

$$\begin{aligned} \Delta P &= KV_0^2\, 7.5\times 10^{-7}\rho,\ \text{psi} \qquad \text{(for any fluid)} \\ \Delta P &= KV_0^2\, 4.2\times 10^{-5},\ \text{psi} \qquad \text{(for oil)} \end{aligned} \tag{8.10a}$$

where K is the effective K-factor based on inlet, local, and exit restrictions of the passage and V_0 is the flow velocity through the orifice, in./s. Velocity $V_0 = V_PA_P/A_0$; therefore,

$$\Delta P = K(V_PA_P/A_0)^2 \times 4.2\times 10^{-5} \tag{8.10b}$$

Now complete the differential equation:

$$-M\frac{dV_P}{dt} = F = A_PKV_0^2\times 4.2\times 10^{-5} = \frac{A_P^3}{A_0^2}K\times 4.2\times 10^{-5}V_P^2 \tag{8.11}$$

For easier manipulation, gather together the constants. Let

$$C_1 = (A_P^3/A_0^2)K\times 4.2\times 10^{-5}$$

Thus,

$$C_1V_P^2 = -M\frac{dV_P}{dt}$$

Rearrange and integrate:

$$\int C_1 dt = \int -M\frac{dV_P}{V_P^2}$$

$$C_1t + C_2 = M/V_P$$

And finally

$$V_P = M/(C_1t + C_2) \tag{8.12}$$

To find C_2, let t = zero, so $V_P = V_i$:

$$C_2 = M/V_i$$

The relationship for piston travel for a given orifice size can be developed with another integration:

$$V_P = dx/dt = M/(C_1 t + C_2)$$

$$\int dx = \int \frac{M}{C_2 + C_1 t} dt$$

$$x = \frac{M}{C_1} \ln(C_2 + C_1 t) + C_3 \tag{8.13}$$

If $x = 0$ at $t = 0$,

$$x = \frac{M}{C_1} \ln (C_2 + C_1 t) - \frac{M}{C_1} \ln C_2$$

Plot graphs to show the general shape of V_P vs. x and x vs. t. In theory, the plunger slows but never stops. For all practical purposes, the motion is stopped when $V_P = 0.1V_i$, so to determine stopping time, use this relationship from Eq. (8.12):

$$t@0.1V_i = \frac{M - 0.1V_i C_2}{0.1V_i C_1} = \frac{0.9MA_0^2}{V_i A_P^3 K \times 0.42 \times 10^{-5}} \tag{8.14}$$

Here is the expression for orifice area, found by solving Eq. (8.14) for A_0:

$$A_0 = \sqrt{\frac{(t@0.1V_i)(V_i A_P^3 K \times 0.42 \times 10^{-5}}{0.9M}} \tag{8.15}$$

which means that for a given desired time to reach a velocity of $0.1V_i$ it requires an orifice area $= A_0$.

A more useful relationship, based on travel x rather than time t, can be developed too. For this, start with a conventional energy balance instead of Newton's law:

$$(E_K)_i - (E_K)_{@x} = \int_0^x F dx \tag{8.16}$$

into which Eqs. (8.2) and (8.10) are substituted, letting V_P represent V_x:

$$\tfrac{1}{2}M(V_i^2 - V_P^2) = \int_0^x \frac{A_P^3 K V_P^2 \times 4.2 \times 10^{-5}}{A_0^2} dx \tag{8.17}$$

Let y = first term. Then

$$\frac{d_y}{dV_P} = -MV_P; \qquad dy = MV_P dV_P$$

Now differentiate Eq. (8.17) and separate the variables:

$$-\frac{dV_P}{V_P} = \frac{A_P^3 K \times 4.2 \times 10^{-5}}{MA_0^2} dx$$

Integrate the expression from V_i to V_P and from 0 to x:

$$\ln \frac{V_i}{V_P} = \frac{A_P^3 K \times 4.2 \times 10^{-5} x}{MA_0^2}$$

$$A_0 = \sqrt{\frac{A_P^3 K \times 4.2 \times 10^{-5} x}{M \ln(V_i/V_P)}} \tag{8.18}$$

which means that for a given design to reach a velocity of V_P at point x in its travel, the orifice area must $= A_0$.

Maximum pressure is at impact, where $V_P = V_i$:

$$\Delta P_{Max} = \frac{A_P^2 K V_i^2 \times 4.2 \times 10^{-5}}{A_0^2} \tag{8.19}$$

Multiple Orifices

The single-orifice design described before is easy to calculate but does not provide optimum deceleration because the back pressure is highest at impact and diminishes rapidly as the plunger slows. One answer is multiple orifices, arranged judiciously along the cylinder wall (Fig. 8.11). The size and spacing of the orifices are determined with an iterative (trial and error) program on a computer. Major manufacturers of shock absorbers now have complex in-house computer programs and can match any load cycle if given sufficient input.

Where the customer is not sure what the energy-absorbing characteristics must be, he or she can order multiple-orifice shock absorbers with manual adjustment. The orifice spacing is still predetermined by computer analysis, but the orifice openings can be modified by the operator within certain limits by moving a sleeve that surrounds the orifices or adjusting each orifice individually.

Where operator adjustment is not desirable, the orifices can be predetermined to match a range of load conditions. Efdyn Corp. (Guthrie, OK) developed a computer program that runs through a wide range of calculations for different masses, different propelling forces (even variable ones), and different velocity conditions.

The computer establishes a pattern of orifices of precise sizes and locations to give the best average performance over the entire expected range. It's called self-adjusting because the flow characteristics of the restrictions automatically match the momentum changes established for all load systems considered within the design range.

Watch Your Assumptions

It's too easy to assume that the pressures increase and decrease linearly, that the load forces are constant, or that all energy is alike.

Ace Controls (Farmington, MI) points out one mistake that inexperienced users make: that of not realizing that a propelling force has a different effect than an inertial force.

As an example, take two systems of identical total energy. First, a load weight of 2000 lb, impacting at an initial velocity of 7.32 ft/s and stopping within 4.876 in., will dissipate 29,736 in.-lb of energy. But so will a system with a load weight of 15,945 lb, a propelling force of 5489 lb, and an initial velocity of 1 ft/s with the same stroke. It's not the same.

The difference is that the first example has 19,985 in.-lb of kinetic energy and 9751 in.-lb of propelling energy (force × stroke), whereas the second example has only 2974 in.-lb of kinetic energy and 26,762 in.-lb of propelling energy. Too high a ratio of propelling energy to kinetic energy makes the shock absorber misbehave.

The problem is that a constant propelling force continues unabated to the end of the stroke and does not diminish with velocity the way an inertial load does. The plunger might ram into the back wall unless the final flow restriction is small enough to resist the large propelling force.

SPECIAL CASE: MULTIPLE ORIFICES (FIG. 8.11)

Ace Controls Co. purposely "manipulated" its computer inputs to see what happens when wrong assumptions are made (Fig. 8.12). The nominal assumptions are as follows: 600 lb. mass; approach velocity, 10 ft/s; zero propelling force; a linear decel-

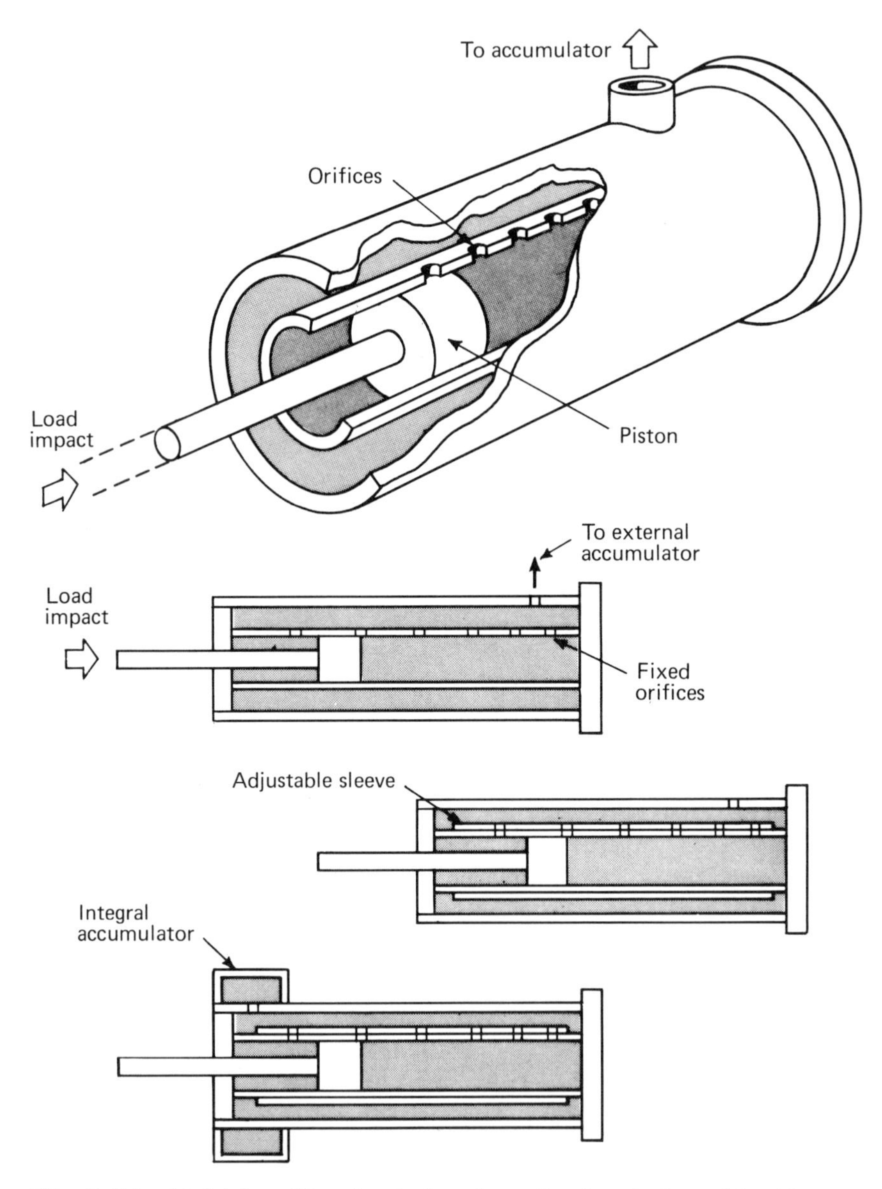

Fig. 8.11 Multiple-orifice shock absorbers: fixed and adjustable. (Courtesy of Ace Controls Co., Farmington, MI.)

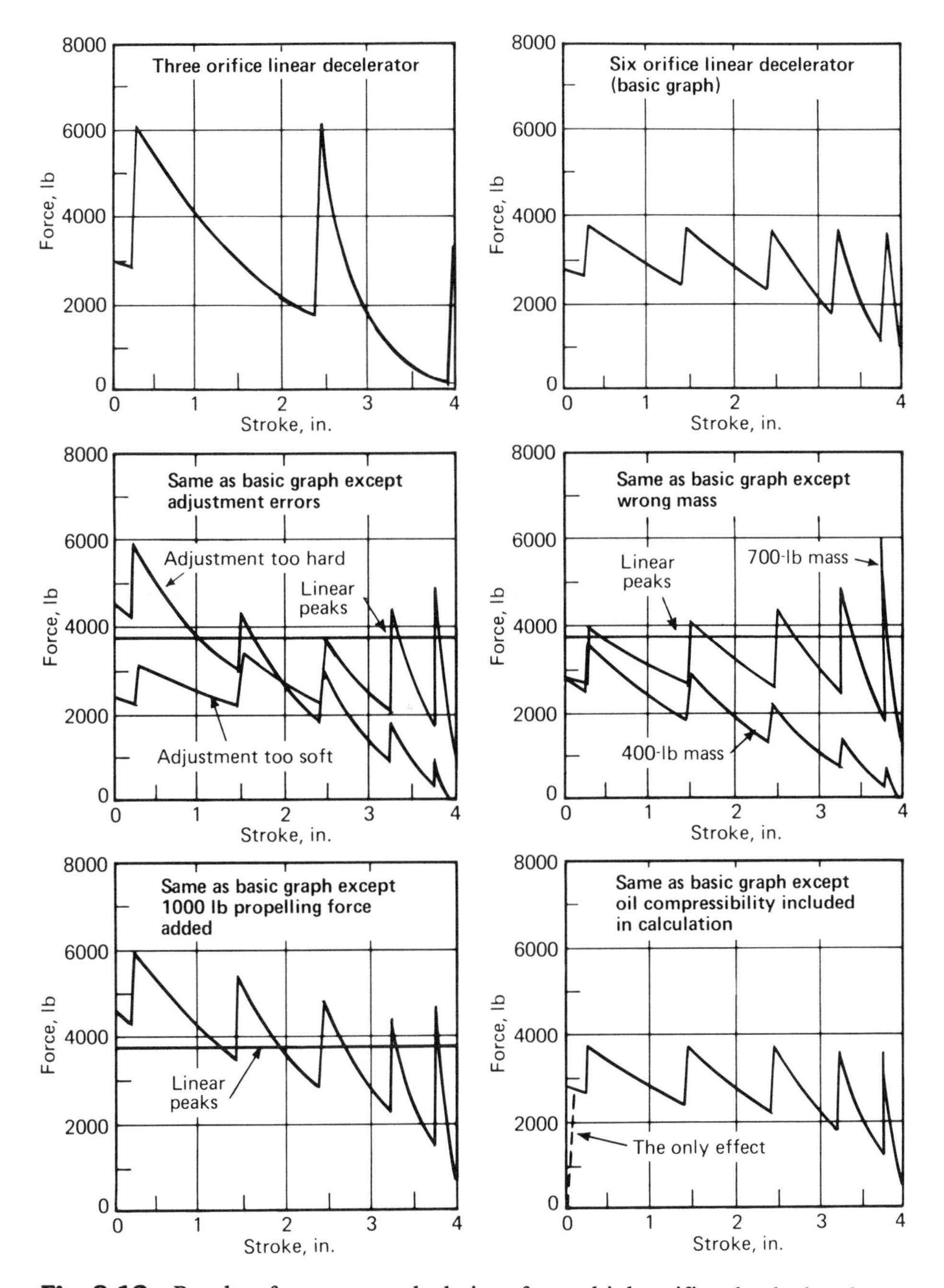

Fig. 8.12 Results of computer calculations for multiple-orifice shock absorber performance. (Courtesy of Ace Controls Co., Farmington, MI.)

erator with a 1-in. bore, 4-in. stroke, and 10-lb-force return spring; and six orifices, with the first orifice ¼-in. away from the piston ring.

Experience has shown that these dimensions will be reasonable for a 600-lb load. When the orifices (or slots) are adjusted later by a mechanism like that shown in Fig. 8.11, the linearity is retained, and only the force levels are altered.

The first design step (before manipulating the values) was to select the proper sizes and spacings of the orifices to hold the series of peak pressures equal. In this way the energy absorption is spread throughout the stroke, approximating an ideal linear decelerator (Fig. 8.13). There always will be peaks and valleys unless an extremely large number of orifices are specified, but six orifices are more than adequate for any normal shock absorber.

The computer does not calculate the orifice dimensions directly, because the equations involve pressure, inertia, force, area, volume, density, and all the complexities of orifice flow. Iteration is the program method used. The computer is instructed to find the orifice sizes and spacings required and to continue to do so automatically until the calculated pressure peaks meet the specifications.

The computer is asked next to plot force, elapsed time, velocity of piston, remaining kinetic energy, and deceleration in Gs.

The graphs in Fig. 8.12 were taken from the X-Y printout of the computer and are fairly close to what you'd see in an oscilloscope.

To see the effects of major changes of inputs or adjustments, they were entered into the computer. For example, the programmer independently increased the load mass to 700 lb, reduced it to 400 lb, added a constant propelling force of 1000 lb, varied fluid compressibility, and reduced the number of orifices to three.

Most interesting is the effect of misadjusting the movable orifice sleeve. This sleeve changes the effective orifice area of each orifice simultaneously and alters the way the decelerator absorbs the energy. A bad adjustment can set up excessive peak

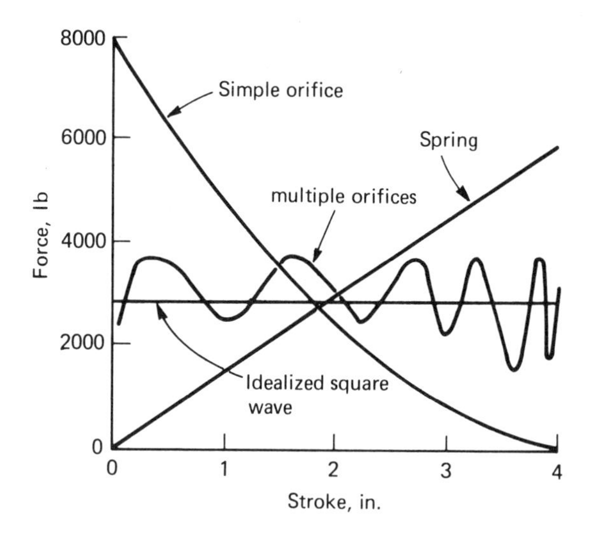

Fig. 8.13 Multiple-orifice shock absorber performance approximates a square wave.

forces. Fortunately the proper adjustment is easily made by observation of the moving mass as it stops.

Past studies showed that oil compressibility, density, and temperature have some effects on performance but not great enough to be considered significant for most applications. The assumptions in the plotted graphs were that temperature stayed near ambient, compressibility was zero, and viscosity was constant. However, when compressibility is considered (the bulk modulus of standard hydraulic oil is 227,000 psi), this new assumption delays the buildup of peak force about 0.1 in. of initial stroke.

A decrease in viscosity caused by temperature increase during cycling requires a slight change of adjustment to the orifice sleeve but otherwise has no substantial effect. The oil density effect, engineering analysis showed, is negligible.

Hard cycling for long periods will raise the temperature of the shock absorber fluid considerably, and this may affect performance somewhat. However, the amount of sleeve adjustment to compensate for this if necessary is small and can be done conveniently in the field during application. A good way to handle that problem is to provide an external accumulator with good thermal capacity, with a double-check valve system to make sure that all of the oil cycles through the accumulator before it returns to the decelerator.

Orifice restrictions rapidly lose their damping effect as the piston slows down, in contrast to laminar restrictions. Either way, the effects can be computed and allowed for. Some shock absorber manufacturers feature sharp-edge orifices to lessen the effect of viscosity (less wall area to introduce wall friction variations). Also, more is known about sharp-edge orifices than about simple drilled holes in literature and ASME standards.

Contoured nozzles smooth out flow at high Reynolds numbers and therefore have higher flow coefficients than sharp-edge orifices. Long orifices introduce wall friction, which means that temperature and viscosity begin to have more influence. Crude openings of any sort introduce wide variations in flow coefficients, but these effects can be compensated for in sizing the openings. That's another good reason for having some adjustment in a shock absorber or linear decelerator.

Annular restrictions, particularly those with laminar flow, are greatly affected by eccentricity. A plunger centered has over twice the flow resistance of a plunger that is completely eccentric in the bore. Centering sleeves are added to the bores of some dashpots.

LIQUID SPRINGS

Oil is compressible (see Chap. 1) and works well as a spring under proper conditions. Any liquid spring (Fig. 8.14) operates from compressibility of the contained fluid. It is the piston rod entering the fluid chamber and not the piston itself that builds up fluid pressure. Restoring force for the direct-compression type equals fluid pressure times cross-sectional area of the piston rod.

The action is elastic: When force is applied, the energy is absorbed, stored, and fully released. A single liquid spring can provide as much "springiness" as 30 coil springs of the same length and diameter or a single large spring many times larger (Fig. 8.15). Some have been designed to reciprocate at 1000 cycles/min. Yet by restricting fluid flow with orifices in the piston, there is a controlled slowdown effect useful for shock absorption, damping, and timing.

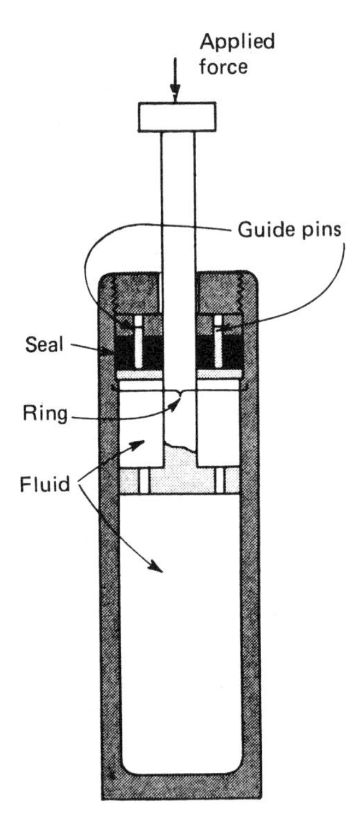

Any liquid spring . . .
operates from compressibility of the contained fluid. It is the piston rod entering the fluid chamber and not the piston itself that builds up fluid pressure. Restoring force for this example (a direct-compression type) equals fluid pressure times cross-section area of piston rod. Orifices or check valves in piston can speed or restrict motion to any desirable degree in either direction—useful when shock-absorbing action is needed.

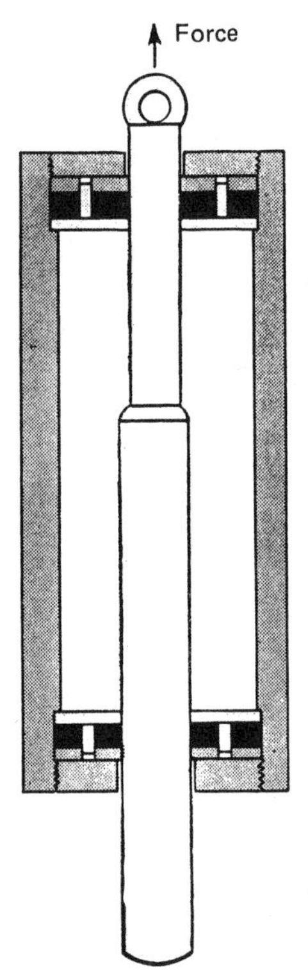

Tension-type . . .
builds fluid pressure when rod moves upward because larger-dia, lower half of rod displaces more fluid than upper half. Restoring force, in downward direction, equals fluid pressure times difference in cross-section areas of rod.

Fig. 8.14 Types of liquid springs.

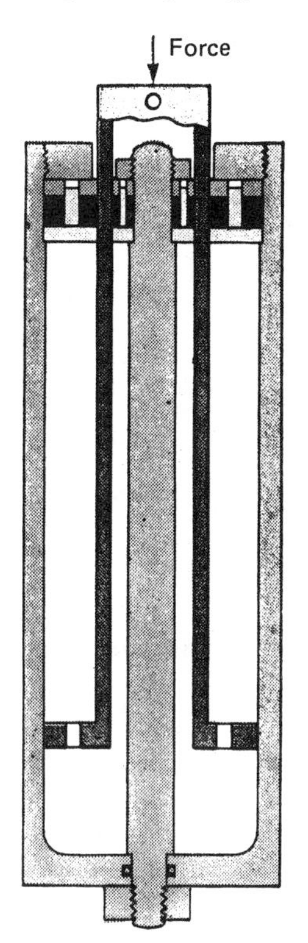

Long-stroke type . . . has hollow shaft to increase structural rigidity. Restoring force equals fluid pressure times cross-section area of hollow-shaft wall. Strokes up to 5 ft are possible.

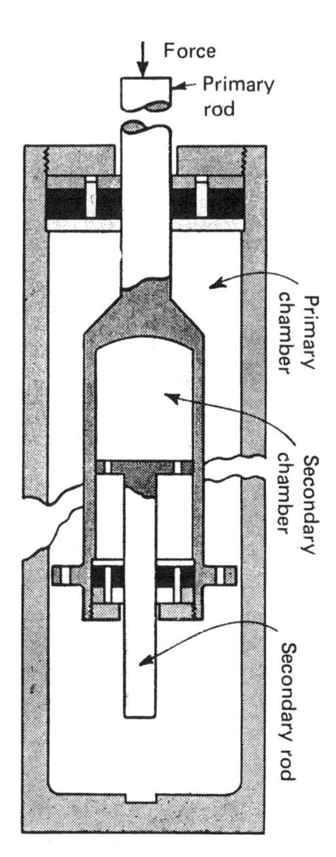

Compound spring . . . has dual spring rate, shown by graph. Zero load finds primary rod in upward position; secondary rod resting on bottom of secondary chamber. As load is applied, primary rod moves downward, compressing fluid in primary chamber; while secondary rod moves small distance into its chamber because of differential pressure against its lower end. It will not float in balance until pressures are about equal in both chambers. When load is increased to change-over point (knee of curve), secondary rod will finally be touching bottom of outer cylinder; still more load mechanically forces this rod into its chamber. Slope of load-vs-travel curves is greater under these conditions because contained volume of secondary chamber is relatively small and pressure rise is rapid for small movements of rod.

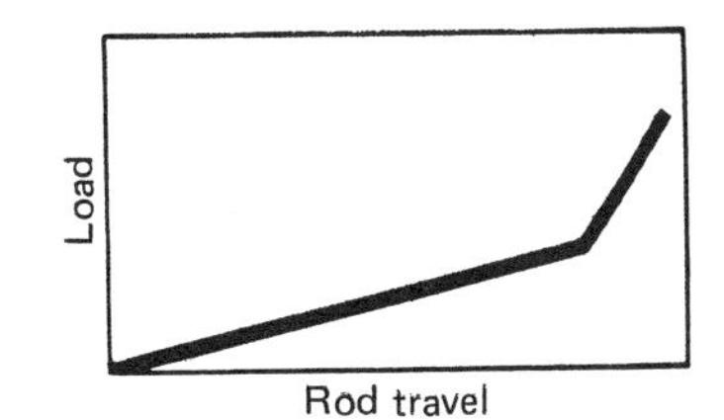

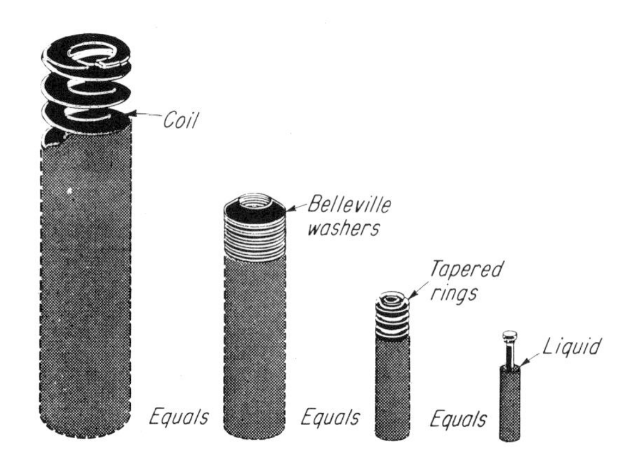

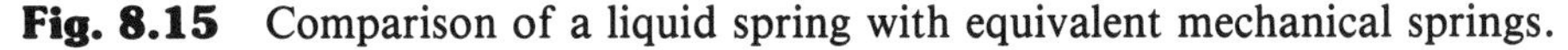
Fig. 8.15 Comparison of a liquid spring with equivalent mechanical springs.

But liquid springs also have some drawbacks too; high costs eliminate their consideration for any but those jobs that lower-priced springs cannot handle. Loads must be high before a liquid spring design is practical, and the resulting high pressures of the compressed fluid must be contained.

Sealing must be absolutely tight, or fluid will be lost and spring characteristics will change. This problem was solved with the "unsupported-area seal" developed by P. W. Bridgman of Harvard in his research on high pressures. With these seals, liquid springs have been run through thousands of cycles without adding fluid. The sealing is so complete that the rod is wiped clean of fluid on each stroke, and reservoirs of grease must sometimes be added for lubrication.

A temperature variation—whether ambient or self-generated—will tend to expand or contract the fluid and change the preload. It does not affect the slope of the load-detection curve, however, because compressibility stays constant over a reasonable temperature range. Natural cooling of the cylinder surfaces will usually prevent the temperatures from rising too high if heat generation is small. If heat is generated rapidly during a work stroke (such as in a shock absorber) but cooling intervals between strokes are long, then the mass of the spring and foundation can act as a temporary heat sink. Helpful is the fact that both the compression and expansion of a fluid are elastic and adiabatic, so that the only energy not restored is frictional energy generated in the rod packing and throttling energy that is purposely generated to absorb shock or slow down movement.

Designing a Liquid Spring

The calculated example (by Lloyd Polentz) in Table 8.3 is for a hypothetical liquid spring with a 50,000-lb load and a 5-in. stroke. Silicone oil is chosen over all other fluids because it is more compressible (Fig. 8.16).

Fluid pressure rises to 50,000 psi, and compression is 18% (82% volume remaining when compressed to 50,000 psi). The rod diameter is 1 in.; therefore the total volume of fluid required is $(5 \times 0.785)/0.18$, or 21.9 in.^{3} Ignoring the volume change of steel parts under stress, satisfactory cylinder dimensions (in inches) are 1.8 ID $\times$ 3 OD $\times$ 8.6 long. Allowing a 3.4-in. total for cylinder ends and seals and 5 in. for the stroke, the total length is 17 in.

For comparison, the equivalent coil spring in Table 8.3 was designed for a

Table 8.3 Liquid Springs Can Outperform Mechanical Springs (See Fig. 8.15)

Performance of Four Typical Spring Types

	Coil	Nested:		Liquid
		Belleville washers	Tapered rings	
Useful range:				
low load	1 oz	20 lb	2 ton	100 lb
high load	10 ton	100 ton	150 ton	200 ton
force vs deflection	low to high	high	high	med. to high
stroke	short to long	short	short to med.	short to long
Damping ability	low	low	low	low to high
Relative cost	low	low	medium	high
An example: for 50,000-lb load, 5-in. stroke:				
Size, in: length	68	37	24	17
dia	11.5	8	5	3

65,000-psi maximum stress, including the Wahl correction factor. The dimensions (in inches) are wire diameter, 3; coil OD, 11.5; and length, 68. There are 19 active turns and 2 inactive ones. The Belleville washers and tapered ring springs were estimated from published design data. All four designs are pictured roughly to scale in Fig. 8.15.

The liquid spring will perform well, even in applications where strokes are repeated frequently. Frictional heat is not likely to be a problem. For example, assume a 5000-lb average frictional load for the hypothetical spring described above. The energy absorbed per stroke is $5000 \times 5/12$ or 2080 ft-lb (2.65 Btu). The assembly weighs about 35 lb, has an average overall specific heat of about 0.15 Btu/lb/F. Therefore the temperature rise per stroke is $2.65/(0.15 \times 35)$ or about 0.5 F—an amount easily dissipated. But a smaller liquid spring under rapidly fluctuating loads may experience higher temperature rise; each design must be evaluated on its own merits.

If heat generation and ambient temperature combine to significantly overheat a liquid spring, then cooling means must be provided, or the spring must be compensated for temperature. One compensating design allows the fluid to expand (or contract) into a constant-pressure reservoir during the no-load condition; controlled valving can prevent backflow during the operating stroke. Another answer is to make the rod stroke adjustable—or design the product to be insensitive to small variations in stroke.

This example of a noncompensated design shows what happens if temperature rise is too high: The coefficient of thermal expansion of a silicone fluid is about 6.5×10^{-4}/F. A 100-F temperature rise in any contained volume corresponds to a volume compression of 6.5% and raises the pressure about 10,000 psi. Such a spring would have poor repeatability.

For improved accuracy, assume isothermal compression for the charging cycle, which is gradual. Assume adiabatic expansion during discharge, which usually is rapid.

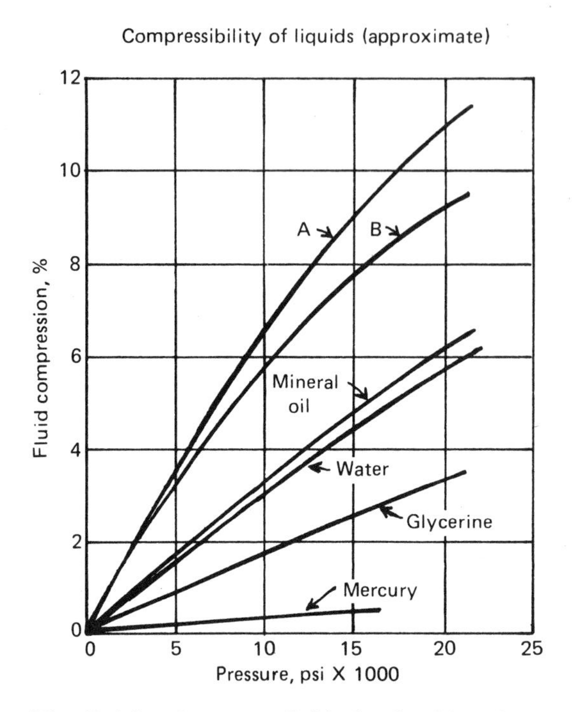

Fig. 8.16 Common fluids for liquid springs are Dow-Corning type F-4029 (curve A) and type 200 (curve B). Each is silicone-based and has a viscosity of 100 cst at 77 F. The other four curves are for reference only—these fluids would not ordinarily be selected.

FLUID HAMMER IN PIPING

This design guide, based on the work of Louis Dodge (New Richmond, Ohio), helps take the bang out of a fluid system by anticipating each pressure-transient problem.

In theory, instant closing of a valve or piston would burst the piping, assuming that flow can be stopped in zero time and that the fluid and pipewall are wholly inelastic. What actually does occur is an elastic phenomenon known as *fluid hammer* (explained later). Its theory and ways to avoid the damage it can cause have been the subject of treatises as far back as 1775.

The scientific approach is awkward when you're designing for production and need quick answers, but with a few simplifying assumptions researchers' findings can be applied easily:

- Single run of pipe (no branches)
- Constant equivalent pipe diameter
- Elastic system (damping neglected)

The simple reservoir-piping-valve circuit shown in Figure 8.17 is our model, and each pressure-transient problem should first be reduced to that form. The accompanying calculation procedures show how to do this. Results will be conservative, because damping in an actual system helps lessen the shock.

What Fluid Hammer Is

Fluid hammer is the term for the overall effect of pressure transients in a piping system. More specifically, it is fluid shock, accompanied by noise, resulting from the opening or closing of a valve or other flow-control component. It includes effects of initial system pressure, fluid inertia, piping dimensions, pipewall elasticity, fluid bulk modulus, and valve operating time (to open or close). A simplified graphic explanation is shown in Fig. 8.18.

During the closing cycle of a valve the kinetic energy of the moving fluid is converted into potential energy. (During the opening cycle the opposite occurs.) Elasticity of fluid and pipewall produces waves of positive and negative pressures which are superimposed upon the original pressure and travel from the valve, through the fluid in the pipe, to the reservoir, and back to the valve. The process would keep repeating indefinitely except that internal friction and other energy losses gradually damp out the waves.

The initial shock of suddenly stopped flow can induce transient pressure changes that exceed the static pressure in both the positive and negative directions. So-called negative pressures in reality cannot be less than absolute zero, but the pipewall contracts during the negative wave, and the effect is the same as a negative pressure.

An analogy to a positive-negative pressure wave is a coil spring that is extended and then released: The stresses reverse from tension to compression. Pipe failure could occur from negative pressure but rarely does, because the pipe is stronger in the inward direction.

Burst piping is not the only damage possible from excessive pressure surges. Any component in the circuit or mounted on the same structure is subject to impact stresses induced by the pressure wave. And even if there is no physical damage, the noise alone can be intolerable.

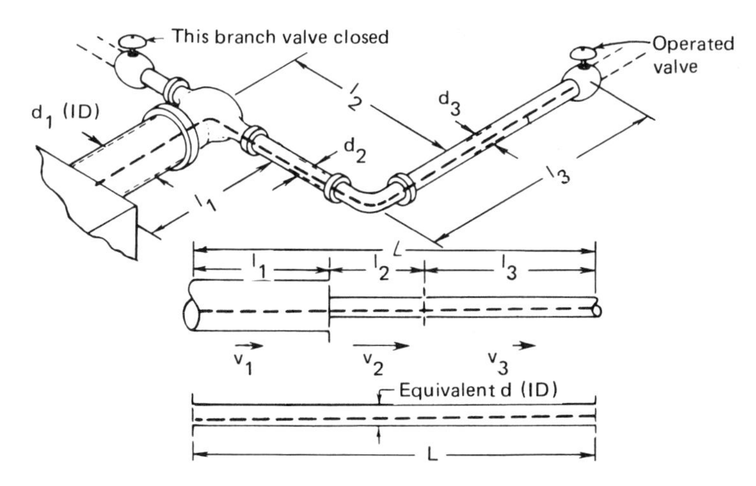

Fig. 8.17 Equivalent single run of piping (for accompanying fluid hammer calculation).

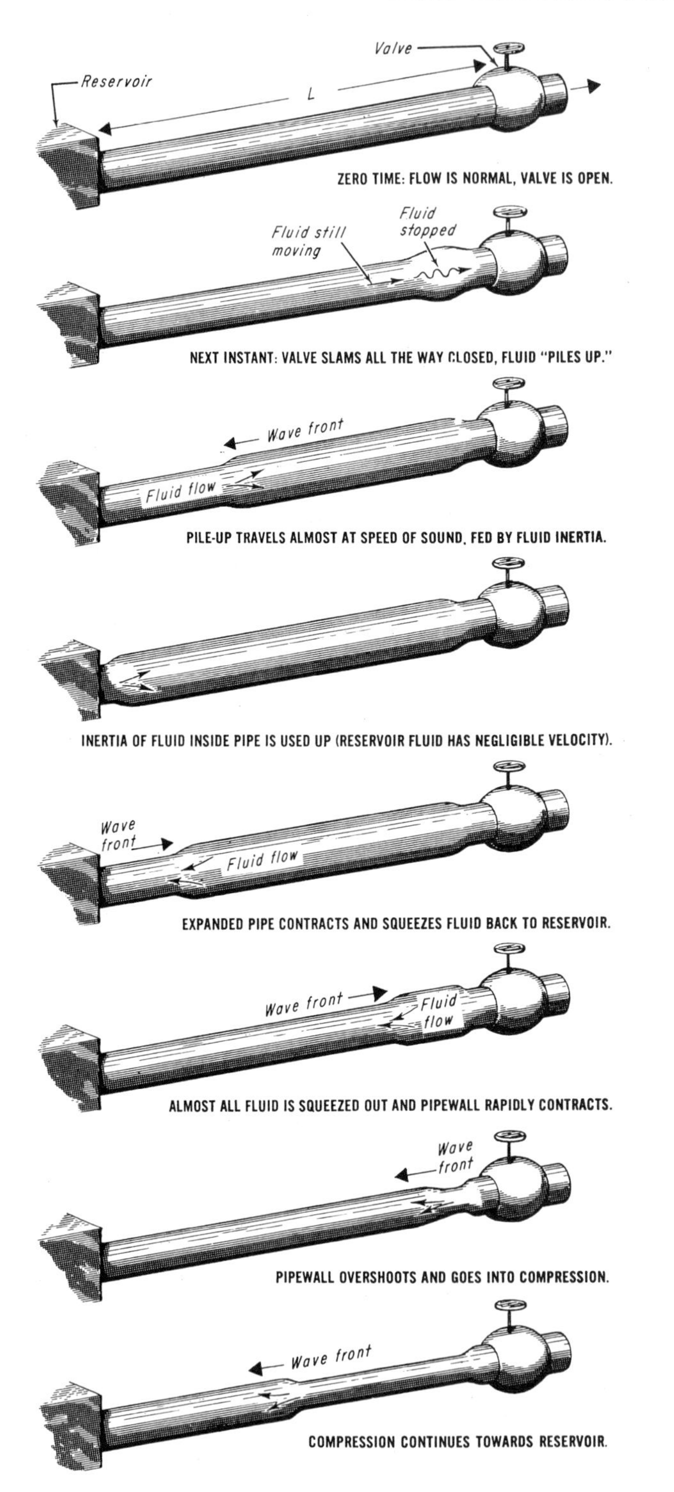

Fig. 8.18 In the pressure-wave example (hypothetical) the pipewall alternately contracts and expands until the energy is dissipated. The entire process can take less than 1 s.

Critical Period

The time T_C of one round trip of the pressure wave from valve to reservoir and back is the critical time (one period) that establishes the severity of the fluid hammer.

If the valve closes within the critical time (called fast closing), the pressure rise is the maximum. It doesn't matter how much faster than T_C the closing time is as long as the fluid is stopped completely before the first pressure wave completes the round trip.

If the valve does not close within critical time T_C, the maximum pressure rise will be less. Special equations are needed.

For valve opening, the same reasoning holds, except that the initial effect is a pressure drop. Figure 8.19 compares the damped pressure waves for both the opening and closing of a valve.

Calculating Fluid Hammer

The basic equations (Table 8.4) are grouped according to speed and direction of valve operation: fast closing, slow closing, fast opening, slow opening. Each pressure-wave equation applies to single run of pipe of constant diameter, so the first step in the mathematical procedure is to convert your piping system to that form. Here is how (also see the completed example, Table 8.5):

Equivalent circuit: For a simple series system with pipes of varying diameter but transporting the same amount of fluid (see Fig. 8.17) you can calculate the average fluid velocity using the general equation (Eq. 1 in Table 8.4) for $V_{F(avg)}$. Note that overall length L in the equation is not modified.

Sharp bends (more abrupt than $5 \times d_i$) and elbows lessen the shock because of

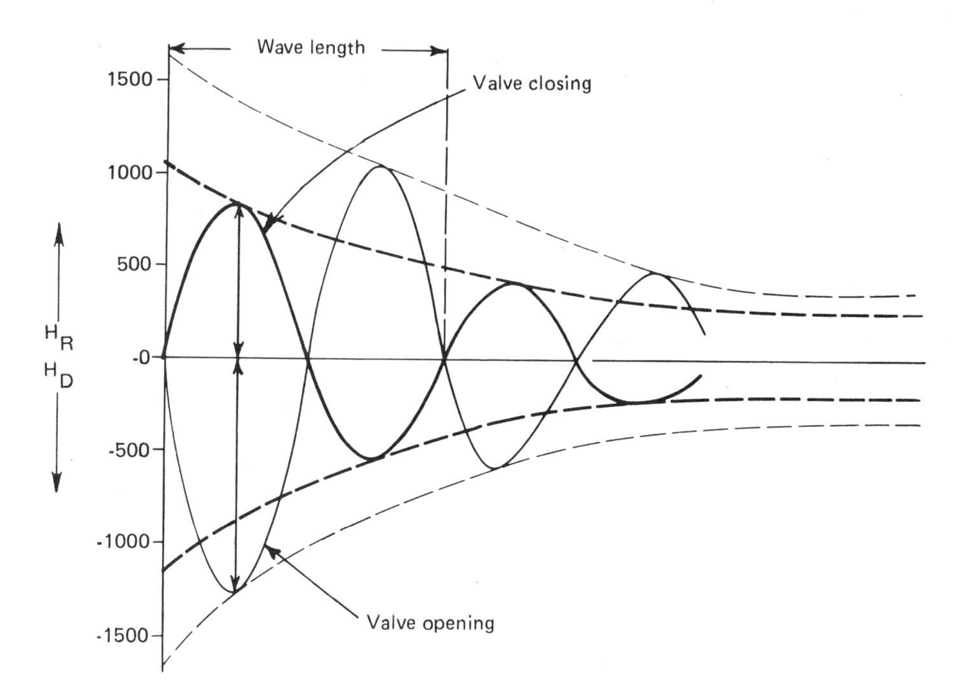

Fig. 8.19 Typical damping of pressure waves after the opening and closing of a valve.

Table 8.4. Simplified Equations for Calculating Fluid Hammer (See Fig. 8.17)

GENERAL

$$V_{F\,(avg)} = \frac{l_1 v_1 + l_2 v_2 + l_3 v_3 + \ldots}{L} \quad (1)$$

where

l_1 = **length of first section of pipe in a series of such sections**

v_1 = **velocity of fluid in first section**

$$V_P = \frac{12}{\sqrt{\frac{\rho}{g}\left(\frac{1}{\beta} + \frac{d_i}{eE}\right)}} \quad (2)$$

T_C = $2L/V_P$ **sec, which is time of pressure-wave round trip** **(3)**

S = $(P + P_R)\,d_i/2e$ **(hoop stress in pipewall)** **(4)**

FAST CLOSING ($T \leq T_C$)

$H_{R\,(max)} = V_P V_F/g$, **ft** **(5)**

$H_R = V_P \Delta V_F/g$ **ft, for fast partial closing** **(6)**

ΔV_F = **initial** V_F **— final** V_F

P_R = **50** V_F, **psi (Simplification of equation 5 for "average" conditions)** **(7)**

SLOW CLOSING ($T > T_C$)

$H_R = M - H - \sqrt{M^2 - m_1^2}$, **ft** **(8)**

$M = m_1 + m_2$

$m_1 = H + H_{R\,(max)}$

$m_2 = V_F^2\,(V_P - 2L/T)^2/2g^2H$

FAST OPENING ($T \leq T_C$)

$H_D = \sqrt{n_F\,(2H + n_F)} - n_F$, **ft** **(9)**

$n_F = H_{R\,(max)}^2/2H$

SLOW OPENING ($T > T_C$)

$H_D = \sqrt{n_S(2H + n_S)} - n_S$, **ft** **(10)**

$n_S = 2V_F^2L^2/g^2T^2H$, **ft**

THE SYMBOLS

- P = system static pressure, psi
- P_R = pressure rise, psi
- H = system static head, ft (H for oil $= 2.52\,P$; H for water $= 2.31\,P$).
- H_R = pressure rise, ft
- H_D = pressure drop, ft
- V_F = velocity of fluid, ft/sec
- V_P = velocity of pressure wave in fluid, ft/sec (depends on pipe dimensions and fluid)
- E = modulus of elasticity of pipewall, lb/in.² (E for steel $= 30 \times 10^6$; copper $= 15 \times 10^6$; cast iron $= 12 \times 10^6$; aluminum $= 10 \times 10^6$; asbestos cement $= 3.4 \times 10^6$)
- β = bulk modulus of elasticity of fluid, lb/in.² (β for water = 294,000; β for oil = 250,000)
- L = pipe length, ft
- d_i = pipe inside diameter, in.
- e = pipewall thickness, in.
- S = hoop (circumferential) stress in pipewall, psi
- T = valve closing or opening time, sec
- T_C = critical period, sec. $T_C = 2L/V_P$
- g = acceleration of gravity, 32.2 ft/sec²
- ρ = fluid density, lb/ft³

TERMINOLOGY

Fluid hammer — total effect of pressure transients. Also called water hammer.

Pressure rise — increment above system pressure.

Pressure surge—peak pressure in absolute units during a pressure rise.

Negative pressure—term for the hypothetical fluid pressure sometimes used to explain pipe contraction resulting from elastic springback of a pipewall after a positive-pressure surge.

their high damping effect—thus for the circuit shown, all the equations are very conservative.

For systems with branches or parallel circuits the conversion to an equivalent single run of constant diameter is difficult or impractical, depending on how much math you want to tackle.

In general, branches add volume and elasticity to the main line. They diminish the shock and decrease the magnitude of the pressure wave. The more branches there are in a circuit, the more damping there is. Hence, complex circuits in most instances will not have a fluid-hammer problem.

Parallel nonidentical circuits are impossible to calculate in a simple procedure because the pressure waves will travel in a different manner in each line and will not return to the valve simultaneously. Parallel identical circuits can be calculated individually and pose no special problem.

Safe pipe size: Eqs. 2 to 10 in Table 8.4 determine maximum surge pressures. For initial sizing of your system piping, use simplified Eq. 7, $P_R = 50V_F$, which is accurate enough to estimate pressure rise. This equation is based on the following: oil weighing 55 lb/ft³; steel pipe; pressure-wave velocity of 4300 ft/s; closing time less than T_C. For better accuracy, use Eq. 5 and the V_P graph, Fig. 8.20.

It is simple to calculate pipe stress from Eq. 4:

$$S = (P + P_R)d_i/2e$$

where S is hoop stress, psi, and $P + P_R$ is calculated surge pressure, psi.

Typical steel tubing has an allowable stress of 85,000 psi. For other materials, use these figures: soft copper, 36,000 psi; half-hard copper, 40,000 psi; annealed stainless steel, 85,000 psi. All are based on temperatures up to 100 F. Always use latest handbook values.

Table 8.5. Completed Example of Fluid Hammer in Oil Line

EXAMPLE .. **Oil line to hydraulic press**

Fluid = oil	Bulk modulus $\beta = 25 \times 10^4$ psi	Density ρ = 55 lb/ft³
Flow = 10 gpm	Velocity V_F = 16.4 ft/sec	System head H = 17,000 ft
Pipe dia, d_i = 0.5 in.	Length L = 40 ft	= 6000 psi
Pipe material = steel	Modulus $E = 30 \times 10^6$ psi	Wall thickness e = 0.11 in.

GENERAL EQUATIONS:

(2) V_P = (read approximate value from Fig 8.20 — Answer: V_P = 4540 ft/sec

(3) $T_C = 2L/V_P = 2 \times 40/4540$ — Answer: T_C = 0.0174 sec

FAST CLOSING ($\leqslant T_C$):

(5) $H_{R(max)} = V_P V_F/g = 4540 \times 16.4/32.2$ — $H_{R(max)}$ = 2300 ft = 875 psi

SLOW CLOSING: (For example, 0.1 sec)

(8) $H_R - M - H - \sqrt{M^2 - m_1^2} = 19{,}407 - 17{,}000 - \sqrt{19{,}407^2 - 19{,}300^2}$ — H_R = 447 ft = 164 psi

$m_1 = H + H_{R(max)} = 17{,}000 + 2300 = 19{,}300$

$m_2 = V_F^2 (V_P - 2L/T)^2/2g^2H = \frac{16.4^2}{34{,}000 \times 32.2^2}\left(4540 - \frac{80}{0.1}\right)^2 = 107$

$M = m_1 + m_2 = 19{,}300 + 107 = 19{,}407$

FAST OPENING ($\leqslant T_C$):

(9) $H_D = \sqrt{n_F(2H + n_F)} - n_F = \sqrt{158(34{,}000 + 158)} - 158$ — H_D = 2163 ft = 800 psi

$n_F = H_{R(max)}^2/2H = 2300^2/34{,}000 = 158$

SLOW OPENING: (For example, 0.1 sec)

(10) $H_D = \sqrt{n_S(2H + n_S)} - n_S = \sqrt{4.9(34{,}000 + 4.9)} - 4.9$ — H_D = 408 ft = 150 psi

$n_S = 2V_F^2L^2/g^2T^2H = 2 \times 16.4^2 \times 40^2/32.2^2 \times 0.1^2 \times 17{,}000 = 4.9$

Note: Closing pressure rise is greater than opening pressure drop in this example. Use 6000 + 875 = 6875 psi for pipe-stress calculations.

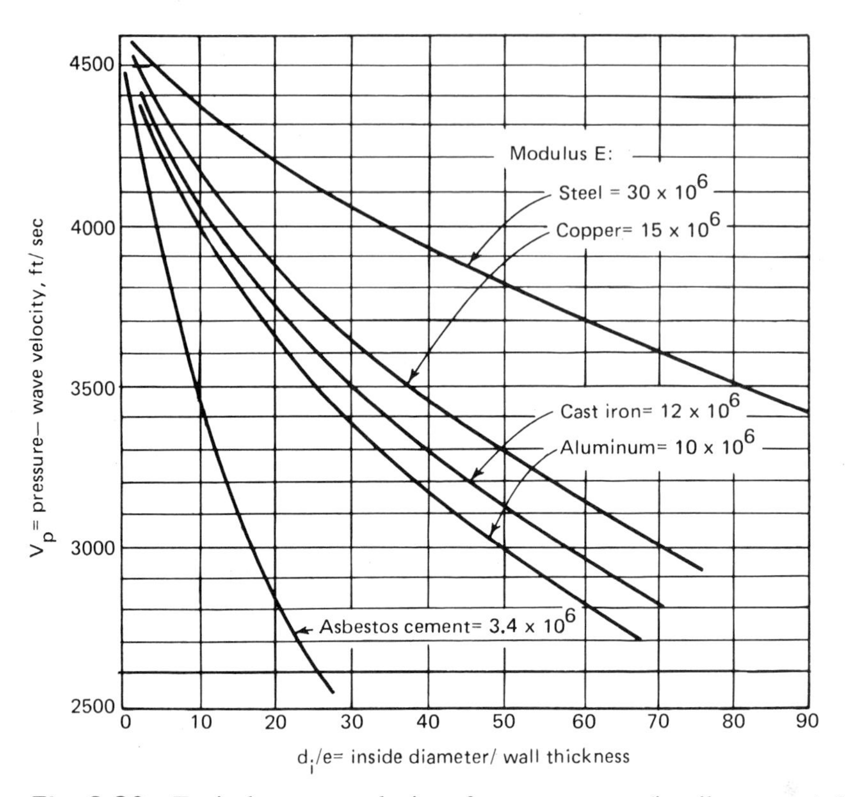

Fig. 8.20 Typical average velocity of pressure wave (in oil or water) for various pipe materials.

But be careful how far you trust simple arithmetic. Remember these facts:

- Stresses and deformations caused by impact pressures are double those caused by slow application of the same pressure.
- Safety factors from 4 to 8 are usually required in rigid piping. Typical recommendations are safety factors of 8 for a working pressure of 0 to 1000 psi, 6 for 1000 to 2500 psi, and 4 for pressures above 2500 psi.

Ordinary surges won't harm the small-diameter tubing and piping common in most hydraulic and pneumatic systems. But the ability to withstand surge goes down as tubing size goes up.

If the calculated surge pressure is too high for the maximum wall thickness available from stock (very likely if the tubing diameter is more than 2 in. and pressures are high), there is a simple answer: Use two or more lines of smaller diameter.

For example, assume that the flow velocity is 20 ft/s through 2-in.-diameter steel tubing with a 0.120-in. wall. Allowable stress is 85,000 psi. Let system static pressure equal 2000 psi.

Using Eq. 7, $P_R = 50 \times 20 = 1000$ psi. Surge pressure $P + P_R = 2000 + 1000 = 3000$ psi. Hoop stress $S = 3000 \times 2.0/(2 \times 0.12) = 25{,}000$ psi. The safety factor at 3000 psi must be 4 or more, so the 85,000 allowable stress is inadequate.

One answer is to select heavier-walled tubing, but a search will reveal that there's not much available in large tubes with heavy walls.

A better answer is to parallel three 1¼-in. lines—they have greater total flow area and also can handle surges better. Hoop stress at 3000 psi for a 1¼-in. tube with a 0.109-in. wall is $3000 \times 1.25/0.218 = 17{,}200$ psi, which gives a safety factor of more than 4. And because the larger flow area reduces velocity V_F, the effective safety factor is even greater. Cost is less too: Three small tubes are cheaper than one large tube, and they are easier to bend and flare because of the smaller outside diameter and thinner wall thickness.

Pressure lines should be rigidly supported at short intervals to prevent vibration and subsequent damage to flared tubing ends, fittings, and seals. Accompanying cylinders, pumps, and fluid motors should be mounted to withstand vibration.

Even if surge damage is unlikely, there still may be a noise problem. Noise depends on the time rate of change of pressure and can occur even at low absolute pressures. Noise is not eliminated merely by making the piping stronger. You must damp or smooth out the pressure wave.

Reducing Fluid Hammer

The discussion will be based on noise, because this is the best indicator of a serious condition. Diminishing noise solves the shock-damage problem. Fast-acting valves are the primary source of fluid-hammer noise, but there are other sources such as cylinders, pumps, and fluid motors. The same remedies apply to all.

There are two basic ways to reduce noise: (1) modify the valve or component to have smooth motion, preferably sinusoidal, within the time allowed for operation; (2) add capacitive volume or elasticity to absorb the pressure waves. Sketched are devices (Fig. 8.21) that reduce noise and shock. Sometimes, of course, you can eliminate the problem by reducing system flow or by increasing pipe diameter to reduce the velocity.

Inexpensive changes to a valve can often smooth out flow. Examples are valve spools tapered to throttle flow, nested springs to control closing force of a solenoid valve, dashpots to reduce end-of-stroke shock, and cams to time valve or pilot operation.

Cylinders should have cushions or dashpots to ensure shockless reversal, particularly if the load has high inertia and is controlled by fast-reverse valves.

For example, suppose a cylinder driving a load of 5000 lb at 1.4 ft/s is controlled by a valve closing in 0.08 s. Normal system pressure is 200 psi, and effective piston area is 12.5 in². Further assume that the frictional load during stopping just balances what's left of the 200 psi after being throttled by the valve. Thus the shock pressure principally will be caused by inertia of the load and can be calculated easily:

$$\text{acceleration } a = dv/dt = 1.4/0.08 = 17.5 \text{ ft/s}^2$$

$$\text{force} = \text{mass} \times a = 5000 \times 17.5/32.2 = 2720 \text{ lb}$$

$$\text{pressure } P = 2720/12.5 = 218 \text{ psi}$$

The actual pressure surge will be smaller than 218 psi because the equations neglected the compressibility of the oil trapped in the cylinder and the elasticity of the metal parts surrounding the oil. However, this is more than offset by the fact that

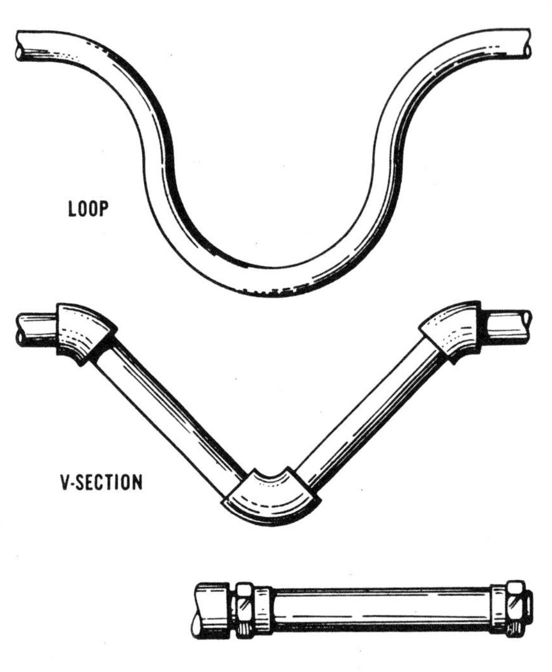

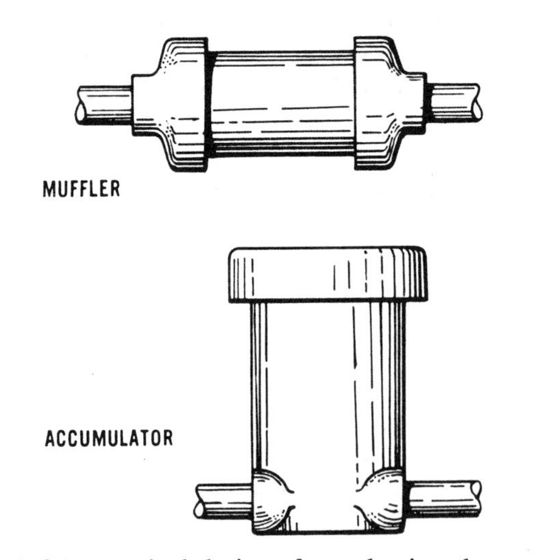

Fig. 8.21 Typical designs for pulsation dampers.

sudden pressures are twice as damaging as gradual pressures. Suggested: Assume an effective peak pressure of 400 psi and allow a safety factor of 8. Better: Add a dashpot or a decelerating valve to reduce the shock.

Where the pulsation is continuous and in resonance with the piping (possible when the source is a pump), then the remedy may be to redesign the pump. Resonance can amplify noise up to annoying or dangerous levels.

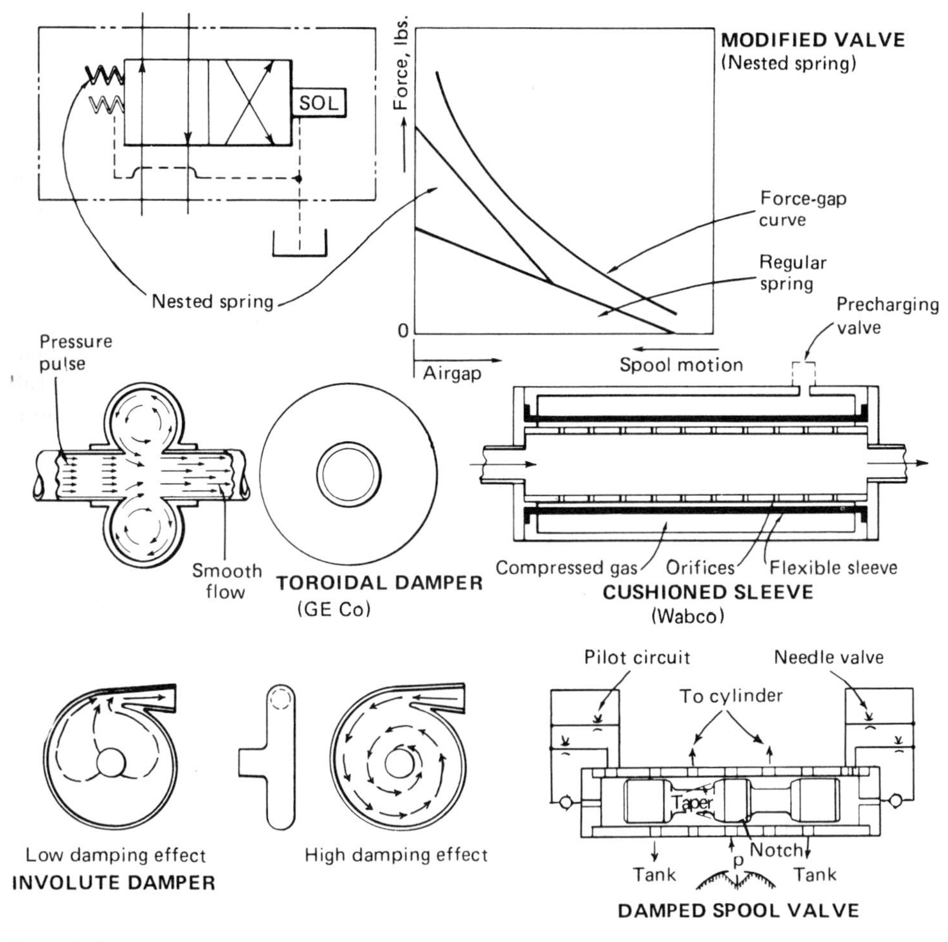

Fig. 8.21 (Continued)

Without altering the valve, cylinder, or pump, you can reduce fluid hammer by adding volume, inertia, or elasticity to the pipeline adjacent to the source of the hammer. Most of the sketches (Fig. 8.21) show such in-the-line pressure-wave absorbers. They are called by many names, including *isolators, capacitive-load chambers, mufflers, surge dampers, suppressors, snubbers,* and *accumulators.* A common name for all of them is *pulsation damper.*

The sketches are self-explanatory, but remember the following:

- A half-loop of copper or steel tubing should have a loop radius equal to three times the tubing outside diameter.
- A muffler can be simply an enlarged section of pipe.
- Spring-type or bladder-type accumulators are good surge absorbers, but they slow down response of the system to a change in pressure.
- Each pulsation damper adds some resistance to flow—negligible in most cases.

- Dampers containing elastomer parts are limited to moderate temperatures.
- Involute and toroidal dampers depend on fluid velocity for the damping effect.

The involute damper has an interesting principle of operation: The velocity of a fluid particle increases toward the center and is inversely proportional to the radius. The result: Centrifugal forces smooth out pulses.

HYDRAULIC ACCUMULATORS

An accumulator holds reserve fluid at system pressure to provide or absorb momentary flow. (Figure 8.22 shows typical examples.) With this extra capacity, the overall system can be designed with smaller pumps without loss of performance. Equations later in the text describe the operation.

The Basic Jobs (Fig. 8.23)

The most obvious application is as an extra volume of oil, teed off the hydraulic supply line and held at system pressure by the force of the compressed gas. In that form, it can serve as an auxiliary power source, pulsation damper, fluid dispenser (discharging a known volume), suction stabilizer for pumps, leakage-makeup source, or thermal expansion volume compensator.

In conjunction with special valving and control, an accumulator can perform special functions. For example:

The *emergency return* of the piston exploits the differential in area between the rod end and blind end of the cylinder to move the piston outward when supply is on and inward when supply is lost. The *expansion tank* absorbs excess volume as the dashpot piston is forced inward by external force, again a phenomenon of the differential area between blind and rod ends.

The *dual-pressure* circuit for hydraulic presses provides maximum flow at low pressure while the ram moves outward initially because the accumulator and both pumps are in action. When the ram strikes the load, high pressure at low flow is sup-

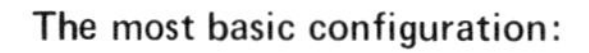

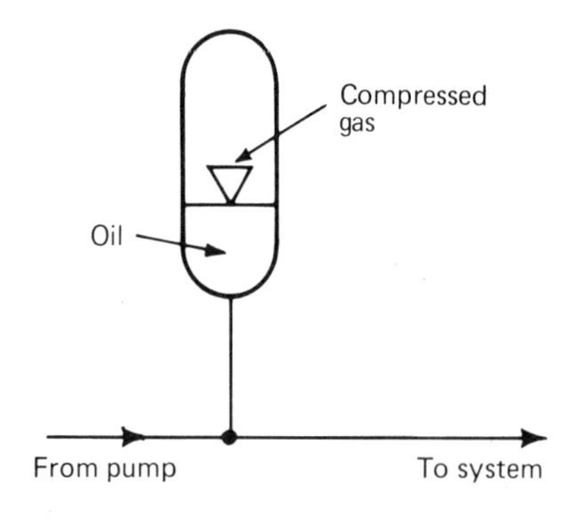

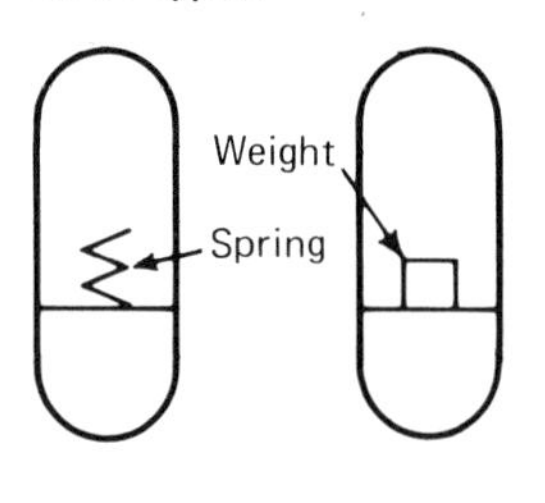

The most basic uses:
Auxiliary power source
Pulsation damper, shock absorber
Fluid dispenser
Pump suction stabilizer
Leakage makeup
Thermal expansion compensator

Fig. 8.22 Most accumulators operate on one of these principles.

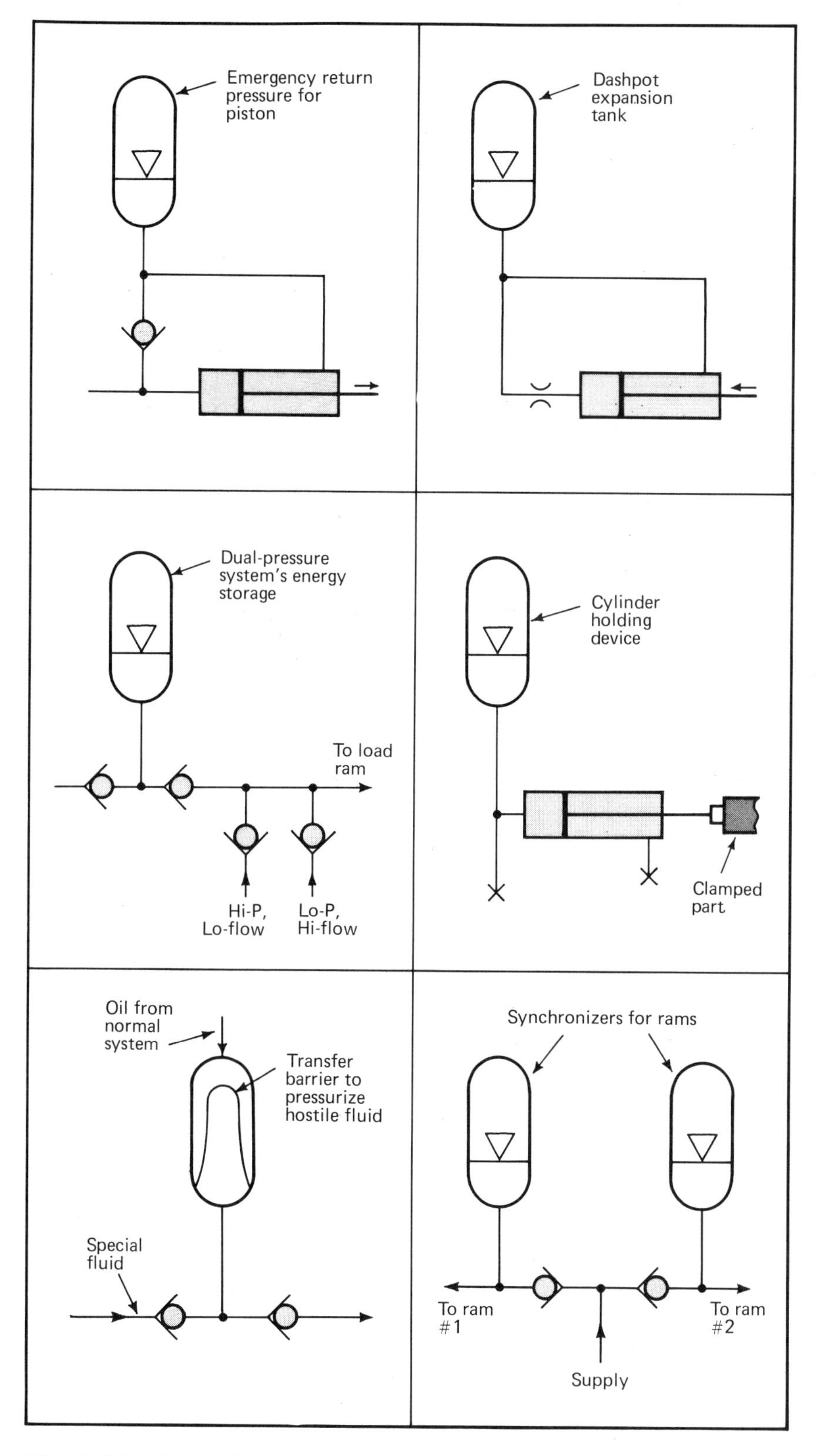

Fig. 8.23 Six applications for accumulators.

plied by the hi-P pump. Replenishment occurs in the return cycle. All is controlled by intricate valving and switching.

The *cylinder holding* device is just a clever way to hold a workpiece for long periods without requiring energy from the main system pump. The *transfer barrier* shows how to pressurize a hostile fluid without compromising the main pumping system.

A modification of this idea is to pressurize a reservoir with air, using a bladder accumulator as the interface. The bladder, full of the working fluid, is confined in an airtight tank, and about 15-psig air pressure is maintained around the bladder. All environmental problems such as rusting and contamination are avoided, and pump-suction cavitation is eliminated (See Chap. 2, Fig. 2.7).

The *synchronizer* stores equal volumes of fluid in twin accumulators and discharges them in parallel under command of valving.

A variation on the accumulator is to use it as a simple oil pump: The gas side is pressurized from a controlled external source such as shop air, and this pressurizes the hydraulic system. It's a compact way to pump enough oil to stroke a cylinder (Fig. 8.24).

For any accumulator application, remember that *stored energy can be dangerous.* Provisions must be made for bleeding off pressure during machine shutdown (Fig. 8.25 shows an example). Also, make sure that all controls are failsafe and that there will be no unexpected machine motions—zero surprises.

Six Accumulator Types

Most common is the bladder type (Fig. 8.26), and most of these discussions are based on it. However, the math applies to all oil-gas (hydropneumatic) accumulators where system oil compresses the trapped gas.

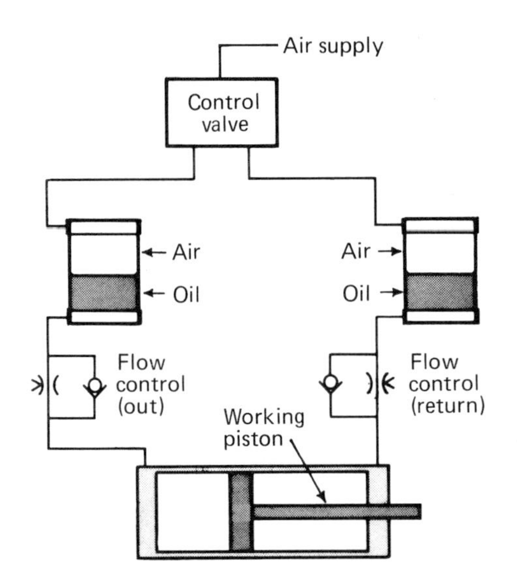

Fig. 8.24 Hydraulic cylinder can be cycled with air pressure using control valves and air-over-oil accumulators.

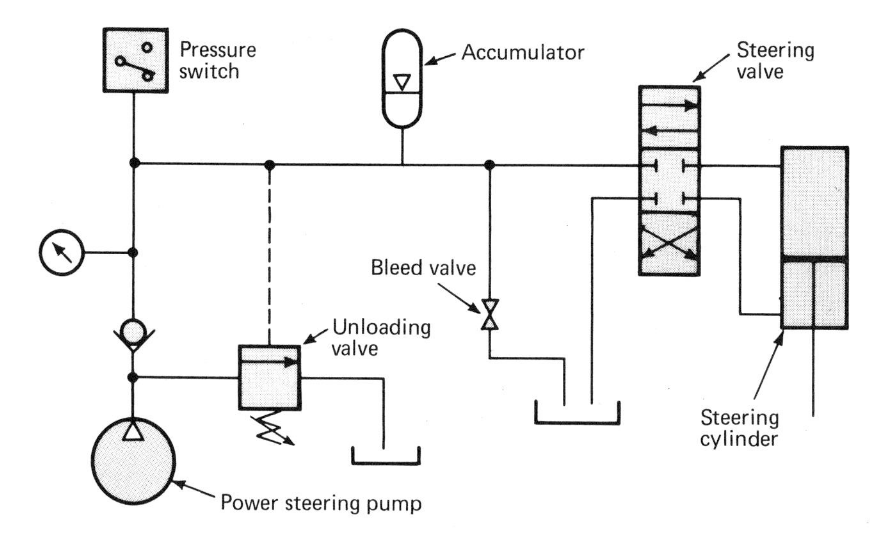

Fig. 8.25 Accumulator provides emergency oil pressure for steering mobile equipment when the pump fails.

The five other types are floating piston, free surface (direct oil-gas interface), spring-loaded piston (gas side open to atmosphere), dual-diameter piston (discussed later), and weighted-piston accumulator (rarely used in industrial systems). There are also variations and combinations; for example, diaphragms are substituted for bladders in some designs. A spherical accumulator has the highest volume-to-weight ratio and is chosen for aircraft. And a free piston can be spring-biased.

The *bladder* is a synthetic-rubber bag charged with compressed air or nitrogen. It has low inertia and can respond very fast—good for absorbing line shocks. Special

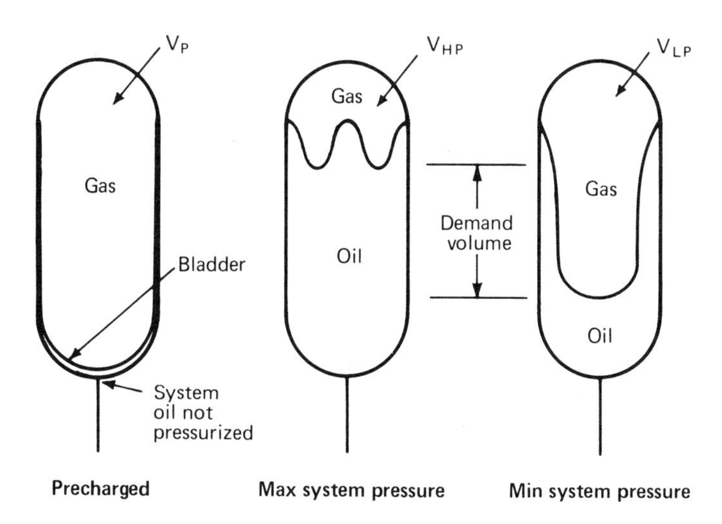

Fig. 8.26 Bladder accumulator model (for the accompanying calculation procedure).

supports or valving prevent the bladder from extruding into the oil port at low oil pressure.

For longest life, the gas bag should be operated in the middle range of its travel: from one-third volume to about two-thirds volume. Details of this are discussed under "demand volume."

The single-diameter *floating piston* is a barrier to keep the system oil from contacting the gas. One advantage of a piston instead of a bladder is that failure is by leakage rather than rupture. Also, higher temperatures can be withstood, depending on the piston seals.

A *dual-diameter piston* accumulator (Fig. 8.27) doubles as an intensifier. Operation is the same as that of the single-diameter accumulator, except that the oil pressure is greater than the gas pressure by the ratio of gas-side piston area to oil-side piston area. The volume of oil flow is less by the same ratio.

Demand Volume Calculations

The demand volume ΔV of an accumulator is the oil that must be discharged into the system during the period of need (Fig. 8.28). The energy comes from the stored gas.

The first step is to establish the precharge pressure, because this determines the

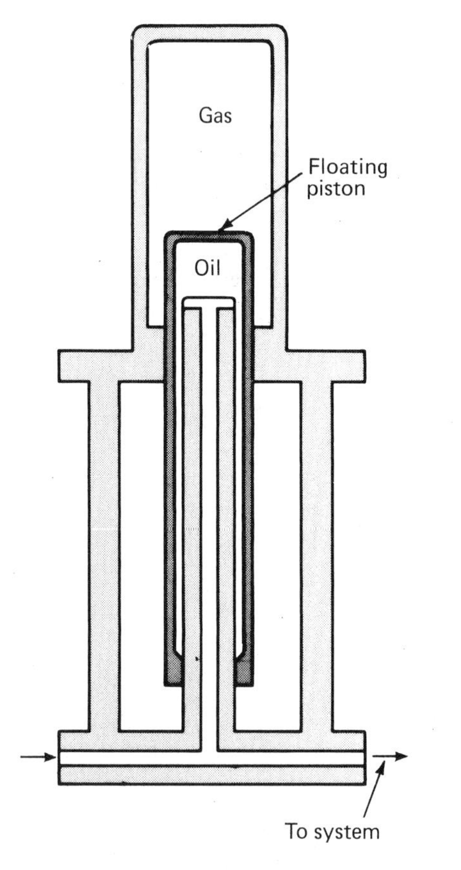

Fig. 8.27 Dual-diameter piston accumulator doubles as an intensifier.

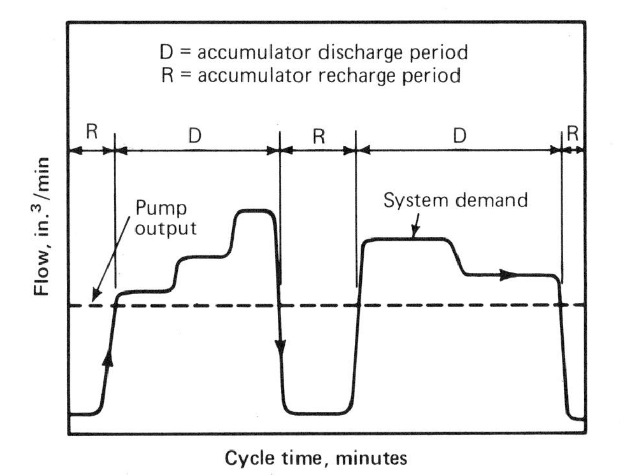

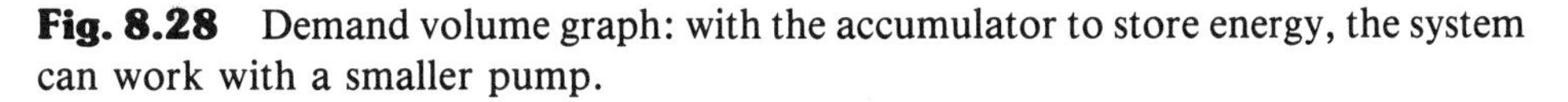

Fig. 8.28 Demand volume graph: with the accumulator to store energy, the system can work with a smaller pump.

total size of the accumulator. As an example, a typical precharge for a 3000-psi hydraulic system is 1000 psi. Then the gas will be compressed to about one-third of its precharged volume, assuming isothermal compression, when system pressure rises to 3000 psi from the unpressurized condition. Engineers at Haskel say, if you specify precharge to high pressure, make sure you don't exceed the pressure of available bottles of nitrogen. At least, you should recommend a gas intensifier.

A 300-psi precharge will result in 10 : 1 compression when system pressure rises to 3000 psi. An extreme case is atmospheric precharge: Here the compression ratio would be 3000/14.7 = 204, and the accumulator overall volume would have to be at least 204 times the volume of the gas at 3000 psi, a size that normally is not practical.

A lot depends on the application. If the system must maintain constant pressure within 5% or 10%, the accumulator gas volume at system pressure should be large in proportion to the oil volume so that the accumulator can discharge considerable liquid without excessive pressure drop. But if all that's needed is a large volume of oil at a rapidly decreasing pressure, gas volume at the highest operating pressure can be small.

The demand volume relationship is

$$\Delta V = V_{LP} - V_{HP} \tag{8.20}$$

See Table 8.6 for symbols. Computing the initial V_{HP} and final V_{LP} gas volumes is simple, using Boyle's gas law:

$$P_{HP}V_{HP}^{n} = P_{LP}V_{LP}^{n} = PV^{n} \tag{8.21}$$

where exponent n = 1 in isothermal expansion and 1.4 to 1.7 (depending on average pressure and temperature) in adiabatic expansion for air or nitrogen. Actual expansion is never wholly one or the other. So for the purpose of this discussion, polytropic expansion is assumed, with n = 1.25. The error is small for low- to moderate-pressure applications using air or nitrogen, where pressure changes are less than 25% during operation.

Table 8.6 Nomenclature of Accumulator Demand-Volume Calculations

Symbols	
System	
P_{LP}	Lowest operating pressure
P_{HP}	Highest operating pressure
x	Ratio $(P_{HP} - P_{LP})/P_{HP}$
ΔV	Demand volume, in.^{3}
Q	Pump capacity, in.^{3}/cycle
s	Piston stroke, in.
d	Piston diameter, in.
N	Piston strokes/cycle
Accumulator	
P_P	Precharge pressure, psi
V_P	Gas precharge, in.^{3}
V_{LP}	Gas volume at low P
V_{HP}	Gas volume at high P
V_{ACC}	Total accumulator volume

For best accuracy, include heat transfer, coefficient of thermal expansion, and variation in gas properties. Remember that heat transfer is more rapid through metal than it is through a bladder.

Pressure requirements are set by the system. To calculate the accumulator size, you must know the maximum pressure and the allowable pressure drop. We'll express it as the ratio x:

$$x = (P_{HP} - P_{LP})/(P_{HP}) \qquad \text{or} \qquad (P_{LP})/(P_{HP}) = 1 - x \tag{8.22}$$

$$V_{HP}/V_{LP} = V_{HP}/(V_{HP} + \Delta V) = (1 - x)^{1/n} \tag{8.23}$$

which is the basic equation relating demand volume and pressure drop, derived from Eqs. (8.20), (8.21), or (8.22). Or

$$V_{HP} = \left[\frac{(P_{LP}/P_{HP})^{1/n}}{1 - (P_{LP}/P_{HP})^{1-n}} \right] \times \Delta V = m\Delta V \tag{8.24}$$

Values of m are plotted in Fig. 8.29.

Accumulator size = gas volume + oil volume. The desired volume ratio of gas to oil must be specified before the combined volume of gas and oil is calculated. As a rule the gas volume is specified as a multiple of the oil volume at the fully charged condition. A typical value for industrial service is 3 : 1. Then the total accumulator volume $V_{ACC} = 1.33V_{HP}$.

The precharge volume V_P can be calculated directly if the precharge pressure, maximum operating pressure, minimum operating pressure, and demand volume are known. The relationship, given in a Greer manual, is

$$V_P = \frac{\Delta V(P_{LP}/P_P)^{1/n}}{1 - (P_{LP}/P_{HP})^{1/n}} \tag{8.25}$$

Before you can specify a demand volume, you have to estimate peak demands of the system. Overall demand volume is the summation of all the component demands in the system. It can best be expressed graphically, as shown in the volume-cycle-time graph (Fig. 8.28). The summation, which includes all working cylinders and the system pump, can be expressed as

$$V = (\pi/4)(d_1^2 s_1 N_1 \ . \ . \ . \ + d_n^2 s_n N_n) - Q \tag{8.26}$$

where d is each piston diameter, s is stroke, N is strokes per accumulator cycle, and Q is pump delivery per cycle.

Example: Compute Total Volume. Assume a bladder-type accumulator with a gas-to-oil volume ratio = 3 : 1 at the fully charged condition. Exponent n = 1.25. System demand volume $\Delta V = 6$ in³ The allowable pressure drop is 5% (x = 0.05). Use Eq. (8.23):

$$V_{HP}/(V_{HP} + 6) = (1 - 0.05)^{1/1.25} = 0.96$$

$$V_{HP} = 0.96(V_{HP} + 6) = 144 \text{ in}^3$$

$$V_{ACC} = 144 + 144/3 = 192 \text{ in}^3$$

Application: Emergency Steering

A loaded off-road vehicle becomes an unguided missile when steering fails. Therefore safety regulations require that an auxiliary steering method be included in the design and that even if the engine fails, the vehicle can be steered to a safe stop (see Figs. 8.25 and 8.30).

If engine failure occurs, system pressure drops, and the unloading valve diverts pump flow to the tank. Pressurized fluid stored in the accumulator forces the check valve closed, and this stored fluid operates the steering cylinder.

That technique is common in the mobile equipment field and in many other in-

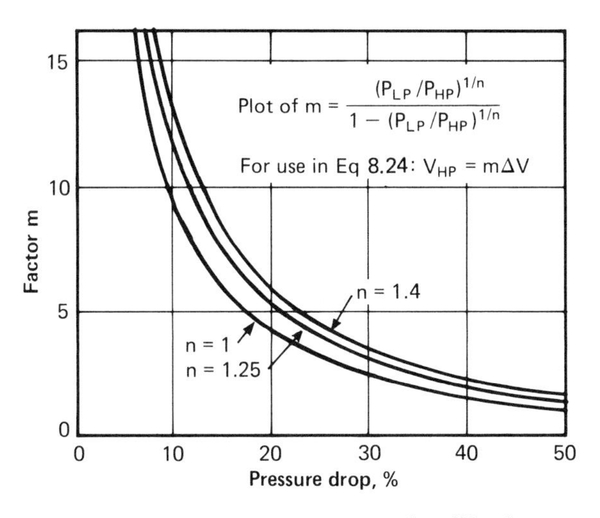

Fig. 8.29 Plot of factor m (to simplify the accompanying accumulator demand-volume calculation procedure).

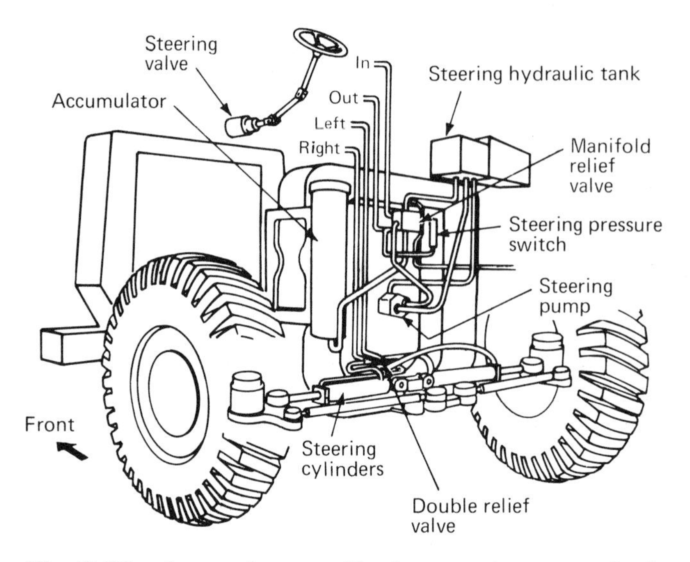

Fig. 8.30 Accumulator application: steering system for heavy-duty off-road vehicle.

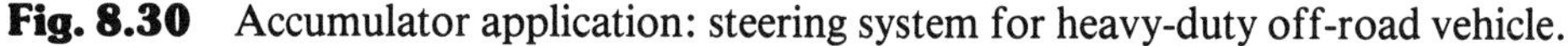

dustries, for example, a steering accumulator in big haulers (Fig. 8.30). The storage capacity is enough for two full steering cycles. In each of these systems, recharge occurs slowly on the next startup.

Another good application for an accumulator is the pressure-loading circuit for a rock crusher (Fig. 8.31). Whenever an uncrushable object starts to jam the rotating bonnet, the loading piston quickly rises to relieve the load, and the accumulator absorbs the sudden flow of fluid.

Three novel applications are hydropneumatic supports (shocks) for trucks (Fig. 8.32), supply source and barrier seal for high water base fluid systems (Fig. 8.33), and rotary shock supports for rail cars (Fig. 8.34).

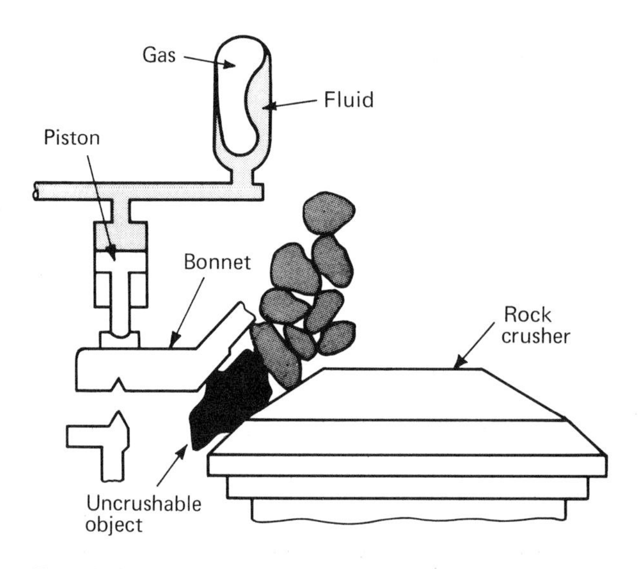

Fig. 8.31 Accumulator application: shock cushion for rock crusher.

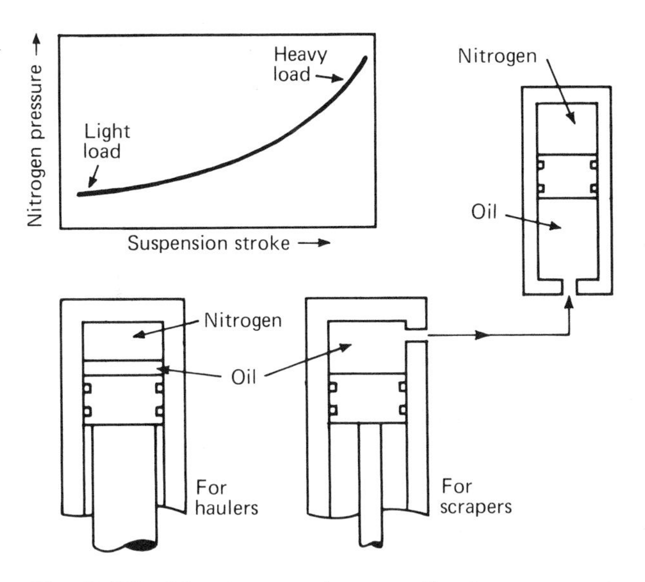

Fig. 8.32 Novel accumulator application: suspension system for heavy earth-moving equipment.

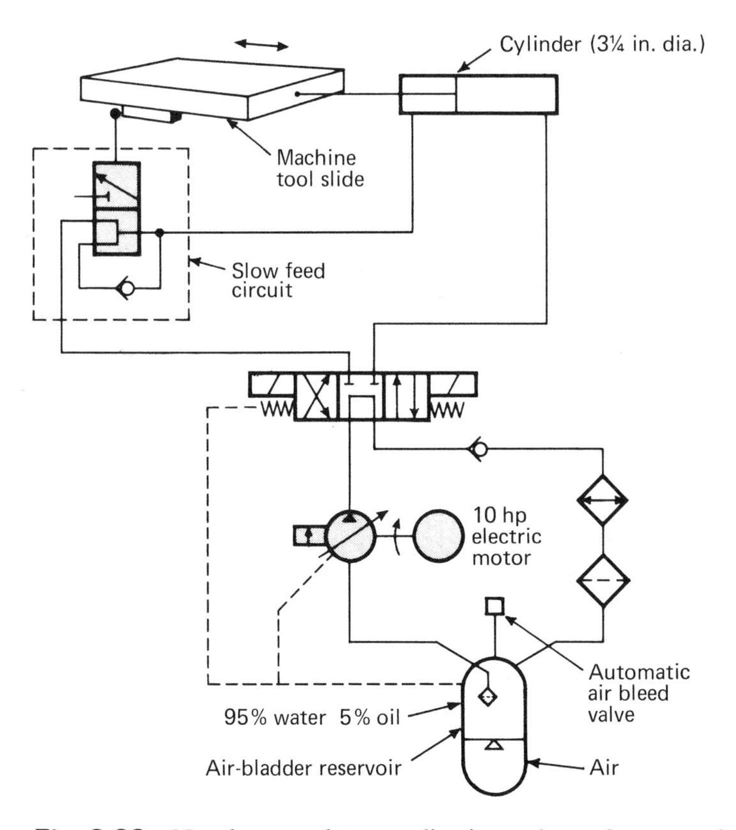

Fig. 8.33 Novel accumulator application: scheme for pressurizing a high water base fluid hydraulic system.

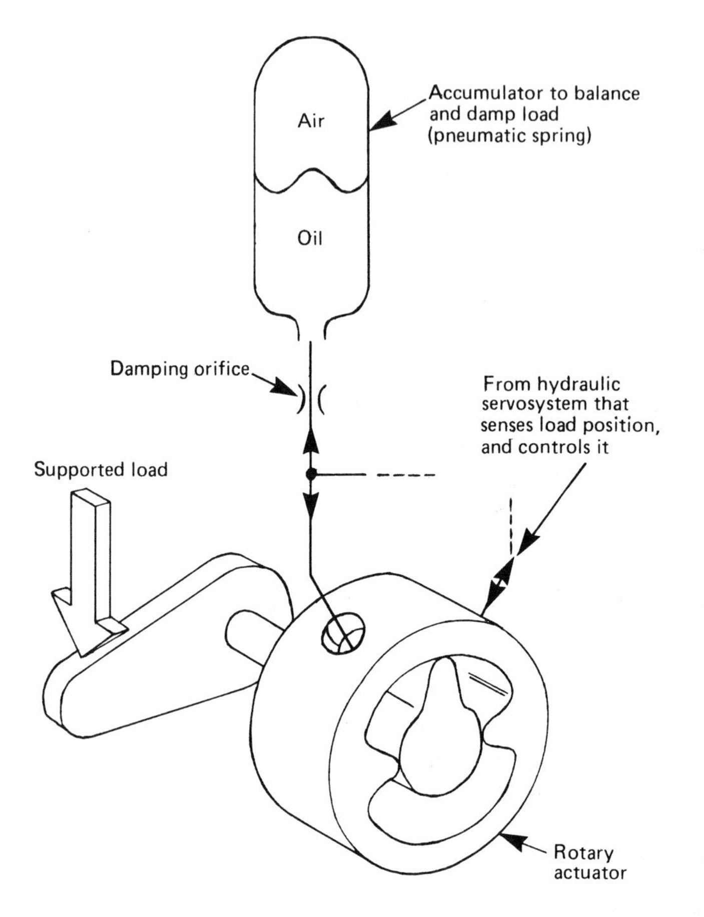

Fig. 8.34 Novel accumulator application: this balanced hydraulic rotary actuator, using an accumulator as a pneumatic spring, not only supports a heavy load but adjusts the height, stabilizes the system, and absorbs shock. The design is for rail cars. (Courtesy of Houdaille Hydraulics, Inc., Buffalo, N.Y.)

Other applications include hydraulic die presses, robots, operators for circuit breakers, hydraulic systems for the plastics manufacturing machinery, nuclear power plants, and machine tools. A typical die press design reduces pumping requirements to 7.5 gpm from 42 gpm by adding an accumulator. It increases cycle speed with less energy.

Three experimental accumulator designs are pictured in Fig. 8.35. Each tackles hard-to-solve problems.

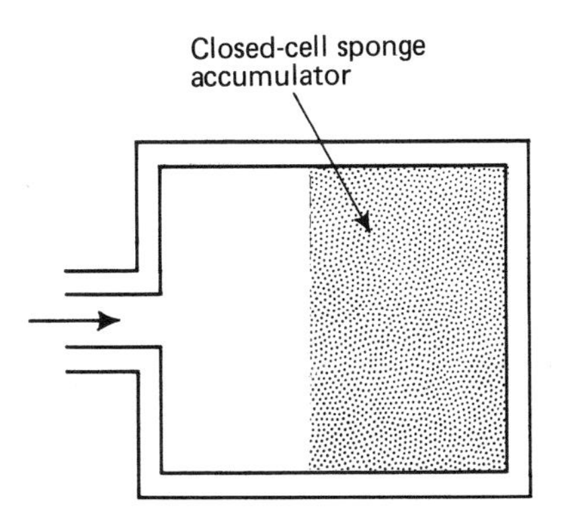

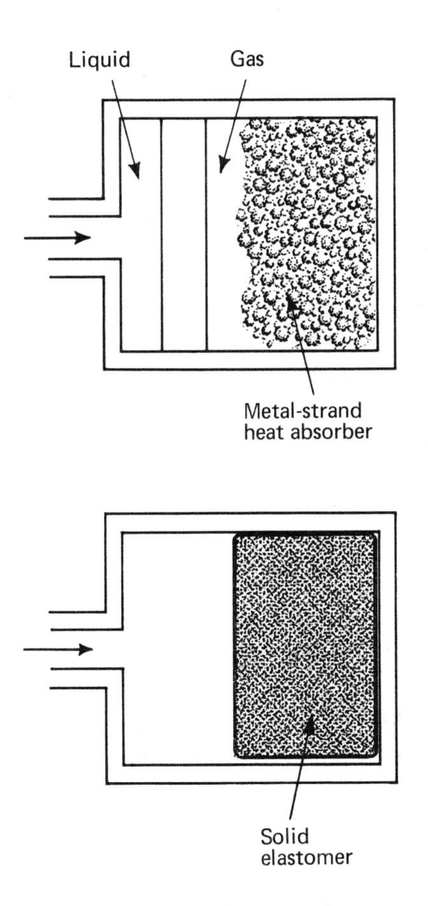

Fig. 8.35 Novel accumulator concepts: closed-cell sponges (to replace gas bag), metal strands (to absorb heat of compression, and solid elastomer (for extreme loads).

9

Hydraulic Flow, Pressure Drop, Response

There's no lack of information on pressure losses in piping, fittings, and restrictions—just difficulty in keeping units straight in calculation and estimating the probable accuracies.

Bernoulli, Poiseuille, Darcy, Weisbach, Moody, Pigott, Reynolds, and other pioneers set down the rules for flow of hydraulic fluids years ago, and the equations haven't changed much. However, only the simplest of the relationships are used at all extensively by fluid power and control engineers.

This chapter covers most types of liquid conductors: piping, hose, fittings, valves, orifices, and clearances.

The final example—response time of a pilot-operated spool valve—introduces electrical and mechanical delays as well.

FLOW RESISTANCE IN PIPE

Tables 9.1 and 9.2 summarize seven variations of flow resistance in pipe, and you can add a few of your own.

It's not necessary to include changes of elevation or even local changes in velocity. The following accurate-enough empirical relationship for flow losses in pipes usually will suffice:

$$H_L = f\frac{L}{D}\frac{V^2}{2g}$$

where head loss depends on pipewall friction factor f, length over diameter ratio, and velocity in ft/sec.

That empirical equation has limitations, of course. Pipe diameter is assumed constant for the length examined to assure constant velocity: And the friction factor is assumed constant at some average value even though it is known to change with temperature, pressure, and variations in pipewall conditions.

Friction factor f (Figs. 9.1A and B) is a function of the Reynolds number and

Table 9.1. Review of Pressure Loss Equations for Piping

General empirical relationships:

All flows, $H_L = f\frac{L}{D}\frac{V^2}{2g}$; laminar only, $H_L = 32\frac{\mu L V}{\varrho g d^2}$

Useful conversions, for oil only ($= 55$ lb/ft^3)

Variables							Equations		
L	V	D	ΔP	Q	f	μ, ν	pressure loss	velocity	flow
ALL FLOWS									
ft	$\frac{ft}{sec}$	in.	psi	$\frac{ft^3}{sec}$	f	–	$\Delta p = f\frac{L}{d}V^2 \times 0.072$	$V = \sqrt{\frac{d\Delta p}{fL}} \times 3.73$	$Q = \sqrt{\frac{d^5\Delta p}{fL}} \times 0.0203$
ft	$\frac{ft}{sec}$	in.	psi	gpm (q)	f	–	$\Delta p = f\frac{L}{d^5}q^2 \times 0.0123$	$V = \sqrt{\frac{d\Delta p}{fL}} \times 3.73$	$q = \sqrt{\frac{d^5\Delta p}{fL}} \times 9.1$
LAMINAR ONLY									
ft	$\frac{ft}{sec}$	in.	psi	$\frac{ft^3}{sec}$	–	cst	$\Delta p = \nu\frac{L}{d^2}V \times 0.0006$	$V = \frac{d^2\Delta p}{\nu L} \times 1670$	$Q = \frac{d^4\Delta p}{\nu L} \times 9.1$
ft	$\frac{ft}{sec}$	in.	psi	gpm	–	cst	$\Delta p = \nu\frac{L}{d^4}q \times 2.45 \times 10^{-4}$	$V = \frac{d^2\Delta p}{\nu L} \times 1670$	$q = \frac{d^4\Delta p}{\nu L} \times 4080$
in.	$\frac{in.}{sec}$	in.	psi	$\frac{in.^3}{sec}$	–	$\frac{lb\text{-}sec}{in.^2}$	$\Delta p = \mu\frac{L_{in.}}{d^4}Q_{in.^3} \times 40.75$	$v = \frac{d^2\Delta p}{\mu L_{in.}} \times 0.0312$	$Q_{in.^3} = \frac{d^4\Delta p}{\mu L_{in.}} \times 0.0245$
ft	$\frac{ft}{sec}$	ft	psf	$\frac{ft^3}{sec}$	–	$\frac{lb\text{-}sec}{ft^2}$	$\Delta P = \mu\frac{L}{D^2}V \times 32$	$V = \frac{D^2\Delta P}{\mu L} \times 0.0312$	$Q = \frac{D^4\Delta P}{\mu L} \times 0.0245$
ft	$\frac{ft}{sec}$	ft	psf	gpm	–	$\frac{lb\text{-}sec}{ft^2}$	$\Delta P = \mu\frac{L}{D^4}q \times 0.091$	$V = \frac{D^2\Delta P}{\mu L} \times 0.0312$	$q = \frac{D^4\Delta P}{\mu L} \times 11$

Table 9.2. Nomenclature for Pressure Loss Equations for Piping

d	= pipe inside dia, in.
D	= pipe inside dia, ft
f	= friction factor, dimensionless
g	= gravity constant, 32.2 ft/sec^2
H_L	= head loss, ft
L	= pipe length, ft
$L_{in.}$	= pipe length, in.
Δp	= pressure loss, psi
ΔP	= pressure loss, psf
q	= flow, gpm
Q	= flow, ft^3/sec
$Q_{in.^3}$	= flow, in.^{3}/sec
R_e	= Reynold's Number $\rho DV/\mu$. Typical units are: $3162\ q/d\nu_{cst}$ gpm, in., centistoke $50.6\ q\gamma/d\mu_{cp}$ gpm, lb/ft^3, in., cp $\gamma DV/g\mu$ lb/ft^3, ft, ft/sec, lb-sec/ft^2
v	= oil velocity, in./sec
V	= oil velocity, ft/sec
μ	= absolute viscosity, lb-sec/ft^2 or lb-sec/in.^{2}
ν	$=\frac{\mu}{\rho}$, kinematic viscosity, ft^2/sec
ρ	= mass density, lb-sec^2/ft^4 = slugs/ft^3
γ	= weight density, lb/ft^3
ϵ/D	= relative roughness of pipewall. Values of ϵ, ft: drawn tubing, 5×10^{-6}; steel or wrought iron, 150×10^{-6}. D = inside dia, ft.

the roughness of the pipewall. When flow is laminar ($R_e<2000$) and isothermal, then

$$f=\frac{64}{R_e}$$

When oil system temperature is variable, as when system components are subject to heating and cooling, the friction factor tends to vary as follows:

$$f=\frac{75}{R_e}$$

For turbulent flow based on smooth steel or brass tubing, an average relationship is

$$f=\frac{0.3164}{(R_e)^{1/4}}$$

Reynolds number

$$R_e=\frac{\rho DV}{\mu}$$

is dimensionless and thus independent of the particular set of units chosen. The nomenclature in Table 9.2 suggests several alternatives. However, it's tough enough even staying within a given set of units.

Viscosity μ usually is the culprit in most wrong calculations, so let's examine that. Absolute viscosity (the one in R_e before) is a direct measure of the fluid shear-

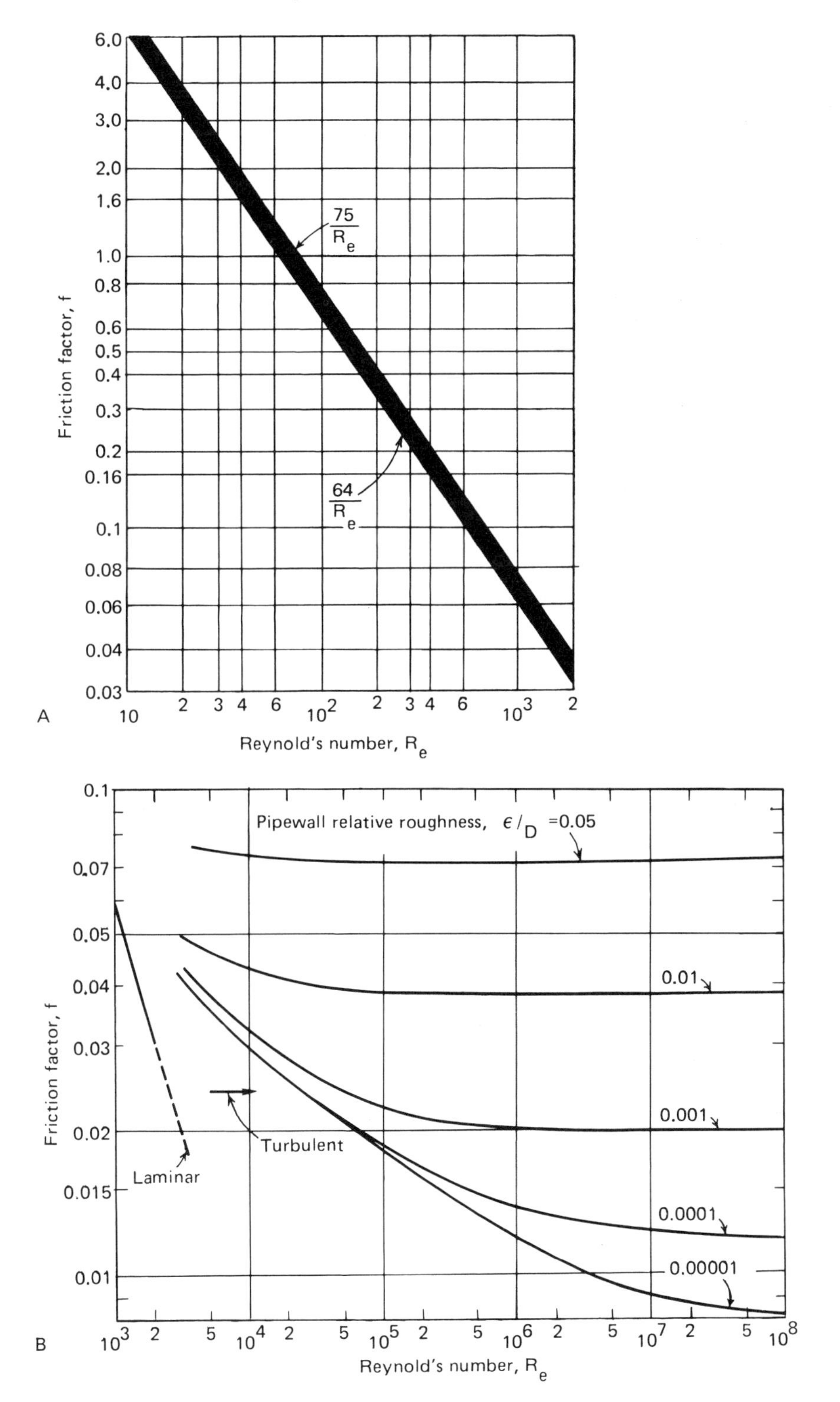

Fig. 9.1 (A) Friction factor for laminar flow in pipe; (B) friction factor for turbulent flow in pipe.

ing forces per unit area, velocity, and layer thickness. Kinematic viscosity includes density and is useful mainly because most viscosity-measuring instruments give readings in kinematic viscosity. You still have to convert to absolute units for many equations. (See Chap. 1 for theory and application.)

Velocity V is important because losses vary as V^2 does. A popular rule of thumb is to size piping to give some specified velocity; what it lacks in accuracy is made up for in convenience. Typical suitable maximum values are indicated in Table 9.3, converted in Fig. 9.2 to gpm for several pipe sizes. That is a good place to start figuring your system losses.

FLOW IN CORRUGATED HOSE

Pressure losses for liquid or gas flow in corrugated hose can be computed by assuming that the corrugations behave as a series of uniformly spaced orifices (Fig. 9.3). Pressure drop is caused by a succession of individual flow expansions.

It's a more valid concept than two previous beliefs, namely (1) the old rule of thumb that a hose behaves as an equivalent duct of three or four times the actual hose length and (2) the assumption that the losses are induced in the valleys of the corrugations and are in some way related to corrugation height—specifically, roughness ratio e/D where e is corrugation height and D is the equivalent diameter of the duct.

The *equivalent duct* concept is misleading because flow through a smooth duct or pipe usually is not fully turbulent, whereas that through a corrugated hose usually is. It is not practical to compare the two flows on this basis unless both are fully turbulent.

Table 9.3 Recommended Maximum Velocities in Piping (See Fig. 9.2)

Piping	Velocity ft/sec	Flow, gpm
Pressure lines,		curve:
800 psi and up	16	A
400 to 800	13	B
up to 400 psi	10	C
Return lines	7	D
Suction lines	5	E

Example: Required flow is 50 gpm, system pressure is 900 psi. What pipe sizes are needed to assure recommended maximum velocities?

On graph at right, draw a horizontal line from 50 gpm, intersecting all curves. Draw vertical lines to establish that:

Pressure lines need 1¼-in. pipe
Return lines need 2-in. pipe
Suction lines need 2-in. pipe

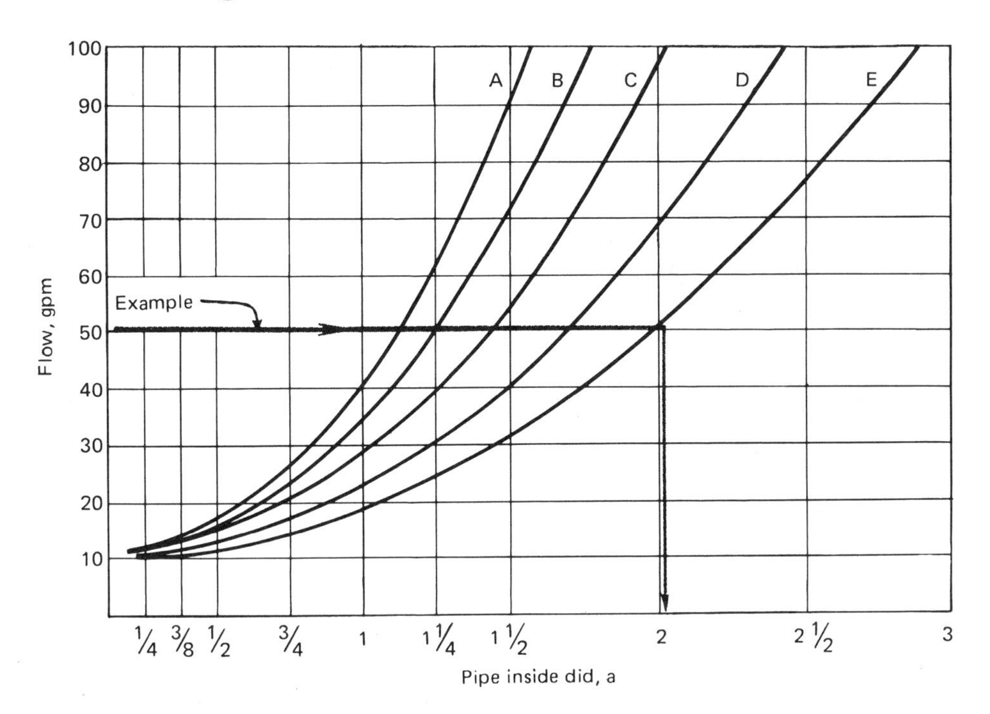

Fig. 9.2 Recommended maximum flow in pipe (see the example in Table 9.3).

Flow through a corrugated hose or bellows becomes turbulent almost immediately upon reaching a Reynolds number of $R_e = DV/n = 2000$ (see the symbols in Table 9.4), and 2000 is exceeded in most applications. Flow through a very smooth pipe may not become fully turbulent until R_e = orders of magnitude higher than 2000.

The *corrugation height* concept is also wrong. In no cases examined during the test program was flow affected by the full corrugation height. The sketch, Fig. 9.3, shows what probably does happen—simple expansion from corrugation to corrugation. The empirical flow equations and charts in this section (from R. C. Hawthorne and H. C. von Helms of Flexonics Co.) have been developed on that basis.

Liquid and Gas

Liquid and gaseous flow are treated alike. The only essential difference is the compressibility of a gas, and this usually can be ignored at subsonic velocities. For example, in uninsulated ducting or piping having a 5-psi pressure drop in a 100-psi line, the calculated pressure drop ignoring compressibility is only 2½% less than that including compressibility.

It is assumed here, in calculating liquid flow, that cavitation does not occur. Cavitation could greatly increase pressure loss, but this did not occur in any of the tests, and it is not likely to occur in any pressurized system. In open discharge at high flow rates, however, there might be cavitation.

Pressure loss in the context of this section is always a loss of static head. Velocity head does not vary unless some of the fluid is lost or altered or the duct diameter is changed.

Laminar flow is excluded from the graphs because the pressure drop in pounds

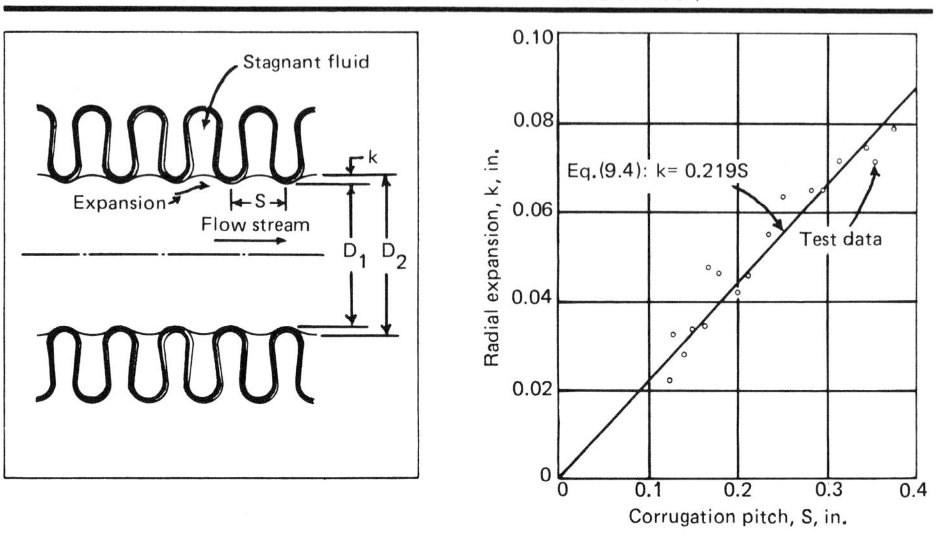

Fig. 9.3 Flow in corrugated hose is a series of expansions.

per square inch is easily solved with this general-relationship Darcy-Weisbach resistance equation:

$$\Delta P = f\frac{L}{D} \times \frac{V^2\rho}{9266} \qquad (9.1)$$

where f = friction factor = $64/R_e$ for laminar flow; L = pipe length, in.; D = pipe or hose inside diameter, in.; V = flow velocity, ft/sec; ρ = weight density, lb/ft^3; and 9266 is a conversion constant.

Table 9.4 Nomenclature for Pressure Loss Equations for Corrugated Hose

C=loss coefficient for straight run
C_B=loss coefficient for bend
$D_{1,2}$=stream diameters, in. (see sketch)
e=surface roughness, in.
f=friction factor=$64/R$ for laminar flow
h_L=head loss, ft
k=expansion of the stream radius, in.
L=conduit length, in.
N=total number of corrugations
n=kinematic viscosity, ft^2/sec
ΔP=pressure loss, psi
r=radius of bend centerline
R_e=Reynold's number
S=corrugation spacing (pitch), in.
V_1=flow velocity, ft/sec at D_1
W=corrugation height, in.
ρ=weight density, lb/ft^3
θ=angle of bend (zero to 180°)

Straight Lengths

This method of calculating pressure drop of turbulent flow in corrugated hose assumes that each convolution is an orifice plate. Loss in contraction of the fluid while passing through the orifice is ignored because all losses eventually appear in the expansion and are evaluated there.

Head loss for a sudden expansion in a straight pipe is expressed as

$$h_L = C\frac{V_1^2}{2g} \tag{9.2}$$

where h_L = head loss, ft; C = loss coefficient = $(1 - D_1^2/D_2^2)^2$, plotted in Fig. 9.4; and g = 32.2 ft/sec². Terms D_1 and D_2 are effective stream diameters, in., defined in the first illustration.

Equation (9.2) can be rewritten in terms of the actual radial expansion k, in., if it is noted that $D_2 = D_1 + 2k$:

$$h_L = \left[1 - \left(\frac{D_1}{D_1 + 2k}\right)^2\right]^2 \times \frac{V_1^2}{2g} \tag{9.3}$$

The empirical part of this development was to find a relationship for radial expansion k. This was done for enough different hose configurations and fluids to establish Fig. 9.3. Note the linearity, which makes this general expression valid:

$$k = 0.219S \tag{9.4}$$

where S is the corrugation spacing or pitch, in.

Equation (9.2), converted to ΔP and with the new expression for C, becomes

$$\Delta P = N\left[1 - \left(\frac{D_1}{D_1 + 0.438S}\right)^2\right]^2 \times \frac{V_1^2\rho}{9266} \tag{9.5}$$

where N = the total number of corrugations or expansions.

Bends and Elbows

Once a suitable friction factor or loss coefficient is established for straight lengths of corrugated hose or bellows, pressure loss in bends and elbows can be handled in the same manner as pressure losses in piping and ducting.

The method is to calculate the pressure loss in the developed length assuming that the bend has been straightened. Then add to this value a loss based on the kinetic energy of the fluid in the bend. The combined expression is

$$\Delta P = N\left[1 - \left(\frac{D_1}{D_1 + 0.438S}\right)^2\right]^2 \times \frac{V_1^2\rho}{9266} + C_B\frac{V_1^2\rho}{9266} \tag{9.6}$$

Values of C_B for 90-deg bends are plotted in Fig. 9.4. Values of pressure drop for other angles have been developed by many experimenters and can be determined by using these approximate relationships:

$$\text{zero to } 90°\text{:} \quad \Delta P_\theta = \Delta P_{90}\left(\frac{\nu}{90}\right)^{1/2}$$

$$90° \text{ to } 180°\text{:} \quad \Delta P_\theta = \Delta P_{90}\left(\frac{\theta}{90}\right)$$

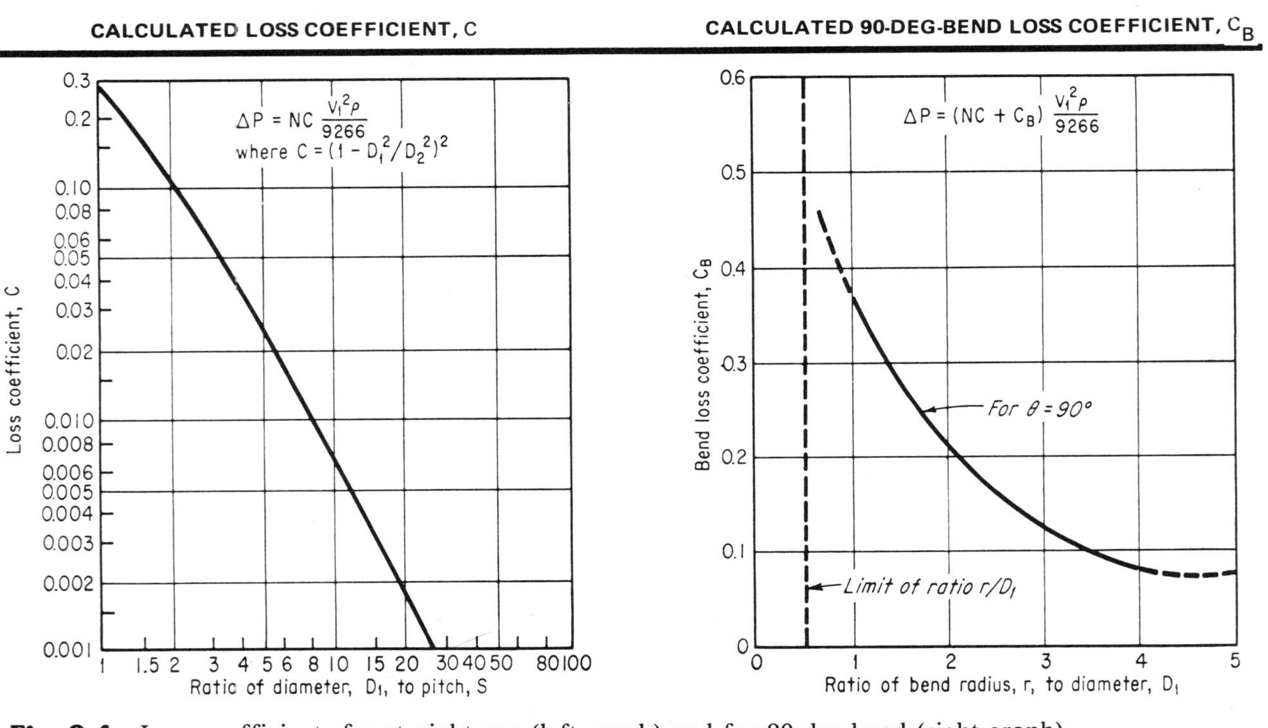

Fig. 9.4 Loss coefficients for straight run (left graph) and for 90-deg bend (right graph).

where ΔP_{90} is computed with the help of the bend-loss graph, Fig. 9.4.

The Tests

Tests were run on straight sections of corrugated hose only. The parameters varied were length, pitch, and height of corrugations; lay of corrugations (annular or helical); Reynolds number (24,000 to 1,300,000); fluid (water and air); and fluid flow. Table 9.5 lists some of the test results.

The water-flow tests were closed-loop, whereas the airflow tests were open-discharge. The change in corrugation spacing (pitch) was accomplished for the 4-in. size by simply extending conventional 4-in.-ID bellows to many different lengths. None of the parameter changes greatly affected the basic relationship of Equation (9.4), namely, $k = 0.219S$.

LOCAL RESISTANCE TO FLOW

Here is a compilation of every type of loss inducer from elbows to filters, with easy-to-use flow coefficients for each.

Local resistances mean unintentional, though unavoidable, loss inducers such as bends, filters, and wide-open stop valves. Many are irregularly shaped obstructions to flow, and the resistance is impossible to calculate accurately.

Basic Equation

Fortunately a simple empirical relationship exists for rough computation of pressure loss for almost any restriction:

$$H_L = KV^2/2g \qquad (9.7)$$

$$\Delta p = KV^2(1.08 \times 10^{-4})\gamma \qquad (9.8)$$

$$\Delta p = K(q^2/d^4)(1.8 \times 10^{-5})\gamma \qquad (9.9)$$

where H_L = head loss, ft; Δp = head loss, psi; γ = weight density, lb/ft^3; V = average velocity, ft/sec (with the exception of section changes); q = flow, gpm; d = diameter, in.; and K (see Tables 9.6 through 9.8) is an empirical constant, different for each type of restriction.

These empirical relationships fall far short of perfection, and a better answer is always possible if you want to break down the flow resistances into entrance losses, friction losses, turbulence losses, velocity head changes, and exit losses. But the K-factor method is much simpler and is accurate enough, particularly when flow is turbulent.

If you want to improve accuracy somewhat, assume that wall friction losses will be higher when flow is laminar. Increase the indicated K values 30% for R_e (Reynolds number) near 2000 and 50% for those near 100.

You can convert K to equivalent pipe length measured in diameters if you divide by friction factor f:

$$L_E/D = K/f \qquad (9.10)$$

where f is a function of Reynolds number R_e for laminar flow and of both R_e and wall roughness for turbulent flow. The $L_E/D = K/f$ relationship is derived from these equations:

Table 9.5. Test Data on Water and Airflow Through Corrugated Hose

METAL-HOSE DIMENSIONS, IN.						WATER-FLOW DATA (closed circuit)		PRESSURE-LOSS CALCULATIONS		
Dia. D_1 Nominal	Actual	Length L	Pitch S	Height W	Number of corrug.	gpm	Reynold's Number R	Loss coeff. C for one corrug. (note)	Test Friction Factor f	Radial flow expansion k, in.
¾	0.765	12	0.125	0.226	96	5.5	2.44x10⁴	0.01225	0.0750	0.0225
1	1.000	20.9	0.128	0.159	164	72.0	2.04×10^5	0.01350	0.1050	0.0318
	1.016	12	0.138	0.239	87	10.0	3.44×10^4	0.01015	0.0745	0.0275
1¼	1.284	12	0.150	0.273	80	12.0	3.18×10^4	0.00927	0.0793	0.0330
1½	1.500	20.3	0.163	0.207	124	216.0	4.02×10^5	0.00724	0.0666	0.0340
	1.517	12	0.200	0.327	60	17.0	3.82×10^4	0.01050	0.0795	0.0420
4	3.990	16	0.252	0.234	64	871.0	7.45×10^5	0.00376	0.0598	0.0640
	3.990	24	0.375	0.213	64	898.0	7.68×10^5	0.00548	0.0584	0.0785
	4.016	3	0.178	0.337	17	942.0	8.00×10^5	0.00198	0.0448	0.0460
	4.016	6	0.353	0.306	17	935.0	7.93×10^5	0.00454	0.0515	0.0710
						AIR-FLOW DATA (Discharge to atmos.)				
2	2.000	36	0.212	....	170	0.2 lb/sec	1.18×10^5	0.00683	0.0647	0.0455
4½	4.460	2.5	0.167	0.375	15	5 lb/sec	1.32×10^6	0.00168	0.0448	0.0470

Note: $C = f\,S/D$ **for one corrugation based on** $h_L = C\frac{V^2}{2g} = f\frac{L}{D}\frac{V^2}{2g}$ **when** $L = S$.

(Courtesy of Flexonics Div., UOP Inc., Bartlett, IL; Boeing Co., Seattle, WA; Martin Co., Orlando FL)

Table 9.6. K-Factor for Pressure Loss Explained

Invented to simplify pressure-loss calculations in ordinary fittings and valves, the K-factor takes some liberties with true energy-balance laws. It neglects changes in velocity, elevation, viscosity, and wall friction. It also discounts the effects of piping roughness preceding and following the fitting.

In fact, the main basis for using K is that observation of numerous test data on many fluids over the years has indicated that pressure losses vary with V^2 of the fluid, give or take 10% on the exponent 2.0. It was logical and useful to express the relationship in this way:

$$\text{head loss } H_L = K\frac{V^2}{2g}$$

$$\text{or:} \qquad \Delta p = KV^2(1.08 \times 10^{-4})\gamma$$

thus tying in nicely with the velocity-head concept in Bernoulli's conservation of energy theorem. Factor K is dimensionless and has a special value for each fitting or restriction. The value tends to decrease somewhat with diameter. For example, a large elbow has a smaller K than a small elbow. Exceptions are sudden enlargements and contractions: The K-factors are less dependent on nominal diameter and thus remain fairly constant, regardless of size.

If the K-factor for a given component is based on actual tests, good accuracy can be expected-assuming your operating conditions are similar to the test.

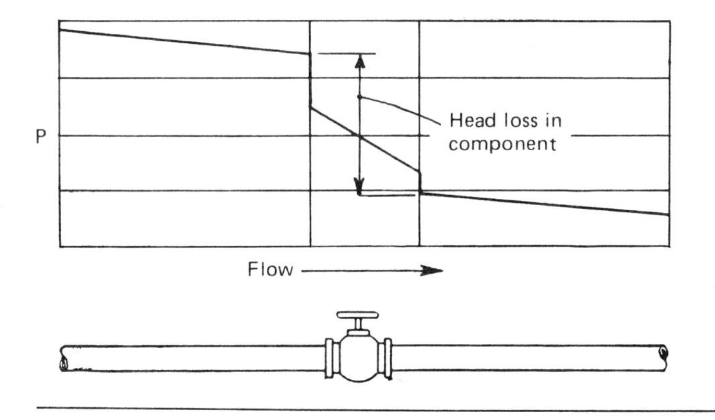

$$H_L = f\frac{L}{D}\frac{V^2}{2g} \text{ for pipe}$$

$$H_L = K\frac{V^2}{2g} \text{ for restrictions} \qquad \text{[Eq. (9.7)]}$$

Where test values of flow resistance are unobtainable, or where values vary with time, such as in filters, rough estimates of the K-factor can be made.

Reynolds number is the key. Coefficient K depends on R_e, and it can be calculated if you know the perimeter or some other critical dimension of the restriction.

For round holes,

$$R_e = \frac{\rho DV}{\mu} = \frac{DV}{\nu} \tag{9.11}$$

Table 9.7 Nomenclature for Pressure Loss Equations for Restrictions

A	= cross-sectional flow area, ft^2
A_O	= open area in coarse mesh or screen, ft^2
A_2	= total effective area of filter, ft^2; $A_2 = A_O +$ solid area
C	= coefficient of pressure loss for filter
d	= flow diameter, in.
D	= flow diameter, ft
f	= friction factor, dimensionless
g	= acceleration of gravity, 32.2 ft/sec^2
H_L	= head loss, ft
K	= K-factor, dimensionless
L	= length, ft
L_E	= equivalent length, ft
L_W	= wetted perimeter, ft
Δp	= pressure drop, psi
ΔP	= pressure drop, psf
q	= flow, gpm
Q	= flow, ft^3/sec
$Q_{in.^3}$	= flow, in.^{3}/sec
R_e	= Reynold's Number, $\rho DV/\mu$.
R_H	= hydraulic radius, ft; $R_H = A/L_W$
V	= velocity, ft/sec
W	= wall thickness of filter element, in.
ρ	= mass density, lb-sec^2/ft^4 or slugs/ft^3
γ	= weight density, lb/ft^3
μ	= absolute viscosity, lb-sec/ft^2
ν	= $\frac{\mu}{\rho}$, kinematic viscosity, ft^2/sec

For holes of noncircular cross section,

$$R_e = \frac{4VR_H}{\nu} \tag{9.12}$$

where R_H = hydraulic radius = A/L_W = (cross-sectional area)/(wetted perimeter of opening).

For coarse-mesh suction filters,

$$R_e = 34 \frac{V_1 d}{(A_O/A_2)\nu} \tag{9.13}$$

where R_e = Reynolds number, dimensionless; μ = absolute viscosity, lb-sec/ft^2; ν = kinematic viscosity, ft^2/sec; V_1 = approach velocity, ft/sec, based on effective cross-sectional area of filter element; d (or D) = mean diameter of flow passages, in. (or ft); and A_O/A_2 = ratio of open flow area to total cross-sectional area of filter element exposed to flow.

Deriving K-Factors

Most K-factors are already derived, from tests by manufacturers over a period of many years. The results, an average of data from many sources, are listed with the sketches of the components and fittings in Figs. 9.5 through 9.11.

K-factors for screwed fittings are 25% higher than those for flanged fittings. Any value given will increase if the fitting is other than clean and conventional.

Valve and fitting K-factors in Table 9.8 are based on Crane Bulletin 409. Valves are assumed fully open (when other than fully open, then the valve function is controlled throttling, the subject of the next section). Flow is assumed turbulent.

Orifices other than ordinary inlet and outlet holes are not covered here, because those are control-type restrictions, also to be covered in the next section.

Strainers and filters are a special breed of restriction and cannot be calculated by analytical means. However, K-factors for strainers and coarse-mesh filters can be simply though crudely estimated by two simple relationships, one for R_e above 400 and one for R_e below 400:

$$R_e > 400: \quad K = 1.3\left(1 - \frac{A_O}{A_2}\right) + \left(\frac{A_2}{A_O} - 1\right)^2$$

Table 9.8 Values of K-Factors for Standard Valves and Fittings

Component	L_E/D (assumes rough surface)	Internal diameters			
		0.5 in.	1 in.	2 in.	4 in.
Gate valve	13	K = 0.36	0.3	0.25	0.21
Swing check	135	3.6	3.0	2.5	2.1
Angle valve	145	3.8	3.3	2.7	2.3
Globe valve	340	9.5	7.9	6.6	5.7
Standard elbows:	Note: Smooth	fittings have much lower values			
90 deg	30 (high)	0.82	0.68	0.58	0.5
45 deg	16	0.43	0.36	0.3	0.26
Long elbow:					
90 deg	20	0.55	0.45	0.38	0.33
Street elbow:					
90 deg	50	1.4	1.2	0.96	0.82
45 deg	26	0.7	0.59	0.5	0.43
Standard Tee:					
through run	20	0.55	0.45	0.38	0.33
through branch	60	1.7	1.4	1.2	1.0
Close return bend	50	1.4	1.2	0.96	0.82

(Courtesy of Crane Co., New York, NY.)

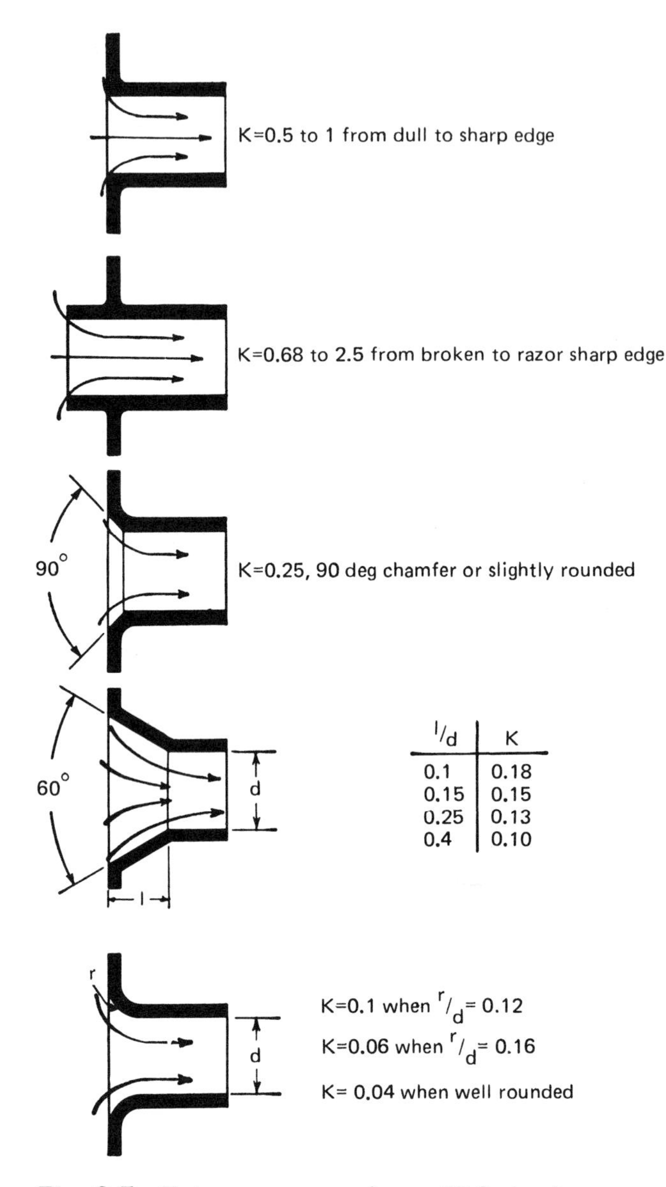

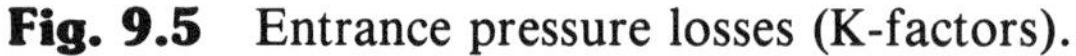

Fig. 9.5 Entrance pressure losses (K-factors).

$$R_e < 400: \quad K = Z \times (K \text{ for } R_e > 400)$$

where K is the value for one thickness of wire mesh or perforated screen and Z is the correction for $R_e < 400$. To calculate flow resistance of several sheets, just add the K-factors. Typical values are plotted in Fig. 9.11.

For fine filters (fiber or sintered metal), the Reynolds number is meaningless because the passages are too small to be considered channels. Pressure drop in psi can be estimated with this empirical relationship:

$$\Delta p = \frac{Q_{in.^3} C \mu W}{A_2}$$

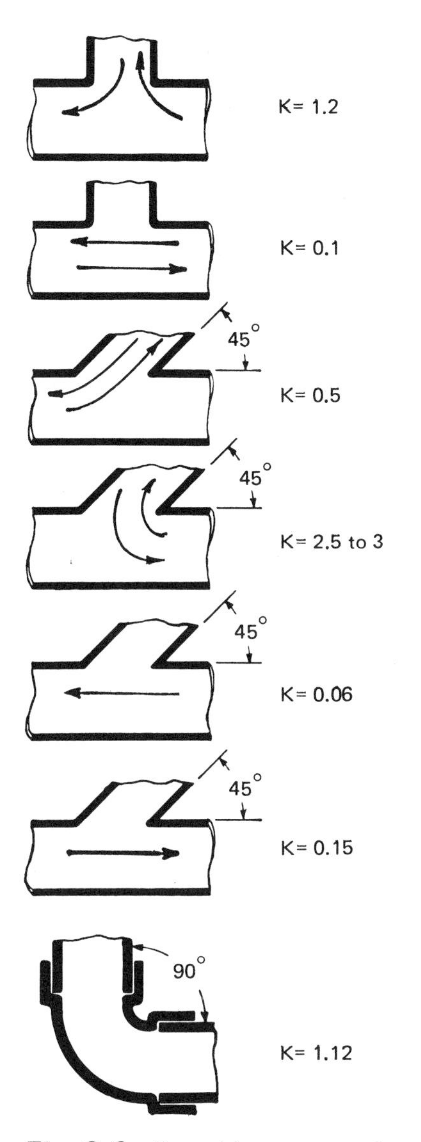

Fig. 9.6 Branching pressure losses (K-factors).

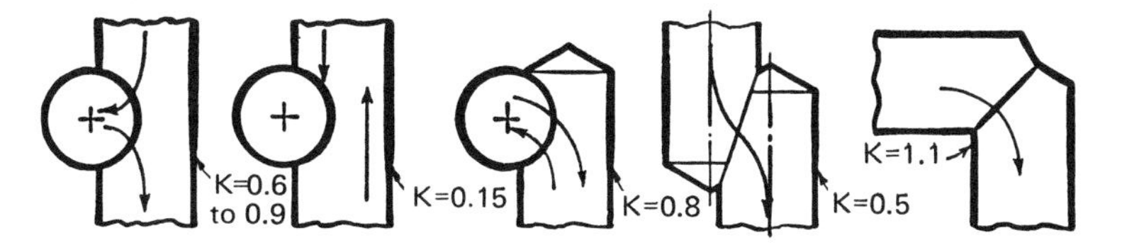

Fig. 9.7 Pressure losses in intersecting passages (K-factors).

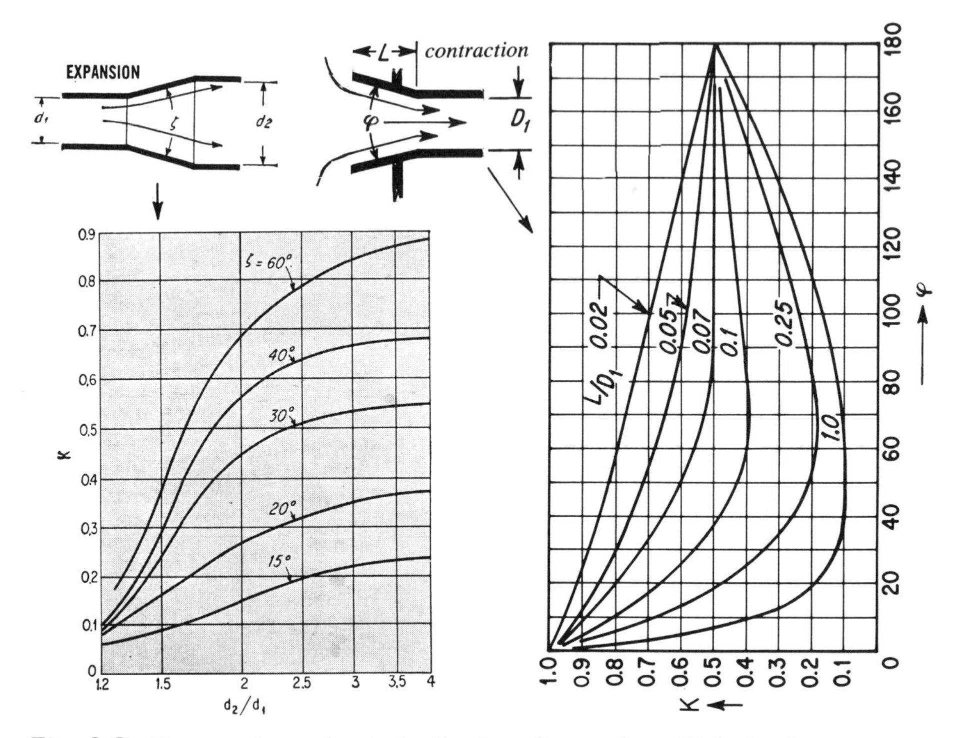

Fig. 9.8 Pressure losses in gradually changing sections (K-factors).

where W = wall thickness in direction of flow, in., and C is a constant peculiar to the filter element. The filter manufacturer runs tests to establish C. Typical values for a 40-micron (nominal) element of impregnated cellulose fiber are as follows:

- For SAE 10 oil at 100 F: $C = 1.22 \times 10^6$
- For SAE 20 oil at 100 F: $C = 1.28 \times 10^6$
- For SAE 30 oil at 100 F: $C = 1.36 \times 10^6$
- For SAE 40 oil at 100 F: $C = 1.48 \times 10^6$

Add 25% to the pressure loss to include the housing, because the C value pertains to the element only. And remember that Δp increases as the filter clogs, and eventually the entire pressure drop occurs across the element.

Validity of Tests

All the K values in this chapter are based on actual tests. The danger is not that the K values are wrong but that they are often misapplied. For one thing, velocity V usually is average velocity, $\frac{1}{2}(V_1 + V_2)$ but sometimes is inlet or outlet velocity, as in Figs. 9.8 and 9.9. Furthermore, slight differences in inlet contour or finish can alter K threefold (Fig. 9.5).

FLUID THROTTLING DEVICES

Planned (and, we assume, desirable) pressure drops include those through orifices, nozzles, control valves, capillaries, clearances, and porous plugs (not filters). See Figs.

9.12 through 9.23 and Tables 9.9 and 9.10 (nomenclature). The coefficients are variable when the restriction is a manipulated one.

Orifices and Nozzles

More is known about orifices and nozzles—perfect ones, that is—than about any other planned restriction in a fluid-power system. ASME specifications describe every dimension, and the flow coefficients are accurate within a few percent. The basic relationship is

$$Q = AC\sqrt{2gH_L} \tag{9.14}$$

where Q = flow, ft³/sec; A = orifice or nozzle area, ft²; H_L = head loss, ft; and C = the flow coefficient, plotted in Figure 9.12. If you want a precise ΔP for a given flow, or vice versa, that's the way to get it. Note that oil temperature and viscosity are not factors.

Not all orifices and nozzles are perfect, however, and many have nonstandard dimensions to solve special problems. Here are the principal types, perfect and otherwise:

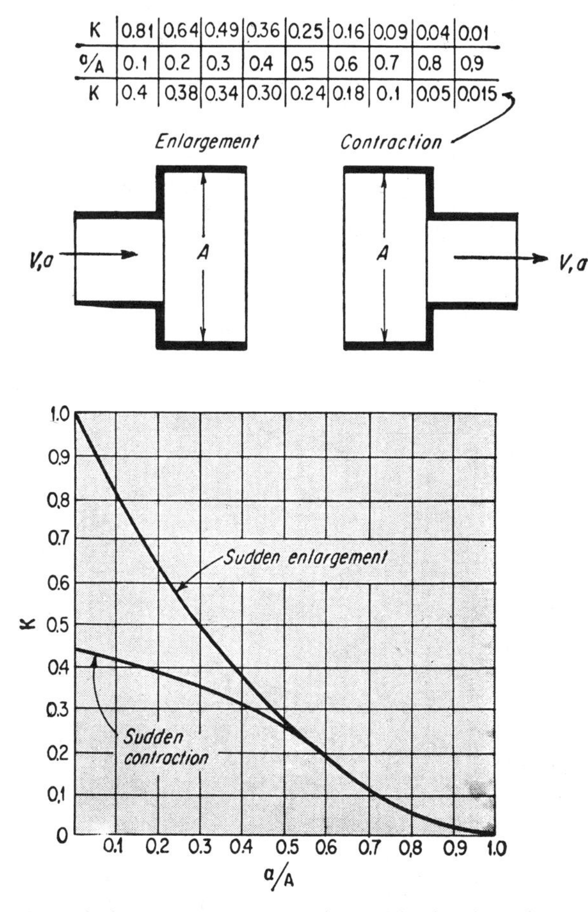

K	0.81	0.64	0.49	0.36	0.25	0.16	0.09	0.04	0.01
a/A	0.1	0.2	0.3	0.4	0.5	0.6	0.7	0.8	0.9
K	0.4	0.38	0.34	0.30	0.24	0.18	0.1	0.05	0.015

Fig. 9.9 Pressure losses in suddenly changing sections (K-factors).

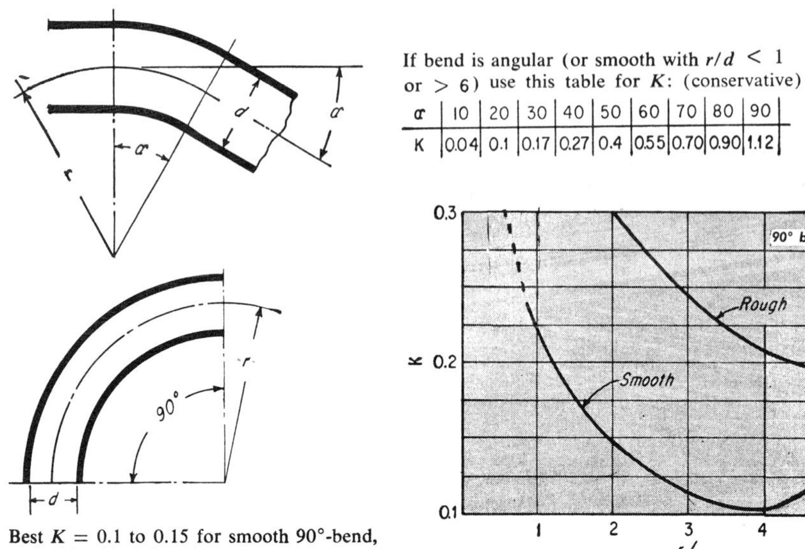

If bend is angular (or smooth with $r/d < 1$ or > 6) use this table for K: (conservative)

α	10	20	30	40	50	60	70	80	90
K	0.04	0.1	0.17	0.27	0.4	0.55	0.70	0.90	1.12

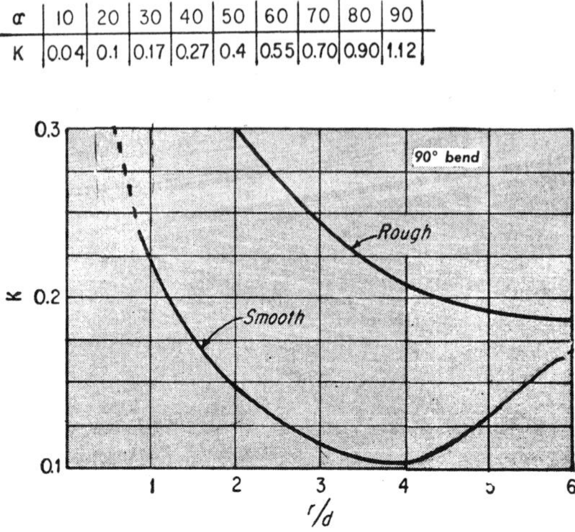

Best $K = 0.1$ to 0.15 for smooth 90°-bend, and r/d at ideal 2 to 4

Fig. 9.10 Pressure losses in bends (K-factors).

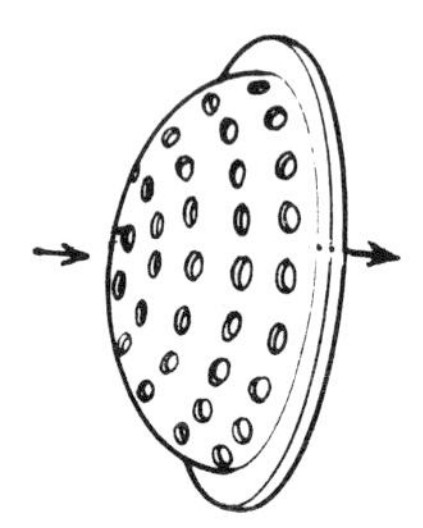

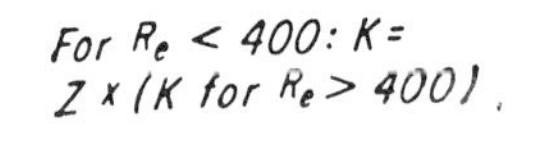

For $R_e < 400$: $K = Z \times (K \text{ for } R_e > 400)$.

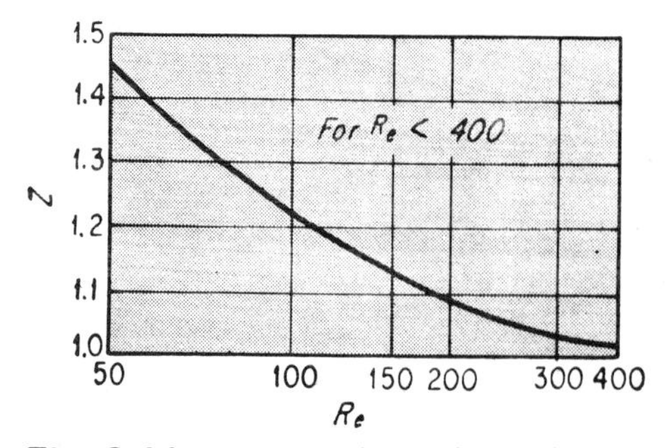

Fig. 9.11 Pressure losses in strainers and coarse filters (K-factors).

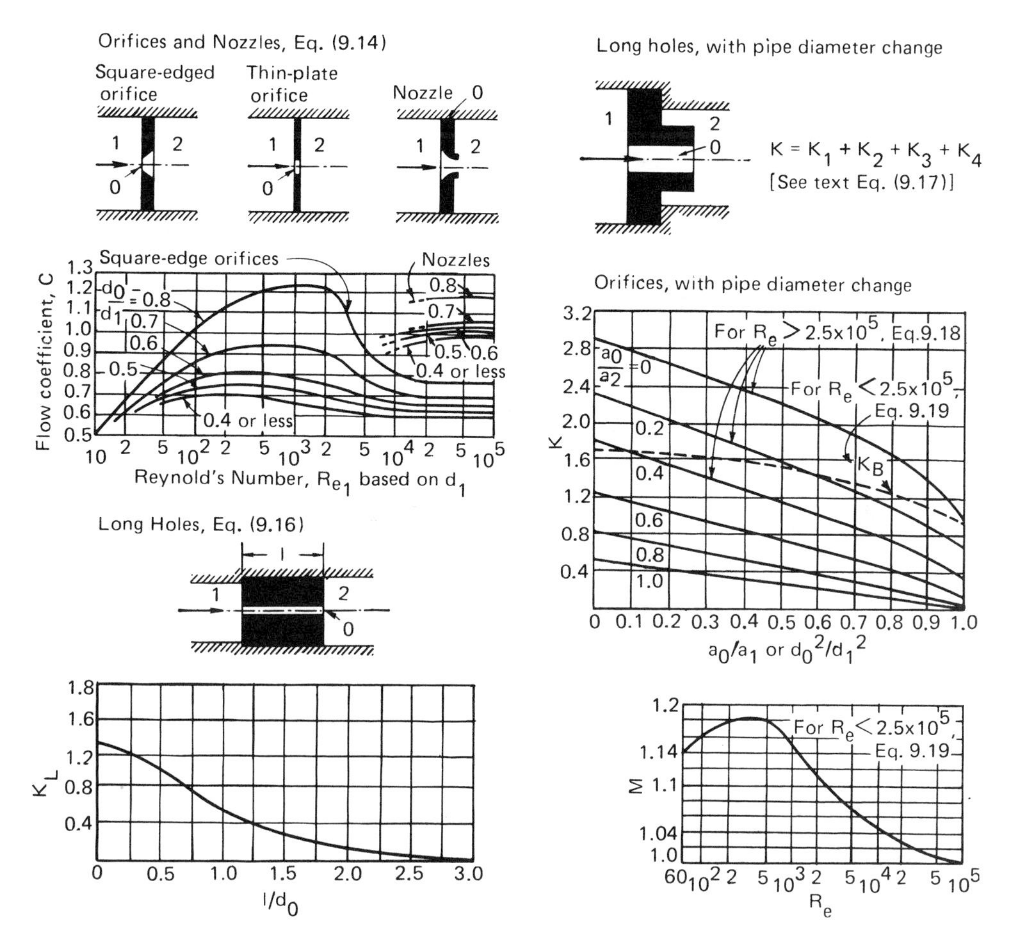

Fig. 9.12 Controlled pressure drops in orifices and nozzles (K-factors and flow coefficients).

- Square-edged orifice (ASME standard)
- Drilled hole in thin plate
- Contoured nozzle
- Drilled hole in plug or thick plate
- Orifice plus change in pipe diameter
- Punched hole or other crude opening

The square-edged orifice flow coefficient is fairly well defined. The dimensions must be sharp and true, however. A square-edged hole in a thin plate has nearly the same flow characteristics as a square-edged orifice. Contoured nozzles smooth out the flow at high Reynolds numbers and therefore have higher flow coefficients than orifices. They are as accurate as square-edged orifices but more expensive. Long orifices such as drilled plugs or plates introduce wall friction losses, and these must be accounted for in any expression for pressure drop.

Punched, cast, or rough-drilled orifices and other crude openings are useful flow

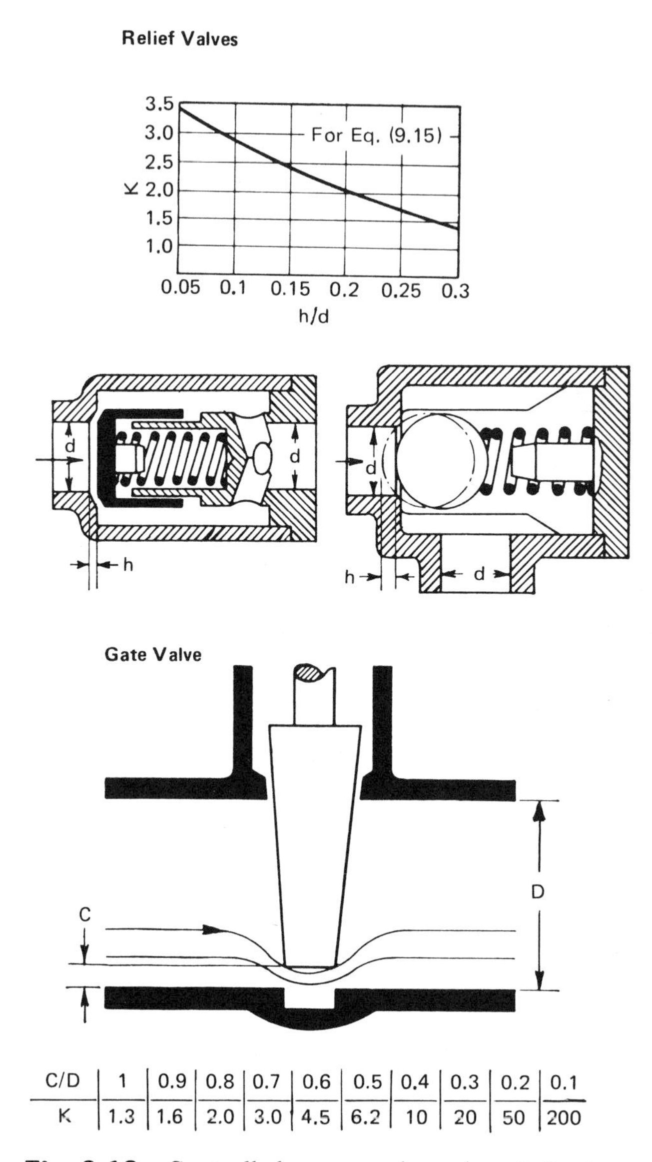

C/D	1	0.9	0.8	0.7	0.6	0.5	0.4	0.3	0.2	0.1
K	1.3	1.6	2.0	3.0	4.5	6.2	10	20	50	200

Fig. 9.13 Controlled pressure drops in relief valves and gate valves (K-factors).

restrictions, but flow might easily differ 20% from what you calculate based on drill or punch size.

Take a 0.25-in. drilled orifice, for example. A 0.01-in. error in the assumed diameter gives an 8% difference in area; a 0.1 error in an assumed 0.8 flow coefficient is equivalent to a 12% change in flow; and any rounding of the hole entrance might boost flow another 10 or 20% because of the nozzle effect.

Where you are not sure about the edge sharpness of a drilled hole, compromise by using a flow coefficient midway between that plotted for orifices and nozzles.

Special Coefficient K. Although flow coefficient C for orifices and nozzles [Eq. (9.14)] is the conventional one to use, expressions involving coefficient K (the K-factor) are available too. These are convenient when other calculations in your hydraulic system also involve K. The general expression, applicable in Figs. 9.12 to 9.15 is

$$H_L = K\frac{V^2}{2g} \tag{9.15}$$

where V is the average velocity, ft/sec, of the fluid just before and just after the restriction, and H_L is head loss in ft. More details were given in Fig. 9.6.

The K-factor is awkward to use if the inlet and outlet have greatly different flow areas or if they are irregular in shape. Assuming a value of K, just which velocity do you select for calculating pressure loss? (Text continues on pg. 231.)

Butterfly Valve

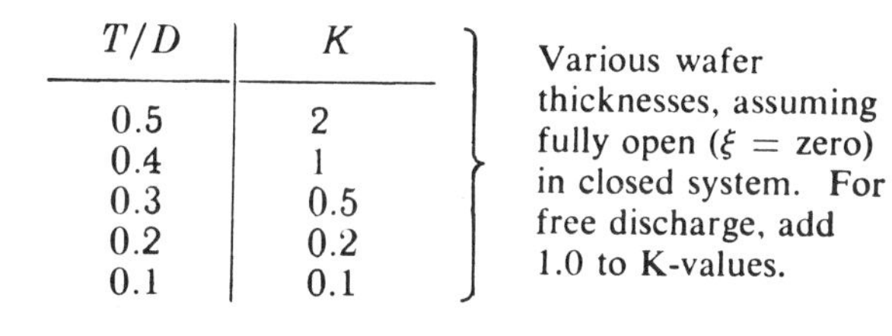

T/D	K
0.5	2
0.4	1
0.3	0.5
0.2	0.2
0.1	0.1

Various wafer thicknesses, assuming fully open (ξ = zero) in closed system. For free discharge, add 1.0 to K-values.

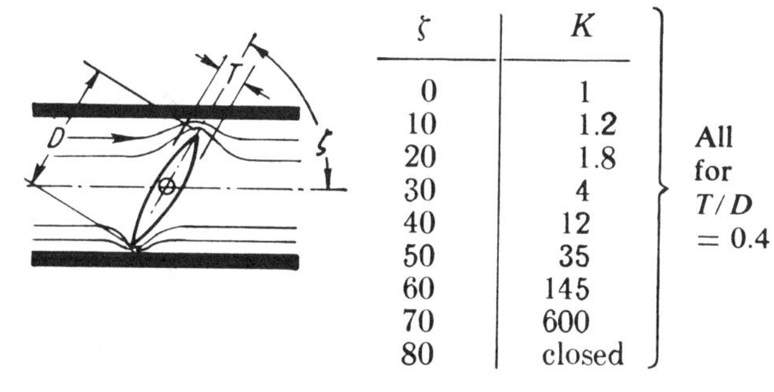

ζ	K
0	1
10	1.2
20	1.8
30	4
40	12
50	35
60	145
70	600
80	closed

All for $T/D = 0.4$

Ball Valve

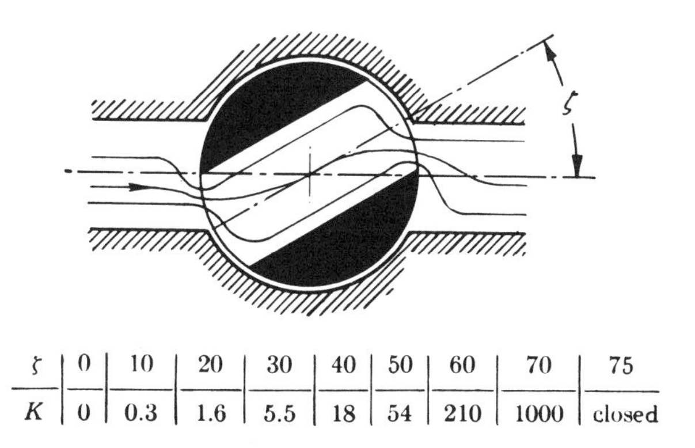

ζ	0	10	20	30	40	50	60	70	75
K	0	0.3	1.6	5.5	18	54	210	1000	closed

Fig. 9.14 Controlled pressure drops in butterfly valves and ball valves (K-factors).

Disk Valve

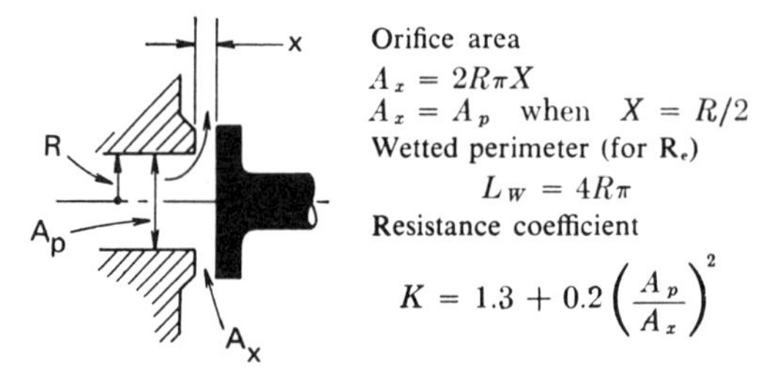

Orifice area

$A_x = 2R\pi X$

$A_x = A_p$ when $X = R/2$

Wetted perimeter (for R_e)

$$L_W = 4R\pi$$

Resistance coefficient

$$K = 1.3 + 0.2\left(\frac{A_p}{A_x}\right)^2$$

Needle Valve

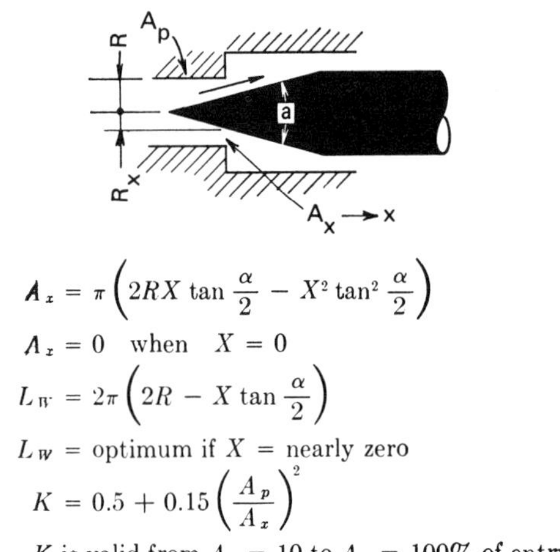

$$A_x = \pi\left(2RX \tan\frac{\alpha}{2} - X^2 \tan^2\frac{\alpha}{2}\right)$$

$A_x = 0$ when $X = 0$

$$L_W = 2\pi\left(2R - X \tan\frac{\alpha}{2}\right)$$

L_W = optimum if X = nearly zero

$$K = 0.5 + 0.15\left(\frac{A_p}{A_x}\right)^2$$

K is valid from $A_x = 10$ to $A_x = 100\%$ of entry area, A_p

Simple Ball Valve

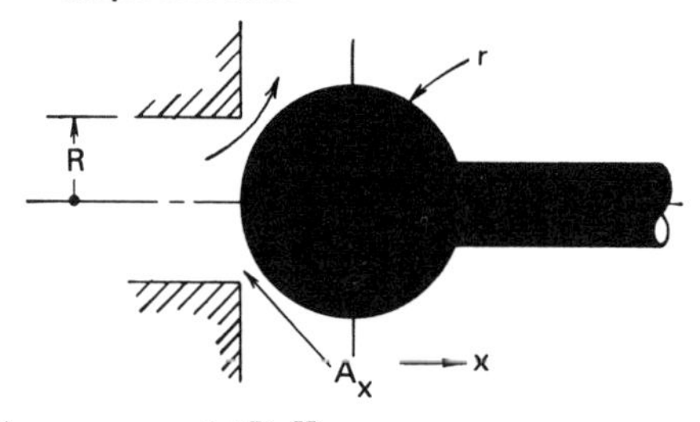

A_x = approx. $1.5R\pi X$

A_x is valid when r = approx. $1.3R$

L_W = approx. $4\pi R$

$$K = 0.5 + 0.15\left(\frac{A_p}{A_x}\right)^2$$

Fig. 9.15 Controlled pressure drops in disc valves, needle valves, and ball valves (K-factors).

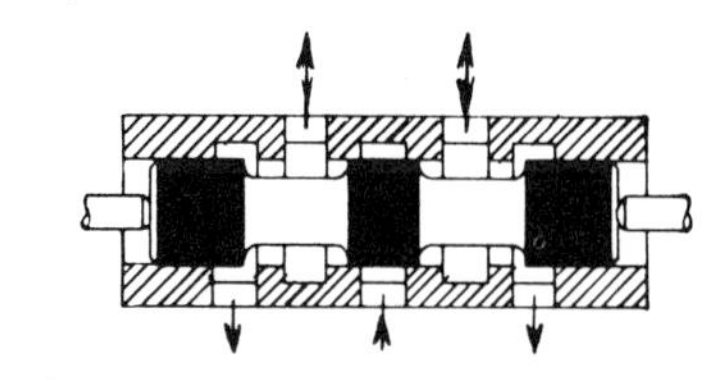

K = 3 to 5.5 at fully open position

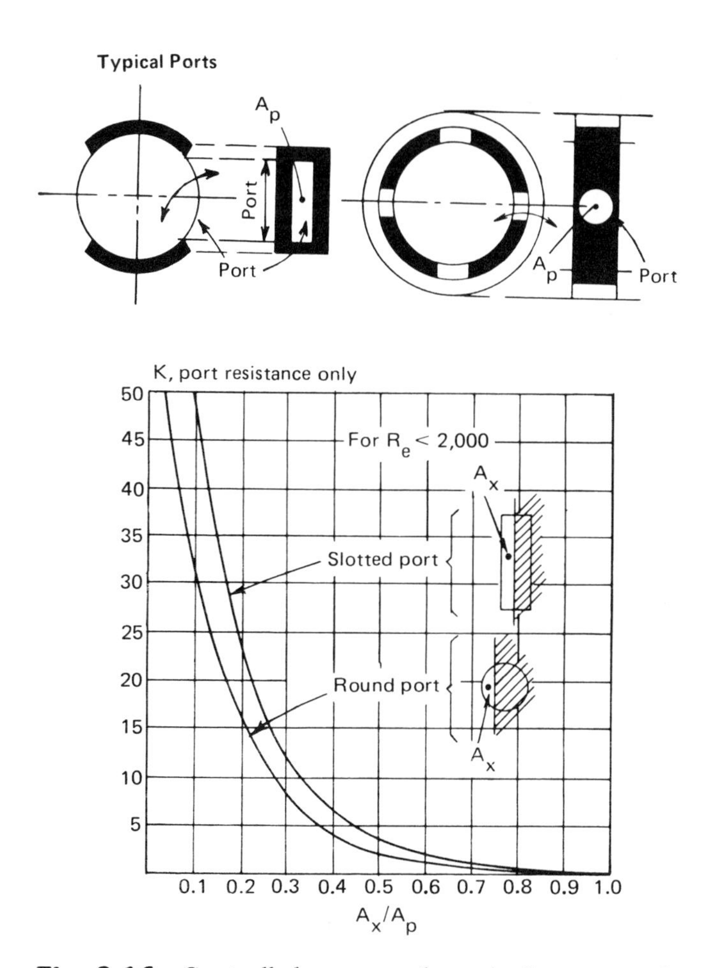

Fig. 9.16 Controlled pressure drops in four-way valves and ports (K-factors).

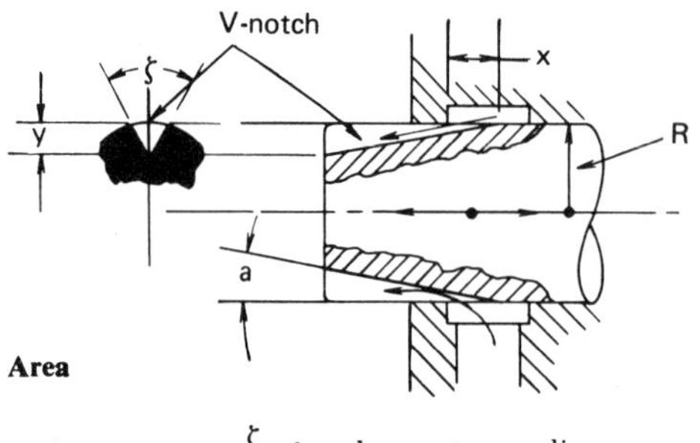

Area

$$A_z = \text{approx. } \frac{\zeta}{2} y^2 \quad \text{where} \quad \zeta = \text{radians}$$

$$y = X \tan \alpha \quad \text{when} \quad \zeta = \frac{\pi}{3} = 60°$$

$$A_z = N \frac{\pi}{6} X^2 \tan^2 \alpha$$

N = number of notches

Wetted perimeter (for R_e)

$$L_W = \pi y \frac{\zeta°}{180°} + 2y$$

$$L_W = (3.05 \tan \alpha) X \quad \text{if} \quad \zeta = 60°$$

Resistance coefficient

K = approx. $400/R_e$ for $R_e < 150$
K = approx. $10/R_e^{0.25}$ for R_e = 150 to 2000

Fig. 9.17 Controlled pressure drops in plunger notches (K-factors).

Table 9.9. Flow Equations for Simple Orifices and Nozzles

Units		Equations based on flow coefficient, C	
Flow	Pressure drop	Flow	Pressure drop (measured across flange taps)
$\frac{ft^3}{sec}$	ft	$Q = d_o^2 C \sqrt{H_L}\, 4.38 \times 10^{-2}$	$H_L = \frac{Q^2}{d_o^4 C^2} 5.21 \times 10^2$
$\frac{lbs}{sec}$	psi	$W = d_o^2 C \sqrt{\gamma \Delta p}\, 5.25 \times 10^{-1}$	$\Delta p = \frac{W^2}{d_o^4 C^2 \gamma} 3.63$
gpm	psi	$q = d_o^2 C \sqrt{\frac{\Delta p}{\gamma}}\, 2.36 \times 10^2$	$\Delta p = \frac{q^2 \gamma}{d_o^4 C^2} 1.8 \times 10^{-5}$

(Courtesy of Crane Co., New York.)

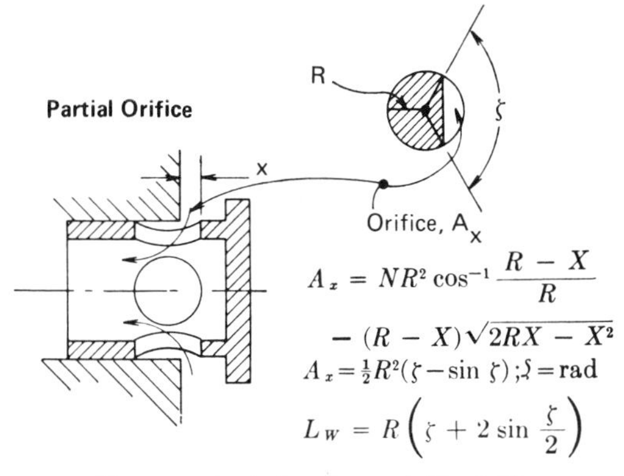

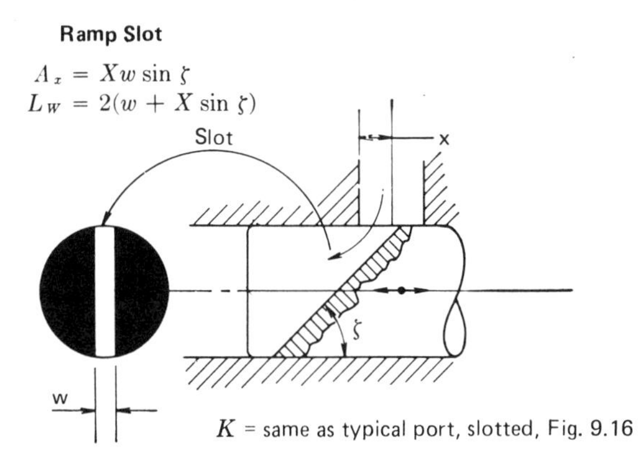

Fig. 9.18 Controlled pressure drops in partial orifices and slots (K-factors).

Velocity for K. In general practice, K is based on the average velocity measured through the restriction. Specifically, note the following.

For orifices and holes, except the nozzles and sharp-edged orifices covered by flow coefficient C, the velocity V is measured through the orifice area A_O.

For valves, and for slots and notches, velocity V is approach velocity, based on inlet port area.

Equation (9.15) applies in every case. Here are the special derivations of graphs and constants used, starting with orifices.

Long-hole Orifices. Friction coefficient f creeps into the calculations when the orifices have significant length (high L/D). If D_O is small in relation to D_1 and D_2, this handy expression for K will suffice:

$$K = 1.5 + K_L + f\frac{L}{D_O} \tag{9.16}$$

where f = the friction factor and K_L is a correction factor based on the L/D_O ratio, plotted in Fig. 9.12.

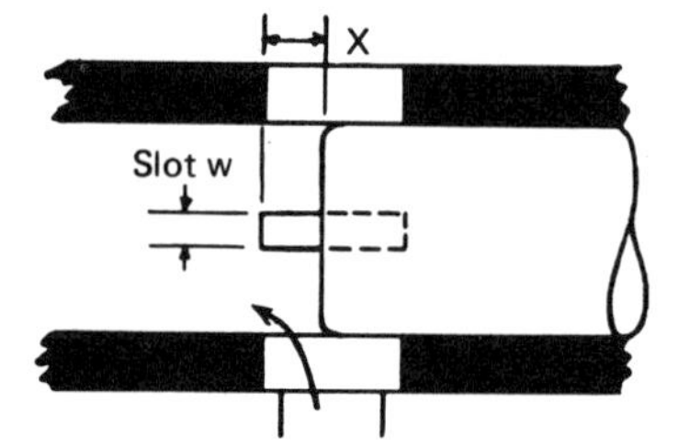

$A_x = NwX$ where N = number of ports
$L_W = 2N\,(w + X)$
K = same as for typical port, slotted

Intersecting Holes

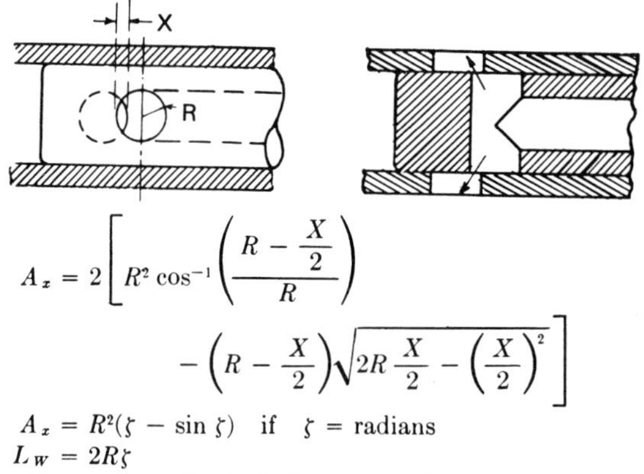

$$A_x = 2\left[R^2 \cos^{-1}\left(\frac{R - \frac{X}{2}}{R}\right) - \left(R - \frac{X}{2}\right)\sqrt{2R\frac{X}{2} - \left(\frac{X}{2}\right)^2}\,\right]$$

$A_x = R^2(\zeta - \sin\zeta)$ if ζ = radians
$L_W = 2R\zeta$
K = same as for typical port, round

Fig. 9.19 Controlled pressure drops in slotted sleeves and intersecting holes (K-factors).

Long Hole, with Pipe Diameter Change. This more general case includes effects of upstream and downstream pipes as well as conventional losses:

Loss at entrance:
$K_1 = 0.5(1 - D_O^2/D_1^2)$
Contraction:
$K_2 = \sqrt{1 - D_O^2/D_1^2} \times K_L(1 - D_O^2/D_2^2)$
Enlargement:
$K_3 = (1 - D_O^2/D_2^2)^2$
Length:
$K_4 = fL/D_O$

The ratio D_O^2/D_1^2 is of course identical with A_O/A_1. The combined terms are

$$K = K_1 + K_2 + K_3 + K_4 \tag{9.17}$$

Note that if the D_O^2/D_1^2 and D_O^2/D_2^2 terms are close to zero, the composite $K = 0.5 + K_L + 1.0 + fL/D_O$, which is identical to Eq. (9.16).

Rotary Slot

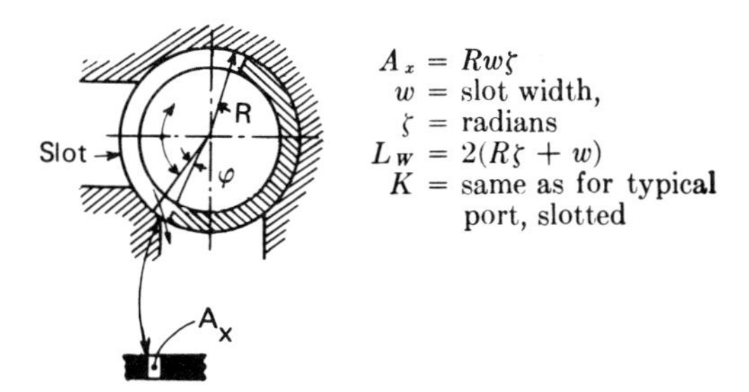

$A_x = Rw\zeta$
w = slot width,
ζ = radians
$L_W = 2(R\zeta + w)$
K = same as for typical port, slotted

Rotary Wedge

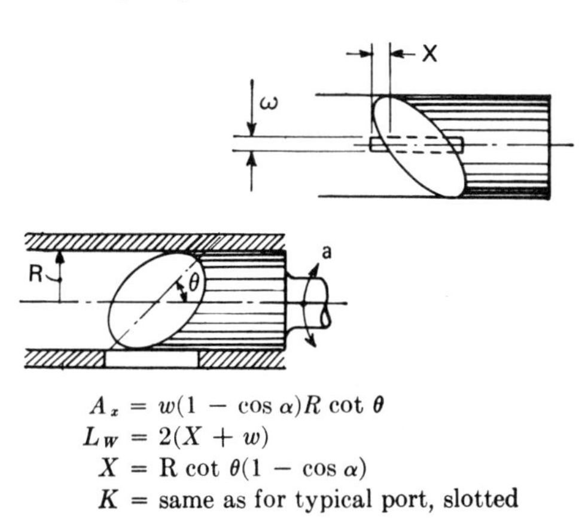

$A_x = w(1 - \cos\alpha)R \cot\theta$
$L_W = 2(X + w)$
$X = \mathrm{R} \cot\theta(1 - \cos\alpha)$
K = same as for typical port, slotted

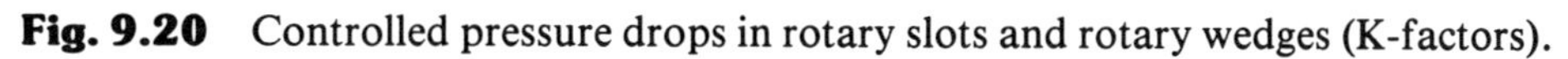
Fig. 9.20 Controlled pressure drops in rotary slots and rotary wedges (K-factors).

Table 9.10 Nomenclature for Pressure Drop Equations for Orifices and Nozzles

a, A	= cross-sectional flow area, in.^{2}, ft^2
C	= flow coefficient
C_{dia}	= diametral clearance
C_{rad}	= radial clearance
d, D	= diameter: in., ft
f	= friction factor, dimensionless
g	= acceleration of gravity, 32.2 ft/sec^2
H_L	= head loss, ft
K	= K-factor, dimensionless
l, L	= length, in., ft
L_W	= wetted perimeter, ft
$\Delta p, P$	= pressure drop: psi, psf
q	= flow, gpm
Q, Q_{in}	= flow: ft^3/sec, in.^{3}/sec
R_e	= Reynold's Number, $\rho DV/\mu$ for round holes and $\frac{4\rho VR_H}{\mu}$ for non-circular openings.
R_H	= hydraulic radius, ft $R_H = A/L_W$
V	= velocity, ft/sec
W	= weight flow, lb/sec
w	= width of slot
ρ	= mass density, lb-sec^2/ft^4
γ	= weight density, lb/ft^3
μ	= absolute viscosity, lb-sec/ft^2
ν	= $\frac{\mu}{\rho}$, kinematic viscosity, ft^2/sec

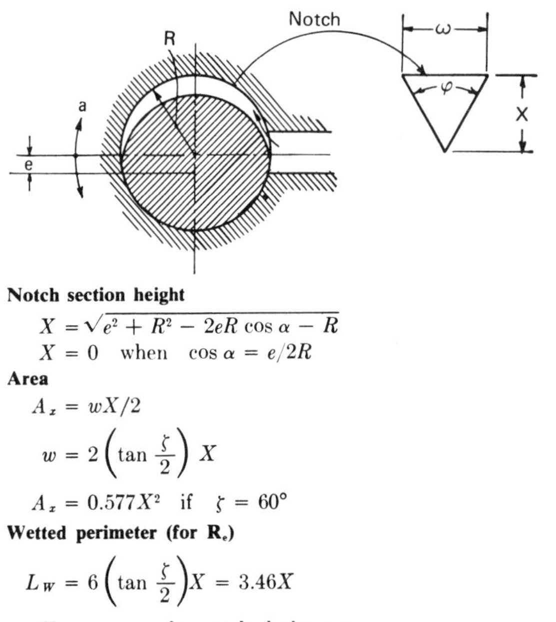

Notch section height

$$X = \sqrt{e^2 + R^2 - 2eR\cos\alpha} - R$$

$$X = 0 \quad \text{when} \quad \cos\alpha = e/2R$$

Area

$$A_z = wX/2$$

$$w = 2\left(\tan\frac{\zeta}{2}\right)X$$

$$A_z = 0.577X^2 \quad \text{if} \quad \zeta = 60^\circ$$

Wetted perimeter (for R_e)

$$L_W = 6\left(\tan\frac{\zeta}{2}\right)X = 3.46X$$

$$K = \text{same as for notched plunger}$$

Fig. 9.21 Controlled pressure drops in rotary notches (K-factors).

Orifices, with Pipe Diameter Change. Overall K-factors are plotted in Fig. 9.12 for various values of D_O^2/D_2^2. The equations are as follows:

When $R_e > 2.5 \times 10^5$:

$$K = (0.707\sqrt{1 - D_O^2/D_1^2} + 1 - D_O^2/D_2^2)^2 \tag{9.18}$$

When $R_e < 2.5 \times 10^5$:

$$K = (K_B/M - D_O^2/D_2^2)^2 \tag{9.19}$$

where K_B and M are indicated on the graphs.

Special cases of Eqs. (9.18) and (9.19) are when D_O^2/D_1^2 and D_O^2/D_2^2 are less than 0.1. Then $K = 2.9$ for $R_e > 2.5 \times 10^5$ and $K = 2.9/M^2$ for $R_e < 2.5 \times 10^5$.

Valves and Ports

The purpose of any flow-control valve is to introduce a flow restriction. The sketched examples (Figs. 9.13 through 9.21) cover almost every type of moving element and port.

The pressure-loss equations are based on the K-factor, Eq. (9.15). The values for K are average ones based on valve sizes from ⅜ to 1½ in. Pertinent test conditions such as range of R_e are indicated on the graphs. Where no conditions are marked, flow can be laminar or turbulent.

For each type of restriction, the pressure loss is for the restriction alone and does not include flow resistances of the valve passages preceding and following the restriction.

Conventional Valves. Relief valves, ball-checks, butterfly valves, four-way spools, ball valves, gate valves, and needle valves are typical valve restrictions. These are accompanied by K-factor graphs for a wide range of openings.

The poppets of spring-loaded relief valves vary their position with flow, and the K-factor varies inversely with the amount of the opening (Fig. 9.13). Just what the final position is for any given flow depends on jet pressures, spring gradient, back pressure, and velocity conversions.

Four-way spools have two ports in series. Throttled flow is in accordance with Fig. 9.16. Wide-open flow is restricted only by the ports, and the K-factor typically is between 3 and 5.5. Closed flow, meaning leakage through clearances, is covered in the section on clearances.

Slots and segments: Even round holes have complicated expressions for flow if they are intersected by a plunger edge or another round port.

The K-factor for any opening is easily estimated. First calculate the cross-section area of the opening, using the equations accompanying each sketch (Figs. 9.17 through 9.21).

Then read the graph of K-factors for slotted and round ports (Fig. 9.16). If the opening is hard to define, take the average K between the slotted and round ports.

Clearances, Capillaries, and Pores. Laminar flow through an annular clearance (Fig. 9.22) varies as the first power of pressure drop but as the third power of the clearance. Thus a 0.001-in. diametral clearance (a normal allowance) passes eight times as much oil as a 0.0005-in. clearance (a close fit) and 1000 times as much oil as a 0.0001-in. clearance (a lapped fit). The latter must be carefully tested under actual operating conditions if you are to be sure that thermal differential expansion doesn't close up the clearance and hang up the plunger. Or you can exploit the thermal expansion to control flow.

K-factors don't apply to clearances (or to any other laminar-type restriction) because Δp is not a function of V^2. The expression covering flow is

$$Q_{in} = \frac{(C_{rad})^3 \ \Delta p d_M \pi}{12 \mu_M l} \tag{9.20}$$

where Q_{in} = flow, in.^{3}/sec; C_{rad} = radial clearance, in.; Δp = psi; d_M/l = ratio of clearance's mean diameter to land length; and μ_M = mean absolute viscosity, lb-sec/in.^{2} The companion expression for Δp is

$$\Delta p = \frac{12 Q_{in} \mu_M l}{(C_{rad})^3 d_M \pi}$$

(A specific example—for the land of a spool valve—is covered in a following section. It includes the effect of an off-center spool.)

Capillaries and small-diameter tubing (Fig. 9.22) have similar expressions for laminar flow:

$$Q_{in} = \frac{d^4 \ \Delta p}{\mu l} 2.45 \times 10^{-2}$$

$$\Delta p = \frac{Q_{in} \mu l}{d^4} 40.75$$

where d is tube inside diameter, in. A variant is a male screw thread in a smooth bore, which forces the fluid to follow a helix.

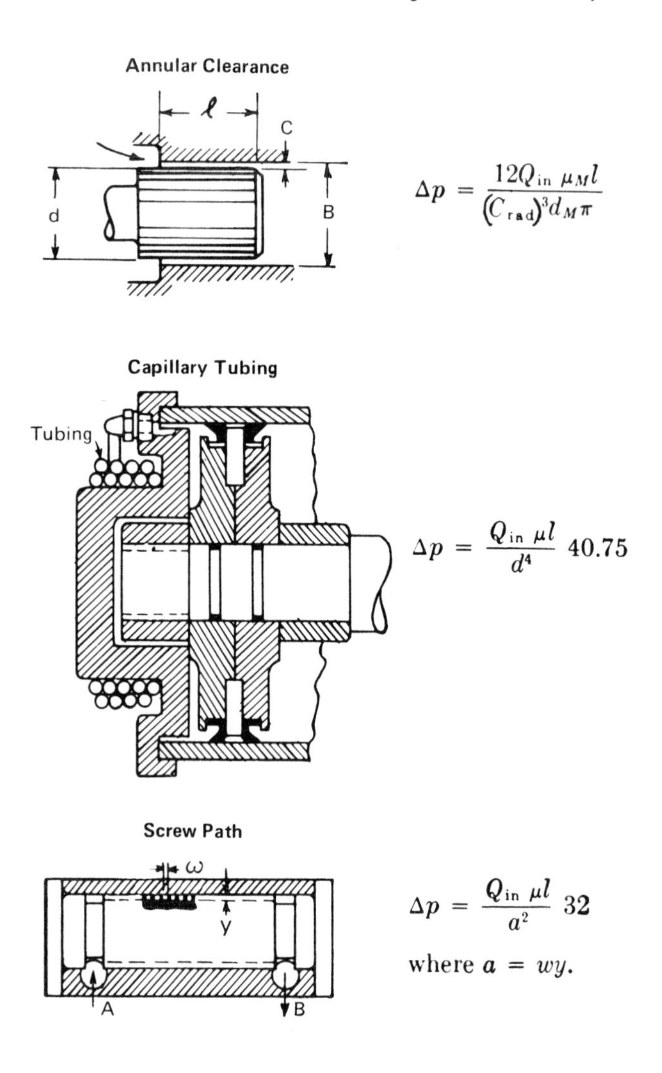

Fig. 9.22 Controlled pressure drops in tightly restricted paths (ΔP). Check Table 9.1 for proper units.

One advantage of using capillaries and annular clearances for flow control is that for any given oil temperature, flow is directly proportional to pressure drop. And oil temperature tends to stay low and constant because the surface-to-volume ratio is favorable for dissipating the heat to atmosphere. Some flow meters are based on capillaries and have bundles of small-diameter tubes in parallel to carry the necessary flow.

Porous plugs of sintered metal (Fig. 9.23) and metal tubes filled with tiny round beads are also used as flow throttlers. The object in each instance is to provide many flow passages and obstructions, assuring laminar flow. A convenient linear relationship between flow and pressure drop is thus achieved. A general expression for pressure drop through a porous plug is

$$\Delta p = CQ_{in}\mu \frac{1}{a}$$

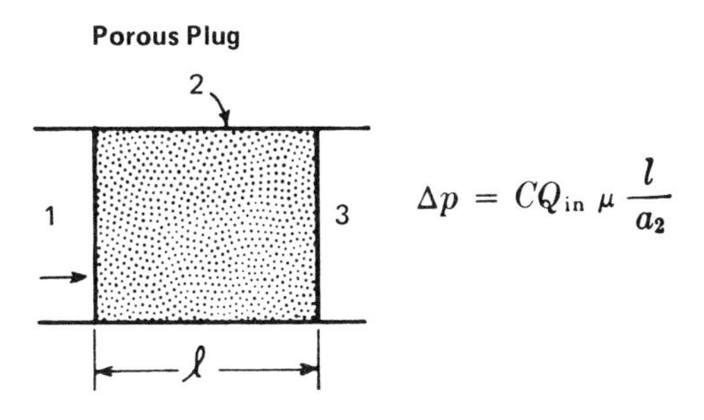

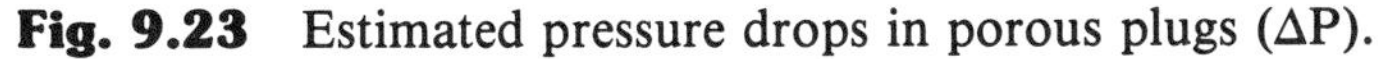
Fig. 9.23 Estimated pressure drops in porous plugs (ΔP).

where C is a constant, different for each material; l = plug length, in.; and a = plug cross-section area, in.[2] Typical value: $C = 1.3 \times 10^6$ for a 40-micron cellulose filter, passing SAE 20 oil at 100 F.

Flow Coefficient C_V

A very common flow coefficient, covered in detail in Chapter 16 (air flow equations) is C_V. Valve manufacturers in particular favor C_V because it is an easy coefficient to assign by test and is readily convertible for use in airflow equations.

Based on water flow, the relationship is $\text{gpm} = C_V\sqrt{\Delta p}$, where p = psi. A restriction has a flow coefficient $C_V = 1.0$ if it will pass 1 gpm of water with a pressure drop of 1 psi.

The C_V coefficients can be combined when valves are in series or parallel (Fig. 9.24). See Chap. 16 for details. The technique is the same as adding electrical resistances in series or parallel.

Flow coefficient C_V can be converted roughly to the K-factor with this expression:

$$K = \frac{1460a^2}{C_V^2}$$

where a = flow area, in.[2]

ERRORLESS ORIFICES

It would be a lot easier to control flow in an orifice or nozzle if the discharge coefficient would hold still. Even with handy charts it's annoying to have to adjust a constant every time the flow or viscosity changes.

What's the problem? Basically, the discharge coefficient of a typical sharp-edge orifice (in a relatively large conduit) drops from about 0.7 to about 0.6 when the Reynolds number increases from 100 to 10,000. Conversely, the discharge coefficient for a typical rounded orifice or nozzle rises from about 0.7 to about 0.9 in the same range. No standard orifice stays at one value.

Answer: Two heads are better than one. Simply by combining a sharp-edge orifice in series or parallel with a rounded orifice (or cone, or nozzle) you practically can cancel out the variation. Fig. 9.25 shows examples of what improvements can be made.

Just putting any sharp-edge orifice in series or parallel with any rounded orifice is not exactly the answer—although it is bound to help. For really constant discharge

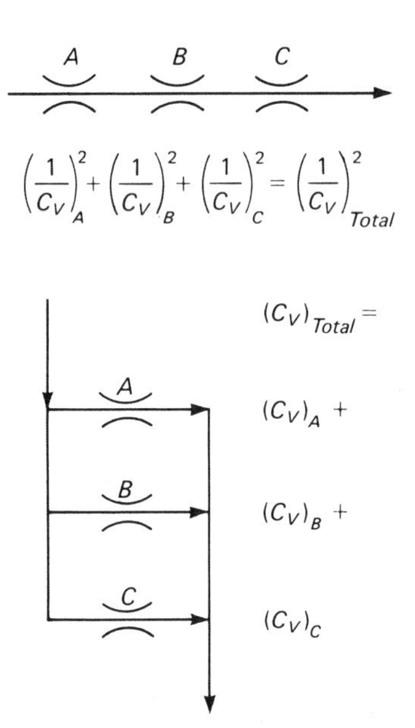

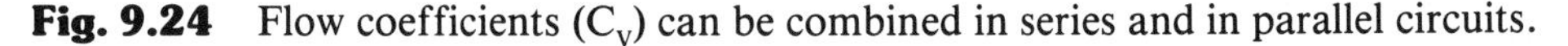

Fig. 9.24 Flow coefficients (C_V) can be combined in series and in parallel circuits.

coefficients you must find the right combination. Fortunately, it can be discovered by calculation using the techniques that follow.

Sizing the Orifices

For best results, limit the study to a few known types. Beveled-inlet orifices are fairly easy to make and have the right characteristics. The values of discharge coefficient C for most sharp-edge orifices are fairly consistent for all sizes, as long as the orifice diameter is small relative to the inlet channel diameter. Above an orifice-channel diameter ratio of 1 : 2, C will tend to be higher for the same Reynolds number. So to be safe, stay below a diameter ratio of 1 : 3 where almost all size effects, including those of inlet and discharge velocity changes caused by crowding, will be eliminated—to within 1% or so.

The first step mathematically is to either set the flows equal if the assumed orifices are in series or set the sum of the flows equal to the main flow if the orifices are in parallel.

If you aim for constant discharge coefficients (within a few percent), then to prove your results you must be able to *measure* the flow with even better accuracy.

LEAKAGE THROUGH ANNULAR CLEARANCES

Suppose you wanted to estimate the leakage in the clearance along the lands of a plunger or spool that fits closely in a sleeve (Fig. 9.26). None of the normal turbulent or transition flow equations apply because the clearance is so small that viscous (laminar) flow prevails . . . as in a capillary.

The first step is to dig up the classic formula for the laminar flow between two close parallel plates of infinite width and adapt it somehow to the annular clearance between a spool and sleeve. The adaption turns out to be not only valid but reason-

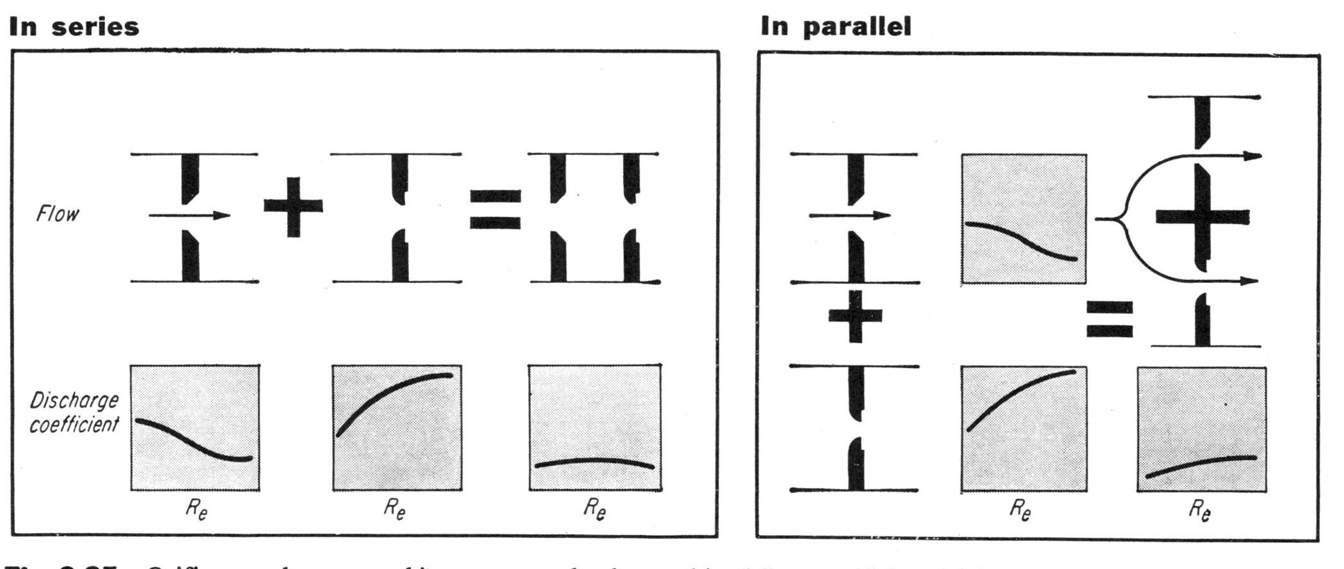

Fig. 9.25 Orifices can be arranged in ways to make the combined flow coefficient fairly constant.

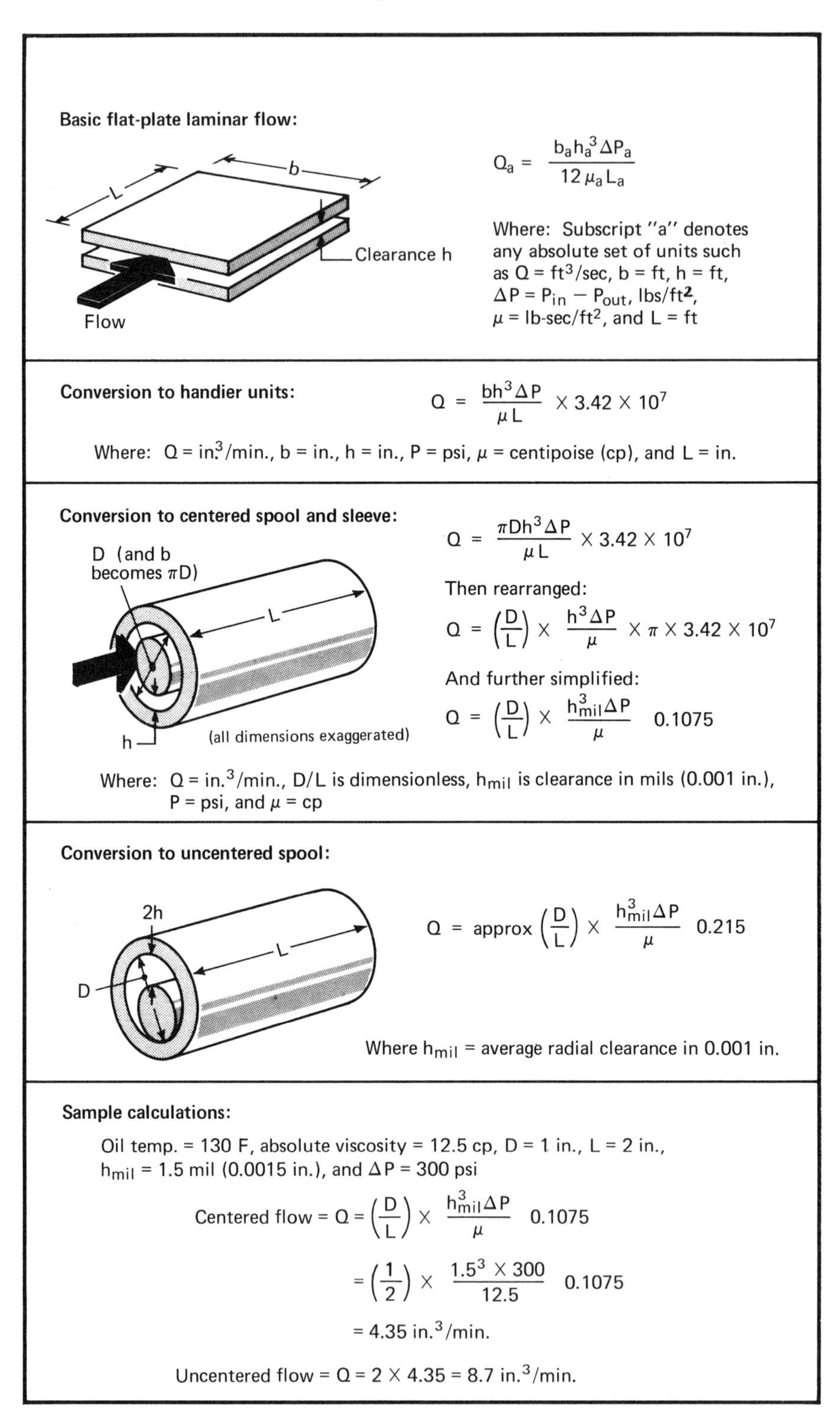

Fig. 9.26 Leakage through spool-to-sleeve clearances.

ably accurate. The clearance we're talking about is in thousandths of an inch, or smaller, which is so small relative to the diameter of the spool that the spool curvature has negligible effect. Sleeve ID is used for all diametral dimensions.

Spool off center: If the spool is assumed centered in the sleeve, there is no further problem other than careful conversion of units. However, the typical spool will be pushed by pressure to one side or the other, and a large increase in flow occurs. Why the increase? Because the laminar flow rate is proportional to the cube of the clearance, and the effect of increasing clearance on one side has much more effect proportionately than the reducing of clearance on the other side. Uncentered flow will be up to 2.5 times centered flow.

The off-centered condition assumed for the accompanying equations and graphs is not the theoretical maximum but allows for the fact that metal-to-metal contact is never achieved, and the spool is never full off-centered. An arbitrary multiplier of 2.0, rather than 2.5, is chosen as the ratio of off-centered flow to centered flow. It's certainly no less accurate than the measurement of clearance and symmetry of the spool.

A convenient graph is Q vs. h for a family of D/L curves, all at an assumed constant value for oil temperature and pressure drop. For a different assumption, use the same curves and just change the scale for Q.

RESPONSE TIME AND FLOW OF A PILOT-OPERATED VALVE

A familiar sight in almost any hydraulic automatic control system is a four-way flow-control valve with a solenoid pilot valve perched on top. There are more of these four-way valves than all other remotely operated flow-control valves combined.

How fast will they respond? It's not just a matter of computing the flow rate through a known restriction, although that is part of the answer. Speed of response also depends on the characteristic force-vs.-airgap curve of the solenoid; the shape, size, clearance, and displacement of each spool; and the fluid viscosity.

The method outlined relates those parameters for the sketched valve, Fig. 9.27, and can be applied to any other pilot-operated spool valve. A special technique is worked out later for a large spool valve actuated by a small auxiliary piston. Symbols and units for all are in Tables 9.11 and 9.12.

Valve in Action

Before applying the response equations, let's establish the sequence of operation (Fig. 9.28). Briefly, here is what happens: The solenoid is energized, the pilot spool moves quickly to the full-open position, and the main spool is shifted at a rate determined by the amont of flow that can move through the pilot ports against these five resisting forces:

- Pilot system back pressure, psi
- Viscous damping force, lb
- Radial jet force, lb
- Axial jet force, lb
- Acceleration force, lb

Starting with the energization of the solenoid, here are the steps to take in calculating individual and total response times.

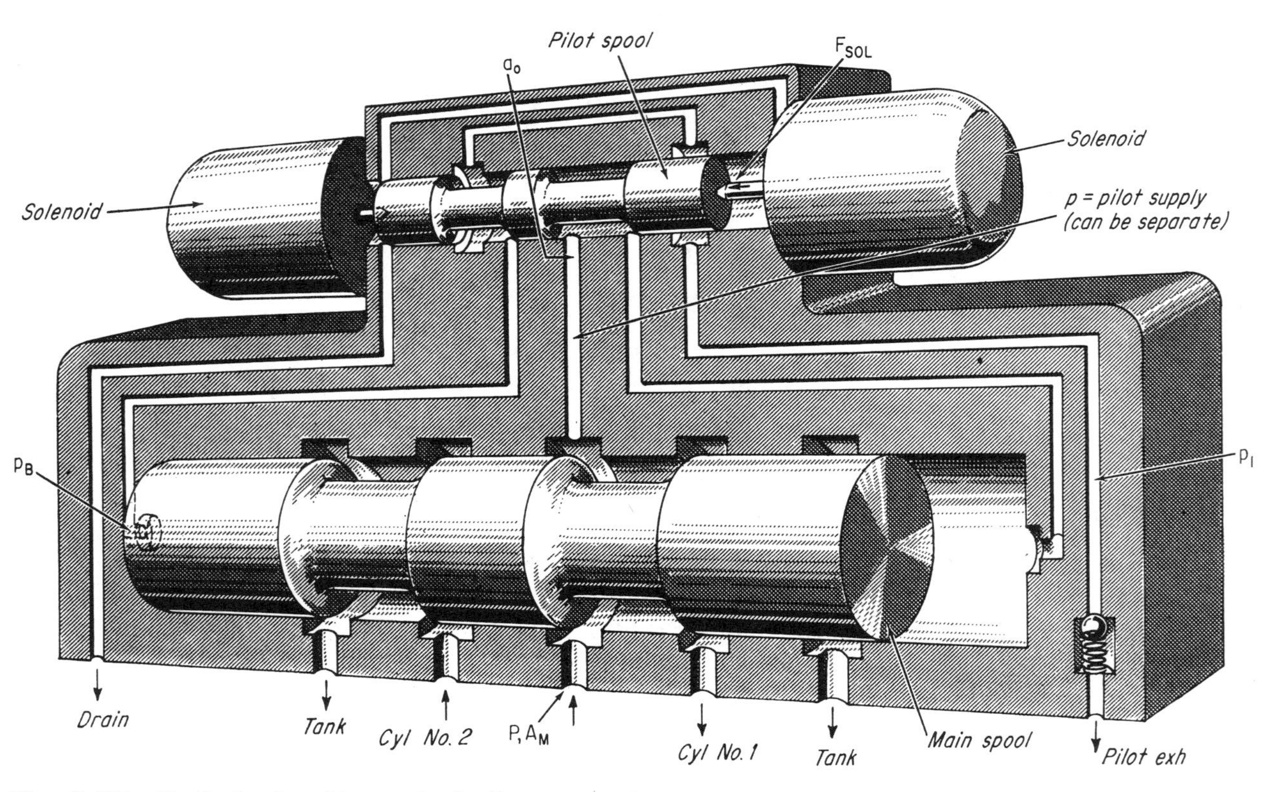

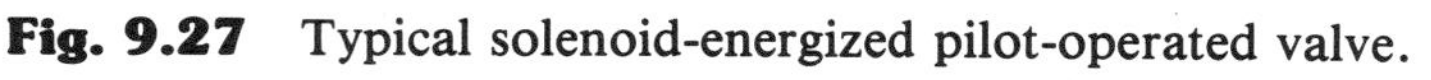
Fig. 9.27 Typical solenoid-energized pilot-operated valve.

Table 9.11 Dimensions and Operating Conditions for Graph in Fig. 9.28

	PILOT SPOOL	MAIN SPOOL
DIAMETER, IN.	$d=0.25$	$D=2.5$
MASS, LB-SEC²/IN.	$m=0.0002$	$M=0.05$
STROKE, IN.	$s=0.375$	$S=1.5$
LAND LENGTH, IN.	.....	$L=6.0$
RADIAL CLEARANCE, IN.	.....	$C=0.0003$
COEFFICIENT OF FRICTION	.....	$F_R=0.04$
SOLENOID FORCE, LB (initial; final)	$F_{SOL}=1$; 8.5	.....
BACK PRESSURE, PSI	.....	$p_B=20$
SUPPLY PRESSURE, PSI	$p=100$	$P=500$
DIFFERENTIAL PRESSURE, PSI	$\Delta p=70$ (approx)	$\Delta P=450$ (approx)
PORT AREA, IN.²	$a_o=0.05$	$A_M=1.2$
FLOW COEFFICIENT	$f=0.6$	$f=0.6$
VISCOSITY, CP	$\mu=80$	$\mu=80$
DENSITY, LB-SEC²/IN.⁴	0.000,085	0.000,085

Solenoid and Pilot

First obtain the force-vs.-airgap curve of the solenoid from the manufacturer. From this you can determine the accelerating force available at each point of the pilot-spool stroke. The approximate linear relationship is

$$F_{SOL} = A + Bx \tag{9.21}$$

where F_{SOL} = solenoid force, lb; A = solenoid force at start of pilot-spool stroke; B = force gradient, lb/in., based on the force-vs.-airgap curve; and x = pilot-spool displacement, in.

Next, calculate the time-displacement curve. Damping forces can be neglected because the stroke is short, velocities are relatively low, oil is of low viscosity, and the pilot spool is lapped to a loose fit. These assumptions are usually valid, and therefore the time vs. pilot-spool displacement can be approximated in this relationship:

$$t = \sqrt{\frac{m}{B}} \ln \left(\frac{x + r + \sqrt{2rx + x^2}}{r} \right) \tag{9.22}$$

were t = time, sec; m = mass of pilot spool, lb-sec²/in.; and r = A/B, the ratio of solenoid starting force to force gradient.

Finally, read the time-displacement graph to find out how long it takes for the spool shift to the new position. The curves in Fig. 9.28 are typical. They are for a ¼-in.-diameter closed-center pilot spool controlling flow to a 2½-in. main spool.

The pilot pressure chosen for this example is 100 psi, which is less than main supply pressure, and a separate supply for the pilot is required. There are no hard and fast rules for establishing pilot pressures, but if possible, keep intensity down to

Table 9.12 Nomenclature for Spool Valve Response Calculations

		PILOT	ACTUATING PISTON	MAIN VALVE
SPOOL DIMENSIONS AND MASS	DIAMETER, IN.	d	d_x	D
	CROSS-SECTIONAL AREA, IN.^{2}	—	a_p	A_s
	MASS, LB-SEC^2/IN.	m	—	M
	STROKE: INTERMEDIATE	x	—	—
	FULL	s	S	S
	ENGAGEMENT (LENGTH IN CONTACT), IN.	—	l_p	—
	LAND LENGTH, IN. (TOTAL)	—	—	L
	SPOOL-TO-BORE RADIAL CLEARANCE, IN.	C	C	C
SOLENOID FORCES	INITIAL, LB	A	—	—
	GRADIENT, LB/IN.	B	—	—
	FINAL, LB	F_{SOL}	—	—
	RATIO, A/B	r	—	—
DRAG FORCES	BACK PRESSURE, PSI	p_B	p_B	p_B
	VISCOUS DRAG, LB (or PSI)	—	d_V	D_V
	RADIAL JET, LB	—	P_{rad}	P_{rad}
	COEFFICIENT OF FRICTION	—	F_R	F_R
	AXIAL JET, LB	—	P_{ax}	P_{ax}
	ACCELERATION FORCE, LB	$F=ma$	—	$F=Ma$
OIL PRESSURE, FLOW, AND PORT SIZE	PRESSURE: SUPPLY, PSI	p	p	P
	PILOT DOWNSTREAM, PSI	p_1	p_1	p_1
	DIFFERENTIAL, PSI	Δp	Δp	ΔP
	PORT AREA, IN.^{2} (EFFECTIVE ORIFICE)	a_o	—	A_M
	FLOW COEFFICIENT (0.55 TO 0.70)	f	f	f
	VISCOSITY, CENTIPOISE	μ	μ	μ
	OIL DENSITY, LB-SEC^2/IN.^{4}	ρ	ρ	ρ
	FLOW RATE, IN.^{3}/SEC	q	q	—
	OIL VELOCITY, IN./SEC (THROUGH PORT)	—	—	V_o
	OIL MASS FLOW, LB-SEC/IN.	—	—	M_f
	OIL-JET DEFLECTION ANGLE, DEG	—	—	α
VALVE RESPONSE	ACCELERATION TIME, SEC	t_a	T_a	T_a
	SHIFTING VELOCITY, IN./SEC	v_p	v	V
	SHIFTING TIME, SEC (AFTER ENERGIZATION)	t	T	T

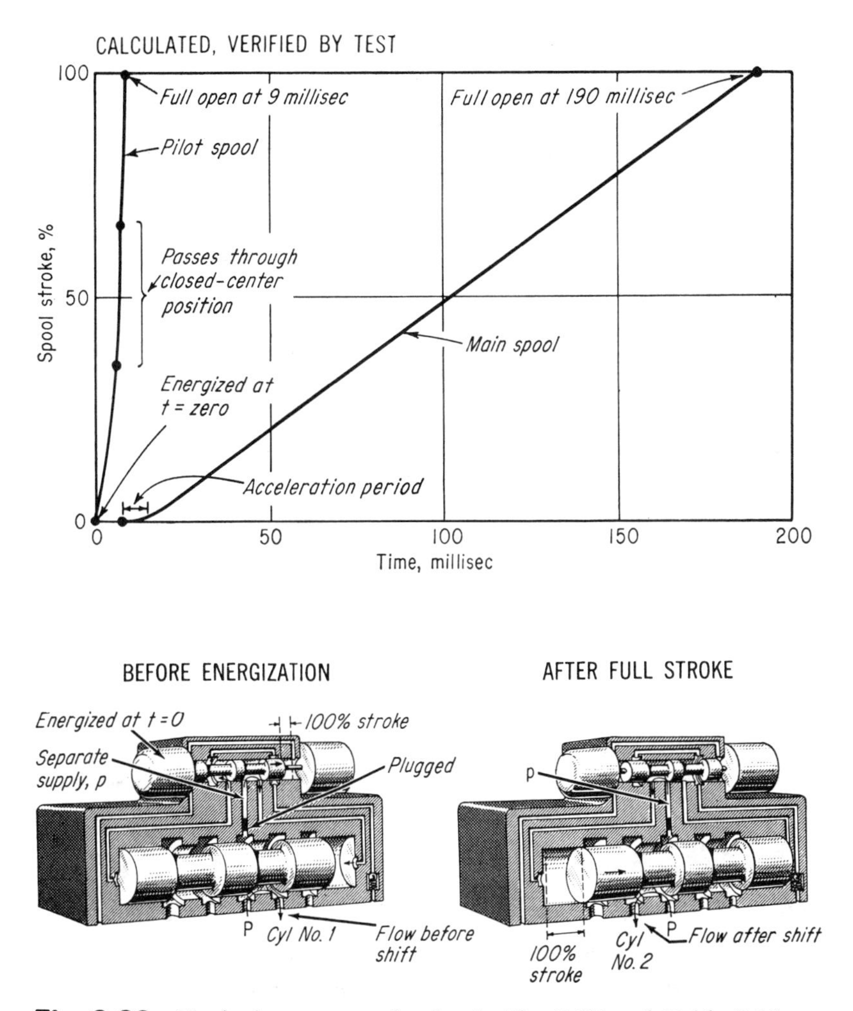

Fig. 9.28 Typical response of valve in Fig. 9.27 and Table 9.11.

a few hundred pounds per square inch if that will do the job, to avoid possible distortion or leakage in the pilot system.

The graph shown reveals this: The pilot spool starts at the far left position, moves a short distance (about ⅛ in.) to the closed center position in about 0.006 sec, and opens to the final position in another 0.003 sec. The average velocity during the last 0.003 sec is 0.125/0.003 = 42 in./sec.

In the open position, the pilot-valve flow rate depends on effective orifice diameter a_O and the pressure drop through the pilot valve:

$$q = fa_O \sqrt{\frac{2\Delta p}{\rho}} \tag{9.23}$$

where q = flow, in.³/sec; f = flow factor, dimensionless, ranging from 0.55 to 0.70 depending on valve type; a_O = cross-sectional area, in.², of the minimum port opening – usually the drilled port hole; $\Delta p = p - p_1$ = differential pressure, psi, measured across the pilot inlet and outlet ports; and ρ = fluid mass density, lb-sec²/in.⁴, normally 0.000085 for oil.

The value of Δp, measured at the pilot spool, depends on the performance of the main spool, including the effects of the back pressure reflected from the pilot exhaust.

Main Spool

The five hydrodynamic resisting forces of back pressure, viscous damping, radial jet, axial jet, and acceleration determine the shifting speed of the main spool.

Some simplifying assumptions must be made; otherwise there is no practical mathematical solution. For one, assume that the back pressure of the pilot system, set by the pilot exhaust valve, is constant. Ignore line resistances, because connecting lines are short. Neglect viscous damping except at the full-velocity portion of the stroke. Then the five dynamic resistances, explained in detail later, can be handled with simple equations.

Pilot-System Back Pressure

Back pressure is usually 5 to 7% of pilot pressure p. Select the higher value if operating pressures are over 200 psi, because it adds a margin of safety that compensates for spool rubbing friction. The friction is from metal-to-metal contact at points where the oil film is partially destroyed.

Above 400-psi operating pressure, a separate pilot supply usually is provided. Pilot pressure in these instances ought to be at least 7% of the main operating pressure to ensure adequate force to move the main spool. For example, when the main operating pressure is 3500 psi, pilot pressure is $0.07 \times 3500 =$ about 250 psi, and back pressure will be $0.07 \times 250 =$ about 20 psi. The back pressure of the main system will range from 50 to 250 psi, depending on the load.

Viscous Damping

For average conditions of constant velocity and constant temperature, the damping force D_V, lb, can be expressed with this approximate relationship (Petroff's law):

$$D_V = \frac{D\pi L V \mu}{C \times 6.9 \times 10^6} \tag{9.24}$$

where D = spool diameter, in.; L = length of the spool lands, in.; V = main spool velocity, in./sec; μ = absolute viscosity, centipoise; and C = spool to bore radial clearance, in.

If temperature varies more than 30 to 50 deg, it is nearly impossible to compute viscous resistance.

Radial Jet Pressure

The spool is forced radially against the far side when oil flow impinges against the near side of the spool (Fig. 9.29). The radial force itself would be of no consequence except that it increases friction in the longitudinal direction in accordance with this expression:

$$P_{ax} = P_{rad} F_R \tag{9.25}$$

where F_R = coefficient of friction and P_{ax} and P_{rad} = force, lb. The value of F_R will range from 0.02 to 0.06.

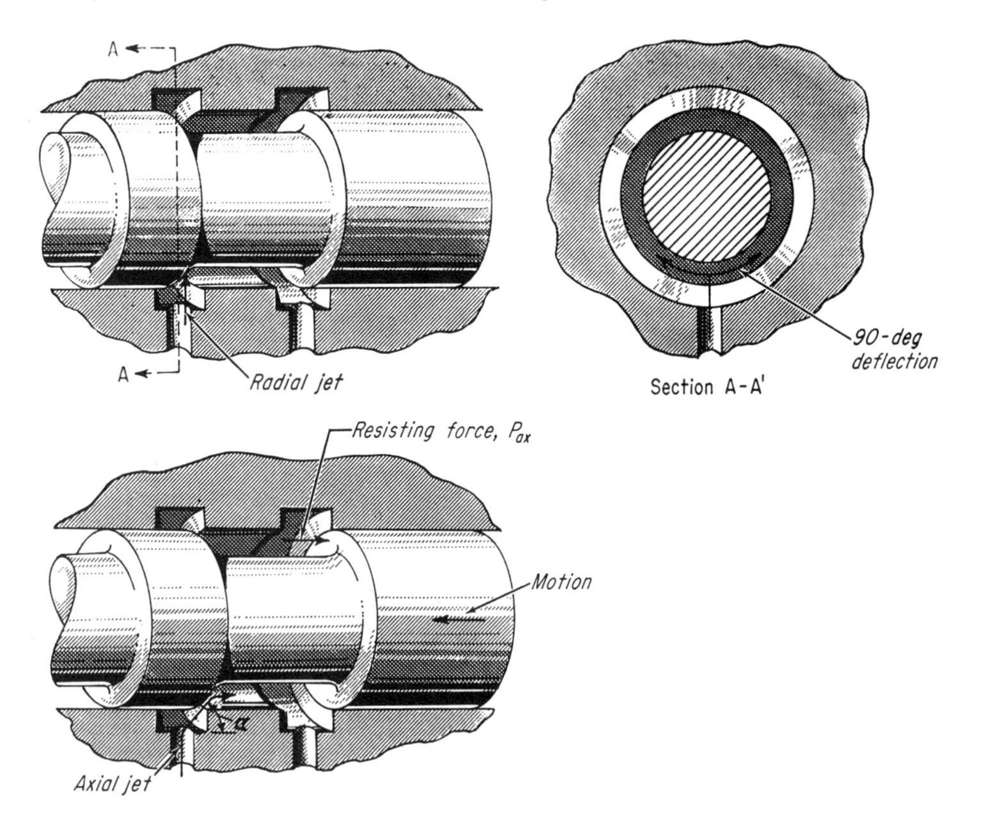

Fig. 9.29 Explanation of jet-force drag.

Formulas for the value of P_{rad} given in Marks' *Standard Handbook for Mechanical Engineers* (McGraw-Hill) and other references depend on the number of degrees the jet is deflected. The deflection is assumed to be 90 deg, giving the equation

$$P_{rad} = M_f V_0 = Q\rho V \tag{9.26}$$

where M_f = mass flow, lb-sec/in.; V_0 = oil velocity through port, in./sec; and V = spool velocity, in./sec.

Then because $V_0 = \sqrt{2\Delta P/\rho}$ and $Q = fA_M\sqrt{2\Delta P/\rho}$, this useful relationship is derived:

$$P_{ax} = 2F_R f A_M \,\Delta P \tag{9.27}$$

where A_M is the port opening of the main valve.

For low system pressures, P_{ax} is small and can usually be neglected. Above 2000 psi, don't ignore P_{ax}.

Axial Jet Pressure

The axial component of the incoming jet of oil will resist or augment spool motion, depending on spool design. Most spool valves have conventional lands, where the jet resists motion during opening of a pressure port and assists motion during closing (Fig. 9.29). To be on the safe side, neglect forces assisting motion.

The jet in effect acts against the spool shoulder away from the port and is effective only in proportion to its axial component, cos α:

$$P_{ax} = 2fA_M \, \Delta P \cos \alpha \tag{9.28}$$

where α normally varies from 70 deg at initial opening to 90 deg at full opening. In calculations, use axial jet pressure during initial opening and the axial component of radial pressure during the remainder of travel.

Mass Acceleration

The force necessary to accelerate the spool is

$$F = Ma \tag{9.29}$$

where M is spool mass, lb-sec²/in., and a is dV/dt, in./sec². This force can be neglected in most instances unless very high speeds are expected.

Spring-Centered Spool

The example thus far has been for a non-spring-centered main spool. A spring-centered spool has the additional resistance of the springs, but otherwise response is calculated in the same way as without the spring. When actuated from the center position, spool travel is one-half that of the free spool, and acceleration is somewhat less. Return speed is faster because of the spring return force.

The total time of spool travel for a spring-centered valve won't differ much from that for a nonspring valve. The performance curves in Fig. 9.30 are from actual

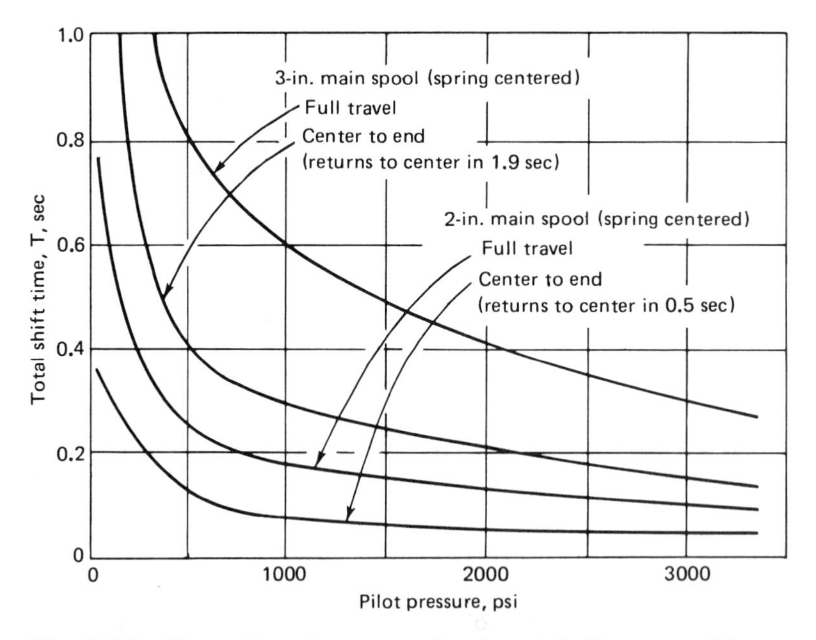

Fig. 9.30 Tests of a spring-centered main spool. (Courtesy of Sperry Vickers, Troy, MI.)

tests made by Vickers on spring-centered valves, showing the effect of pilot pressure on valve response. The 3-in. and 2-in. notations in this case refer to port size rather than spool diameter. Note that the total shift times are of the same order of magnitude as the times for the non-spring-centered valve of Fig. 9.28.

SPECIAL CASE: PISTON-OPERATED SPOOL

Some pilot-operated valves have main spools too large to shift quickly with small pilot-oil flows unless special actuators are added. With special actuators such as the small-diameter pistons shown in Fig. 9.31, an ordinary solenoid pilot is plenty large enough to shift very large main spools.

The actuating piston size must be carefully selected to give optimum response. If the diameter d_x is too large, the spool velocity will be low. If the diameter is too small, spool friction and back pressure will be too hard to overcome.

Compute the best size by writing a differential equation containing velocity and area of the actuating piston and setting the first derivative of velocity vs. area equal to zero—the peak value of velocity. Start with the pilot flow [Eq. (9.23)]:

$$q = fa_O \sqrt{\frac{2\Delta p}{\rho}} \tag{9.30}$$

Substitute constant k_1 for $fa_O\sqrt{2/\rho}$ and νa_p for q. Term ν is actuating piston velocity and a_p is its cross-sectional area. Equation (9.23) becomes

$$va_p = k_1\sqrt{\Delta p} \tag{9.31}$$

Pressure drop $\Delta p = p - p_1 = p - p_B - (P_{ax} + \Sigma\, D_\nu)/a_p$, which is pilot pressure p less the aggregate of the hydrodynamic resistances discussed before. To simplify, set $k_2 = p - p_B$ and $k_3 = P_{ax} + \Sigma\, D_\nu$. Then

$$va_p = k_1\sqrt{k_2 - k_3/a_p}$$

or

$$v = k_1 \sqrt{\frac{k_2}{a_p^2} - \frac{k_3}{a_p^3}}$$

$$= \sqrt{\frac{k_1^2\, k_2}{a_p^2} - \frac{k_1^2\, k_3}{a_p^3}} \tag{9.32}$$

The first derivative is

$$\frac{dv}{da_p} = \frac{1}{2} \frac{\left(\dfrac{3k_1^2\, k_3}{a_p^4} - \dfrac{2k_1^2\, k_2}{a_p^3}\right)}{\sqrt{\dfrac{k_1^2\, k_2}{a_p^2} - \dfrac{k_1^2\, k_3}{a_p^3}}}$$

Let the first derivative = zero, and multiply both sides of the equation by the denominator:

$$0 = \frac{3k_1^2\, k_3}{a_p^4} - \frac{2k_1^2\, k_2}{a_p^3}$$

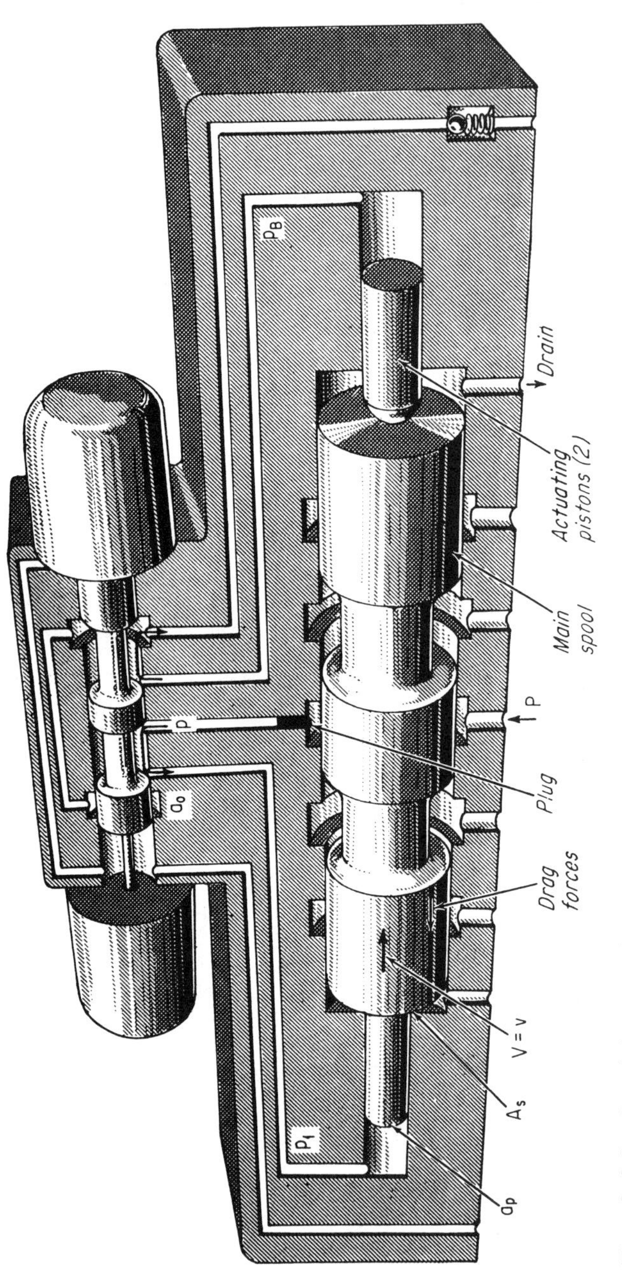

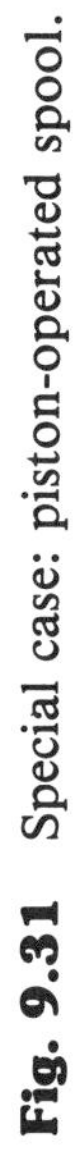

Fig. 9.31 Special case: piston-operated spool.

Solve for optimum area a_p:

$$a_p = \frac{3k_3}{2k_2} = \frac{3(P_{ax} + \Sigma D_V)}{2(p - p_B)} \quad (9.33)$$

Some approximations must be made before a_p can be computed. For instance, a_p appears as a piston diameter in the expression for damping force, Eq. (9.24), thus changing Eq. (9.33) into quartic form with four roots. Accuracy is not improved enough to warrant such a rigorous approach.

The simplification is to use trial and error. First assume a value for the actuating piston diameter, and then substitute it in the viscous-drag equation (following) and Eq. (9.33) to see how close you've guessed.

Viscous drag includes the main spool and the actuating piston, and Eq. (9.24) still applies. The combined expression can be simplified to

$$\text{total } D_V = \frac{4VA_S\,\mu}{C \times 6.9 \times 10^6 \times d_x} \times \left(\frac{DL}{d_x} + 2S\right) \quad (9.34)$$

where D_V = damping force, lb; V = piston and main-spool velocity, in./sec; A_S and $a_p = \pi D^2/4$ and $\pi d_x^2/4$, the cross-sectional areas of the main spool and the actuating piston, in.^{2}; C = radial clearance of piston and spool (assumed same), in.; L = combined lengths of lands of main spool, in.; and S = main spool stroke, in. Term 2S is for two actuating pistons, and S is assumed equal to the average piston engagement l_p, in. The third example at the end of the next section uses this equation.

VALVE CALCULATION SAMPLES

Velocity of Main Spool

Dimensions, operating pressures, and performance are given in Table 9.11. Pilot supply pressure is 100 psi; main supply pressure is 500 psi.

The procedure is to compute maximum velocity first and then the acceleration time. The final computation is for total time.

The forces acting upon the main spool at maximum velocity are pilot back pressure p_B, viscous damping force D_V, and radial jet force P_{rad}. From Eq. (9.27), $P_{ax} = 2 \times 0.04 \times 0.6 \times 1.2 \times 450 = 26$ lb; $26/4.9 = 5.3$ psi. Provisionally, estimate that D_V is equivalent to 3.2 psi and $p_B = 20$ psi. The combined hydrodynamic resistance is $5.3 + 3.2 + 20 = 28.5$ psi.

Hence, pilot-valve pressure differential $\Delta p = 100 - 28.5 = 71.5$ psi, and the flow rate is

$$q = 0.6 \times 0.05\,\sqrt{2 \times 71.5/0.000085}$$
$$= 40 \text{ in.}^3/\text{sec}$$

The maximum velocity of the main spool, $V = 40/4.9 = 8.2$ in./sec. With V known, $D_V = (2.5\pi 6 \times 8.2 \times 80)/(0.0003 \times 6.9 \times 10^6) = 3.05$ psi. The provisional estimate of $D_V = 3.2$ was close enough, and recalculation is not necessary.

The forces acting upon the spool during acceleration are p_B, P_{ax}, D_V, and F, where F = Ma. Assuming a mean value for initial port opening $A_M = 0.4$ in.^{2}, then from Eq. (9.28), $P_{ax} = 2 \times 0.6 \times 0.4 \times 450 \times 0.26 = 56$ lb; $56/4.9 = 11.4$ psi, where $\alpha = 75$ deg; and cos $\alpha = 0.26$. Viscous drag will be the average: $D_V = 3.2/2 = 1.6$ psi. Back pressure is still $p_B = 20$ psi, so the total is $11.4 + 1.6 + 20 = 33$ psi.

Therefore accelerating pressure = $100 - 33 = 67$ psi; $67 \times 4.9 = 328$ lb. From Eq. (9.29), acceleration time $t_a = MV/F = 0.05 \times 8.2/328 = 0.0013$.

The displacement of the main spool during the acceleration period is negligible, being less than 1% of the total stroke. The total stroke of 1.5 in. takes $1.5/8.2 = 0.182$ sec. The time interval from energization of the solenoid to completion of the main valve stroke $T \cong 0.190$ sec.

Higher Pilot Pressure

Dimensions and main operating pressure are unchanged from the first example. However, pilot pressure is now made equal to main operating pressure: 500 psi. Much higher flow will pass through the pilot valve port. Here are the resulting response figures:

Maximum velocity period: $P_{ax} = 5.3$ psi (same as before); $D_V = 7.7$ psi (higher estimate, proportional to anticipated velocity); $p_B = 20.0$ psi (same). The total is 33.0 psi.

New $\Delta p = 467$ psi, and $q = 4.7\sqrt{467} = 102$ in.^{3}/sec; $V = 102/4.9 = 20.8$ in./sec; $D_V = 1.82 \times 20.8 = 37.8$ lb $= 7.75$ psi (proves out assumption).

Acceleration period: $P_{ax} = 11.4$ psi (same); $p_B = 20.0$ psi (same); $D_V = 3.6$ psi (increased). Total is 35.0 psi. Accelerating force $F = 465 \times 4.9 = 2280$ lb.

Accelerating time $t_a = (0.05 \times 20.8)/2280 = 0.0005$ sec. The 1.5-in. stroke takes $1.5/20.8 = 0.072$ sec. Total time $T = 0.081$ sec.

The flow rate of the pilot oil is more important than pressure intensity in obtaining a fast-acting valve. A slightly larger pilot valve and enlarged pilot porting have a marked effect on the operational speed of the main valve.

Note that increasing the pilot pressure fivefold, from 100 to 500 psi, only doubles the speed of response, from 0.19 to 0.08 sec. Increasing the port area can result in a nearly proportional gain in speed, and no additional pressure is necessary. Costs of producing a ⅜-in. pilot spool are not much greater than those for a ¼-in. spool, and the increase in capacity is 50% without the additional heat losses entailed by an increase in pressure.

Special Piston Actuators

Dimensions and operating pressures are unchanged from the first example. However, a small actuating piston is placed at each end of the main spool to increase the longitudinal velocity for a given pilot-fluid flow rate. Trial and error would normally be used to calculate the most effective diameter for the actuating piston. To simplify this example, the guess is $d_x = 1.4$ in., which is fairly close to the known optimum diameter [in parentheses after the calculated value, from Eq. (9.34)]:

$$\Sigma D_V = \frac{8.2 \times 4.9 \times 4 \times 80}{0.0003 \times 6.9 \times 10^6 \times 1.4} \times \left(\frac{2.5 \times 6}{1.4} + 2 \times 1.5 \right)$$

$$= \frac{12{,}800}{2075 \times 1.4}\left(\frac{15}{1.4}+3\right) = 60.5 \text{ lb (59.0 lb optimum)}$$

Introducing the value for ΣD_V in Eq. (9.33),

$$a_p = \frac{3(26+60.5)}{2(100-20)} = 1.62 \text{ in.}^2 \ (1.59 \text{ in.}^2)$$

$$d = 1.44 \text{ in. } (1.425 \text{ in.})$$

With optimum $a_p = 1.59$ in.^{2}, piston velocity using Eqs. (9.30) to (9.32) is

$$v_p = 4.6\sqrt{\frac{80}{2.53} - \frac{26+59}{4.0}}$$

$$= 15.0 \text{ in./sec.}$$

Total time T = 0.100 + 0.009 = 0.109 sec. Using the pilot pressure of the second example (p = 500 psi), V = 20.8 in./sec, and d = 1.06 in. Then

$$\Sigma D_V = \frac{20.8 \times 4.9 \times 4.80}{0.0003 \times 6.9 \times 10^6 \times 1.06} \times \left(\frac{15}{1.06}+3\right) = 258 \text{ lb}$$

$$a_p = \frac{3(26+258)}{2(500-20)} = 0.886 \text{ in.}^2$$

$$d_x = 1.06 \text{ in.}$$

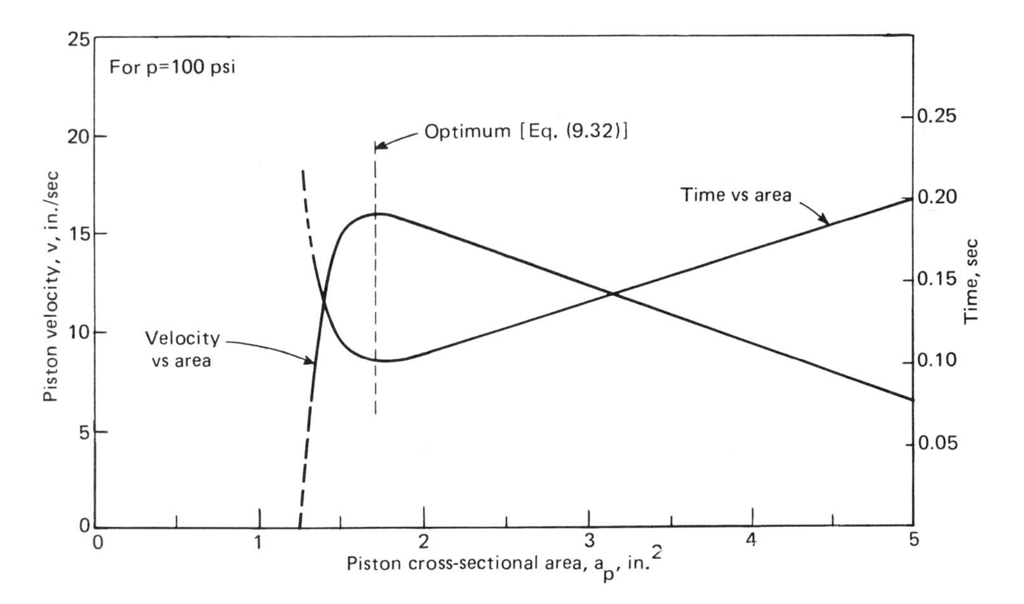

Fig. 9.32 Effects of varying the diameter of the actuating piston in Fig. 9.31.

PILOT PRESSURE p, PSI	MAIN VALVE DIAMETER, IN.	MAXIMUM VALVE VELOCITY, WITHOUT PISTON IN./SEC.	TOTAL SHIFT TIME T, SEC	MAXIMUM VALVE VELOCITY, WITH PISTON IN./SEC	PISTON DIAMETER, IN.	TOTAL SHIFT TIME T, SEC
100	2.50	8.2	0.190	15.0	1.425	0.109
500	2.50	20.8	0.081	55.0	1.06	0.036

Table 9.13. Summary of Effects of Adding Actuating Pistons to Main Spool of Pilot Valve

$$v_p = 4.6\sqrt{\frac{480}{0.784} - \frac{284}{0.61}}$$

$$= 55 \text{ in./sec}$$

$$T = 0.036 \text{ sec}$$

The results are summarized in Table 9.13 and Fig. 9.32, using Eq. (9.32), depicting the effect of actuating-piston area upon spool shifting velocity and shifting time.

10

Hydraulic Applications

The strength of hydraulics is its sustained high force and torque in close quarters. Most applications are chosen because the jobs cannot be done as conveniently—or at all—with electrical electromechanical actuators and drives (which otherwise might prove cheaper and would eliminate the possibility of oil leaks).

Hardly anybody denies that hydrostatic drives and actuators belong in earth-moving equipment (Figs. 10.1 and 10.2), aircraft, certain heavy-duty robots, and other applications where sustained high torque and force are needed in tight spaces. Even some railroad equipment is a natural application for fluid power (Fig. 10.3).

It's the in-plant applications that are in contention. Many manufacturing plants are specifying electrical drive units because they are believed quieter, leak no oil, are more adaptable to electronic analog and digital control, and allegedly cost less. Even sawmills and other nonplant operations have been leaning toward electrical drives.

However, not everybody accepts or believes those limitations. The examples given here show where the strengths lie in fluid power—and why it continues to grow.

Details of the pumps, motors, cylinders, actuators, drives, controls, and accessories mentioned in the applications are covered in other chapters.

This chapter will concentrate on reasons for picking hydraulic systems over electrical and electromechanical systems, showing successful examples. The greatest competition is between electric and hydrostatic drives, so the emphasis will be there.

HYDROSTATIC DRIVES: APPLICATION TIPS

Fluid motors have inherently low inertia compared with most electrical motors because the energy is put into the fluid back at the pump; all that the fluid motor has to do is turn. Also, all fixed-displacement hydrostatic motors (unlike electrical motors) can withstand stall loads for long periods without overheating and can produce high torque in cramped spaces—ideal for most off-road equipment, aircraft, and some robots.

Operation at full speed and load is less efficient than with electric motors, so an efficiency and cost study must be made to determine the best economies over the

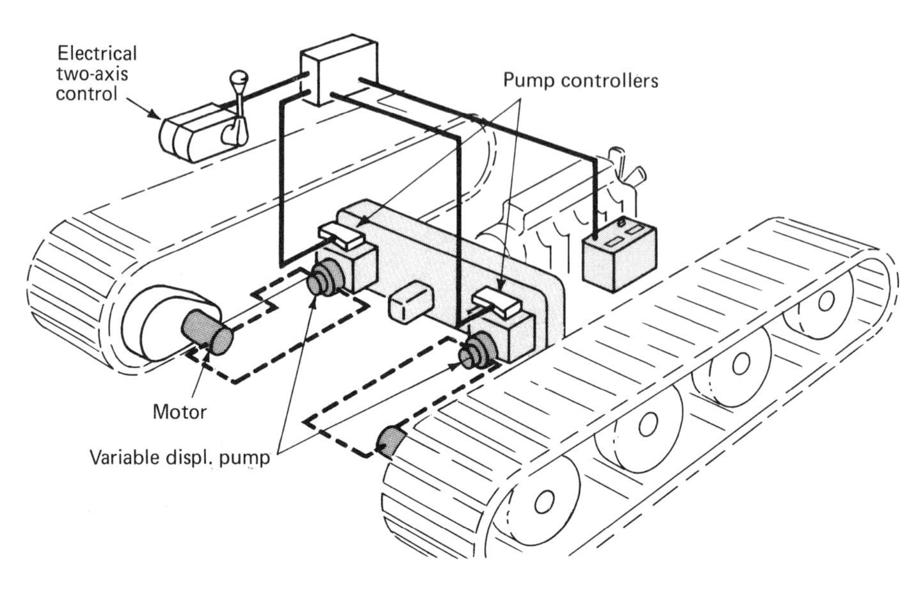

Fig. 10.1 Heavy-duty earth-moving vehicles are a natural application for hydrostatic pump-motor drives.

long haul. If the heavy loads are at low speeds or at stall, hydraulics can prove to be the most economical.

Feed drives for machines in a factory are a special case: They use a moderate amount of power in traverse, but the prime need is for tight control of position and velocity. It can be done electrically or hydraulically.

Fluid resonance is a problem that is special to hydraulic systems. The fluid is compressible and transmits or reflects pulses at speeds and frequencies that can in-

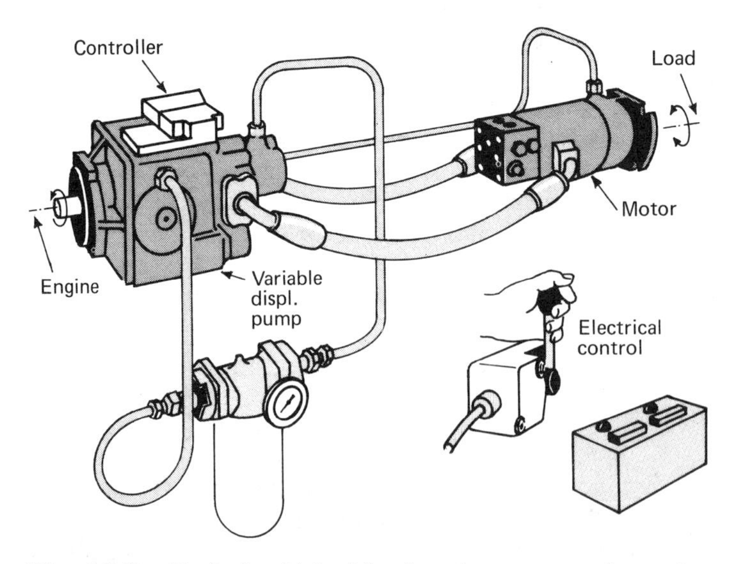

Fig. 10.2 Typical vehicle drive has the pump at the engine and the motor on an axle or wheel.

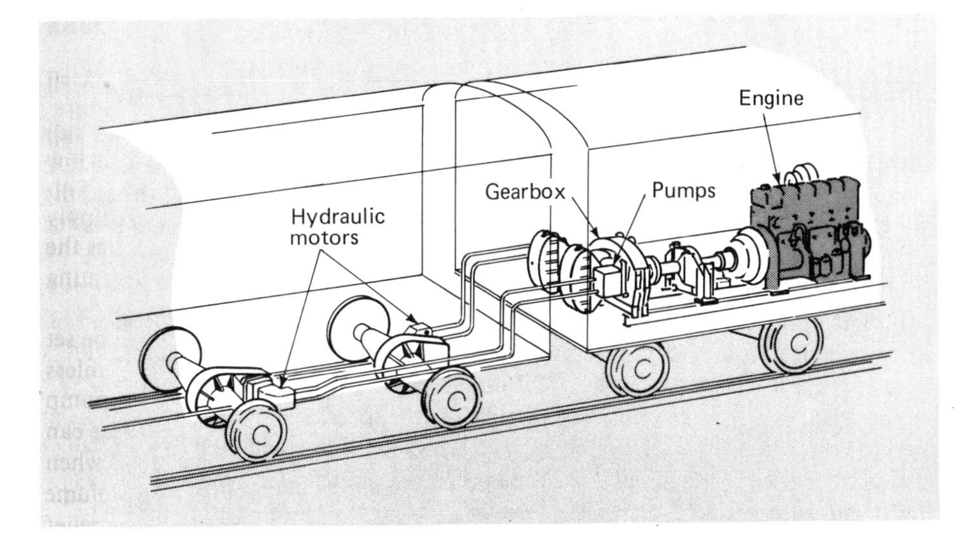

Fig. 10.3 Experimental locomotive is powered with a hydrostatic drive.

terfere with control unless the system is carefully designed. Also, normal pulsations from pumps have been known to trigger false speed signals in pickup tachometers. Most of these problems are solved with adjustments in the controlling electronics, in the hydraulic circuits, and in the internal clearances and leakages of the motors.

Categories of Drives

Fluid motor drive systems can be broadly categorized by their primary purpose: controlling *power, torque, velocity,* or *position.* These four basic categories include a wide variety of individual configurations, some of which operate in more than one mode.

Much development effort has gone into the control of *power-type* hydrostatic drives, particularly the closed-circuit configuration, and attention is usually focused upon the pumps in these systems. The pump (and its pressure, flow, or load compensator) determines the basic characteristics of the drive; the motor is merely the output device, selected to match the load speed and torque requirements to the flow and pressure capabilities of the pump. Power transmission efficiency and smoothness are the main concerns, along with noise, weight, and size per hp and, of course, cost. Power level is generally moderate to high, so efficiency is important.

Drives controlling *torque, velocity,* or *position* tend to be vastly different from power drives and are generally much lower in power level. It is often possible to ignore efficiency until all other performance factors have been examined. At low power levels it is very hard to justify additional cost or complexity to gain efficiency because the amount of energy that can be saved is still fairly small.

In contrast to higher powered hydrostatic drives, control of torque, velocity, or position places the most stringent requirements on the motor rather than the pump. Ultimately, it is the motor's ability to respond faithfully to pressure and flow commands that establishes the control accuracy of a system. The task of the pump in these systems is simply to provide constant pressure or flow to the valves that control the motor.

Generally, it is best to operate high-accuracy systems from a constant pressure source. In particular, position and low-speed (feed) drives operate efficiently and well in this way, because the pump's dynamic response can be isolated from the motors.

However, in applications with a very wide speed range, constant pressure can become very inefficient at the high-speed end where the load torque is often quite low and could be handled with low pressure. Here it is advantageous to add a volume-limiting control to the pump (or use the entire pump output) and let the supply pressure fall to the actual load pressure in the full-speed condition. This enables the system to operate with minimum energy dissipation throughout most of the operating cycle.

In high-speed drives (such as machine spindles) the speed should always be set by an adjustable volume pump, rather than by a restrictive flow control valve, unless the speed regulation is really critical. This minimizes heat generation as well as pump noise and wear. If the drive must also operate at low speed, a flow control valve can be used for only the critical lower portion of the speed range and be bypassed when it is not needed. The pump should be of the pressure-compensated type with a volume limiter. If not compensated, the pump's delivery pressure must be wasted over a relief valve.

Any variable-volume pump (noncompensated) can be adjusted for a flow rate slightly above the no-load flow plus the full-load slip flow of both pump and motor, so that the relief valve can hold full pressure under load. This is expeditious for systems that do not have to be changed often but can be a nuisance for frequently varied conditions. In addition, it is very probable that the pump flow will be set far above the needed amount by a careless operator. This will, of course, produce rapid heating of system fluid, which is a particularly severe problem in closed circuits having a very small reservoir.

Torque control systems can take many forms depending on the specific needs of the job. There are dynamic torque loads, such as web tensioning roll drives, and static torque loads, such as power torque wrenches (or "nut runners"). A vast difference exists in the motor requirements for these very unlike conditions. Motor torque variations can be effectively masked by the load inertia and compliance in most dynamic drives. However, the opposite occurs in static or low-speed drives where motor torque ripple, stiction, and leakage variations can combine to excite the load inertia and compliance into oscillatory movement and large inconsistencies in the stall torque.

Energy-Saving Tips

Many manufacturers recommend special energy-saving circuits, devices, and procedures. Figures 10.4 through 10.6 show the savings in dollars possible.

Oilgear/Hydura (Milwaukee, WI) recommends regenerative braking. By careful design of the system, and with special settings for relief valves, it is possible to let the kinetic energy of the braked load be absorbed by the driving hydraulic motor, which in turn drives the pump as a motor. If the prime mover is an ac electric motor, it can act as an induction generator and feed electric power back through the lines.

Auto Specialties Mfg. Co. (St. Josephs, MI) suggests that energy can be saved by putting clutches in series with those pumps, motors or drives that can be conveniently idled during off-load cycles (Fig. 10.7).

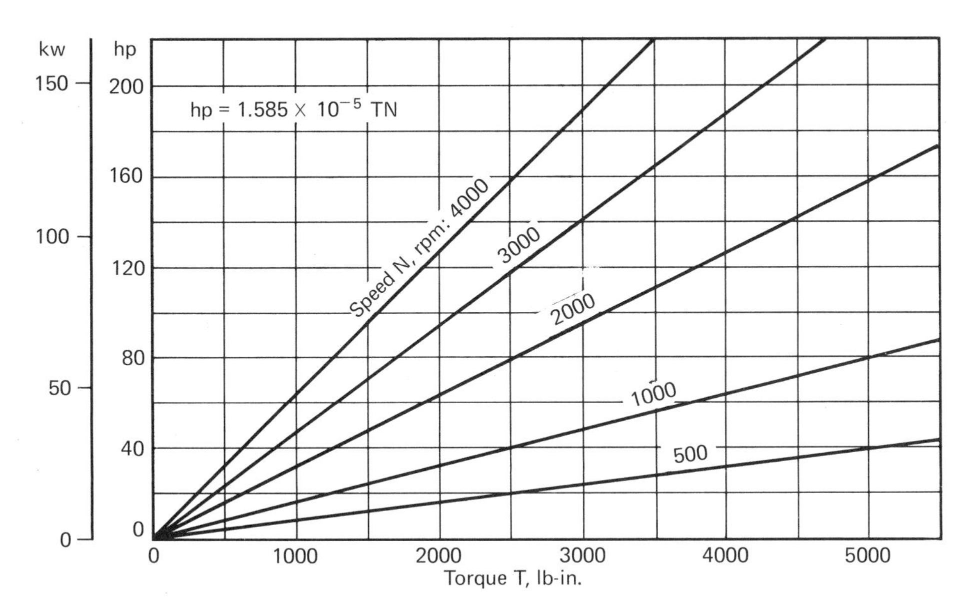

Fig. 10.4 Torque and speed convert to horsepower, which eats up dollars (see Fig. 10.6).

Load-sensing hydrostatic drives also save energy. For example, Abex (Columbus, OH) offers a horsepower summation control (Fig. 10.8) that limits the total power delivered to all the functions driven by a diesel engine. Diesel engine speed is a good indicator of power used, because the engine slows as load builds up. A signal from the engine speed governor is sufficient to modulate a cone-and-seat summation valve and thus send limiting signals to the pressure compensator on the hydrostatic pump.

Eaton/Char-Lynn (Eden Prairie, MN) has developed special priority bypass circuits for its gerotor-type steering pumps, designed to send unused pressurized hydraulic fluid to cylinders and motors elsewhere on the vehicle. Energy is saved by not throttling excess flow to tank, yet the important steering function maintains first priority (Fig. 10.9).

MACHINE SLIDE DRIVES

Designers of slide drives for machining or assembly stations in factories are awakening to the fact that hydrostatic motors of the right type and with the right position sensors (precision limit valves) cost less and perform as well as or better than electromechanical slide drives.

In Fig. 10.10 the shaft output to the ball lead screw is 1200 lb-in. during closed-loop final feed, at feed rates from 0 rpm to 60 rpm. Torque during open-loop rapid traverse is 120 lb-in., from 0 rpm to 400 rpm.

The hydrostatic motors depicted in the schematic displace 3 in^3/rev and 10 in^3/rev, respectively, and are of a special low-internal-leakage design to enable them to accurately move the lead screw and hold the final position. The designs shown can turn as slowly as 1 rpm or less, at full torque. (Motor details were disclosed in Chap. 6.)

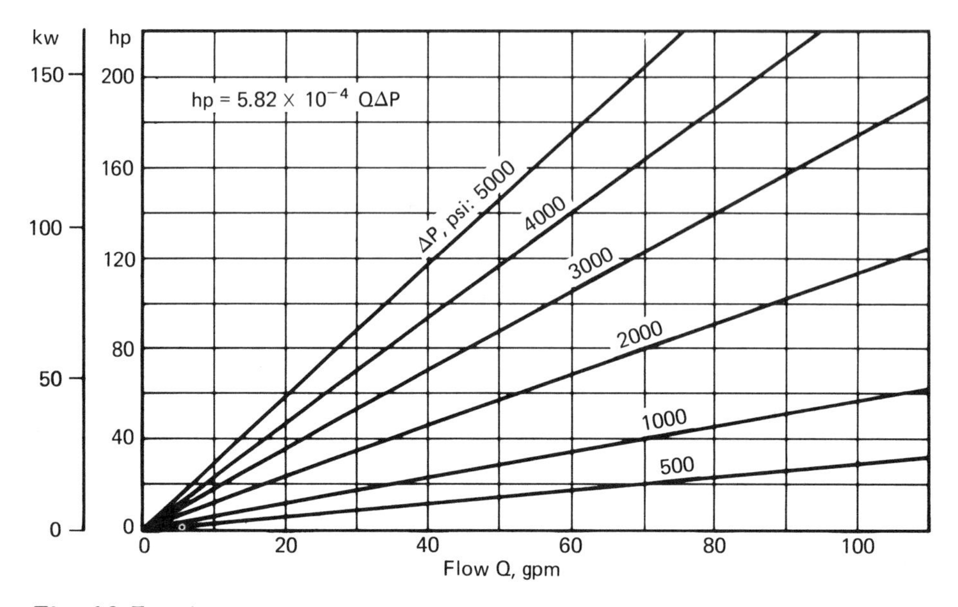

Fig. 10.5 Flow and pressure changes also convert to horsepower (and kilowatts).

There is a simple way to estimate the minimum operating speed of any fluid motor without elaborate testing: just compute the ratio of the internal leakage (at stall) to the displacement per revolution. This ratio, expressed as in^3/min divided by in^3/rev, has the dimensions of revolutions per minute.

The significance is that a motor operating unloaded at precisely that shaft speed can just develop full torque when stalled if the flow is held constant. If the flow is

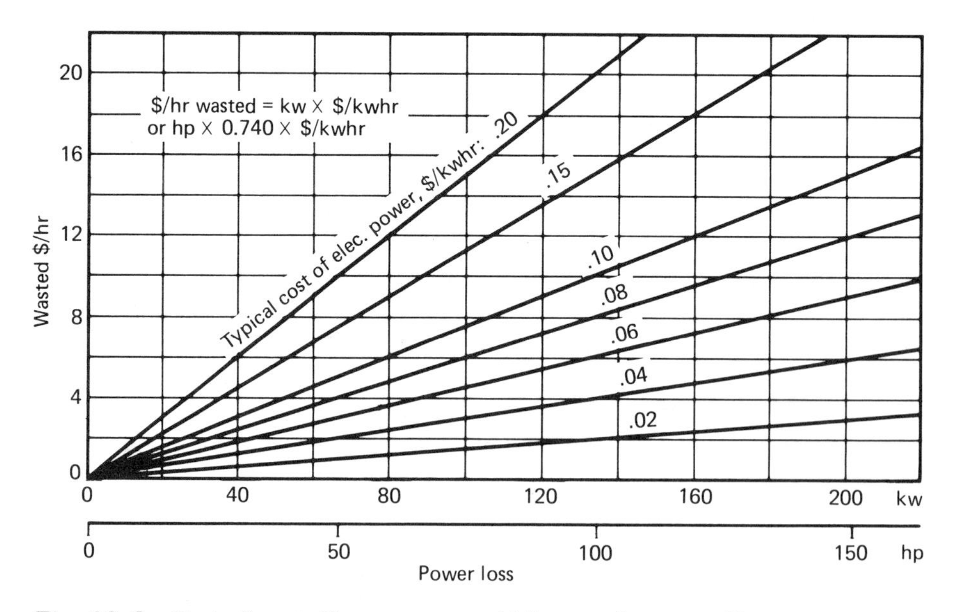

Fig. 10.6 Cost of wasted horsepower and kilowatts for any utility rate can be quickly estimated.

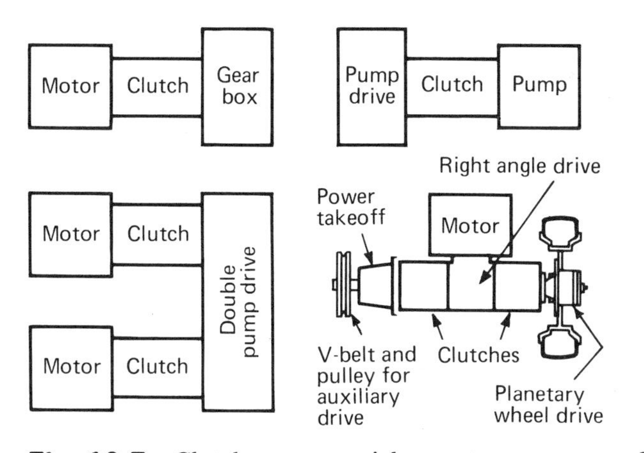

Fig. 10.7 Clutches are a quick way to save energy by uncoupling unneeded accessories.

reduced, the unloaded speed is proportionally reduced also, and the motor can now be stalled with proportionally less torque.

Reducing the rpm below the calculated value can be done only until the motor's friction torque exceeds the now reduced torque capability at stall. When this happens, the motor simply stops rotating. Just a little above that speed, the motor can turn continuously but has essentially zero torque capability.

What is needed for best low-speed performance is lowest leakage and largest displacement, to produce the lowest calculated "minimum speed." The motor must also have uniformly low leakage, or speed variations will occur under constant torque load. Similarly, the motor must have uniform friction and constant displacement characteristics, or torque and speed ripple are induced. The lower the speed, the more exaggerated this ripple becomes, until "cogging" (intermittent rotation) begins.

Certain types of motors have typically higher leakage rates than others. Gear motors, whether of the spur, internal, or orbiting gear (gerotor) type, are examples.

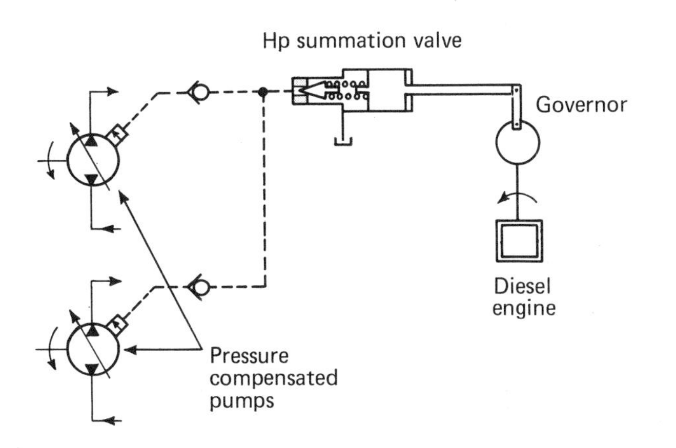

Fig. 10.8 Load-sensing hydrostatic drive automatically reduces load when diesel engine slows. (Courtesy of Abex, Columbus, OH.)

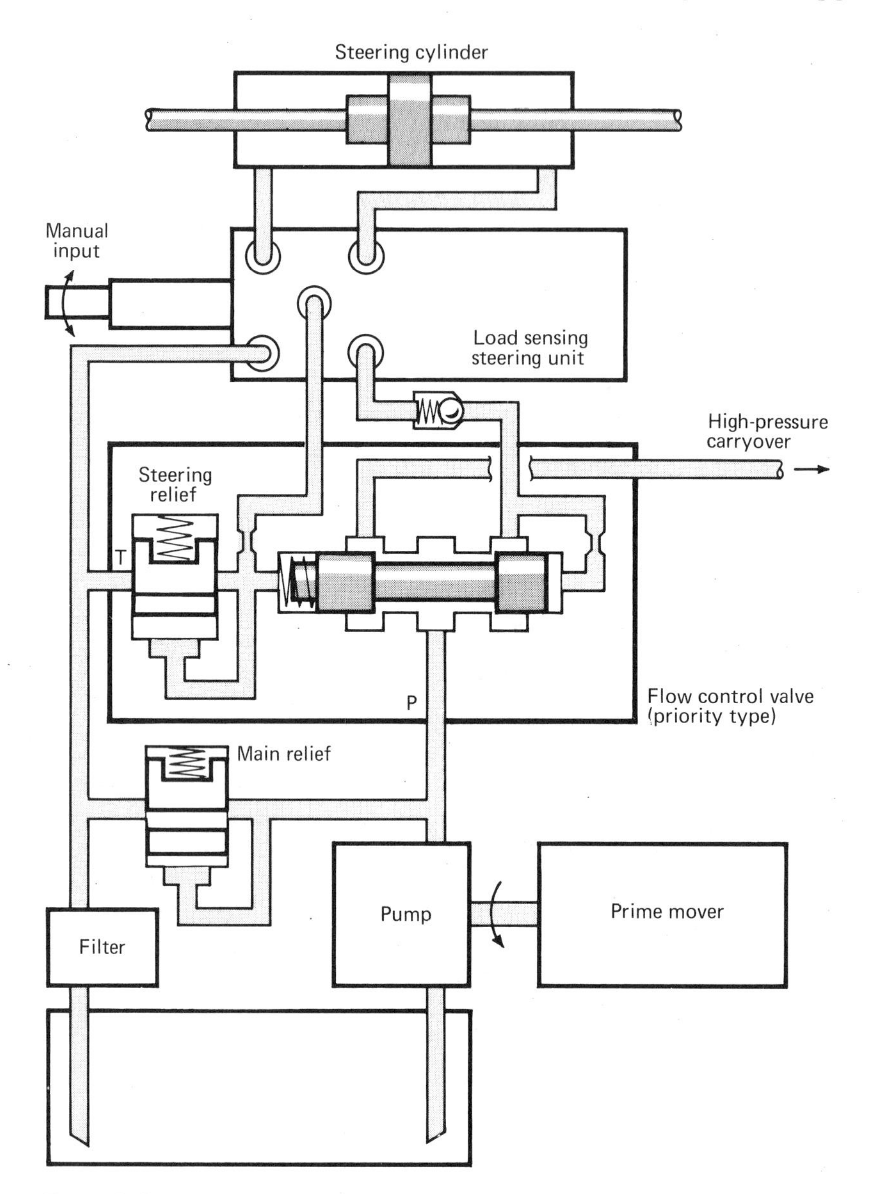

Fig. 10.9 Priority bypass circuit in steering system makes sure that steering cylinder gets fluid first.

Piston motors, whether of the ball or cylindrical type, can be made with low internal leakage because of circular sealing surfaces.

Cross-port leakage—sometimes necessary for fluid slippage to avoid damage from pressure peaks within the motor—spoils the motor's slow-speed controllability. Zero-lap valve ports, for minimum cross-port leakage, are best for low-speed control, but care is needed to avoid sudden pressure surges, such as blocking of flow from the motor. High-leakage motors are better in that respect.

In the slide drive schematic (Fig. 10.10) accuracies are ±0.001 in. (position), ±1 rpm (speed), and ±1 lb-ft (torque), all without the usual shot-pins, ratchets, pawls, solid stops and—best of all—without electronics.

What about speed control between points? Without electronics it's not practical to modulate speed continuously at every point, but for slide drives that's not needed. Basically, the table should move quickly and smoothly to the final position and then come to a quick, smooth stop. All of that can be done with simple flow control valves for open-loop traverse, plus a final limit valve with an arm long enough to close off the valve ports gradually. The limit valve must have zero lap for accurate final cutoff.

REGENERATIVE DRIVES

Hydraulic power drives for machinery and vehicles can be made more efficient by storing energy from one part of a cycle (braking) to be used later in another (acceleration). Tyrone Hydraulics (Corinth, MS) completed such a design (Fig. 10.11). It's for a sawmill carriage drive and uses only fixed-displacement pumps and motors.

The carriage hauls the log back and forth past the saw. It not only transports the log but indexes it to a specific size. The carriages weigh from 2000 lb to 150,000

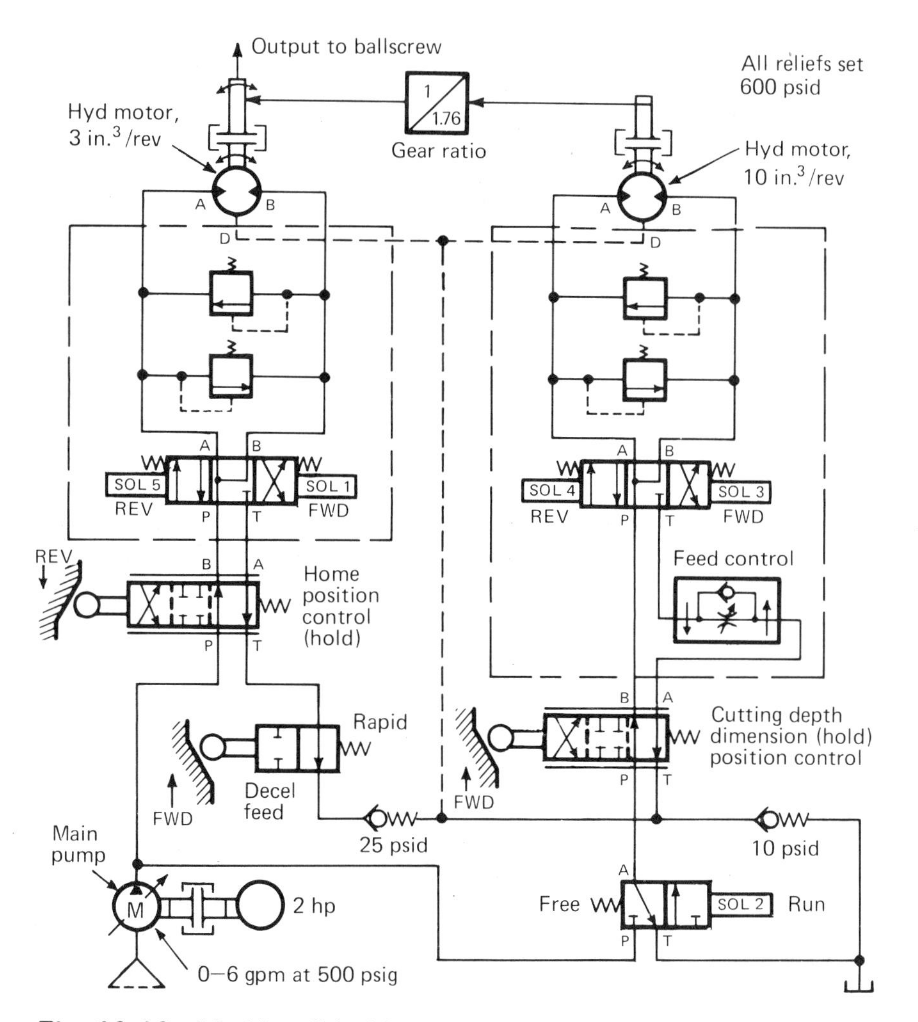

Fig. 10.10 Machine slide drive under hydraulic control competes well with electrical types. (Courtesy of Nutron Corp., Hingham, MA.)

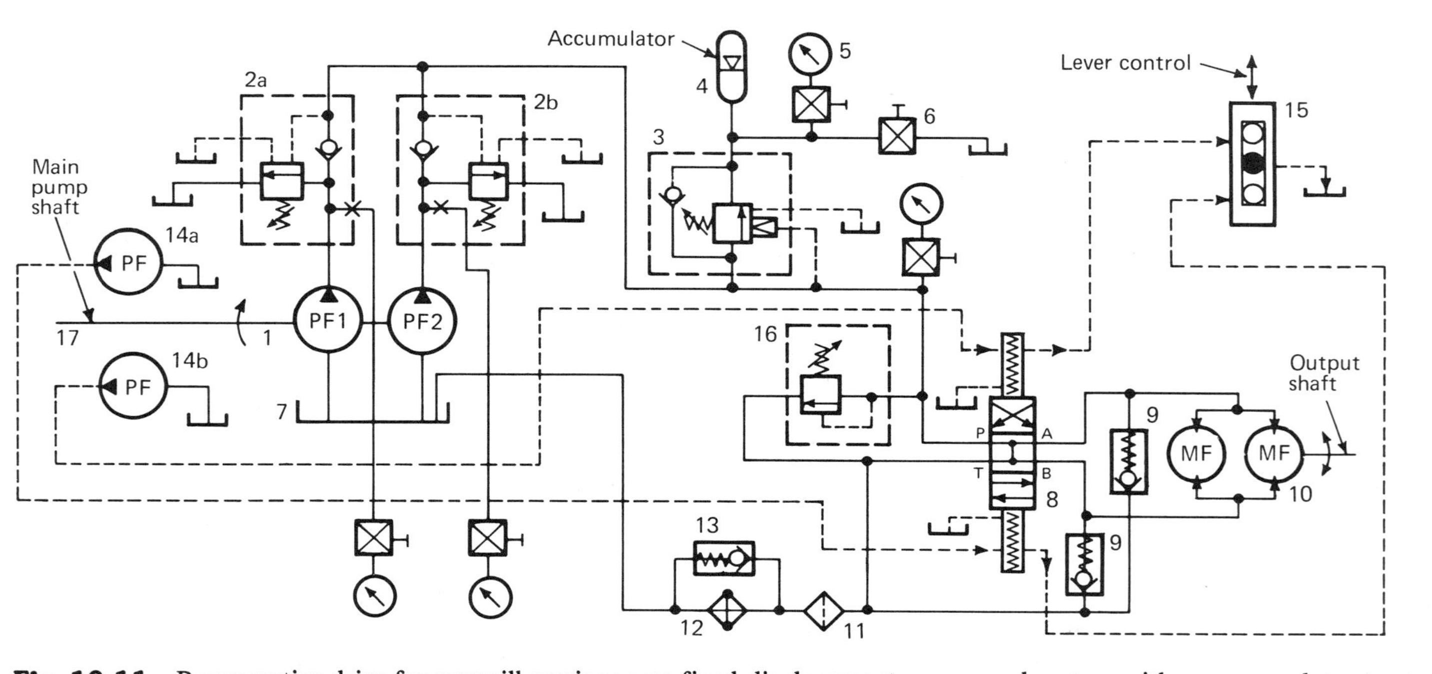

Fig. 10.11 Regenerative drive for sawmill carriage uses fixed-displacement pumps and motors with an accumulator to store the hydraulic energy. (Courtesy of Tyrone Hydraulics, Inc., Corinth, MS.)

lb and operate from 750 ft/min to 900 ft/min over strokes from 15 ft to 20 ft. All of this takes place 20 times per minute, which adds up to a lot of acceleration, deceleration, and high traverse speeds.

A typical 15,000-lb carriage producing a dozen boards a minute takes an average input of 200 hp, peaking at about 375 hp. A 200-hp electric motor is sufficient to drive the pumps because the peak load occurs for only a second and will not overheat the motor.

Considerable additional saving in input horsepower could have been effected by incorporating adjustable flow hydrostatic pumps, but the initial costs would have been up to 25% higher than for the fixed-displacement design chosen. Furthermore, the regenerative feature would not be there, and the braking energy would be wasted.

Electrical drives also were considered, but these are more expensive and require maintenance personnel that most mills do not have. Some mills use electrical drives anyway and take advantage of a different regenerative feature: to let braking return electrical current to the power line.

The hydraulic pump unit and/or pressure storage system (accumulator) shown in Fig. 10.11 supplies oil under pressure through the control valve (item 8) to the double hydraulic motor (item 10). The motor drives a rope drum through a roller chain reduction to give the desired carriage speed. A lever-operated control (item 15) feeds the log at a speed proportional to lever movement, forward and back. The center position is neutral, allowing the oil to circulate to tank at very low pressure. Storage system oil stays at high pressure.

The regenerative system unloads pump section 2 (PF 2) when system pressure is over 900 psi. Pump section 1 (PF 1) continues to deliver oil to the motor and/or storage system until system pressure reaches 1500 psi, at which time it is unloaded. Now both pump sections are unloaded, requiring virtually no input power.

The storage system (and pump section 1 if required) delivers oil to the motor during acceleration and stores oil pumped by the motor when stopping the carriage. During the acceleration to top speed, system pressure decreases because stored oil is being used with both pump sections unloaded.

When system pressure drops to 1250 psi, pump section 1 will cut in and continue to accelerate the carriage. As the carriage speed increases, system pressure again decreases, and when it drops to 850 psi, pump section 2 cuts in to reach top speed. When at top speed, the storage system valve (item 3) closes, and both pump sections move the carriage at top speed and low pressure.

During carriage braking, system pressure climbs rapidly, both pump sections unload, and the oil pumped by the motors while stopping the carriage is stored to be used for the next carriage acceleration. Note that no power is required from the pump sections to stop the carriage, resulting in substantial power savings and low system heat generation. A relief valve (item 16) set at 1600 psi protects the system.

Performance of the regenerative carriage drive was compared with a conventional carriage drive. This test was conducted under controlled laboratory conditions using a flywheel inertia load to simulate a very heavy carriage. A Honeywell Visicorder System recorded input amperes, pump pressure, system pressure, and storage system pressure vs. time. The power consumption for the regenerative system was over 40% less than the power consumption for the conventional system.

There was no substantial difference in the cycle time for the same load and pressure conditions. This demonstrates that a slight variation in system pressure is not a significant factor during acceleration and braking. Control characteristics were almost identical for the two drives. Control was improved in the regenerative system because of less mechanical shock.

LIFTING-CYLINDER CIRCUIT

Engineers at Moog Inc., East Aurora, NY, figured a way to save pump energy in the hydraulic system for a lift truck. They simply ported the static load pressure into special pilot circuits that help exploit the stored energy in the high-pressure fluid for the down stroke of the lift (Fig. 10.12).

The idea is simple enough, but it had never been applied to a lift truck. The justification is up to a 20% saving in power consumption over comparable proportional remote-controlled warehouse vehicles. Moog developed a special modular valve package that contains all of the necessary controls. All the truck designer has to do is supply the pump-reservoir package and the output load cylinder.

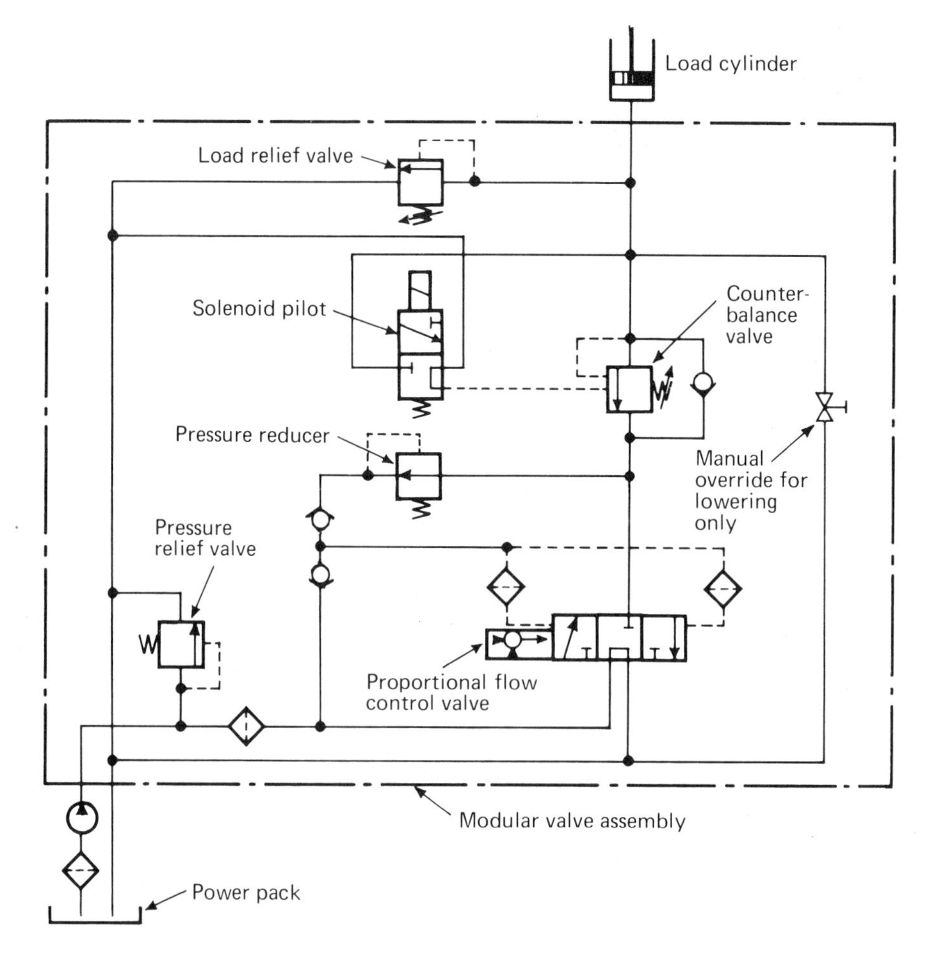

Fig. 10.12 Lift-truck lifting circuit exploits weight of load to pressurize and control the down stroke.

Starting with the raise cycle of the lift truck, the pump supplies pressure directly to the pilot stage of the three-way proportional open center flow control valve. Application of an electrical signal to the valve shifts the spool proportionally, closing the open center return port and raising system pressure. Simultaneously, system pressure is metered by the opposite three-way spool lobe to raise the load cylinder via the counterbalance valve. The pump relief and load relief valves protect the circuit from overpressure. Load-raising velocity is a function of input electrical signal.

The lowering cycle is unique. A constant voltage (12 or 24 Vdc) electrical signal is applied to the counterbalance pilot-stage solenoid coincident with the application of an opposite polarity command signal to the three-way control valve. Energizing the pilot-stage solenoid results in full opening of the counterbalance valve. Load pressure is then applied to the pilot and second stage of the proportional valve. Displacement of the proportional valve spool in response to the command signal opens the load port to tank. Load flow is therefore metered to tank, permitting lowering of the load. Similar to load raising, lowering velocity is proportional to the "lower" signal.

The interesting part is that during lowering, the electric motor is shut off at the pump, and the pilot pressure to the flow control valve now must come from the load cylinder through certain controlling circuits. The pressure downstream of the counterbalance valve is tapped for this purpose and fed through a pressure-reducing valve, half of a two-way check, and into the pilot stage of the flow control valve. The purpose of the pressure reducer is to control the minimum load-lowering velocity resulting from load flow through the first stage of the proportional valve with the second stage spool at null. A lower pilot setting reduces load flow through the first stage while a higher pressure increases first stage flow. The pressure reducer holds the pilot pressure to whatever value is decided upon by the customer. This particular one was set at 600 psi.

The load can be stopped at any point by returning the three-way spool to null and simultaneously deenergizing the counterbalance valve. In the event of an electrical failure, the load can be readily lowered with the manual override valve coming directly from the load cylinder. The valve can be placed where convenient.

SPLIT-FLOW PUMP FOR DOZER

The concept of split flow, where a single pump has multiple outlets serving different functions (Figs. 10.13 and 10.14) can reduce the pump size and the sizes of the driven motors and cylinders on off-road equipment.

A rubber-tired tractor bulldozer, four-wheel drive, will demonstrate the method nicely (Fig. 10.15). Engineers at Dynex prepared this example from typical data. Let's assume that it has these functions:

1. Propel vehicle
2. Operate dozer
3. Drive winch

The *propel* function is to be handled by a hydraulic motor on each wheel. Performance of an individual wheel must be as follows: maximum speed, 59.2 rpm; torque at maximum speed, 30,000 lb-in.; speed at maximum torque, 29.6 rpm; and

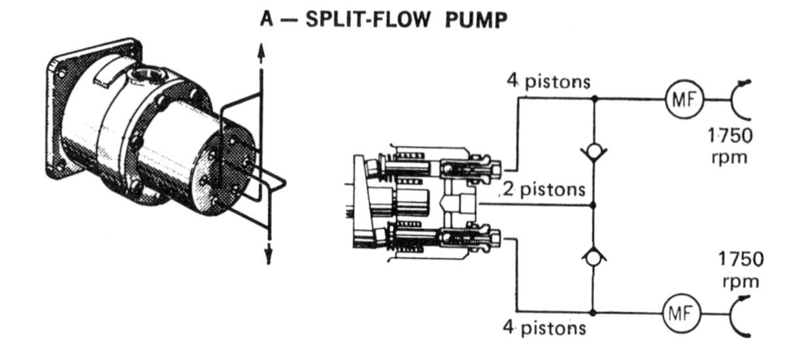

B — STEP-VARIABLE MOTOR

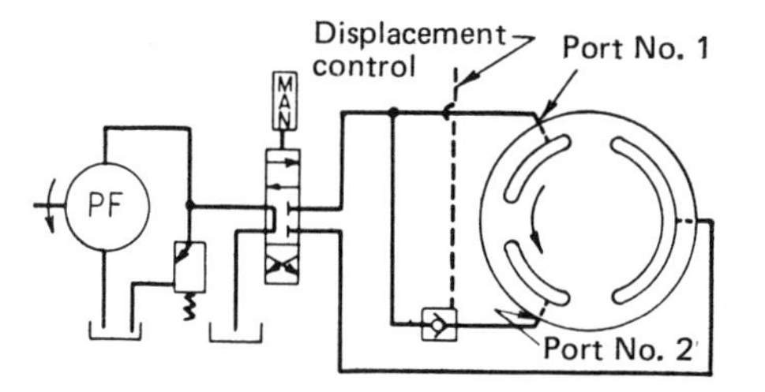

Split flow means that the individual pistons or groups of pistons are made to discharge into independent lines, free to develop different pressures depending on the loads. **Step variable** means that at least one pressure port of the motor has been divided in two, and by selection of valving the incoming fluid can enter one or both halves. If both halves are pressurized, fluid enters during full displacement of the motor (piston type) as it turns, and normal speed results. If only one half is pressurized, fluid enters during only half of the displacement and the motor has to run twice as fast to handle the fixed flow from the pump. Efficiency is less at partial displacement because the intake is blocked before the end of the stroke, but many applications can justify this to save a second pump or a speed changer.

Fig. 10.13 Split-flow pump and step-variable motor for tractor-dozer help reduce sizes of the hydraulic components. The overall design is described in the accompanying text and figures. (Courtesy of Dynex/Rivett, Pewaukee, WI.)

maximum torque at low speed, 74,500 lb-in. Tractor speed must be adjustable in two ways: for overall forward or reverse motion and for turning, where the outside wheels turn at a faster rate than the inside ones.

The *dozer* blade will be operated with two cylinders. Each must produce a maximum force of 10,000 lb at a maximum speed of 10 in./sec.

The winch will be turned with a single hydraulic motor. Maximum line pull must be 20,000 lb, maximum line speed 280 ft/min at 20,000 lb, maximum drum torque 200,000 lb-in., and maximum drum speed 53.5 rpm.

Normally, the *winch* will not have to operate at full speed while the vehicle is moving. Also, the winch motor and dozer cylinders will not have to operate together. The dozer and propel functions will be simultaneous.

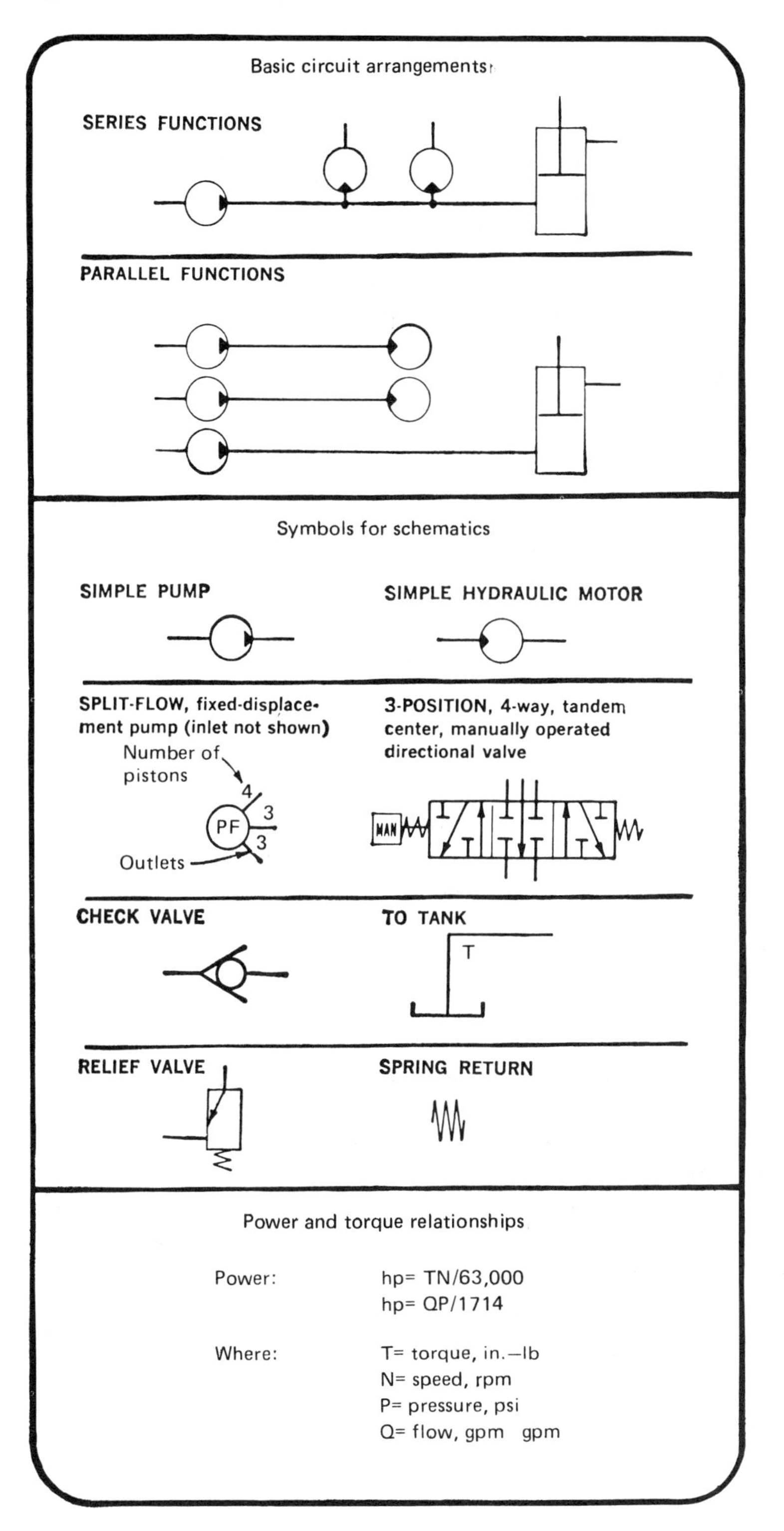

Fig. 10.14 Nomenclature for design of tractor-dozer.

Fig. 10.15 Tractor-dozer used for design example.

Ten Steps in Design

Step 1—Output Requirements. *Propel:* Condition I (high speed) is where wheel speed = 59.2 rpm and torque = 30,000 lb-in.; condition II (full torque) is where wheel speed = 29.6 rpm and torque = 74,500 lb-in.; and condition III (full torque and full speed), although it doesn't occur in practice, is where speed = 59.2 rpm and torque = 74,500 lb-in. The motor must be able to handle all three.

Propel power can be calculated for the three conditions using the conventional torque-speed formula:

$$\text{hp} = \frac{TN}{63{,}000}$$

For condition I, hp = 30,000 × 59.2/63,000 = 28.2; for II, hp = 74,500 × 29.6/63,000 = 35; for III, hp = 74,500 × 59.2/63,000 = 70.

The wheel motors will be geared down to wheel speed. For this example, 3000-rpm motors of the step-variable type are selected. Each motor operates at either of two displacements. At maximum vehicle loads, the higher displacement is chosen to provide maximum torque (low speed); at light loads, where higher speed is desired, the lower displacement is chosen (reduced torque).

Motor specifications are as follows: maximum displacement, 2.1 in^3/revolution; rated pressure, 6000 psi; rated speed, 3000 rpm; power output at rated speed and pressure, 90.5 hp; and torque at rated pressure, 1900 lb-in.

Gear reduction between motor and wheel = (output torque required)/(input torque × gear reduction efficiency). Assume efficiency = 92%, and calculate gear reduction GR for maximum torque:

$$\text{GR} = \frac{74{,}500}{1900 \times 0.92} = 42.6:1$$

Therefore, maximum motor speed = $59.2 \times 42.6 = 2520$ rpm, and motor speed at full torque = $29.6 \times 42.6 = 1260$ rpm. Required oil flow for four motors (each with 2.1-in^3 displacement) at 1260 rpm = about 50 gpm (12.5 gpm/motor), including leakage allowance. Power output is about 35 hp for one motor, 140 hp for four.

Dozer: The cylinder specifications are as follows: Maximum force = 10,000 lb/cylinder, and maximum speed = 10 in./sec. Assume 3500-psi maximum system operating pressure. Area A = 10,000/3500 = 2.86 in^2 (about a 2-in. bore). Based on a 2-in. bore, at 3200 psi, required flow to each cylinder = $3.141 \times 10 \times 60/231 = 8.15$ gpm, or 16.3 gpm for both. Power to cylinders is computed with a conventional flow-pressure formula:

$$\begin{aligned} hp &= gpm \times psi/1714 \\ &= 16.3 \times 3200/1714 = 30.4 \text{ hp} \end{aligned}$$

Winch: The drum horsepower = TN/63,000 = $200{,}000 \times 53.5/63{,}000 = 170$ hp. Motor horsepower must be that much or greater.

The motor selected has these specifications: displacement = 6 in^3/revolution; rated pressure = 6000 psi; rated speed = 2500 rpm; torque at rated pressure = 5500 lb-in.; and output at rated speed and pressure = 218 hp.

Gear reduction GR between the hydraulic motor and the winch drum, based on maximum motor torque, is

$$GR = \frac{200{,}000}{5500 \times 0.92} = 39.5 : 1$$

Therefore, maximum motor speed = $53.5 \times 39.5 = 2110$ rpm, and approximately 57.2 gpm will be required.

Step 2—Categorize Outputs.

propel = rotary
dozer = linear
winch = rotary

Step 3—Determine the Total Number of Simultaneous Functions.

A. Propel motors and dozer cylinder
B. Propel motors (at slow speed) and winch

For A, maximum flow = 50 + 16.3 = 66.3 gpm. Maximum propel motor pressure = 6000 psi. Maximum dozer cylinder pressure = 3200 psi.

Step 4—Determine the Number of Series, Nonsimultaneous Functions. Do it for dozer, propel, and winch circuits.

Step 5—Determine the Number of Parallel Simultaneous Functions. Do it for propel and dozer circuits.

Step 6—Establish Priority. Propel and dozer have priority over the winch.

Step 7—Size the Piping and Valving. See Fig. 10.16 for normal functions. Each of

A — Piping

BRANCH OF CIRCUIT	PROPEL MOTOR	DOZER CYLINDER	WINCH MOTOR
Max flow, gpm	12.5	16.3	57.2
Max pressure psi	6000	3200	6000
Tube size, in.	¾	¾	1½
Tube material, ASTM	4130	4130	4130
Tube wall in.	0.120	0.109	0.250

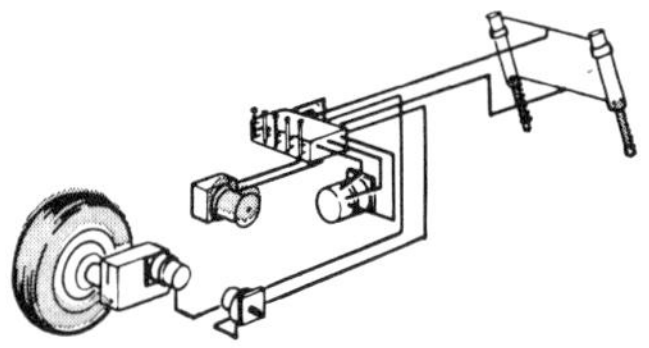

B — Valving

FUNCTION	TYPE OF VALVE
Step variable selector	3-way, 2 position
Propel directional	4-way, 3 position, tandem center
Winch directional	4-way, 3 position, tandem center
Dozer directional	4-way, 4 position

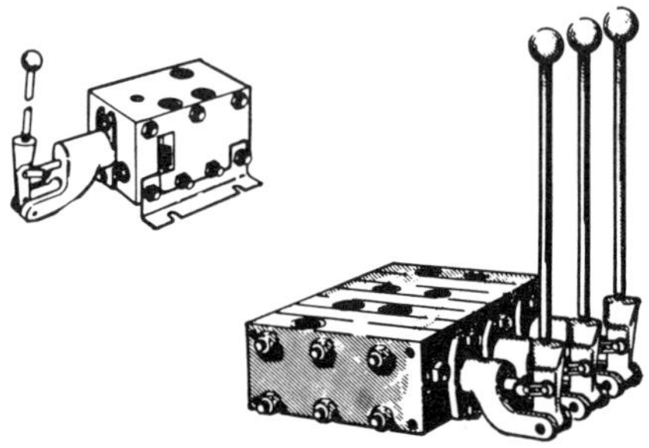

Fig. 10.16 Piping and valving for tractor-dozer example, step 7.

the valves incorporates additional functions: The step variable selector valve has a built-in check valve; the propel directional valve and winch directional valve have built-in relief valves and motor overload valves; the dozer directional valve has a built-in relief valve and a fourth position called "float." In float position, all ports (pressure, cylinder port 1, cylinder port 2, and tank) are interconnected, thus allowing the dozer blade to move up or down as the ground contour varies.

Step 8—Add Up the Simultaneous Power Requirements.

horsepower for propel and dozer = 205.4 hp

horsepower for winch = 200 hp

Because the propel-dozer functions do not operate at the same time as the winch, the prime mover power needs to provide only 205.4 hp.

Step 9—Finish the Circuit Layouts, Specify the Central Hydraulic System. To provide the independent simultaneous flow to each of the four propel motors plus the dozer cylinder, choose two split-flow pumps (independent outlet ports). Each pump will be split into three independent flows. Two pumps rated for 33.15 gpm at 6000 psi will provide the oil.

When steering the vehicle, additional flow is required by the outside wheels. The circuit is designed to ensure flow from three pump pistons to each motor. Four pistons from one split flow pump are connected through check valves to all four motors. In this way flow will go to the motors with the least resistance.

One of the objectives is to make use of all or part of the oil from the propel-dozer circuits for the winch circuit. So if the outlet series ports of the propel and dozer valves are connected into the winch circuit, the winch circuit is inoperative only when both the propel *and* the dozer are operating. But when only the propel function is in operation, the winch is able to operate slowly but at full torque.

At this point, an adjustment of the winch gear ratio GR might be desirable. In the example, we initially based the winch gear ratio on torque. Now, because we have a known gpm available for the winch motor (from propel and dozer circuits when not used), we can base the gear ratio on motor speed resulting from the available gpm. Flow from propel and dozer circuit = 66.3 gpm, winch motor speed = 2450 rpm, and required winch drum speed = 53.5 rpm. Thus, GR = 2450/53.5 = 45.8 : 1.

With the proposed circuit, the winch gear reduction should be increased from 39.5 : 1 to 45.8 : 1. Winch circuit pressure becomes 5100 instead of 6000 psi for the required drum torque. Winch tubing can be reduced to 0.219 in. wall thickness.

Step 10 – Select a Prime Mover. Based on a maximum pump output horsepower of 205.4, an engine of approximately 230 hp is required. The final design is shown in Fig. 10.17.

SUPERPRESSURE FLUID POWER

The art of designing for pressures far above 10,000 psi – and even up to 500,000 psi – holds a special fascination.

Beeswax and die lubricant at up to 200,000 psi experimentally squeezed metal billets successfully into wire at Western Electric's lab in Princeton, NJ.

Water jets at up to 100,000 psi have been cutting thin metal, bricks, and assorted materials in labs in the United States, Canada, Russia, and Europe for many years. Water-jet mining of coal is being actively developed in Russia and China, and tunnel-digging is not far behind.

Clothing and upholstery researchers have discovered that high-pressure jets of water can cut through multiple layers of cloth or plastic and can be computer-controlled to follow complex patterns. Canadian paper mills already are using such jets at 50,000 psi to slit newsprint in a process for the recovery of defective rolls, and it's proving cheaper than knives.

Isostatic pressing of metal powder is becoming commonplace. Industrial diamonds are being formed in hydraulically powered presses all over the world.

Hydroforming of large plates, autofrettaging of gun barrels, and testing of high-pressure cylinders and housings are additional uses for superpressures.

Even plumbing manufacturers are in the act. At least one is hydrostatically expanding metal tubing against reinforced dies to create instant faucets, exploiting pressures up to 40,000 psi.

Making Putty of Metal

Fundamental research on materials, particularly brittle materials, has uncovered great new ways to make things. The father of it all was P. W. Bridgman, Harvard Univ., who received the Nobel prize for his work in high pressure. He discovered the underlying principles of superhydrostatics and developed much of the equipment.

He proved that metals and ceramics become increasingly ductile as ambient pressure is boosted to 100,000 psi and more. The higher the pressure, the greater the

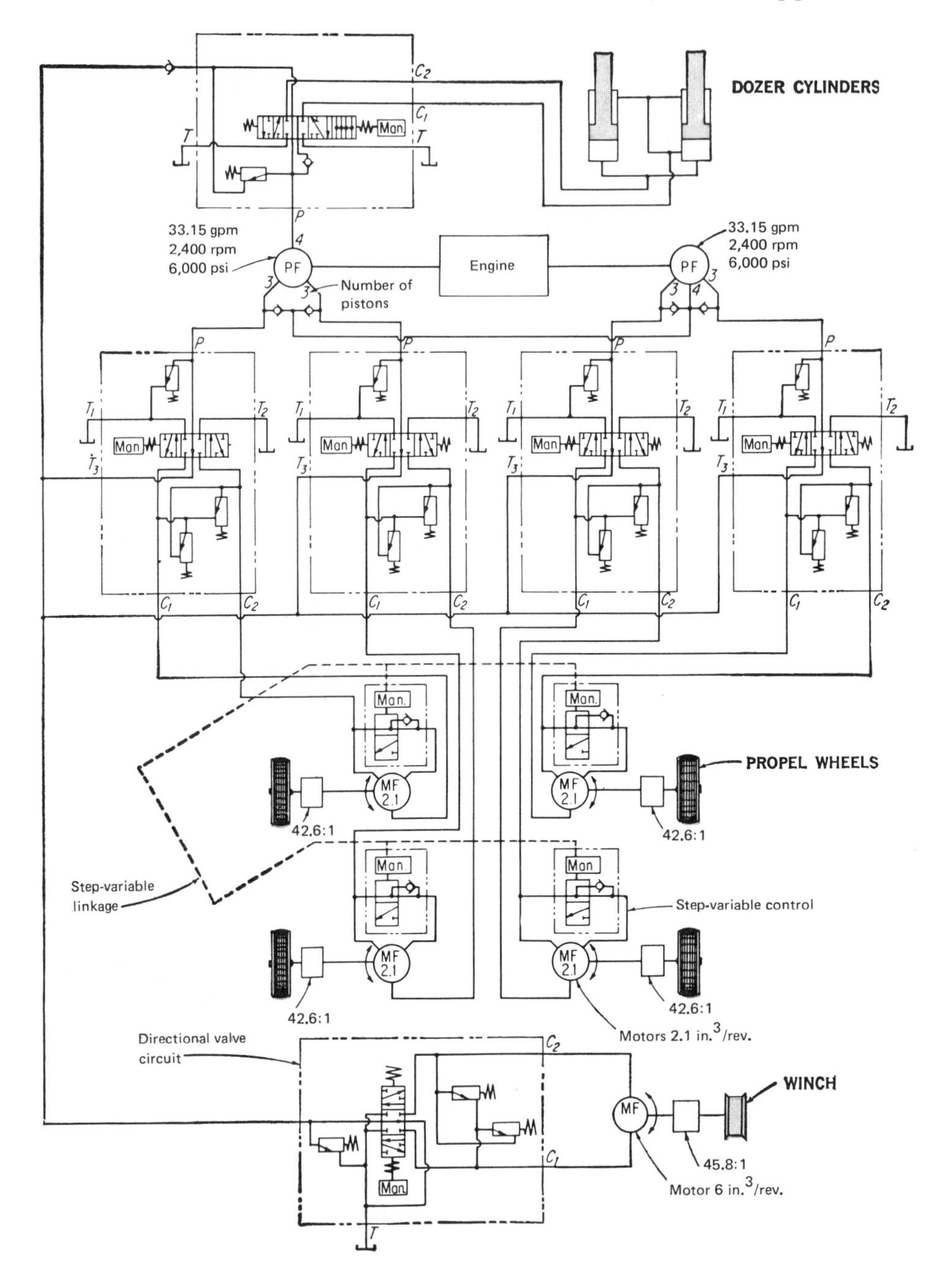

Fig. 10.17 Final tractor-dozer circuits (Courtesy of Dynex/Rivett, Pewaukee, WI.)

effect. Mild steel responds well at 200,000 psi. Some other materials require much higher pressures.

Even high-alloy gears and high-speed twist drills can be extruded if the pressure is high enough, but not yet economically. Dissimilar materials often can be bonded if joined in a high-temperature superpressure environment.

Special Kinds of Fluids

Ordinary hydraulic oil will not work at superhigh pressures. Only special liquids can stand the gaff. Diesters are useful up to 200,000 psi or more. Normal pentane, isopentane, gasoline, some silicones, and certain other liquids can be used at pressures to 500,000 psi and above.

Gases sometimes are selected as the medium for pressures from 70,000 to 500,000 psi. Not all gases have that range, however. Argon becomes solid at 180,000 psi and 80 F.

For pressures above 500,000 psi, certain solids are plastic enough to serve as a pressure transmitter. Examples are pyrophilite, lead, and indium. Potassium works well at cryogenic temperatures.

Water, Too

Water is a special case. It will turn to ice at about 160,000 psi at room temperature. But it is inexpensive and relatively inert and, therefore, is chosen for high-pressure cutting jets. The only problem seems to be wet sheets: Upholstery and clothing manufacturers find it somewhat inconvenient to dry out cut cloth after it is neatly sliced with water jets. The dyes run at the edges, and mildew spoils some pieces.

Industrial experimenters are moving ahead seriously with plans to exploit water jets for cutting metal commercially. Bendix Research Labs, Southfield, MI, has cut thin-gage metal, 1-in. plywood, ½-in. plastic sheet, firebrick, glass, rocks, coal, and stacks of woven fabrics.

The pressure source can be a specially built continuous-flow system that delivers upward of 1.4 gpm of water or other fluid at pressures from 10,000 to 100,000 psi. Jet velocities range from 1000 to 3000 ft/sec and nozzle sizes from 0.025 to 0.005 in. diameter.

METALWORKING UNDER PRESSURE (MUP)

The potential for hydrostatic working of metals under extreme pressures is vast, and researchers worldwide are competing strongly to move from lab to shop with various techniques (Fig. 10.18).

Metal forming and bending under high pressure, including some versions of explosive forming, are part of an old art. Hydrostatic extrusion from high pressure to atmospheric pressure is just emerging as a technology. Extrusion from high pressure to a somewhat lower intermediate pressure, to keep the rod always under compression, is in its infancy.

The whole concept has been dubbed MUP, for metalworking under pressure. Hybrid hydrostatic processes based on older metalworking methods have been around a long time. Typical are forming-die, blind-die, and combination-die processes, deep-shell drawing, and tube-flanging.

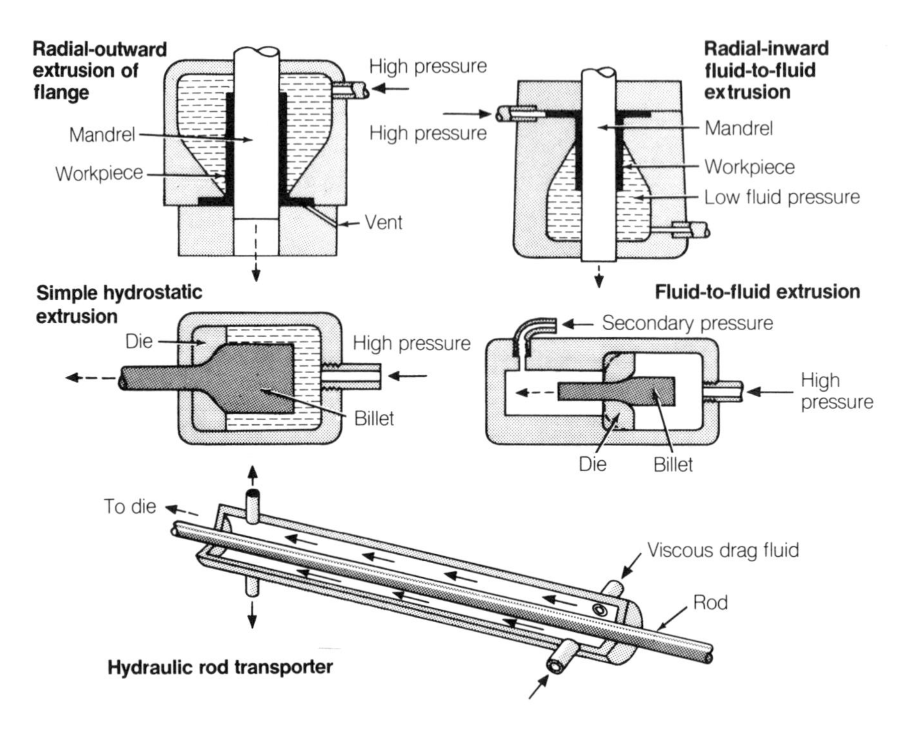

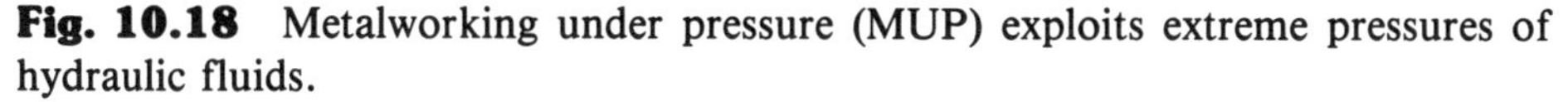
Fig. 10.18 Metalworking under pressure (MUP) exploits extreme pressures of hydraulic fluids.

Western Electric developed several of these techniques for commercial use. It used a hydrostatic-forming-die method to make coaxial connectors at less than a tenth of the former cost. The company also has made cores for electrical relays with hydrostatic blind dies and coaxial jacks with combination dies. The coaxial jacks can be spat out at the rate of 10 a minute, including extrusion, cutoff, bulging, and hole punching, in one operation.

A new technique is hydrostatic deep-shell drawing. Western Electric does radial-inward extrusion of cylindrical shells to unprecedented length-diameter ratios. It also flanges rectangular shells with variable cross section and wall thickness and forms ultra-thin-wall (0.010-in.) shells by friction-aided extrusion under pressure. A related area is tube flanging under pressure, with a very hard (Rockwell B-90) copper tubing that remains ductile. The pressures needed to extrude these hollow forms range upward to about 400,000 psi.

American Standard's faucet-making process, called Hydramold, is a form of bulging. Pressure inside the tube forces it radially outward into a faucet-shaped die. Simultaneously, the two ends of the tube are pushed inward axially by mechanical means, assisting the bulging.

Hydrostatic Wire Extrusion

One outstanding potential commercial application for MUP seems to be in the hydrostatic extrusion of billets into wire (Fig. 10.19). The billets or rods are not forced mechanically through the die, as in conventional techniques, but are extruded by extreme hydrostatic pressure and made to slip through the die on a film of fluid.

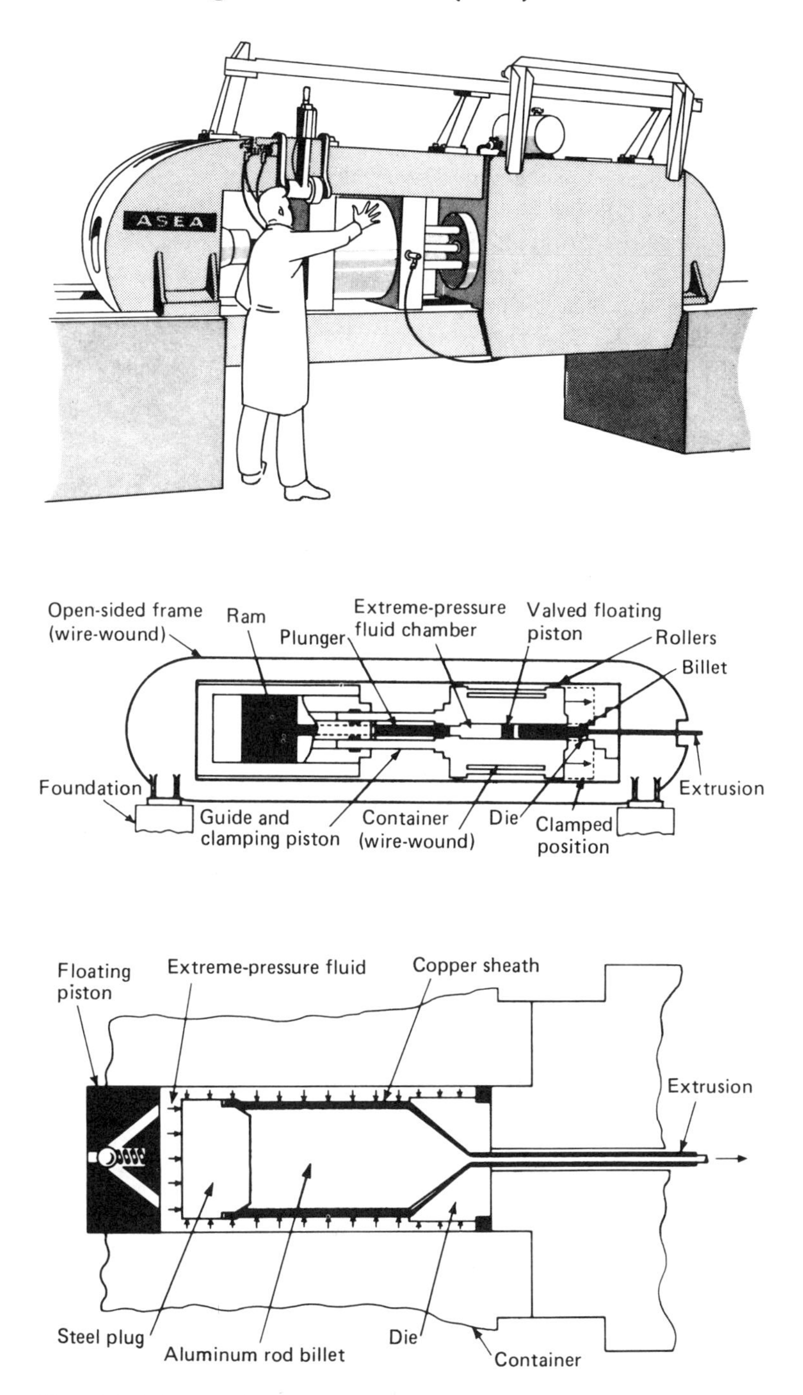

Fig. 10.19 Hydrostatic extrusion of aluminum billets directly into wire is possible with the apparatus in these three drawings. In the top drawing, the operator has his hand on the ram intensifier assembly. The middle drawing shows details of the ram and the extreme-pressure chamber. The bottom drawing is a close-up of an aluminum billet being squeezed through a die. The billet can be 40 in. long; the plunger insertion creates pressures to 290,000 psi (ASEA, Houston, TX).

Battelle Memorial Institute, Columbus, OH, has been successful in drawing 0.018-in.-diameter beryllium wire at 2000 ft/min through a die, using a combination of hydrostatic extrusion and mechanical pulling. That compares with 150 ft/min for conventional drawing.

The process was able to reduce the wire diameter to 0.005 in. in three passes at 325 F. Hydrostatic pressure around the wire was built up to about 200,000 psi with a modified platen press for the experiments. Several kinds of die lubricant were tried successfully: acidless stearine, triaryl phosphate, paraffinic resin, and castor oil. The work was conducted under contract to the U.S. Air Force.

Other successful Battelle MUP projects include tight bending of titanium and extrusion of profiled tubing from alloy steel, maraging steel, and aluminum.

Pressure-to-Pressure Extrusion

The ultimate in cold working of brittle materials is to surround the entire extrusion process with a superpressure environment. Much experimental work has been done, but no process is yet in commercial operation.

The simplest concept is a hydrostatic extruder that necks down the billet at one pressure—say 400,000 psi—and delivers the extruded shape continuously into a slightly lower-pressure environment, called the receiver.

The advantage is that incipient cracks caused by the extrusion are suppressed by the extreme pressures surrounding the billet and the extruded shape. Extensive tests have proved it to be so, but nobody seems to know exactly why. Even tungsten rod can be extruded through a die without cracking.

The receiver pressure must be held at a sufficient differential below the primary pressure to allow extrusion to occur. One way to regulate this pressure is an auxiliary billet used as a plug between the receiver and atmosphere. The plug will extrude at the desired receiver pressure, thus holding it constant. It is a lab technique only. Another regulator concept is a capillary tube to bleed off pressure.

Hot Gas Bonding

Many normally unbondable materials can be joined with a combination of superpressure and high temperature.

Commonly, such equipment operates at pressures to 30,000 psi and at temperatures to 3000 F with inert gas. Kobe Steel Ltd., Kobe, Japan, reports it has autoclaves pressurized to 100,000 psi. At least a dozen U.S. companies make autoclaves, some up to 200,000 psi.

For even higher pressures, special electrically heated capsules can be inserted in modified hydrostatic chambers, where small-scale tests can be run on hot-gas bonding and diffusion up to over 400,000 psi. Designs for this vary, but the principle is as follows:

The capsule is a miniature oven, laced with electric heaters in several zones. A hollow in the capsule contains the experiment. A jacket of inert porous material is placed around the capsule, and the package is inserted in the chamber.

The electrical leads are brought out through seals in the chamber. Inert gas introduced into the jacket serves three functions: pressurization, thermal insulation, and electrical insulation.

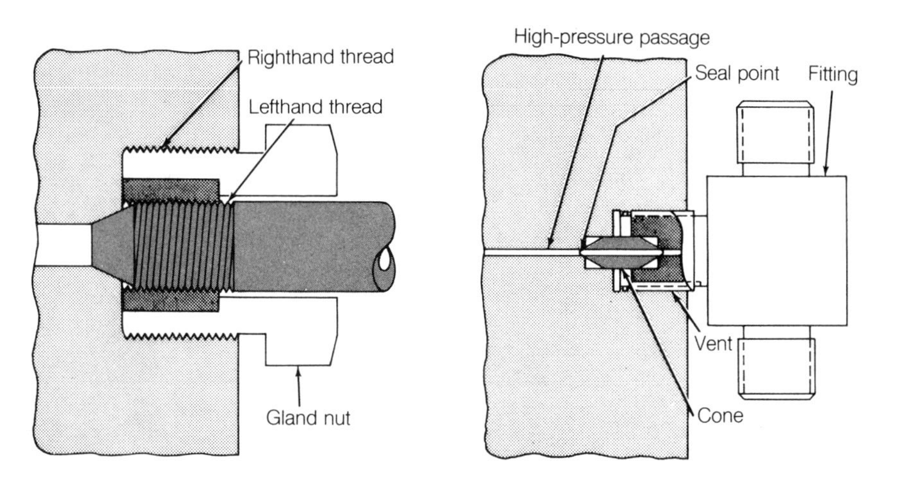

Fig. 10.20 Extreme-pressure fittings: The left-hand drawing shows reverse threading to ensure that the tube tightens when the gland does; the right-hand drawing shows a replaceable cone insert.

Basic Equipment Needed

Pumps and intensifiers for liquid and gas are the first requirement, and standard items up to 200,000 psi are offered. The first commercial 200,000-psi intensifier was designed and built by Harwood Engineering in about 1948. The Russians developed a 20-kilobar (290,000-psi) compressor/pump for gases or liquids. Special units (nonstandard) in the United States go up to over 550,000 psi. However, most listings are under 100,000 psi.

Tubing is specially made for this service. The manufacturer starts with special-grade tubing of normal commercial size and then reduces its diameter through progressive dies until the desired extremely small inside diameter is achieved. It takes special skill to reduce the diameter uniformly without causing surface cracks (sunbursts). Concentric tubes also are used.

Valves are available up to 200,000 psi, and even higher for custom applications. Most fittings are metal-to-metal (Fig. 10.20). One design has a right-hand thread on the outer element and a left-hand thread on the inner. The combination makes the inner thread self-tightening when the outer thread (gland) is wrenched down. Above 125,000 psi, a conical insert may be introduced. It is a replaceable element that helps extend the life of the more expensive body portions of the valves.

11

Pneumatic Systems and Components

The primary emphasis is on low- to moderate-pressure systems used in manufacturing, process control, vehicle control, machine control, materials handling, road repair, shop maintenance, engine starting, energy storage, ordnance, ship control, mining, and railroad braking.

Some data on high-pressure air and other gases are included to help in calculating storage volumes and wall stresses in high-pressure containers. For more detailed properties, see any good handbook—such as Ingersoll Rand's *Compressed Air and Gas Data.*

The final sections of this chapter cover low-pressure systems for shop air and instrument control, plus unusual applications of air.

SUMMARY OF AIR PROPERTIES

The accompanying graphs (Figs. 11.1 through 11.5) are approximations based on assorted data. The values above 5000 psi were taken from early tests by Amsterdam Institute in Holland. More exact values at certain lower pressures are given in succeeding chapters where needed in calculations.

Compressibility Factor (Fig. 11.2)

A compressibility correction factor Z inserted in the ideal equation of state $Pv = RT$ will make the equation valid for a real gas. Thus, $Pv = ZRT$, where P is pressure in lb/ft^2, v is specific volume in ft^3/lb, R is a perfect-gas constant (53.3 for air), and T is temperature in deg Rankine (absolute).

Factor Z, plotted for real air, depends on temperature, pressure, and the nature of the gas. If the value is known for each condition, then perfect gas relations apply. At points where Z is unity, a real gas behaves as a perfect gas.

A way to express the deviation of air from the perfect state is as a percentage of ideal compressibility: $100(Z - 1)/1$. The next plot (Fig. 11.3), obtained from source data separate from those for factor Z, gives the deviation in the range from 0 to 5000

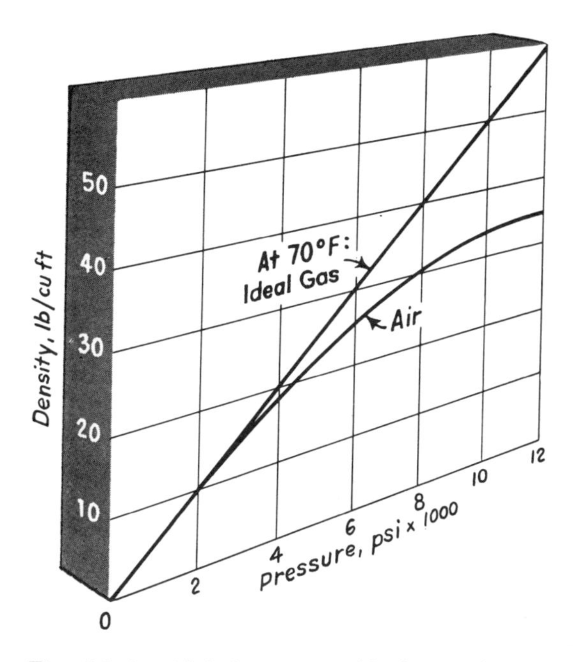

Fig. 11.1 Air behaves as an ideal gas at low to moderate pressures and temperatures.

psia, −100 to +250 F. A positive value indicates greater specific volume than that for a perfect gas at the same temperature and pressure.

Specific-Heat Coefficients (Fig. 11.3)

Individual values for coefficients c_v (at constant volume) and c_p (at constant pressure) are plotted. The ratio $k = c_v/c_p$ is important to thermodynamic calculations and is affected considerably by changes in temperature and pressure. In general, k increases directly with pressure but inversely with temperature.

Velocity of Sound (Fig. 11.4)

For a perfect gas,the velocity of sound $a_0 = \sqrt{kgRT}$, where g is 32.2 ft/sec²; the other units are already defined. For air, the velocity of sound is 1126 ft/sec at 68 F, 14.7 psia. For low pressures (less than 100 psia) this velocity is relatively independent of pressure (changes in density are compensated by changes in elasticity) and is roughly proportional to the square root of temperature. In round figures $a_0 = 49\sqrt{T}$. Example: At 70 F, 14.7 psia, $a_0 = 49\sqrt{530} = 1128$ ft/sec.

At higher air pressures, the approximate equation does not apply because the velocity is not constant at all pressures for each temperature. The greatest variation is for −100 F air: Sound velocity at 5000 psia is 1801 ft/sec, which is almost twice the value of 928 ft/sec based on $49\sqrt{T}$.

Viscosity (Fig. 11.5)

Absolute viscosity μ is a measure of the force F to shear a 1-ft cube of fluid at a specified velocity V: $\mu = (F/A, \text{lb/ft}^2) \div (V/H, \text{ft/sec-ft}) = FH/VA$, lb-sec/ft². Alternate

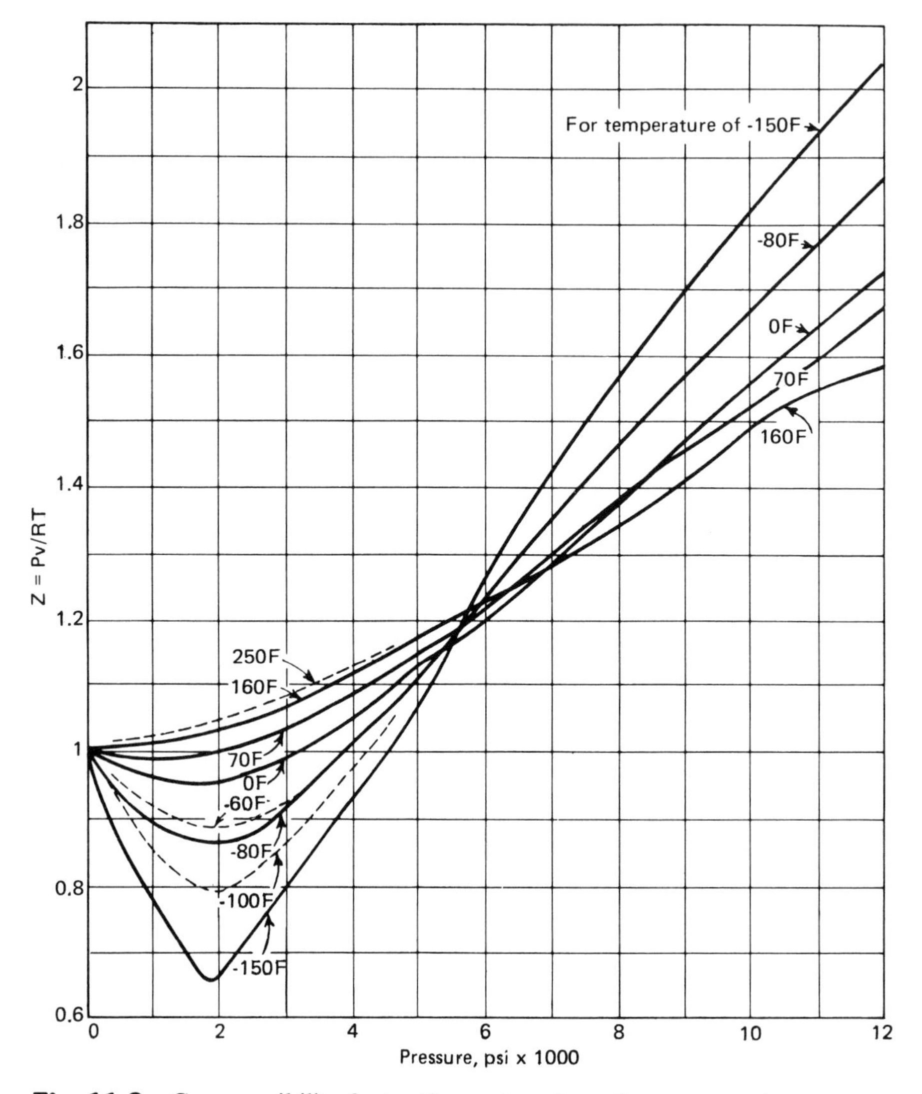

Fig. 11.2 Compressibility factor Z, as a function of pressure and temperature, for air.

units, slug/ft, are obtained by replacing force F with its equivalent in mass and acceleration units. For even different units, $lb\text{-}sec/in^2$, see Table 11.1.

Kinematic viscosity ν is absolute viscosity, slug/ft-sec, divided by mass density, $slug/ft^3$, giving units of ft^2/sec.

An Example

Find the properties of air at 3000 psi, 160 F. Compressibility factor Z is approximately 1.064; therefore the deviation is plus 6.4%. Mass density = P/ZRT = (144 × 3000)/(1.064 × 53.3 × 620) = about 12.3 lb/ft^3 or 0.381 $slug/ft^3$. Specific heat $c_p = 0.287$ btu/lb/F; $c_v = 0.179$. Ratio $k = 0.287/0.179 = 1.61$. Velocity of sound = 1433 ft/sec.

Kinematic viscosity is 14×10^{-7} or 0.0000014 ft²/sec. Absolute viscosity = $(14 \times 10^{-7}) \times 0.381 = 5.33 \times 10^{-7}$ lb-sec/ft². In metric units, absolute viscosity is $5.33 \times 10^{-7} \times 47{,}880 = 0.0255$ centipoise.

PRESSURIZED CONTAINERS (FIG. 11.6)

With good reasons, the Bureau of Explosives developed the ICC regulations for shipment of pressurized containers. Make no mistake!

Just filling a tank or bottle with compressed gas and integrating it into the design of a machine is not too difficult. What's hard is doing it cheaply and safely. You've got to understand three disciplines: gas physics, stress analysis, and safety engineering. Of the three, safety engineering exerts the greatest influence. If you cannot satisfy

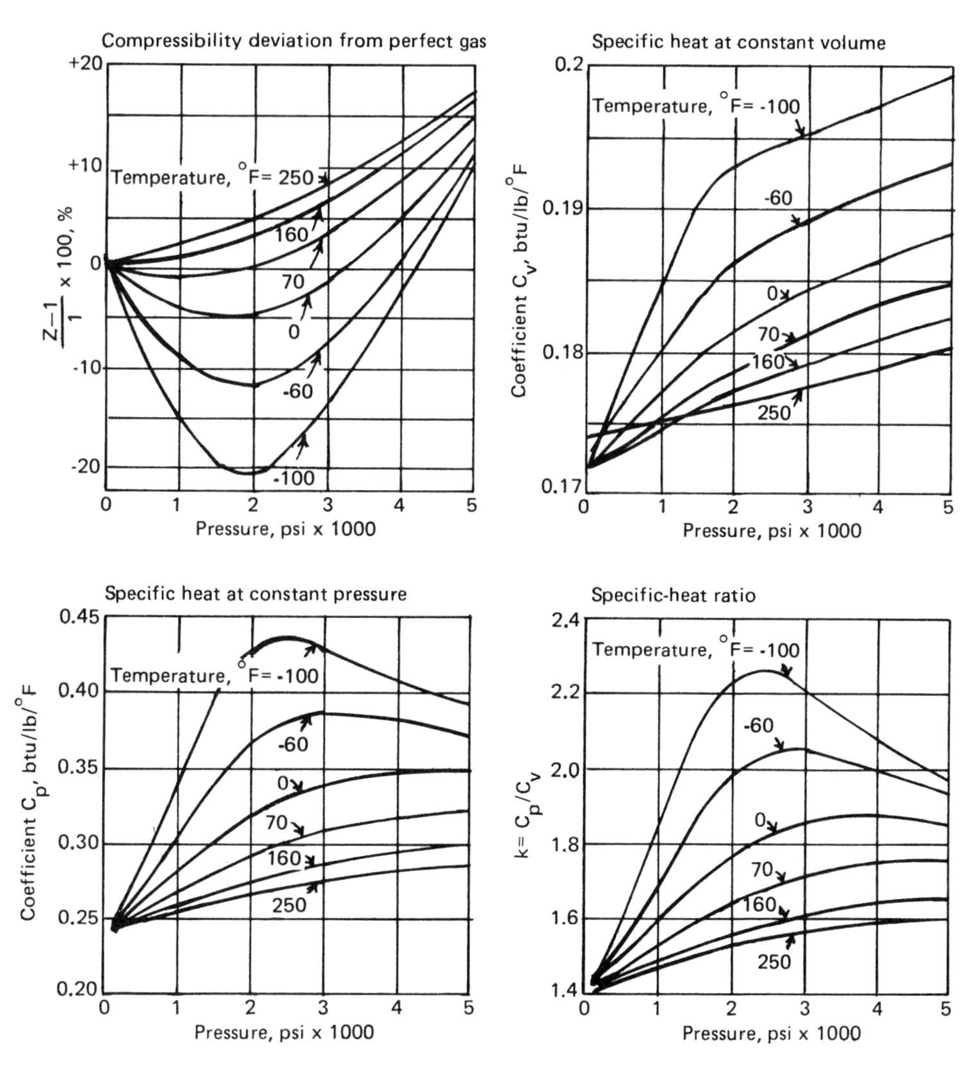

Fig. 11.3 Effects of temperature and pressure on compressibility and specific heat coefficients.

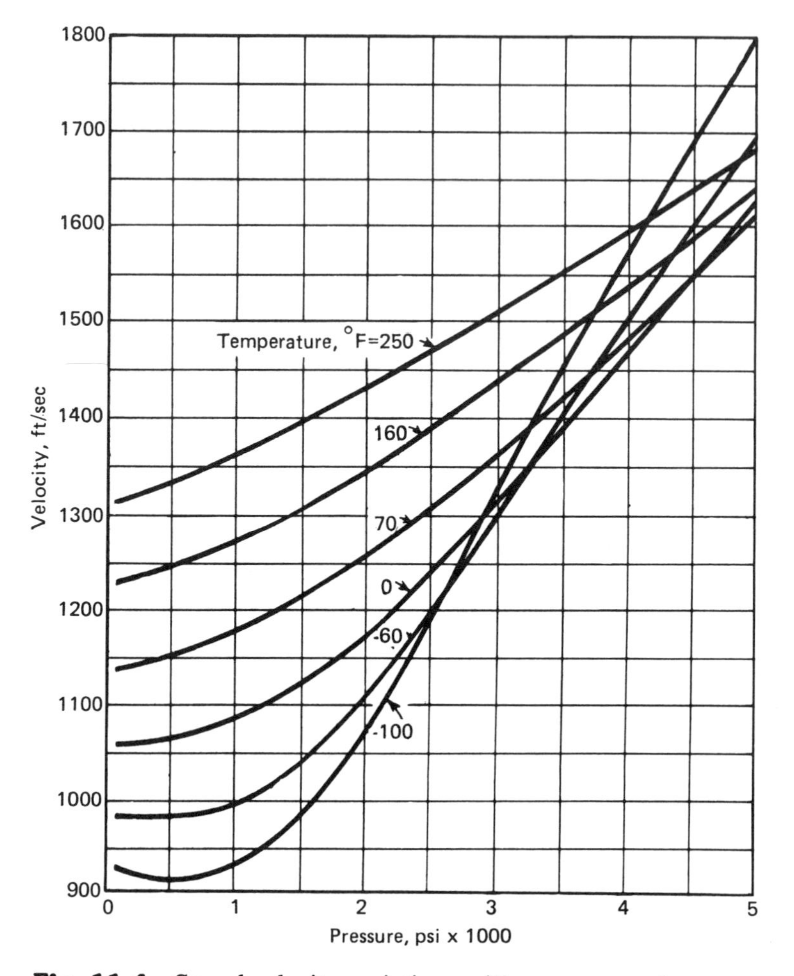

Fig. 11.4 Sound velocity variations with pressure and temperature, for air.

state, national, and industrial safety regulations, you may end up with a machine that is uninsurable and which nobody will buy.

Iron bottles of laughing gas, imported from London in 1870, were the first gas cylinders in the United States. By 1911, Pittsburghers were bottling oxygen and acetylene in tanks of "tough, pliable steel . . . tested to at least twice the working pressure."

Today, a surprising variety of machines are bottle-fed from one type of pressurized gas container or another.

For instance, helium compressors for some nuclear reactors have hydrostatic helium gas bearings that lift the heavy rotors friction-free until the rotating speed generates a supporting gas film hydrodynamically. Gas bottles supply the pressure for that initial lift. Diesel, gasoline, and jet engines have bottles of compressed gas for powering the pneumatic cranking motor. Metalworking shops have tanks of compressed nitrogen to accelerate the high-velocity pistons of certain impact machines. Airbrakes on trains and buses always need pressurized tanks. Refrigeration systems are pressurized. Even industrial hydraulic systems have compressed-gas chambers if accumulators are used or if the reservoir is pressurized with inert gas.

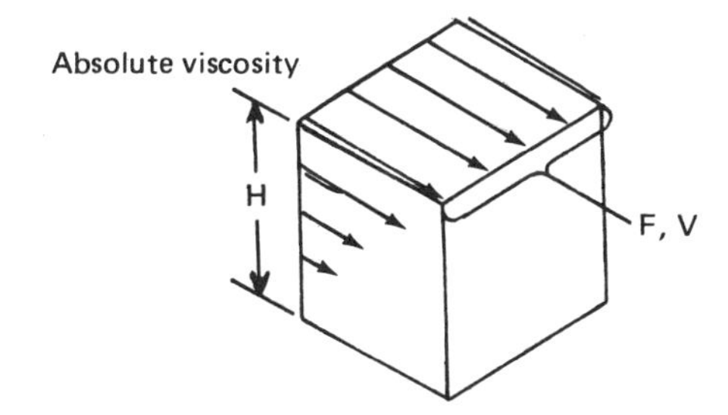

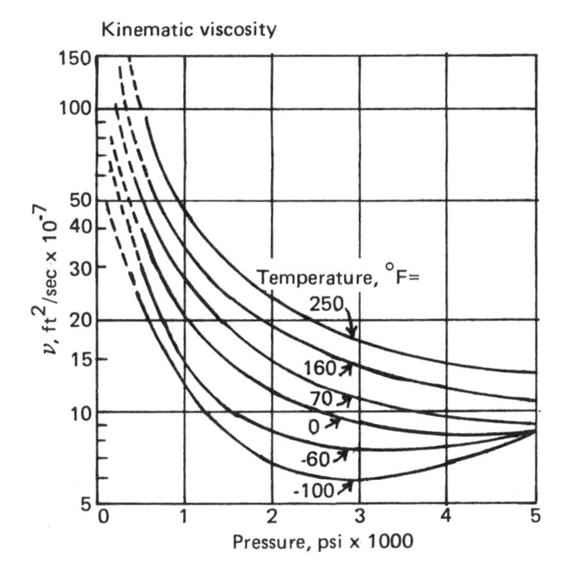

Fig. 11.5 Absolute viscosity (drawing) is a physical measurement. Kinematic viscosity (graph) is a defined quantity, the ratio of absolute viscosity to mass density (also see Chap. 1).

This chapter concentrates on containers small enough to mount on a machine—up to about 10 gal. Site-mounted tanks are discussed briefly; the principles are the same. The chief problem always is how to get as much usable pressurized liquid or gas (or both) as possible into the container without creating a hazard.

Table 11.1 Viscosity and Density of Air at Standard Conditions

Viscosity (absolute)	2.6×10^{-9} lb_F -s/in.² at 68 F 3.3×10^{-9} lb_F -s/in.² at 100 F
Density (lb mass)	0.0752 lb/ft³ at 14.7 psia, 68 F 0.0709 lb/ft³ at 14.7 psia, 100 F

Note: standard cfm (scfm) $= \text{actual cfm} \dfrac{P_{abs}}{14.7} \times \dfrac{528}{T, {}^{\circ}R}$

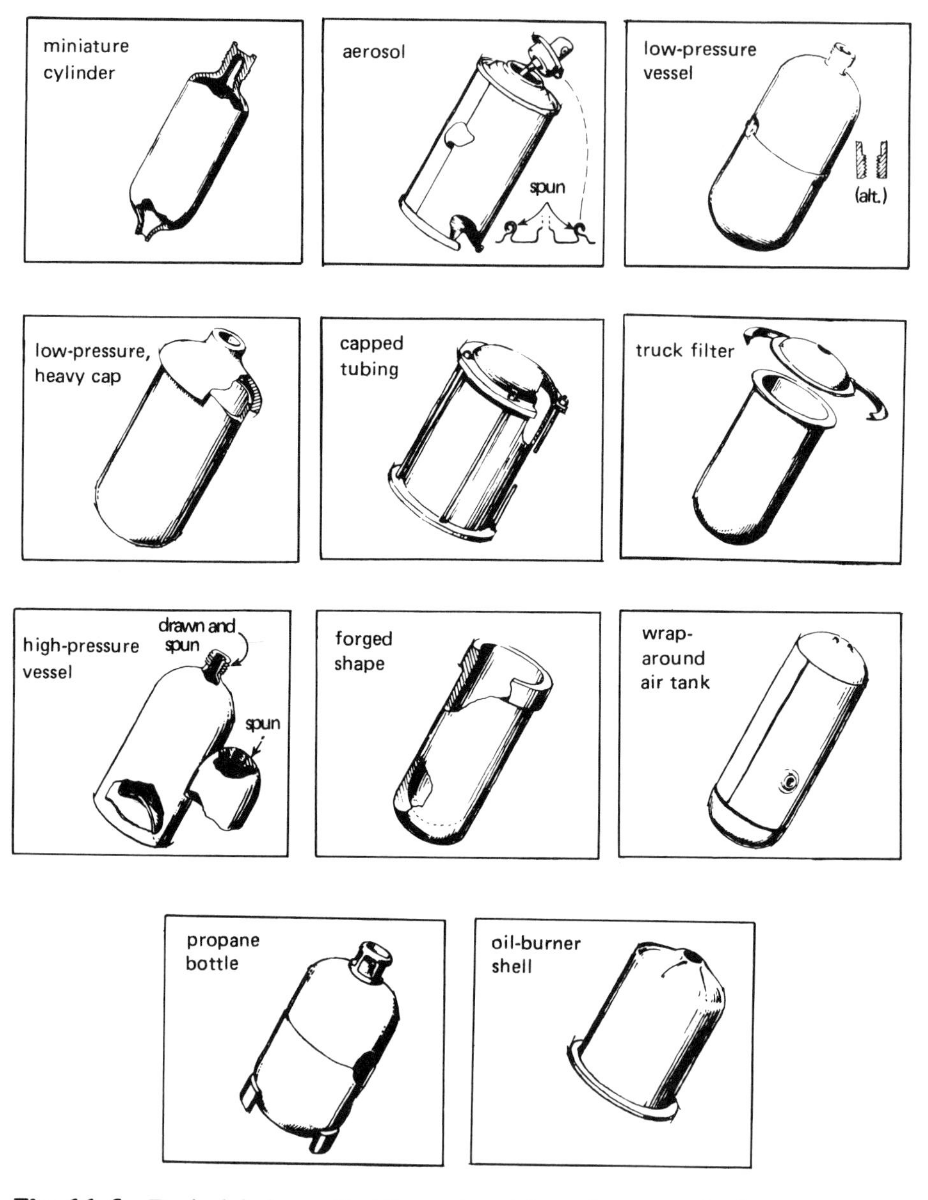

Fig. 11.6 Typical designs of bottles and containers for pressurized gases and liquids.

Thar She Blows

What constitutes a hazard is carefully, perhaps arbitrarily, spelled out by various regulatory agencies. The Interstate Commerce Commission (ICC), Coast Guard, and Federal Aviation Agency (FAA) took early interest, and any container shipped loaded was fair game to them. On-site pressurized tanks are scrutinized by state agencies, backed up by ASME boiler and unfired pressure vessel codes.

Even the aerosol cans that squirt your shaving cream are designed in strictest accordance with carefully written safety standards. And the Compressed Gas Assn in New York is a good source of information on working pressures of various gases.

The dangers are greater than you might realize. Some special bottles have test pressures up to 43,000 psi! Even an aerosol can of shaving cream (less than 100 psi) can make an awful mess in your suitcase if punctured. A cartridge of carbon dioxide at 838 psi and 70 F will reach 2265 psi if the temperature is increased to only 130 F. A 1-in. diameter CO_2 cylinder for life belts (1000 psi) can soar 200 ft in the air when punctured. Oxygen cylinders in military aircraft are wrapped with steel wire to prevent shrapnel effects should the cylinder be punctured by a bullet.

Obviously this is a field to be conservative in. For one thing, learn and accept recommended safety factors. And second, keep the tanks away from heat. Don't leave aerosols lying around in the hot sun. Never scratch or dent a high-pressure container. Rules of thumb are given in Table 11.2.

The rest of the text will help you design, select, and apply small pressurized containers with a minimum of trial, error, and danger.

Table 11.2 Safety Rules for Pressurized Containers

1. Gas containers are marked to note contents in accordance with Booklet Z48.1-1954 of ASA (American Standards Assn). Do not substitute other gases.

2. Purge before filling any small cylinder from a large one, and then only with the help of an expert on gas characteristics, filling densities, and safety.

3. Detect leaks with soapy water, never with a flame.

4. Empty any leaking or damaged cylinders, and those that have been exposed to fire. Close the valve, replace any protective caps or outlet plugs, and consult the supplier. DO NOT SHIP.

5. Keep below 120°F. Shield against direct sun's rays, never use a flame to melt accumulated ice or snow, never weld to cylinder even if it is empty.

6. Avoid extreme cold on high-pressure cylinders because the alloys may lose shock resistance.

7. Store cylinders in dry, cool, well-ventilated area. Avoid damp places, salt air, corrosive atmosphere.

8. Keep volatile gases away from oil, gasoline, and waste. Oxygen should be stored at least 20 ft from gasoline or fuel gas. LPG gases are not allowed in certain tunnels.

9. Open pressurized containers slowly. Do not tamper with the safety devices—your life may depend on them.

10. Tighten connections between tanks firmly but do not force; they are designed to be made up easily.

11. Store cylinders of argon, nitrogen, and oxygen in an upright position to make vapor venting possible.

12. Provide gas masks or self-contained breathing apparatus for handlers of poisonous gases, with appropriate and prominent instructions.

13. Use old cylinders first, because of outdating, possible corrosion, or aging of steel, affecting flexibility.

(continued)

Table 11.2 (Continued)

14. Avoid electrical currents through container, such as accidental grounding near arc welders.

15. Avoid mechanical damage, because small scratches and dents create high-stress points that later might fail. In other words, use tanks as containers, not as rollers or supports. Protect them from loose or falling objects by suitable covers.

16. Move cylinders with fork truck or other proper lifts, never by dragging. Do not grasp by caps or valve handles.

17. Build safe stands for topheavy cylinders.

18. Retest all pressure containers periodically in accordance with applicable regulations, some of which are mandatory.

19. Avoid horse play, and jet-dusting of clothing. High-pressure gases are as lethal as a bullet.

Gas or Liquid Containers

A typical small pressurized vessel contains gas, or gas and liquid. Ideally, the liquid is a puddle of liquefied gas, chosen to have the desired vapor pressure at the expected operating temperature (see Tables 11.3 and 11.4). As the vessel is depleted, the liquefied gas vaporizes and keeps the pressure up until nothing but gas remains. Then pressure drops rapidly.

Aerosols, which are the commonest and cheapest pressurized containers, are based on that principle. Thus they maintain constant pressure throughout their life and are relatively safe at all ordinary temperatures. At elevated temperatures, however, vapor pressure increases dangerously. Also, check government standards to learn which fluids are banned for environmental reasons.

Many gases, including nitrogen and oxygen, cannot be liquefied at room temperature. These are stored in one of two ways: as a cryogenic liquid or as a gas at

Table 11.3 Vapor Pressures of Typical Liquefied Gases

	At these temperatures, deg F						
	−35	**0**	**35**	**70**	**105**	**140**	**175**
Freon 12	8 in. Hg vacuum	9	33	70	126	206	320
nitrous oxide	165	295	475	745	1500	2450	3450
propane	3	24	57	109	185	290	435
carbon dioxide	145	294	512	835	1580	2530	3520
ammonia	5 in. Hg vacuum	16	52	114	214	365	579

Above cylinder pressures assume cylinders charged to recommended filling densities.

Table 11.4 Volume-Pressure Relationships of Gases and Vapors

	Perfect gas	Air	Acetylene	Ammonia	Argon	Carbon dioxide	Freon 12	Helium	Nitrogen	Nitrous oxide	Oxygen	Propane
Chemical symbol	none	none	C_2H_2	NH_3	A	CO_2	none	He	N	N_2O	O_2	C_3H_8
ICC regulation:	none	varies	8; 8AL	4A480	varies	3A1800 3E	9; 41; 4B225	varies	varies	3A1800; 3E	varies	4A150; 41; 9; 4B225
Liquid* filling density (typical %)	none	none	acetone used as filler	54	none	68	119	none	none	68	none	42
Volume, in.^{3}, to hold one ft^3 of free gas at 70 F and at various psi	221.5 at 100 psi; 14 at 1800 psi	221 at 100 psi; 13.97 at 1800 psi	15 at 250 psi	2.22 at 114 psi	219.7 at 100 psi; 13.22 at 1800 psi	4.65 at 835 psi	7.74 at 70 psi	222.3 at 100 psi; 14.87 at 1800 psi	221.3 at 100 psi; 14.13 at 1800 psi	4.66 at 745 psi;	219.5 at 100 psi; 13.19 at 1800 psi	7.68 at 109 psi
Expansion ratio, %:: volume free/ volume compressed	7.8; 123.5 (for conditions above)	7.82; 123.7	115.2	778.3	7.86; 130.7	371.6	223.2	7.77; 116.2	7.81; 122.3	370.8	7.87; 131	225
Comments	none	none	dangerous to load	unpleasant odor	inert	high expansion; cools	good mixer	leak detector	inert	harmless to foods	dangerous with oil	gaseous fuel

*Filling density = percent by weight of liquefied gas actually in container at 70 F, related to weight of water the container will hold.

high pressure but at room temperature. Only the latter is within the context of this chapter. Cryogenic liquids such as liquid oxygen (LOX) are a subject in themselves.

Vessels for plain liquids such as hydraulic oil and water may also operate under pressure. Examples are fuel filters built to withstand fuel pump pressure, cylinders for piston-type accumulators, and heat-exchanger drums. However, the dangers are less severe because the likelihood of explosive expansion is less—except in the case of water expanding into steam.

Working Pressures

The following values are based on early standards: Always use the latest revisions in any design.

Aerosol cans for paint and insecticides rarely exceed 100 psi. Shop air goes up to 130 psi, as does city water pressure. Propane and butane cylinders usually are charged up to 240 psi. Low-pressure vessels according to ICC definitions have less than 1000-psi fill pressure, and most are less than 500 psi. Nitrous-oxide pressurization in food aerosols goes as high as 1087 psi at 98 F. Sunken ocean buoys must withstand external pressures of over 7000 psi, which means you must either make the chamber strong or prefill it with high-pressure gas. If there were a practical way to equalize external water pressure with internal gas pressure as the chamber was lowered, its wall could be thinner.

Test or proof pressure under ICC procedures, for conventional high-pressure cylinders from about 1000 psi and above, is 5/3 working pressure. Low-pressure cylinders are tested at twice working pressure. The ICC performs some tests in water jackets, though not all tests are run that way. A jacketed test usually is applied to one sample out of 200 low-pressure vessels and measures the total or plastic expansion and permanent expansion that occurs when test pressure is applied internally. Permanent expansion allowed is 10% of total expansion.

The ICC demands that pressurized cylinders be stress-relieved and flexible, meaning that yield strength must not be more than 67% of tensile strength. This lowers the absolute yield strength but eliminates variations due to work hardening.

ASME tests differ from those of the ICC. Test pressures are 1.5 times working pressure, and the vessels are struck with hammer blows, making the tests more severe and thus more conservative. Burst strength generally is specified over 3 times working pressure.

On low-pressure cylinders with pressures below 200 psi, such as ordinary fire extinguishers, UL specifies burst strength 6 times working pressure and for cartridges, 5 times. Actual testing is at 3 times working pressure. Stress relieving afterwards sometimes is required.

Container Styles

There is nothing proprietary about good container design, and the most economical method of manufacturing becomes the key consideration. Quantities naturally play their part. Here we shall primarily discuss containers under 10 gal, which is 2300 in^3, 83½ lb of water, or a container about 12 in. in diameter by 24 in. high.

Choose a standard container, preferably one suited for a higher pressure and larger than your need. It often will prove less costly than a precise fit and will be safer.

Most containers are cylindrical because that shape is easiest to make. However, a sphere can withstand higher pressures and contains more volume for a given total wall area.

An aluminum sphere 13-in. OD with 0.5-in. wall can withstand 7500 psi. Some are applied as deep-sea submerged buoys. A compromise here could be a sphere of thinner wall pressurized internally with inert gas at half the external test pressure. The wall stresses would be halved but would go from tension to compression during descent.

For practical purposes, three classifications cover almost any type of cylindrical pressurized container: miniature cylinders, low-pressure vessels, and high-pressure vessels. The terms container, cylinder, vessel, bottle, flask, tank, and tube mean essentially the same thing: a pressurized reservoir of some sort.

Miniature Cylinders. Not actually cylinders but steel bottles, they come in assorted sizes from 8 g (about ½-in.^{3} content) to 35 g (about 6¼ in. long and 1-in. OD). They are mass-produced with highly automated tooling and usually filled with CO_2. For sizes greater than standard, you can manifold the cylinders together.

Miniature cylinders are punctured to release the contents and are not reusable. They are manufactured by deep drawing and are seamless. The filling end is sealed in any of several ways, including a welded-in or a spun-in pierce disc or a staked end. Manufacture to other than standard designs would be uneconomical.

Low-Pressure Vessels. In sizes from ⅓ to 2 lb of water (1 lb of water occupies about 27.7 in.^{3}), lock-seam construction is chosen. Aerosol cans are made that way. The amount of charging gas in each is small, keeping pressures below 100 psi.

For fairly large sizes, consider using beer kegs or carbide drums. Or try wraparounds. A number of air tanks for compressors, reservoirs for truck brakes, etc., are made by wrapping a sheet around (see sketches), welding them longitudinally, and adding two elliptical heads. In standard sizes, these are made with elaborate jigging and welded with special submerged-arc fixtures. The spuds (attached threaded bosses for the container valves) are projection welded or welded metal projections tapped.

We now see more low-pressure containers made of two drawn halves overlapped and brazed or girth welded and with the spuds welded on. Because long-stroke drawing presses needed are comparatively slow and tooling is costly, most containers over 6 in. long are centrally joined to avoid using a deep shell.

Some capped containers are made of tubing with two end caps. These are cheap to tool but in quantity cost more than joining two halves.

Tubing is useful in refrigerant lines, where screens holding decontaminants are placed in lengths of copper or stainless tubing and the ends spun.

Aluminum float cylinders are sometimes impact-extruded. The finished container, good for holding LPG (propane) for welding or camp stoves, is shipped under special permit and has a rupture disc safety release.

High-Pressure Vessels. If a container is not too large, the ICC favors seamless construction. Spun tubing is limited by the diameter of tubing available (about 16 in.) but can be quite long when used for gas storage. ASME-designed vessels such as accumulators are welded and may have caps held by bolts. Larger high-pressure vessels, 7 to 54 in. in diameter, may call for forgings or perhaps shell castings with heavy walls.

Spun tubing is used in varying quantities and outlet configurations. In large quantities, such as for life-raft cylinders, drawn shells with the outlet end spun in are used. The bottom is closed; the spun wall of the outlet end will be considerably thicker than the side wall, handy for threads and O-ring grooves. Half-pound carbon dioxide cylinders (21 in^3) and some larger fire extinguishers are standard.

If one end is to be completely open, or at least have a large opening, drawn cylinders may be preferred over tubing. One company supplies drawn tanks to 100-gal capacity. Necking-in a drawn cylinder in a press is not easy or cheap due to possible bulging. To reduce a 4-in.-diameter shell to 2.4 in. on the end (40% reduction) calls for three necking operations and probably two anneals with appropriate tooling.

Both spun and forged containers may have heavy walls and substantially heavier ends, adequate to receive male threaded valves, drain plugs, etc. Drawn cylinders have extruded ends for the same purpose, and necking down adds some thickness to the end. Bottoms remain at the original stock thickness, of course.

Where pistons are to be installed, there may be problems in sealing. The ID of seamless tubing is rather inaccurate, and hot spun tubing, though more accurate, is not smooth. Welded tubing, resized, comes closer. Tolerances on drawn shells are often tighter: a 0.006- to 0.010-in. spread on 3- and 5-in.-diameter shells, not counting out-of-roundness. However, a slight draft is needed—in the order of 0.002 in., closed end smaller, for each 3 in. of length, to help in stripping parts off punches without scratching. It might be necessary to grind or hone. Heat treating, of course, increases out-of-roundness.

For small-lot production, such as special valves or machine hydraulic cylinders, we still find uses for tubing lengths and caps that are gasketed and held by through bolts. An advantage is that they can be completely disassembled.

Stand-up cylinders are a separate problem. To get a drawn or spun cylinder with a rounded end to stand upright, rings and collars can be welded on if the tank if intended for low pressures, like those for propane. Where welding is not permitted and adhesives are not sufficiently reliable, the bottom can be pushed back into the body. On a drawn cylinder, a rather generous internal radius is needed so the die post will be substantial. Some spun and forged cylinders have almost flat bottoms.

Correct Wall Thickness

The one place where error can spell disaster is in the design of the container wall. You must consider all parameters and allow for every possible abuse. Table 11.5 lists typical causes of failures. Fortunately there are equations for about every situation, and most of them appear in the ASME Code for Unfired Pressure Vessels, available from the American Society of Mechanical Engineers, New York. Special equations used by the pressurized-container industry are summarized in Table 11.6.

Proper specifications for construction should consider working pressure ranges and possible surges; working temperature ranges; description of contents including recommendations on whether the vessel should be lined with rubber, glass, alloy metal, or painted to prevent corrosion; gas volumes; information affecting metal properties, including welding; possible explosion of lubricating oil vapors in air tanks; and vibration, particularly if a compressor is mounted on the tank.

Stress on the ends of the container varies with shape. A flat end is under the greatest stress. The closer the end comes to being a hemisphere, the less it will expand

Table 11.5 Typical Causes of Pressurized Container Failure

1. Design errors (see text for recommended practices)
2. Failure to inspect or retest properly
3. Overpressurizing (or excessive vacuum)
4. Corrosion or erosion of the metal
5. Malfunction of automatic devices permitting overpressure
6. Excessive heat or excessive cold
7. Water hammer or carryover of system material
8. Ventline not functioning, or not used

under pressure. Frequently, cylinder ends are shaped with internal or external elliptical shapes, usually external. In testing at twice working pressure, such ends tend to expand permanently and may lead to rejections. If a shallow elliptical head is designed on a cylinder that is only two or three times as long as its diameter, the problem is greater than on a longer cylinder. In a long cylinder, sidewall expansion helps take up out-of-roundness. The ICC does permit prestressing to within 10% of proof pressure, which helps reduce the degree and percent of permanent expansion.

This stress on cylinderical walls must be contained. Steel seems to be the best answer. Aside from aerosols, there are few aluminum cylinders in the United States. Reinforced fiberglass containers, although very dependable, are not widely used where lowest cost is a requirement.

Low-carbon steels are figured to a maximum stress of 24,000 psi, multiplied by joint efficiency if less than unity. Certain steels usually sold as having 50,000-psi minimum yield are figured at 35,000 psi. These steels often are preferred on a strength-to-cost ratio and give lighter cylinders.

High-pressure cylinders usually are made of alloy steels such as AISI-4130 (chrome molybdenum), heat treated and drawn, and are figured at 70,000 psi.

Valves and Caps

Safety valves may be spring-loaded, frangible-disc, or pin type (for medical gases). An ICC-approved safety valve must be used on cylinders over 4½ in. in diameter and over 12 in. long exclusive of neck. For CO_2, N_2O, and gases at 1800 psi or over, relief valves are required. They may be located within the release valves. Safety releases in lower-pressure cylinders permit 10% more fill.

Aerosol valves often are spun over a curl on the can outlet. Releases for powder fire extinguishers are male projections screwed into spuds with an O-ring above the threads. Propane and refrigerant tanks have similar valves. Miniature cylinders may have threads below the pierce end and are screwed into releases with central piercing pins. On sophisticated military bottles, the pins may be actuated by an electric current. Larger high-pressure bottles may be filled through a valve screwed into threads cut in an extruded bottle neck. Generally, O-ring construction is simple and effective. It is also used on drain plugs.

There are so many types of valves, spring valves, valves supported by gas pres-

Table 11.6 Calculations for Stress and Wall Thickness of Pressurized Containers

The stress on the wall of a sphere or an infinitely long tube can be figured, but most cylinders are of irregular shape. The stress we seek to measure is at working pressure (or at required pressure for temperature-sensitive gases) in temperature ranges from −20 to 130 F, without special allowances for bumping, corrosion, and aging of steel. Hence retesting is called for on a regular basis.

Sidewall strength is the criterion. The minimum yield strength of materials used in a stress-relieved and flexible condition determines the minimum wall thickness of a cylindrical shape that contains fluid under pressure. Failure rarely will occur at the ends if they are hemispherical or at least elliptical.

S = working stress, psi
t = minimum wall thickness, in.
P = proof-test pressure, psi
D = outside diameter, in.
d = inside diameter, in.
R = inside radius, d/2, in.

Applicable equations.

- ICC (based on the Bach Clavarino formula)

$$S = \frac{P\,(1.4D^2 + 0.4d^2)}{D^2 + d^2}$$

$$S/P = \frac{1.3\,(D/d)^2 + 0.4}{D/d - 1.0}$$

Poisson's ratio is assumed = 0.3.

- **ASME code (Barlow formula modified by Boardman)**

$$S = \frac{P\,(R + 0.6t)}{t}$$

$$S/P = \frac{0.6\,(D/d) + 0.4}{D/d - 1.0}$$

This is not used for very heavy small vessels, where t is greater than $R/2$. From that formula is derived the following, where inside diameter is known:

$$t = \frac{PR}{S - 0.6P}$$

Thus for a 2.8-in. ID cylinder at 3000 psi proof pressure, and 70,000 psi allowed for S, $t = (3000 \times 1.4)/(70{,}000 - 1800) = 0.0615$-in. minimum wall.

In the event outside diameter is governing, the formula is:

$$t = \frac{PD}{2S + 0.8P}$$

For a 2.87 OD cylinder this is $(3000 \times 2.87)/(141{,}400) = 0.0605$ minimum t. Drawing strip would be ordered 0.072 plus or minus 0.003 in.,

to allow for material tolerance, a degree of thinning in drawing, and a slight margin for decarburization.

sures, cylinder packings, valves with replaceable puncture discs for release or safety, and special explosion-protected designs for missiles that much ingenuity with little standardization goes into containers and controls. The desirable standard sizes for fittings are suggested by the Compressed Gas Assn.

WHAT'S IN THE AIR

One of the changing scenes is industrial pneumatics. Like it or not, the air you rely upon to operate valves, cylinders, motors, and controls is being examined critically by everybody in the industry and will come out purer if not better. The motivation is chiefly from two sources: OSHA and EPA. OSHA wants pure breathing air in plants, and EPA wants clean streams.

You might be misled into thinking it's just a local problem with each plant owner. Not so. If the changes described come about as expected, different kinds of components will be required to run on the new air.

No longer will moist, oily air be a fact of life in plant pneumatic systems. Each machine or component will be required to operate with dryer and cleaner air and probably will have to operate nonlubricated or with minimal lubrication. Thick mists of oil no longer will be tolerated, nor will be excessive forced lube. Instead, built-in lubrication might be expected for each cylinder or motor, and it must not carry over into the atmosphere.

It used to be enough to aim the exhaust away from the operator but not now. Any air released into a breathing atmosphere must be purer than the air in a pine forest and no noisier than a swamp full of crickets.Tricks like exhausting air into passages within the body of the machine are clever, but it still must be pure. Where no exhaust at all is allowed, as in some woodworking plants, the only answer is to pipe it away.

Some designers advocate closed-circuit pneumatic systems with 140 psig for operation and 70 psig for return. It eliminates exhaust problems, but there aren't many companies actually doing it.

Varnishing, silting, and gumming of valves must be avoided somehow. Mill scale in plant piping will find its way into valves and cylinders, and there's no easy way to filter a system at enough spots to catch it all. The answer is for plant engineers to demand scale-free pipe, flushed clean and capped. To do otherwise asks for trouble. Many plants already specify stainless steel piping and avoid the problem.

At the moment there is no rush to further this revolution. The change is too drastic to handle quickly or arbitrarily, as OSHA is finding out. Filter-regulator-lubricator packages (see the section following) still are depended on to condition air coming into your pneumatic power system in most instances, because that's the prevalent method and everyone is familiar with it.

Yet nobody denies the inevitability of some sort of change. We talked with designers of compressors, valves, cylinders, filters, regulators, lubricators, and related components. None deny that the average plant air system is far from satisfactory. They point out that everything would be OK if only maintenance were done correctly.

But plant maintenance is not easy. Compressors generate oil fumes, poorly tended desiccant driers send debris downstream, regulators are set too high and waste air, filters are the wrong choice or are allowed to overload, lubricators are left unattended and either go dry or deliver excessive oil, and valves exhaust oily air to atmosphere.

There are many tales of woe. An air tool operator, bothered by the oil mist exhausting from his tool, adjusts the lubricator but manages only to turn it off. Everybody is happy because the mist is gone. The air tool supplier is happy because he will sell twice as many replacement parts.

A new maintenance man has forgotten to refill those lubricators that are more than 2 ft above eye level and almost always forgets those next to the roof. Many lubricators go bone dry, and if it weren't for the wet oily air coming into the system from the old compressor, nothing would work right. The filters in another system are changed only when air can't flow through them anymore.

Often the filter-regulator-lubricator package is hidden from view in a cabinet. Others are behind machines or behind high ductwork. Even if the operator sees them (binoculars help), he or she might not be able to tell if they are full or empty because of accumulated dirt on the outside of the transparent bowl. If the bowl is metal, or just covered with a perforated metal protector, the problem is intensified.

A great deal of engineering work has gone into developing reliable filters, regulators, lubricators, dryers, compressors, and the other hardware in air systems, but the ultimate user is not as reliable as the designs. The really sophisticated users—such as auto manufacturers, major food processors, and chemical plants—have perfected the maintenance as much as is practical, but even they aren't completely happy.

The New Pneumatics

Logic suggests that an ideal system should not require irksome if not impossible maintenance. We talked to plant engineers at several modern plants to see how they solved the problem for themselves. They offered a consistent story, tempered only by how recently the plant was built. Newer plants were able to incorporate more efficient systems from the start; older plants could only make adjustments.

High on the list were these concepts: air cylinders with built-in lubrication, coalescing filters, refrigerant dryers, exhaust muffler-reclassifiers (to diminish noise and catch oil mist), and oil-free air valves and cylinders wherever possible. A lot of the technical details on the various devices are given in Chaps.

FILTERS, REGULATORS, LUBRICATORS (FRLs)

Probably the most common solution to the final conditioning of in-plant air is the FRL—the filter-regulator-lubricator. The makers and users of FRLs are not totally satisfied with the ways they are applied, but nobody seems to have a better low-cost answer.

The typical FRL is a packaged unit containing all three functions: filtration, pressure regulation, and in-line mist-type lubrication. Half a million are sold every year to OEM and plant buyers, who seem happy enough. So who's kidding whom?

Researchers explored the topic from every angle and discovered this: The functions certainly must be fulfilled in some manner or other for any pneumatic system, and for the average designer of equipment the easiest and least costly way is a packaged FRL unit. The package costs as little as $50 to $150 and promises a lot. Typical sizes range from ¼-in. to 1-in. pipe diameter, with the most popular size being ½ in. Beyond 1 in., the functions usually are supplied separately and not as an FRL unit.

Many companies offer filter-regulator-lubricators. Also, privately labeled versions are sold by big distributors such as Grainger and even by mail order houses such as Sears. A typical example of an FRL unit is shown in Fig. 11.7.

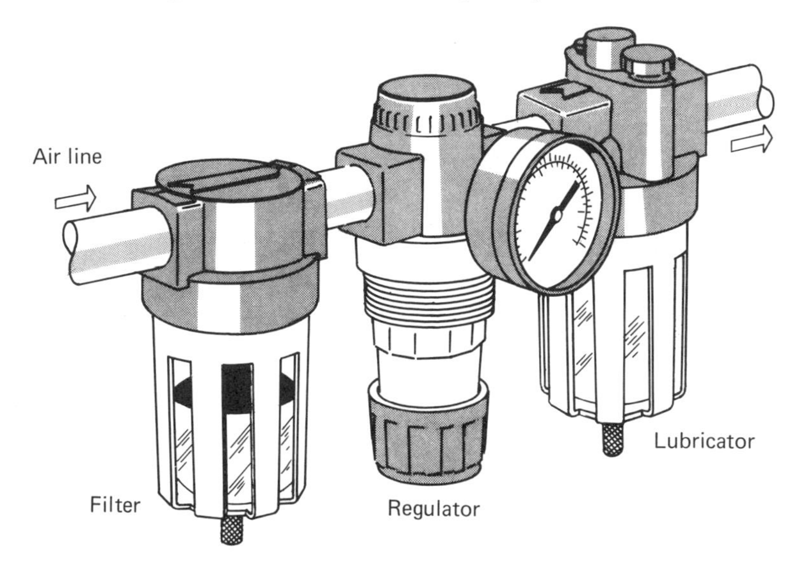

Fig. 11.7 Typical filter-regulator-lubricator for pneumatic systems. (Courtesy of Master Pneumatic-Detroit, Detroit, MI.)

The basic reason for the combination is that pneumatic cylinders and motors, including air tools, share a common need. Each requires air that is relatively free of water, rust, dirt, and other carryovers from the shop air system.

Furthermore, the pressure level always should match the need so as not to waste energy. And proper lubricant must be furnished at each friction point. A FRL unit can do all of these, and the chief limitation fundamentally is the fact that the lubricating portion is directly in the air line and therefore a prisoner of the vagaries of pneumatic systems in general.

The other functions—filtering and regulating—are easier to solve and, judging from the many responses we received from suppliers and users of pneumatic equipment, create no insurmountable problems if the guidelines discussed in this chapter are accepted and followed. Let's take the functions separately, pinpoint the problems, and offer some solutions.

Filtering Comes First

Two requirements stand uppermost in the minds of the users: The filters must be large enough to operate without attention for long periods of time, and the drains must be very convenient to operate or must be automatic.

Solutions include solenoid drain valves, remotely operated; pressure-actuated drain valves; float-operated drain valves; or simply a flexible poppet valve that can be opened by pushing it with a stick or finger in any direction.

Apparently many maintenance people are notorious for neglecting any filter that is not directly in line of sight (some units are hidden) or that is not easy to drain (nobody wants to look for a special tool or in some cases a union pipefitter). The reluctance is understandable, but the results of poor maintenance are high costs in worn or damaged equipment. A filter bowl that fills with water ceases filtering and instead will deliver dirty water downstream.

Master Pneumatic-Detroit (Sterling Heights, MI) recommends automatic poppet-type drains with poppet-type pilot valves, energized with air from the pneumatic system. Areas should be large enough to preclude any chance of jamming or clogging. Poppet valves have the best record of tolerance to debris and resistance to jamming.

The water must be removed or it collects in piping, forms puddles, washes out lube, forms emulsions with oil, and accelerates rusting. It accumulates particles of rust along with the water and oil, making a great lapping compound to wear out cylinders and valves.

In addition to filtering, water must be gotten rid of by sloping the air mains always downward (Fig. 11.8), putting automatic drains at the end of the headers, taking off air from the tops of the headers, and adding coolers, separators (centrifugal action), and strainers.

Oil is not as readily centrifuged out or blown off a surface as water is, because oil spreads and clings whereas water coalesces and washes off. Also, water is chemically predictable, and oil is not.

The other requirements—proper filtering media, pressure rating, flow rating, pipe size (Fig. 11.9), housing and media materials, temperatures, and so forth—are under control of the original specifier, just as if he or she bought the filter separately. For example, avoid dissimilar metals because they can start electrolytic corrosion. Choose whatever micrometer rating gives the protection needed.

Some specials are not available in FRL packages. Fisher Body Div., GM Corp. (Warren, MI), pointed out that for the heavy-duty service required in-plant, Fisher engineers chose to designate their own special, large filters housed in brass. They are expensive but do the job.

They added several other requirements: metal bowls, 40-micrometer filtering media, no coalescing (see Chap. 17), no desiccant dryers, and no synthetic lubricants in compressors upstream (to avoid possible damage to seals). The idea was to spread maintenance periods as far apart as possible, even to the point of accepting some water and contaminants. Fisher tests each air-using component on simulated shop air, including rust.

Where cleaner air is a necessity, then Wilkerson Corp. takes the opposite ap-

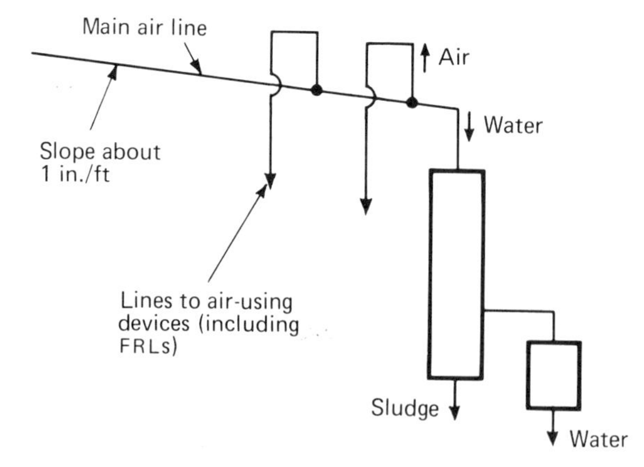

Fig. 11.8 Industrial pneumatic system must be designed to eliminate water and sludge to protect instruments and devices.

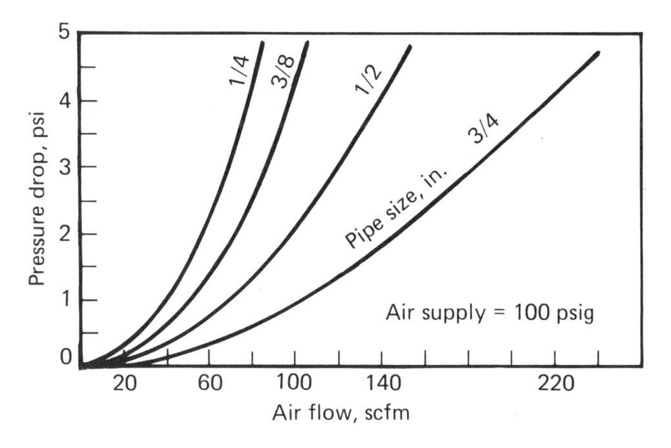

Fig. 11.9 Typical flows and pressure drops through clean filters.

proach: Why accept 50 micrometer contamination when for relatively little more money you can buy a 5-micrometer filter that will have the same service life? The choice depends on whether or not the system can thrive on dirt. If it can, let it. Otherwise specify fine filtration.

Filtration of superfine particles and aerosols (less than 1 micrometer) using coalescing (agglomerating) filters is not necessary for most cylinders and valves but is needed for some air bearings, gaging, and instruments.

The phenomenon of coalescence (accumulating tiny droplets of mist until they become big drops and fall by gravity) is useful in another way. Some coalescing filters are applied as reclassifiers in air exhaust tanks at the outlet of cylinders. They agglomerate oil mist into liquid drops and collect them to prevent contamination of the atmosphere.

Pressure Regulation

Not many engineers voiced concern here. The typical pneumatic system does not require special performance beyond that provided by FRL units. Engineers at Master Pneumatic-Detroit offered good advice on how to specify a regulator:

If the load is a cylinder, then quick inlet flow and rapid relieving (bleedoff) usually are more important than controlled pressure during piston movement. However, if the load is a torque motor tool, pressure control is more important because the tool must not overtorque or undertorque critical bolts on automotive or aircraft assemblies.

The internal design of a regulator is a science in itself (whether of the diaphragm or piston type, with or without air-domes, and so forth) and is covered in Chap. 8. The important application information is the actual expected airflows and pressures at each point in the load cycle: Give those to the regulator experts.

If special regulator performance is required, then the FRL functions should be ordered separately. Fisher Body, for example, specifies large regulators not available in any FRL package.

Lubricators Misunderstood

Most of the slings and arrows have been against mist lubrication. It wasn't that anyone argued against the theory of airborne transport of finely dispersed lubricant for eventual reclassification (return to liquid form) at the precise point needed. That's a great idea, and it works when the flow conditions are correct. Norgren Co. (Littleton, CO) pioneered it three decades ago.

Two major categories cover most mist-type designs: differential type and capillary type (Fig. 11.10). The *differential* type relies on a pressure difference (between the surface of the oil in the reservoir and the injection point of the oil in the airstream) to lift the oil. A restriction such as an orifice or venturi in the airstream is sufficient. It creates an aspiration vacuum.

There are various proprietary designs of the restriction, and some of them adjust flow area or bypass flow automatically to hold the oil-air ratio fairly constant over a wide range of flows. Study them before you choose.

The *capillary* type relies on wicking to lift oil from the reservoir to the airstream. A sintered bronze porous rod is typical. The airstream sweeps off oil as a function of velocity, and some models will work in either direction of flow.

Nobody doubts that the mist is created. The argument concerns what happens to the mist afterwards. We've summarized the many comments received from users and suppliers of mist-type lubricators, which are the type used in most FRLs:

Airflow rate critically affects the amount of lubricant delivered and doesn't necessarily maintain the right proportion of oil to air; not every air-using device downstream of the lubricator has the same need for oil, and some differ dramatically; and the oil mist and droplets generated by the lubricator do not reliably follow all the twists and branches of the piping, fittings, and valving leading to the cylinders and motors.

Reservoir bowls are relatively small and sometimes are hard to reach and fill. If the oil-flow adjustment is too easy to make, unauthorized persons are likely to alter the adjustment to turn off the flow entirely. It's a great temptation for an operator to stop an oil drip that way.

The bowls often are made transparent for visibility, but the polycarbonate plastic usually chosen for the job is sensitive to certain chemicals and vapor found in some manufacturing areas. The bowls can craze and crack while under air pressure.

There are ways to engineer around those lubricator problems. For instance, perforated metal guards around polycarbonate bowls will safeguard personnel. Sight

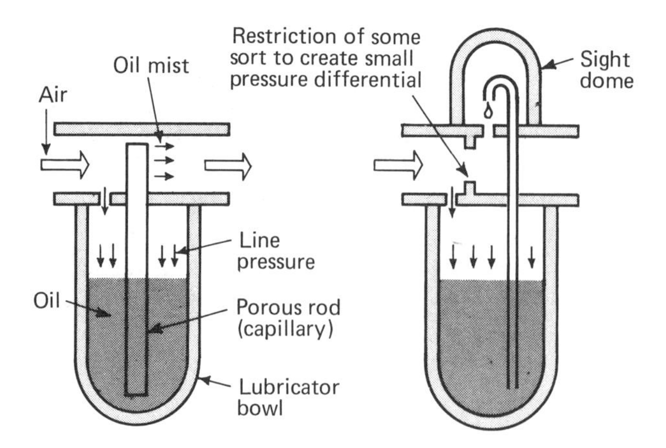

Fig. 11.10 Two mist-lube concepts: capillary (wicking) and differential pressure.

gages installed on metal bowls can eliminate polycarbonate entirely if desired. Also, there are new plastic alloys, such as GE's Xenoy 3500, that resist chemicals. Remote large reservoirs can feed individual lubricators and reduce frequency of filling.

Oil-flow adjustments can be controlled in various ways, such as placing a dummy adjustment screw over the real screw to foil unauthorized tampering. Wilkerson offers a "variable venturi" automatic oil metering adjustment that is in effect an internal elastomeric element that flexes under varying flow demands to allow a bypass of air.

Studies have been made to establish the ideal ratio of oil to air; one was by Boeing Aircraft (Seattle, WA). Manufacturing and engineering people there determined that one drop of oil for every 20 ft^3 of air was satisfactory for most air tool applications. A typical tool operating with a ¼-in. inlet consumes about 20 scfm, so one drop a minute is the desired first-try setting. Final adjustment depends on other factors such as tool size and load. Just make sure oil is not dripping from the cylinder or motor exhaust and that the tool is not running dry.

An inescapable conclusion, however, is that the best applications are where the airflows are totally predictable and reasonably uniform, the downstream devices are like Siamese twins or at least close together, and the net flow is always toward and through the air-using device. Probably the majority of applications on well-designed equipment meet those criteria. The difficulties seem to relate mostly to nonpredictable shop-type applications.

The Myth of the Mist. Some suppliers and users of FRLs take a strong position against oil mist as an air line lubricant. For example, Master Pneumatic-Detroit questions the belief that oil can be broken up into minute particles, each of which is lighter than the supporting air at the moment of introduction into the air line, and stay that way. Quoting a vp: "If each particle is to remain suspended in the airstream through the connecting nipple, lengths of pipe, and through the hose to the valve, cylinder or tool, why should it not continue to float right on through the operating device and out the exhaust port to the atmosphere? If it remains suspended, without wetting out on a surface to create an oil film, how can it possibly provide lubrication?"

There are certain exceptions to the argument. One is in bearing applications that use the air strictly as a transportation medium at low pressure (a few psi) and low flow (1 cfm or less). Under these conditions, some of the oil can remain airborne. The minute droplets can be made to wet-out if the air velocity is increased with a nozzle just before the bearing. The process is called reclassification and occurs when the minute droplets of lubricating oil impinge on the final surface (Fig. 11.11).

It doesn't work that way for compressed air in a pneumatic power system. The velocities are so high that the minute droplets will wet-out at many surfaces along the way. The company ran tests (which anyone can readily repeat) to prove this theory. Briefly, connect the outlet of a mist-type lubricator to a 6-ft length of clean and dry clear plastic tubing and connect that to a completely clean and dry nongoverned tool. Turn on the air pressure; then watch and listen.

The tool will rotate but initially receives no oil. The oil will begin to coat the inside of the tubing next to the lubricator and gradually work its way along the tube until it reaches the air tool. The air tool speed will increase the moment the visible oil reaches it, but not before. The mist, therefore, has done nothing: The liquid oil has done the lubricating.

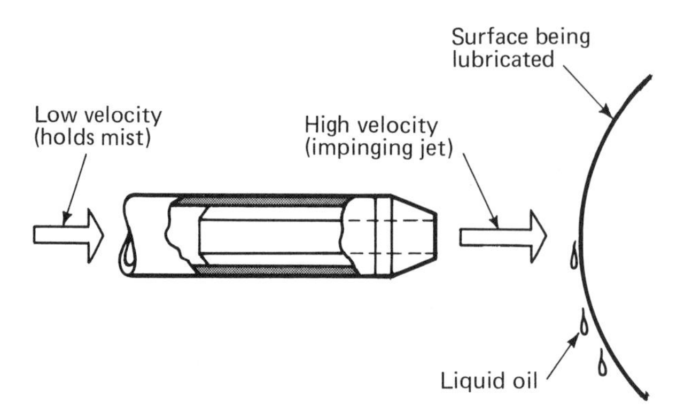

Fig. 11.11 Reclassifier nozzle converts oil mist to liquid by impinging it at high velocity. (Courtesy of C. A. Norgren Co., Inc., Littleton, CO.)

If that's true, then these assumptions logically follow: The lubricant is dragged along the wall of the tubing by the friction of the air on its surface; high-velocity air will move the oil film faster than low-velocity air will; the oil will move more readily on downhill slopes than on uphill slopes; vertical rises will inhibit oil passage; and low spots in piping loops will puddle oil.

For example, suppose a lubricator is placed upstream of the four-way directional control valve feeding a reversing cylinder. It will deliver oil as mist or liquid to the valve, and the valve will pass the oil through to the line connecting the valve to the cylinder ports.

The problem is this: The air velocity in the line feeding compressed air to the pressurized end of the cylinder is likely to be less than that of the unrestricted, expanding air exhausted from the same port when the cylinder reverses. It's a simple thing, but the net flow of oil toward the cylinder might be zero because of it. The valve will be flooded with oil, and the cylinder will run dry. It actually happens.

It can be solved (as can many of the other problems associated with in-line lubricators) but requires clear understanding of a lot of complicated flow technology, according to experts.

Parker Hannifin (Otsego, MI), suggests this: Install quick exhaust valves right next to the cylinder so that the exhaust flow does not travel against the incoming oil film. Lubrication is easier if the flow of oil-carrying air is always into the cylinder.

Another way to limit exhaust velocity is to put a flow control valve in the line near the cylinder port such that the full inrush velocity is allowed, but exhaust flow is restricted (Figs. 11.12 and 11.13).

Alternative Lubrication. Manufacturing managers in major automotive assembly plants are concerned about downtime and seem to favor direct lubrication over in-line mist lubrication. Some, including Fisher Body, are developing proprietary prelubed cylinders. Others buy them from cylinder suppliers. Still others choose central lube systems to inject lubricant under pressure directly to the friction points.

A method between in-line and centralized pressure lubrication is pulse lubrication. Here a lubricator unit is piped into the air line, but the oil is not injected at this point. Instead, air pressure operates a small spring-returned piston inside the lubricator and pumps a charge of oil into a length of tubing that ends at the cylinder point to be lubricated.

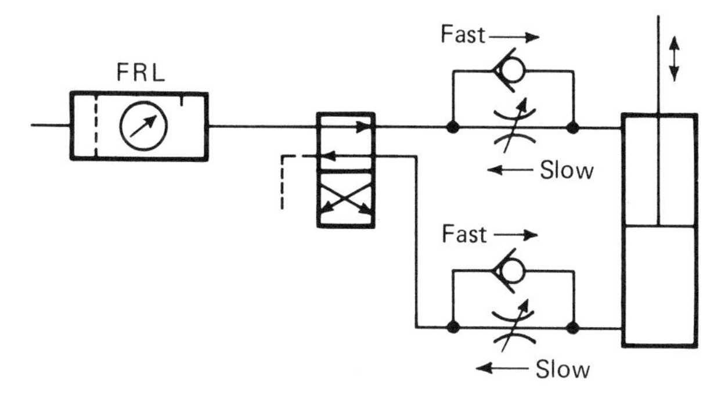

Fig. 11.12 Lubricant can be helped to flow into a reversing cylinder if the exhaust velocity is kept low (see Fig. 11.13).

The air pressure must fluctuate from about atmospheric pressure up to over 45 psi in order to pulse the oil piston and allow it to return. The disadvantage is that every lubrication point needs a length of tubing, and the machine begins to resemble a spaghetti dinner. One answer here is to run the lubrication tubing inside the air hose, right up to the point that needs oil.

If one charge of oil each cycle is too great, the pulses can be held off by a pneumatic or electric timer. A pulse after every 10 cycles or any large number may be programmed.

The same principle is followed in a line of multiple-piston lubricators made by Master Pneumatic-Detroit, called SERV-OIL (Fig. 11.14). They are stacked air-operated oil injectors and rely on normal fluctuations in air pressure (at least 45-psi variation) to power the pistons.

Overall Shop Air System

Don't forget the effects of one part of a system on all the rest. A prime example is the shop air compressor (see Chap. 12) with its accessory equipment and often strange lubricants. The coolers, separators, and dryers can be analyzed readily. But what about the lubricants?

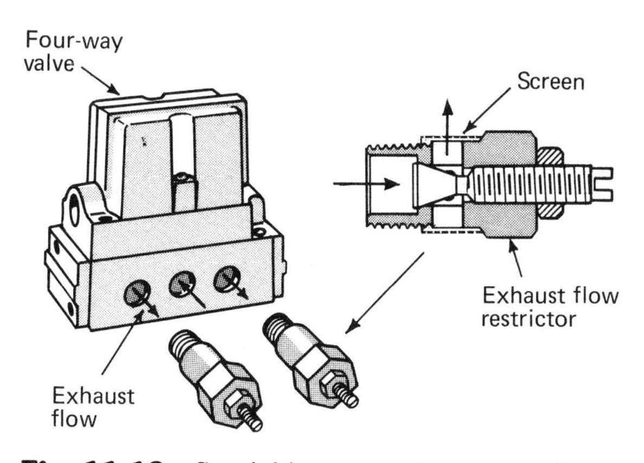

Fig. 11.13 Special insert restrictors can slow down cylinder exhaust any desired amount. (Courtesy of Parker Hannifin Corp., Cleveland, OH.)

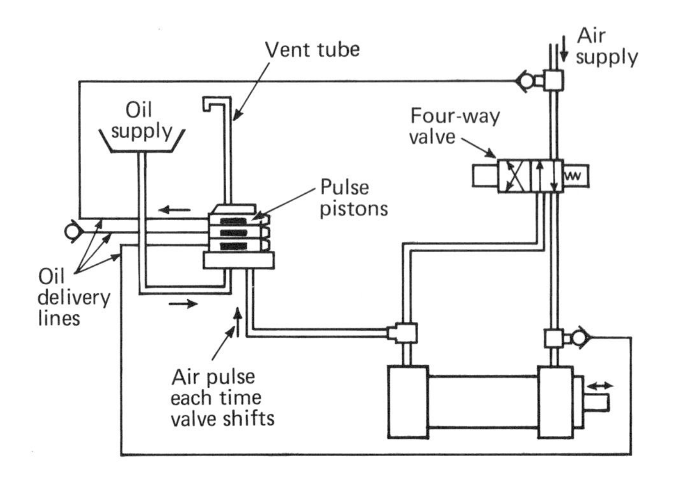

Fig. 11.14 Air-powered lubricator is driven by the cycling action of cylinder air lines, feeding many points. (Courtesy of Master Pneumatic-Detroit, Detroit, MI.)

Many modern compressors operate on synthetic oils because they have strong advantages over petroleum oils: better life, better performance, lower friction, higher operating temperatures. However, the effects on seals and plastic housings (particularly polycarbonate bowls) downstream are uncertain at best, and unless great care is taken throughout the entire shop air system, some device or other will prove incompatible with the chemistry of the lubricant.

The problem is serious enough to cause many manufacturers of air-using devices to include warnings in their literature. Table 11.7 shows a typical label attached to devices sold by Wabco (Lexington, KY).

One answer might be a new chemical-resistant transparent material called Xenoy 3500 by GE Plastics (Pittsfield, MA). It's a physical blend of polycarbonate (which is amorphous) and polyester (which is crystalline). It has 8000-psi tensile and 12,000-psi

Table 11.7 Warning Label for Lubricants Used in Pneumatic Systems (Courtesy of Wabco Contracting and Mining Group, Peoria, IL.)

Use of noncompatible lubricants voids warranty
Many synthetic oils and additives are being promoted for their superior performance features without warning customers of the serious compatibility problems which may develop from their presence in even small quantities
It is an industry-wide problem. Review the document developed by the National Fluid Power Assoc.: *Recommended guidelines for the use of synthetic lubricants in pneumatic fluid power systems (NFPA/T1.9.2-1978)*
Selection of oils compatible with Buna-N, neoprene, urethane, silicone and Hytrel is critical. Only petroleum-based non-detergent oils without synthetic additives and with analine points between 180° and 210° are acceptable

flexural strength and low water absorption and can be injection molded. In one test, a bowl was cycled continuously from 0 to 300 psi with a fluid mixture of 50% carbon tetrachloride and 50% trichloroethane; the polycarbonate failed immediately; Xenoy survived 90 cycles. As a scientific check, bowls were tested on pure water: Both the Xenoy and polycarbonate exceeded 1000 psi without any failures.

On the other side of the coin is burned petroleum oil residue. Compressors using old-fashioned oil create particles of oxidized lubricant that get carried downstream. They have no lubricating qualities and only clog filters and cause some valves to stick. The answer in any case is to know what lubricating system the user has and design for it.

Another inherited problem is loose parts floating downstream and damaging other devices. Fisher Body insists that every in-line valve have no internal parts that will blow downstream when broken and that every adjustment be permanent. Jam nuts are used on variable orifices.

Be careful about mixing one concept with another. For example, prelubed cylinders will be washed out if in-line lubricant is allowed to flow through them.

REALLY DRY CONTROL AIR

"Your living room might be dryer than the Sahara," say the ads. True! But even Sahara air is too wet for some pneumatic controls (Fig. 11.15).

Remember that dry air at 275 F still has moisture and might start condensing out at 200 F or 70 F. It is essential to know what the true moisture content is at every expected pressure and temperature, so that proper dewpoints can be preestablished. The accompanying charts (Figs. 11.16 and 11.17) are guides.

The whole problem hinges on dewpoint, or temperature of condensation. In any pneumatic system that is sensitive to moisture, you've got to remove enough of the water (techniques follow) to lower the dewpoint safely below the minimum temperature the system will reach. Minus 100 F is called for in some process applications. Minus 40 F is a common specification for outdoor systems of all types in cold climates. Indoor systems can get away with +35 F at line pressure in many instances.

The first step in setting up your own requirements for dry air (or any gas) is to establish what the lowest system temperature will be. After you've listed the ambients and low-point temperatures throughout the system, check to make sure you've considered these:

- Cooling effect of adiabatic expansion
- Piping exposed to outside air
- Spots adjacent to colder equipment.

Then you've got to decide how much moisture, if any, the system can stand. Some systems can tolerate wet air at certain local points, on the assumption that the air will heat up and its moisture evaporate before it reaches any of the sensitive regulators or controls. But this approach is risky.

Water in air lines can cause corrosion, rusting, and scaling of metals; blistering of paint; removal of lubricant; and (when solidified as ice) blocking or rupturing of conduits. In most cases moisture is intolerable, and that is the assumption we'll follow here.

Start with the compressed air supply. Let's look at a typical air-supply system. The compressor delivers 75 to 120 psig, or a ratio of absolute pressures ranging from

Fig. 11.15 Instrument air that seems dry under one set of conditions often proves wet when the conditions are changed.

6 : 1 to 9 : 1. Outlet temperature is from 250 to 450 F, and there is practically no condensation in this part of the air system.

However, the aftercooler drops the compressed air below its dewpoint, and the trouble begins. Figure 11.18 depicts a complete compressed air supply system from the compressor to your inlet piping connection. Hypothetical values have been assigned to give you a feel for the quantities involved. The dryer shown in the figure can be of any type capable of −40 F dewpoints.

Drying the Air

Drying the air simply means getting out the water (Fig. 11.19 and Table 11.8, Drying Techniques). Some can be separated out mechanically by baffles or centrifugal action. Fine filters take out some more. Cooling coils can condense moisture at dewpoints 5 to 15 F above the cooling water temperature. The cooling water can be precooled with refrigerants, or the refrigerant itself can be pumped through cooling coils to condense the moisture in the air system directly (sketches).

But none of these conventional direct-cooling techniques can be used below

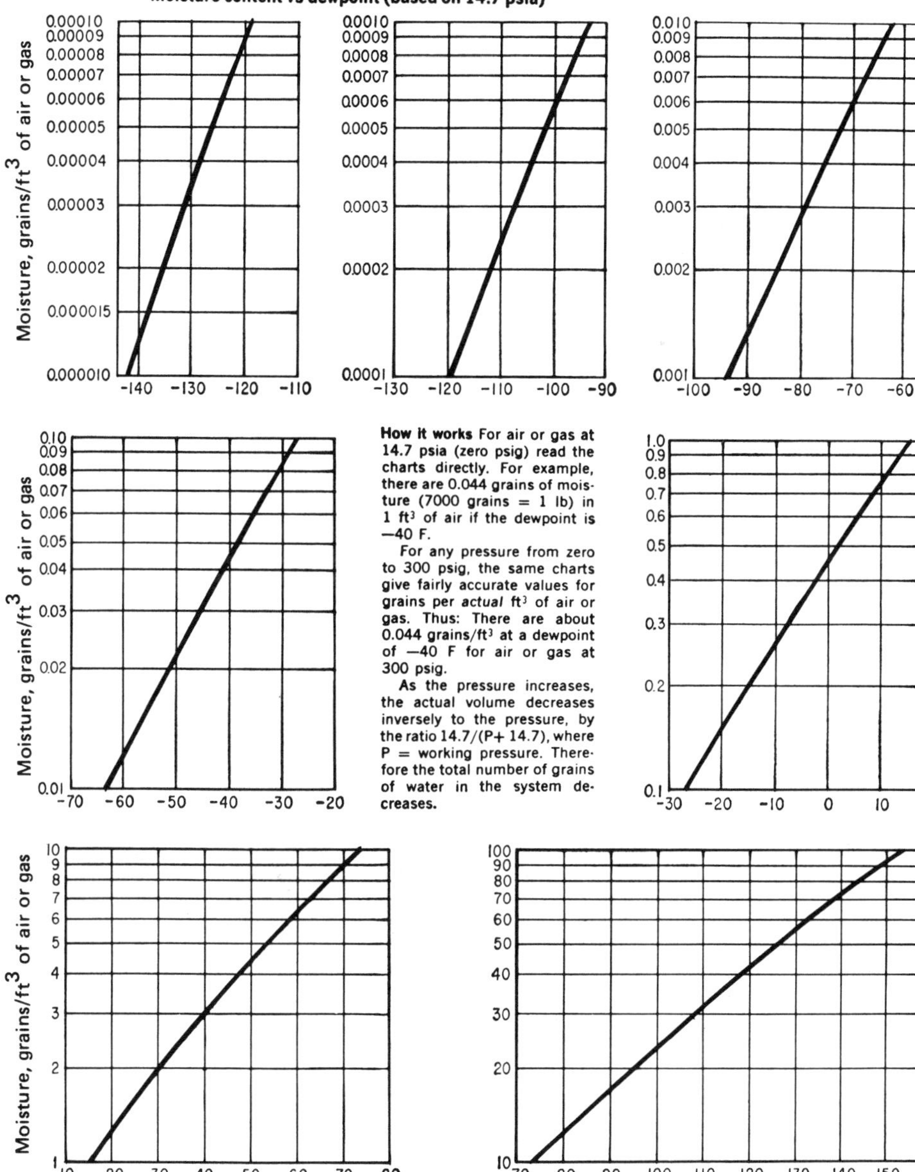

Fig. 11.16 The actual water content of air at any dewpoint can be read directly from these charts.

about 35 F because the moisture condensed in the air system will freeze. For dewpoints lower than 35 or 45 F, chemical or desiccant techniques are applied.

Chemical

One noncooling air-drying system available is based on a deliquescent or absorption-type drying agent. The first sketch in Fig. 11.19 (sketch A) is of such a system. The chamber is filled with sodium chloride, with a small quantity of calcium and magnesium chloride and calcium phosphate. An alternate charging chemical is urea.

These chemicals are salts and absorb the moisture from the air which is passed through them. The driving force that removes the moisture is the differential in vapor pressure of the water vapor in the air vs. that of the water in solution. The practical limit is not much less than a 40 F dewpoint at line pressure, although 30 F is possible. Because these salts go into solution and are drained from the vessel, periodic recharging is necessary.

Desiccant. Unlike absorptive or deliquescent chemicals, a desiccant is *ad*sorptive and

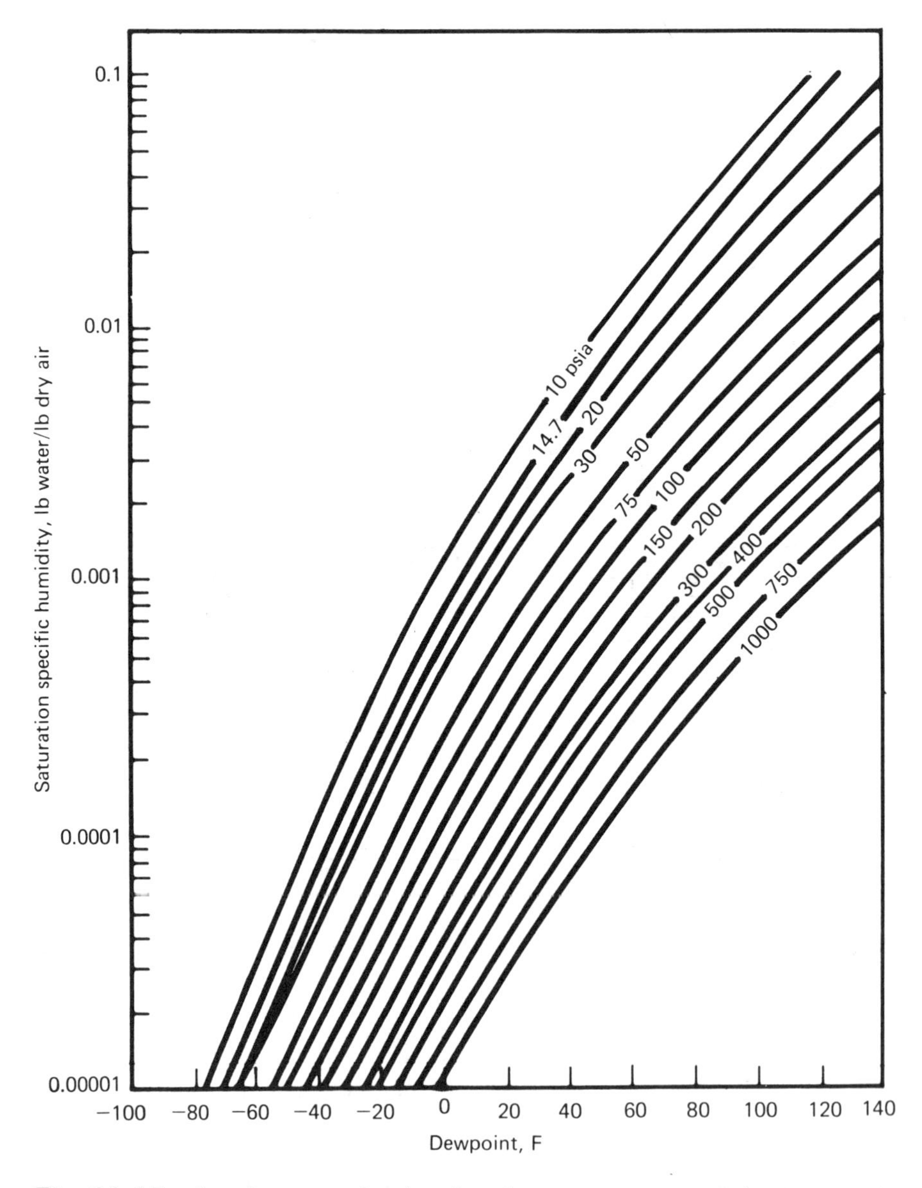

Fig. 11.17 Another way of estimating the water content of air. (From *Chemical Engineering Magazine*).

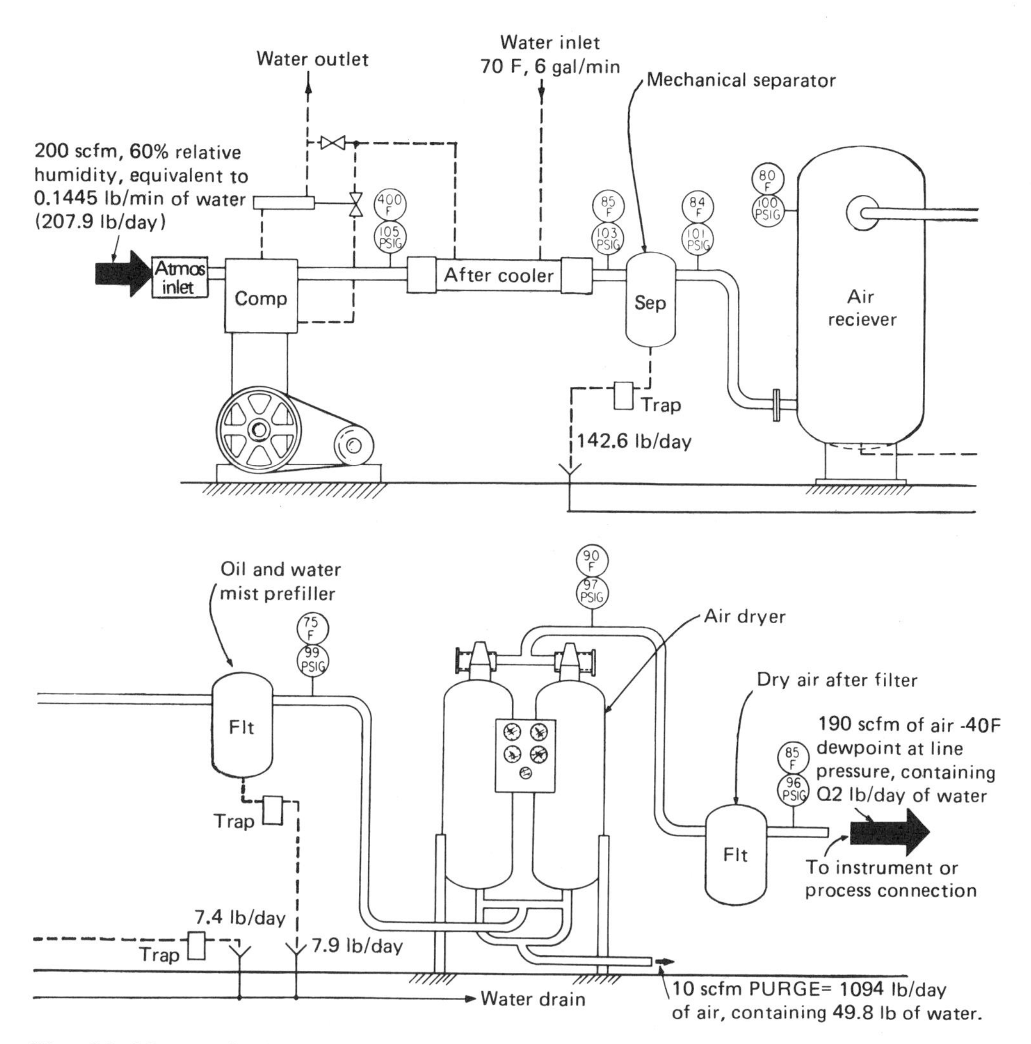

Fig. 11.18 Typical compressed air supply system to show techniques for drying air.

does not form a solution or cause a phase change. Instead, it adsorbs moisture on its surface and holds it as a mono- or bimolecular film. By optical examination, you cannot determine if the substance is wet.

Typical desiccants are silica gel, activated alumina, and molecular sieves of various types, all of which have enormous surface-to-mass ratios. Desiccants can be regenerated by heat or by purging a small portion of the dry air over the desiccant. Let's look at both methods: heat regenerated and so-called pressure-swing regenerated.

Heat-Regenerated Desiccant. Sketches E to H in Drying Techniques, Fig. 11.19, illustrate the four similar but distinct concepts. In each case, one chamber does the normal work of drying the airstream, while the alternate chamber is being regenerated by heat. Then the chambers are switched. It takes 48 hr for the heating cycle, and the resulting line-pressure dewpoint will be from zero down to -40 or -50 F.

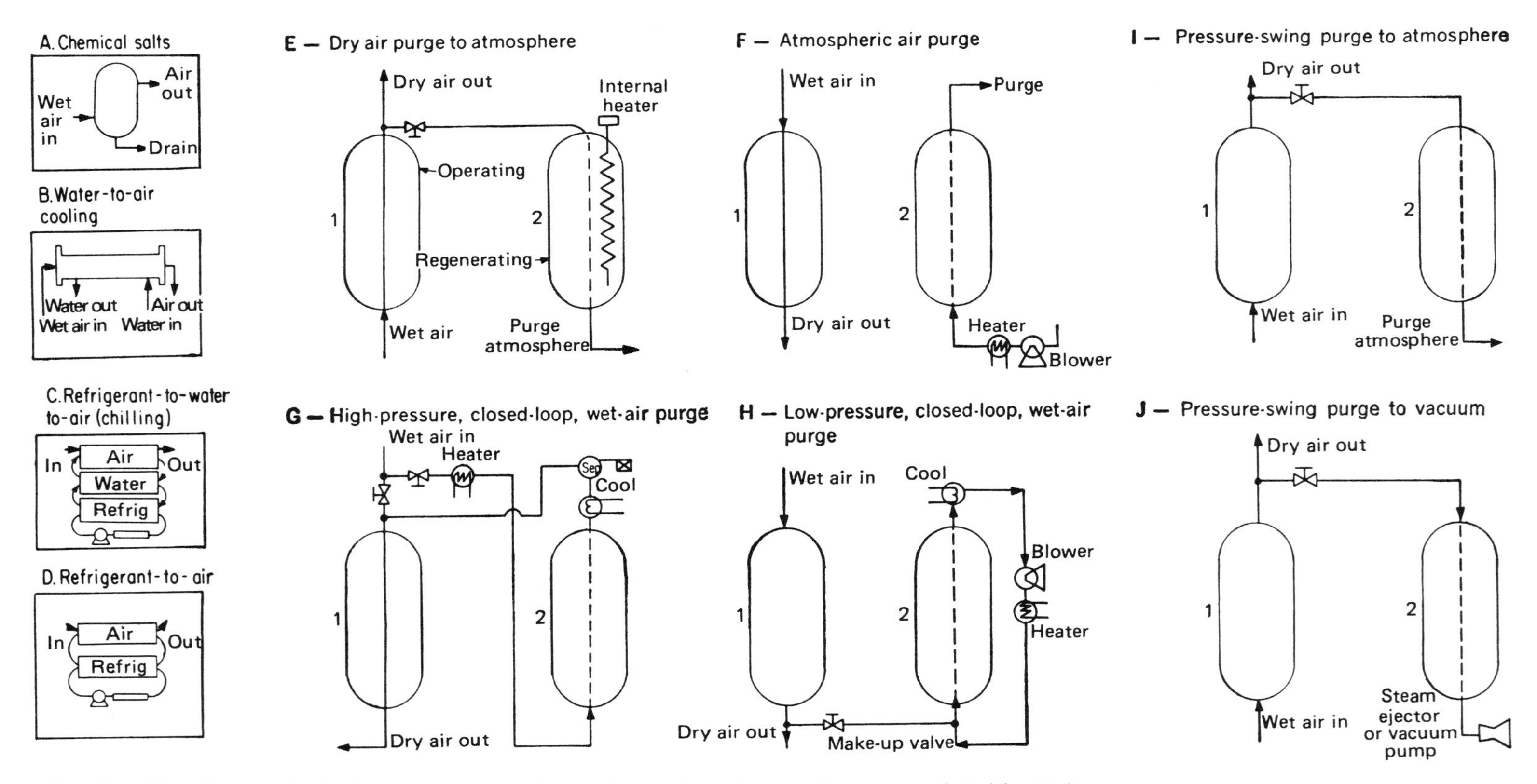

Fig. 11.19 Ten methods for removing moisture from air. Also see the text and Table 11.8.

Table 11.8 Performance and Applications for Ten Air-Drying Techniques (Also See Fig. 11.9)

		Coolants			Heat-regenerated desiccant				Pressure-swing-dried desiccant	
	Chemical salts	Water to air	Refrig to water to air	Refrig to air	Dry air purge to atmos	Atmos air purge	Hi-P closed loop wet air purge	Lo-P closed loop wet air purge	Purge to atmos	Purge to vacuum
	A	B	C	D	E	F	G	H	I	J
Typical performance range: Inlet flow, scfm (based on 100 F, 100 psig)	0-5000	0-10,000	10,000-50,000	0-1000	0-1000	300-5000	500-30,000	500-10,000	0-5000	200-20,000
Inlet temp, °F	50-100	300-450	80-200	50-150	60-130	60-130	60-130	60-130	60-120	60-120
Inlet press, psig	0-150	50-300	50-300	50-300	50-5000	20-5000	50-1000	3-50	60-5000	60-5000
Outlet dewpoint, °F at line press	+30 and up	+50 to 180	+50 to 180	+35 and up	−60 and up	−60 and up	−60 and up	−60 and up	−40 and below	−40 and below
Utilities required	none	water	water and electricity	electricity	electricity, steam, and either process or external air or gas				elec (control), proc. air or gas.	elec (control), steam, proc. air or gas.
Applications	Drying high dewpoint air	Aftercooler for high dewpoint compressor	Large flow, high dewpoint	Low flow, moderate dewpoint	Instrument air, moderate flow, low dewpoint	Instr. and proc. air, med. to high flow, low dewpoint		Proc. air high purity, med. to high flow, low dew.	Low to med. flow, high pressure, low dewpoint.	Med. to high flow, high pressure, low dewpoint.

In each sketch, dried, heated air or gas is forced through the wet desiccant to reactivate it. Where the dried air comes from and how it is pressurized and heated are different in each of the four systems. Details are shown in the sketches.

Sketch E: Dry air purge to atmosphere. A portion of the mainline dried air is bled off at reduced pressure to carry out the water vapor. The desiccant primarily is heated by radiation, which drives off the adsorbed water. When the desiccant is dried, the chambers are switched by valving (not shown in the sketches). The other chamber is dried in the same way.

Sketch F: Atmospheric air purge. Ambient air is pulled into the regeneration loop by a low-pressure blower and then pushed through a heater, heated, forced through the wet desiccant chamber, and exhausted back to atmosphere. Regeneration is by convective heat transfer. Because the regeneration air is heated to sufficiently high temperatures, the humidity of the inlet air is relatively unimportant.

Sketch G: High-pressure, closed-loop, wet air purge. Some wet air or gas (conveniently at the highest pressure in the system) is bypassed, heated, and passed through the wet desiccant. The air is then cooled to condense the moisture it picked up and is reintroduced at a lower pressure point in the wet air line. The lower pressure is produced by partially throttling the main airflow during regeneration.

Sketch H: Low-pressure closed-loop wet air purge. Here, the regeneration circuit is an independent loop, except for system air that is bled in to keep the loop charged. A blower circulates the loop air through a heater, into the wet desiccant chamber, and finally to a cooler. The cooler condenses the moisture picked up from the desiccant.

Pressure-Swing Regenerated Desiccant. This technique eliminates the need for heat, substituting mainline dry air throttled down to atmospheric pressure or partial vacuum to reduce the relative humidity. Thus, it has a greatly increased capacity for holding moisture and easily dries out the wet desiccant. The throttled air used in the regeneration is exhausted to atmosphere (purged). The cycle is repeated every few minutes, compared with hours for the heat-regeneration cycles.

At 100 psi, about 15% purge air is typical for sketch I, 3% for J.

Sketch I (purge to atmosphere) and sketch J (purge to vacuum) differ only in the level of the expansion of the mainline dry gas. The pressure-swing dryer produces very low dewpoints—minus 100 F is not unusual.

How to Choose. The charts in Fig. 11.16 provide the needed psychrometric information. The example following, although simple, lists the key steps. The economics are more difficult of decision than the engineering, however, because they involve costs of utilities such as compressed air, steam, electricity, and even process gas. Whether you select electric heaters in conjunction with a motor-driven blower (both are electric power users) or elect to purge air or gas from the main stream (it can take 1 to 25% of the mainline flow, according to the required dewpoints) depends in large measure on the relative costs of these utilities.

Example: Heat-Regenerated Dryer. Problem: Determine the weight of water per hour to be eliminated from 25 scfm of air (14.7 psia and 70 F reference), actual inlet conditions of 90 psig and 90 F, saturated with moisture, to give outlet air with a dewpoint of −40 F.

The first step is to calculate the actual cfm:

$$\text{acfm} = \text{scfm} \times \frac{P_1T_2}{P_2T_1} = 25 \times \frac{14.7 \times 550}{104.7 \times 530} = 3.65$$

Water load (to be removed by desiccant):

$$W = \text{acfm} \times w \times t$$

where W is lb water to be adsorbed by desiccant per hour of drying, w = lb water in 1 ft^3 of air saturated at 90 psig and 90 F, and t = drying time, minutes. Thus,

$$W = 3.65 \times 0.00234 \text{ (see Fig. 11.16)} \times 60 = 0.445 \text{ lb of water per hour}$$

Assuming the total drying time per cycle is 4–6 hr, the desiccant mass must be sufficient to adsorb 1.78–2.67 lb before regeneration, while holding a dewpoint of −40 F or better.

At this point, the dryer manufacturer must be contacted. The amount of desiccant needed depends on many variables: type of desiccant, reactivation temperature, nature of contaminants if any, desired duration between desiccant changes, and the required size and shape of the desiccant bed.

Example: Adapt Fig. 11.16 to 100 psig. *Problem:* The dewpoint at 14.7 psia is −70 F; what is the dewpoint at 100 psig?

(1) On curves read 0.006 grains/ft^3 at −70 F and 14.7 psia; (2) multiply 0.006 $\times (100 + 14.7)/14.7 = 0.049$ grains/ft^3; (3) on curves read −38 F dewpoint at 0.049 grains/ft^3.

UNUSUAL PNEUMATIC SYSTEMS

Two pneumatic systems are selected as different from the run-of-the-mill types: (1) dual-pressure systems and (2) ram-air systems. There are hundreds more, but the two selected are a good indication of the variety possible.

Dual-Pressure Systems

Dual-pressure systems have important advantages in economy and performance. For example, Wabco Fluid Power (Lexington, KY) offers a two-tank reduction system (Fig. 11.20) for petroleum drill rigs. The first tank is at 175 to 250 psi, and the second is at 125 to 150 psi.

The higher pressure squeezes out most of the moisture, which is automatically removed with a drain valve. The air entering the regulator is saturated, but as soon as the pressure is reduced, the relative humidity drops, and very little condensation will occur from that point on. As an example, reduction of saturated air to 100 psi from 200 psi lowers relative humidity to 53% from 100%.

Freezing still might be a problem with outside rigs, so there is another trick: an antifreezer. These devices are tanks of alcohol with large porous wicks extending upward into the airstream to allow the air to pick up vapor. Depending on the circumstances, the antifreezer can be placed at the compressor inlet or downstream. Do not specify permanent-type antifreeze, however, because vaporization is an important characteristic.

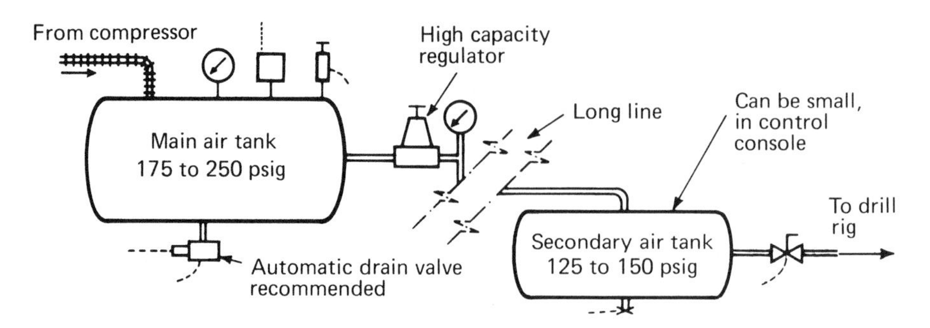

Fig. 11.20 Dual-pressure compressed air system removes much moisture by compressing to high pressure in the first tank. Expansion to lower operating pressure reduces the relative humidity. (Courtesy of Wabco Contracting and Mining Group, Peoria, IL.)

Another dual-pressure concept is the air amplifier or intensifier. Haskel Engineering and Supply Co. (Burbank, CA) has an "air doubler" (Fig. 11.21) which has a spool-type piston that takes input pressure and automatically doubles it. If the downstream pressure decreases momentarily, the device reacts to maintain the twice-input level. It's not designed for continuous high flow (it exhausts a portion of the input) but is ideal for occasional use of high pressure.

Ram-Air Cooling

Fast aircraft today use a combination of ram-air and bleed-air to power the entire environmental system. Hamilton-Standard, a div. of United Aircraft (Windsor Locks, CT) is one of the companies that has designed and built this type of system (Fig. 11.22).

It is called bootstrap because it takes its energy from the air itself and doesn't need electrical power. According to its designers, the system is unique in that the turbine, compressor, and a booster-circulating fan are combined on one shaft, and the whole package, including control, is integrated.

Operation of the system is basically simple: Hot bleed-air at up to 600 F from

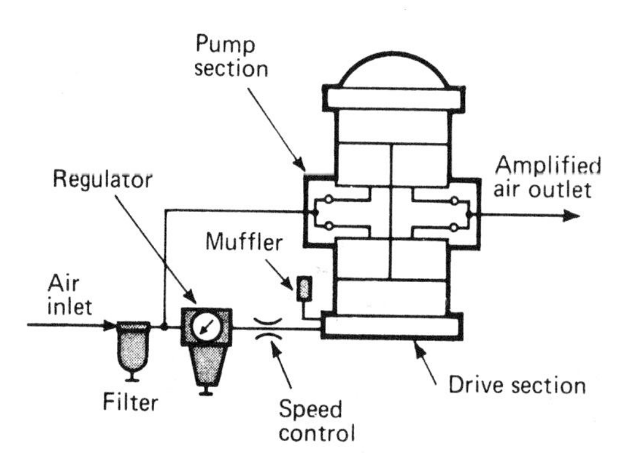

Fig. 11.21 Pressure doubler is an air-to-air intensifier with a 2 : 1 ratio. (Courtesy of Haskel, Inc., Burbank, CA.)

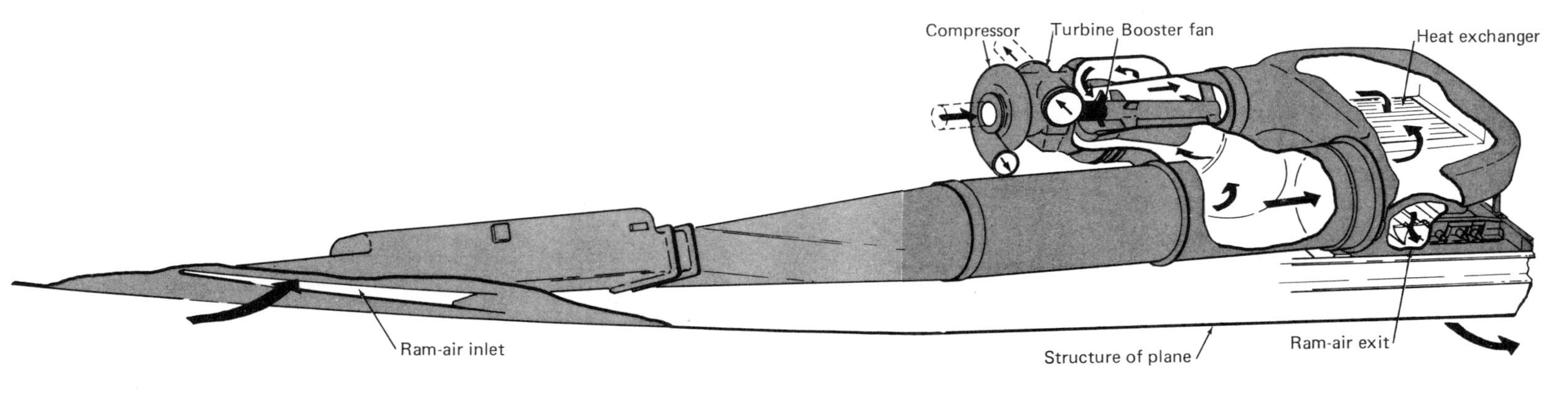

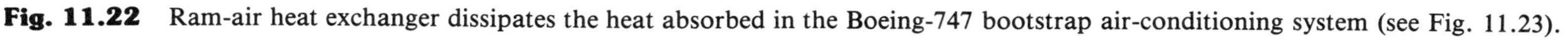

Fig. 11.22 Ram-air heat exchanger dissipates the heat absorbed in the Boeing-747 bootstrap air-conditioning system (see Fig. 11.23).

midstage or final-stage locations on the plane's main engine compressors supplies the stored energy. This air is cooled in a ram-air heat exchanger (Fig. 11.23), is compressed, then cooled again, and finally expanded through the turbine to lower its final temperature to less than ambient.

The turbine-expander does three jobs: It drives the compressor and a special booster fan and while so doing expands the hot, high-pressure air down to cooling temperature. The operating speed is 37,000 rpm.

Ram-air scooped up by the plane as it streaks through the sky takes heat out of the exchanger. The exchanger is made of aluminum, assembled with a fluxless brazing technique in a high-temperature vacuum furnace. Lack of flux eliminates the possibility of later corrosion caused by salt deposits from conventional molten-salt brazing operations.

The booster fan moves the air through the heat exchanger whenever the air speed of the plane is insufficient to ram the air through or when the plane engines are idling.

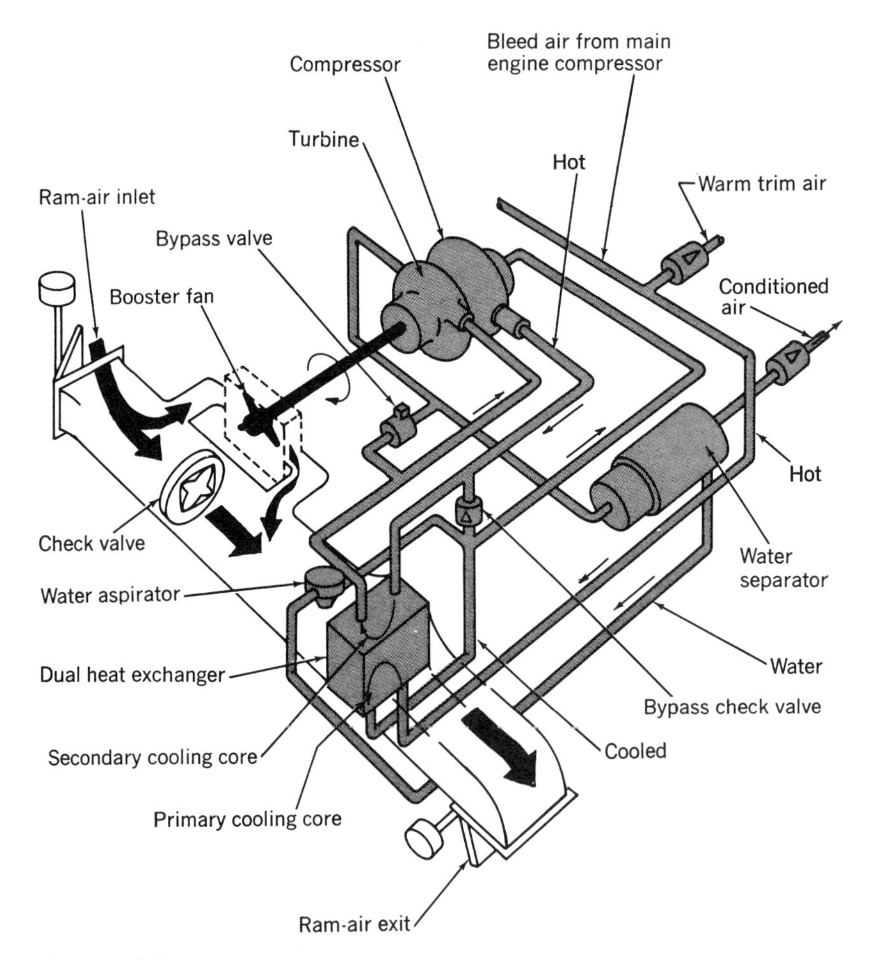

Fig. 11.23 Bootstrap air-conditioning package for Boeing 747 has a turbine, compressor, and booster-circulation fan on the same shaft. Pressurized air powers the system (see Fig. 11.22).

What a breeze. Each package has a capacity of about 72 tons of refrigeration. Combined, this performance is equivalent to the cooling of 36 five-room houses during a summer heat wave. Airflow is 8000 cfm, which is enough to change the cabin air once every 210 sec. In a typical cruise condition, air entering each bootstrap package at 380 F is continuously cooled to 62 F. At high altitudes (above 30,000 ft) the bootstraps are not needed because ram-air comes in cold enough to provide the air conditioning.

The accessories are fairly elaborate. The ram-air flow is controlled with flapper valves. Turbine-compressor-stage flows are monitored by thermal sensors to prevent excessive temperatures. Condensed moisture is removed with water separators that work by a combination of coalescing through fabric, cyclone centrifuging, and aspiration.

The aspirator is powered by bleed-air, which forces the water into the ram-air stream ahead of the exchanger so that it enhances the cooling effect. The water is evaporated as it passes through and goes overboard.

12

Compressors, Blowers, Fans

The distinction is not always well defined, but for the typical pneumatic power and control applications—including shop air systems—the following definitions will suffice: A *compressor* produces substantial working pressure (typically 100 psig) at low to moderate flows to run powered equipment and to pressurize control systems; a *blower* produces light working pressures (2 to 15 psig, for example) for auxiliary low-energy functions such as paper handling, low-force clamping, combustion, and ventilation; and *fans* normally provide cooling or ventilation at less than 1 psig.

Figure 12.1 shows most of the types to be discussed, with a hint of the flow ranges encountered. This chapter will concentrate in four areas: (1) general compression theory, (2) packaged positive-displacement air compressors, (3) regenerative blowers, and (4) conventional low-pressure fans.

GENERAL COMPRESSION THEORY

Operating characteristics of all types of compressors and blowers can be compared by the combined use of two parameters—specific speed and adiabatic head—each a function of the geometry of the machine. Specific speed (Fig. 12.2) is a dimensionless ratio first developed for liquid pumps (explained in Chap. 3) and then applied to compressible fluids:

$$N = nQ^{1/2}/(gh)^{3/4}$$

For liquids, h is simply pressure × density. But for compressible fluids, h must be integrated for density change in the compression process, using the thermodynamic description of a reversible adiabatic process, and becomes the "adiabatic head" H:

$$H = \frac{R(T+460)}{[(K-1)/K]M} \left[\left(\frac{P_2}{P_1}\right)^{(K-1)/K} - 1\right]$$

or

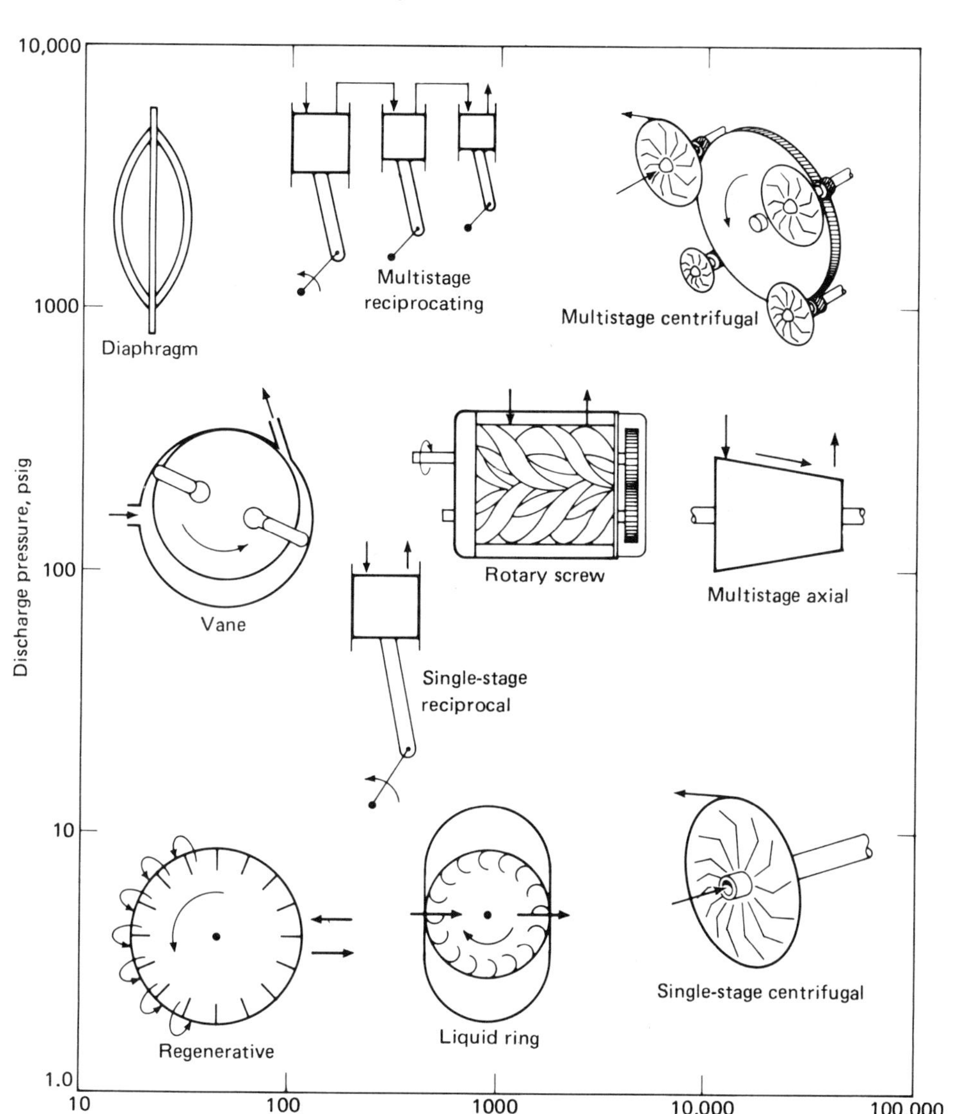

Fig. 12.1 Flow and pressure ranges are characteristics of compressor design and geometry.

$$\frac{HM}{T+460}=f\left(K,\frac{P_2}{P_1}\right)$$

This relationship, presented graphically in the adiabatic-head curves (Fig. 12.3) expresses compressor performance in terms of pressure ratio of the machine, fluid inlet temperature, molecular weight, and ratio of specific heats of the fluid (Tables 12.1 and 12.2). The relationship can be used to determine the adiabatic head needed for a given application or to evaluate a specific compressor from catalog performance data.

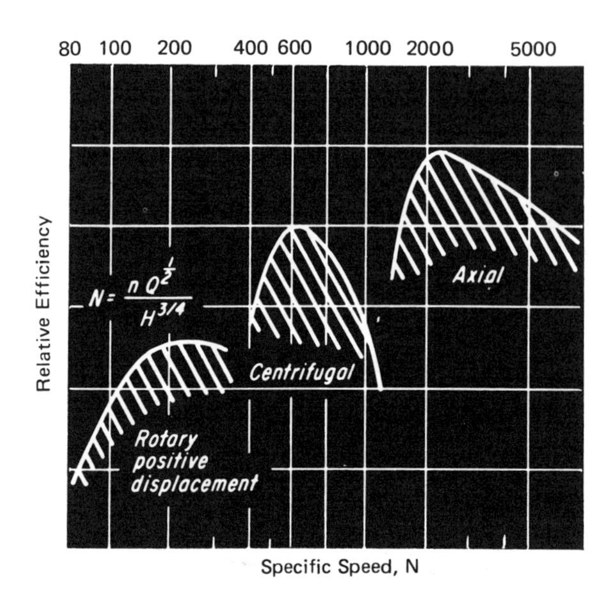

Fig. 12.2 Specific-speed ratio compares the efficiency of three types of compressors for any given speed, volume, and adiabatic head.

In a centrifugal or axial-flow compressor, H is determined by the speed, impeller diameter, and vane geometry. In a positive-displacement machine, H is determined by the volume change in the fluid space during the compression stroke. Thus, this factor is a useful constant in evaluating the capability of any compressor under given operating conditions.

Further, the adiabatic-head concept helps evaluate the effect of varying the fluid properties, inlet pressure, and inlet temperature on compressor performance. For a given machine;

- An increase in inlet temperature decreases the pressure ratio.
- A change in inlet pressure produces a proportional change in discharge pressure.
- A change to a gas with higher molecular weight increases the pressure ratio.

Since this relationship applies to both positive-displacement and turbocompressors, it offers a direct method of comparing performance in a specific application. When specific speed is plotted as

$$N = nQ^{1/2}/H^{3/4}$$

as shown in the diagrams, the relative efficiency of positive-displacement, centrifugal, and axial-flow machines is readily evaluated for a specific speed, flow, and head.

In using this method to select a machine for a specific application, the first step is to define the system requirements:

- Molecular weight and ratio of specific heats for the gas
- Pressure ratio required, including system losses
- Inlet gas temperature

- Flow rate

Next determine the optimum type of compressor by these steps:

- Calculate adiabatic head H from curve
- Decide on speed
- Calculate specific speed
- Select compressor type from chart

If the chart indicates positive displacement, use the pressure ratio to narrow the choice.

While pressure ratio, flow, and speed are the primary factors in selecting a compressor, the decision must also consider pressure losses in the system and uniformity of load; corrosiveness, dirt, and moisture content of fluid; and service life and maintenance problems.

A compressor system must provide the required flow and pressure at the discharge of a pneumatic system, which includes pipe, valves, and fittings. Methods for estimating system pressure losses are described in Chap. 16.

Provision should also be made for more flow capacity for potential growth. Additional components are almost invariably added to any shipped system, so an

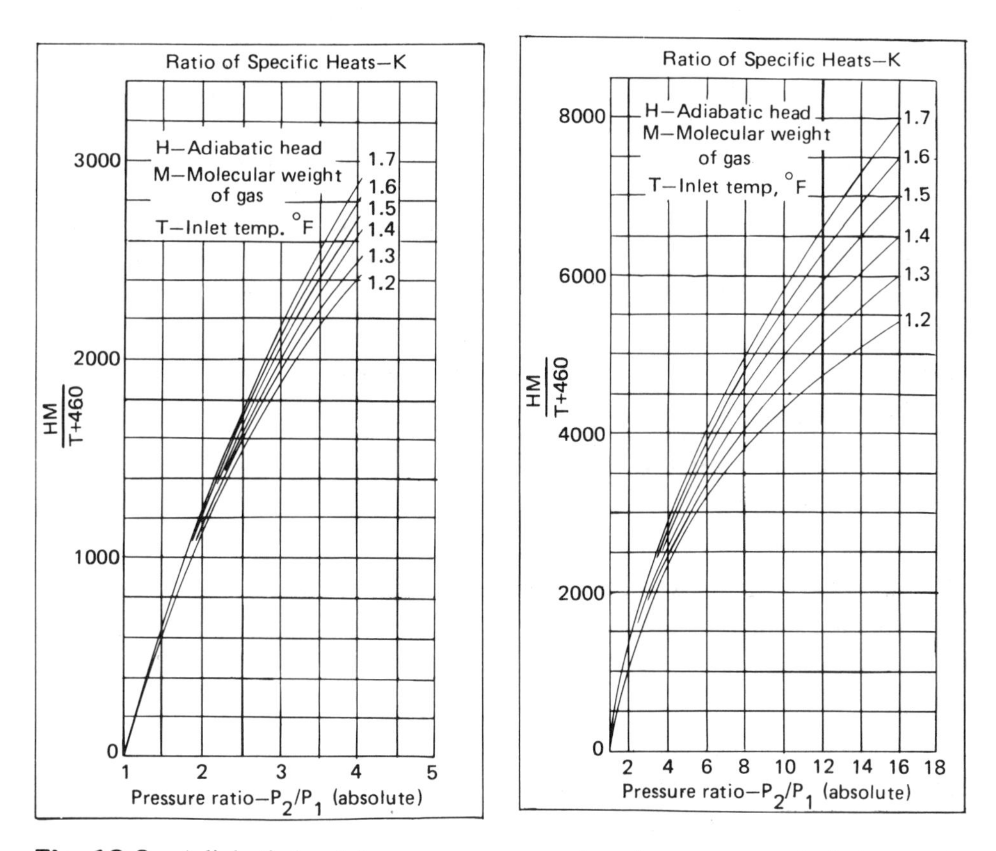

Fig. 12.3 Adiabatic head for a compressor, when the pressure ratio is known, is computed from the values read on the vertical scale.

Table 12.1 Gas Properties for Computation of Adiabatic Head

	Molecular weight	Ratio of specific heats[a]
Hydrogen	2	1.4
Helium	4	1.7
Methane	16	1.3
Ammonia	17	1.3
Steam	18	1.3
Nitrogen	28	1.4
Carbon monoxide	28	1.3
Air	29	1.4
Oxygen	32	1.4
Carbon dioxide	44	1.3
Freon 12	121	1.1

[a]Values are for input temperatures in the normal ambient range.

allowance of at least 25% over immediate needs, depending on the situation, will call for little increase in investment. Also, compressor performance decreases with age and use, primarily from wear of vanes, valves, or piston-rings. This shows up as a decrease in discharge pressure, rather than a decrease in flow.

While there is no hard-and-fast rule, in general, the system should be designed for one additional pressure ratio. For example, a system discharging at 100 psi (gage) with atmospheric inlet (a pressure ratio of approximately 8) would need a pressure

Table 12.2 Nomenclature for Calculation of Adiabatic Head and Specific Speed

g = gravitational acceleration
h = pressure head, ft
H = adiabatic head, ft
K = ratio of specific heats of gas for ambient temperatures
M = molecular weight of gas
n = rotational speed, rpm
N = specific speed
P_1 = compressor inlet pressure, psia
P_2 = compressor outlet pressure, psia
R = gas constant, 1544 ft lb/lb/deg Rankine
T = inlet temperature, °F
Q = volume, cfm

ratio of 9 and a discharge pressure of approximately 120 psi. Again, the additional investment is minor.

For a system with an intermittent load—the more common case—the compressor capacity should be less than the system capacity, with the difference made up by an accumulator in the system. If some variation in supply pressure is acceptable, a relatively small compressor can be used with a large accumulator. If the supply is used for long periods of time with only short idle periods, compressor capacity should nearly equal the supply. If a relatively constant discharge pressure is required, use a small accumulator and a large compressor. For regulated pressure, use an oversize compressor with a pressure regulator.

Example: Fan Adiabatic Head

Suppose we need 50 cfm of room air at 2 psi and plan to use an ac universal motor at 20,000 rpm with a life of 1000 to 2000 hr. The molecular weight of air is 29, the temperature is 70 F (530 R), the ratio of specific heat is 1.4. The pressure ratio (absolute pressure) is 16.7/14.7 or 1.136. From the adiabatic-head curve (Fig. 12.3) HM/T is 200, and H is $200 \times 530/29$ or 3680.

Substituting this value for H in the equation for specific speed gives $N = 20{,}000 (50)^{1/2}/3680^{3/4}$ or 300. Turning to the chart for specific speed (Fig. 12.2), this value falls between the displacement and centrifugal types. A centrifugal type can be made at relatively low cost from aluminum stampings, while a positive-displacement compressor needs more parts and closer tolerances and therefore costs more.

If the application were industrial, requiring a service life of 5000 or 10,000 hr, problems of brush wear and bearing life in a universal motor could be eliminated with a 3600-rpm induction motor. At this lower motor speed the specific speed would be 55, and the choice would now swing to a vane-type compressor.

Example: 85-psi CO_2 Compressor

Consider a machine to compress 400 cfm of carbon dioxide from 5 psi and 100 F to 85 psi, with a compressor rotating at 3600 rpm. The molecular weight of carbon dioxide is 44 and the ratio of specific heats is 1.3. The pressure ratio (absolute) is 5. Entering the adiabatic-head curve from the pressure ratio, the value of 3000 is read on the ordinate. Adiabatic head H is $3000 \times 560/44 = 38{,}200$. Specific speed is $3600(400)^{1/2}/38{,}200^{3/4}$ or 26, at the low end of the positive compressor range. Since the pressure ratio is 5, either a piston or vane type can be considered.

The flow would have to be increased by a factor of 500 or the rpm multiplied by 10 before a centrifugal compressor would be effective.

Performance data in catalogs are generally given in terms of air at room temperature and atmospheric pressure. The adiabatic-head chart can be used to find the equivalent performance for air instead of carbon dioxide. Adiabatic head H will be the same, 38,200. The molecular weight of air is 29, the inlet temperature 520 R, the inlet pressure 14.7 psi, and the ratio of specific heats 1.4. Therefore,

$$HM/T = 38{,}200 \times 29/520 = 2130$$

Entering the chart for adiabatic head at 2130 and K at 1.4, the pressure ratio is 3.2, and the outlet pressure is 47.0 psia. Therefore, a compressor designed to deliver 400 cfm of air at 47 psia with 60 F inlet air temperature would compress 400 cfm of carbon dioxide from 20 psia to 100 psia at an inlet temperature of 100 F.

Example: 75-psi Helium Compressor

Take an unusual example: helium with a molecular weight of 4 and a ratio of specific heats of 1.66. Assume the compressor is to take helium at 80 F and 50 psig and deliver 75 psig. The pressure ratio (absolute) is 1.39. From the adiabatic-head chart, the vertical scale value is 520. With T = 540 R, H is calculated as 70,200.

Translating these values to air as a base (T = 520 R, K = 1.4, and M = 29) and using the value of H calculated for helium, the new ordinate is

$$70{,}200 \times 28.95/520 = 3910$$

This value, read down from the curve for K = 1.4, gives a pressure ratio of 6.8. Therefore, a compressor designed to deliver a pressure ratio of 6.2 in air with an inlet temperature of 60 F would compress the same volume of helium to a pressure ratio of 1.39 with an inlet temperature of 80 F and inlet pressure of 50 psig.

For Specialized Requirements

Other requirements include corrosive gases, wet gases, high temperatures, high pressures or extremely low pressures, as well as gases of differing physical properties. For high pressures and very low vacuums, a range of heavy-duty high-pressure compressors and close-clearance high-suction compressors is available. For corrosive applications, the compressor surfaces in contact with fluid may be glass, stainless steel, or graphite.

Wet gases, and gases other than air, influence selection. Compression of wet gases leads to a more nearly isothermal compression with a consequent change in compressor performance; lubrication may present problems not present with dry gases. Corrosion-resistant construction may be needed for high-temperature gases, or the choice may be between an expensive specialized compressor or the use of a cooler to decrease the inlet temperature. But these specialized compressor conditions often present problems that only the compressor manufacturer can solve.

Compressors—In Series or Parallel

When additional flow or pressure is needed beyond the capacity of an available compressor, within certain limitations compressors can be operated in parallel or in series. In series, matching problems arise in positive-displacement machines—piston, rotary vane, rotary lobe.

If two of these compressors are operated in series, the discharge flow rate of the first must match inlet flow rate of the second. If not carefully matched, a series system may show an appreciable vacuum or excessive pressure between the two compressors rather than the intermediate pressure desired. Mismatch may also introduce seal problems in the second compressor, since inlet and discharge pressure, and the compressor-case pressure, will be higher than design values.

With turbocompressors in series, a flow mismatch will cause surging of one or the other and flow instability.

Positive-displacement compressors can be operated in parallel if the discharge pressures of the two machines are about the same; flow through one compressor has little or no effect upon the other. However, turbocompressors operated in parallel do present a matching problem; one compressor will frequently pick up the entire flow while the second idles or becomes unstable. Consequently, an automatic flow-control system is required.

Turbo and positive-displacement compressors can be combined; a turbocompressor can be used as a supercharger ahead of a positive-displacement machine or vice versa. Or a turbocompressor could be used in parallel with a positive-displacement compressor as a flow booster.

PACKAGED COMPRESSORS

A compressed air package is to pneumatics what a pump-reservoir unit is to hydraulics. Here's a way to match the compressor to your application needs.

The custom or "dedicated" compressor packages we'll be talking about are only part of the overall scene. Ingersoll-Rand's air compressor group (Charlotte, NC) says that a large share of the market is for packaged utility compressors, marketed and serviced directly by the compressor manufacturers. In these applications, thermodynamic theory is of little concern to the buyer. Compressors under 100 hp often are bought out of a catalog.

Just the same, a summary of thermodynamic relationships is included in graphs and boxes accompanying this text. Though theoretical, they are useful in making initial estimates of temperature and power. A thorough text, called *Compressed Air and Gas Data*, is published by Ingersoll Rand (Washington, NJ).

Flow is most important. An application goes nowhere until accurate figures for flow are obtained. It's not enough to make conservative (high) estimates, because a system designed for high flow might not work well at low flow. For example, regenerative desiccant dryers included in some systems will not perform properly unless the loads are at least 50% of the rated values. Also, a large motor-compressor system operating at 20% to 50% load costs a lot more to buy, run, and maintain than a smaller motor-compressor system running at 75% to 100% load.

A worse mistake is to specify oversized dynamic-type (centrifugal or axial) compressors which are designed to run continuously. During periods of low demand, these machines must be operated in the relief or bypass mode to avoid surge (flow reversals).

The best way to predict flow is to review carefully all the potential uses of the compressed air produced by the system to verify the true demands. Inexperienced engineers will add up catalog values of pneumatic cylinder or air motor displacements as well as valve flow coefficients (C_V) and then convert them to estimated free air, actual cfm, or whatever. They will fail to recognize the most useful values of all:

- Average airflow required
- Maximum/minimum pressure (peaks and valleys)
- Maximum/minimum airflow
- Quality of air (humidity, oil content, contamination)
- Duty cycle (times at each flow and pressure)
- Proper units of measurement (psig vs. psia, true mass flow, etc.)

That last item—true mass flow—is too often ignored or misused. The dangers are these: So-called actual cfm (acfm) might be based on nonstandard conditions, thus misleading the compressor application engineer. Free cfm (fcfm) is another measurement to avoid because it's not specific.

Some small compressor manufacturers exploit nonspecific terminology to exaggerate equipment performance. The figures can be very inaccurate if they are based on throttled inlet, high-temperature discharge, or other conditions not consistent with standard measure. Using theoretical displacement of cylinders or air motors is misleading too because it ignores volumetric efficiency and leakage inherent in compression machinery.

Only mass flow, or a conversion of mass flow to standard cfm (scfm), should be used to determine the size (volume) of a compressor. It's a bother to convert flow data into cfm at standard conditions (14.696 psia and 60 F), but it's worth it.

With good flow and pressure data, a packager can determine if the job can be done with a single compressor discharging directly into the system or if multiple units or special packages are needed.

Pressure generally is not a problem because the demand is known fairly accurately ahead of time. An exception is where the peaks are of short duration; it might pay to specify a lower-pressure compressor for the major duty, with a small booster compressor or intensifier to handle the peak loads.

Average air consumption is important to know. A pressure tank or receiver (reservoir) of the correct size will store air during idling and can discharge it during peak load demand. The compressor should be large enough to compress the average flow but shouldn't have to handle the peak loads. The most compact design is where the peak load is shared jointly by the pressure tank and the compressor—both working at maximum capacity.

On-off operation of the compressor is another possibility. The pressure tank can be pumped up quickly and the compressor shut off. It takes a higher-flow compressor than would be required in the averaging duty mentioned above, but the compressor and drive motor will benefit from long cooldown between pumping periods. Components can be designed for high peak-load operation without danger of overheating. If the shutoff periods are long enough, significant energy savings result, compared with a constantly idling motor-compressor.

Control of Capacity

The capacity of a compressor can be readily adjusted during operation by these simple methods: suction or discharge unloading valves that can be held to prevent compression, clearance-pocket control to add effective volume to the cylinder of a reciprocating compressor and lower its volumetric efficiency, on-off control of the driving motor to shut down the compressor whenever cycle pressure is up, and bypass of compressor discharge.

Clearance-pocket control is done in discrete steps by adding fixed volumes through valving and normally is not infinitely variable.

On-off control is satisfactory unless the motor-compressor is cycled too frequently. Once in 6 min is considered acceptable. The way to increase intervals between cycles is to set a wide range between shutoff and turn-on pressures—say, 20 psi in a 100-psig system.

Bypass of compressor discharge partially unloads the compressor, and a check valve holds the receiver pressure up, but the compressor is still running and is still compressing air.

A combination of several of the above control methods can save energy. For example, stop-start control can be added to inlet-unloading control, with a time-delay relay to prevent too rapid cycling.

Rotary-screw compressor flow can be reduced by beginning the compression at a later point along the axis. It's done by blocking off some of the inlet porting at one end of the casing with an integral sliding or rotary valve. It decreases compressor efficiency but saves energy by compressing less air.

The ultimate control of a positive-displacement machine is to vary the speed of the driving motor to match the compressed air usage, but the costs are higher and the control more complicated.

Work of Compression

The work of any stage of a reciprocating compressor can be approximated by assuming adiabatic compression, according to most authorities (see Table 12.3). The compression cycle is too fast for any significant amount of local cooling, so isothermal compression or anything near it cannot occur.

In any event, friction and other losses not related to compression are so high that it is hardly worthwhile to derive the exact polytropic equation. (See the follow-

Table 12.3 Equations for Calculation of Simple Compressor Performance

---

Adiabatic relationships

$$\frac{P_2}{P_1} = \left(\frac{V_1}{V_2}\right)^k ; \quad \frac{P_2}{P_1} = \left(\frac{T_2}{T_1}\right)^{\frac{k}{k-1}} ; \quad \frac{T_2}{T_1} = \left(\frac{P_2}{P_1}\right)^{\frac{k-1}{k}}$$

$$\text{Ideal } \frac{P_2}{P_1} \text{ per stage} = \sqrt[s]{\frac{P_F}{P_1}} \quad \text{with perfect intercooling}$$

$$\text{Adia. hp per stage} = \frac{144}{33{,}000} P_1 Q_1 \frac{k}{k-1}\left[\left(\frac{P_2}{P_1}\right)^{\frac{k-1}{k}} - 1\right]$$

Conversion to standard flow (14.7 psia, 60 F)

$$\text{scfm} = \text{acfm}\left(\frac{\text{actual } P}{14.7}\right)\left(\frac{520}{\text{actual } ^\circ\text{R}}\right)$$

$$\text{based on } \frac{P_1 V_1}{T_1} = \frac{P_2 V_2}{T_2}$$

Symbols

P = psia; V = ft^3; T = °R; k = c_p/c_v = 1.4 for air; s = number of stages; Q = cfm (or ft^3/min); Q_1 = cfm at inlet conditions; scfm = flow converted to standard conditions; acfm = actual cfm; subscript 1 = inlet and 2 = discharge of stage; F = final discharge

---

ing section on efficiency.) Typical overall efficiency for small reciprocating compressors can be 60%.

Motor power (brake hp) can be roughly estimated by calculating ideal adiabatic compression power (gas hp) from Table 12.3 and adding the appropriate mechanical losses from vee-belts, bearings, and so forth.

One quick method is to allow 40% of the calculated gas hp in machines below 5 hp, 30% for units up to 25 hp, and 20% for larger compressors.

For instance, a 100-cfm, 14.7-psia inlet compressor discharging at 100 psig would require 18 gas hp in the equation but actually requires a 25-hp motor. A general fact is that bigger compressors are more efficient.

If more accurate calculation is required, then be prepared for a lot of hard work. Here are some of the tools you'll need: charts of temperature, entropy, and humidity; tables of constants for deviations from a perfect gas; and extensive details on compressor design, including clearances, materials of construction, side loading, lubrication, cooling, and so forth. Only the compressor manufacturers are likely to get into that much analysis.

Efficiency a Problem

Compression is not efficient by any measure, since a lot of the work of compression converts directly into heat, which then must be removed. The available energy content of compressed air or gas at room temperature is only a fraction of the work energy that went into compressing it.

Hypothetical example: A control-air system that operates on 100-psig dry air taken from a 120-psig receiver. It has a two-stage compressor driven by an ac motor through pulleys; included are intercooling, aftercooling, regenerative desiccant dryers, pressure regulators, and an automatic drain.

Efficiencies might be roughly as follows: motor (electrical), 85%; motor and pulley (mechanical and windage), 90%; compression (including suction and volumetric), 90%; compressor (friction and mechanical), 90%; and flow efficiency (including all pressure drop and purging losses), 90%.

The best that can be done is to reduce losses by some of these good design techniques:

- Lowered slip leakage flow from discharge back to inlet during compression
- Multiple staging with intercooling to avoid extremes in temperature and to lower the work of compression
- Exploitation of waste heat for other processes
- Utilization of the compressed air or gas at the highest level of temperature to take advantage of the increased volume
- Extreme conservation (compress only what you need)

Rotary screw compressors have a special problem in that the compression takes place gradually along the axes of the meshing screws, culminating in discharge when the compressed pocket of air reaches the end of the casing.

If the pressure in the discharge line is different from the pressure in the pocket of air when they interface, there is a sudden expansion or contraction of the compressed air. Such compressors are carefully designed to match the system; otherwise, efficiency will suffer considerably.

Reciprocating compressors don't have this problem because the discharge valves are loaded against system air pressure, and they do not have to open until cylinder pressure is up to system pressure.

Rotaries have another special problem: rotor-to-rotor clearance. To avoid backflow leakage (slip), the rotor surfaces should almost touch. The most efficient answer is to inject oil or water between the active surfaces and let the rotors function like lubricated gear teeth in mesh. It not only closes clearances and reduces slip but eliminates some of the need for precise timing gears and quiets and cools the compressor. Most rotary screw compressors are built to be operated "flooded" (lubricated).

The penalty is that the oil or water must be removed at the output, requiring elaborate separators and dryers downstream. The alternative is to run them dry and maintain close rotor clearances (as small as 0.0005 in.) with precision timing gears. The gears increase power requirements by as much as 10%.

What's in the Air

The design of a compressor package is strongly influenced by how much moisture, oil, and contamination the final air system can tolerate.

Pure, dry air is needed for instruments, breathing apparatus, processes, and other applications where liquids and contaminants are unwelcome. (An example for moisture-free air is given at the end of this section.) At the other extreme are applications that can tolerate wet, dirty air.

Subcategories of each are legion, and the compressor package must be designed carefully to meet special demands. For instance, some applications work well with dewpoints close to atmosphere, while others demand dewpoints down to −100 F.

Refrigerated dryers can lower the dewpoint to about 33 F, but below that condensate freezes inside coils and blocks flow of compressed air. For really low dewpoints, where condensation must not occur at any operating level of temperature and pressure, chemical (absorptive) or dessicant (adsorptive) dryers are recommended. See Chap. 11, for example.

Many desiccant dryers have twin towers or beds so that while one bed is drying the compressed air, the other bed is being purged of moisture by blowing low-pressure dry air through it and then venting the moist air to the atmosphere.

The actual moisture content of saturated air can be quickly read from charts. More total moisture can be retained at high temperature than at low temperature, and pressure will squeeze out moisture.

Don't be misled by air that seems bone dry at 275 F; it might start condensing out moisture at 200 F or at any other temperature. Also, realize that a pound of air saturated at 15 psia and 100 F can hold 10 times as much moisture as a pound of air at 100 psia and the same temperature.

That phenomenon can be exploited as follows: Compress the air and store it at a much higher pressure than you intend to use. Drain off the condensed water. Then, when the air is expanded to the operating pressure, it will have a much lower relative humidity.

Oily air is a special problem. One answer is to specify compressors that operate without lubrication. Dynamic compressors such as centrifugals and axials meet that requirement, but they are available only in large sizes (see Fig. 12.1). Reciprocating units can be made to operate lubeless with special designs and are widely used for supplying instrument-quality air. The industry calls a reciprocating compressor *oil-free* if a crosshead seal separates crankcase from piston; it's called *oil-less* if the crankcase is dry (Fig. 12.4).

How much is too much? Any contamination that's discharged from the compressor, no matter how small the amount, will accumulate. If it's oil, sooner or later it will deposit on downstream surfaces, unless removed with good separators and filters that are properly maintained.

Too-pure air can be a problem also. Make sure that downstream equipment can function on pure, dry, unlubricated air. If not, specify oil lubricators to meter precise, small amounts into the airstream or directly lubricate the friction points (Chap. 11).

Typical Application

A customer of Pressure Specialties Inc. (Hackensack, NJ) specified four identical compressor packages to be shipped to the Middle East for mounting on water-treatment equipment. Each was to supply 10 scfm of oil-free dry air to 100 psig.

The first proposal by PSI was as the customer implied: four separate but identical units. Each consisted of an oil-free compressor (crosshead seal between crankcase and piston), air-cooled aftercooler, regenerative desiccant dryer, and applicable filters, drains, valves, and accessories.

The customer quickly replied, "We budgeted only 35% of your bid. What can you do to reduce costs?" With a further look, PSI discovered the following:

- Only two of the units were to be on line at one time.
- Each unit had a peak demand of 10 scfm, and maximum pressure was only 50 psig. Average demand was only 3 to 5 scfm.
- Ambient conditions were extraordinary: Delivered air had to be dry at ambients from 40 F to 140 F.

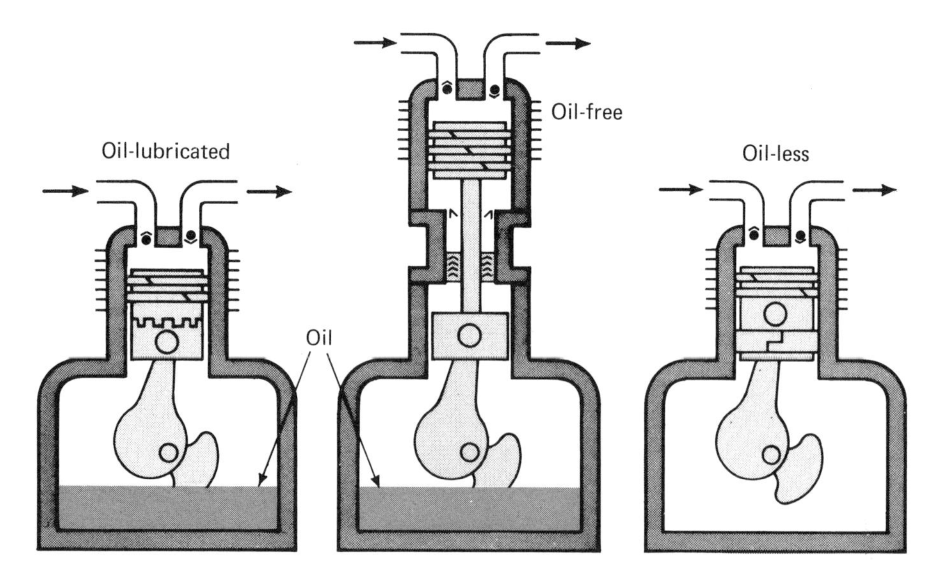

Fig. 12.4 Oil-free and oil-less air compressors are distinct from oil-lubricated designs. Differences are shown here graphically. (Courtesy of ITT/Pneumotive, Monroe, La.)

- It wasn't instrument air they needed but air dry enough to prevent accumulation of moisture.
- Cooling water was available but didn't have a quality good enough to circulate through the compressor cooling passages directly.

The final package (Fig. 12.5) has the following features:

- A single, base-mounted unit with two compressors in duplex arrangement, each featuring an air-cooled, single-stage oil-free compressor (Corken Co.) rated at 10 scfm and 75 psig
- Electric motors designed for mill and chemical duty, rated 25% oversize to allow for excessive ambient temperatures
- Individual aftercoolers, shell-and-tube design, water-cooled
- 120-gal receiver (tank) for the cooled air and its condensate, with an automatic condensate drain
- A refrigerated dryer downstream of the receiver, 50% oversize to ensure 35 F dewpoint or lower at all operating pressures and flows
- Duplex control to automatically alternate the compressors (to take turns being the "lead" during normal load cycling)
- Entire system—including piping, wiring, gages, valves, and controls—mounted on a single steel baseplate, strong enough for export crating
- Only three field connections: cooling water in-and-out, compressed air out, and electrical to control center

The system was built and tested, and it met all air demands, including 100% spare capacity. And, because of the packaging, the costs for startup and the period of training were a fraction of what was originally expected.

REGENERATIVE BLOWERS

There's a region of flow and pressure from about 10 to 1000 cfm and from about ½ psi to 8 psi (Fig. 12.1) that most off-the-shelf fans and compressors don't satisfy. Ordinary fans can supply the flow range but are limited in pressure. Small vane compressors can economically produce the pressure range but are limited in flow. Turbocompressors have plenty of pressure range but are impractical at anything but substantial flows. Rotary lobe blowers have plenty of flow range but aren't economical at extremely low pressures.

This is where regenerative blowers are becoming increasingly helpful. They naturally fit into the 0- to 400-cfm and 0- to 4-psi ranges and by putting them in series and parallel can reach the 1000-cfm and 8-psi ranges. The same blowers also will pull a vacuum of the same head.

Many companies offers regenerative blowers as stock items: Rotron (Woodstock, NY), Gast (Benton Harbor, MI), Siemens (Iselin, NJ), and Spencer (Windsor, CT) are examples. Gast introduced a design that has a separate torus on each face of the impeller (Fig. 12.6, second drawing). Each face builds up about 2 psi, and a crossover port stages them in series if 4 psi is wanted. The alternative is to run them in parallel at 2-psi head, doubling flow.

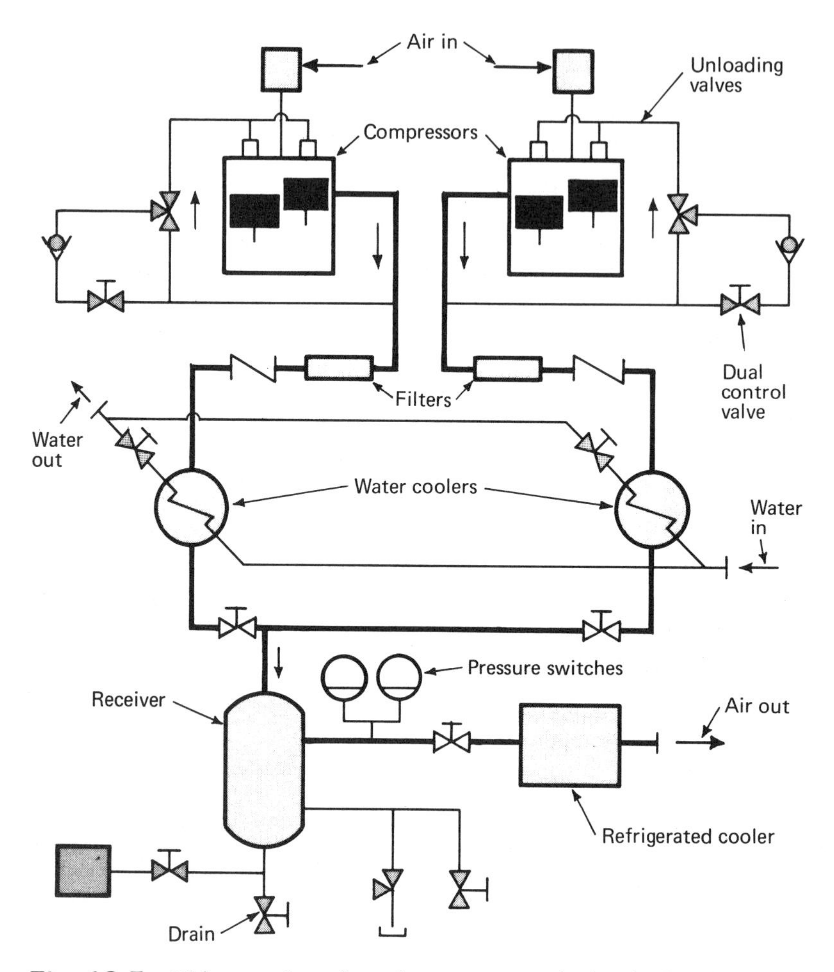

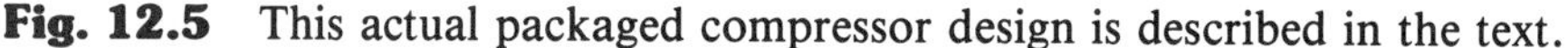

Fig. 12.5 This actual packaged compressor design is described in the text.

The Principle

A regenerative blower operates on a principle that has been described variously as viscous, drag, turbine pumping, forced ring-vortexing, and centrifugal spiraling. Some are called side-channel blowers and others peripheral. In any case, the pumping chamber is a torus around the periphery of the housing. The impeller induces vortices of flow that spiral through a curved tunnel—the torus—from inlet to outlet (drawing). Regenerative pumps (Chap. 3) operate in a similar way.

Each spiral is a stage of compression, however slight, and the effect builds up regeneratively. A barrier section between inlet and outlet in the torus housing keeps the higher pressure outlet from leaking back into the inlet. Details of different companies' designs differ considerably.

Rotron engineers describe the operation of their design (Fig. 12.6, drawing on the left) as follows: When the impeller revolves, air between the vanes of the impeller is driven by centrifugal force from the vane root outward to the vane tip, into the annular area between the vane and the housing and back to the vane root again. This circular motion, coupled with rotation of the impeller, forces the air into a spiralling

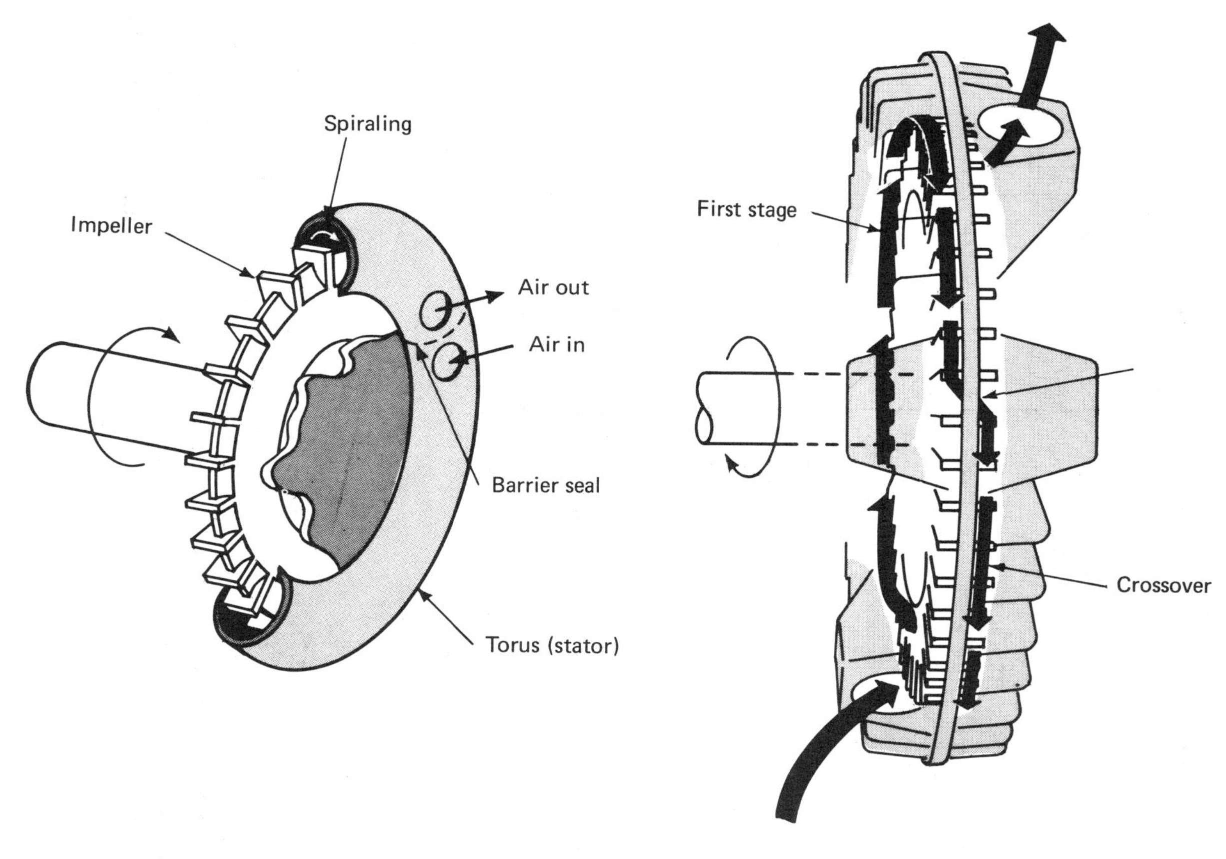

Fig. 12.6 Regenerative principle is indicated in these artist's conceptions of blowers. The first resembles Rotron's Cyclonair; the second resembles Gast's RAM-II.

forward motion. The repeated passing of the air through the vanes transfers energy to the air.

Pressure rise from inlet to outlet typically is up to 4 psi in a single stage rotating at 3600 rpm at zero flow. At zero head, the flow is up to 220 cfm. Larger flows are possible.

The pressure rise can be doubled by running two blowers in series; the flow can be doubled by running two in parallel. There is no theoretical flow limit—it's just a function of how large a blower you can buy or how many you want to put in parallel.

The efficiency of a regenerative blower is only between 40 and 50%, compared with 60 to 70% for turbo types and 70 to 80% for rotary lobes. That's because the pressure buildup is by a combination of viscous shear, turbulence, forced convection, and vortexing. A good explanation of what really happens is not readily available, but it works. The concept becomes useful when it has a lower total lifetime cost than another method.

Noise can be a problem because of a whine generated in the internal stripper (barrier seal). However, with proper silencers on inlet and exhaust the applicable Walsh-Healy standards are readily met.

One type of silencer is based on the old Helmholtz resonator principle. The embodiment is a tuned acoustical filter containing precise multiple holes and a resonance chamber. Acoustical foam is added within the chamber for additional silencing. Most of the silencing occurs at the inlet. The outlet resonator has less effect. For maximum quiet the final design and tuning of the chambers should be done on blowers mounted on actual or similar systems.

Typical applications include the following: ventilation systems for computers and other electronic equipment, suction and pressurization for stacked sheets, vacuum and airblast for cleaning, pump priming, mist spraying, holding devices, cushions of air for floating paper or other flat material, air-supported structures, fluidic supply air, breathing support systems, and bubblers for waste treatment or plating operations. Each of these applications takes a little more pressure than an ordinary fan will supply but doesn't need a positive-displacement compressor or a high-flow centrifugal.

LOW-PRESSURE FANS

Low-pressure fans include centrifugal (squirrel-cage), mixed-flow, and axial types—along with variations (Fig. 12.7). One variation is the transverse-flow fan (Fig. 12.8), which resembles the squirrel-cage type but sucks in air along the length instead of at the end; a vortexing action develops static pressures higher than those for ordinary centrifugals. Also, the transverse design is not limited in length: Air enters freely along the side. The following data, however, pertain only to standard centrifugals, mixed-flow, and axial types.

Fan Impeller Selection

When low-pressure cooling or ventilating airflow and head requirements are known, it's possible to deduce the approximate type and shape of fan impellers from simple theoretical relationships (see key derivations of equations at the end of this section). Then the final search through manufacturers' catalogs is greatly simplified. The method was developed by engineers at Torrington Manuf. Co. (Torrington, CT).

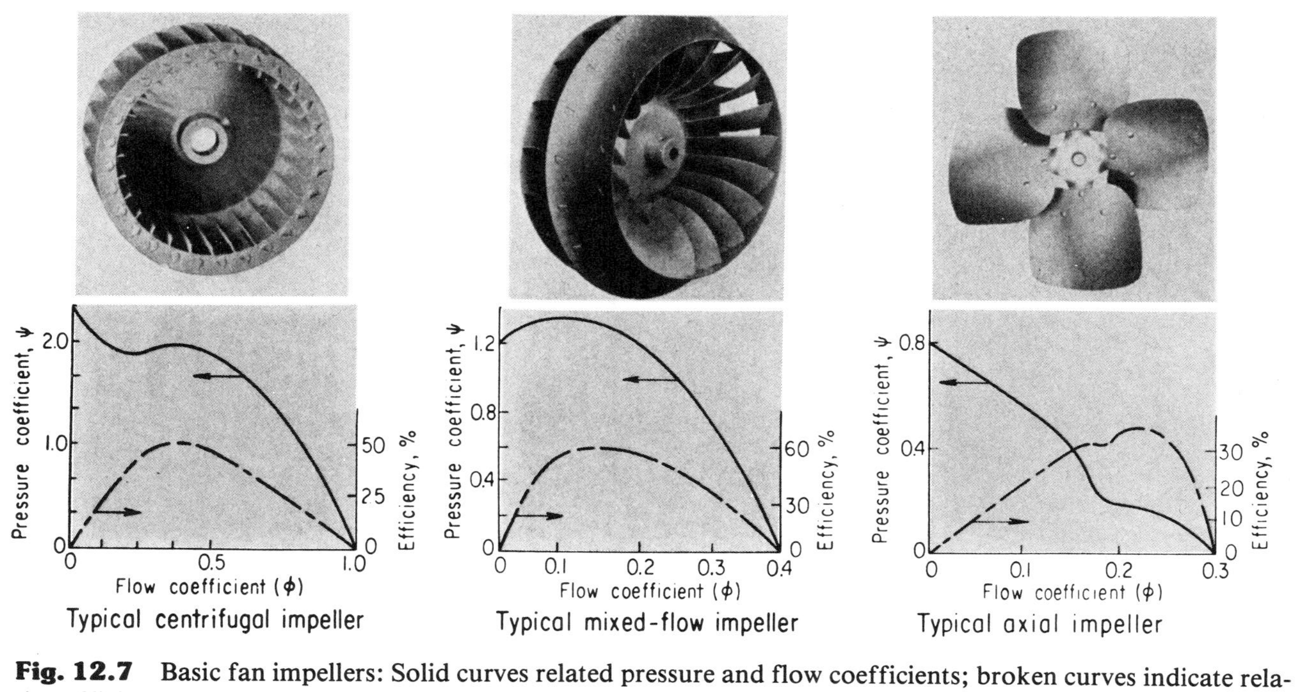

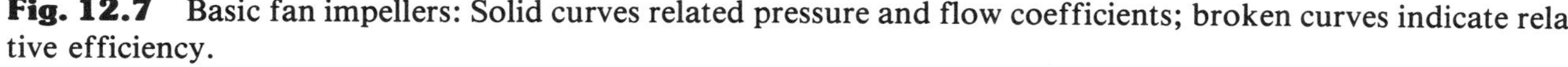

Fig. 12.7 Basic fan impellers: Solid curves related pressure and flow coefficients; broken curves indicate relative efficiency.

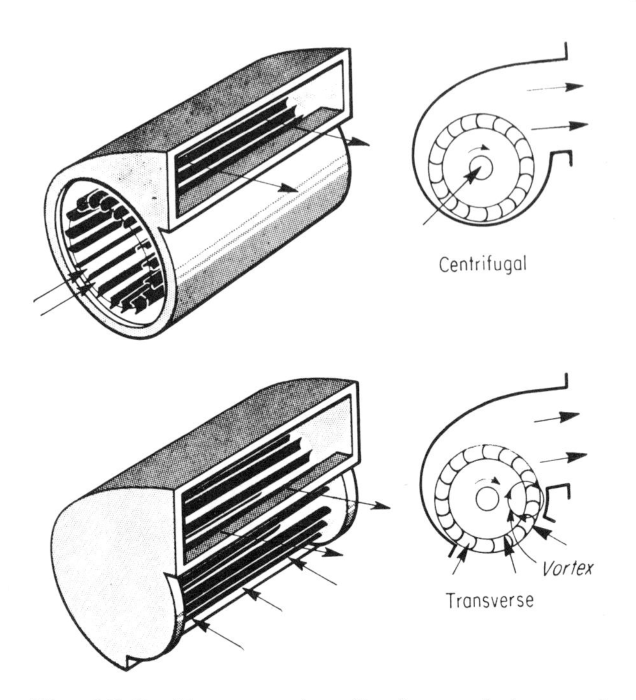

Fig. 12.8 Transverse impeller is a variation on the centrifugal type.

The most frequently used commercial types are forward-curved centrifugal, mixed-flow, and propeller-type axial impellers. Forward-curved centrifugal impellers are divided into two subgroups—normal blade area, where the total blade surface is about equal to the peripheral surface of the impeller, and high blade area, where the total blade surface is 30 to 40% greater than the peripheral area. This division is important because of the effect of blade surface on pressure and flow coefficients. Propeller fans are subdivided into groups with high, medium, and low pressure coefficients. Vane-axial and tube-axial impellers are usually designed to customers' specifications and so should be selected by working directly with the manufacturer. See Table 12.4.

The first step is to list your application requirements: airflow Q, in cfm; static pressure p_s, in inches of water; and impeller speed N, in rpm. The airflow rate is determined by established heat-transfer rates and the unit performance required for the application. Static pressure is determined by the air-path configuration, coil resistance, filters, louvers, and other elements in the flow system. Impeller speed is determined by the choice of motor (standard four- and six-pole shaded-pole and split-capacitor types are available) and also by the drive—direct or belt.

Calculate Specific Speed. Use the equation

$$N_s = \frac{NQ^{1/2}}{p_s^{3/4}}$$

or find the specific speed using the nomograph in Fig. 12.9.

The meaning of specific speed is explained earlier in this chapter and in Chap. 3.

Check the Tables. Table 12.5 shows the type (or types) of impellers that operate at high efficiency at the calculated specific speed. The thickest portion of the bar indicates the point of peak efficiency. If more than one impeller type is appropriate, base your final selection on such other factors as cost, size and shape of space available, and nature of flow path. Generally, one type of impeller will be a clear choice for a particular application.

Find the Right Standard Diameter. Using the appropriate pressure coefficient ψ from Table 12.5, calculate the theoretical impeller diameter from this equation:

$$D = \frac{1.53 \times 10^4}{N} \sqrt{\frac{p_s}{\psi}}$$

Consult a catalog to find a standard diameter that most closely matches the theoretical diameter you have calculated. Use this value for the remaining steps.

Find Other Dimensions, If Needed. *Propeller type:* Find the pitch angle by substituting in this equation (see Table 12.5 for flow coefficient ϕ):

$$\alpha = \sin^{-1} \frac{350Q}{\phi_1 N D^3}$$

Forward-curved centrifugal or mixed-flow impeller: Find the impeller width:

$$W = \frac{175Q}{\phi_1 N D^2}$$

Radial-bladed or backward-curved centrifugal impeller: Consult a manufacturer's catalog for additional information, because they may be carried as special products.

Table 12.4 Flow and Pressure Coefficients for Various Fan Impeller Types

	Type		Pressure coef ψ	Flow coef ϕ	Typical applications
Centrifugal impellers	Forward normal curved	High blade area	1.5–2.0	0.20–0.75	Room air conditioners Central air conditioning and heating Electronic-equipment cooling
		Normal blade area	1.0–2.0	0.15–0.65	
	Radial blade		1.0–1.4	0.002–0.09	Washing machines Clothes dryers Assorted household appliances
	Backward curved		0.60–1.10	0.09–0.30	Industrial exhaust systems Electronic-equipment cooling
Mixed-flow impellers			0.60–1.10	0.10–0.25	Room air conditioners Roof ventilators
Axial impellers	Vane axial		0.20–0.60	0.10–0.40	Aircraft air conditioning Pressurizing aircraft compartments Shipboard air conditioning
	Tube axial		0.10–0.30	0.10–0.40	Electronic-equipment cooling Aircraft air conditioning
	Propeller	High ψ	0.15–0.20	0.25–0.40	Air conditioning condensing units Light industrial exhaust fans
		Med ψ	0.10–0.15	0.15–0.35	Unit and space heaters, Air circulators
		Low ψ	0.05–0.10	0.10–0.25	Assorted portable appliances

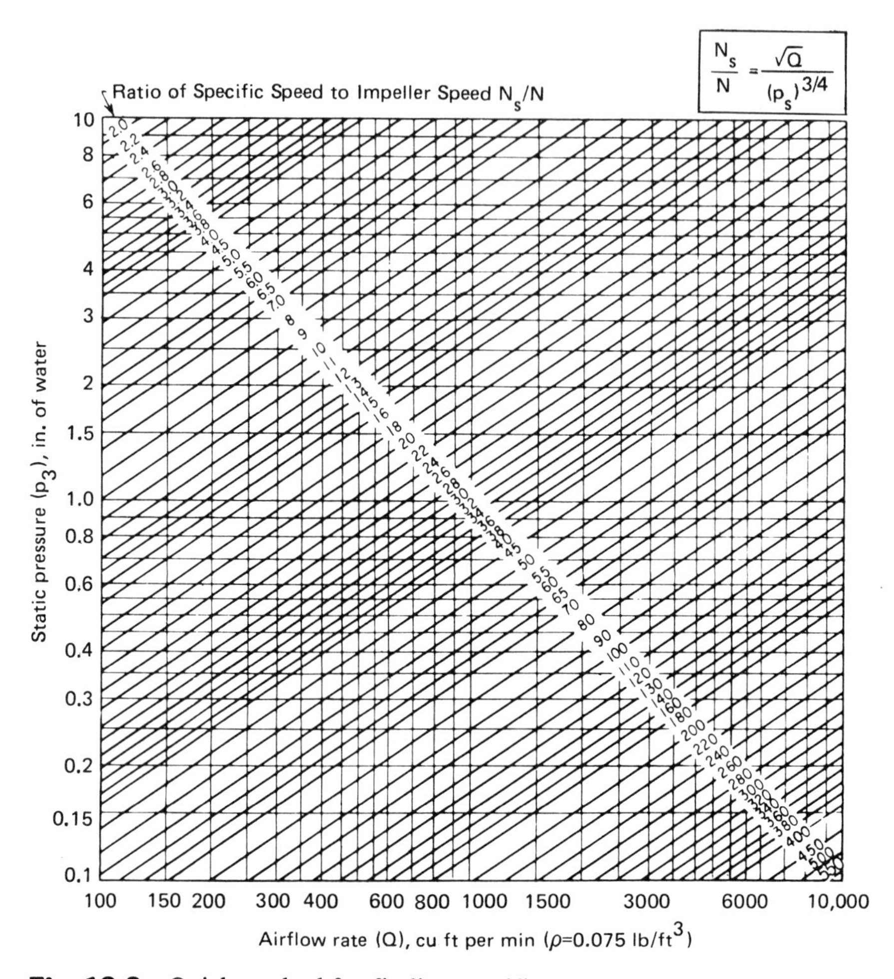

Fig. 12.9 Quick method for finding specific speed: Multiply the diagonal column (N_S/N) by the impeller speed (N).

Make the Final Choice. Consult a manufacturer's catalog for the final selection. It should be easy because type and approximate size are known.

Other Application Requirements. Although this selection method has been presented with airflow rate, static pressure, and rotational speed as the application requirements, it is not limited to impeller selection where they are the only known quantities. For example, if you must use either a particular type of impeller or a particular impeller diameter and you must find the best impeller, the selection procedure described above can still be used.

If diameter is known, for instance, the equation for D can be rearranged to solve for one of the other factors.

Even where there are too many unknowns, you can accurately estimate a solution by varying a particular parameter within reasonable limits and calculating the other parameters until a satisfactory impeller is found. After a little practice, solutions can be found in this manner almost as quickly as if all the requirements were given.

Table 12.5 Performance Data for Various Fan Impeller Types

Type			Specific speed $N_s \times 1000$ (Peak efficiency at widest part of bar)	Typical ψ	Typical ϕ
Centrifugal impellers	Forward curved	High blade area		1.70	0.254
		Normal blade area		1.40	0.176
	Radial blade			1.0 – 1.4	0.002–0.09
	Backward curved			0.60–1.10	0.09–0.30
Mixed–flow impellers				1.00	0.112
Axial impellers	Vane axial			*	*
	Tube axial			*	*
	Propeller	High ψ		0.175	0.263
		Med ψ		0.125	0.250
		Low ψ		0.075	0.238
*Wide deviation possible within range			6 8 10 20 30 40 60 80 100 200 300		

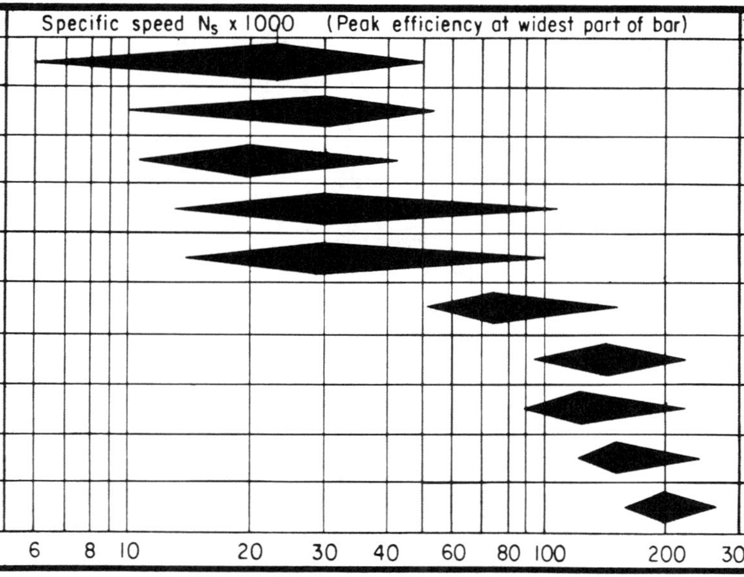

Examples.

Example I: Assume the following requirements:

$Q = 500$ cfm
$p_s = 0.10$ in. of water
$N = 1550$ rpm

To find the impeller type required, calculate the specific speed:

$$N_s = \frac{NQ^{1/2}}{p_s^{3/4}} = \frac{(1550)(500)^{1/2}}{(0.10)^{3/4}} = 194{,}500$$

In Table 12.5, the most efficient impeller for a specific speed of 194,500 is a propeller with a low pressure coefficient. The table gives 0.075 as the typical pressure coefficient for the low-ψ propeller. The diameter is then calculated:

$$D = \frac{1.53 \times 10^4}{N} \sqrt{\frac{p_s}{\psi}}$$

$$D = \frac{1.53 \times 10^4}{1550} \sqrt{\frac{0.10}{0.075}}$$

$D = 11.4$ in.

To find the fan pitch angle, find a standard diameter closest to 11.4 in. One catalog lists a standard fan with an 11-in. diameter.

The Table gives 0.238 as the typical flow coefficient ϕ for the low-ψ propeller. The pitch angle α is then:

$$\alpha = \sin^{-1} \frac{350Q}{\phi_1 ND^3}$$

$$= \sin^{-1} \left[\frac{(350)(500)}{(0.238)(1550)(11)^3} \right]$$

$$= \sin^{-1}(0.357)$$

$\alpha = 21°$

The catalog lists performance data for 16° and 24° pitch angles for the 11-in.-diameter fan. The 24° fan delivers 660 cfm at 0.10 in. of water and 1550 rpm, which is higher air delivery than necessary. With the 16° pitch angle, the fan delivers about 515 cfm at 0.10 in. of water and 1550 rpm. This is very close to the requirements, so final selection is an 11-in.-diameter, low-ψ propeller fan with a 16° pitch angle.

Example II: Assume the following requirements:

$Q = 100$ cfm
$p_s = 0.25$ in. of water
$N = 1000$ rpm

The specific speed is

$$N_s = \frac{1000(100)^{1/2}}{(.25)^{3/4}} = 28{,}200$$

Table 12.5 indicates that any of four impeller types may be efficiently applied:

- Forward-curved centrifugal impeller of high blade area
- Forward-curved centrifugal impeller of normal blade area
- Backward-curved centrifugal impeller
- Mixed-flow impeller

Typical pressure coefficients from Table 12.5 for these types of impellers start from 1.7 for high-blade-area forward-curved centrifugals and drop down to 1.0 for mixed-flow impellers.

Because diameter is an inverse function of pressure coefficient for a fixed speed and a fixed static pressure, it is evident that for a higher pressure coefficient the impeller diameter will be smaller. Thus, a forward-curved centrifugal impeller of high blade area will require the smallest impeller diameter in this particular application and is the most desirable.

The required diameter will then be

$$D = \frac{1.53 \times 10^4}{1000}\sqrt{\frac{.25}{1.7}} = 5.85 \text{ in.}$$

The closest diameter listed in the manufacturer's catalog for a standard forward-curved centrifugal blower wheel of high blade area is D = 5-19/32 in.

Table 12.5 indicates 0.25 for a typical value of ϕ for a forward-curved centrifugal impeller of high blade area, resulting in a blade width of

$$W = \frac{(175)(100)}{(.25)(1000)(5.59)^2}$$

$$W = 2.25 \text{ in.}$$

The closest standard width listed in the catalog is W = 2-16/32 in.

The catalog performance curve for this particular impeller shows it will deliver 98 cfm against a static pressure of 0.25 in. of water when operated at 1000 rpm. A slight increase in speed, such as from 1000 to 1020 rpm, will raise the airflow rate to at least the 100 cfm originally required. This increase in speed is well within normal motor capabilities, so the final selection is a forward-curved centrifugal impeller of high blade area, 5-19/32 in. in diameter and 2-1/2 in. wide.

Derivations of Key Fan Equations

Symbols.

N_s	specific speed
N	impeller speed, rpm
Q	airflow rate, cfm
p_s	static pressure, in. water
ψ	pressure coefficient
ϕ	flow coefficient
g	acceleration of gravity, ft/sec/sec
ρ	weight density of air, lb/ft^3
U_t	blade tip velocity, ft/sec

D	impeller diameter, in.
V_0	airflow velocity, ft/sec
ΔA	cross section of airflow, ft^2
α	fan pitch angle, deg
K	constant for given class of fans
W	width of impeller, in.

Pressure and Flow Coefficient. Pressure coefficient ψ, a dimensionless quantity, is the ratio of potential energy developed by an impeller to pressure energy corresponding to blade-tip velocity:

$$\psi = \frac{2gp_s}{\rho U_t^2} \tag{12.1}$$

It can also be expressed as

$$\psi = \frac{2.35 \times 10^8 p_s}{N^2 D^2} \tag{12.2}$$

By rearranging, an equation for impeller diameter becomes

$$D = \frac{1.53 \times 10^4}{N} \sqrt{\frac{p_s}{\psi}} \tag{12.3}$$

Flow coefficient ϕ, also a dimensionless quantity, is the ratio of airflow velocity to blade-tip velocity:

$$\phi = \frac{V_0}{U_t} \tag{12.4}$$

To maintain a uniform approach for all axial- and centrifugal-impeller types, ΔA is defined as $\pi D^2/4$. Therefore the flow coefficient can be expressed as

$$\phi = \frac{700Q}{ND^3} \tag{12.5}$$

where $Q = \Delta A \times V_0$.

Blade Pitch Angle and Impeller Width. Pitch angle can be determined by

$$\sin \alpha = K\phi$$

This equation is empirical and holds true only for flow rates from 70 to 100% of maximum flow. But for this range of flow rates, propeller fans are most efficient and are most effectively used. If, in the equation

$$\frac{\sin \alpha_1}{\phi_1} = \frac{\sin \alpha_2}{\phi_2} \tag{12.6}$$

the ratio $\sin \alpha_1/\psi_1$ for a particular fan is known, the sine of the pitch angle α_2 of another fan of the same class can be found by finding its flow coefficient ϕ_2. If, for example, a pitch angle of 30° is assigned to α_1, then $\sin \alpha_1 = 0.5$. In Table 12.5, flow coefficients are given out for fans of 30° pitch in each class of propeller fan. Therefore, pitch angle for any fan can be calculated from the application requirements, diameter, and flow coefficient by substituting the value of ϕ from Eq. (12.5) for ϕ_2 in Eq. (12.6):

$$\sin \alpha_2 = \frac{350Q}{\phi_1 ND^3} \tag{12.7}$$

or

$$\alpha = \sin^{-1} \frac{350Q}{\phi ND^3} \tag{12.8}$$

For centrifugal and mixed-flow impellers, the width of the impeller must be known. The ratio of width to diameter is proportional to the flow coefficient:

$$\frac{W}{D} = K\phi \tag{12.9}$$

where K is constant for a given impeller design, $W/D = 0.25$ is assumed, and corresponding values of ϕ_1 are determined in the table for each class of impellers.

Substituting Eq. (12.5) in Eq. (12.9) gives

$$W = \frac{175Q}{\phi ND^2} \tag{12.10}$$

TURBOS, INTENSIFIERS, INDUCERS

Although not classic compressors, blowers, or fans, the engine turbocharger (Fig. 12.10), the air-to-air intensifier (Fig. 12.11), and the inducer (Fig. 12.12) are interesting examples of the air-moving art.

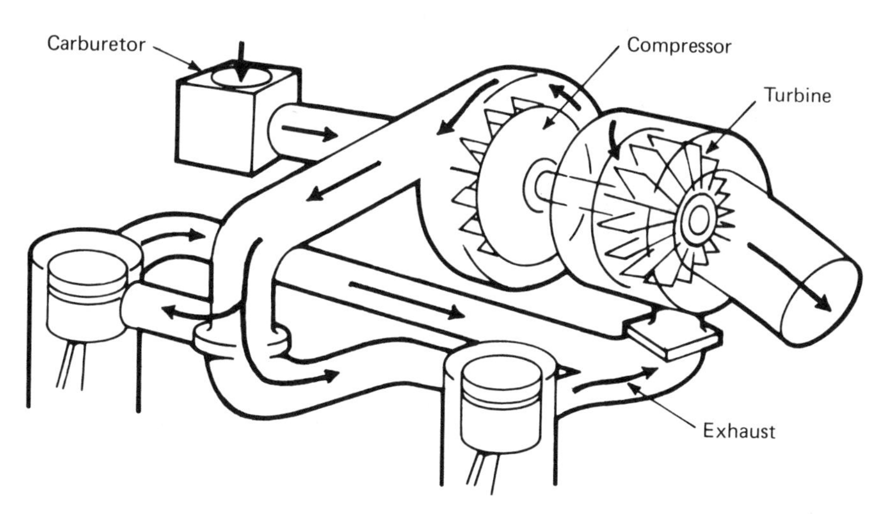

Fig. 12.10 Automotive supercharger compressor is powered by an exhaust-driven turbine.

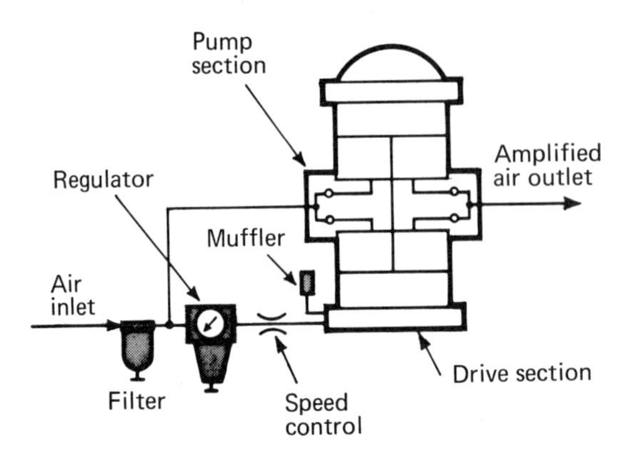

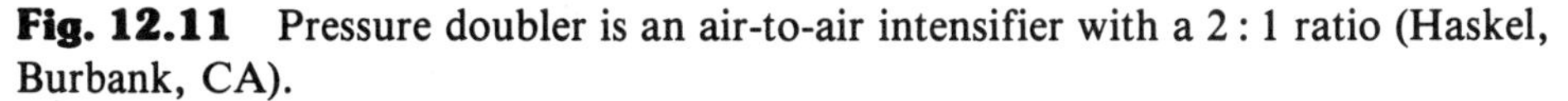

Fig. 12.11 Pressure doubler is an air-to-air intensifier with a 2 : 1 ratio (Haskel, Burbank, CA).

The turbocharger increases the pressure of engine inlet air by exploiting the pressure and flow energy of the exhaust to spin a high-speed turbocompressor.

The air-to-air intensifier takes part of the flow of ordinary shop-pressure air (100 psig) and uses it to pump up another portion of the flow to much higher pressure (say 200 psig).

The inducer injects high-pressure air into an axially aimed slot, creating sonic flow at low pressure along the inner wall of the pipe. The high-velocity flow induces large volumes of atmospheric air into the stream, amplifying the flow by as much as 25 : 1.

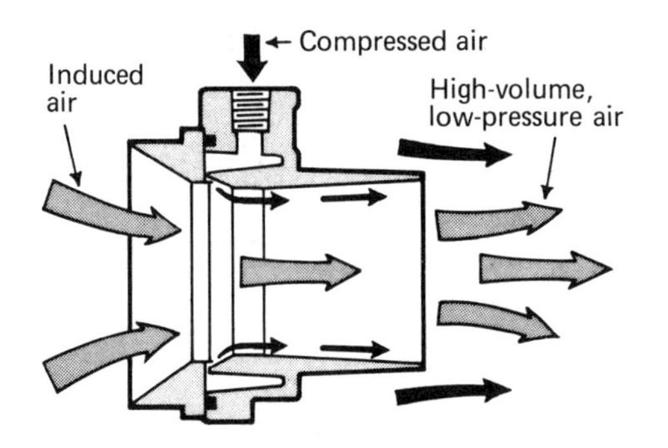

Fig. 12.12 High-volume induced flow results from small-volume flow of compressed air (Vortec, Cincinnati, OH).

13

Pneumatic Valves and Controls

This chapter will concentrate on the basics of directional valves and pressure regulators, which are the most numerous of pneumatic controls for machines and systems.

Additional application and performance details are covered in Chaps. 14 through 16.

Also, see Chap. 4 for details on a wider selection of valve types.

DIRECTIONAL CONTROL VALVES

Dozens of manufacturers make pneumatic directional control valves. They cover the gamut of spool, poppet, plate, plug, ball, shuttle, and simple check (a two-port directional). A few examples are shown in Figs. 13.1 through 13.3.

They all have roughly the same purpose: to direct or prevent flow through selected passages. They shift quickly from position to position, whether under solenoid, pneumatic, mechanical, or manual control. The valves aren't intended as throttlers or regulators. A typical application is to pressurize cylinders or air motors sequentially or to provide the "logic" to trigger the final power valves.

Figure 13.4 illustrates typical passage configurations available for spool valves, but by no means do they cover the dozens of proprietary products sold. For example, rotary selector valves are offered that have a single input and dozens of outputs, chiefly for test instrumentation. Various choices exist in rotary and linear slide-plate valves and rotary plug and ball valves, and there is a plethora of miniature pneumatic logic valves (Chap. 14).

It's been common practice for pneumatic valve manufacturers to customize designs to solve specific user problems, followed up by substantial support such as replacement parts and engineering advice. Current industry design standards have been flexible enough to accommodate this relationship, and big users like the car manufacturers write design and purchasing specs to ensure performance and interchangeability from their standpoints.

These old relationships will persist because of enormous numbers of installed

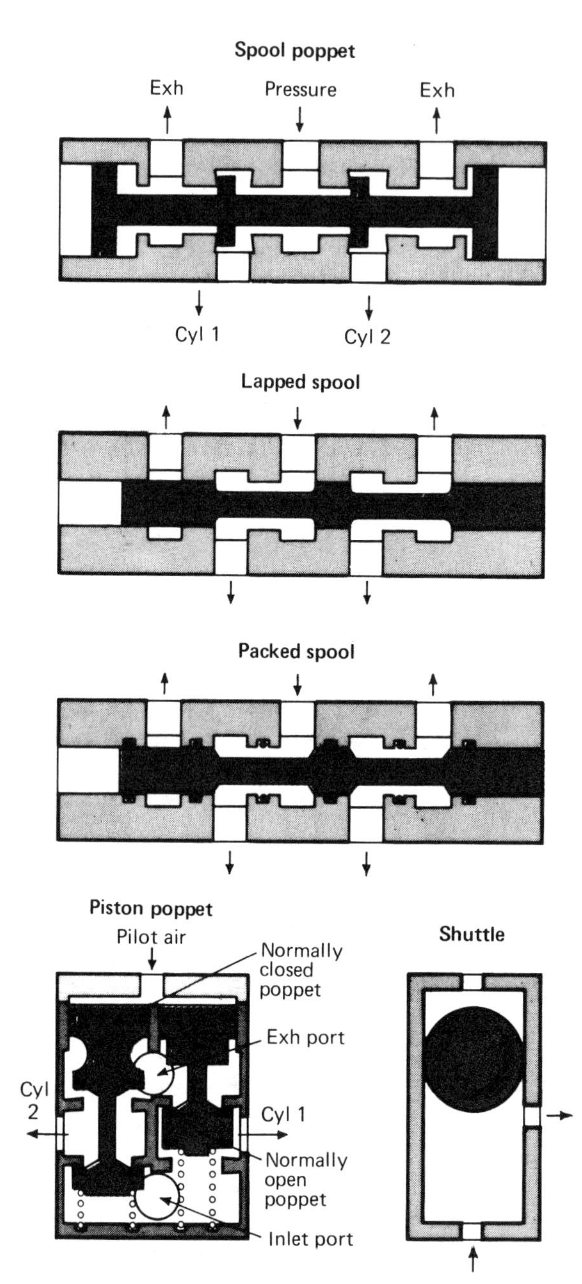

Fig. 13.1 Most valves have one or more of these typical operating elements.

valves and backup parts and decades of successful experience, but there are strong winds of change, namely, the possibility of worldwide interchangeability via ISO (International Standards Organization), pressure from the Federal Trade Commission (FTC) to alter the way voluntary industry standards are written, an increase of liability litigation in all engineering fields, demands for more efficient performance to conserve energy, and last but not least, the need for a universal way to diagram complex pneumatic circuits so they can be designed, built, and maintained by ordinary people.

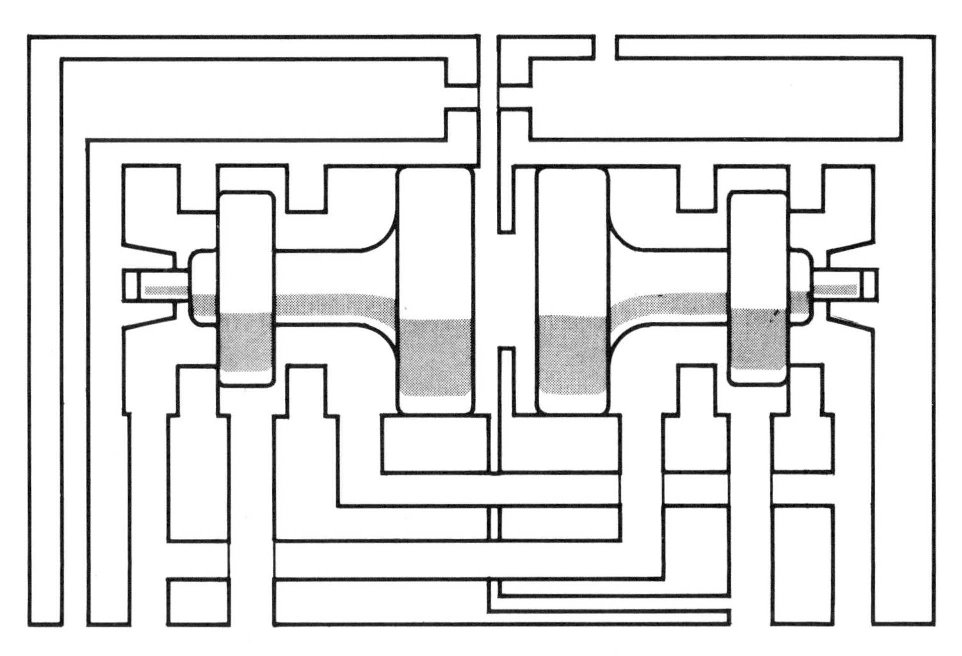

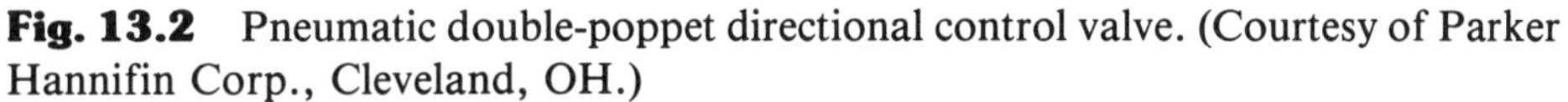

Fig. 13.2 Pneumatic double-poppet directional control valve. (Courtesy of Parker Hannifin Corp., Cleveland, OH.)

Many such problems are faced by all segments of fluid power, but the air valve people seem to be facing them all at once.

International Interfaces

Over 35 countries, including the United States, reached consensus on a set of mounting surfaces (Fig. 13.5) for five-port directional air control valves; a preliminary standard was introduced as ISO 5599/1-1978. If universally accepted and adopted, it would for the first time make possible the mounting of valves from many manufacturers on the same subplate.

There was some controversy. For one, there were U.S. standards and specs similar to the ISO standard but not interchangeable (the ANSI interface standard and

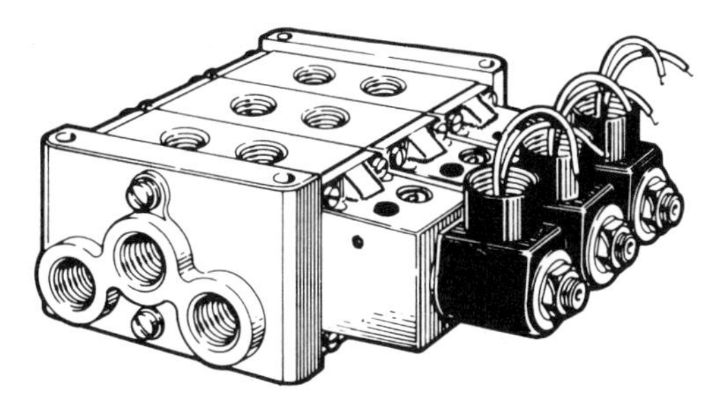

Fig. 13.3 Stacked single-solenoid-operated spring-return valve with common inlet and exhaust ports. (Courtesy of C. A. Norgren Co., Inc., Littleton, OH.)

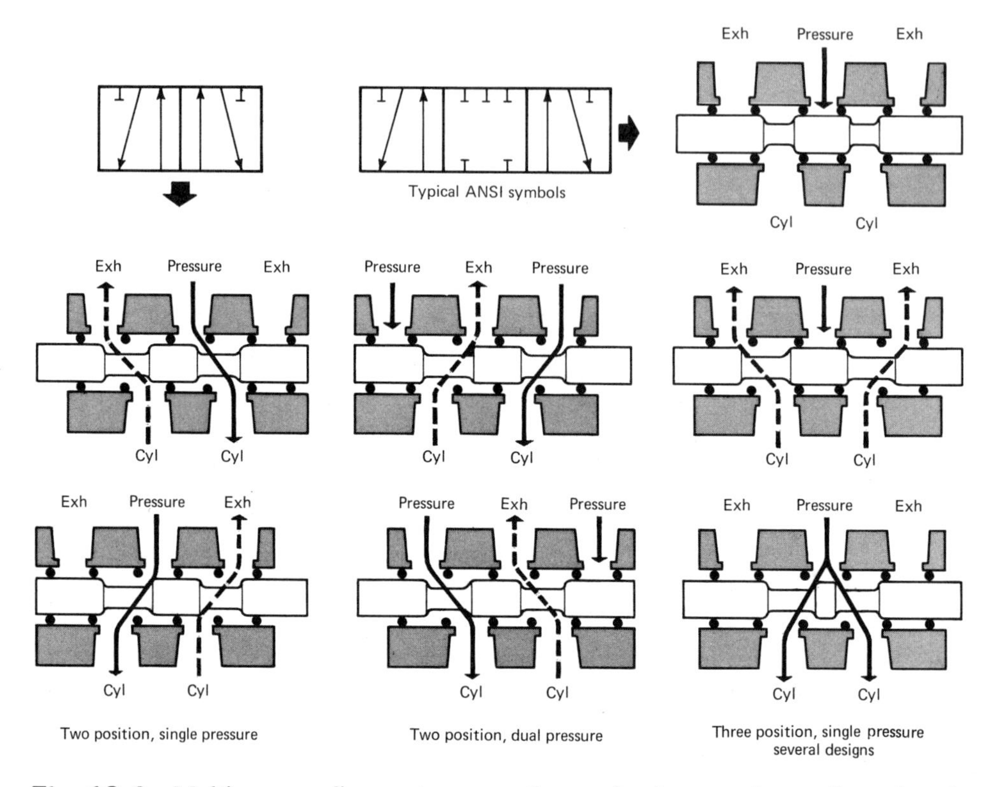

Fig. 13.4 Multipurpose five-port pneumatic spool valves can be configured and pressurized in numerous ways to control extension and retraction strokes of power cylinders. (Courtesy of C. A. Norgren Co., Inc., Littleton, Oh.)

the Ford interface spec shown in Fig. 13.5). Valve makers building to one or both were not anxious to scrap dies and begin again.

The interfaces look similar, but there are important differences. In essence, the ISO interface is a tight pattern that provides maximum flow passage with minimum bolt spacing and minimum weight of metal. As presently conceived, it readily accommodates spool valves but is snug for some poppet and diaphragm designs. Advocates claim that the majority of industrial pneumatic directional valves are (or could be) spool type.

The reasons for the pattern are partly engineering, partly marketing, partly political. It was the consensus of the countries working on the standard that no single country should benefit more than any other from the new interface; it must be different from any existing standard. Furthermore, the interface must be based on the best engineering judgment of all groups and must be practical and inexpensive to manufacture.

The electrical options in the ISO standard are not included here because details are still being developed as the book goes to press.

The Europeans so far have shown greater interest in using the new standard than have U.S. companies. It seems to match reasonably well what they're already doing.

Another challenge is the fact that if a large number of valve designs will fit on the same standard baseplate (that's the purpose of the ISO work), then new variations in performance might result in unexpected flows and motions.

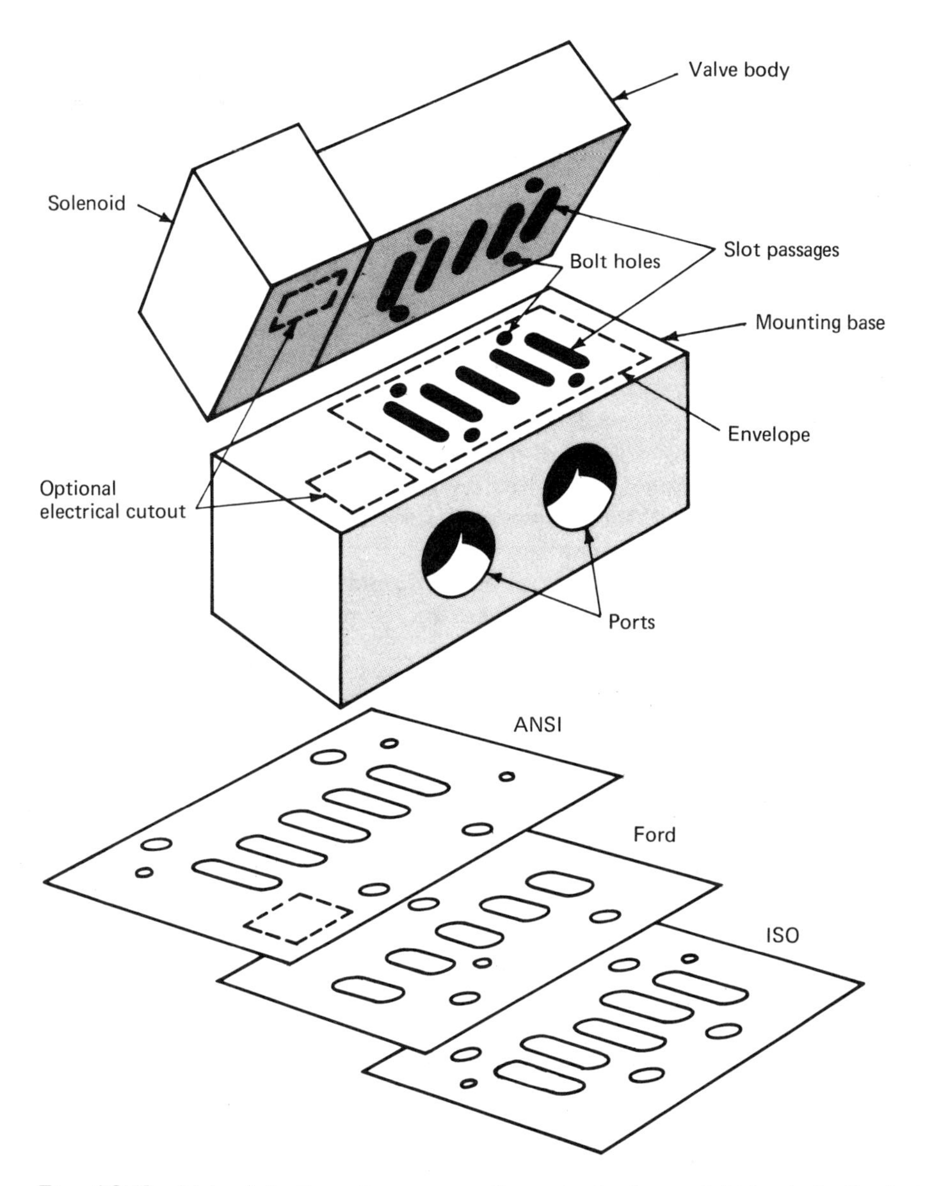

Fig. 13.5 Valve interfaces for pneumatic control valves might look similar but are not necessarily interchangeable among contending standards. These conceptual drawings show differences among three early standards: ANSI, Ford, and ISO (see the text).

If ports aren't designated uniformly, a pneumatic press might close when the operator expected it to open. Also, some U.S. valves have very high C_V (flow) coefficients, yet the valve body will fit onto the same interface as another valve of much lower C_V. The change in machine operating speed could be unsafe if valves were changed.

The same problems of interchangeability exist even with ANSI and proprietary interfaces, but there are fewer opportunities for errors because there are fewer valve designs to choose among. Anyway, good design practice is necessary whatever set of standards is used.

As another example, a five-port four-way valve can be converted to a dual-pressure valve by reassigning the supply port to exhaust and pressurizing the exhaust ports separately (Fig. 13.6). It's a good way to conserve energy by reducing pressure levels wherever possible but could present an interchangeability problem.

One consensus is this: An ISO interface standard is a good idea for international trade, even if it might not be adopted locally in each country. So it is predicted that ISO 5599 will be accepted eventually, and its success will depend on its merits.

APPLICATION AND PERFORMANCE

In any air valve, flow is hard to calculate, costs are illusive, and control is tricky. Yet no other working fluid is as safe and as versatile in as many environments. Temperature and radiation have no harmful effect, water hammer doesn't exist, and electrical shorts in an air line are impossible. Furthermore, high-pressure air in a container stores more energy per pound than a storage battery and can sustain a pneumatic system during temporary outage of the main air supply. Also, air is easy to work with so long as you don't have to calculate anything.

There's no mystery to air valves if the basics are understood. The proprietary features that make every valve look different are not as important as the fundamental design of the plunger and passages.

Key operation and design parameters are pressure level, flows (load vs. leakage or crossover), clearances, cycle speed, percent time in each position, and porting arrangement (including manifolding).

Also important are operator force, shock and vibration tolerance, corrosion resistance, plunger friction, need for lubricant, tolerance of contamination, life, detenting ability, and pressure balance. Not all of these are important in every application. The art of tying it all together is called system design, and if this isn't done thoughtfully, every good feature built into the valve is diminished or lost.

The biggest saving is not to pressurize the air in the first place, or at least to pick the lowest possible level. It may take 100 psig to extend a cylinder as fast as you would like but considerably less pressure to return it. Some companies offer modular valve packages with built-in regulators that automatically select the correct pressure or meter-out restriction for the motion or action desired.

Watch out for sticking valves: They can interrupt sequences or can waste large volumes of air if stuck in an intermediate position. Contamination is a big part of the problem. Poppet valves have somewhat more tolerance to dirt than do spool valves, but then again, any valve can stick.

Hesitation is not just a dirt problem. Here are many other ways that hesitation or lockup can occur: low solenoid power (or voltage fluctuations), wrong spring return force (too high), clogged vent holes or pilot exhaust holes, tight or gummy seals, wrong pressure signals (too weak), and insufficient or wrong lube.

Lubrication of valves ranges from zero to excess, but there are OSHA and dollar incentives to lubricate precisely. Where in-line lubricators (Chap. 11) are used, the net airflow through a valve including crossover leakage affects lube delivered.

One culprit is so-called "varnish." It's caused by the drying of oil and the separation of chemical additives from the lubricant. As the oil dries, varnish forms on the lands of a lapped spool valve, slowing down or stopping movement. It also damages elastomer seals on packed valves.The cure usually is to add fresh lubrication to absorb varnish.

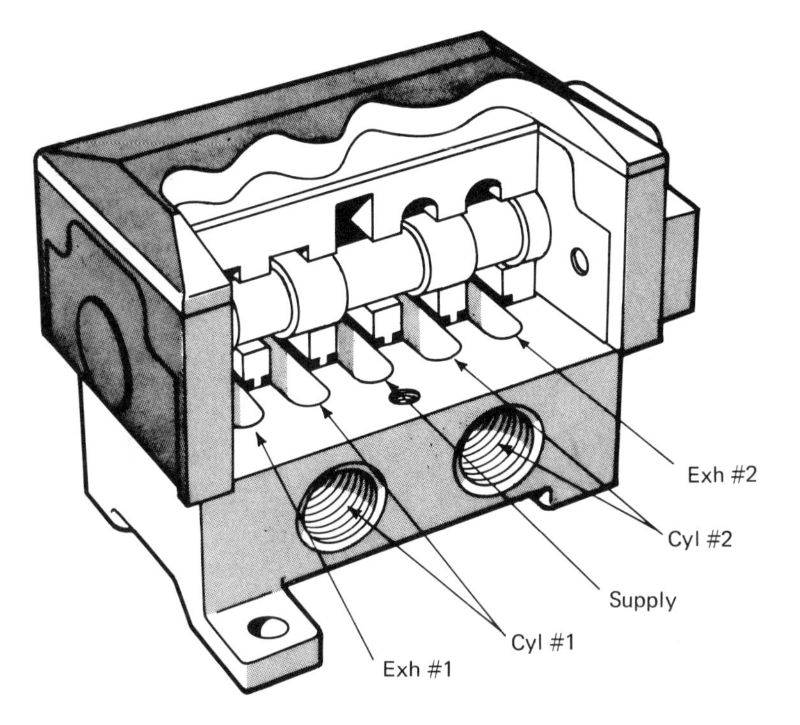

Fig. 13.6 This five-port four-way valve (only two ports visible) can be piped up various ways to the supply, exhaust, and cylinder lines. For example, each exhaust port can become a supply port, allowing two different supply pressures (Alkon, Wayne, NJ).

Burned oil is a worse problem. Some plant compressors reach temperatures of up to 400 F—above the combustion point of most lubricating oils—creating fumes that deposit onto valve parts and do not readily dissolve in lubricating oil. Special high-temperature oils solve this problem.

Crossover Leakage

Leakage occurring while an air valve spool or poppet moves end to end can be costly. There's always a running controversy over which type of valve leaks the most. The fundamental valve choices are seating-action (poppet types in general), shearing-action (spool, plate, plug, or ball), and combinations.

Seating-action valves shift flow by moving the poppets off one set of seats onto another. The motion is short, and the flow change is quick. However, during excursion all poppet surfaces are momentarily unseated, and some blow-by (crossover flow) to the exhaust port or between active ports occurs. There are poppet-valve designs that close at crossover, but they require separately moving poppets.

Shearing-action valves (such as spools) have land surfaces available to close off the exhaust or any other port during intermediate travel, thus preventing blow-by. The penalty is that valve travel is longer than that for poppet valves of the same flow rating. Furthermore, unless shearing-action valves are designed for zero clearance metal-to-metal sealing, they will have either leakage paths (packless spool valves) or dynamic seals subject to friction and wear (packed valves). (See Fig. 13.1 for examples.) An advantage of lapped packless spool valves is that they can operate without lube.

Five of the key parameters come into play simultaneously: pressure, flow, clearance, speed, and percent time in each position. For example, a valve that shifts flow once per hour and takes but 10 msec to transfer seat-to-seat won't leak much long-term even if open to exhaust. It is much more important that sealing be good at both ends of the travel, because that's where the valve spends 3,599,990 out of every 3,600,000 msec. On the other hand, if the valve cycles once per second and takes 500 msec to transfer, high crossover leakage will result.

Let's put some numbers on the problem (see Fig. 13.7 and Table 13.1). A 1-in.-diameter packless (lapped) spool with a 1-in.-long land, an average radial clearance

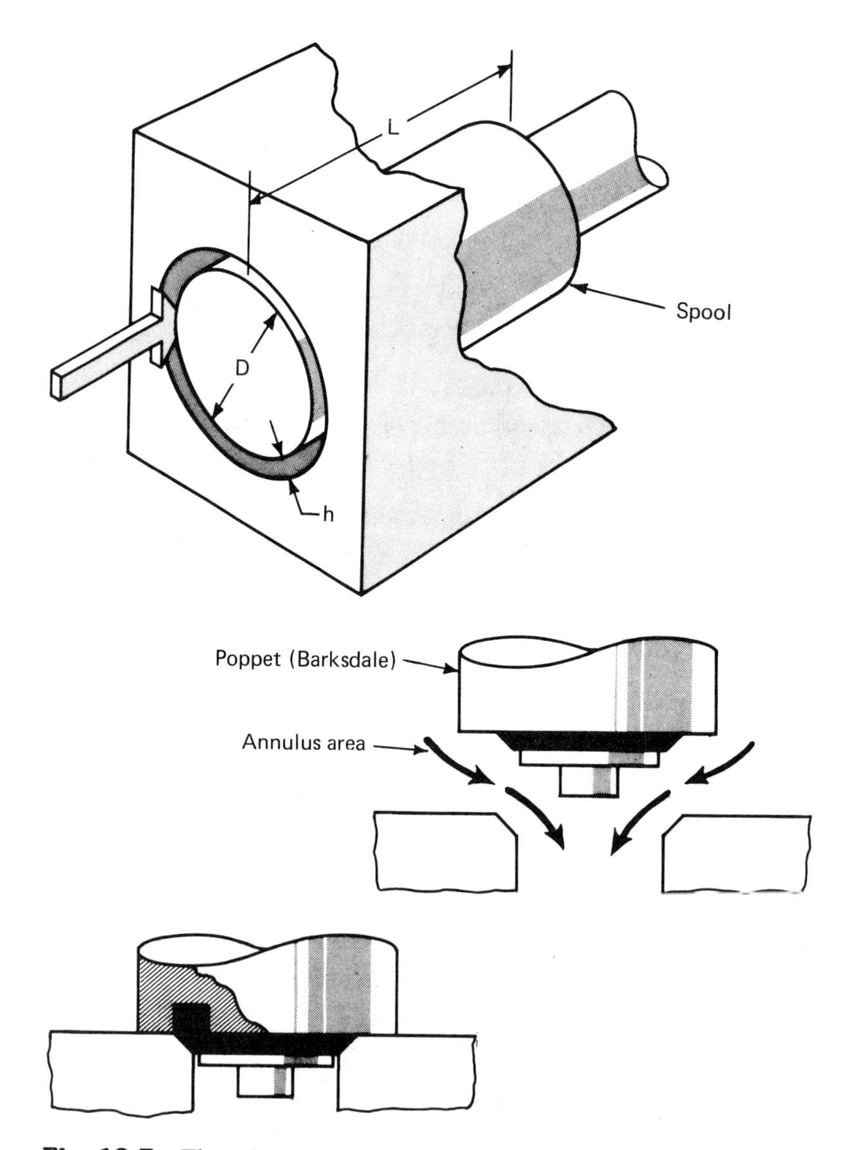

Fig. 13.7 Flow through annulus of spool or poppet can be calculated with simple equations given in the text.

Table 13.1 Properties of Standard Air

Viscosity (absolute)	2.6×10^{-9} lb_F-s/in.^{2} at 68 F 3.3×10^{-9} lb_F-s/in.^{2} at 100 F
Density (lb mass)	0.0752 lb/ft^3 at 14.7 psia, 68 F 0.0709 lb/ft^3 at 14.7 psia, 100 F
Note: standard cfm (scfm)	$= \text{actual cfm} \dfrac{P_{abs}}{14.7} \times \dfrac{528}{T, ^\circ R}$

of 0.001 in. (exaggerated), a pressure of 100 psig (114.7 psia), and a temperature of 68 F (528 R) will pass laminar flow in accordance with this modification of the flat plate flow equation:

$$Q = \frac{\pi D}{L} \times \frac{h^3 \Delta P}{12\mu}$$

where Q is in^3/sec at average density; D is nominal spool diameter, in.; L is axial flow path, in.; h is radial clearance (spool centered), in.; ΔP is pressure drop, psi; and μ is absolute viscosity, lb-sec/in^2 (about 2.6×10^{-9} for air at 68 F assuming constant with compressibility). Note: An uncentered spool will have over twice the leakage, but we'll assume a centered spool:

$$Q = \frac{\pi 1}{1} \times \frac{(0.001)^3 \times 100}{12 \times 2.6 \times 10^{-9}} = 10.07 \text{ in}^3/\text{sec}$$

$$= 10.07 \frac{60}{1728} = 0.35 \text{ cfm (say 50 psig)}$$

$$= 0.35 \times \frac{50 + 14.7}{14.7} = 1.54 \text{ scfm}$$

$$= 1.54 \times 60 \times 24 \times 365 = 809{,}424 \text{ ft}^3/\text{yr}$$

But that's an exaggeration, because a lapped spool can have far tighter clearance than 0.001 in. Let $h = 0.0001$. Then

$$Q = 1.54 \times \left(\frac{0.0001}{0.001}\right)^3 = 0.00154 \text{ scfm}$$

That's $0.00154 \times 60 \times 24 \times 365$ or 809 ft^3/yr. At 25¢/1000 ft^3, the cost is $809 \times 0.25/1000 = \$0.20$/yr. And if the clearance is 0.00001, Q becomes 0.809 ft^3/yr, a negligible amount.

Many exceptions and variations exist. If the land length available for sealing is only 0.1 in., leakage will be at least 10 times greater. Scratches or wear cause leaks too.

Now try a short-stroke (0.2-in.) poppet (Fig. 13.7) with a 1-in.-diameter seat, an effective area (annulus) at half-stroke of $0.1\pi D = 0.314$ in^2, and 100-psi blow-by to exhaust. Blow-by flow (sonic) will be approximately

$$W = \frac{0.5318 P_1 A}{\sqrt{T_1}}$$

where W is lb/sec; P_1 is upstream pressure, psi; A is flow area, in² and T_1 is upstream temperature, 528 R.

$$W = \frac{0.5318 \times 114.7 \times 0.314}{\sqrt{528}}$$

$$W = 0.834 \text{ lb/sec} = 50 \text{ lb/min}$$

$$Q = 50/0.0752 = 664 \text{ scfm}$$

where 0.0752 is ft³/lb of air at standard conditions.

As a hypothetical example, assume that the effective valve travel time during crossover is 10 msec. The maximum theoretical loss of air at crossover is $664 \times 0.010/60$ or 0.11 ft³, say 0.1 ft³. If the valve operates once a minute for a solid year, that's $0.1 \times 60 \times 24 \times 365 = 53{,}560$ ft³/yr. If compressed air costs $0.25/1000 standard ft³ (scf), each valve wastes $52{,}560 \times 0.25/1000 = \$13.10$/yr.

But this, too, is an exaggeration. There are dozens of ways to modify and interpret the calculations. Here are a few:

- None of the crossover flows will be quite as high as calculated, because there will be local flow restrictions, inertia, and other impedances.
- If the average year-round air usage of each valve is 10 scfm, then 0.1 scfm becomes only 1.0% loss, which is relatively small. If average usage is 1 scfm, then 0.1 scfm loss is 10%, which is considerable.
- If valve excursion time is 1 msec instead of 10 msec, then the air loss per valve is minor (5256 ft³/yr = $1.31).
- If valve operation is once/hr for one shift only, five days/week, then even a 10-msec crossover leak will be tolerable: $52{,}560 \times (1/60) \times (8/24) \times (5/7) =$ 208.6 ft³ or $0.05/yr. For that matter, even a 500-msec leak would cost but $2.50/yr.
- If a plant has 1000 valves, multiply all these figures by 1000. For instance, the $13.10 becomes $13,100.

AIR PRESSURE REGULATORS

An air regulator (Fig. 13.8) is a balanced system within itself, and all the major problems such as stability, response, hysteresis, chatter, drift, and hunt have been sweated out and solved by the manufacturers. For most applications the regulator can be assumed to respond instantly and smoothly throughout its range.

The best indicator of performance is the flow curve, also called the regulation curve (Fig. 13.9). The effect of flow on output pressure is by far the greatest consideration, thus submerging lesser effects of sensitivity, hunt, drift, and transients. Even negative droop, where increased flow raises the controlled output pressure, creates no problem, thanks to inherent damping in the regulator valves and passages and pressure losses in the subsequent system. The manufacturer always makes sure the critical frequencies are much greater than any frequencies encountered in actual operation.

Most pneumatic systems are based on simple flow and pressure relationships, particularly in cylinder applications. The major need is to provide high flow with good regulation. Where complex transients must be controlled, or where extreme sensitivity is paramount, special instrument-type regulators are available.

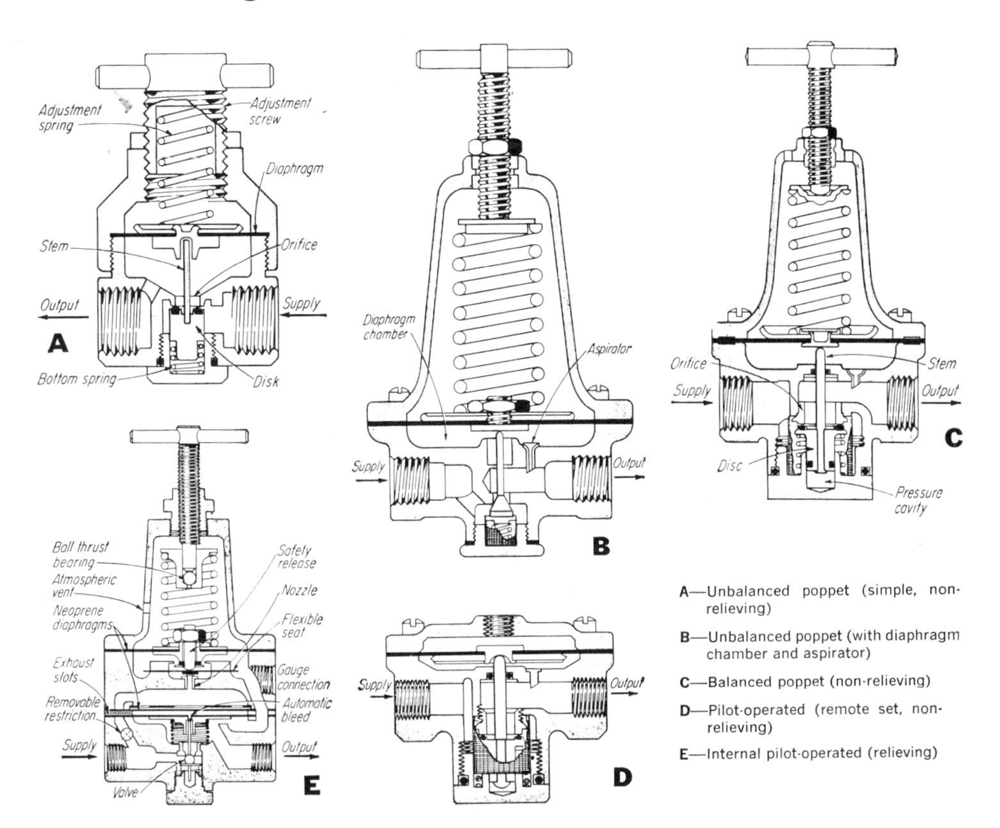

Fig. 13.8 Most pneumatic pressure regulators are similar to one of these five designs.

Five typical types of air regulators are shown in Fig. 13.8. Briefly, they operate as follows.

Unbalanced Poppet (A)

This simplest form of pressure regulator is called unbalanced because the disc (poppet) is exposed to unequal forces: upstream pressure on one side and reduced pressure on the other.

In operation, supply pressure enters the inlet port at the right and flows up around the disc. By turning the adjustment screw clockwise, the adjustment spring is compressed, forcing the diaphragm to move downward. The diaphragm pushes the stem down, and the disk moves away from the orifice. Flow through the orifice causes the downstream pressure to build up and act on the underside of the diaphragm to balance the force of the adjustment spring. Therefore, the disc throttles the orifice to restrict the flow and produce the desired pressure. Generally a pressure gage, not shown, is connected to the outlet port to assist in this adjustment. The bottom spring primarily assures closure of the regulator. It also combines with other forces, including the net upward force on the disc caused by the supply pressure, to complete the dynamic force balance that opposes the force of the adjustment spring.

A number of pressure ranges are available for this and each type of regulator. Typical ranges in most catalogs are 1–25, 3–50, 10–125 psig, etc. The regulator is the

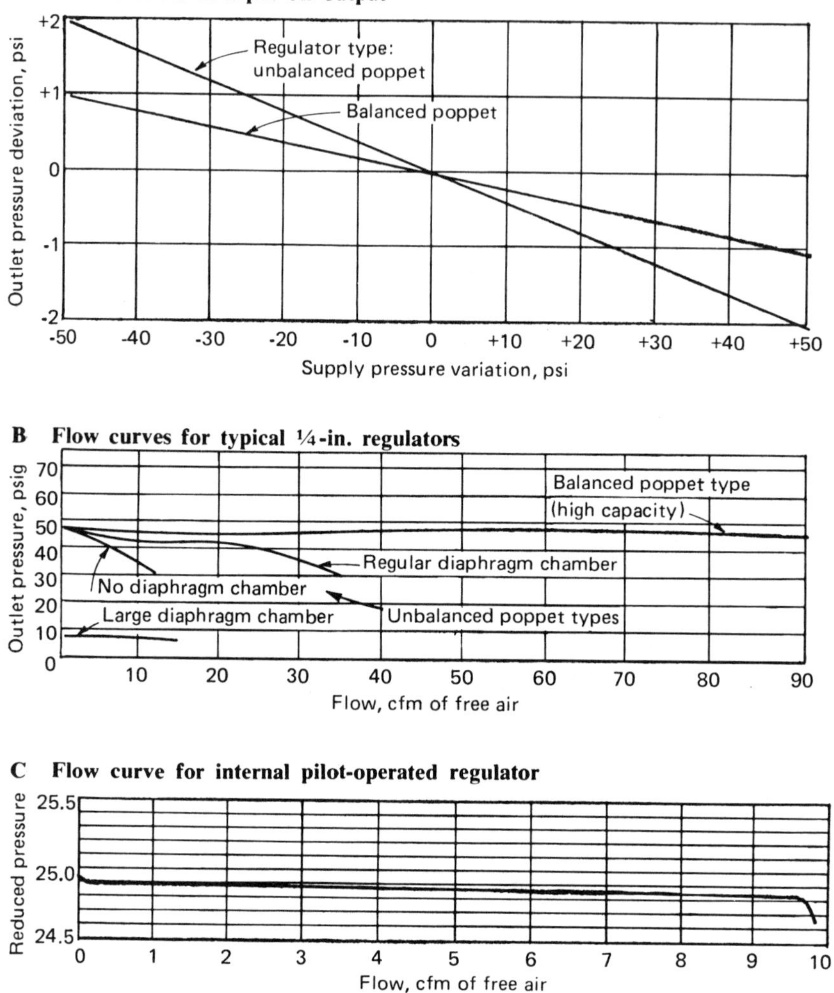

Fig. 13.9 Pressure regulator performance can be as precise as required, depending on design features.

same in each instance: The only difference is in the rate, lb/in., of the adjustment spring. With a lower rate (softer) spring, a reduced-pressure change of a given magnitude (psi) will produce a greater movement of the diaphragm and disc assembly. The droop is minimized, while the response and sensitivity are improved. Select the lowest range that will still yield the maximum pressure setting you need.

Regulators can function as shutoff valves too. Generally, regulators are designed so that there is no loading on the upper spring when the adjustment screw is backed off all the way; the bottom spring, assisted by supply pressure, closes the valve. Conversely, they may be adjusted to produce pressures below the minimum lower limit, such as a 5-psi setting on a 10–125-psi regulator. However, the ease of setting and degree of regulation are not the best.

Unbalanced poppet regulators are furnished either with or without an integral relief. Figure 13.8, item A, shows a nonrelieving type. To make it relieving, a small hole, called the relief orifice, is drilled in the center of the diaphragm assembly. Under normal conditions this hole is closed by the end of the stem. However, if an overpressure is created in the downstream line, it will close the disc against the primary orifice and lift the diaphragm away from the stem. The overpressure then is bled through the relief seat into the spring cage and out to atmosphere through the vent hole.

Once the overpressure is exhausted, the relief orifice reseats on top of the stem. The relief feature is of significant advantage in closed circuits such as those for cylinders. Here overpressure, caused by either external loading of the cylinder or by thermal expansion of the air, automatically will be relieved to atmosphere. Also, when a relieving type of regulator is used, the reduced pressure readily may be adjusted downward without having to bleed the line or energize the circuit to reduce the pressure.

Unbalanced Poppet with Diaphragm Chamber (B)

It has a chamber which isolates the diaphragm from the flow passageway, eliminating the abrasive effects of the air. More important, the diaphragm chamber, in combination with the aspiration tube (usually standard), makes for less droop. Increased airflow across the mouth of the tube slightly reduces the pressure in the diaphragm cavity. The diaphragm deflects downward, forcing the disc a little farther away from the orifice, producing a slightly higher reduced pressure. The increased pressure setting has a compensating effect on the pressure drop downstream of the regulator that is caused by the higher flow.

In addition to the aspirator effect, this regulator has a much larger diaphragm area, assuring more movement of the disc with a given change in reduced pressure because of the greater resultant force. The larger the diaphragm is with respect to the orifice, the greater will be the response speed and sensitivity. The flow graphs show that the droop of this regulator is significantly improved over that of a smaller regulator of the same size orifice. A more dramatic example is the very large diaphragm regulator at the bottom of the graph in Fig. 13.9.

Atmospheric bleed is incorporated in some models, through a small hole in the diaphragm. This further reduces droop and increases sensitivity because the disc never completely throttles the orifice. Therefore, the initial droop necessary to break the disc away from the orifice is eliminated.

Balanced Poppet (C)

It has the same general internal construction as the other regulators except for a much larger orifice, to allow for greater flow. Stability is still good because the disc is balanced—exposed to the same pressure on both top and bottom surfaces. In this design, both ends of the disc are subjected to reduced pressure, bled into the pressure-tight cavity in the bottom plug through the annular space between the stem and disc. The adverse effects of supply and reduced-pressure fluctuations on the disc are canceled out by balance of forces. Therefore, sensitivity and response are greatly improved.

The balanced poppet type of regulator is picked for some of the most rigorous applications: for example, the cylinder-operated spot-welding guns used in the automotive industry. A double-acting cylinder is operated up to a rate exceeding 200 cycles/min. Therefore, the regulator must open and close 400 times/min, with sufficient capacity to let the air flow into the hose and cylinder in the very short time permitted. In addition, it must have sensitivity and fast response to level off the pressure and shut off the flow before it overshoots.

Remote-Set, Pilot-Operated (D)

You can connect a ⅛-in. pilot line to the main regulator from a convenient point where another regulator can produce the control signal.

The remote-set regulators differ from the integral types only in that the top spring assembly has been replaced with a shorter, pressure-tight bonnet to receive the control signal. The regulator works against a force created by air pressure instead of by a spring. The air chamber essentially is an air spring.

Droop is significantly reduced because the mechanical spring rate has been eliminated. In spring-set regulators, as the flow increases, the spring length increases due to diaphragm deflection, but the adjustment spring force decreases. Consequently, the disc opening is restricted, and the reduced pressure falls off. An air spring, however, maintains a constant force on the upper side of the diaphragm because the control signal is held at a constant pressure by the setting regulator.

Internal Pilot-Operated (E)

Here, the setting regulator is put in the same housing as the main line regulator. It operates on the same force-balance principle as the remote type, except that a portion of the supply pressure bleeds into the cavity over the lower diaphragm and escapes up through the nozzle. As the flexible seat is opened by an increase in air pressure on the upper diaphragm, the pressure above the lower diaphragm is reduced, which causes the disc (valve) to approach the primary orifice, reducing the flow and thus the pressure. A bleed-type relief seat is incorporated by venting up through the center layer of the diaphragm, which is made like a sandwich with a porous member in the center. A safety-type relief valve, located above the pilot mechanism, will rapidly dump if triggered by an exceptionally high overpressure.

The performance is nearly perfect (see Fig. 13.9) with very low droop and excellent sensitivity. This explains its use on instrumentation and gaging work where close regulation is essential. Its cost is two to three times that of the balanced poppet type.

A modification to these basic types of regulators are the so-called piston-type regulators where a piston is used in place of the diaphragm (or diaphragms, in the case of the internal pilot-operated regulator). The pistons are free-floating and have the advantage (at additional cost) of eliminating the restrictive effect of diaphragms.

14

Pneumatic Logic

Pneumatic logic is simply the use of air valves and other pneumatic devices to perform purely control functions. They don't do any work but transmit YES or NO signals in the control circuit, just as electrical relays do in a switching circuit or transistors do in a microcomputer. See Figs. 14.1 through 14.4 for examples of *logic* modes.

There are two types of pneumatic logic devices: moving-part air valves (usually tiny spool or poppet valves) and no-moving-part fluid amplifiers (miniature devices that exploit the dynamics of tiny streams of air or other fluids to trigger YES or NO signals).

The moving-part valves predominate in industry; fluid amplifiers are used only in special circumstances (see the final section of this chapter).

MOVING-PART PNEUMATIC LOGIC

Miniature pneumatic control valves of every description have appeared from time to time all over the world. These so-called logic valves (Figs. 14.5 through 14.9) are not asked to take on brute tasks. They aren't even required to act as pilots for working valves and cylinders. They are assigned exclusively to the functions of sensing, interpreting, and decision making.

Thus they fill the gap in control hardware where the system is too complex for ordinary bulky control valves yet not complex enough to warrant investment in an electronic digital computer, or even microcomputer chips with special power supplies. And in many applications, air is without doubt more reliable than electricity.

The Essence of Control

Pneumatic control consists of information (sensing) and decision (logic) and does not include power elements such as power valves, cylinders, and motors. However, the final element in each control branch is a power valve (mounted on the machine), and the essence of the control system is to instruct the valve spool which way to move.

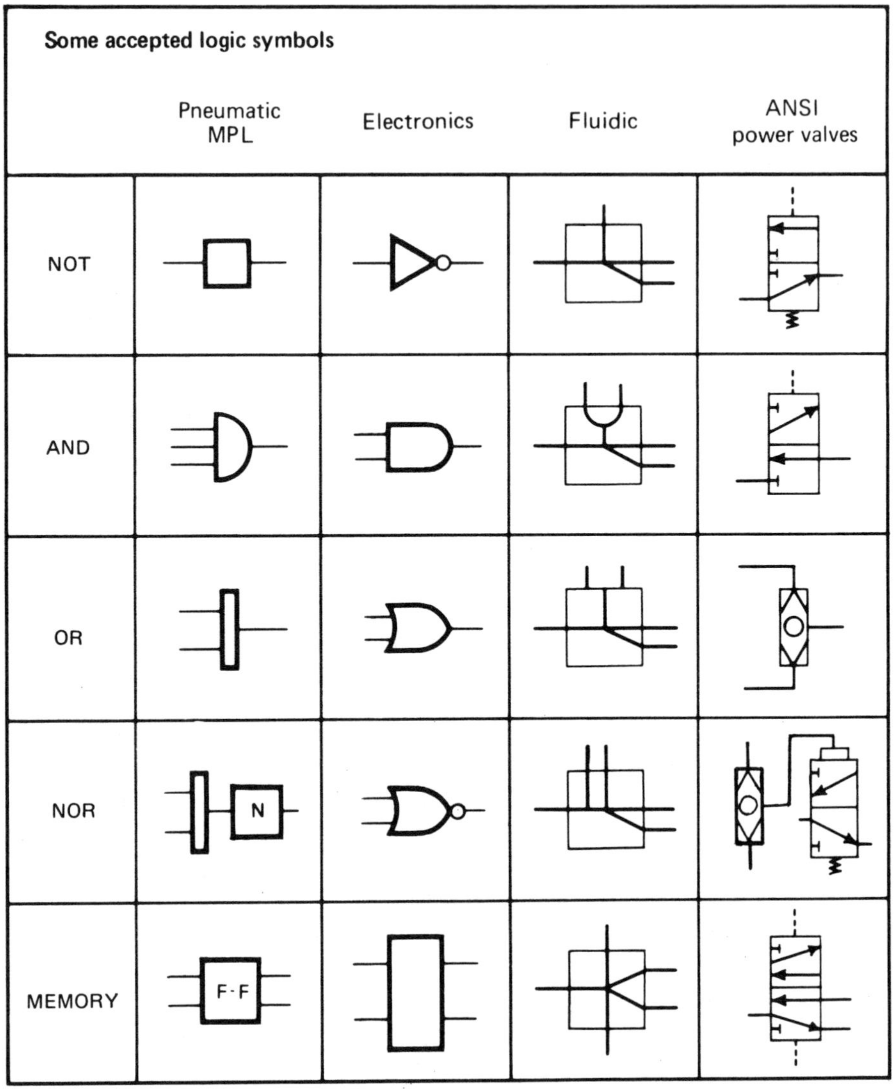

Some accepted logic symbols

	Pneumatic MPL	Electronics	Fluidic	ANSI power valves
NOT				
AND				
OR				
NOR	N			
MEMORY	F-F			

Fig. 14.1 Pneumatic logic concepts are patterned after electronic logic.

The information devices—such as limit valves, pushbuttons, and special transducers for sensing pressure, temperature, velocity, force, and even proximity—are separate.

The decision devices in most instances are mounted together in a protective cabinet. Troubleshooting of shuttle valves, timers, pneumatic relays, and any other logic elements are thus localized and convenient. Typical logic is a five-port multipurpose valve (Fig. 14.5, item A). It can be ported in various ways to give two-way, three-way, four-way, and five-way action.

Less typical is the Dutch-designed "wheatcake" logic concept (Fig. 14.5, item B). It essentially is a stacked diaphragm valve without connecting stems. The only mechanical communication from one diaphragm to the next is a spacer disc that nearly fills the void between the diaphragms. The concept was developed at TNO-Delft, the Netherlands, and first built by Sempress.

Each void is an input chamber, identical to every other but energized separately. When any void receives an input signal, it swells the surrounding diaphragms, which push up and down against the rest of the stack.

The highest spacer comes to a stop at the top button; the lowest spacer forces the bottom button down and closes off the output to exhaust.

It's a NOR-logic device. It produces an output until it receives a signal from an input. Combinations of NORs provide any decision-making function, and that sort of logic is used widely in control systems.

Another unusual pneumatic logic valve concept is based on a diaphragm (Fig. 14.5, item H, and Fig. 14.6). It's a board made in two halves, each molded with all passages complete. The operating pressures are only a few psi, and the flexing is small. The devices nearly qualify as fluidic devices—meaning "no moving parts."

There are as many variations on the principles sketched in Figs. 14.5 through 14.9 as there are companies in the business. For example, Telemecanique has modular poppet-type valves and relays that snap into racks and have snap-in accessories that change the circuit instantly. One shown (Fig. 14.7) has the circuit printed on the

Logic Term Symbol (Based on NEMA)	Word Description
AND	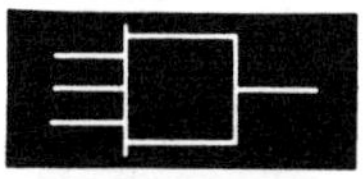 All inputs must be present to produce an output
OR	 Any input will produce an output
NOT	 Absence of input produces an output
MEMORY	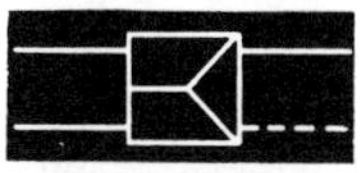 See variations in table of typical variations
DELAY	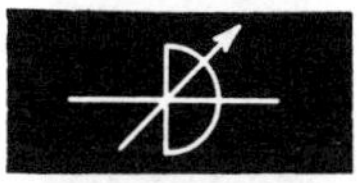 See variations in table of typical variations
Amplifier	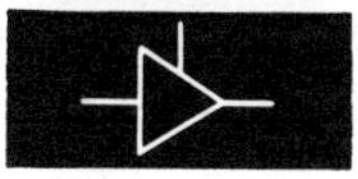 Amplifies logic-level input to give solenoid-level output. Power source goes to top terminal
Original input	 Converts conventional signal from contact-type devices into low-v signal for logic devices

Fig. 14.2 Alternative symbols from NEMA help explain the logic functions of Fig. 14.1.

Three MEMORYs

a basic MEMORY

Off-return type. Retains output corresponding to input last energized. Returns to off state if power is interrupted

ON-OFF-NOT type

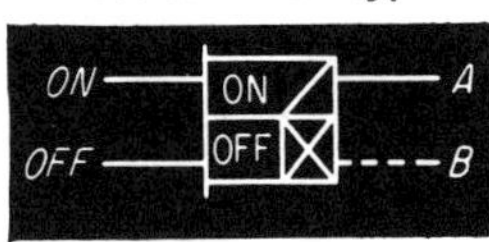

Output A will correspond to the input. Output B has a NOT function, and will always be opposite to A. If both inputs are energized (on plus off), A goes off, B goes on

FLIP-FLOPS

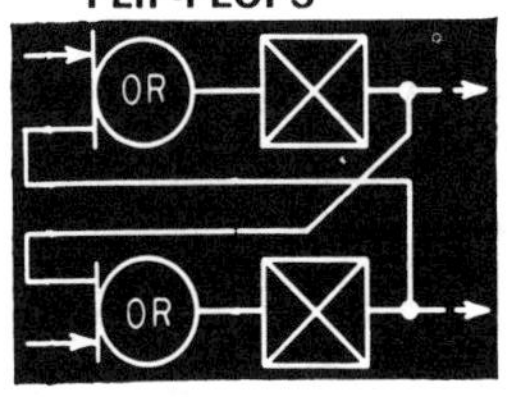

Energizing one of the inputs, even momentarily, will turn on the corresponding output and turn off the other output

Three DELAYs

DELAYs

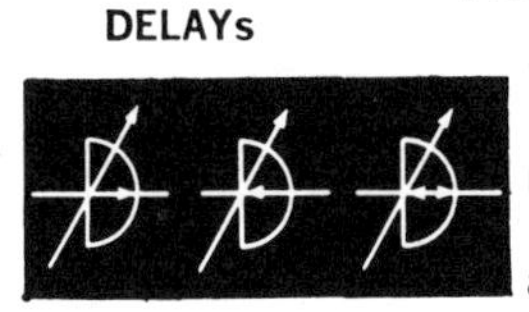

Each is adjustable. First has an output after an input. Second has a loss of output after an input. Third has an output after an imput, and a loss of output after a loss of input

Fig. 14.3 Typical variations from NEMA show how logic functions are created.

face, and when the accessory is remounted end-for-end, it changes the circuit and the diagram.

Accepted Standards

Diagramming of moving parts fluid control is already an American national standard: ANSI-B93, 38-1976. Portions of this standard have worked their way into some manufacturers' catalogs, though there are many more manufacturers who chose to develop their own proprietary diagrams. Unfortunately, not many of the proprietary techniques agree with each other, so much logic design work becomes a personal relationship between the supplier and his customers. We'll limit this discussion to the ANSI method, although others have value too.

Basically, the ANSI standard separates the control devices from the output devices (see Figs. 14.10 through 14.12). The supply air and the inputs are generally to the left and the output devices to the right.

Two methods are depicted: the *attached* method (Fig. 14.10), which resembles an electronic logic diagram with complete ANDs, ORs, and NOTs, and the *detached* method (Fig. 14.11), which resembles the standard electrical ladder diagram widely used in electromechanical relay circuits and in some programmable controllers. Attached in this sense means that the symbol represents a complete valve, with all the

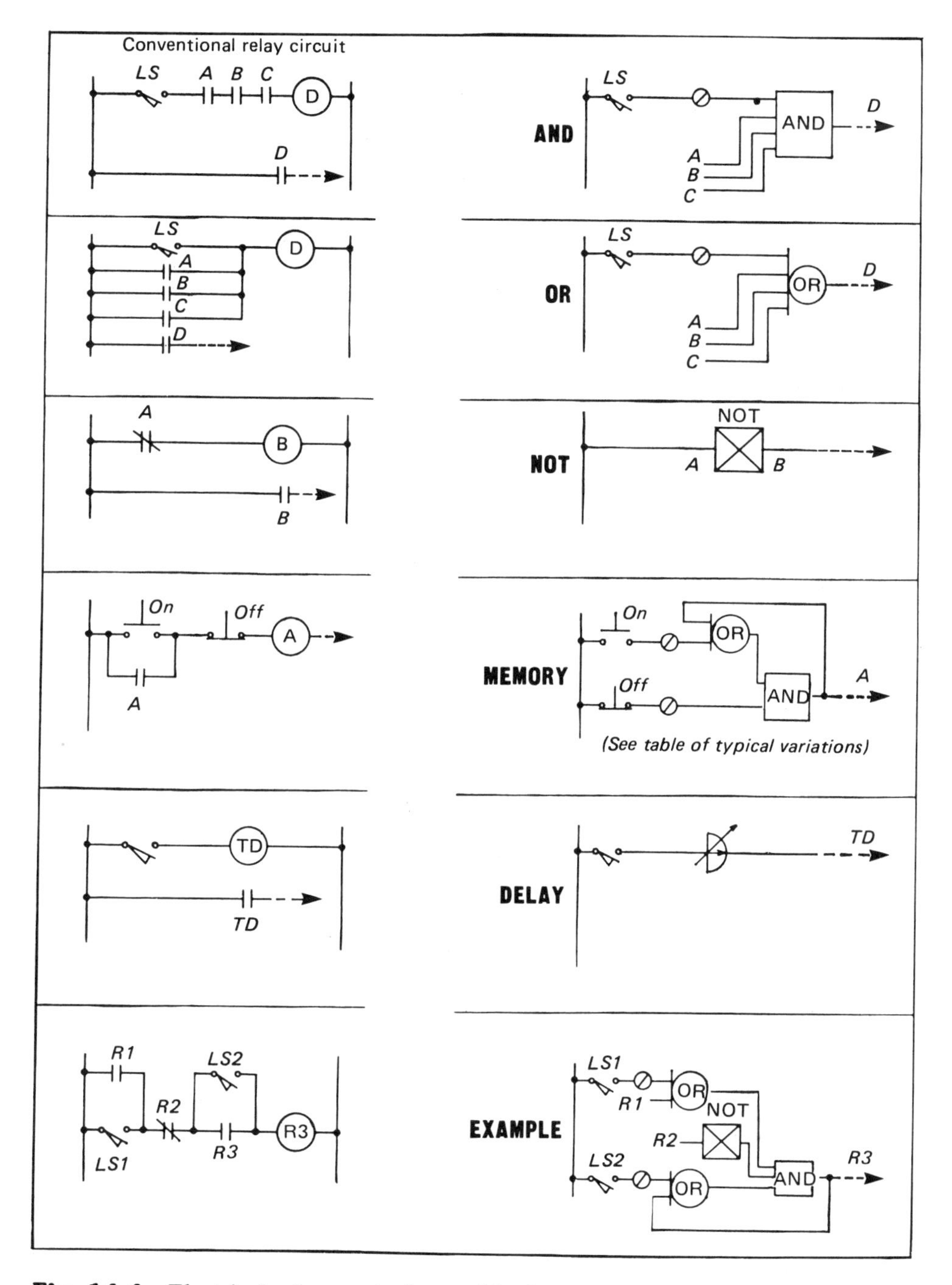

Fig. 14.4 Electrical relay equivalents of logic symbols.

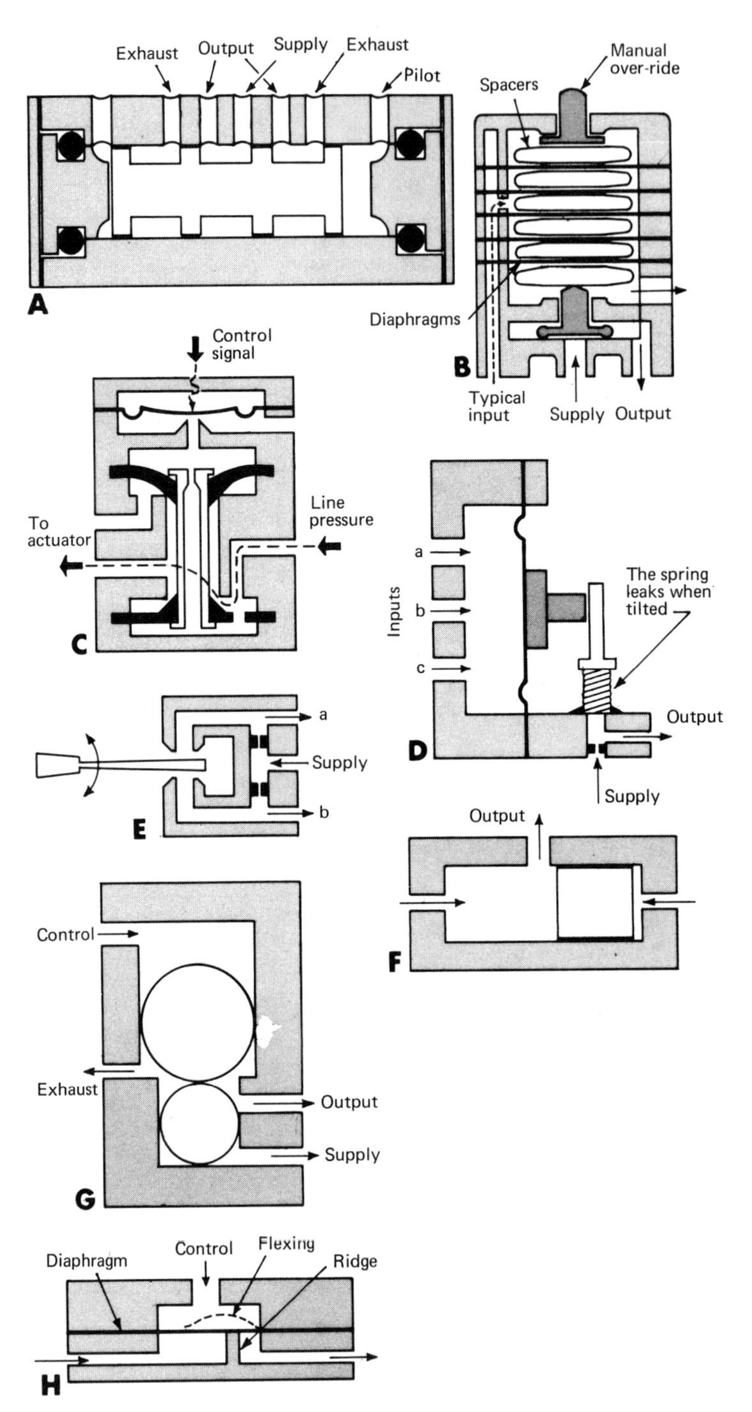

Fig. 14.5 Some of the hardware for creating pneumatic logic.

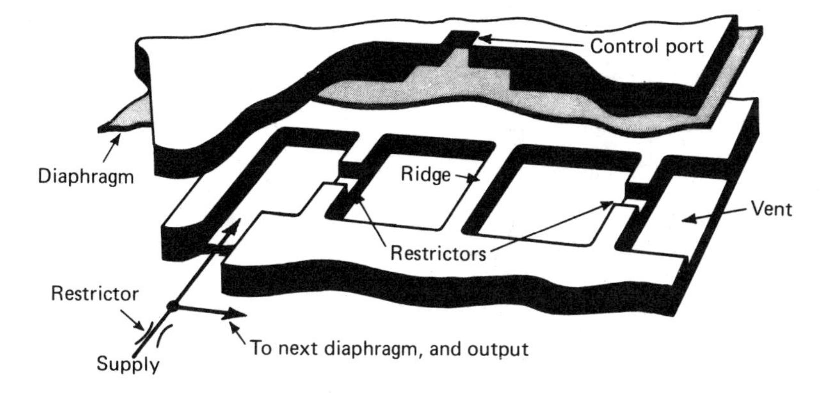

Fig. 14.6 A simple diaphragm becomes a logic valve: output occurs if control port pressure forces the diaphragm over the ridge. (Courtesy of Double A Products Co., Manchester, MI.)

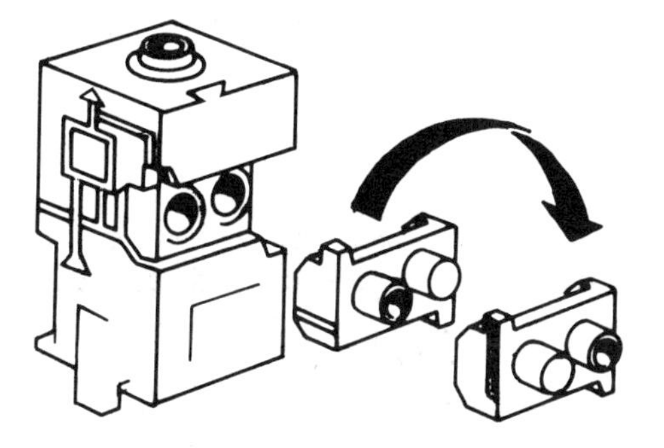

Fig. 14.7 Modular pneumatic relay changes the logic mode if the element is flipped. (Courtesy of Telemechanique, Inc., Arlington Heights, IL.)

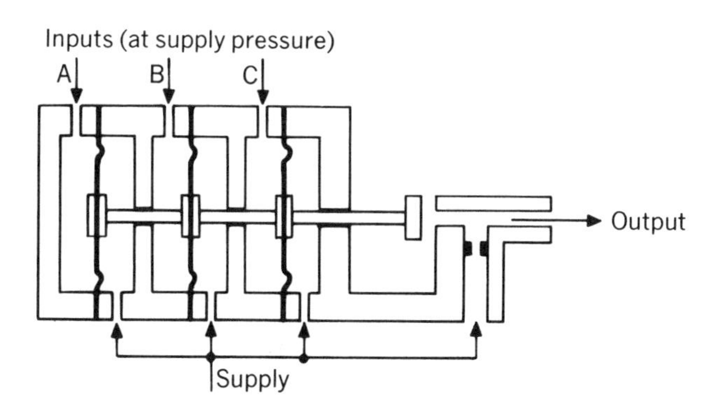

Fig. 14.8 Stacked diaphragm AND logic valve has output only if A, B, and C are pressurized.

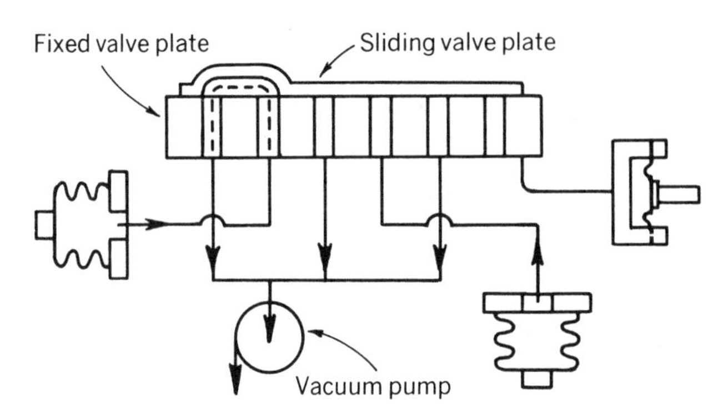

Fig. 14.9 This vacuum-actuated device is not exactly a logic valve, but it could be. Atmospheric pressure operates the bellows.

parts attached within the symbol. Detached means that the functions of the valve are drawn separated from the valve pilot, at any point in the schematic where needed.

All resemble well-documented electrical diagramming standards but with important differences. For one, air valves when open pass air, but switches when open stop electricity. So a new language had to be invented: A valve is not called open or closed but rather passing or nonpassing, as shown in Fig. 14.12.

A lesser problem is exhaust: Electricity needs a return path, whereas air can be exhausted at any valve. So the air schematics do not have to show a return path for the exhausted air.

Gravity is exploited in many of the symbols to differentiate between a valve that is held actuated before the cycle starts, as opposed to one that is not held actuated. In the symbols, if the valve "arm" hangs down, it is not held actuated; if it seems to be defying gravity by hanging upward, it is assumed held actuated.

Multiple ports of relay valves are represented by vertical parallel lines, exactly as in an electrical ladder diagram. A gap means nonpassing (the designated port is closed to flow). A slash mark means passing.

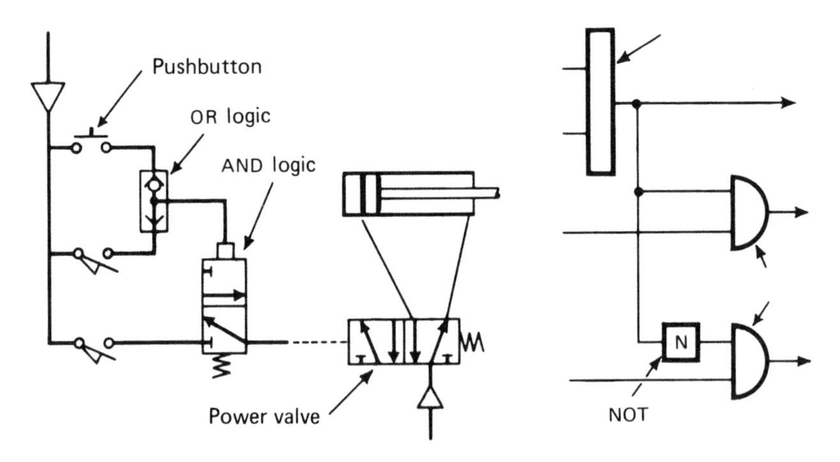

Fig. 14.10 Attached method of diagramming, where each logic valve or function is kept intact (two different circuits are shown).

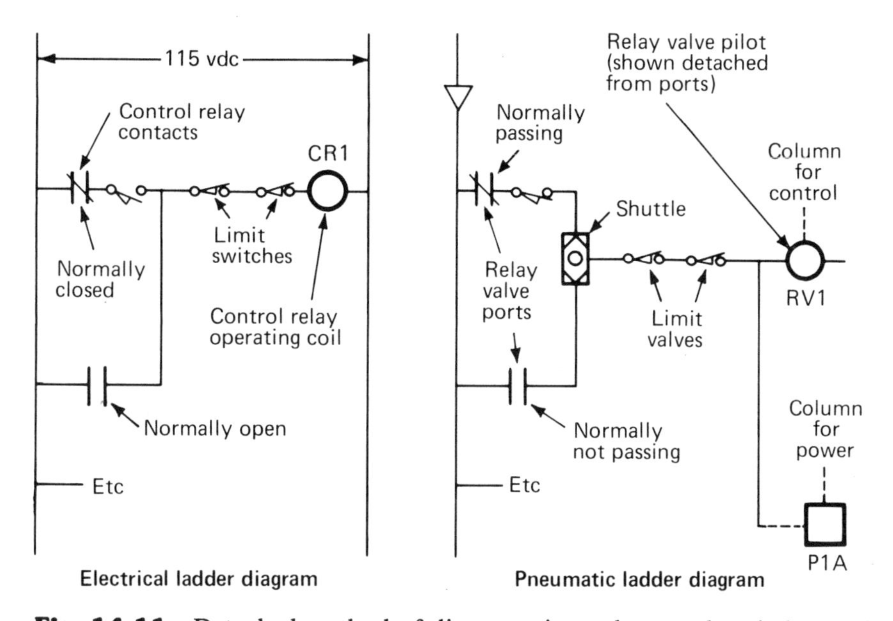

Fig. 14.11 Detached method of diagramming, where each switch or valve function is detached and placed in the circuit where action takes place.

One difficulty is to show a port normally passing (meaning it is passing when not energized by the control) but forced nonpassing at the start of the cycle because of pressurization at the control point. The answer? Show the passing slash mark broken and the halves pushed outside the gap, suggesting that the pressure signal pushed them there.

A selector valve can be shown in two ways: as a rotary valve (which it often is) or as a matrix. The matrix symbol has the additional advantage of being able to represent various combinations of passing and nonpassing by simply placing the "×" symbol at the desired ports and positions.

Where the air circuits must be fairly complex (as in controlling an assembly machine or a transfer line), the detached-type diagramming is preferred by many engineers because each rung of the ladder can be a complete function and fewer vertical cross-connections result. But therein lies a controversy.

The Controversy

The attached-method devotees do not always see eye to eye with the detached-method (ladder diagram) devotees. The reason seems to be a difference in the applications sought.

Doig Associates (Highland, MI), ladder diagram experts, explain it this way: In the attached method, every function built into the logic valve is shown attached to the valve, even if some of the functions are piped into other circuits at different portions or even on different pages of the schematic.

If the circuit is at all complex, the outcome is a plethora of crisscrossing lines, seemingly with everything connected to everything else. Only if the circuit is relatively simple will the attached method be convenient. However, many companies use the attached approach, and it seems to work well.

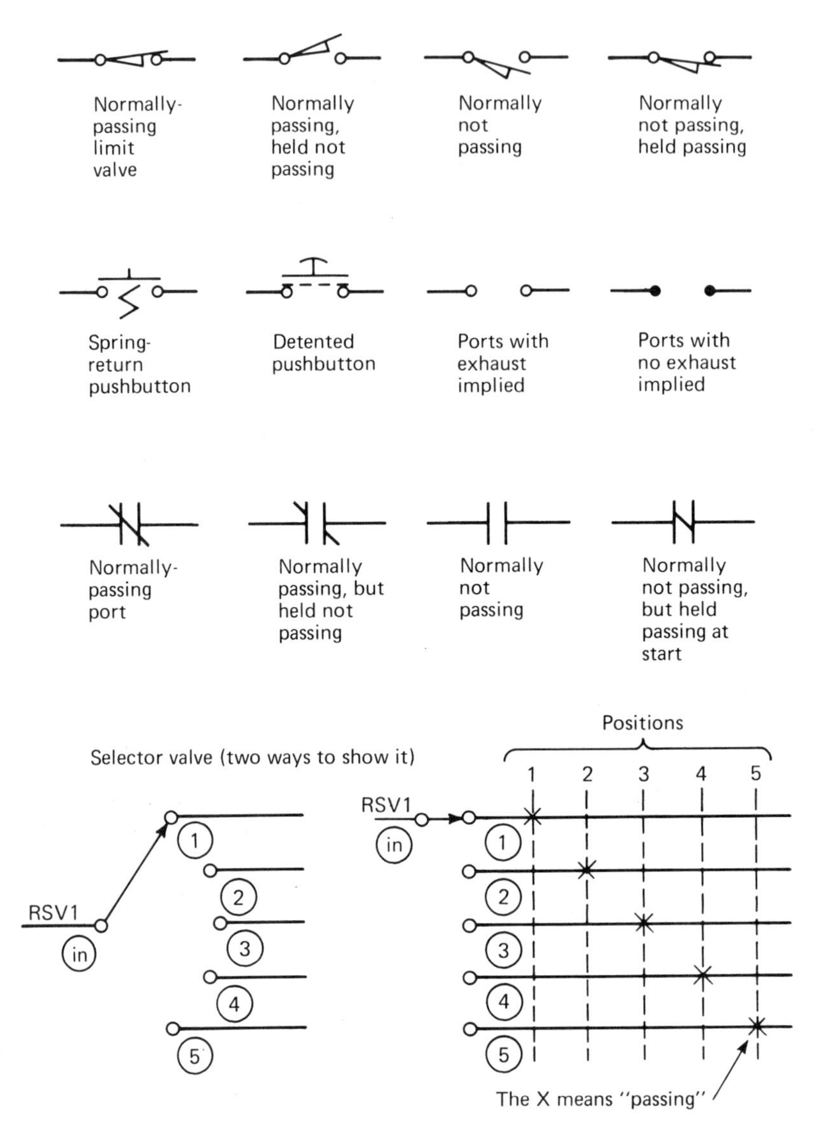

Fig. 14.12 Pneumatic circuit elements resemble electrical circuit elements in this NFPA/ANSI pneumatic MPL (moving-part logic) standard diagram.

The detached method doesn't try to depict every function of a valve in a single symbol but sorts out the functions one by one, as simple symbols, and places them precisely in the schematic where the function takes place. Each symbol is suitably coded with the name of the valve it's part of, and it takes only a quick scan of any schematic to find related symbols.

Does it sound complicated or confusing? Well, Doig says the electrical industry has been using this detached-symbols ladder diagramming method with outstanding success for over 50 years, and ANSI pneumatic control standard B93.38-1976 with detached symbols is practically a copy of it.

To see how it works, we'll go through a hypothetical 20-line (rung) circuit as an example (Fig. 14.13). It was designed by engineers at Dynamco (Dallas, TX) and is based on the ANSI standard.

The Ladder Diagram

The basic logic control valve in this system (Fig. 14.13) is of the spool type, four-way, five-port, similar to Fig. 14.5, item A, but the concept will work with any valve.

The company has grouped several valves in a module called Program-Air™. The valves are called relays because they are triggered by pilot air signals and change the flow conditions in one or more controlled flow passages.

Within the module are three precision, double-pilot, detented pneumatic relays mounted to an integrated circuit base assembly. The base is a sandwich of aluminum plates.

The top plate provides mounting surface for the relays and communicates the ports of these relays into the baseplate. The back plate is the connector plate, with barbed tube fittings for convenient connection to the pneumatic power devices and interlocks of the machine being controlled. Between the top plate and the back plate, a series of air passageways interconnect the relays in a definite sequence.

The assembled module delivers three pneumatic output signals in sequence: event 1, event 2, and event 3. Here is how the unit operates in the circuit:

When a 50-psig supply pressure is applied to port N (line 1), there will be no output signal or pressure present from any of the other ports as long as all three relays have their spools in the B position (meaning "initial" or "back home"). Alternate flow paths RV1, RV2, and RV3 are nonpassing as shown (lines 2, 8, 14). A momentary start signal at port O (line 1) pressurizes the A pilot of RV1 through the module, and since there is no signal present at RV1B (line 7), RV1's spool shifts to the A position (meaning "away"), making RV1's alternate flowpath (line 2) passing and its normal flow path (line 9) nonpassing. This provides a continuous pressure signal at port M (line 2) and port 5 of RV3 (line 3). Signal M is used to turn off the previous event 3. With RV3's normal flow path (line 3) passing, a pressure signal is also present at ports 1 and B (lines 3 and 5).

The pressure signal at port 1 starts event 1 by energizing the pilot of the power device that is to be actuated during event 1. The signal at port B supplies the interlock that determines when event 1 is completed. As soon as the interlock selected (limit valve, pushbutton valve, time delay, or other) determines that event 1 has been completed, the interlock provides the flow path for signal B to pass to port C (line 5). The signal port at port C now pressurizes the A pilot of RV2, causing its alternate flow path (line 8) to become passing and its normal flow path (line 15) to become nonpassing.

This provides a pressure to the B pilot of RV1 through S1 and to port 5 of RV1. Since the spool of RV1 was in the A position during event 1, event 2 cannot be started until RV1's spool returns to the B position. With a signal at RV1B (line 7) and no signal at RV1A (line 1), RV1 shifts to the B position, making its alternate flow path (line 2) nonpassing and thereby exhausting the pressure signal at ports M, B, and 1, which terminates event 1. The RV1 normal flow path (line 9) becomes passing which presents a constant pressure signal at ports 2 and F (lines 9 and 11).

The signal at port 2 starts event 2, and the signal at port F supplies the interlock used to indicate the completion of event 2. When the interlock is completed, it passes

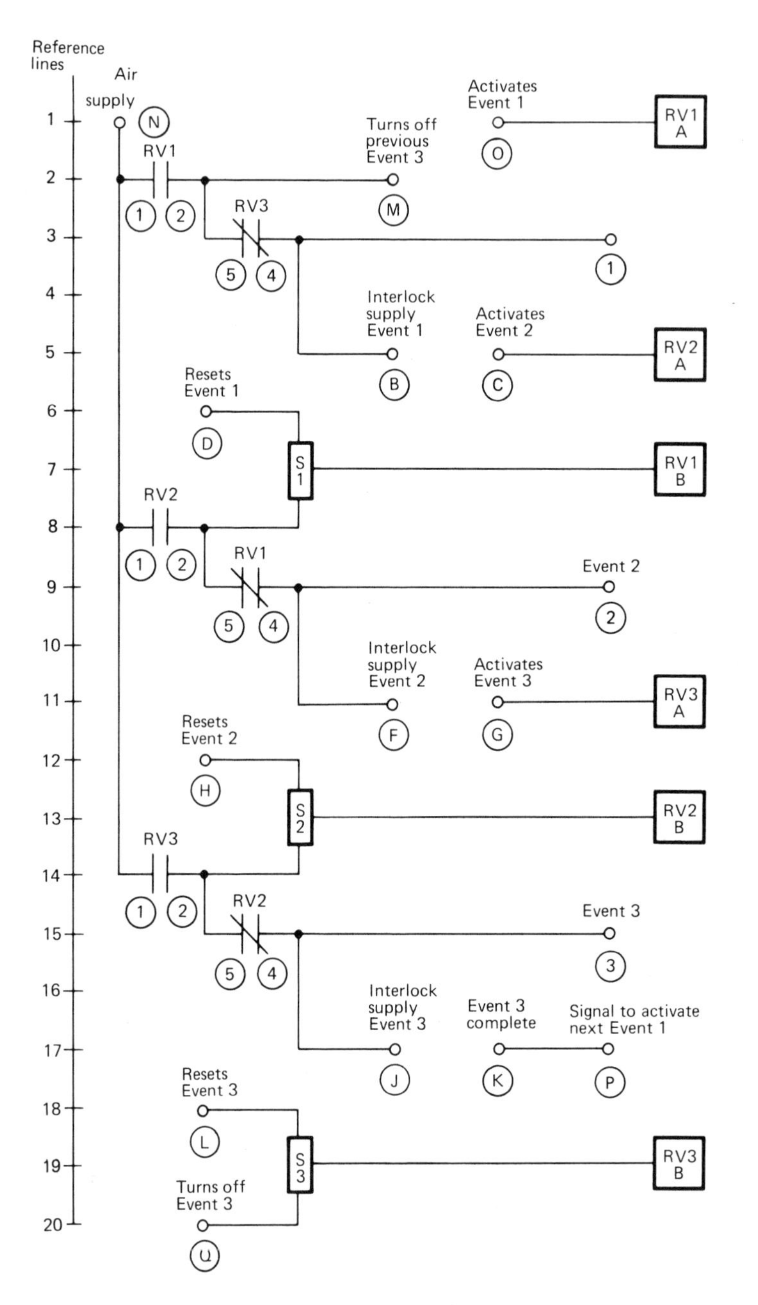

Fig. 14.13 Hypothetical example of ladder diagramming with detached-type symbols. See the text for a detailed explanation. (Courtesy of Dynamco, Dallas, TX.)

the signal at port F to port G, which pressurizes the A pilot of RV3. The shifting of the spool to the A position in RV3 makes its normal flow path (line 3) nonpassing and the alternate flow path (line 14) passing. This provides a pressure signal to the B pilot of RV2 through S2 and to port 5 of RV2. Since RV2 was in the A position during event 2, event 3 cannot be started until RV2 returns to the B position. With a signal at RV2B (line 13) the spool shifts to the B position, causing RV2's alternate flow path (line 8) to become nonpassing and its normal flow path (line 15) to become passing. This provides a continuous pressure signal at port 3 to start event 3 and at the J port to supply the interlock for event 3.

As soon as the interlock for event 3 is completed, it passes the signal from port J to port K, indicating that event 3 is complete. Port K through the module provides output P, which is connected to the O port of the next module. Reset ports D, H, and L (lines 6, 12, and 18) allow the relays to be reset into the B position. Note: Pressure at port N must be removed before the relays can be reset.

FLUID AMPLIFIERS

A fluid amplifier (also called a flueric amplifier or fluidic amplifier) is a pneumatic logic element that has no moving parts and gives its YES or NO signal by exploiting the dynamics of a fluid.

The theory of the device is that if a stream of air or any fluid can be deflected or disturbed by a weaker signal (the control), and the diversion or disturbance of flow can be readily detected or utilized, it's an amplifier.

Anyway, many novel designs have been created to exploit the phenomenon (Figs. 14.14 and 14.15), and although big commercial applications never materialized, many important special applications are successful. A simple valve application (Fig. 14.16) shows the principles.

The absence of mechanical moving parts gives the fluid amplifier unique advantages (Table 14.1, *truth table*).

A well-chosen fluidic device can tolerate almost any environmental abuse. You can make the amplifiers out of nearly any material, can stack them like pancakes to eliminate interconnecting tubing, and can mount them anywhere.

The material can be ceramic, glass, high-strength plastic, Photoceram (Corning), stainless steel, beryllium copper, or any rigid substance. Elements can be made by casting, stamping, molding, machining, or etching. They can be stacked loosely and then bonded or clamped mechanically. Bonding is harder because of the need to avoid distorting or contaminating the critical edges and passageways, plus the need to achieve a perfect seal between passages. But, once together, the elements stay together.

General Electric manufacturing people wanted to see just how rugged fluid amplifiers are. So they abused fluid amplifiers with extreme temperatures, oily atmospheres, wet steam, and abrasive Arizona road dusts, inside and out. No harm came to the parts, nor did operation suffer.

Even so, some materials and some precision passage designs do not tolerate oil and dust as well as others. Where dust or oil canot be tolerated, fine filters (about 5 microns) may be used.

DIGITAL (bi-stable)

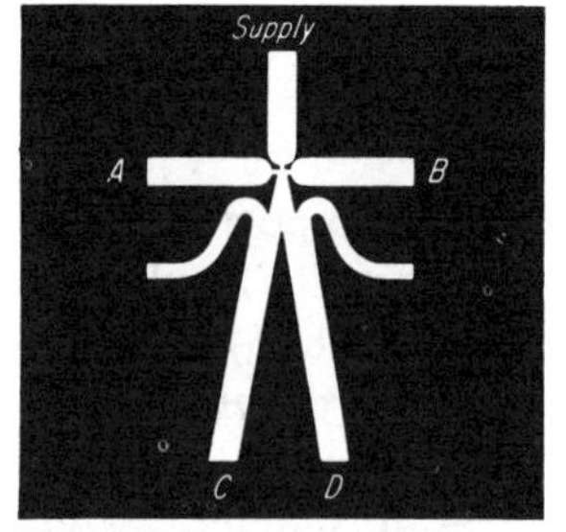

Wall-effect flip-flop is one of the pioneer devices. The supply jet will tend to hug one wall (or the other), thus emerging from outlet **C** (or **D**). Once established, the jet continues through the one outlet until a control signal at A or B deflects the jet to the opposite outlet. The remaining two outlets are vents.

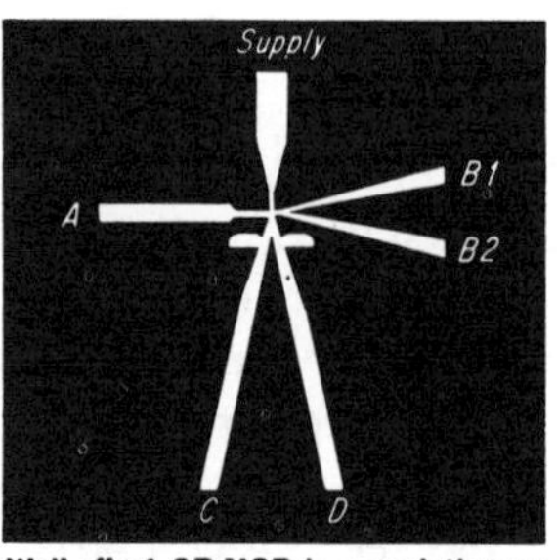

Wall-effect OR-NOR is a variation on the flip-flop. Control ports **B**[1] and **B**[2] constitute the **OR** effect, in that either one **OR** the other can switch over the device. If neither **B**[1] **NOR B**[2] are energized, this produces an output at **D.** This example is only one of the many logic functions that can be performed by fluidics.

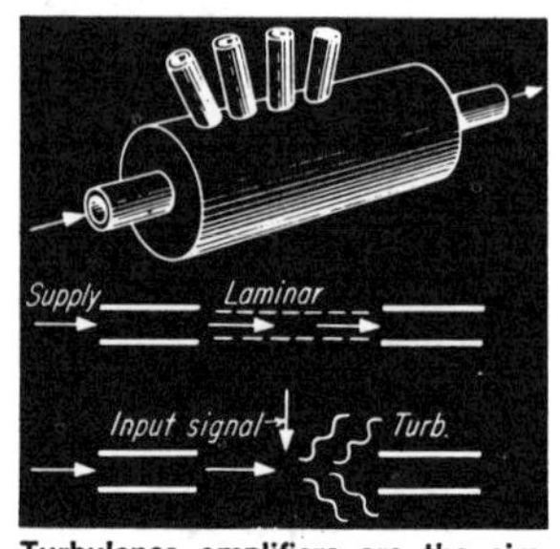

Turbulence amplifiers are the simplest in concept. The supply jet is aimed at the output and will continue to reach it until the laminar characteristic of the jet is disturbed. It takes only a whisper of air to disturb the jet, thus destroying the output signal. The logic mode is **NOT** or **NOR** (no output when any input).

ANALOG (proportional

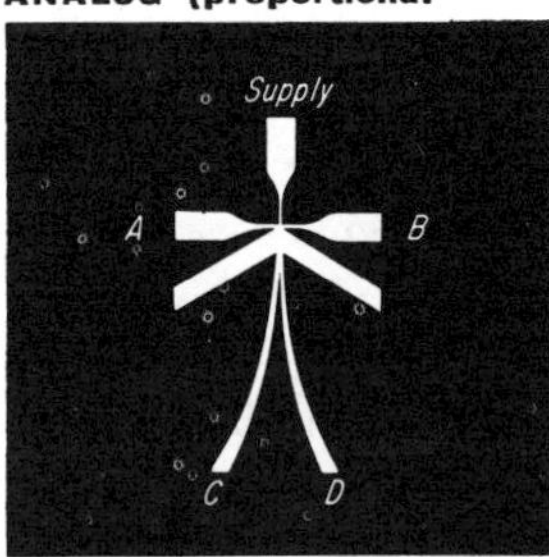

Momentum exchange fluid amplifiers depend on the interaction of control jets A and B with the main supply jet, which aims a portion of the jet to one output or the other (**C** or **D**). The other legs are the vents, as in the wall-effect digital amplifiers. The appearance of the proportional and digital devices are similar.

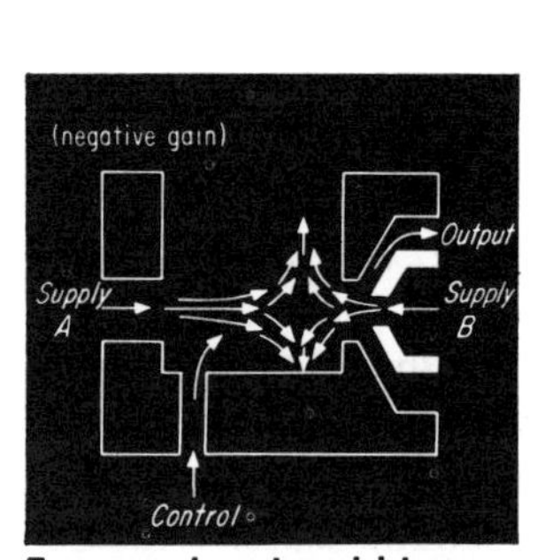

Transverse impact modulators can vary output by moving the area of impact axially with a force balance between the opposing jets **A** and **B.** The control jet deflects the supply-**A** stream and tends to negate its effect, thus moving the impact area to the left, reducing output. The device therefore has negative gain.

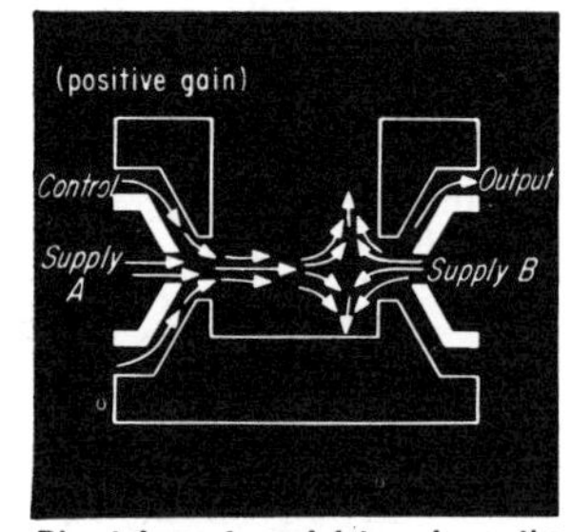

Direct impact modulators have the same basic function as the transverse variety, but use the control jet to strengthen, rather than to weaken, the supply-**A** jet. There are many performance charts plotted to show where the spread-out region occurs, but it is practically impossible so far to compute it mathematically.

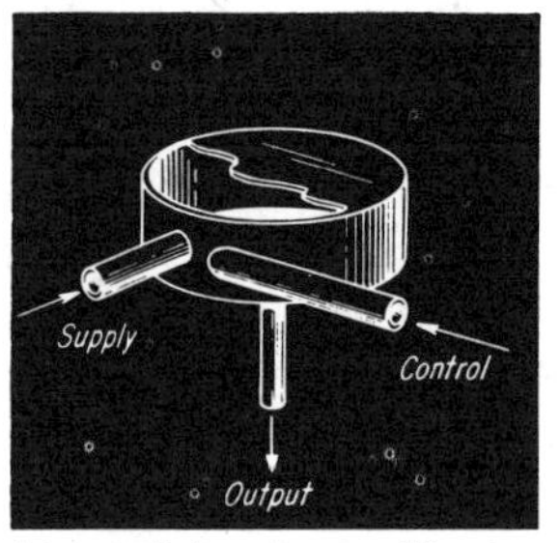

Jet-controlled vortex amplifier has least resistance to flow when the control jet is not operating, because the flow goes directly from supply to output without vortexing. The control jet causes the supply jet to vortex, thus lengthening the effective flow path in proportion to the amount of control flow.

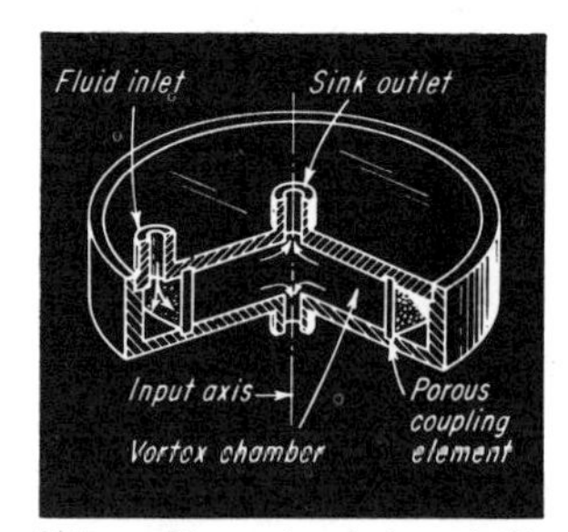

Rate-sensitive vortex device actually rotates in operation, imparting vortex-producing forces which serve to lengthen the effective flow path from supply to output. The principle is similar to that in the jet-controlled version. The rate-sensitive vortex is used in some experimental missiles and aircraft.

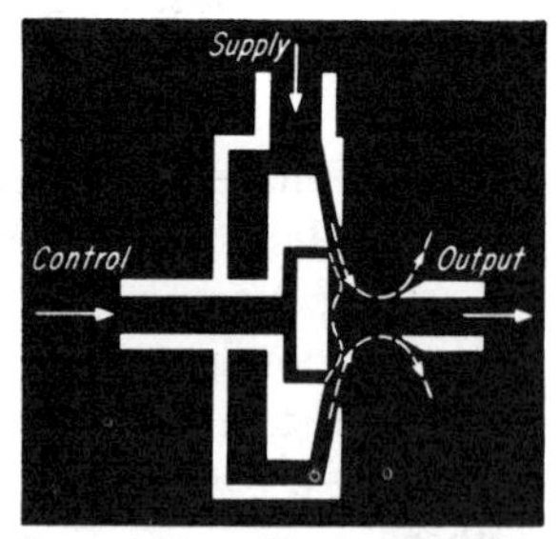

Focused-jet amplifier is designed to aim the supply jet directly into the output pipe except when a control signal tends to deflect the jet away from the target. In some versions, there can be several output pipes clustered in the target area, with several control inputs to steer the main jet selectively.

Fig. 14.14 Fluidic logic has no moving parts. These diagrams explain the principles.

Fig. 14.15 Fluid amplifier operates on tiny jets of air. (Courtesy of Bowles Engineering, Silver Spring, MD.)

Applications

Already identified as applications for fluidic devices are temperature, pressure, and speed controls on gas turbines; proximity sensors of all kinds (for pneumatic instruments, electric relays, liquid-level sensors, thread detectors in looms, air gaging, parts counters, seismic mass detectors, machine position controls, etc.); counting and timing circuits for missiles and projectiles.

Also recommended are process controls where electricity is not allowed; bootstrap flow instruments that use the process fluid itself; diversion valves for process fluids, cryogenic fuels, and hot metal vapors; respiratory instruments; artificial organs, using body fluids; sequencing controls for machines and processes where conventional electromechanical controls cannot be used, cannot stand up under abuse, or are not reliable enough; short-time controls for projectiles or mobile machines in which a bottle of air or gas is the only supply; flight controls for aircraft; steering and intertial controls for missiles and rockets; and proportional amplifiers for process and air-conditioning controls.

Proximity Sensors. The back pressure of a blocked air jet has always been a good way to sense the proximity of a surface. Some can sense objects several inches away; others can even detect the open-pore surface of a sponge. Many exploit fluid amplifiers to pick up the pressure change that occurs when the nozzle reaches a predetermined closeness to the surface. Examples of jet sensing are shown in Figs. 14.17 through 14.19.

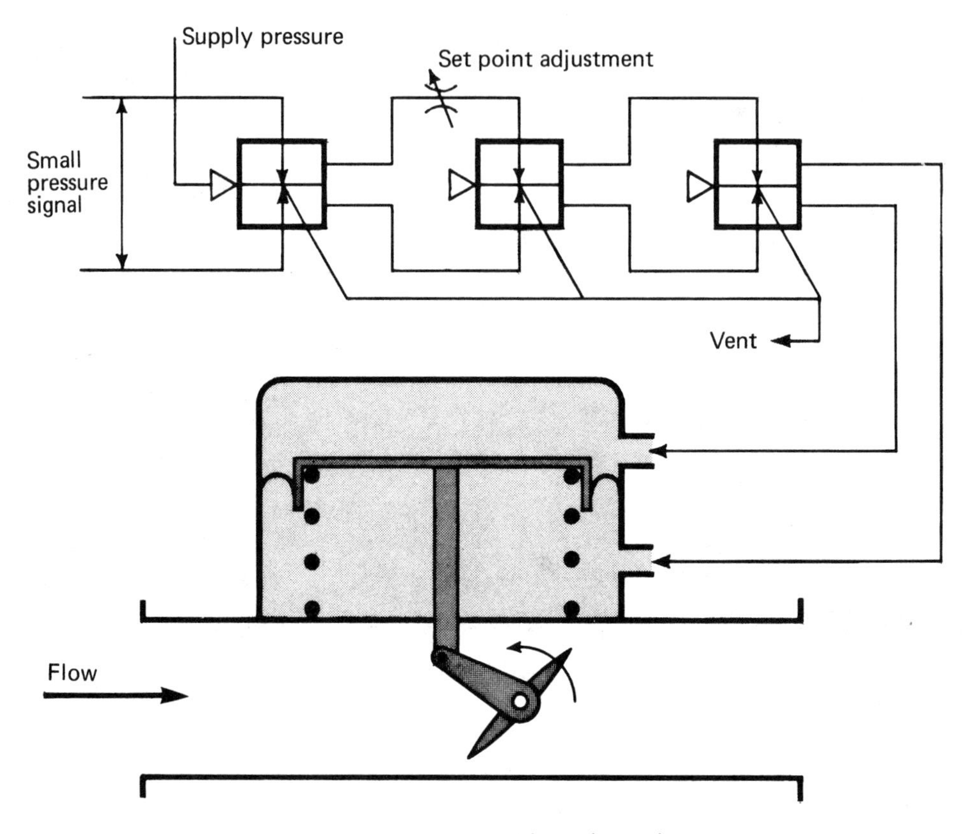

Fig. 14.16 Fluid amplifiers can be staged in series to increase output pressure to operate a flow valve. (Courtesy of AiResearch, Div. of Garrett Corp., Phoenix, AZ.)

Stacked Fluidic Timers. Six of the seven qualities in the truth table (Table 14.1) make fluid amplifiers a natural choice for projectile fuse-timer controls: stackability, absence of moving parts, freedom from electricity, carry-along gas-bottle or ram-air supply, tolerance to shock and vibration, and speed. Each of these is essential when a timer is inside a projectile and is shot out of a gun.

General Electric's R & D Center developed an experimental model of such a timer (Fig. 14.20) for Picatinny Arsenal.

About 450 beryllium-copper laminations are etched with patterns of fluid amplifiers and associated circuits. No other parts are needed except a punched card that is inserted in a slot in the timer before the projectile is fired. Instructions on this card preset the interval at any value from 0 to 280 sec.

In principle, this tiny device, 1 × ¾ × ¾ in., counts the pulses from a built-in fluidic oscillator, which has a known fixed frequency. Thus, a given count takes a known time interval. Adjustability is in the presetting of the number of counts before the timer produces an output (at the right end of the drawing).

A separate supply bottle, occupying several cubic inches, can be used for the motive gas. After the projectile is launched, ram air can take over.

How it times: Gas is fed to the timer at two points: at the power supply connection (lower left in drawing) and at the code-preset supply connection (upper right in

Table 14.1 Fluid Amplifier Truth Table Shows When to Choose (or Not to Choose) a Fluidic Device

Do you need a device that.....	Fluid amplifiers	Spools, poppets	Solid-state integrated circuits
...is environment-proof?			
Can be designed to operate at extremely high temperature.	✓		
Can be made tolerant of any atmosphere.	✓	✓	✓
Is unhurt by nuclear radiation.	✓	✓	
Is unhurt by heavy vibration or shock.	✓		✓
...has no moving parts? No stiction, hysteresis, dead zone, or jamming. Also, no mechanical blocking, so no fluid hammer.	✓		✓
...is tiny, stackable, monolithic? Whole circuits can be made in one integrated block, all permanently sealed, and with no moving parts.	✓		✓
...can be supplied from any fluid source? Air, gas, water, oil, or process fluids will work. Even the water rushing by the hull of a boat, or the air slipping past an airfoil, can be exploited.	✓		
...needs no electricity? And is not affected in performance by radio or electrical interference.	✓	✓	
...is responsive to extremely small inputs? Breaths of air; proximity of anything, such as tiny threads, specks, liquid surfaces, and air bubbles; motion of housing (fluid inertia effect); shock waves; fluid disturbances; spark discharges; sound waves; controlled vibration; localized heat.	✓		✓
...is fast (millisec or better)?	✓		✓
Or do you seek a device that.....			
...has high energy output? Easily transduced to mechanical movement.		✓	
...is widely available from many sources?			
For proportional control.		Some	✓
For on-off control.		✓	✓
...is ultrafast (microsec)?			✓
...can be shut off individually when not in use? Power requirement drops drastically when device is switched off.		✓	✓

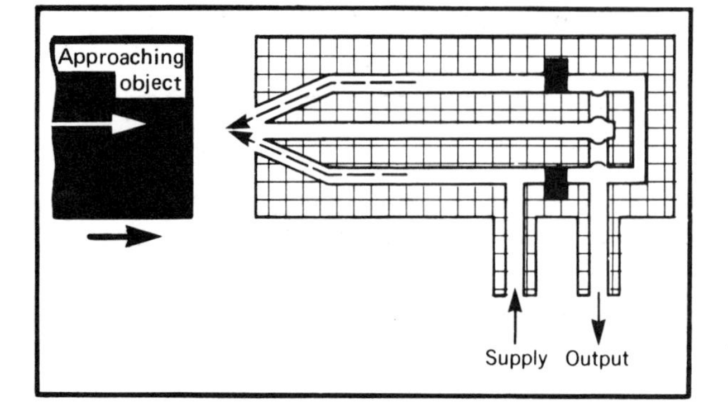

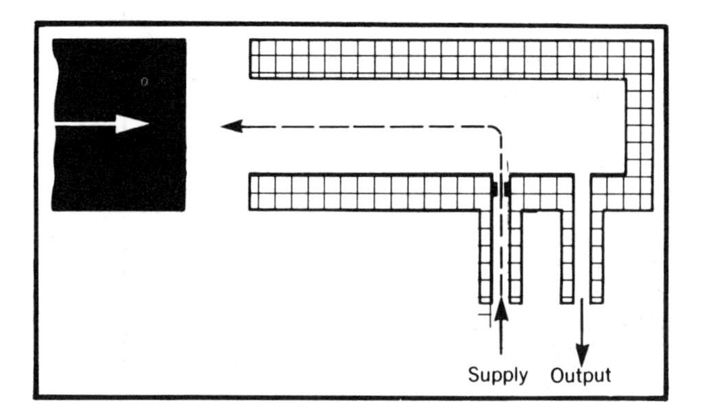

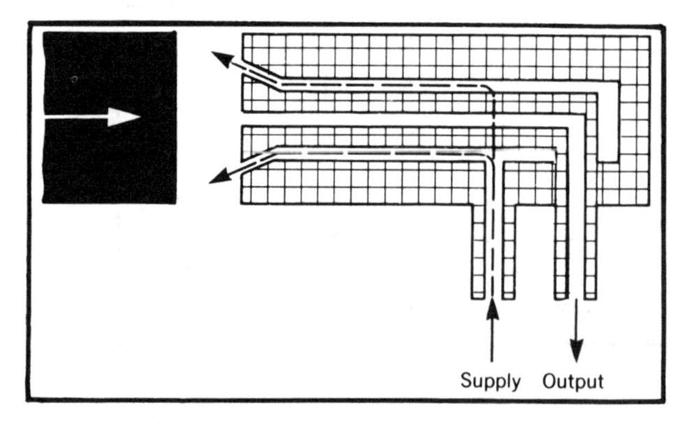

Fig. 14.17 Fluidic (air jet) proximity sensing: converging-cone type (top drawing), back-pressure sensor (middle), and diverging cone (bottom).

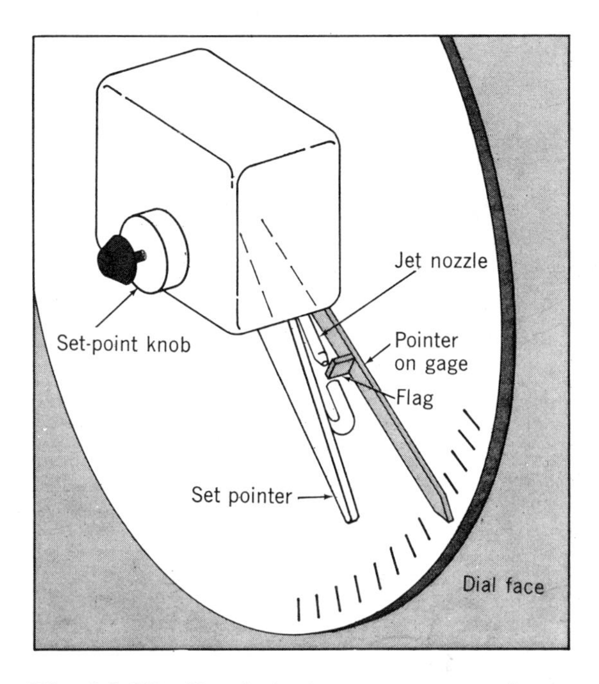

Fig. 14.18 Proximity jet sensor mounted on a set pointer can detect a flag on the pressure-gage pointer. (Courtesy of Martin-Decker Co., Santa Ana, CA.)

drawing). The card, coded for each timing period selected, presets the circuits to transmit the pulses through requisite elements of the stack. It has punch-holes that admit gas to certain passages, while blank areas block flow to other passages.

The oscillator is tuned to about 500 cps as a convenient basic frequency. The counters are versions of wall-attachment fluid amplifiers. Every time it receives a pulse, a counter flips from one output to the other, thus splitting the frequency at a given output leg.

Counting is in binary form—zeros and ones rather than, say, decimals. One leg is considered the *zero* binary; the other, the *one* binary. Each output feeds the next stage's input; the final stage will not deliver its output until every preceding stage has operated.

The drawing (Fig. 14.20) shows only the concept of the timer. The actual prototype has 18 counting stages, each of about seven laminations etched with many fluid-amplifier elements and connecting passages. Laminations are 0.002 or 0.004 in. thick, depending on the function. The nozzles are about 0.009 in. wide. Also included are two oscillator stages and related sonic feedback loops.

Rotary-spray etching of the laminations holds tolerances in the passages to 0.0001 in. The laminations are not bonded in any way, merely squeezed together mechanically and slipped into a housing that holds them in compression.

Operating pressure can be varied. GE and Picatinny Arsenal ran their tests at supply pressures from 1.8 to 20 psi. The timer is adjustable in increments of 0.002 sec. A stop watch was pneumatically coupled to the timer prototype during tests. When the start button was pushed, the watch triggered the gas supply to the oscillator; when the count was completed, the output pressure stopped the watch.

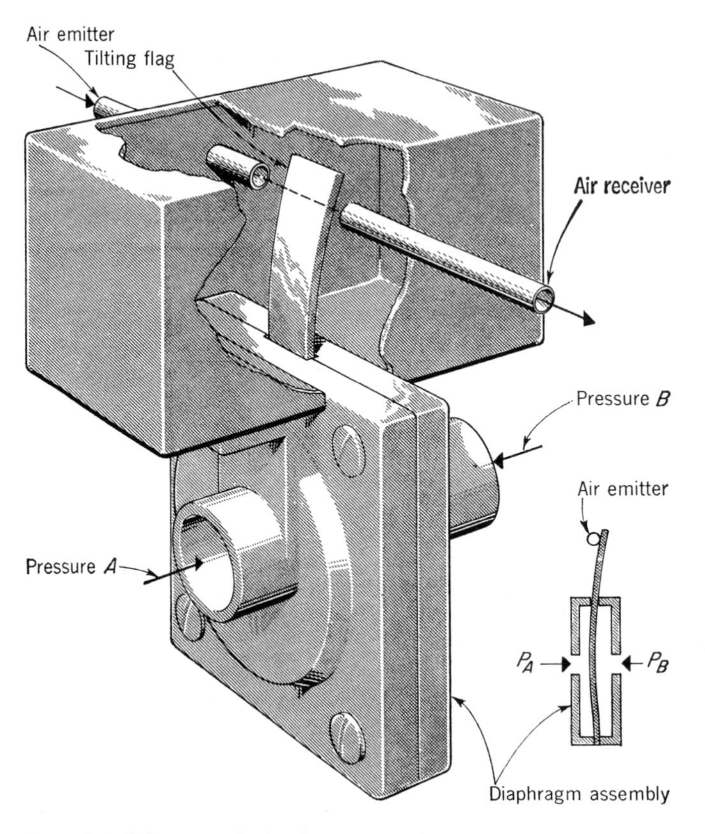

Fig. 14.19 Proximity jet sensor detects operation of a differential-pressure gage. (Courtesy of Gem Sensor, Div. Transamerica Delaval, Plainsville, CT.)

Accuracy of the fluidic timer proved to be 99.5% throughout the pressure range, at constant temperature. It dropped as low as 93% when temperature was varied between −65 F and +165 F.

System Control

Industrial machine control activity is nil, but fluidic systems are in planes, missiles, nuclear plants, and climate-controlled buildings.

There are only a handful of companies building non-moving-parts fluidic devices and systems, and these few are concentrating in areas where nothing else works as well or at all: aircraft gas turbines, for instance, and respiratory systems, textile thread gaging, air conditioning, controls for hazardous areas such as processing and nuclear, and proximity sensing in place of mechanical, electrical, or optical devices.

It's not surprising that most applications for fluidics are in the sensing area. What more direct way is there to sense a hole in a plate, tool in a chuck, edge of a fabric, clearance of a thread passing through a tube, tolerance of a close-fitting part, gust of air, slight vacuum, or protrusion on a red-hot part than with fluidics? All are standard applications today. For example, GE Co. (Schenectady, NY) has sold thousands of fluidic wire and thread gages.

Hot Temperature Sensing. Gas turbine designers need fast, accurate, and reliable sensors to monitor gas pressure, temperature, and flow under the worst possible conditions of vibration, shock, and thermal extremes. Major aircraft and aerospace manufacturers with government support have spent millions to develop fluidic controls to solve some of these problems, and the research is going on just as strongly today as ever.

Why? Because the potential reliability of fluidics is superb, and most other methods involve mechanical or electronic parts, subject to wear or damage. Most designs are still in prototype, but thorough testing has proved the worth of the concept.

The choice seems to be between capillaries and oscillators (Fig. 14.21). A capillary (top schematic) restricts gas flow in a linear manner with temperature, based on a laminar flow phenomena. An oscillator (bottom schematic) pulses at a rate determined by temperature.

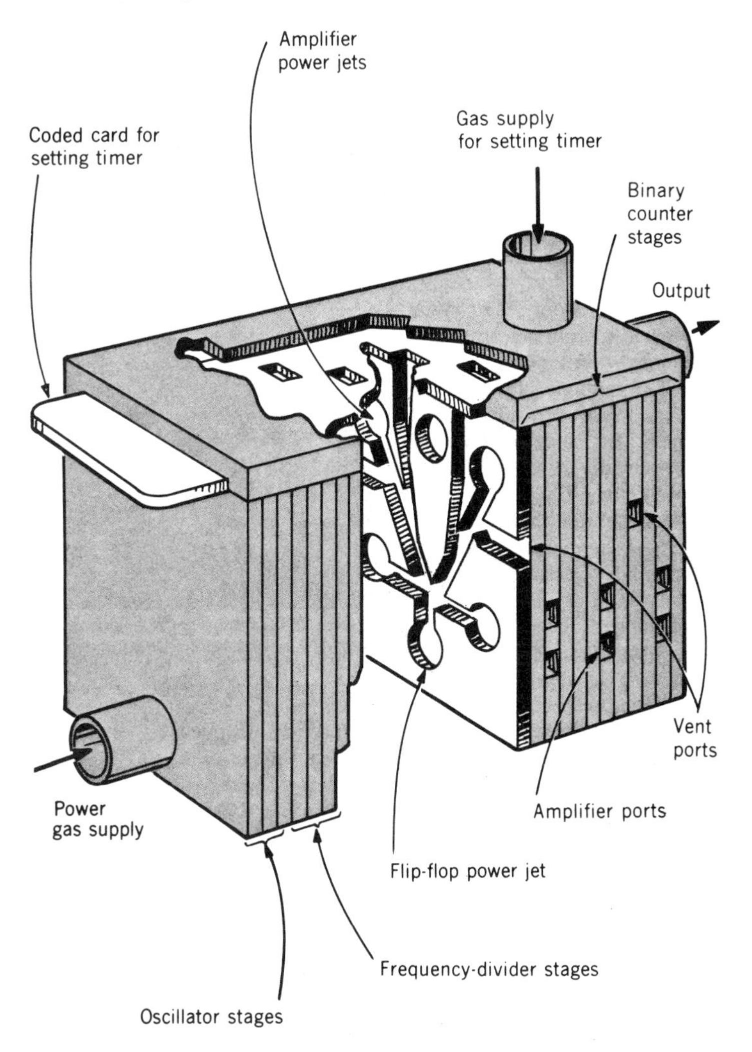

Fig. 14.20 Stacked fluid amplifiers provide compact circuits for a timer designed to ride inside projectiles. (Drawing is conceptual.)

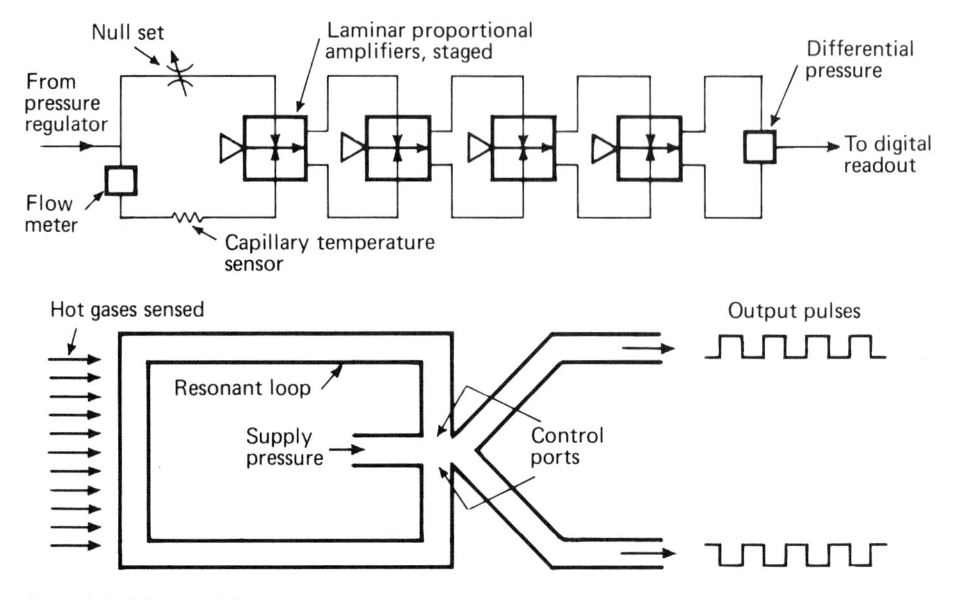

Fig. 14.21 Fluidic temperature sensors by Harry Diamond Lab, Adelphi, MD (top schematic) and AiResearch, Div. of Garrett Corp., Phoenix, AZ (bottom).

In an oscillator, gas viscosity increases slightly with temperature, and for a given flow there is a measurable pressure drop nearly linear with temperature. The problem in the past was that the pressure drop was too small to measure accurately. With fluidic sensors working in a Wheatstone bridge configuration, however, the drop can be sensed with good accuracy, and many experiments have proved the principles.

The Wheatstone bridge method, by the way, is used by Johnson Controls (Milwaukee, WI) for sensing temperature in thousands of commercial air-conditioning systems. The sensing leg of the bridge is a capillary, called a fluidic resistor. Fluidic impact modulators (Fig. 14.14) are in the bridge-balancing circuit.

AiResearch-Garrett (Phoenix, AZ) has also developed a laminar-flow fluidic exhaust temperature sensor (as part of a complex all-fluidic pressure, speed, and temperature control for a 60-kW single-stage gas turbine). It's a bridge network but matches a vortex element against an orifice, using the difference in response as a measure of sensed temperature. The gain of the unit at the set point is about 10 in. of water differential pressure per 100 F. The signal is further amplified fluidically to about 200 in. water per 100 F.

An oscillator-type temperature sensor works on a different principle. AiResearch-Garrett tells of three basic configurations: cavity oscillators, feedback oscillators, and sonic oscillators. The sonic type is a bistable fluid amplifier with a passage connecting the control ports. The supply flow will initially latch to one output leg or the other, and the resulting rarefaction wave propagates through the passage at sonic velocity to the opposite control port, switching the flow to the opposite leg. The frequency of oscillation is proportional to the square root of the absolute temperature of the gas.

Pressure and Speed Controls. Bendix (South Bend, IN) developed a compressor bleed valve control system (Fig. 14.22) for small gas turbines. It is needed only at low speeds where the first stages of the compressor tend to go into stall because of the reduced relative flow capacity of the later stages. The valve allows air to escape from a mid-compressor stage to help maintain sufficient flow in the forward stages.

A fluidically computed fraction of compressor discharge pressure is applied against a fluidically computed pressure based on midstage and fan-duct pressures. The resulting differential pressure across the control legs of the fluid amplifier guides the output to either the valve control diaphragm or to exhaust. The bleed valve is forced open or closed to maintain a predetermined limiting relationship among these pressures.

A combined speed, pressure, temperature, and fuel flow fluidic control (Fig. 14.23) was tested by AiResearch (also see temperature sensors in Figs. 14.20 and 14.21). It exploits fluid amplifiers in numerous ways and is a good example of what can be done. The gas turbine instrumented for the tests is an AiResearch Model GTPR36-61 60 kW recuperated unit owned by the army.

Compressor discharge air is bled into the fluidic system pressure regulator and provides the only supply pressure. Vent flow from the fluidic circuits is exhausted through a unique converging-diverging nozzle that establishes sonic flow at all altitudes and holds the vent pressure fairly constant at about 19 psia. This constant pressure is used to bias the regulator, thus holding output pressure constant despite changes in altitude. Bendix has a similar vent nozzle.

Primary control is of fuel input. Engine speed is sensed by an interrupted-jet pneumatic pulse generator, with output frequency amplified and rectified by a series of fluid amplifiers. The pressure output of these stages is compared with a pneumatic pressure reference representing the desired engine speed.

Exhaust gas temperature is monitored to prevent overtemperature by limiting the fuel flow. This feature is extremely important because the turbine can be damaged during fast startup or under certain overload conditions. The sensor (described before)

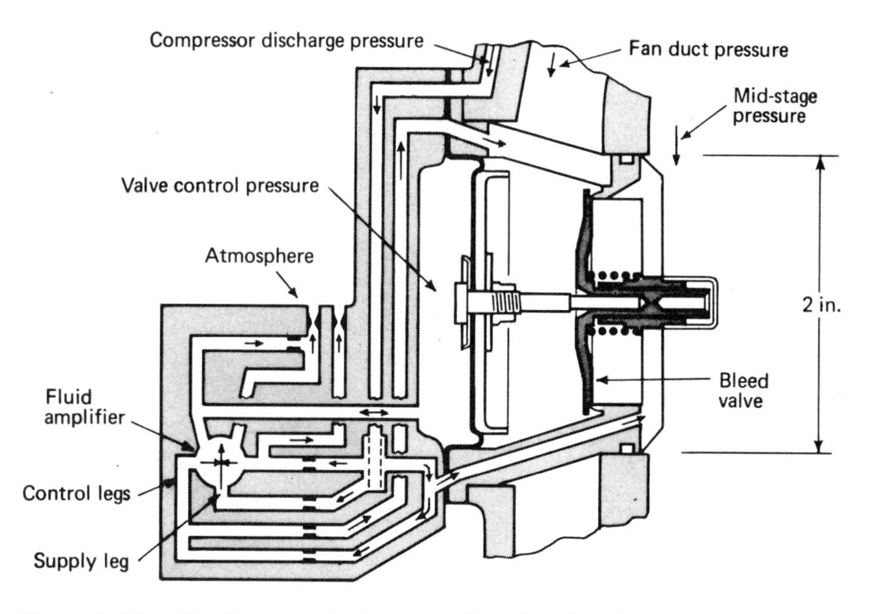

Fig. 14.22 Fluidic control of a gas turbine fan-duct bleed valve. (Courtesy of Bendix Corp., Lewisburg, WV.)

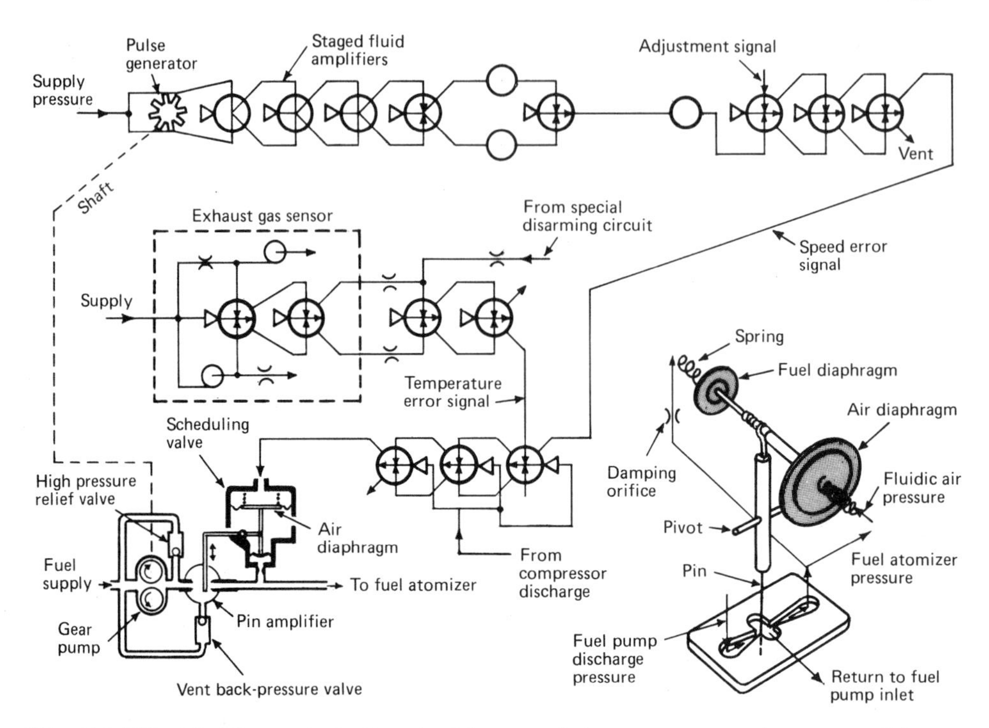

Fig. 14.23 Fluidic control of gas turbine speed, temperature, and pressure. (Courtesy of AiResearch, Div. of Garrett Corp., Phoenix, AZ.)

produces a differential pressure that is fluidically amplified and summed with the speed error signal and with a pressure proportional to compressor discharge. The final signal is pressure to a fluidic pin amplifier (patented by AiResearch) that modulates fuel directly.

The pin amplifier (Fig. 14.23) basically interposes a pin in the nozzle section of a fluid amplifier, thus allowing or diverting the flow depending on pin position. The pin is moved around a pivot point with fluidic air pressure against a diaphragm. There is extremely low hysteresis, high gain, and linear response. Accuracy of the fluidic system is very good, holding set point within 1.7%, compared with 3% for conventional proportional gas turbine controls.

Air Velocity and Flow. Airstream pressure drop or oscillation is the key to direct fluidic measurement of flow velocity. Two oscillating types are being developed for gas turbines. One is a fluid amplifier with flow feedback to set up oscillations. The other is a passage with an obstruction that sets up trailing vortices (Fig. 14.24).

The trailing-vortices design relies on sensitive fluid amplifiers to detect the vortices. In an AiResearch version a piezoceramic transducer converts the pressure pulses to electrical pulses for counting and readout.

Fluidic Ear. A subminiature relative of sonar—an acoustic ear—proved itself more versatile than either air jets or electronic sensors for detecting objects and then measuring their dimensions (Fig. 14.25).

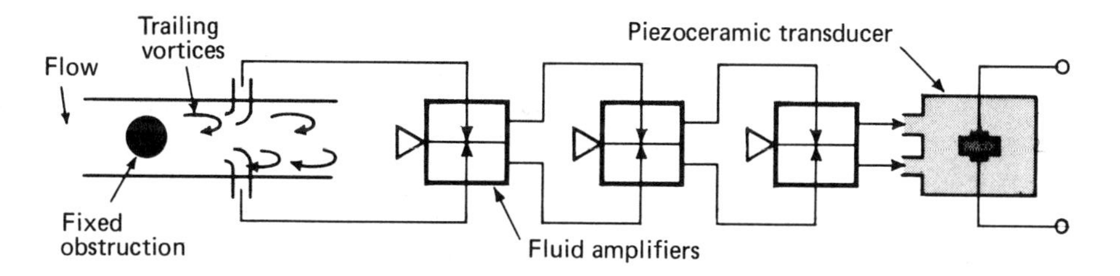

Fig. 14.24 Fluidic sensing of flow by counting frequency of vortices. (Courtesy of AiResearch, Div. of Garrett Corp., Phoenix, AZ.)

The device was designed by Pitney-Bowes, Inc. (Stamford, CT). It's an ultrasonic no-moving-parts fluidic whistle in combination with a tuned, sound-sensitive amplifier. A 50-kHz signal is generated in the whistle. This signal is a purely acoustic wave—not in any way a jet of air. In fact, no airflow needs to be received by the ear, and the designers even interposed a "muff" of 0.0005-in.-thick plastic to keep the ear dust-free. The acoustic signal can travel right through the membrane.

The only air jet needed is in the transmitter, which creates the acoustic wave. This jet flows directly to discharge: The acoustic wave proceeds at right angles from the jet. It is a whistle phenomenon based on the edgetone effect, in which the jet oscillates laterally approximately one-half wavelength between the edges.

Blowing on the acoustic wave will not disturb it, a feature that distinguishes it from the conventional air-jet sensor. The sound-sensitive receiver (fluidic ear) is an exponential horn coupled to a fluidic amplifier. The horn collects the acoustic wave and concentrates it at a control point in the cavity of the amplifier.

No extra noise: The receiver is directional and will be affected only by an acoustic wave or reflected wave aimed at the horn. The amplifier is designed to be sensitive only to 50 kHz, so it effectively filters out extraneous noise, which usually falls outside this region.

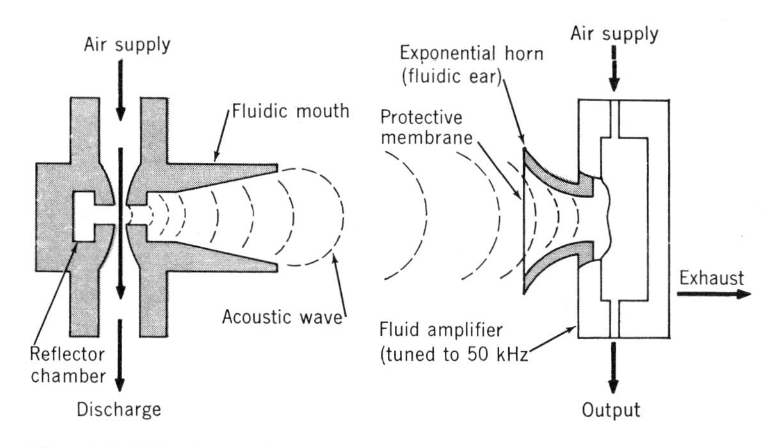

Fig. 14.25 Acoustic wave sensor (fluidic ear) detects the presence or absence of obstructions. (Courtesy of Pitney Bowes, Inc., Stamford, CT.)

The fluidic amplifier into which the wave is fed has output only in the absence of the wave, because the wave destroys the output. Thus it is a NOT-LOGIC device.

In applications where limit switches, photoelectric sensors, or conventional air-jet sensors prove unsatisfactory, the fluidic ear serves as a good presence detector. It is not affected by vibration, high ambient temperatures (when fabricated from special materials), or radio frequency interference, and it has an exceptionally high degree of immunity from contaminated environments.

15

Pneumatic Cylinders, Cushions, Actuators

There are many physical similarities among pneumatic actuators, hydraulic actuators, and shock absorbers—both the linear and rotary types. But the chapters on these are kept separate in this book because the application problems are considerably different (see Chaps. 5, 6, and 8).

Air is highly compressible; oil is not. Air is a poor lubricant; oil is a good lubricant. Air is not a fire hazard; oil is often a fire hazard. Air is expensive to compress; oil is easy to pressurize.

Where the applications are similar, the design information will appear in more than one chapter. If you don't find it in this chapter, look in one of the others.

PNEUMATIC CYLINDERS

The simplest air component is a cylinder, and engineers claim to know how to use it . . . but do they? Too many engineers we questioned said, "It's all explained in the catalogs." Yet manufacturers admit that users frequently ignore one or more rules such as in-line load forces, oil-free exhaust, no-sag rods, properly keyed mounts, and bounce-free cushioning.

The basic errors in application add up to over 30—not necessarily all by any single user. See the checkoff list (Table 15.1) and typical problems (Fig. 15.1). Some of the solutions are pictured in Fig. 15.2.

Off-Axis Loading

A leading mistake is to forget that pneumatic cylinders are push-pull devices and are not load guides. Sideloads and bending moments should be taken care of separately, or at least designed into the cylinder structure beforehand.

The ideal application has the cylinder, the rod, and the load mounted in line and rigidly constrained to stay in line. If the load moves otherwise in relation to the

Table 15.1 Typical Air Cylinder Problems and Solutions

Category	Problems	Solutions
Off-axis loading	Side loads; cocking; buckling	Heavier bearing; stop tubes
Misjudged load conditions	Mass; acceleration; extreme environment	Resize cyl; increase pressure; call manuf.
Cushioning	Shock; bounce; noise	Cush.; valv.; muffler
Speed, pressure, position	Too slow; too fast; sudden peaks; inaccuracy; drift	Meter-in; meter-out; position feedback; indexing cyl's; hyd. control
Lubrication	Too little, too much; wear; incompatibility	Positive lube; self-lube; oil-less
Contamination	Water; oil; dirt	Separators; dryers; filters
Mounting and stress	Thread stress; thermal expansion; sag	Rolled threads; shoulders; proper keys and mounts
Efficiency and conservation	Leakage; wasted pressure; dead volume	Seals; dual pressure; integral valving

piston rod, the art and science of vector diagramming must be brought into play, and the fun begins.

Every manufacturer has a variety of mountings and accessories, including clevis (hinge) mountings, universal rod eyes, lugs, trunnions, keys, and stop tubes to accommodate any load path and keep the rod in line with the load—or almost always.

The nature of the load is oft forgotten, according to experts. It is easy to forget the friction or inertia of a load—until the piston stalls or smashes into the head. Vertical loads and accelerating forces are many times misjudged. The changes in friction, flow, and alignment occasioned by ambient variations are another source for error.

Mounting and Stress

A cylinder is a structure and needs stress analysis as does any dynamically loaded member. Most manufacturers have done the analyses beforehand and offer application guidelines in their catalogs. Use them.

Always draw load vectors, moment diagrams, and rod travel curves before settling on cylinder mounts (Figs. 15.3 and 15.4) and attachments. Misalignment and sideloads can drastically shorten the life of a cylinder.

A fully extended rod is subject to high moments at the rod-piston joint when sideloads are high. Even the weight of the rod can cause problems. Specify stop-tubes

(Fig. 15.2) to ensure sufficient supporting moment without overstressing the rod bearings or piston. Ask for hollow rods if extreme weight reduction is needed.

A retracting rod suffers most when the load path is parallel with but offset from the axis of the rod path. The sideload gets more severe the closer it approaches the cylinder, where there is less and less rod length to flex.

Sagging rods are sometimes a problem, but not necessarily because of stress. If the rod is long and must be kept straight, it can be prestressed to stay horizontal at full extension. Miller, for example, offers such a rod in its catalog.

Rod buckling is a problem only if manufacturers' clear instructions in manuals and catalogs are ignored. If in doubt, pick next largest size rod.

Another possible problem is noncenterline mounting. Unless a cylinder is mounted along the axis of the load (centerline mounting), a bending moment results, equal to (load force) × (offset of axis from mounting surface). A short cylinder suffers most because the balancing moment arm (cylinder length) is short and the bolt tension as a result is high.

Take great care in selecting mounting styles, pivots, rods, extra supports—whatever is necessary.

Examples: Make sure that the clevis on the cap end is in the same motion plane as the clevis on the rod end. Avoid self-aligning mounts for trunnions, because trunnions are designed for shear and not bending. Prevent bending of threaded piston-to-rod or load-to-rod joints by providing shoulders to tighten them against.

A cylinder itself can provide some self-alignment if the rod head is mounted to

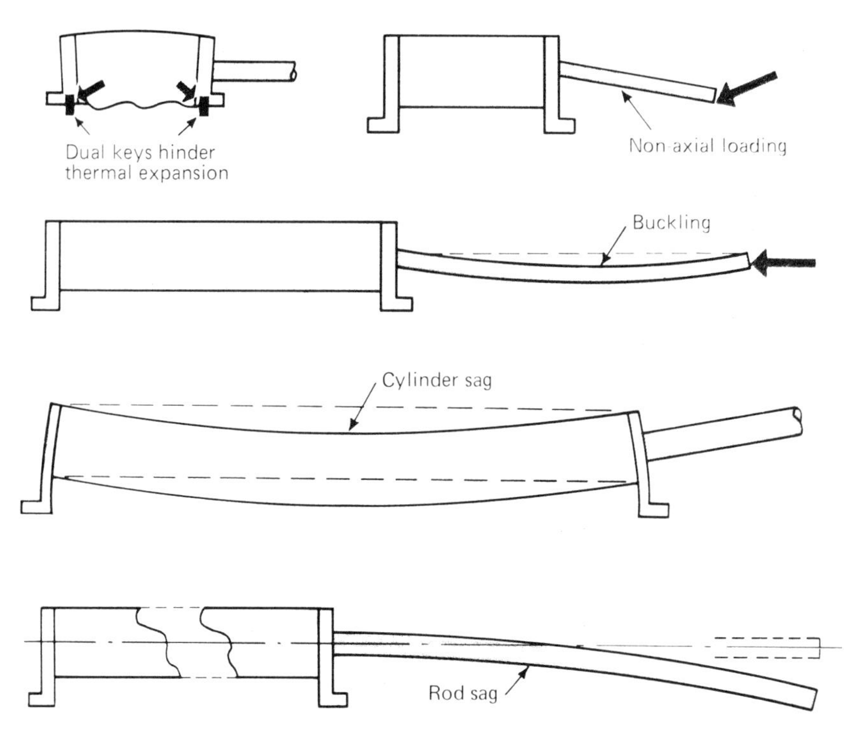

Fig. 15.1 Typical problems of misapplication experienced by users of pneumatic cylinders.

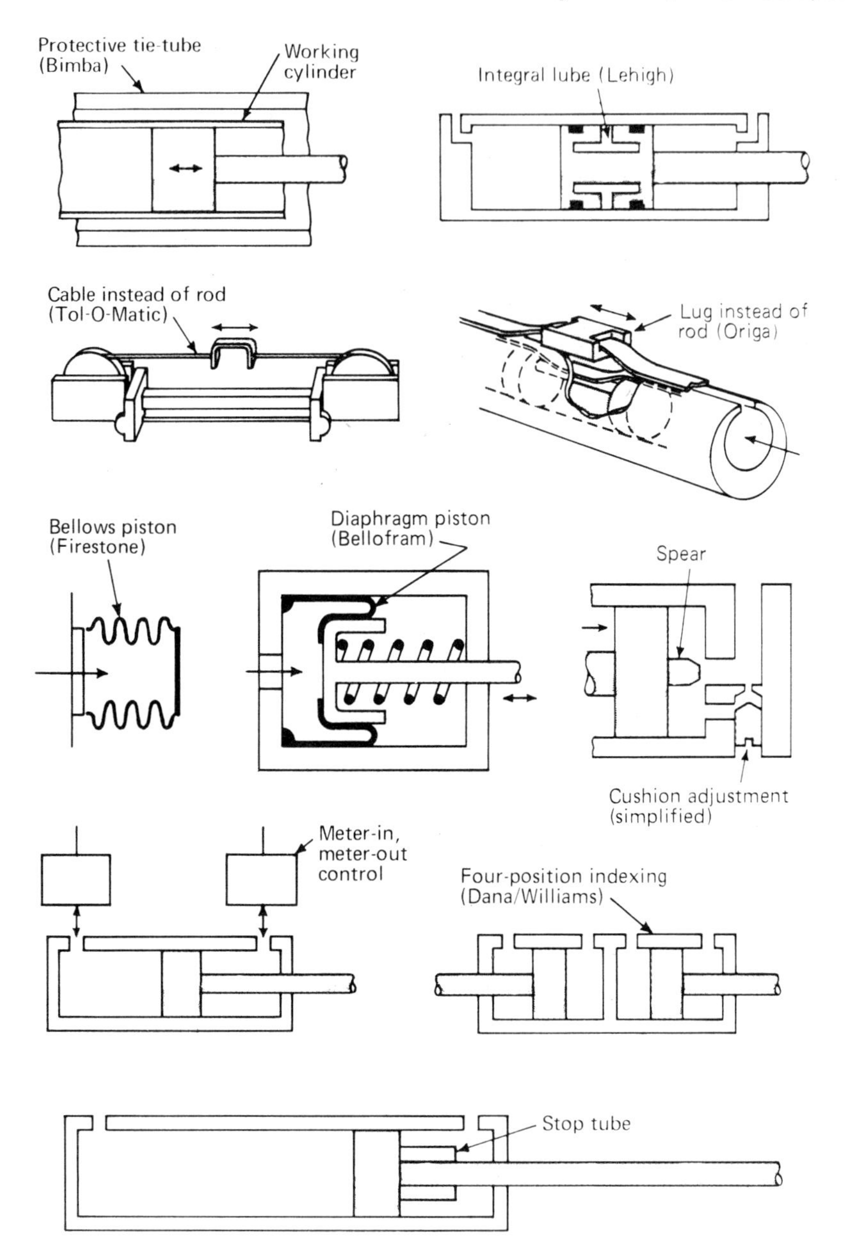

Fig. 15.2 Special cylinders solve specific problems such as physical abuse, lack of lube, cramped space, bounce, and sideloads on rod.

float. It can be done by drilling the mounting hole in the cylinder lug oversize relative to the positioning dowel in the baseplate.

Shock loads can be taken in part by shear keys, pins, or plates in the cylinder mounts (drawings). Just be sure not to put them at both ends of the cylinder (they restrict thermal expansion) or at diagonally opposite corners (distortion under load will occur).

Put the keys or pins at the end which exploits the cylinder elasticity the most (Fig. 15.5). If the shock load is likely to be in compression (upper view), put the key or keys at the rod end. If the shock load will be in tension, put the key or keys at the cap (blind) end.

Pressure shock loads have another effect; sudden mechanical expansion of the cylinder barrel. The stresses may be well within the design limits for ultimate strength, but the seal clearances might briefly open up enough to cause blow-by or mechanical damage to the seals. If there is danger of this, pick a higher-pressure cylinder (called derating).

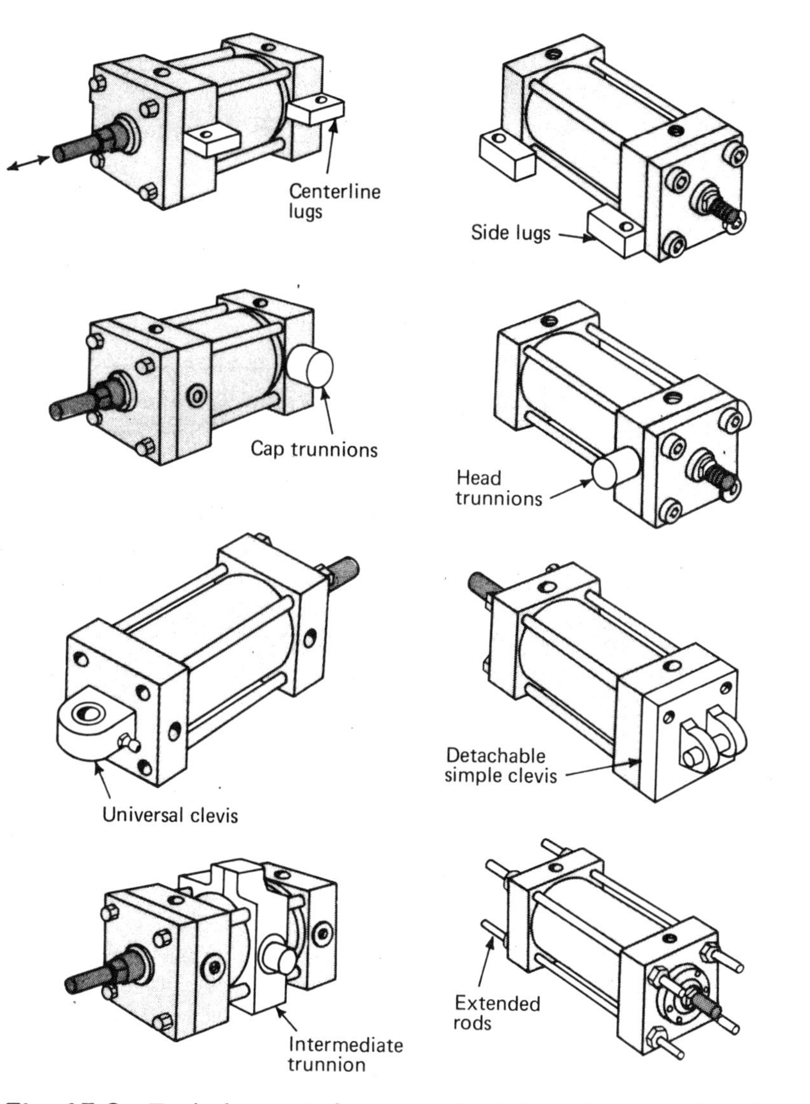

Fig. 15.3 Typical mounts for square-head tie-rod pneumatic cylinders. Also see Fig. 15.4.

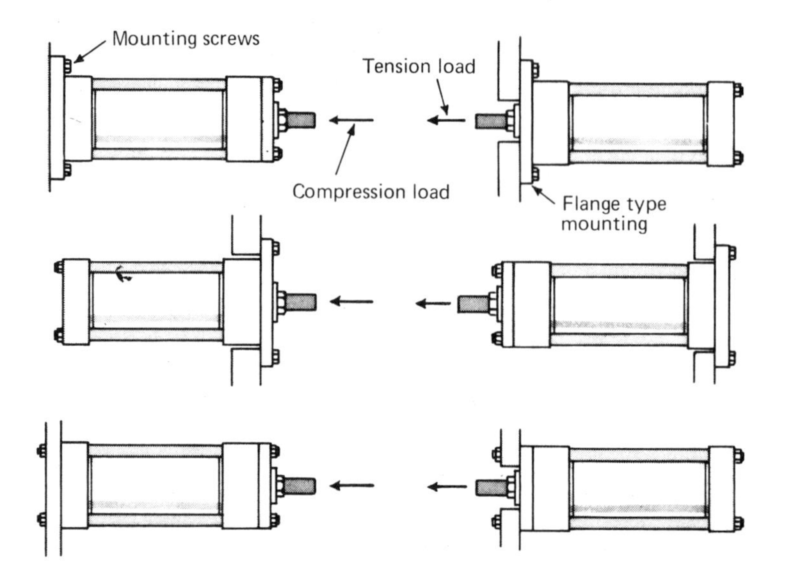

Fig. 15.4 Mounting screws for flange-type mounts are protected from tension stresses when arranged as shown. (Courtesy of Schrader-Bellows Div., Akron, OH.)

Position Feedback

The position of any cylinder can be simply controlled by a limit switch in the rod path or by magnetic or electronic sensors mounted on the cylinder itself to detect piston position. Some have a flexible magnet in the piston assembly to give the sensor something to detect. In any event, the position signal can open or close a valve to slow or stop the piston at any point.

For more sophisticated control of position, a pneumatic feedback servo can be applied. SMC Pneumatics offers one (Fig. 15.6), and the operator can call for any piston position by setting pilot pressure remotely. It works like this:

With zero pressure to the pilot, supply air to the spool pressurizes both diaphragms. However, pressure unbalance occurs because of air escaping through the nozzle, and the valve spool moves to the right. Supply pressure is directed to the rod end of the pneumatic cylinder, moving the power piston to the left (retraction). A neutral position is reached when the diaphragm pressures balance and the valve spool is recentered.

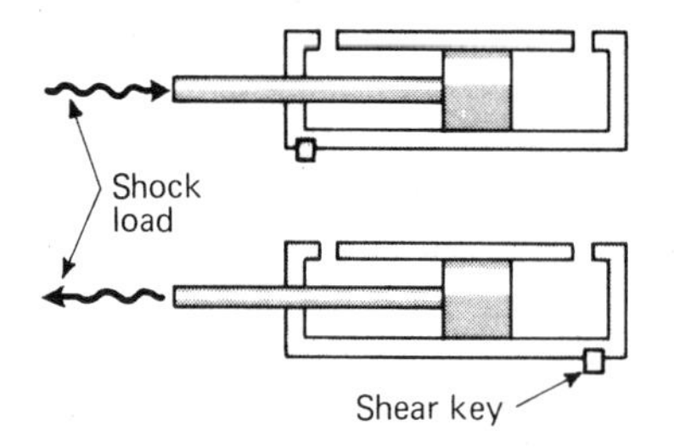

Fig. 15.5 Shear keys absorb shock best when they exploit cylinder elasticity as shown.

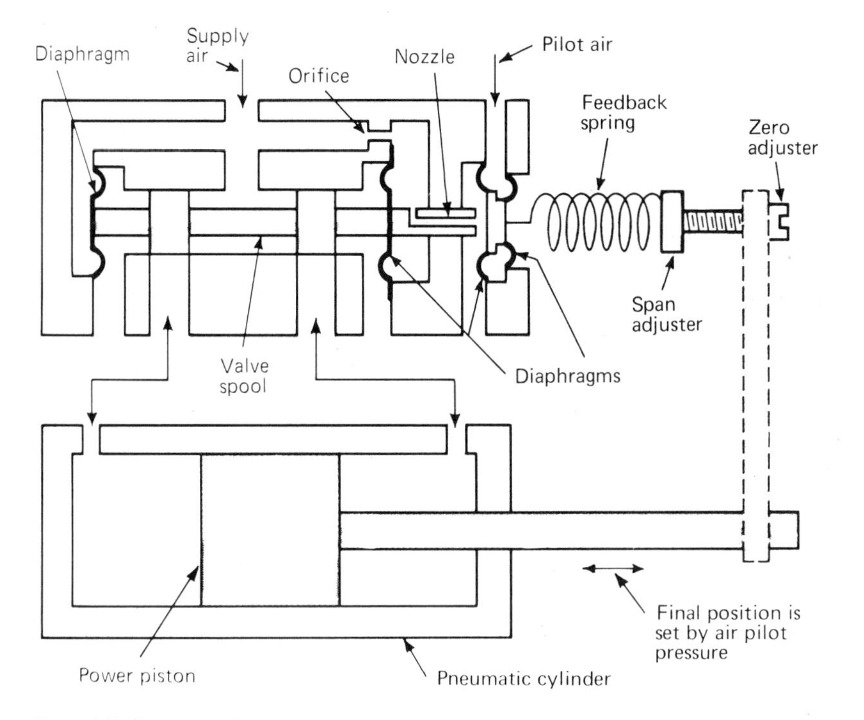

Fig. 15.6 Position feedback from the pneumatic cylinder adjusts the valve to control piston motion and the final stopping point (Courtesy of SMC Pneumatics, Inc., Indianapolis, IN.)

When pilot air pressure is introduced, the pilot diaphragm moves toward the nozzle, restricting flow. Back pressure increases against the valve spool, moving it to the left. Now, the power piston is pressurized to move to the right (extend). Movement continues until the feedback spring force moves the pilot diaphragm to the right enough to rebalance the valve spool at the neutral position. The final position of the power piston is a function of pilot pressure.

Speed, Pressure, Position

Control of air cylinders is harder than control of oil cylinders because air is compressible. It expands fast but builds up cylinder pressure gradually, compared with oil, which usually is flow-limited by pump capacity but builds up cylinder pressure fast.

Air piston speed is a result of three forces: driving pressure, back pressure, and load forces, including friction and inertia. None alone can control speed. Calculation of speed and acceleration, even given all the conditions, is uncertain.

A good first step is to put a flow control valve in the exhaust (meter-out). This will help establish back pressure. Driving pressure can be set with a pressure regulator, which will maintain value even if flow varies.

Meter-out control can be a fixed restriction, a needle valve, a flow control valve, or a servovalve with position feedback. There are plenty of examples from one extreme to the other.

Some manufacturers offer packaged cylinder-with-control units that are ready to hook up to an air supply and run. Miller, for example, calls its package a linear air motor.

Another approach is do-it-yourself. Suppose you want two adjustable rates of speed: fast for traverse and slow for end of stroke. Two flow control valves will do it, and a simple way to switch from one flow rate to the other is with a three-way limit valve and a check valve (Fig. 15.7). Piston motion is mildly restricted by the fast flow valve until the rod actuates the limit valve and diverts cylinder exhaust through the slow flow valve. Return stroke (retraction) in this example is unrestricted.

The most precise slowdown is with hydraulics. An example is the Hydro-Check (Fig. 15.8) manufactured by Schrader-Bellows. The moving piston rod of the pneumatic cylinder comes in contact with an adjustable stop on the hydraulic cylinder piston rod.

A self-contained hydraulic circuit with adjustable restriction slows the hydraulic piston to a rate set by the control knob. The restraint can be in either direction or in both directions.

Lubrication

A perfect answer so far eludes the industry. The heavy users of air cylinders impose strict regulations on themselves and on their suppliers to avoid certain hazards: starved cylinders, excessive lube, incompatible lube. Some choose to design and build their own cylinders, with prelube for nonlubricated service.

Three levels of lube (or nonlube) exist: lubricated (lubricant added during course of operation), prelubricated (lubricant installed at time of assembly and sufficient to last for an extended period), and nonlubricated (no lubricant installed or added).

The first two provide reliable lubrication for at least millions of cycles; the last is a question mark. The name of the game is compromise.

The crudest lubricating technique is to rely on the normal water and oil content of shop air, and this method actually is accepted in some plants.

The most direct technique is to provide lubrication integral with the cylinder (Figs. 15.9 and 15.10). This includes miniature reservoirs within moving pistons, lubricated wicks, stored grease in grooves, and direct lubrication under pressure at each friction point.

The most controversial yet the most widely accepted way to lubricate cylinders is with atomizing-type lubricators in the compressed air supply lines. Dozens of com-

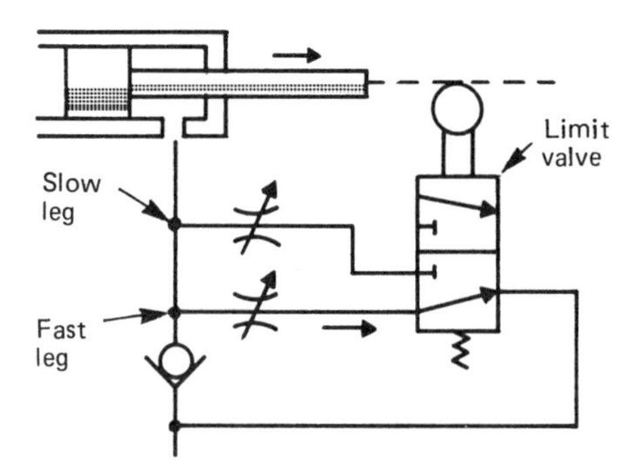

Fig. 15.7 Cylinder slows down after the limit valve is actuated.

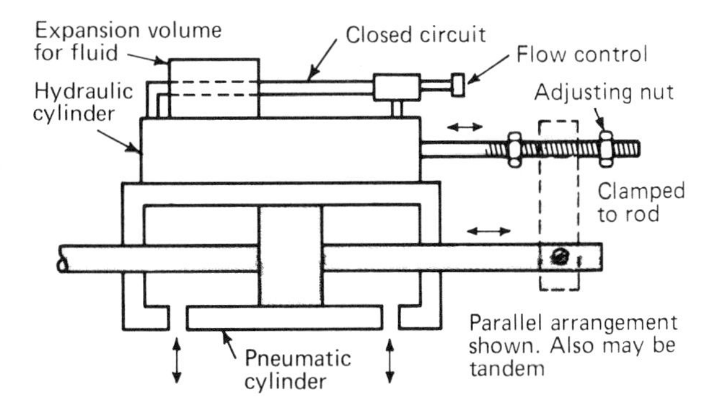

Fig. 15.8 Hydraulic retarding circuit with a flow control valve stops the pneumatic cylinder. (Courtesy of Schrader Bellows Div., Akron, OH.)

panies, particularly the manufacturers of FRLs (filter-regulator-lubricators) have developed ingenious designs to go about this. (See Chap. 11.)

The controversy arises partly from the difficulties in getting maintenance people to keep the lubricators properly filled with oil and adjusted in a complex factory environment. An empty lubricator doesn't lubricate. Or even if it works, will some cylinders get excess oil, and others go dry?

A second problem is washout of prelube when the air-line lubricant passes through the cylinder. If the lubricator runs dry afterwards, it leaves the cylinder without lubricant.

Another controversial point is mist carryover. Some major manufacturing plants refuse to use in-line lubricators in certain areas because even small amounts of oil mist if allowed to exhaust will adversely affect operations elsewhere.

In any case, if the oil is introduced as mist, it must be reclassified back to liquid at the cylinder. Most air-powered devices do this automatically as the mist enters with the air. However, some of the mist goes straight through and out the exhaust and must or should be collected.

Slippery Seals

Elastomer seals and wear rings with built-in lubrication qualities are reducing the need for oil or grease in many light-duty applications of air cylinders. But only certain elastomers have the combination of slipperiness, resilience, mechanical strength, wear resistance, and temperature tolerance to meet tough industry requirements. Also, the metallurgy, hardness, and finish of the mating cylinder bores are critical for proper performance and life. (See Chaps. 19 and 20.)

Fluorinated elastomer is one type of slippery seal. Buna-N compounded with dry moly lubricant (molybdenum disulfide) is another.

A different way to apply solid lubricant is to impregnate it in the piston or in the cylinder barrel material itself. An example is an air cylinder made of glass-fiber reinforced epoxy resin containing a homogeneous dispersion of low-friction additives (solid lubricants). Another is a Teflon-impregnated aluminum cylinder tube and a reinforced Teflon piston wear ring. The rod gland can have an oil-filled sintered bronze bushing or be chemically etched to hold moly grease.

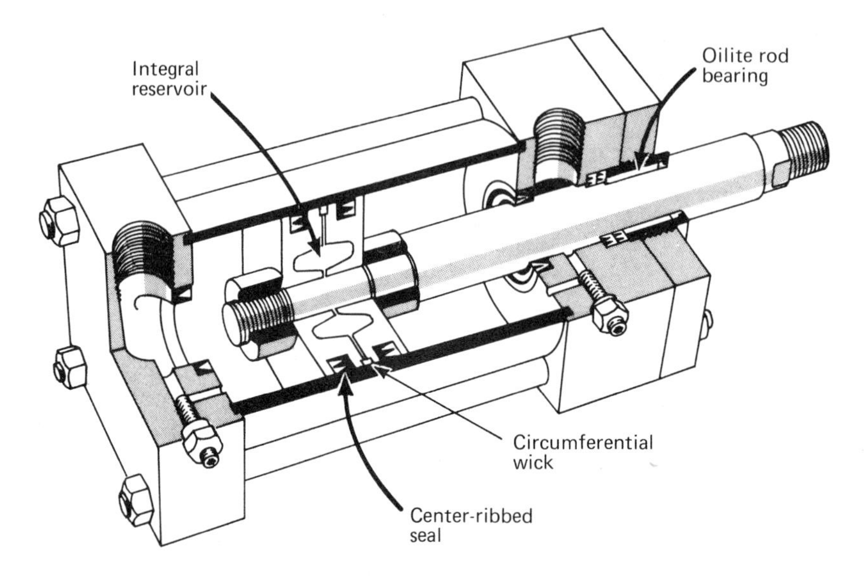

Fig. 15.9 Integral oil reservoir supplies lubricant to the pneumatic cylinder. (Courtesy of Lehigh Fluid Power, Lambertville, NJ.)

Totally nonlubricated air cylinders are not recommended by most experts except in very special situations. The problems are as follows: Surface finishes must be excellent, tolerances on sizing must be precise, the air or gas used must be free of contaminants, and the operating speeds, pressures, and sideloads must be carefully considered and totally predictable.

Grease cups are a reasonable answer if the other methods fail. Some designs automatically dispense the grease by spring force, pneumatic piston, or chemical action. The biggest problem is remembering to check the results frequently enough. Failing in that, expect to (a) starve the cylinder or lubricant or (b) clog the air passages with excess grease.

Whatever lube is chosen, make sure that the seals are compatible with it. There's a lot of strange chemistry around, and seals swell or deteriorate if not matched to the lube.

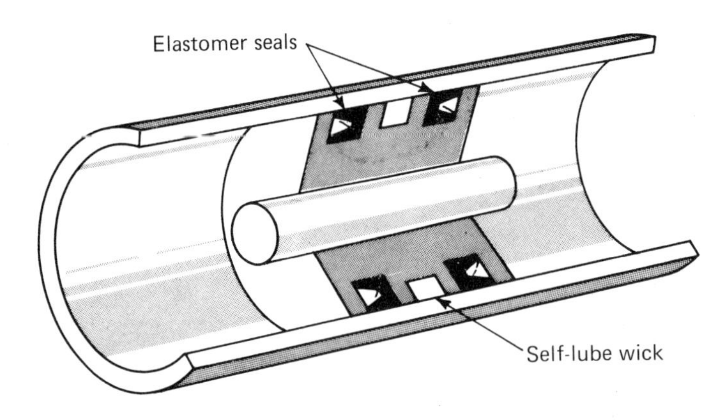

Fig. 15.10 Oil-impregnated wick lubricates the pneumatic cylinder.

And while you're at it, make sure that no other harmful contaminants get into the cylinders. Shop air lines are collecting points for a lot of water, scale, and burned lube from the compressors. Proper air systems have pipes that slope down about 1 in. per 10 ft and have drop-legs (tee-offs) from the top of the line to pull off air for cylinder use without dragging water and dirt with it (see Chap. 11).

Follow that with filters and regulators as required, and top it off by cleaning up cylinder exhaust and quieting it down before discharging it back to the human environment.

Special Designs

Cylinder manufacturers seem willing to build custom designs, or at least that's the claim in their catalogs and literature. We studied the variety of designs already available and concluded that there must be a cylinder type to solve almost any known problem. The illustrations in Figs. 15.11 through 15.14 show a few.

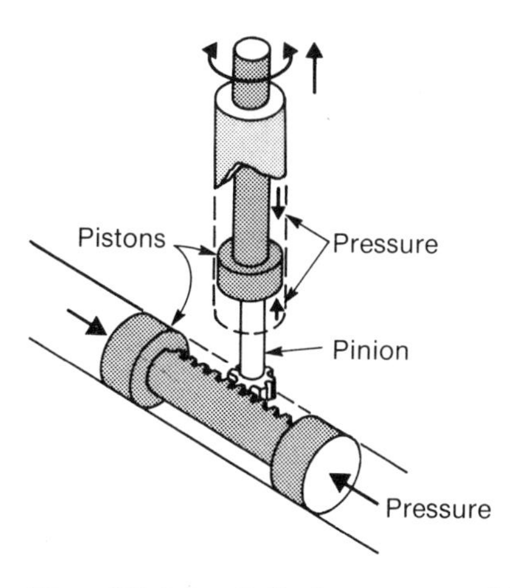

Fig. 15.11 Cylinder-gear combination creates both linear and rotary motion. (Courtesy of PHD, Fort Wayne, IN.)

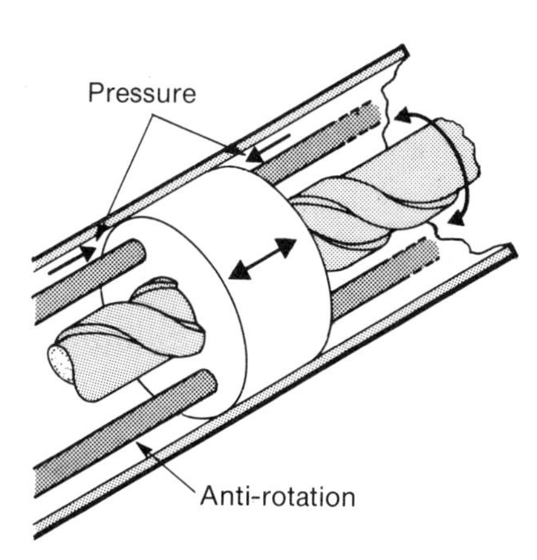

Fig. 15.12 Rotary rod motion can be generated with a linear piston.

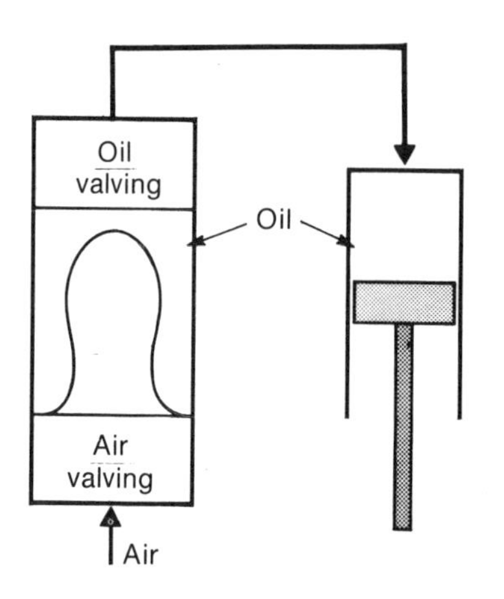

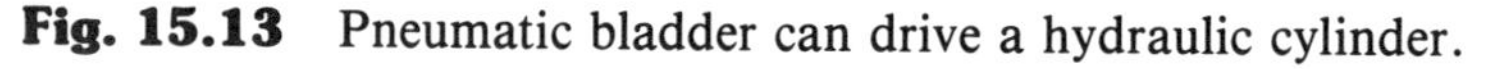

Fig. 15.13 Pneumatic bladder can drive a hydraulic cylinder.

In addition, there are tiny cylinders such as a ¼-in.-diameter model by SMC. And there's the Adjust-a-Stroke air cylinder by Flair-Line, which has floating heads inside the cylinder; they can be moved hydraulically to shorten or lengthen the effective length of the cylinder. Each floating head in the Flair-Line cylinder is energized and held with oil pressure fed in through the fixed heads. The pneumatic piston between them is energized through its own hollow piston rod.

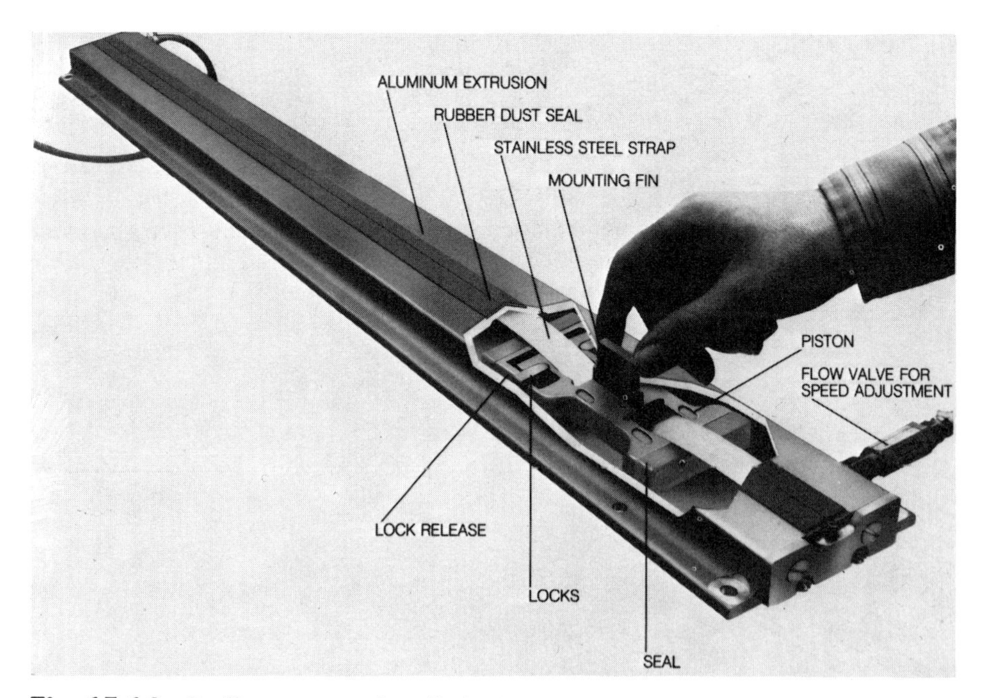

Fig. 15.14 Rodless pneumatic cylinder has a fin to carry load. A stainless strap seals in the compressed air when forced against the slot in the extrusion. (Courtesy of Abex/Hansen, Gurnee, IL.)

AIR-SPRING CAM FOLLOWER

This application example (Fig. 15.15) combines pneumatics and mechanisms to solve a machine motion problem.

Mechanical cams are great for positive motion and accurate positioning but are unforgiving when something jams. The answer, developed by engineers at Android Corp. (Auburn Heights, MI), is a cam-driven motion where the follower is held against the cam by an adjustable air spring.

A big advantage, besides eliminating "crunching" of parts during a jam, is force adjustment. By reducing air pressure in the spring cylinder, the operator can run the machine (a precision parts assembler in this instance) through its paces under reduced load, or even manually, for checking purposes. When full force is required to advance the machine table, pressure can be adjusted upward.

The air circuit design was developed by Doig Associates (Bloomfield Hills, MI). Air is supplied to a pair of valves in tandem via a regulator, which has the ability to relieve downstream pressure almost as fast as it can supply air. This keeps the air pressure in the blind end of the air spring virtually constant.

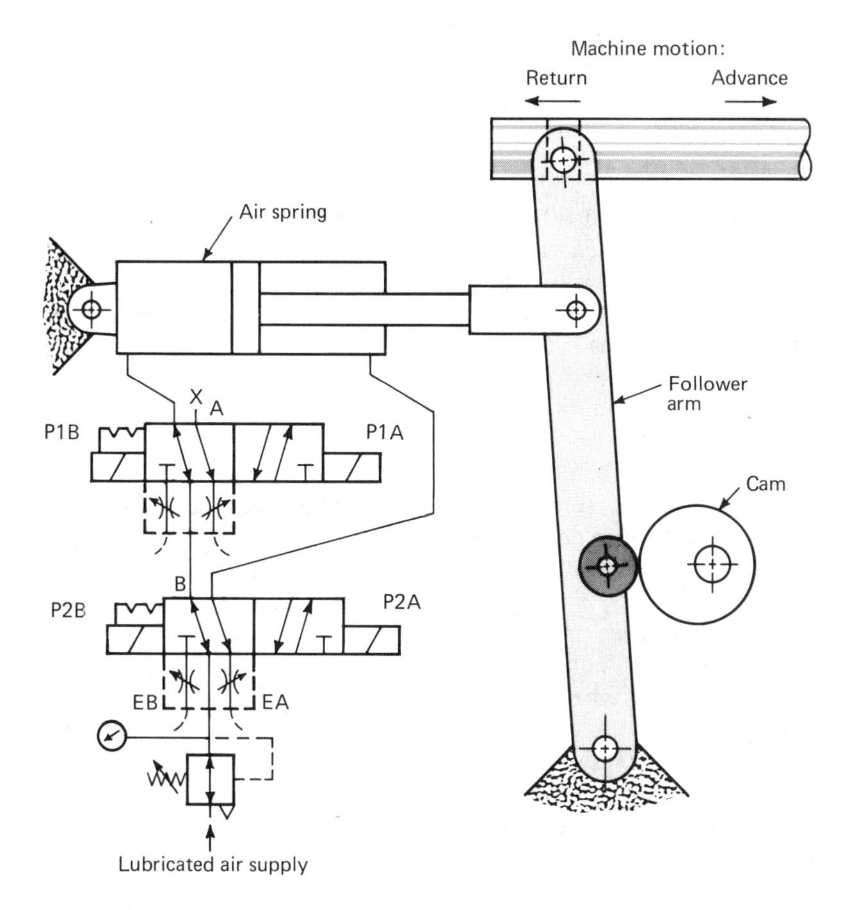

Fig. 15.15 Pneumatic cylinder used as an air spring keeps a steady load on the follower arm throughout the cam cycle. (Courtesy of Doig Associates, Bloomfield Hills, MI.)

Doig says that several previous problems were solved with the new design. Air consumption is one. In conventional air-cylinder applications the working cylinder is filled and emptied each cycle, whereas in this new design it is only charged and exhausted a few cubic inches—the amount needed to follow the short excursion of the cam.

Lubrication was feared to be another problem. How could lube be carried into the cylinder with such a small volume of air? The answer was simple: The lack of airflow prevented washout of the prelube, and the cylinders are working fine. Just to be sure, the air regulators were placed above cylinder level so that oil fog would plate out on the walls of the tubing and thereby run downhill to the cylinders.

Safety was an early worry, because of the energy stored in air that could cause unscheduled motion of the machine. The solution was the circuit shown. Direct-solenoid-operated detented valves in compact modular array ensure that each main spool will remain where it is placed and that no electrical or air failure can cause valve motion. There are no mechanical springs or air pilots.

Flow valves control the speed of piston movement except in the unusual instance when air pressure is purposely removed from both ends of the cylinder for repairs or adjustment. The cylinder, however, must be repressurized in a specific manual sequence, starting at the return position, at which instant the piston will strike the blind end harmlessly.

CUSHIONING AND SAFETY

Pneumatics is faster and can be more dangerous than hydraulics because air expands rapidly and carries piston and load with it. Stopping the load with simple cylinder cushions alone is tricky because air is highly compressible.

If the cushioning air pressure is not exactly correct, the piston will do one of several of these: slam into the head, not go anyplace, move backwards, or bounce back and forth near the end of stroke.

If the load conditions—including friction—change after the system is built and shipped, all the good design work might be for naught.

Moderate-duty cushioning can be handled with built-in adjustable restrictions and elastomeric bumpers (Fig. 15.16). Heavy-duty cushioning is another matter, and equipment designers often choose one of these alternatives: let the load absorb the shock external to the cylinder, mount a hydraulic decelerator integral with or mounted next to the pneumatic cylinder, or incorporate feedback control to automatically adjust driving and cushioning air pressures with flow control valves.

For moderate applications, adjustable internal air cushions are offered by most manufacturers. A typical spear type is shown simplified in Fig. 15.2. (For examples of other types, such as stepped, tapered, orificed, or spring energized, see Chap. 5.)

To make sure the spear stays centered, a floating bushing can be incorporated in the opening. To speed up the return stroke after cushioning, an integral check valve can be added in parallel with the cushion adjustment. Relief valves can be specified to limit the pressure extremes.

Parker Hannifin has combined a check valve function with its floating bushing by leaving grooves in the lip seal on the side facing the piston. During return, the air easily passes around the seal and pressurizes the full face of the piston.

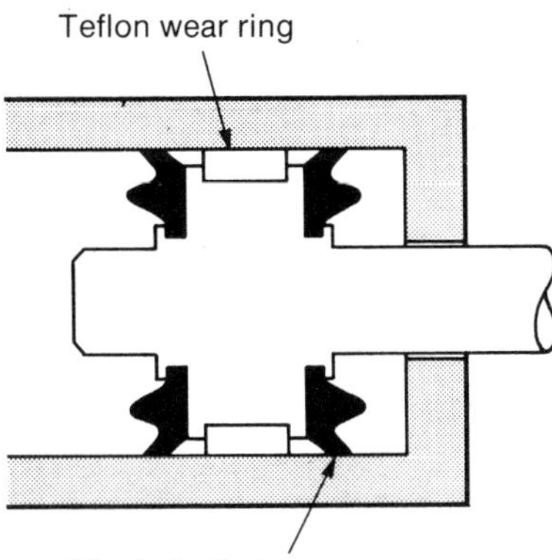

Fig. 15.16 Elastic bumpers on the pneumatic cylinder act as quiet shock absorbers at the end of stroke. (Courtesy of K. C. Mosier Co., Inc., Dayton, OH.)

Big cylinders have the greatest need for careful cushioning, and they have the most room to work with. Small cylinders (less than a few inches in diameter) often have fixed restrictions such as drilled orifices—or no special air cushioning at all.

AIR DASHPOT CALCULATIONS

Nondimensional parameters simplify calculations for the orifice area, maximum pressure, and stroke time of simple snubber-type dashpots (Fig. 15.17). Ordinary air-compression equations are nonlinear, and even with graphs to help you they take time and are prone to errors.

A method developed by Tom Carey and Ted Hadeler at Walter Kidde (Belleville, NJ) overcomes the difficulties by substituting dimensionless ratios for each of the key parameters. For instance, pressure is replaced by the dimensionless ratio P/P_i, which is actual pressure vs. initial pressure. And piston velocity becomes v/v_i, which is actual vs. initial velocity.

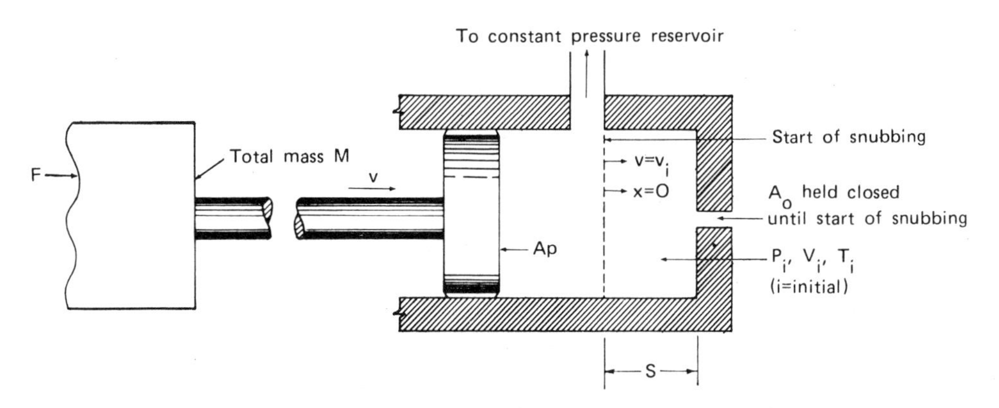

Fig. 15.17 Model of pneumatic snubber, used for accompanying calculations.

Eight of these dimensionless ratios, plus four simple parametric relationships (Table 15.2), and the charts in Fig. 15.18, enable you to quickly calculate orifice size for any given combinations of initial pressure, load mass, external constant force, snubber length and diameter, initial and final velocities, initial temperature, and gas constant.

Assumptions

The method has been validated by actual tests and will work for any snubber from pea-size to several feet in diameter. However, it must be used within these limitations:

- Only the piston-orifice type of dashpot is considered.
- The piston is slowed throughout the stroke and is firmly stopped when it reaches the end of the stroke. It does not rebound before the end, or oscillate or bounce, or increase in velocity during the stroke because of excessive external force.
- Friction is zero.
- The external force is constant. If the force will vary, then you must work out solutions for several force levels (say at zero, midpoint, and maximum) and select the calculated orifice size that is the best under the circumstances.
- Air, hydrogen, nitrogen, oxygen, and any other gas with a specific-heat ratio of about $k = 1.4$ can be computed with this method. Gases with other than $k = 1.4$ cannot.
- Contained air or gas is ideal.
- Compression is adiabatic.
- Flow through the bleed orifice is critical. This is a good assumption except when dashpot initial pressure = atmospheric, such as in snubbers for screen doors.

In summary, this is the problem that is solved: A load mass with a constant external force is slowed by a pneumatic snubber until it bottoms at full stroke. Velocity is positive throughout the stroke but reaches a minimum value at, or near, impact.

The accompanying charts give nondimensional velocity at the point of impact, as a function of nondimensional external force, pressure-volume stored energy of the snubber cylinder, and weight flow of the air or gas.

No data for nondimensional time are given because the choice of parameter ($K_t = v_i t/S$) proved fortunate in that it held a constant value of 0.95 ± 0.15 for every combination of variables Thus, within reasonable limits of accuracy, there was no need to graph the minor differences.

The maximum value of pressure occurs just before impact at the end of the stroke, and these points are recorded in Fig. 15.19. It is rarely necessary to know the intermediate points from P_i = initial to P_{max} = final, because cylinder wall dimensions are based on P_{max}.

How the Data Were Obtained

Dimensionless equations 11 through 17 comprised the mathematical model for a computer simulation. The dimensionless parameters were varied, and the output was recorded and plotted (Figs. 15.18 and 15.19).

Table 15.2 Nomenclature for Air Snubber Calculations

SYMBOLS

a = acceleration, in/sec^2

A_o = orifice area, in.^{2}

A_p = piston area, in.^{2}

C_D = orifice discharge coefficient (dimensionless)

F = constant force, lb (includes force of atmospheric pressure)

g = gravitational constant, in/sec^2

k = ratio of specific heats (dimensionless)

m = weight of gas in actuator, lb

M = total load mass, lb-sec^2/in. (includes piston and rod)

P = snubber pressure, psia

R = gas constant, in. lb/lb $-°R$

S = snubber active length, in.

t = time, sec

T = gas temperature, $°R$

V = snubber volume, in.^{3}

v = piston velocity, in/sec

w = orifice flow rate, lb/sec

x = piston travel, in.

ρ = gas density, lb/in.^{3}

SUBSCRIPTS

i = initial (at start of snubbing action) such as in v_i

p = piston

o = orifice

I—Basic equations

Force and motion

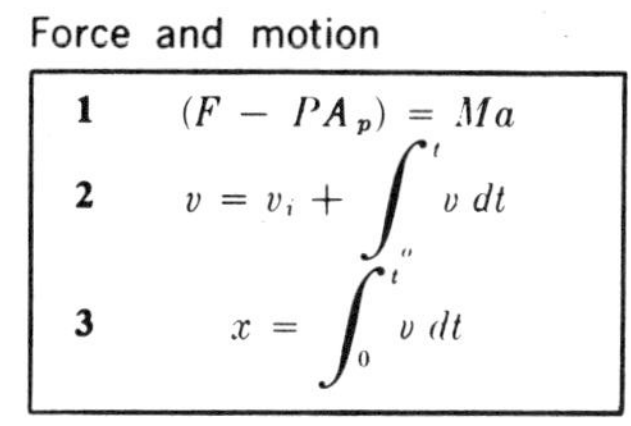

1 $$(F - PA_p) = Ma$$

2 $$v = v_i + \int_0^t v\,dt$$

3 $$x = \int_0^t v\,dt$$

Perfect gas law

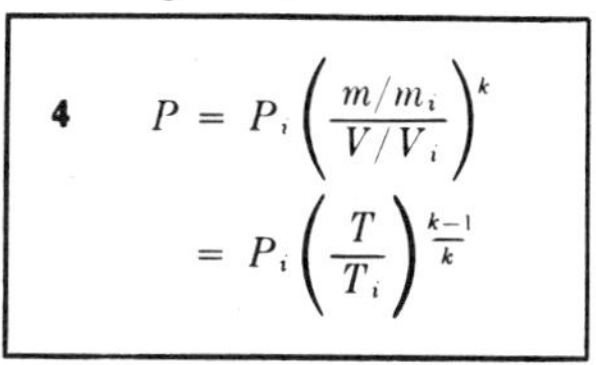

4 $$P = P_i\left(\frac{m/m_i}{V/V_i}\right)^k = P_i\left(\frac{T}{T_i}\right)^{\frac{k-1}{k}}$$

Gas flow through bleed orifice

5 $$w = C_D P A_o \sqrt{\frac{kg}{RT}\left(\frac{2}{k+1}\right)^{\frac{k+1}{k-1}}}$$

II—Dimensionless parameters

6	$K_E = \frac{P_i V_i}{M v_i^2}$	stored energy / kinetic energy
7	$K_F = \frac{F}{P_i A_p}$	constant external force / initial pressure force
8	$K_P = \frac{P}{P_i}$	pressure / initial pressure
9	$K_w = \frac{w_i}{\rho A_p v_i}$	initial orifice flow / initial displacement flow
10	$K_x = \frac{x}{S}$	piston travel / snubber length
11	$K_v = \frac{v}{v_i}$	piston velocity / initial piston velocity
12	$K_a = \frac{Ma}{P_i A_p}$	acceleration force / initial pressure force
13	$K_t = \frac{v_i t}{S}$	travel at initial velocity / snubber length

III—Parametric relationships

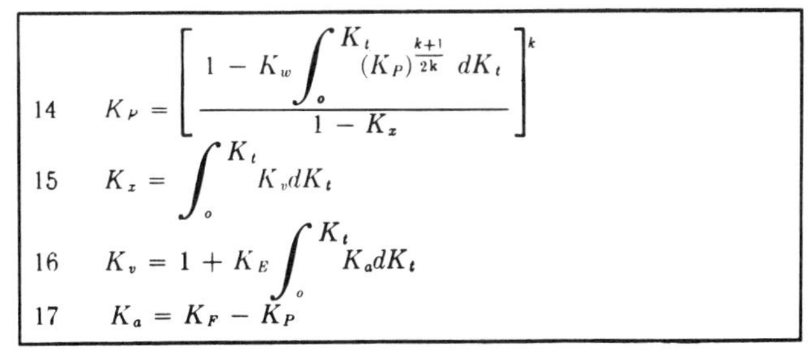

14 $$K_P = \left[\frac{1 - K_w \int_0^{K_t} (K_P)^{\frac{k+1}{2k}}\, dK_t}{1 - K_x}\right]^k$$

15 $$K_x = \int_0^{K_t} K_v dK_t$$

16 $$K_v = 1 + K_E \int_0^{K_t} K_a dK_t$$

17 $$K_a = K_F - K_P$$

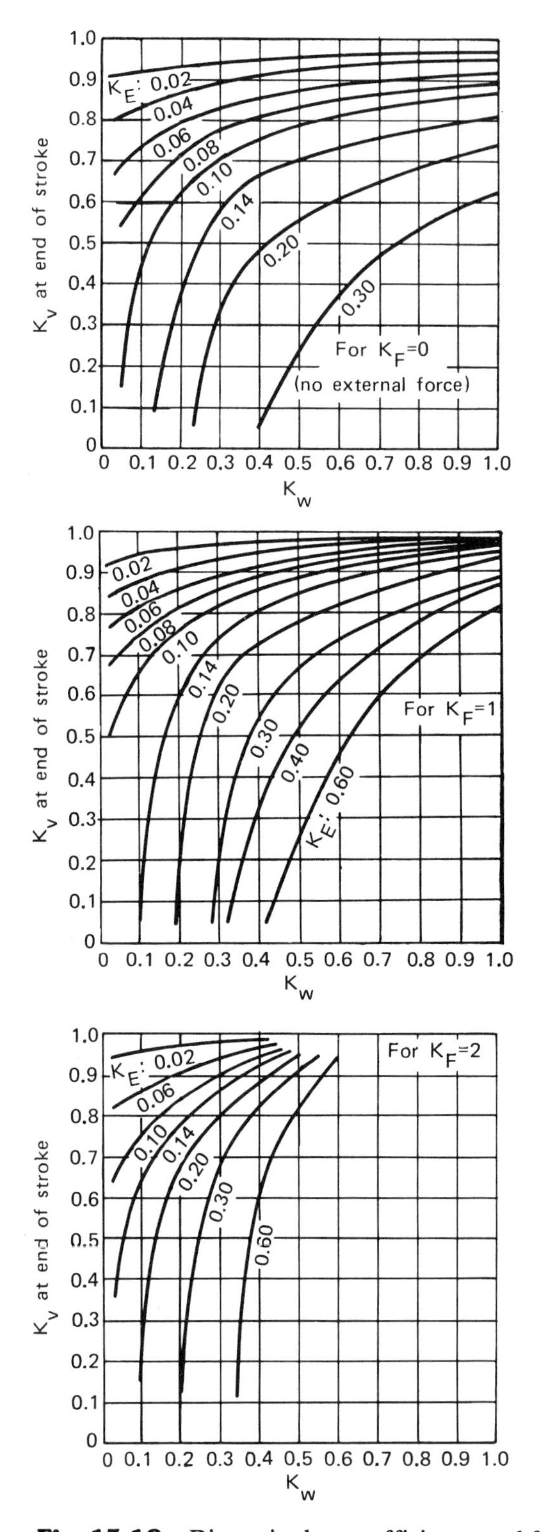

Fig. 15.18 Dimensionless coefficients used for estimating impact velocity vs. orifice flow of the pneumatic snubber in Fig. 15.17.

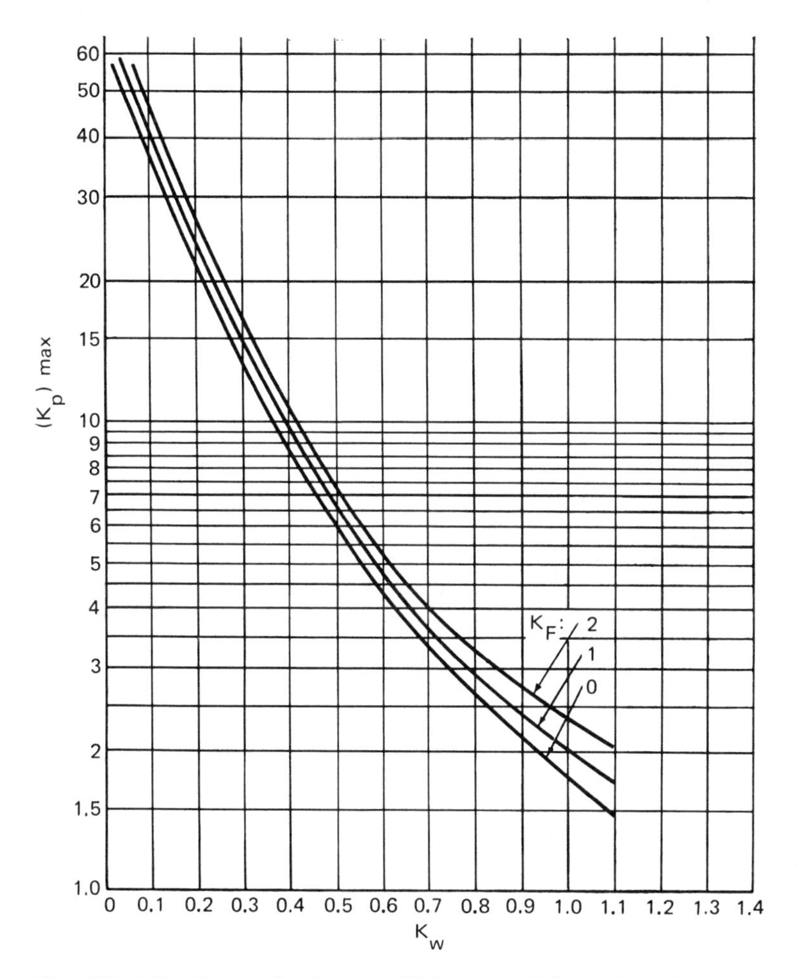

Fig. 15.19 Dimensionless coefficients used for estimating the maximum pressure at impact of the pneumatic snubber in Fig. 15.17.

Because the plotted parameters are dimensionless, each point represents an unlimited number of solutions which can be converted to your actual conditions by simple algebra.

Calculated Example: 3-in³ Snubber

What is required orifice area A_O when $M = 0.1$ lb-sec²/in., $P_i = 100$ psia, $A_p = 3$ in², $V_i = 3$ in³, $S = 1$ in., $v_i = 100$ in./sec, v at end of travel (impact) = 20 in./sec, $F = 150$ lb, $T_i = 530°$ R, $R = 639.6$ in.-lb/lb-°R (air), and $C_D = 0.9$?

Also determine peak actuator pressure, peak acceleration (negative), and approximate time of stroke.

Solution

From equations 6, 7, and 11, $K_E = (100 \times 3)/(0.1 \times 100^2) = 0.3$; $K_F = 150/100 \times 3 = 0.5$; K_V at end of stroke $= 20/100 = 0.2$.

Now find the value for K_W at $K_E = 0.3$ and $K_V = 0.2$ (Fig. 15.18). Note that there is no chart for $K_F = 0.5$. Therefore you must interpolate between the charts for $F = 0$ and $F = 1$. Do it as follows: Prepare a chart of K_W vs. K_F as in Fig. 15.20 by reading values of K_W from each plot in Fig. 15.18. Read off $K_W = 0.375$ from Fig. 15.20.

Solve for true flow w_i in equation 9: $w_i = K_W \rho A_p v_i = 0.375(P_i/RT_i) \times 3 \times 100 = 0.0332$ lb/sec. Solve for A_O in equation 5, initial flow:

$$A_O = \frac{w_i}{P_i C_D \sqrt{\frac{kg}{RT_i}\left(\frac{2}{k+1}\right)^{k+1/k-1}}}$$

$$= 0.016 \text{ in}^2$$

Read off $K_P(\max)$ from Fig. 15.19, knowing that $K_F = 0.5$, $K_W = 0.375$: $K_P(\max) = 10.2$. Then true P_{max} (at end of stroke) $= 10.2 \times 100 = 1020$ psi, from equation 8. $K_a(\max) = K_F - K_P(\max) = 0.5 - 10.2 = -9.7$, from equation 17. Thus $a_{max} = (P_i A_p/M)(-9.7) = -29{,}100$ in./sec², from equation 12.

Travel time, based on K_t = about 0.95, is $t = (S/v_i)0.95 = 0.0095$ sec. A slight increase in K_t can be expected in a system with friction.

STORED-AIR ACTUATORS

The method of dimensionless ratios used to compute air dashpots in the previous section will work also for the reverse action: stored air to drive an actuator (Fig. 15.21).

Again, absolute pressure ratio is substituted for absolute pressure, volume ratio for volume, and even synthetic ratios for real time.

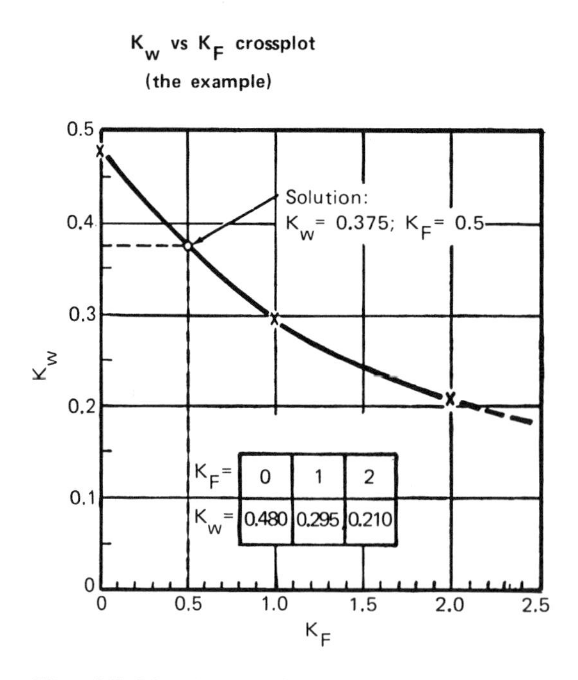

Fig. 15.20 Sample interpolation of typical data points from Fig. 15.19.

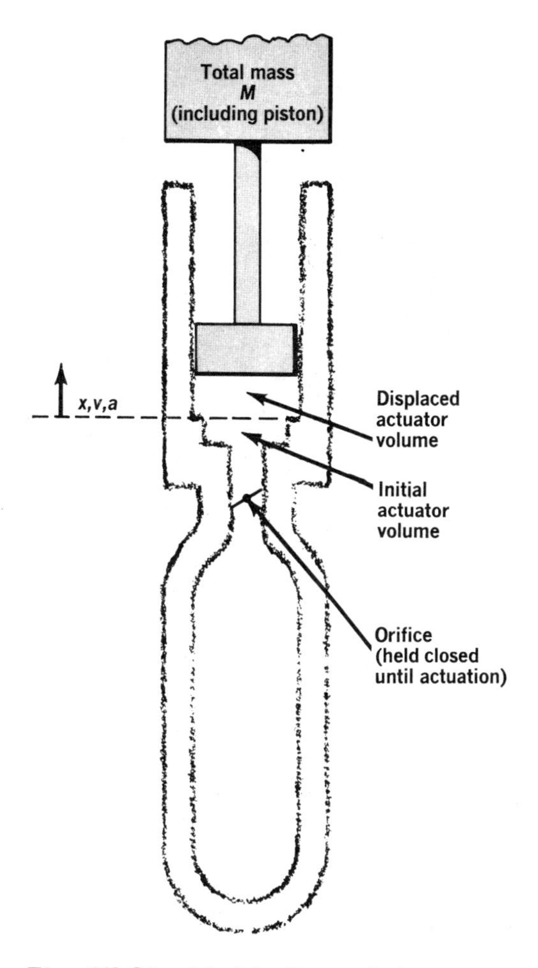

Fig. 15.21 Model of stored-air actuator, used for accompanying calculations.

With the new ratios, called dimensionless parameters, one family of graphs will suffice for all sizes, pressures, weights, and so forth (Fig. 15.22).

One exception: The ratio of storage volume (volume in bottle) to initial actuator volume K_{V_s} has a marked secondary effect on overall performance, so all the graphs are plotted assuming a single value of that ratio: 50. If you want to design your actuator with any other ratio, you'll have to set up the equations again on a computer. There are a few other assumptions:

- Friction is zero.
- System is adiabatic.
- No external forces act on load or piston.
- Working fluid is an ideal gas.
- Gas specific heat ratio is 1.4.
- Gas volume in actuator initially is negligible (no appreciable energy).

Parametric Graphs (dimensionless)

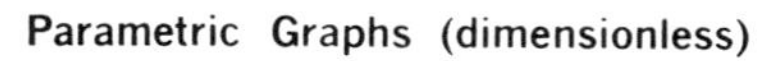

1. Storage pressure vs time

K_{Ps}

1.0
0.8
0.6
0.4
0.2
0

0.1 0.2 0.3 0.4 0.5

K_t

$K_{Vs} = 50$

$K_Q = 20, 40, 60, 80$ or 100
(same plot applies)

2. Max actuator pressure vs volume displacement

K_{Pa} max

1.0
0.8
0.6
0.4
0.2
0

20 40 60 80 100

K_Q

$K_{Vs} = 50$

3. Piston travel vs time

K_x

100
80
60
40
20
0

0.1 0.2 0.3 0.4 0.5

K_t

K_Q
100
80
60
40
20

$K_{Vs} = 50$

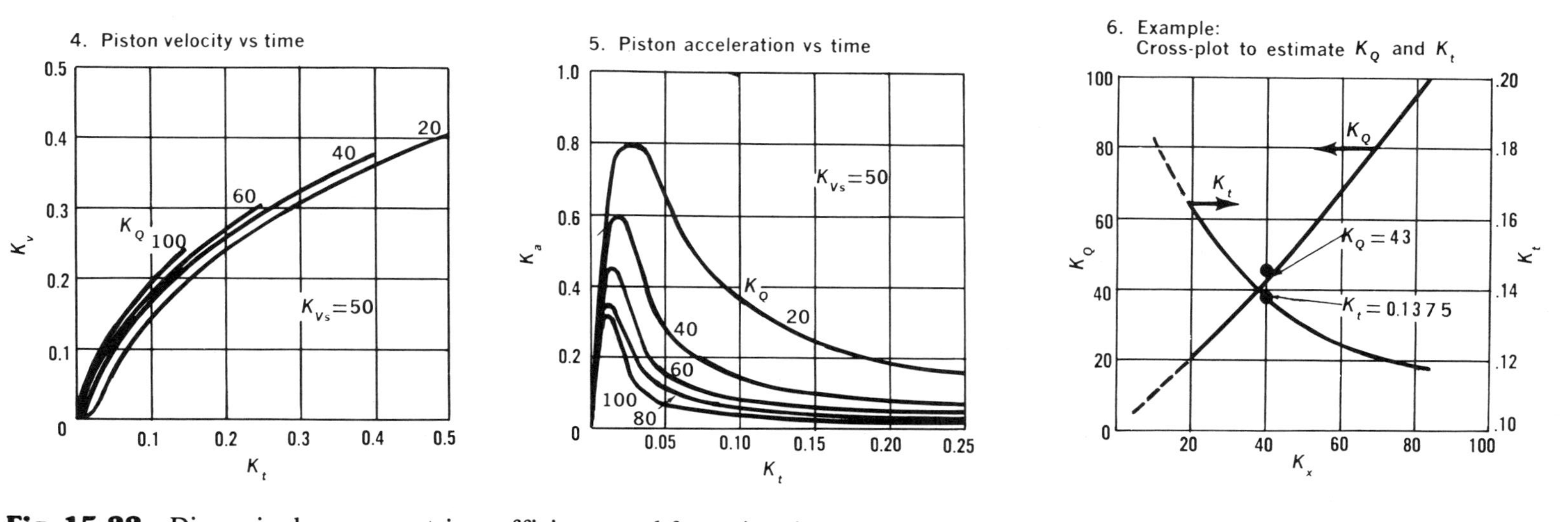

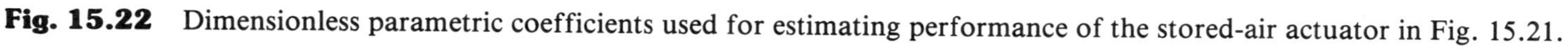

Fig. 15.22 Dimensionless parametric coefficients used for estimating performance of the stored-air actuator in Fig. 15.21.

Standard Equations

The symbols, equation boxes, and example (Tables 15.3 through 15.5) summarize the math better than words. The equations are standard for computing force and acceleration when gas expands through an orifice from a pressure chamber into an actuator volume and moves a piston.

The dimensionless parameters, although arbitrary, were picked after careful study and experiment. All graphs have been checked by tests to validate the method.

AIR MOTORS AND ROTARY ACTUATORS

The basic advantages of air motors are low weight per horsepower, small size, indifference to overload and stall, cool operation, the availability of low-cost speed controls, use of air rather than oil, and relative indifference to dirty atmospheres. The disadvantages as compared to electric power are lower efficiency, higher first cost (nearly double the equivalent electric motor), and a maintenance cost which will be approximately the same and perhaps more expensive than electric motor maintenance.

Poor efficiency is the result of combining the cost of compressing air with the inefficiencies in expanding it during the work cycle. Efficiency overall will be lower than 20% in many instances.

Most air motors are the vane type but can be piston (Fig. 15.23), gear, and gerotor types as well. (See Chap. 6 for drawings, general theory, and design of fluid motors; most of it applies to air motors, too.)

The chief difference is that air is highly compressible and oil is not. True air motor equations are the inverse of compressor equations (see Chap. 12).

An ideal air motor would expand the compressed air to atmospheric pressure within the motor. However, physical limitations in eccentricity and number of vanes restrict the compression ratio severely.

Most applications of air motors and actuators are where the special performance qualities in the first paragraph are essential and efficiency is not a problem. Examples are certain flight-surface actuators on aircraft, rotary air tools in manufacturing plants, air clamping devices, and rotary stirrers or actuators in hazardous environments.

Table 15.3 Nomenclature for Stored-Air Actuator Equations

a = acceleration, in./sec^2
A_o = orifice area, in.^{2}
A_p = piston area, in.^{2}
g = gravitational constant, in./sec^2
k = ratio of specific heats (dimensionless)
K = parametric symbol (see dimensionless parameters below)
M = total load mass, lb-sec^2/in. (includes piston and rod)
P = actuator pressure, psia
dp/dt = rate of pressure change, psia/sec
Q = rate of actuator to critical orifice flow (dimensionless)
R = gas constant, in. lb/lb-°R
t = time, sec
T = gas temperature, °R
V = air volume, in.^{3}
v = piston velocity, in./sec
w = orifice flow rate, lb/sec
x = piston travel, in.
ρ = gas density, lb/in.^{3}

SUBSCRIPTS

a = actuator volume
i = initial (at start of launching action) such as in v_i
p = piston
o = orifice
s = storage volume

Table 15.4 Basic Equations for Stored-Air Actuator Design

Basic equations

1. $(P_s)_i V_s = P_s V_s + P_a V_a + \frac{k-1}{2} M v^2$

2. $dP_s/dt = -\frac{kR}{V_s} T_s w$

3. $dP_a/dt = \frac{k}{V_a}\left(RT_s w - A_p P_a \frac{dx}{dt}\right)$

4. $w = Q P_s A_o \sqrt{\frac{kg}{RT_s}\left(\frac{2}{k+1}\right)^{\frac{k+1}{k-1}}}$

5a. $Q = 1.00 \quad \text{for} \quad 0 \leqq \frac{P_a}{P_s} \leqq \left(\frac{2}{k+1}\right)^{\frac{k}{k-1}}$

5b. $Q = \sqrt{\frac{\left(\frac{2}{k-1}\right)}{\left(\frac{2}{k+1}\right)^{\frac{k+1}{k-1}}}} \times \sqrt{\left(\frac{P_a}{P_s}\right)^{\frac{2}{k}} - \left(\frac{P_a}{P_s}\right)^{\frac{k+1}{k}}}$

for

$\left(\frac{2}{k+1}\right)^{\frac{k}{k-1}} < \frac{P_a}{P_s} \leqq 1.00$

6. $Ma = P_a A_p$

7. $V_a = (V_a)_i + A_p x$

8. $\left(\frac{T_s}{(T_s)_i}\right) = \left(\frac{P_s}{(P_s)_i}\right)^{\frac{k-1}{k}}$

9. $P = \rho RT$

Dimensionless parameters

10. $K_{Ps} = \frac{P_s}{(P_s)_i}$ — $\frac{\text{storage pressure}}{\text{initial storage pressure}}$

11. $K_{Pa} = \frac{P_a}{(P_s)_i}$ — $\frac{\text{actuator pressure}}{\text{initial storage pressure}}$

12. $K_{Vs} = \frac{V_s}{(V_a)_i}$ — $\frac{\text{storage volume}}{\text{initial actuator volume}}$

13. $K_Q = \frac{A_p \sqrt{\frac{2(P_s)_i V_s}{M(k-1)}}}{w_i/(\rho_s)_i}$ — $\frac{\text{max theo. piston displacement rate}}{\text{initial critical gas flow}}$

14. $K_t = \frac{t}{(\rho_s)_i V_s / w_i}$ — $\frac{\text{actual time}}{\text{storage discharge time at initial flow rate}}$

15. $K_x = \frac{x}{(V_a)_i / A_p}$ — $\frac{\text{piston travel}}{\text{travel to displace initial actuator volume}}$

16. $K_v = \frac{v}{\sqrt{\frac{2(P_s)_i V_s}{M(k-1)}}}$ — $\frac{\text{actual velocity}}{\text{max theo. velocity for infinite expansion}}$

17. $K_a = \frac{a}{(P_s)_i A_p / M}$ — $\frac{\text{actual acceleration}}{\text{acceleration caused by initial storage } P \text{ acting on piston}}$

Parametric relationships

18. $\frac{dK_{Ps}}{dK_t} = -kQ(K_{Ps})^{\frac{3k-1}{2k}}$

19. $\frac{dK_{Pa}}{dK_t} = kK_{Vs}\left(\frac{Q(K_{Ps})^{\frac{3k-1}{2k}} - K_Q K_{Pa} K_v}{1 + K_x}\right)$

20a. $Q = 1.00 \quad \text{for} \quad 0 < \left(\frac{K_{Pa}}{K_{Ps}}\right) < \left(\frac{2}{k+1}\right)^{\frac{k}{k-1}}$

20b. $Q = \sqrt{\frac{\left(\frac{2}{k-1}\right)}{\left(\frac{2}{k+1}\right)^{\frac{k+1}{k-1}}}} \times \sqrt{\left(\frac{K_{Pa}}{K_{Ps}}\right)^{\frac{2}{k}} - \left(\frac{K_{Pa}}{K_{Ps}}\right)^{\frac{k+1}{k}}}$

for $\left(\frac{2}{k+1}\right)^{\frac{k}{k-1}} < \left(\frac{K_{Pa}}{K_{Ps}}\right) \leqq 1.00$

21. $\frac{dK_x}{dK_t} = K_{Vs} K_Q K_v$

22. $K_v = \sqrt{1 - K_{Ps} + \frac{K_{Pa}}{K_{Vs}}(1 + K_x)}$

23. $K_a = K_{Pa}$

Table 15.5 Sample Calculation for Stored-Air Actuator

Example:

1000 psi storage, 1 sq in. piston.
The wanted values

1) Orifice area, A_o
2) Time of power stroke, t
3) Actuator pressure peak, $P_a(max)$
4) Acceleration peak, $a(max)$
5) Acceleration at the end of power stroke, $a_{x=32}$
6) Storage pressure at the end of power stroke, $(P_s)_{x=32}$

[Known]

Mass $M = 0.1$ lb-sec²/in.
Piston area $A_p = 1.0$ in.²
$(P_s)_i = 1000$ psia
$V_s = 40$ in.³
$v = 300$ in./sec at end of power stroke
$x = 32$ in.
$(T_s)_i = 530$°R
$R = 639.6$ in.-lb/lb-°R (air)

Solution:

From Eq(16), $K_v = \dfrac{300}{\sqrt{\dfrac{(2)\,(1000)\,(40)}{(0.1)\,(0.4)}}} = 0.212$

Use this value for K_v, make up a table of values (below) for K_t and K_x at four values of K_Q: 20, 40, 60, 100. Use graphs 3 and 4.

K_Q	20	40	60	100
K_t	0.165	0.140	0.127	0.117
K_x	19.0	37.5	54.0	83.5

From Eq(12), $(V_a)_i = \dfrac{V_s}{K_{Vs}} = \dfrac{40}{50} = 0.8$ in.³

From Eq(15), $K_x = \dfrac{(32)(1)}{0.8} = 40$

Knowing K_x, find the values of K_Q and K_t that correspond, by plotting K_Q and K_t vs K_x from data in Fig. 15.22 (graph 6). From the graph, for $K_x = 40$, $K_Q = 43$, and $K_t = 0.1375$.

From Eq(13)

$$\frac{w_i}{(\rho_s)_i} = \frac{(1)\sqrt{\dfrac{(2)(1000)(40)}{(0.1)(0.4)}}}{43} = 32.9 \text{ in.}^3/\text{sec}$$

From Eq(4)and(9)

$$\frac{w_i}{(\rho_s)_i} = A_o\sqrt{kgR(T_s)_i\left(\frac{2}{k+1}\right)^{\frac{k+1}{k-1}}}$$

$$A_o = \frac{32.9}{\sqrt{(1.4)(386)(639.6)(530)\left(\dfrac{2}{2.4}\right)^6}}$$

$= 0.00421$ in.², orifice area

From Eq(14)

$$t = \frac{(0.1375)(40)}{(32.9)}$$

$= 0.167$ sec, time of power stroke

From Graph 2, for $K_Q = 43$, $(K_{Pa})_{max} = 0.58$.
From Eq(11)

$(P_a)_{max} = (0.58)(1000)$
$= 580$ psi, actuator pressure peak

Peak acceleration is found from the above value of $(P_a)_{max}$ or from graph 5. Using Eq(6):

$$(a)_{max} = \frac{(580)(1)}{(0.1)}$$

$= 5800$ in./sec², acceleration peak

The acceleration at the end of the power stroke is found from graph 5, for $K_t = 0.1375$, $K_Q = 43$

$$(K_a)_{x=32} = 0.10$$

From Eq(17)

$$(a)_{x=32} = (0.10)\frac{(1000)(1)}{(0.1)}$$

$= 1000$ in./sec², acceleration at end of power stroke

Storage pressure at end of stroke, from graph 1, for $K_t = 0.1375$:

$$(K_{Ps})_{x=32} = 0.87$$

From Eq(10)

$(P_s)_{x=32} = (0.87)(1000)$
$= 870$ psi, storage pressure at end of stroke

Fig. 15.23 Typical radial-piston air motor. (Courtesy of Fenner America, Middletown, CT.)

16

Pneumatic Flow, Pressure Drop, Response

Flow of compressible fluids (air and gases) through piping, valves, and restrictions is much more difficult to compute than is flow of oil or water through the same restrictions. Air is even friskier than electricity when it comes to pinning down performance. It changes shape, velocity, direction, density, viscosity, and relative humidity at the drop of a Fahrenheit or a Pascal.

As a consequence, the vast majority of airflow calculations through restrictions are approximations, often based on oil flow equations. Not only that, there are nearly as many such approximations as there are basic industries.

The process industries have several at least; the steam power people have a few; the HVAC (heating, ventilating, and air-conditioning) companies have their own; the gas pipeline experts have at least one; and many major labs have special equations.

The most popular standard in the fluid power and control industry is the C_V flow coefficient, described later. Table 16.1 lists it along with eight other common flow equations. Additional text and tables describe them in detail and add a few more. See Fig. 16.1 for a common test method.

The newest and probably the most accurate method is the sonic conductance test, described at the end of the chapter. It is considered too sophisticated by many small companies to apply but may yet become a standard.

Not discussed in this chapter is air leakage flow through annular clearances; that topic is explored thoroughly in Chap. 8.

BACKGROUND ON ALL METHODS

There are so many methods that it becomes a challenge to relate coefficients from one with coefficients from another. So, short of asking for actual test data from every manufacturer, how does a system designer mix and match valves, fittings, piping, and other restrictions from a variety of sources, new and old?

One easy answer—unfortunately practiced widely—is to make all the valves and piping the same size as the port on the final power element. For example, if the air

Table 16.1 Comparisons Among Nine Typical Airflow Equations

	Typical equations	Flow coefficients defined
1	$Q = 33 D_O^2 P_U \sqrt{R(R^{0.43} - R^{0.71})}$	D_O = equivalent sharp-edge orifice (C_D = 0.6) $= \sqrt{\dfrac{Q}{33 P_U \sqrt{R(R^{0.43} - R^{0.71})}}}$
2	$Q = \dfrac{963}{60} C_V \sqrt{\dfrac{\Delta P (P_U + P_D)}{G T_U}}$	C_V = valve coefficient $= \dfrac{60Q}{963} \sqrt{\dfrac{G T_U}{\Delta P (P_U + P_D)}}$
3	$Q = \dfrac{1360}{60} C_V \sqrt{\dfrac{\Delta P \times P_U}{G T_U}}$	C_V = flow coefficient $= \dfrac{60Q}{1360} \sqrt{\dfrac{G T_U}{\Delta P \times P_U}}$
4	$Q = \dfrac{1350}{60} C_V \sqrt{\dfrac{\Delta P \times P_D}{G T_U}}$	C_V = ANSI/NFPA coefficient $= \dfrac{60Q}{1350} \sqrt{\dfrac{G T_U}{\Delta P \times P_D}}$
5	$Q = \dfrac{5180}{60} C_V \sqrt{\dfrac{P_U^2 - P_D^2}{M T_U}}$	C_V = valve-flow coefficient $= \dfrac{60Q}{5180} \sqrt{\dfrac{M T_U}{P_U^2 - P_D^2}}$
6	$Q = \dfrac{963}{60} C_V \sqrt{\dfrac{P_U^2 - P_D^2}{G T_U}}$	C_V = flow coefficient (Eq 2 = Eq 6) $= \dfrac{60Q}{963} \sqrt{\dfrac{G T_U}{P_U^2 - P_D^2}}$
7	$Q = \dfrac{2.32^{0.443}}{60} C_G \dfrac{\Delta P^{0.443} \times P_U^{0.6}}{\sqrt{G T_U / 520}}$	C_G = gas-flow coefficient $= \dfrac{60Q}{(2.32)^{0.443}} \times \dfrac{\sqrt{G T_U/520}}{\Delta P^{0.443} \times P_U^{0.6}}$
8	$Q = F P_U \sqrt{4/3} \sqrt{1 - R^2}$	F = NBS flow factor #1 $= \dfrac{Q}{P_U \sqrt{4/3}} \sqrt{\dfrac{1}{1 - R^2}}$
9	$Q = F P_U \sqrt{8/5} \sqrt{R(1-R)(3-R)}$	F = NBS flow factor #2 $= \dfrac{Q}{P_U \sqrt{8/5}} \sqrt{\dfrac{1}{R(1-R)(3-R)}}$

Symbols

D_O	=	diameter of equivalent sharp-edge orifice, in.
C_V, C_G, F	=	typical flow coefficients
G	=	specific gravity, gas/air (= 1 for air)
M	=	molecular wgt, lb_m (= 29 for air)
P	=	pressure, psia (subscript D = downstream, U = upstream)
ΔP	=	pressure drop, $P_U - P_D$, psi
Q	=	air flow, scfm at 14.7 psia, 68 F
R	=	pressure ratio, P_D/P_U
T	=	absolute temperature, $R = F + 460$

Approximate conversions among Eqs 1, 4 and 9

$D_O = 0.236\sqrt{C_V} = 0.316\sqrt{F}$; $C_V = 18 D_O^2 = 1.8F$; $F = 10 D_O^2 = 0.556 C_V$

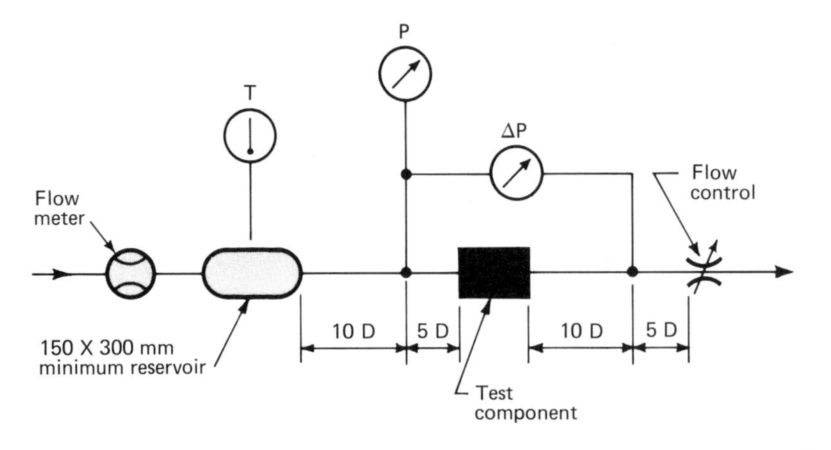

Fig. 16.1 Standard test setup for airflow coefficients. Flow meter may be placed upstream or downstream. (Courtesy of the National Fluid Power Association, Milwaukee, WI.)

cylinder port is 1.0 in., make all the control valves 1.0 in. It's a conservative solution if you don't mind an occasional valve 10 times larger than needed or a pipeline so large in diameter that it acts as a plenum and slows down response to a pressure change.

Another problem is that actual valve passage dimensions often have little relation to the so-called nominal port diameter.

A better answer is to calculate how much airflow is truly required to operate the cylinder and pick a valve (and a coefficient) that fits the need. At least several valve manufacturers offer guides to do it. A 1.0-in. valve can cost six times as much as an ⅛-in. valve, so a little homework might pay off. The hardest part is converting among the many published flow coefficients. Quick conversion equations would be a blessing: The accompanying tables and figures address the problem.

C_V and D_O Coefficients

The C_V flow coefficient is a clear winner in extent of application. In one form or another, in the United States and in Europe, C_V is the most common way to represent the nominal flow capacity of air valves. Equivalent orifice D_O is very similar in concept but is not as widely accepted.

Most of the member companies of the NFPA (National Fluid Power Assoc.) are committed to the C_V flow coefficient and are supporting special industry-wide tests to ascertain the level of accuracy and reliability.

Air valves (and liquid valves) rated by either of these techniques are assumed to have a fixed flow passage in each valve position. By definition, an opening or restriction has a C_V coefficient of 1.0 if it will pass 1.0 gpm of water at 1.0-psi pressure drop. The tests may be run on any fluid, including air, but they are often run on water. The basic equation is

$$\text{flow, gpm} = C_V(\Delta P)^{0.5}$$

where ΔP is psi.

Once the value of C_V or D_O has been calculated, the water flow equation is not used again because you need compressible-flow equations for air. The coefficient stays the same, however, and the airflow estimate is fairly accurate if pressure drop is less than 15% of inlet pressure, so that density does not change too much. Above that the actual flow will be 5 to 20% more than the calculated flow. At pressure drops beyond the critical pressure ratio, $P_2/P_1 = 0.533$ for air, sonic throttling occurs, and volume flow will not increase further.

The method thus is accurate at very low pressure drop, very conservative at moderate pressure drop, and not applicable beyond critical pressure drop.

A Practical Answer. Since there's nothing better than C_V on the horizon, valve companies have developed precise ways of reducing the unpreciseness. Numatics, Inc. (Highlands, MI), for example, ran extensive tests (25,000 data points, 1500 photos) on air valves and cylinders to develop a family of C_V coefficients and cylinder response times for a wide variety of pressures and loads. Within the ranges specified, the accuracy is said to be good. It's all explained in their company manual *Practical Air Valve Sizing*, authored by Henry Fleischer, Director of Engineering, and Paul Tallant, Senior Research Engineer. It's based on equation 4 in Table 16.1 and the graph in Fig. 16.2. (Note: symbol *P* in table is psia, but in graph is psig.)

It includes a series of worked-out examples for single-acting and double-acting air cylinders. Another feature is a series of charts and graphs, including nomographs, covering the company's entire air valve line as they perform with actual air cylinders in the range from 1½- to 14-in. bore. If you know the load and stroke of your air cylinder, the total time of stroke can be picked off the graphs for a variety of valving arrangements, pressures, and effective orifice sizes (C_V coefficients).

The cylinder loading resulting from internal friction and exhaust back pressure is accounted for in the manual's performance curves. Most cylinders have about 15% internal friction.

Six steps enable you to quickly size a system containing a valve, cylinder, piping, and fittings:

1. Determine the cylinder bore size as a result of the given load conditions and available pressure.

2. Pick a performance graph representing the bore size and intended pressure to be used and from it extricate the required overall system flow coefficient C_V to meet the load and time specifications. For dual pressure applications, to obtain minimum size devices, choose pressures such that the required system C_V for extend is essentially equal to the system C_V for retract.

3. Select a valve having a C_V slightly greater than the system C_V arrived at.

4. Choose a pipe size based on the standard port tapping of the valve. Then figure individual C_Vs for the fittings and pipe lengths and the total piping C_V.

5. Knowing the pipe size, obtain the cylinder C_V.

6. Combine all the individual C_Vs to figure the system C_V and check it against what you first assumed. If it is higher, your work's done. If lower, go to the next larger pipe size before trying a larger valve.

Series or Parallel Restrictions. Although there is the perennial question whether or not C_V factors can be added truly in series and parallel, Fleischer and Tallant de-

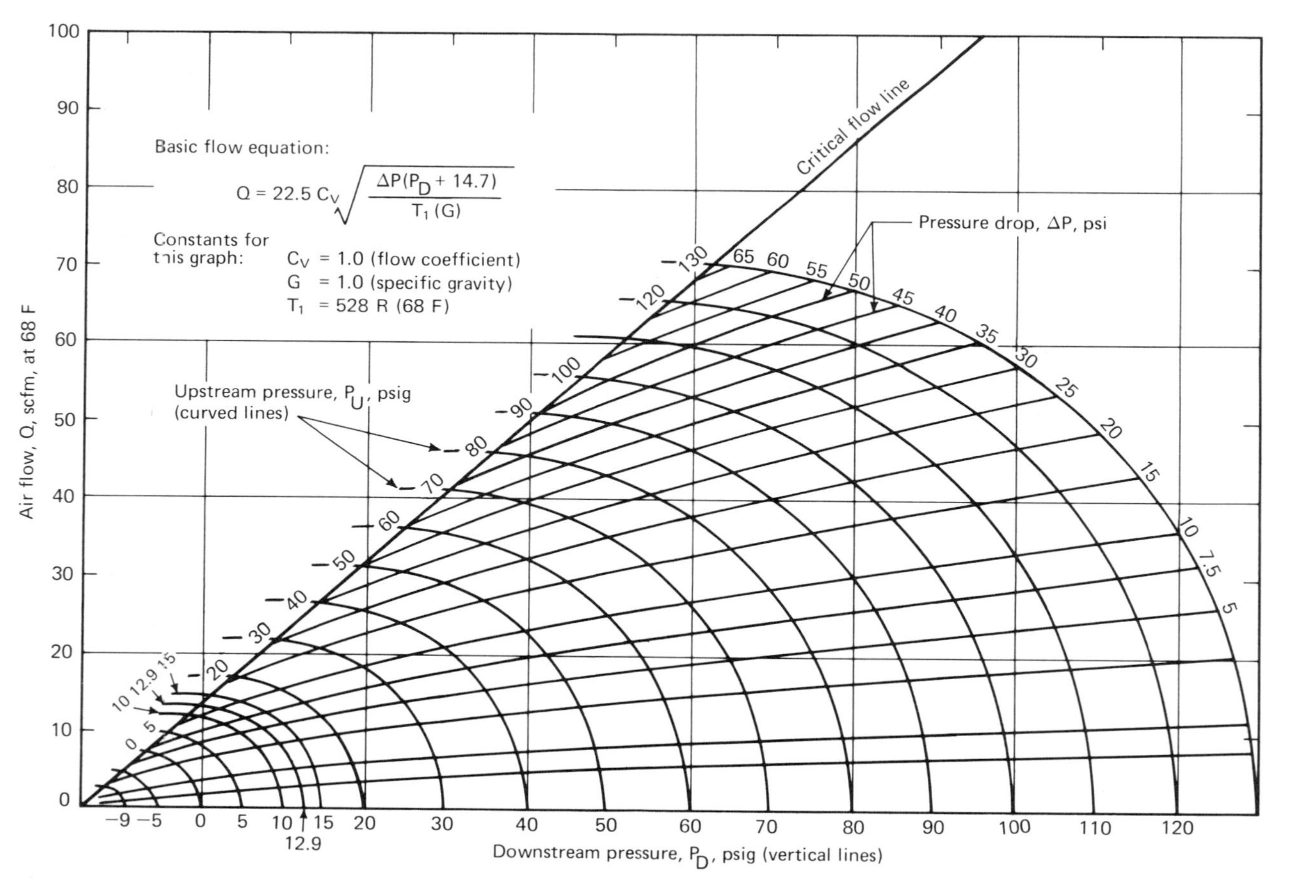

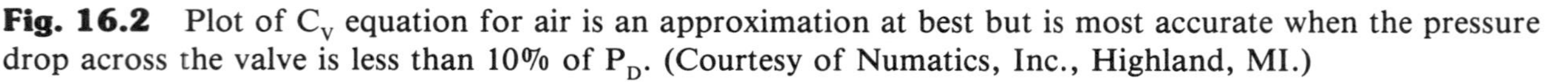

Fig. 16.2 Plot of C_V equation for air is an approximation at best but is most accurate when the pressure drop across the valve is less than 10% of P_D. (Courtesy of Numatics, Inc., Highland, MI.)

termined from their tests that reasonable accuracy will result if the following simple relationships (Fig. 16.3) are exploited:

For series valves and restrictions: $(1/C_V)^2$ for the system equals the sum of all the individual $(1/C_V)^2$s of the devices. It doesn't matter in which order the devices are placed or whether they are valves, orifices, fittings, or tubing. Each has been assigned values of C_V in the manual.

For parallel valves and restrictions: C_V for the system equals the sum of the individual C_Vs of the devices. For combinations of series and parallel, combine one section at a time and then combine the results. The manual has a nomograph for combining any pairs in series, and this is quicker than the math if you have a lot of trial and error.

A Few Tips. Most of the other air-valve companies also have developed techniques for working with C_V coefficients. Typical examples are Airmatic/Snap-Tite, Alkon, Automatic Valve Corp., C. A. Norgren, Parker Hannifin, Ross Operating Valve, Schrader Bellows/Scovill, Skinner/Honeywell, and Wabco/American Standard. Here are some of their ideas and shortcuts:

A valve that has a C_V of 1.0 will pass 10 scfm with 1.0-psi pressure drop if air inlet pressure is 90 psig and temperature is ambient. Flow will increase to about 30 scfm for a pressure drop of 10 psi. Norgren (Littleton, CO) suggests memorizing that last set of figures: 10 psi, 30 scfm, 90 psig. Note that they are multiples of 3.

Air valves and liquid valves rated by this technique are assumed to have in each valve position a fixed flow passage equivalent to a sharp-edge orifice. The C_V coef-

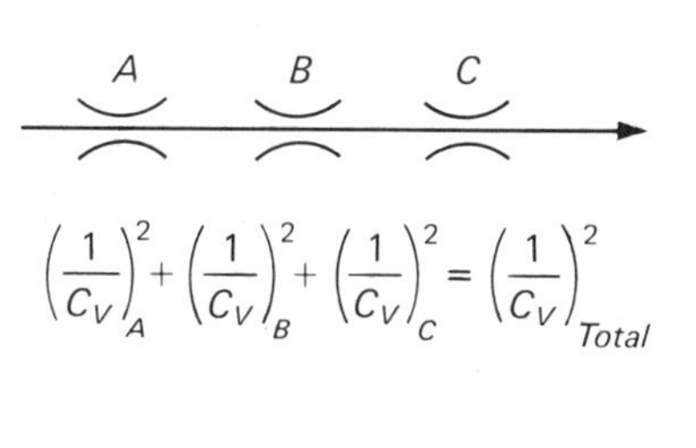

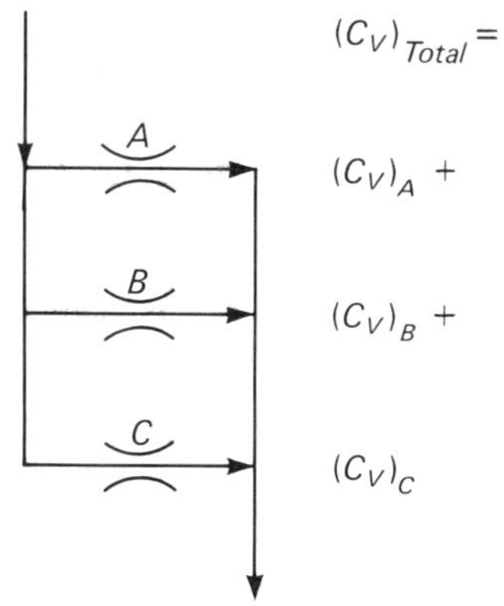

Fig. 16.3 C_V coefficients (Fig. 16.1 and Table 16.1) can be easily handled for valves or restrictions in series or parallel.

ficient for any valve thus represents an equivalent orifice, nothing more. In fact, $C_V = 18D_O^2$, where D_O is an equivalent orifice.

The fact that the true flow path is a tortuous combination of series and/or parallel restrictions is generally ignored because it is impossible to calculate.

Even though the method is approximate, it is usually as accurate as the test data in a typical engineering laboratory. Even an apprentice mathematician must realize that if pressure is observed with 98% accuracy (that's optimistic) and flow with 95% accuracy (also optimistic), the data on those combined measurements alone aren't going to be much better than 93% accurate. Then throw in variations in valve machining, distortions of flow path because of passage shape, faulty measurement of temperature, and severe effects of adjacent fittings and piping, and you realize that it is very doubtful that the final coefficient will be even 85% reliable in practice.

Old NBS Flow Factor (Fig. 16.4)

Gas or airflow through a pneumatic component may be fairly approximated with this relationship:

$$\text{scfm} = \text{constant} \times \sqrt{(P_U^2 - P_D^2)}$$

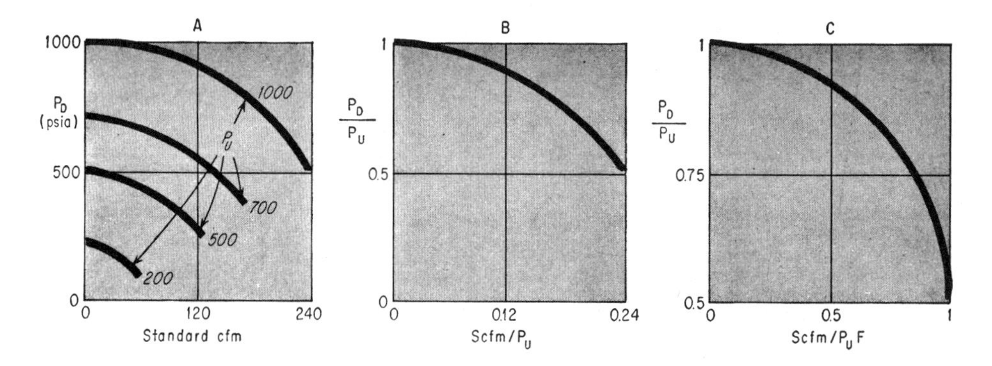

Gas or airflow through a pneumatic component may be fairly approximated with this relationship:

$\text{scfm} = \text{constant} \times \sqrt{(P^2_U - P^2_D)}$

where the constant depends on passage configuration, friction, and inlet temperature; P_U and P_D are upstream and downstream pressures, psia. Scfm is actual flow in cfm converted to "standard" conditions—14.7 psia, 68 F. Although the equation does not represent isentropic flow, and assumes considerable loss in stagnation pressure, its convenient mathematical form simplifies manipulation and the error will probably not exceed 5 to 10%. Test results can be plotted in three ways. Curves in *A* are flow data in scfm for a typical component. Curve *B* is from the same data converted to single curve by dividing by upstream pressure. Curve *C* is from the same data, further converted to "normalized" form by dividing by a defined flow factor, *F*, that is a measurable quantity.

Flow factor, *F*, is the ratio

$$\frac{\text{flow (scfm)}}{P_U \text{ (psia)}} \text{ when } \frac{P_D}{P_U} = 0.5$$

The constant in the flow equation equals $F \times \sqrt{4/3}$, therefore the normalized equation for curve *C* becomes

$$\frac{\text{scfm}}{P_U \times F} = \sqrt{(4/3)} \times \sqrt{(1 - (P_D/P_U)^2)}.$$

or

$$Q = P_U \times F \sqrt{(4/3)} \times \sqrt{(1 - r^2)}$$

This variation of the NBS equation has proven to be a better approximation of actual flow:

$$Q = P_U F \sqrt{(8/5)} \sqrt{[r(1-r)(3-r)]}$$

It is the mean between the $\sqrt{(1-r^2)}$ relationship and adiabatic expansion.

Fig. 16.4 Flow factor F, developed by the National Bureau of Standards, fairly approximates the actual flow of air or gas.

where the constant depends on passage configuration, friction, inlet temperature and other factors: P_U and P_D are upstream and downstream pressures, psia. Scfm is actual flow in cfm converted to "standard" conditions—14.7 psia, 68 F. Although the equation does not represent isentropic flow and assumes considerable loss in stagnation pressure, its convenient mathematical form simplifies manipulation, and the error will probably not exceed 5 to 10%.

Call $P_D/P_U = R$, and pull P_U out of the radical, and the equation becomes

$$Q = P_U F\sqrt{(4/3)} \times \sqrt{(1 - R^2)}$$

For somewhat better accuracy,

$$Q = P_U F\sqrt{(8/5)[R(1 - R)(3 - R)]}$$

It is the mean between the $\sqrt{(1 - R^2)}$ relationship and adiabatic expansion.

Flow factor F rarely is used today, except for development work in some companies. The constant normally is not calculated from a water-flow test; it is based on air-test data. However, a fairly accurate value for F can be derived if the equivalent orifice diameter D_O is known. The converting is done by combining equation 9 (for F) and equation 1 (for D_O) from the airflow equations (Table 16.1) and expressing F in terms of D_O:

$$F = 26.1D_O^2\sqrt{(R^{0.43} - R^{0.71})} \div \sqrt{[(1 - R)(3 - R)]}$$

This complex form has very little practical value, but fortunately the total value of the terms under the square root is fairly constant for a wide range of R. (For $R = 1$, $F = 9.99D_O^2$; for $R = 0.9$, $F = 9.5D_O^2$; for $R = 0.5$, $F = 8.45D_O^2$.) Tests show that a workable equation is $F = 10D_O^2$ for low pressure drops.

The K-Factor

Invented a long time ago to simplify pressure-loss calculations in ordinary fittings and valves, the K-factor takes some liberties with true energy-balance laws. It neglects changes in velocity, elevation, viscosity, and wall friction. It also discounts the effects of piping roughness preceding and following the fitting.

In fact, the main basis for using K originally was that observation of numerous test data on many fluids over the years indicated that pressure losses vary with V^2 of the fluid, give or take 10% on the exponent 2. It was logical and useful to express the relationship in this way:

$$\text{head loss } H_L = KV^2/2g, \text{ ft}$$

or

$$\Delta P = KV^2(1.08 \times 10^{-4})\rho, \text{ psi}$$

thus tying in nicely with the velocity-head concept in Bernoulli's conservation of energy theorem. Factor K in the equation is dimensionless and has a special value for each fitting or restriction. Velocity V is ft/sec average, and density ρ is lb_m/ft^3.

Factor K tends to decrease somewhat with diameter. For example, a large elbow has a smaller K than a small elbow. Exceptions are sudden enlargements and contractions: The K-factors are less dependent on nominal diameter and thus remain fairly constant, regardless of size.

If the K-factor for a given component is based on actual tests, good accuracy can be expected—assuming your operating conditions are similar to the test.

The K-factor for air normally is based on tests. In theory K can be converted mathematically to any other coefficient, but the variables don't cancel out as nicely as they do for D_O or C_V, and the conversion "constant" doesn't stay constant. For example, the value for K contains the pressure ratio. For that reason it is not included in the accompanying box of typical equations and coefficients (Table 16.1).

The K-factor for oil flow is much easier to work with. See Chap. 9 for details.

Sonic Conductance (Fig. 16.5)

Potentially the most accurate method for determining airflow through valves and restrictions is the most recent entry: the sonic conductance method developed in Europe and discussed as a possible ISO/NFPA/ANSI document (ISO is the International Standards Organization, NFPA is the National Fluid Power Association, and ANSI is the American National Standards Institute.) The method is not in the tabulations (Table 16.1). However, the concept is important.

In essence, it first establishes by test what the airflow will be through a valve or other restriction at the critical pressure ratio—called the "critical flow" or "choked flow" condition, occurring at about $P_D/P_U = 0.5$ or less. Then it assumes that flow through the subsonic region, if plotted against P_D/P_U, will diminish along an elliptical path until flow equals zero at $P_D/P_U = 1.0$. The special graphs explain it (Fig. 16.5).

The sonic conductance technique resembles the old NBS flow factor method (explained before) except that the flow in NBS does not follow a pure ellipse. The difference is not great.

However, sonic conductance differs fundamentally from all the C_V coefficients because C_V normally is established at low pressure drops, whereas sonic conductance is established at the critical pressure drop.

In theory, sonic conductance values throughout the range should be consistently more accurate than C_V values because the flows are known at both ends of the curve—at P_D/P_U = critical and at P_D/P_U = zero—while C_V values are known only at P_D/P_U = low subsonic.

The big problem with the sonic conductance method is that it takes an enormous air supply to run tests on big valves. For example, sonic flow through a typical large-body 1-in. valve at P_U = 100 psig is over 800 scfm. If the valve is 2 in., the flow could be over 3000 scfm. And so on. Experimenters in the United States have discovered that the typical lab air system can't cope with it.

However, it is possible to run special transient tests using tanks of compressed air or gas as the source and continuously record the changing pressures, flows, and temperatures for a long enough period (a few seconds) to establish equivalent flow coefficients. Steady-state flow can be computed from transient data, even though the math is complicated.

Transient tests of valve response times are proving more repeatable than steady-state flow tests, probably because a transient test setup (Fig. 16.6) is compact and can be closely controlled.

Transient tests establish the time intervals from the instant an electrical or fluid signal is given to the valve under test until the valve has opened and passed sufficient

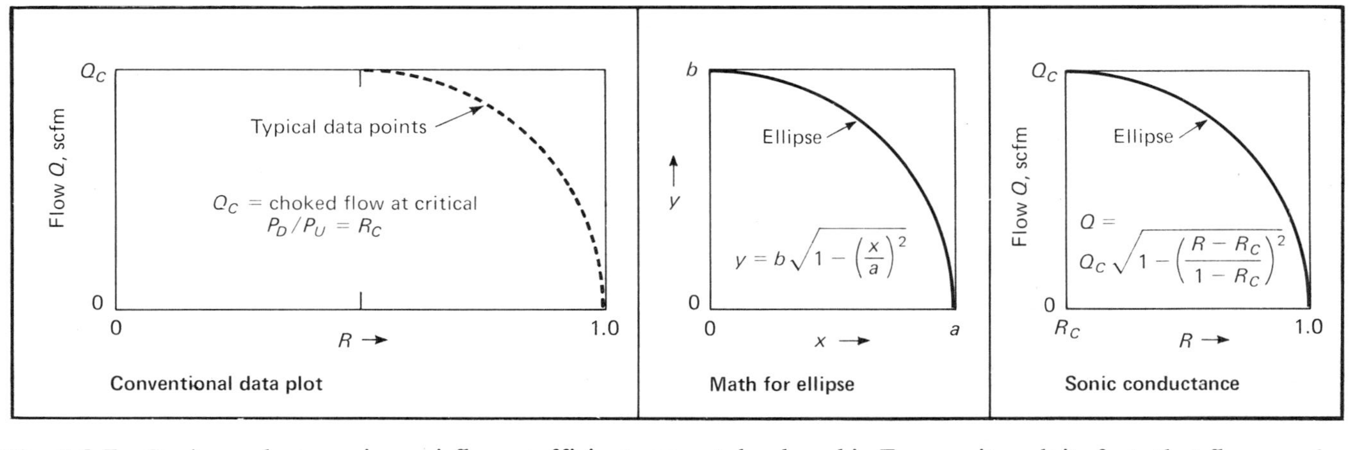

Fig. 16.5 Sonic conductance is an airflow coefficient concept developed in Europe; it exploits facts that flow graphs tend to be elliptical and that choked flow is the maximum obtainable.

Table 16.2 Pressure Loss Through a Restriction Is the Opposite of the Drag of an Automobile, but the Equations Are Similar

---

The mathematics of aerodynamic drag

Aerodynamic drag force on a moving vehicle is directly proportional to the product of the maximum cross-sectional area (frontal area) and the dynamic air pressure, modified of course by the empirical drag coefficient.

One extreme example is a square flat plate perpendicular to flow: the drag coefficient is about 1.2, meaning that the drag force is more than the product of plate face area times dynamic air pressure. The other extreme is an airfoil, typically with a drag coefficient of 0.05. Low-drag cars range from 0.3 to 0.5.

All the following equations assume wind-tunnel conditions: that is, only the air flow and not rolling resistance is creating drag force. It also is assumed that the size of the vehicle has no effect on the drag coefficient. The coefficient of a large car will be identical to that of a small car if the shapes are identical. The actual drag force is proportional to frontal area.

Drag equation: $F_D = C_D(0.5\ \ell\ V^2)A_F = C_D P_D A_F$

Where: F_D = aero drag force, lb; C_D = dimensionless drag coefficient, based on test; ℓ = air density, slugs/ft^3 (= 0.002378 for standard air at 14.7 psi, 60 F); V = air velocity, ft/sec (= 1.4667 × mph); A_F = max frontal cross-sectional area, ft^2; P_D = dynamic pressure = $0.5\ell V^2$, lb/ft^2 = $0.001189V^2$ in standard air.

Aero hp = $F_D V/550 = C_D P_D A_F V/550 = 2.162 \times 10^{-6} V^3 C_D A_F$ for standard air.

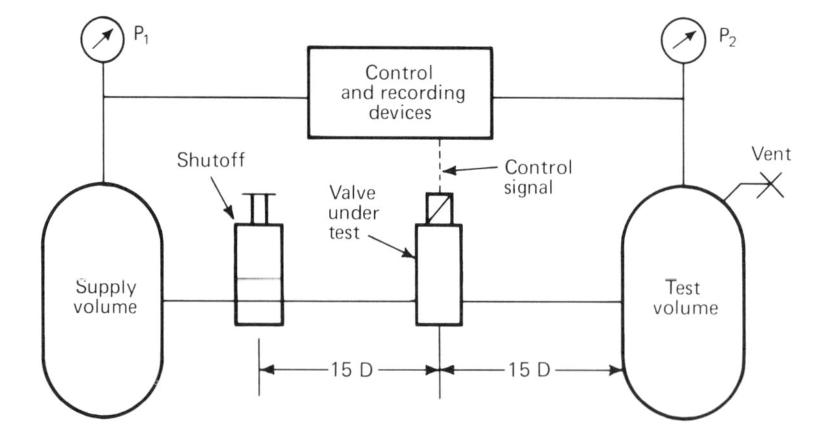

Fig. 16.6 Test setup for transient flow through air valves. It is possible but difficult to use the transient data to estimate steady-state sonic conductance. (Courtesy of Schrader Bellows Div., Akron, OH.)

flow volume to pressurize the downstream line. By definition, *reaction interval* is the time to change pressure (increase or decrease) 10% in response to an input signal; *response time* is the time to change pressure 90%. The difference is called *flow interval*, or the time to change pressure from 10% to 90%.

At least several manufacturers are generating full-range transient data with reasonably simple instrumentation, calculating low-head flow accurately, and deducing choked flow mathematically. In theory, transient tests for a few seconds duration should uncover all the needed input information. Automatic Valve Corp. (Novi, MI) is doing it and reports success.

Aerodynamic Drag

The mathematics of aerodynamic drag force on a moving vehicle is similar in concept to the math of airflow through a valve or restriction. That is, the dynamic pressure is a function of velocity squared. Anyway, Table 16.2 is a brief summary of the math of car drag coefficients.

17

Hydraulic and Pneumatic Filters

In fluid power systems a filter's chief function is to keep the working fluid (oil or air) free of contamination. Each kind of fluid has peculiarities, so this introduction will examine the natures of the fluids as well as the needs of the systems. Hydraulic filters will be treated separately from pneumatic filters, and the available filtering media for both will be discussed independently at the end.

The points to consider are the following:

- Is the fluid a liquid or gas? (A filter of given rating can generally remove much smaller particles from the latter.)
- Is it "nominal" or "absolute" filtration you need—will an occasional oversize particle harm the system?
- Can the system tolerate coarse particles during initial startup? (Some filters improve with time onstream—a porous cake of contaminant particles builds up on surface of filter medium.)
- What type of dirt will be filtered, and how does it compare with the arbitrary "typical" particles specified by test standards? Remember: Some gelatinous particles might deform enough to squeeze through tight openings.
- Is the fluid's dirt concentration large enough to build a filter cake? And will this cake formation pass liquid or block it?
- Will built-in dirt and other loose particles found in some off-the-shelf filters be damaging to system? How about those filter media that occasionally "flake" or break off particles, adding them to the stream?
- Will any single filter have large enough surface or volume to remove solids at the rate required?
- What about total system pressure—and allowable pressure differential across the filter? Compressibility of media? Flow velocity? Vibration in service?
- Will the fluid and filter be compatible at all operating temperatures? If some

corrosion is allowable, how much will this shorten the onstream time of the filter?

Follow these three general steps in choosing a filter:

- First, know your fluid system. Not just the fluid's trade name and expected flow but something of its chemical and physical nature and behavior under various temperatures and pressures.
- Second, evaluate available filtering-media types that might perform well with the fluid under question.
- Finally, consult filter manufacturers to decide on specific filter design. (The last step is not discussed here—it will be wiser to benefit directly from the specialized experience of each manufacturer.)

The first step—knowing the fluid—is not difficult if fluid has been analyzed by you or the manufacturer. This chapter is limited to engineering fluids carrying dirt particles smaller than 200 microns in average diameter. Here are basic facts concerning such fluids:

Gases are usually easier to filter than liquids. From gas, under certain conditions, some filter media will remove particles one-half to one-tenth the size of particles, removable from liquid. One explanation: The lack of lubricity of dry gases plus their generation of electrostatic charges helps hold particles that would otherwise slip through.

Dirt carried by fluid (gas or liquid) can be a mixture of fine and coarse, compressible and noncompressible, slimy and crystalline solids. Filtering is easiest when most particles are coarse, noncompressible, and crystalline. The fine, compressible, and slimy solids tend to pack together and restrict fluid flow drastically.

Concentration of dirt (volume of dirt vs. volume of fluid) must be known, because any filter has a dirt-capacity limit. Remember this: If flow is 1 gpm and concentration is 1%, then dirt-flow alone is 2.3 in^3/min. If the dirt is compressible, then only a thin filter cake is allowable—a fraction of an inch. Coarse, noncompressible dirt can build up to many inches without excessive pressure drop across the filter medium.

Details on choice of filtering media (the basic porous or woven materials that do the filtering) are given at the end of the chapter.

HYDRAULIC FILTERS

The art of hydraulic fluid filtration is sufficiently advanced to practically ensure clean systems and reduced wear—if the designers and users of the systems want to make the effort.

There are, however, a confusing multitude of options (see Fig. 17.1 for a sampling). Therefore, many designers and buyers of machinery and vehicles perceive the art as too sophisticated and/or expensive to meet practical needs.

As a consequence, the filtering strategy selected by many OEM designers is the cheapest that will keep the customer's equipment running without too many problems between maintenance periods. Wear life of the pumps, valves, cylinders, motors, actuators, and related hardware is a secondary consideration, so long as the life exceeds maintenance periods. That philosophy can be proved expensive in the long term (U.S. Army studies put annual losses due to wear at $100 billion).

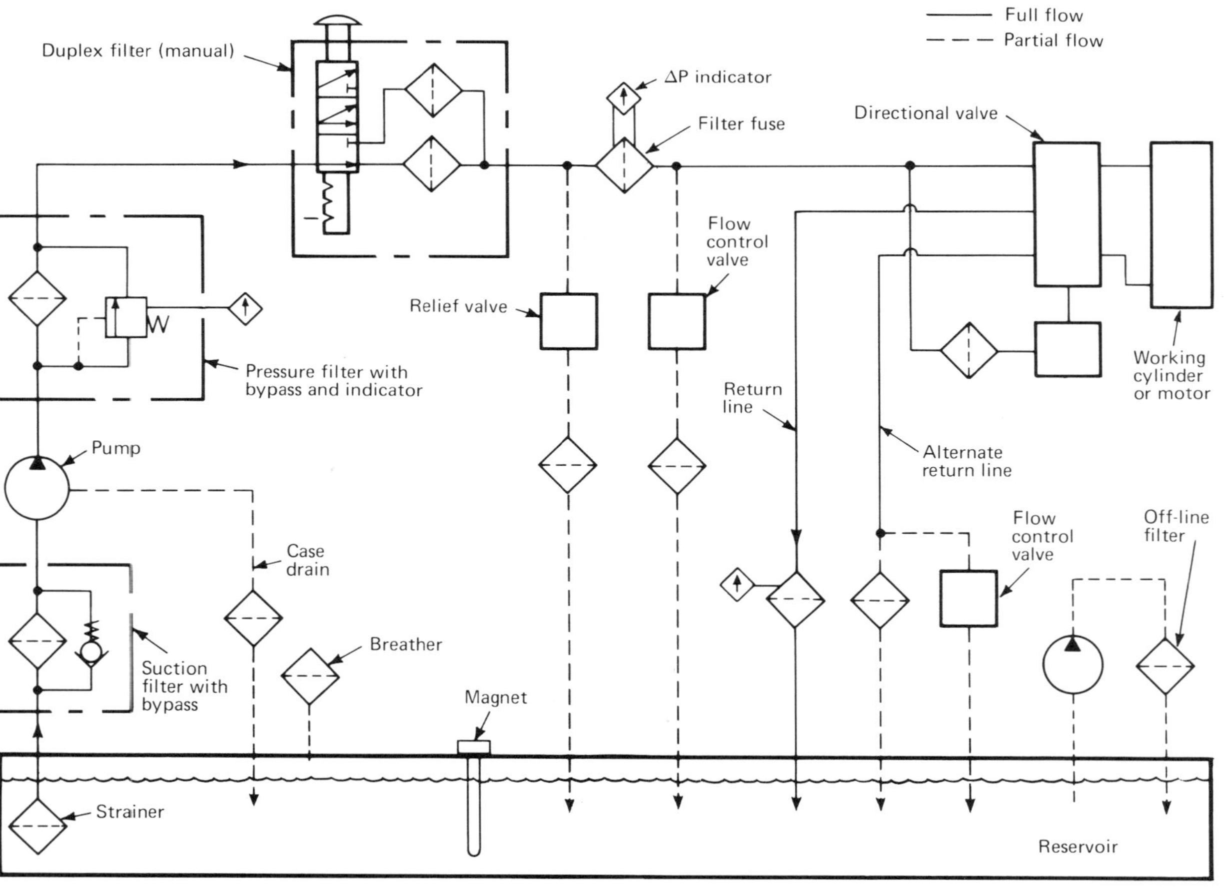

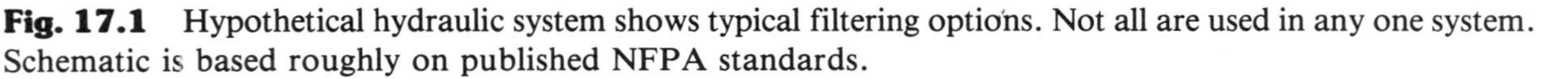

Fig. 17.1 Hypothetical hydraulic system shows typical filtering options. Not all are used in any one system. Schematic is based roughly on published NFPA standards.

Choices are a reflection of the application. A bulldozer's hydraulic system must make it through the day, regardless of the amount of dirt ingested; therefore, coarse filters with automatic bypassing are very tempting. In contrast, aircraft and precision machines are more sensitive to dirt and need finer filters to protect servovalves and other tight-clearance hydraulic controls from silting and jamming. Wear alone is less important and is no grounds for grounding.

Processes with downstream components that will be destroyed if debris gets into them need special assurances: One is a nonbypass filter that can block full system pressure (it's called a dirt fuse by Pall); another is a duplex filter that always has a clean element in parallel for quick transfer. In either instance, a "condition" indicator such as a gage or pop-up can warn the operator.

For systems with constant reversals of flow, there is the bidirectional filter. It contains automatic valving to redirect flow always to the working surface of the filter.

Maintenance on workhorse equipment often is extremely bad—consisting sometimes of calling the maintenance department (if there is one) only when the machine stops working. A lazy operator's solution might be to yank out the element and not replace it, or—if the filter is protected with a built-in valve to prevent operation within an empty housing—just to poke a hole through the filter element with a screwdriver, creating a permanent bypass. With near-sighted customers like that, it's better to build in an automatic bypass.

A possible counterattack for the filter manufacturer is to state in writing the operating and maintenance conditions on which the warranty is based. However, nobody wants to be that arbitrary because sales might drop. A gentler approach is to offer "extended" warranty for good maintenance.

It's all unbelievable and sad when you consider that 70% of system failures are attributable to dirty oil. At least examine the alternatives, even if electing not to invest—yet.

Dirt Is the Enemy (Fig. 17.2)

It's well known that particles greater than 25 μm can jam pumps, valves, and motors. Particles less than 25 μm, and more specifically between 0.5 μm and 5 μm (Fig. 17.3), cause degradation failure. These fine abrasive particles, called silt, enter clearances between moving parts. If the particles are about the same dimension as the clearance (thickness of an oil film, for example), they can score and abrade the surfaces. Wear generates more contaminants, increases the leakage clearances, lowers efficiency, and generates heat. The higher the pressure, the greater the problem.

Dramatic proof of the existence and generation of silt was documented in an SAE paper and rediscussed in a technical filter guide by Pall Corp. (Glen Cove, NY). A traditional 25-micron nominal filter was replaced by a 3-micron absolute filter in a vehicle undergoing a 3-month test. In 20 min of running, the total count of contaminant particles 5 micron and larger in the hydraulic system was reduced 100-fold. In 300 more hours, the count was reduced another 10-fold. Then, the 3-micron absolute filter was replaced with a new 25-micron nominal element, and within 100 hr the count *increased* 300-fold (Fig. 17.4).

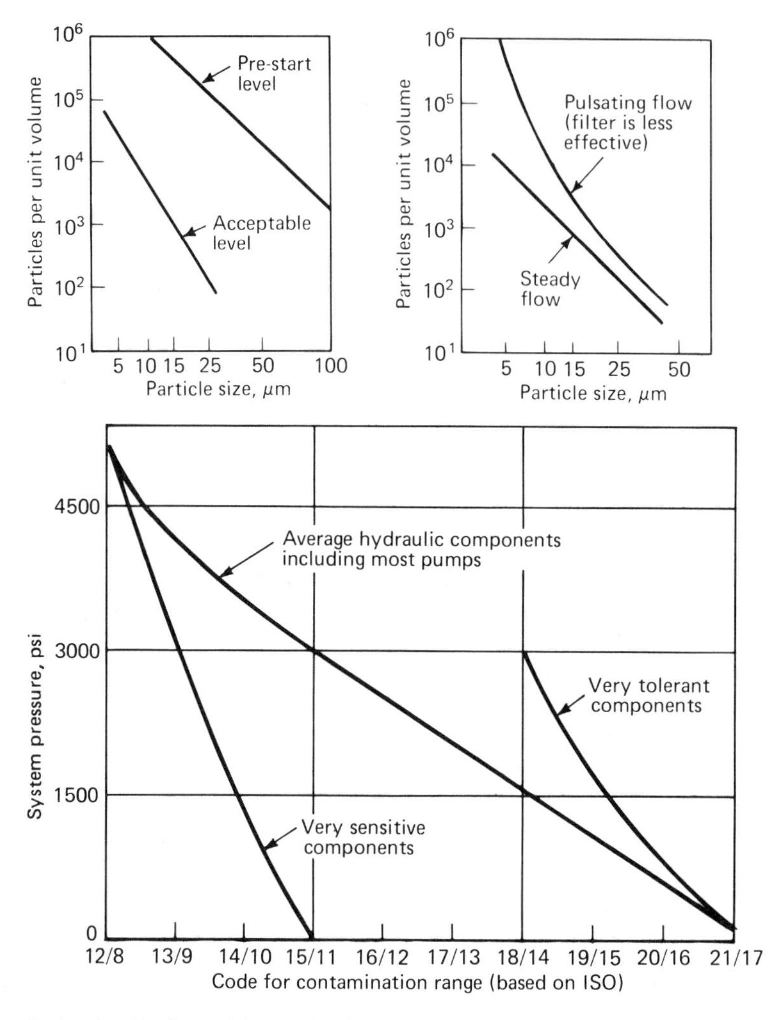

Explanation: Numbers to left and right of slashes are essentially powers of 2. The number of particles per ml are $2^{(N-1)}$ min and 2^N max, where N is the code number. Numbers to left of slashes are for particles $>5\ \mu m$, and to right $>15\ \mu m$.

Example: Code 14/10 means: 8192 to 16,384 particles $>5\ \mu m$; and 512 to 1024 $>15\ \mu m$.

The language of dirt

Defining contaminant dimensions and qualifying the filtering media to remove certain kinds and sizes of particles from oil is close to a black art. Everybody has a definition for a 5-micrometer filter or a 15-micrometer filter or whatever, but they contradict each other. The two most vocal schools of thought are the "absolutists" and the "nominalists".

In between are the ordinary people who refuse to get excited one way or the other. Anyway, most users tend to take so-called absolute and nominal ratings with a grain of salt, and rely instead on field data. If the filter works, who cares what the nominal or absolute rating is?

The accompanying graphs and jargon have been developed to explain contamination ranges in actual systems. The most widely accepted include: 1) The Beta ratio developed at OSU; and 2) log-log plots of particle count per milliliter versus particle size in micrometers.

The Beta ratio of an operating filter during steady-state-flow test is simply the count upstream divided by the count downstream of fine test dust, based on any selected particle size:

$$\beta X\mu m = \frac{\text{particles} > X\mu m \text{ upstream}}{\text{particles} > X\mu m \text{ downstream}}$$

where $X\mu m$ is particle size in micrometers. A ratio of 1.0 means that no particles are stopped. A ratio of 75:1 means that 75 particles are stopped for every one that gets through, an efficiency of 100 - (100/75) or 98.7%. Most filters with absolute ratings have Beta ratios of over 75:1.

The Beta ratios are plotted and manipulated in many ways to show separation efficiency, apparent capacity, actual capacity, and filter life profile—all for any chosen particle sizes within the capability of the instruments.

Experts agree that Beta ratios are practical, and are using them or plan to. One warning: the values measured with a clean filter are not always the same as those measured with a dirty filter. Be sure to ask for Beta values taken at flows and pressure losses expected in actual service. Don't accept data taken at 10 psid if your system will experience 150 psid.

Log-log plots of counts vs size are straight-line graphs, and are easy to read. The samples shown are typical, but are not meant to define any particular system. The first shows the spread between a system before and after it is properly flushed. The second shows the orders-of-magnitude difference in filtering effectiveness between a steady and pulsating flow system. For initial data, all systems should be run at no-load conditions until the contaminant level stabilizes.

The acceptable contamination level depends on the type of system and the application. Data have been accumulated for several years by various system users, hydraulic component manufacturers (pumps, valves and motors), and filter manufacturers. Sperry Vickers (Troy, MI) took some of these data and analyzed them to determine the maximum acceptable particle counts for several different kinds of hydraulic systems (final graph). Note that high-pressure systems need cleaner oil than low pressure systems do; it's because the forces are greater and the clearances are less.

Fig. 17.2 The understanding of contamination in oil starts here: definitions of terms—such as the beta ratio—and a method for determining how dirty the oil is.

Absolute vs. Nominal

The absolute rating is based on the maximum pore size of the filtering medium; the nominal rating (not an industry standard) is based on the average size of the particles stopped by the filtering medium.

The term "nominal" is widely used by many filter manufacturers but is not definable. The NFPA has deprecated the term in its glossary. Each manufacturer has his own method of determining the so-called nominal rating, based on some idealized performance, probably under constant-flow conditions, but there is no standard test.

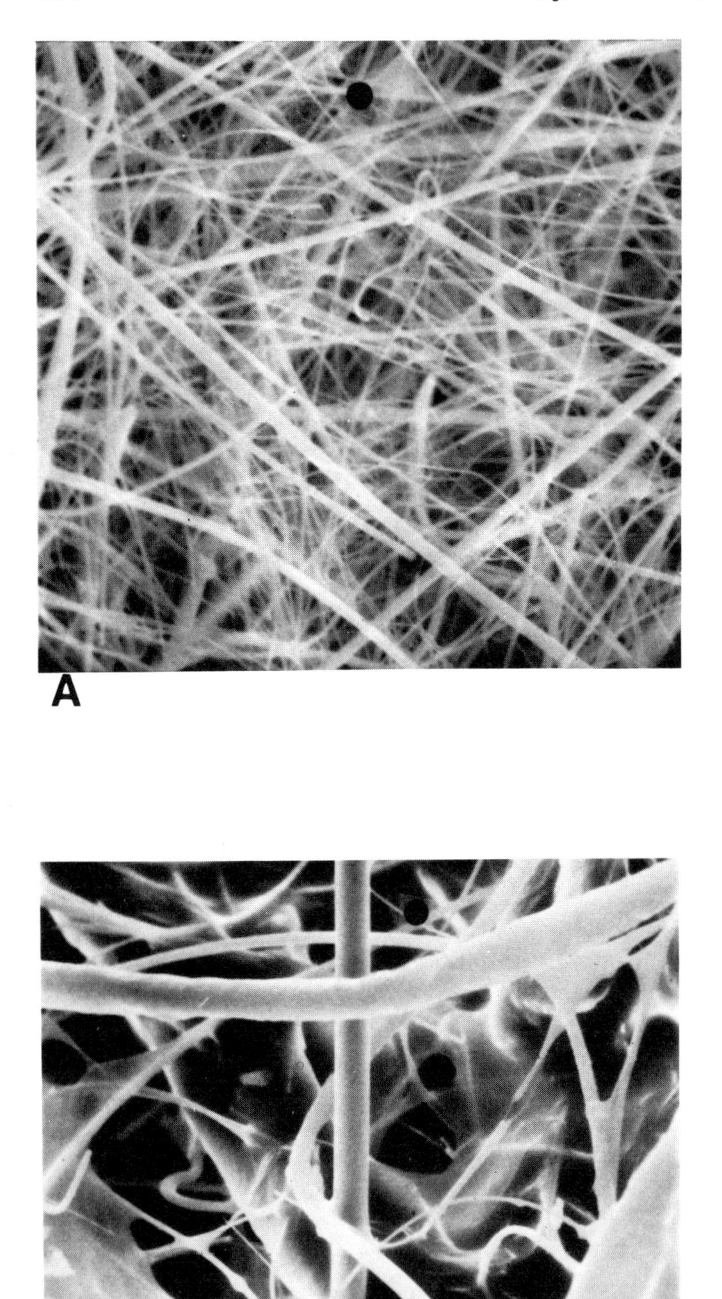
A
B

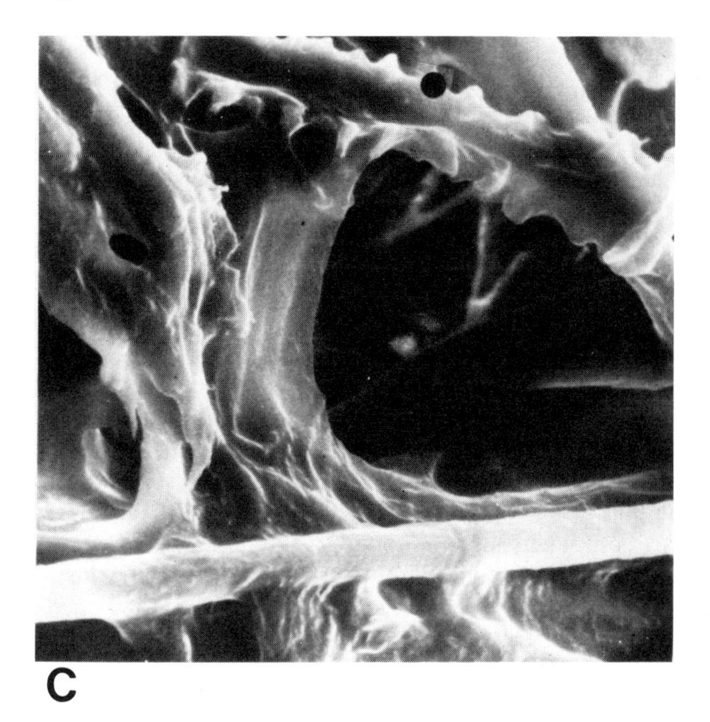

Fig. 17.3 Absolute 3-μm filter paper in photo A (Courtesy of Pall Corp., Glen Cove, NY) has tighter spacing than the nominal 3-μm paper in photo B and the nominal 10-μm paper in photo C. All photos were taken at the same magnification by a scanning electron microscope. The superimposed dot represents a 3-μm dirt particle that the filters are supposed to block.

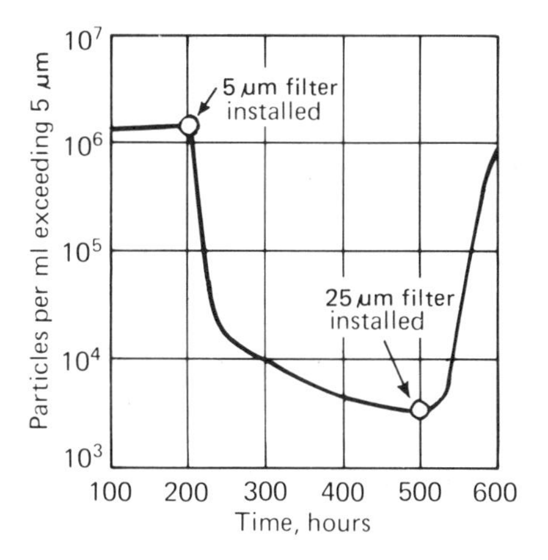

Fig. 17.4 Actual test of the hydraulic system in a vehicle shows the dramatic effect of installing, then removing, a finer filter. (Courtesy of Pall Corp., Glen Cove, NY)

The nominal rating usually is a small fraction of the absolute rating. Pall, for example, rates its 3-micron absolute element as 0.45-micron nominal.

Typically, each 10 gal of oil filtered to 10-micron nominal contains from 1 to 3 billion silt particles (Fig. 17.5). A 3-micron absolute filter removes enough of them to extend valve and pump life from 4 to 100 times, according to Pall.

A true test of filter rating should be variable-flow, not constant-flow. Constant-flow is a more practical test but might allow easy capture of particulate matter on the element face, artificially improving the micron rating. A variable-flow test (and actual operation later) will dislodge many of the particles. Tests at OSU (Oklahoma State Univ.) and elsewhere confirm this.

Assembling Superclean Systems

In some aerospace and other critical applications, we've gone past the point where there is a choice of coarse vs. fine filtration. Only fine filtration, 3 micron and better, will positively eliminate particles that cause wear, erosion, plugging, and jamming of precision fluid power components. The required level of cleanliness is much stricter than that in the best fluids delivered to the aircraft or equipment manufacturer.

Even making the filters is difficult—you need clean-room conditions. For example, Circle Seal's Microporous Filter Div. (Anaheim, CA) makes and qualifies wire mesh and composite filters for hydraulic use down to 3 micron.

In-process inspections had proved self-defeating because they exposed the filters and parts to atmospheric contamination that was greater than the test requirement. The first culprit was found to be a lubricant used in the original drawing of the fine stainless steel wire: It collected airborne contaminants. Special solvents, followed by normalizing heat-treating in a reducing atmosphere, eliminated the problem. The resulting surface ash was relatively easy to remove by ultrasonics. But the ultrasonics also dislodged particles from rough-machined surfaces elsewhere in the test system, and these surfaces had to be electropolished and passivated ahead of time to prevent it.

At final assembly time, it was found impractical to conventionally fill the oil tanks of the system. A superfilter on a portable cart that could be brought right to the oil tank was the only answer. After that, elaborate machine flushing procedures were shown to be essential to make sure that hidden internal contamination didn't negate all the previous work.

Cincinnati Milacron came to similar conclusions. The company has been a leader for years in the application of superfiltration techniques to servo-controlled machine tools. It made a design study to determine how long it would take to pay off the extra cost for fine filtration and discovered it was less than 1 month! This is based on the fact that servovalve and servomotor life is increased fourfold if critical-size contaminants are removed. Critical-size contaminants are those that approximate the size of the operating clearances (minimum hydrodynamic film) in moving parts such as valve spools and rotors. The company estimates that a quarter of a teaspoon of dirt per 55-gal drum of system oil is all that can be tolerated.

All of this sounds like a clarion call for superfiltration, but who wants to assemble a system under clean-room conditions? Prefilter all oil? Never add new oil to a sensitive hydraulic system directly? Specify superflushing at high Reynolds number? Monitor and maintain the system frequently? Yet that's what is required.

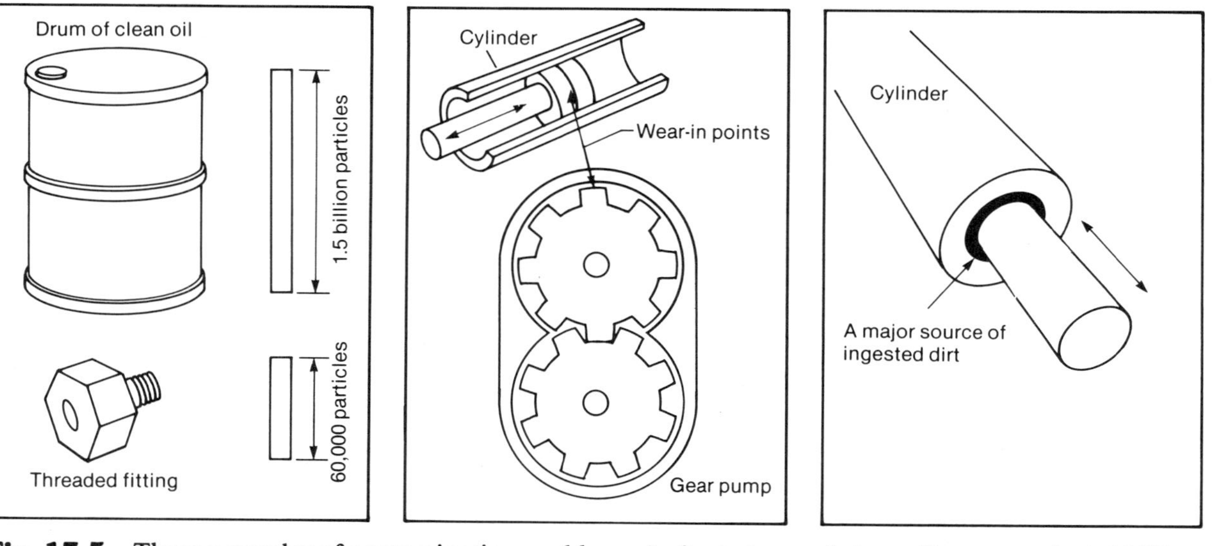

Fig. 17.5 Three examples of contamination problems. Left: A drum of clean oil can contain 1.5 billion particles; a fitting, 60,000. Middle: Contaminants are generated by piston rings and gears wearing in. Right: Ingested dirt at rod seals is a major source of contamination.

Startup failures of both new and overhauled systems often are contaminant-caused catastrophic failures. Contaminants built into each component making up the hydraulic system, and contaminants generated in assembling the components and systems, are significant contributors.

System flushing is a good first step—if done right. But don't forget Reynolds number, $\rho DV/\mu$, where ρ is density, D is pipe diameter (or some other representative dimension), V is velocity and μ is viscosity (see Chap. 1).

Reynolds number for hydraulic oil is eight times higher at 180 F than it is at 75 F. Flushing at 75 F for hours won't dislodge everything: A lot more dirt will show up after the system operates at 180 F in the customer's plant.

Also, if the system or vehicle is shipped after final flush, contaminants can be dislodged by mechanical abuse in the shipping process itself. Field operation can have the same effect, and for a safe flushing procedure, you should vibrate the system at least as vigorously as you expect it to shake in operation.

Steps to Improve the Odds

The cleanliness level can be fairly accurately defined, and there are reliable design and test procedures: It boils down to four steps:

- Determine the maximum level of contamination the system can live with on a continuous basis without damaging pumps, valves, and motors.
- Design and test a filtering system that can maintain it.
- Deliver the system to the customer at least as clean as the promise.
- Make sure he/she keeps the faith.

Each of those four steps is a career in itself. The engineering tools are available: automatic particle counters to measure the dirt level in a flowing stream, radiation-type monitors to detect abrasive wear of pump and valve surfaces (Fig. 17.6), and multipass lab test rigs to simulate actual system flow and contamination levels (Fig. 17.7).

One answer is to do the filtering off-line. That is, simply recirculate all the oil in the reservoir through a special filtering system either continuously or periodically. The advantage is that flow is constant and maintenance is easy. The disadvantage is that it takes considerable time to completely cycle all the oil; it doesn't protect against sudden ingestion of dirt.

What about atmospheric dirt? Air breather filters on top of reservoirs are one answer, but they should be carefully chosen. It's pointless to filter oil to 5 μm inside a system while allowing 50-μm particles to be sucked into the reservoir from the atmosphere every time a cylinder cycles.

Cylinder rod seals (see Chap. 20) also are a great ingester of dirt. They have to ride on an oil film to prevent friction, and this creates an entry path for dirt. OSU tests show that rod seals ingress about six particles greater than 10-μm size for each square inch of swept rod area. Wear of seals or wipers can increase the ingression rate considerably. In bad ambient conditions a 2-in. rod in a 4-in. bore, cycling at 36 ft/min, can ingress 16,000 particles larger than 10 μm every minute, and the ingression rate increases greatly with time.

Even in-shop assembly is fraught with hazard. Experts say, Don't try to clean components or systems after they are fully assembled. That's like trying to wash your

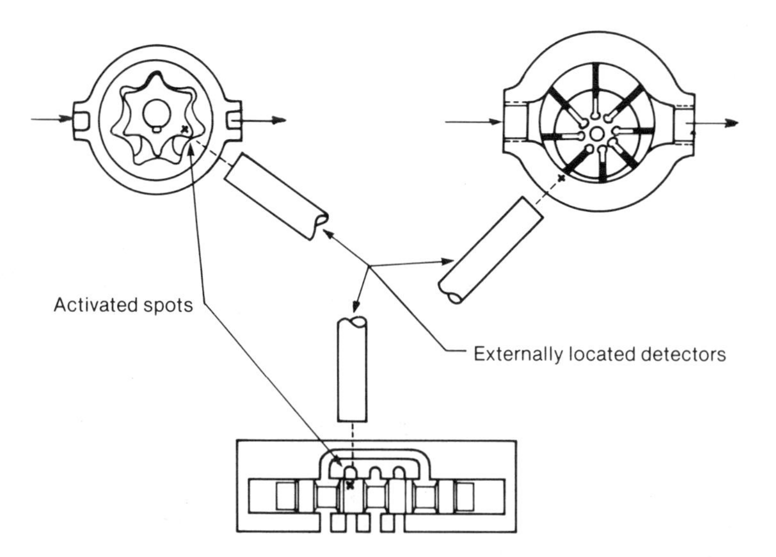

Fig. 17.6 Wear patterns can be detected by irradiating the points of interest, and monitoring them with radiation detectors. (Courtesy of Fram Corp., East Providence, RI.)

underwear by taking a shower with all your clothes on. They further note that there are five sources of dirt in hydraulic systems: built-in dirt, generated dirt, drawn-in (ingested) dirt, repair or rework dirt, and escaped dirt. Here are design tips for all of them:

- Avoid components with built-in dirt traps—you'll never win.
- Clean the components beforehand in one or several of these ways: soap and water, acid cleaner, alkaline cleaner, solvent, ultrasonics, and mechanical cleaning.

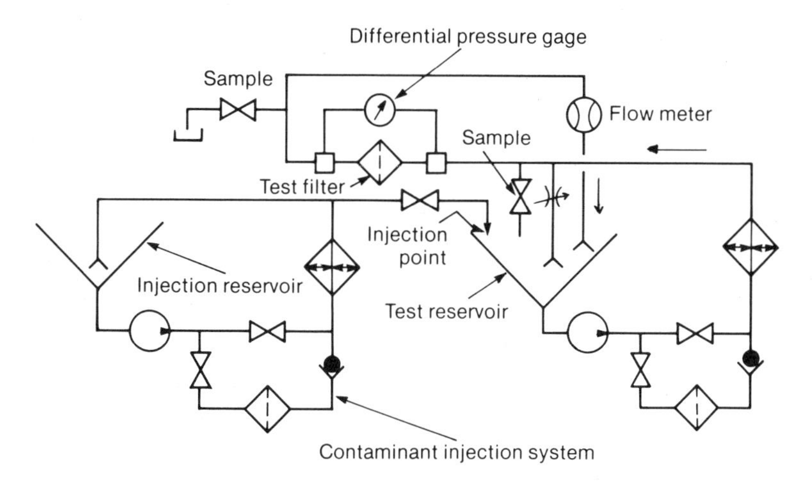

Fig. 17.7 Multipass filter test developed at OSU is accepted worldwide, but not every company has the test rig.

- Use rust preventative for steel or iron parts, but make sure it's clean.
- Flex the hoses to release dirt that might be trapped. One manufacturer uses 50 flex cycles.
- Don't trust that a new barrel has clean oil (Fig. 17.5). Even a clean 55-gal barrel of oil can add 1½ billion particles 10 μm or larger to your system. Filter the oil before using it.
- Do you think straight-thread O-ring fittings are clean? Not so. Tests at Pall show that it is not unusual to add 60,000 particles greater than 5 μm each time you assemble one. Pipethreads are worse, and sometimes they add a little Teflon tape as well.

There is a way to get all that dirt out: Hook up a filter cart ("kidney machine") with high-performance fine filters. In 15 min or less, a good flush is possible. Also, it's a good way to fill the reservoir (Fig. 17.8).

Another warning: Remember that some gear pump manufacturers achieve close clearances by machining the pump housing undersize and letting the gears cut the housing during startup! It's a neat trick but generates metal particles. Piston rings also generate particles as they wear in (Fig. 17.5).

Every Kind of Test

There are national and international standards for classifying and measuring contaminants (size or weight), acid content, flow, collapse pressure, structural integrity, filtering capacity, and pressure drop.

Particles can be either counted or weighed, but counting is preferred because of the availability of accurate automatic particle counters. The counted sizes can be classified per existing standards, such as National Fluid Power Association (NFPA) recommended standard T2.9.3 R1-1979, American National Standards Institute (ANSI) B93.30-1973, National Aerospace Standard (NAS) 1638, Military Standard (MIL-STD) 1246A, and International Standards Organization (ISO) 4402.

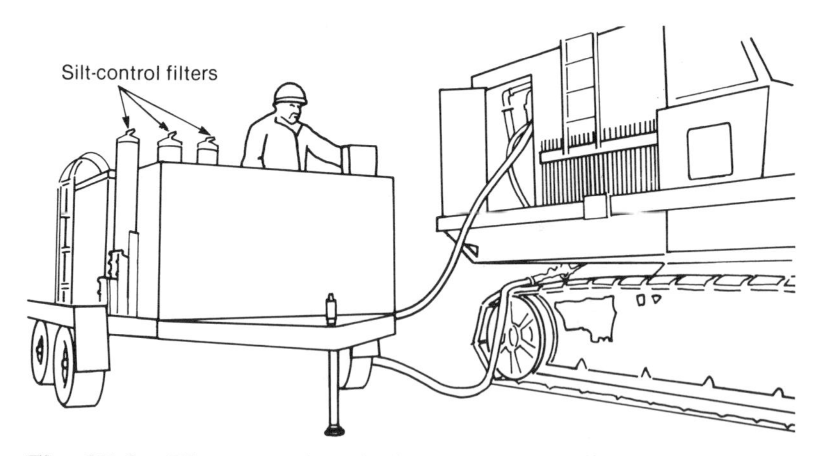

Fig. 17.8 Filter cart solves dual problem of massive flushing of system and refilling with oil that is known to be clean. (Courtesy of Pall Corp., Glen Cove, NY.)

Measurement and test of liquid contamination is more difficult. The water content may be found by a distillation technique issued by American Society for Testing and Materials: ASTM D-95-62. The acid content is measured in terms of neutralization number using ASTM D974.

Some older tests are still in use: permeability of AC test dust measured in seconds/1000 cc of reference fluid, bubble point to measure maximum pore size, mean flow pore size, DOP (dioctylphthalate) efficiency, filter capacity, and air flow resistance.

But a better and more universal test is the Multipass method for evaluating the filtration performance of a fine hydraulic fluid power filter element (Fig. 17.7). It was developed over a decade ago under the direction of Dr. E. C. Fitch at the Fluid Power Research Center, Oklahoma State University (OSU) and now is an accepted national and international standard: NFPA T3.10.8.8-1973, ANSI B93.31-1973, and ISO DIS 4572. Pall Corporation was the first filter manufacturer to build a multipass rig.

Now this multipass test method is the only industrially accepted means for filter element evaluation. It measures contaminant separation, retention, and pressure loss. In essence, the same contaminated fluid is circulated continuously through the test filter, and additional contaminant is added continuously to the stream to maintain the original concentration.

Pressure drop across the filter vs. time is recorded. At a predetermined drop, fluid samples are taken upstream and downstream of the filter. Using an automatic particle counter calibrated per ANSI B93-28-1973, the samples are analyzed for the number of particles per milliliter greater than selected sizes. These counts are used to calculate the "filtration ratios" (called beta ratios). It's all explained in Fig. 17.2.

Materials and Structure

With the advent of synthetic fluids, water-based fluids, and other chemical combinations, filter materials become touchy. Cellulose, for example, swells in water and closes off pores. Here are some other observations:

Microfiberglass media are state-of-the-art and doing very well in regular and fine filtration. They include fine enough fibers for silt control (5 μm and smaller particles), are not too expensive, and are made to be disposable. The disadvantage is low strength, but with epoxy bonding the problem is eliminated. The epoxy withstands synthetic fluids well.

Gaskets are not as good as O-rings in filter assemblies that experience flow surges and other abuse. An O-ring seals better as pressure is increased. A gasket is more likely to leak or bypass.

Five international test standards, developed at OSU, evaluate the structural integrity of the filter itself: fabrication integrity ISO 2942, collapse-burst ISO 2941, material compatibility ISO 2943, end load ISO 3723, and flow fatigue ISO 3724.

Fabrication integrity is a bubble point test that searches for defects, not for pore size. Collapse-burst applies high differential pressures such as those caused by contaminant loading, cold startups, and flow surges. The test is run by adding contaminant upstream of the elements at rated flow until the required pressure drop is exceeded or rupture occurs.

To test for material compatibility, the element is immersed for 72 hr in system fluid 15 C higher than rated maximum temperature. This test is followed by an end load test and a collapse-burst test. Flow fatigue is performed on an element that is loaded with contaminant to yield a specified pressure drop. Flow is cycled between zero and rated flow.

Another fatigue test, though not an international standard, is NFPA recommended standard T2.6.1-1974. It's a statistical method for estimating probability of failure caused by pressure cycling, based on actual data for the materials of construction but not based on finished components. It's controversial but probably is the best way to predict fatigue failure without testing to destruction (see Chap. 22).

Failsafe filters: Many filter manufacturers offer built-in valves and special safety features that protect the system in the event of careless maintenance or some unusual failure upstream of the filter.

The "dirt fuse" (really a "block") offered by Pall is one example. It is designed to continue functioning until the entire system pressure—up to 6000 psi—appears as pressure loss across the filter element. The filter will not fail but merely stops the flow.

Internal check valves, reliefs, bypasses, reverse-flow valves, and shutoffs are available in many filter models from various manufacturers. Some are designed to prevent backflushing filter dirt into the mainstream when flow reversal or bypass occurs. Others automatically close off flow ports when the filter element is removed to prevent loss of fluid or ingestion of dirt during routine maintenance. Still others offer differential pressure indication—even remotely—to warn that filter life is nearing its end.

One warning: The built-in valves are not precision instruments and have reasonable tolerances on operating points. If you rely on them, incorporate a safety margin.

Another warning: Don't overdo the operating differential pressure loss. A differential of 85 psid at 20 gpm wastes 1 hp at the pump.

Disposable vs. Cleanable

Metal-mesh hydraulic filter cartridges (Fig. 17.9) were heralded in the late 1950s as the best answer for reliable filtration where pressures and temperatures were high and

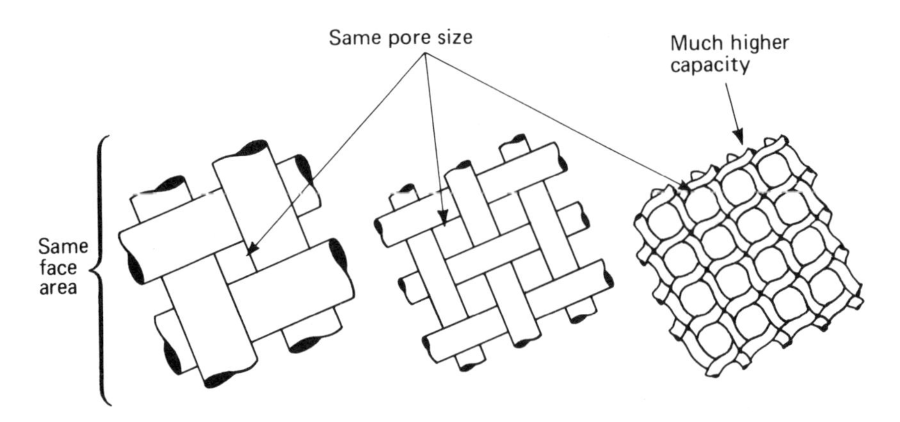

Fig. 17.9 Dirt capacity of woven wire mesh is affected by wire size: Each sample above has the same pore size but not the same capacity.

where cleanability was considered important. But today there are disposable high-performance paper cartridges made of inorganic fibers impregnated with epoxy resin. Pall Corporation offers one called Ultipor (Fig. 17.10).

There is some difference of opinion as to which is best, so we'll quote both sides. Pall will take the side of disposables (though they make cleanables too): Pall estimates that in 4000 hr of service on a typical aircraft, a cleanable filter might be cleaned three times, whereas a high-capacity disposable filter needs to be replaced only once. The cost of cleaning a metal-mesh cartridge in airline shops is nearly as much as the cost of a disposable cartridge, so it's cheaper to use disposables.

A further reason for choosing disposables, says Pall, is the extreme difficulty of setting up proper cleaning facilities in the field. Techniques perfected in labs often prove unreliable in the field. One of the difficulties is the removal of accumulated "varnish" that holds contaminant particles with tenacity. Only special solvents will remove the varnish. Another difficulty is avoiding transfer of the invisible contaminants to the downstream face of the element, where they later can be dislodged during system startup, with potentially catastrophic results.

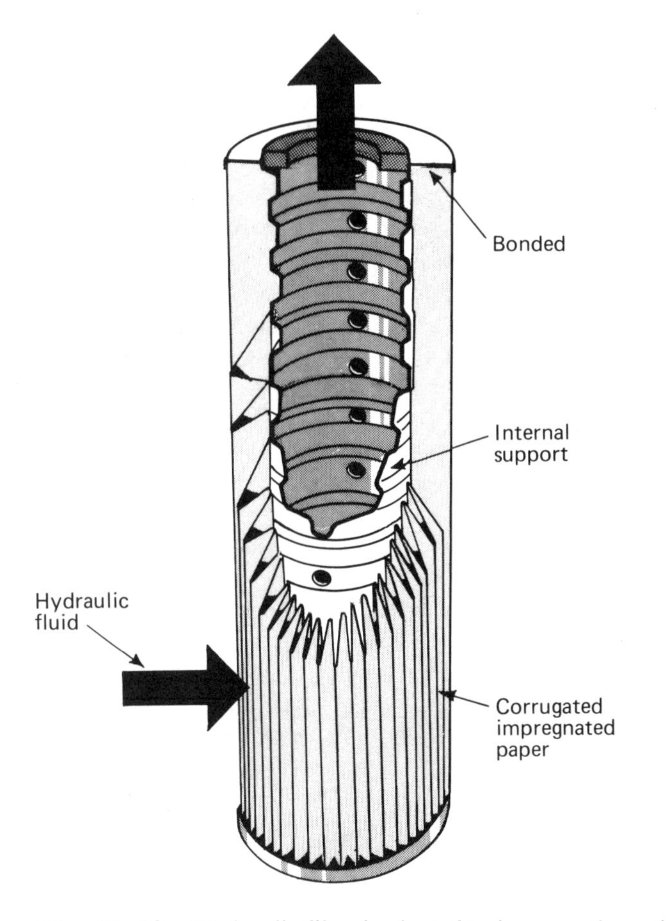

Fig. 17.10 Hydraulic filter is pleated to increase the active area and bonded to metal end caps for strength. (Courtesy of Pall Corp., Glen Cove, NY.)

A counter argument is offered by HR Filters Co., which developed a metal felt cleanable element (Fig. 17.11). The company claims that a simple air backflush with household detergent will suffice in most situations. For highly contaminated filters with special cleaning problems, a more elaborate procedure is necessary. Equipment then required includes two stainless steel pans, a hot plate, a water flush nozzle, an air nozzle, and assorted auxiliaries. The procedure involves a sequence of air cleaning, boiling with two mild chemicals, and water flushes.

PNEUMATIC FILTERS

The bulk of discussion on conventional air filters is in Chap. 11. There, an entire shop air system including filters, regulators, and lubricators is described in detail.

Left to discuss are two important and unusual examples of the air filtering art: the coalescing (droplet-agglomerating) filter and the dynamic air filter for gas turbine inlets.

Coalescing Air Filters

Coalesce means "to grow together." In air filter terms, it means the agglomeration of minute liquid droplets into larger ones, so they can be more readily captured. The minute droplets are among the pollutants in shop air systems, originating primarily in the plant compressor. They don't usually cause damage to air tools, but they can spoil pneumatic logic performance.

Wilkerson Corp. (Englewood, CO) categorized for us the approximate sizes of oil and water particulates in compressed air and other gases. For example, vapors are 0.001 to 0.01 micron (a micron, or more correctly a micrometer, is 1 millionth of a meter, or about 0.00004 in.). Other examples are the following: fumes, 0.001 to 0.1; fogs, 1.0 to 50; mists, 50 to 5000; and sprays 500 micron and larger. The general name for all of them, plus other particulates in air, is *aerosol*. An exception is vapor, which really is in solution with the air as a molecule and is not a particle or droplet.

To rid the air of such aerosols requires a combination of filtering phenomena: direct impaction, inertial impaction, diffusion (Brownian movement), and coalescence. The first three actually enhance the last and most important: coalescence.

Coalescing action works best at low air velocities and at room temperature. Furthermore, the droplets coalesce best if they have high surface tension, are clean, and have high differences in density and high interfacial tensions among the different liquids.

Not all of these conditions exist in typical airstreams, so geometry of the filter becomes important. For example, baffles and centrifugal flow inducers will increase impingement and coalescence. Permanent electrostatic charges (or similar phenomena) in the fibrous filter bed may help, too.

Other mechanical and chemical requirements for effective coalescence include a high ratio of pores to strands in the filtering media (to allow airflow without reducing surface area), mechanical strength (to prevent media breakdown and migration), and chemical inertness. One idea, developed initially by the British and incorporated in the Wilkerson units, is borosilicate glass strands with a mean diameter of about 0.5 μm.

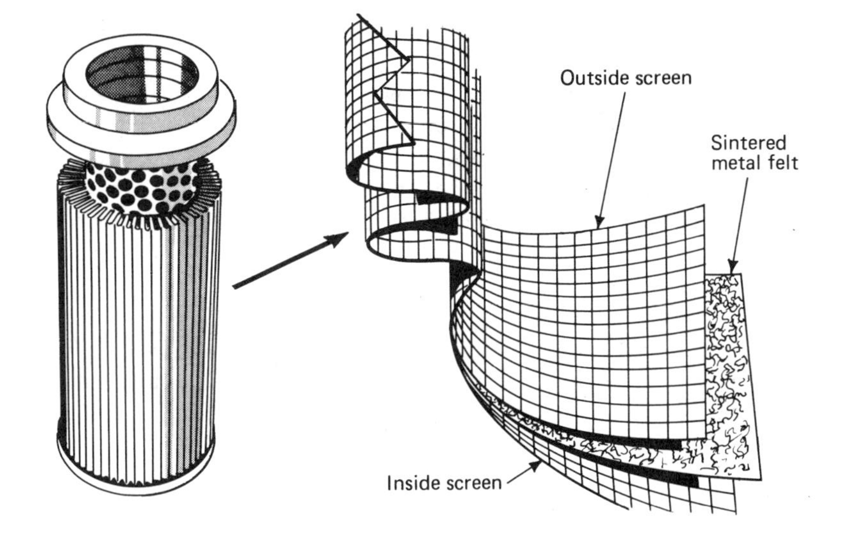

Fig. 17.11 Sintered metal felt is made of fibers about 6 in. long and less than 10 μm in diameter. Pore size is much smaller than that for woven wire mesh. (Courtesy of HR Filters, Div. Textron, Pacoima, CA.)

Coalescing Air Filter Design. The following details relate directly to the air filter cartridge made by Wilkerson, called MICROalescer (Fig. 17.12). The others are similar in principle, although each claims various operating or cost advantages.

Flow in the Wilkerson unit is radially outward through successive layers that have different functions. The pressurized air or gas flows in from the top and passes first through the impregnated wound-ribbon paper core.This core screens out larger dirt particles, much like a conventional air filter. If large amounts of dirt are expected, the company recommends that you specify a regular 5-micron filter upstream of the coalescing filter.

The second layer contains coalescing material: microscopically fine fibers of borosilicate. They trap subminiature particles of dirt and entrain and coalesce liquid water and oil aerosols (microscopic droplets suspended in the air). The company claims that an extremely high percentage (over 99.99%) of submicroscopic dirt particles are trapped and that nearly all of the liquid aerosols coalesce (agglomerate) into droplets heavy enough to drop into the filter sump. The pores are smallest at the ID and become progressively larger towards the OD, to allow for droplet agglomeration.

The final layer is a plastic open-pore foam cover that allows the air to pass through but prevents carryover of liquid. The droplets run to the bottom, inside the foam, then work their way over the bottom lip, and fall into the sump.

All of the oil and water aerosols are continuously removed, *without* saturating the fibers. In contrast, ordinary mist-removing filters are like wicks: They absorb the liquid, eventually saturate, and begin passing liquid into the exit airstream.

Other particulates are stopped, too. Even bacteria can be trapped during initial passage through the coalescing layer, and 99.99% of solid particulates 0.01 micron in size are trapped radially further outward by inertial impaction. The absolute rating of the element is said to be 0.03 micron, based on extensive testing done in the company's lab and at independent Hauser Lab in Boulder, CO. The pore size is not anywhere near as small as 0.03 micron: Brownian motion (diffusion) causes the submicron particles to eventually impact against the fibers, where they are held through some attraction phenomenon not yet fully understood.

In contrast to the aerosols, water vapor and oil vapor are not coalesced, because they are in solution. The water-vapor dewpoint (condensation temperature) of the air does not change during passage through the filter. If the air has 100% relative humidity when it comes in, it will be equally saturated when it comes out. This is important to understand because any cooling of saturated air during passage through, or downstream of, the filter will create liquid water mist or droplets.

The uninterrupted service life of coalescing-type filters is much longer than for absorbent-type mist removers, because the coalescing droplets can be conveniently drained while in use, whereas absorbed moisture cannot. Many users claim 1 or 2 years of normal service life for coalescing filters.

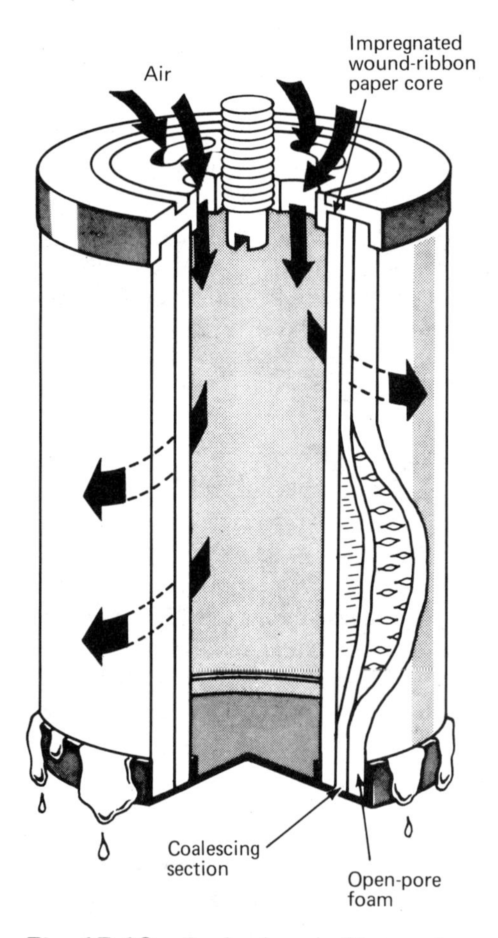

Fig. 17.12 Coalescing air filter agglomerates tiny liquid particles until the particles become large enough to be captured. (Courtesy of Wilkerson Corp., Englewood, CO.)

GAS TURBINE CENTRIFUGAL FILTERS

There's no practical way to squeeze high volumes of dirty air through cramped conventional barrier filters, as aircraft gas turbine designers have found out. A better method is to separate the dirt from the air centrifugally (Fig. 17.13), as is done in fly-ash removal systems in many power plants.

Not just aircraft but any vehicle or machine that has a high-flow gas turbine or other engine for power can benefit from centrifugal separators. Examples are long-haul trucks, off-the-road vehicles, marine engines, and stationary power generators.

Silencers are necessary companions to separators when the applications are in human earshot.

Clean and Dry

Ingested particles from about 20 μm (0.0008 in.) and up in size are harmful to gas turbines. Fortunately this size particle is massive enough to separate centrifugally from the airstream if the air can be spun rapidly. Water droplets also are removed by this action.

Mechanically rotated separators are impractical in the limited space available at the inlet of most gas turbines, but vortex separators with no moving parts will work if designed carefully. The Pall design, trademarked *Centrisep,* is only 2¼ in. deep (Fig. 17.14).

Four elements are needed in the separator portion (Fig. 17.15): one or more flowpipes in parallel to handle all of the incoming air from atmosphere, fixed vortex-generating vanes in each flowpipe to impart spin to the air, an outlet tube from each

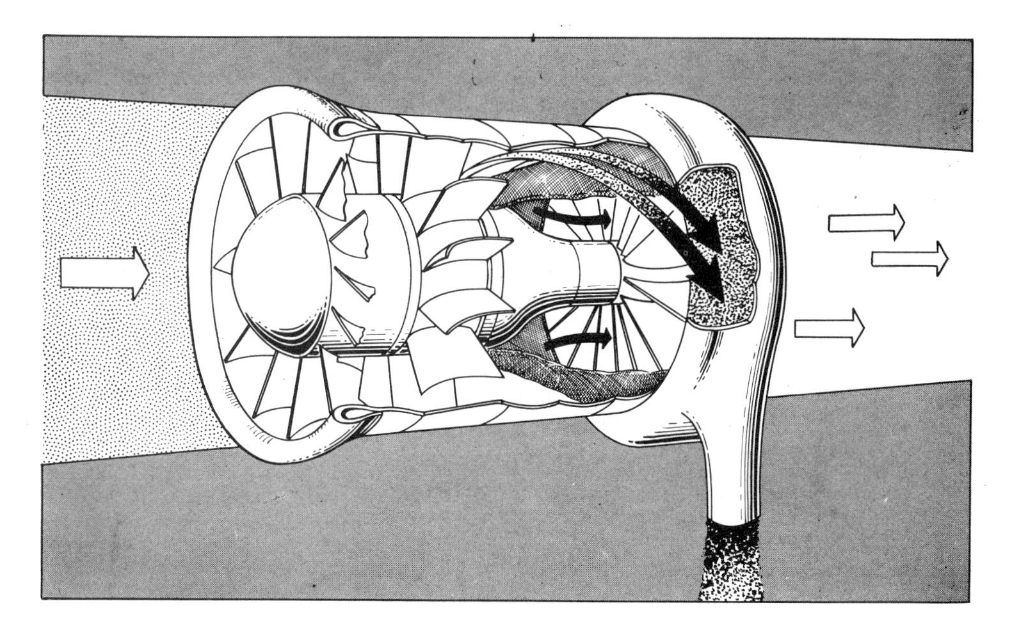

Fig. 17.13 High-speed rotor in cyclone-type dust separator is able to throw out particles as small as 5 μm. (Courtesy of Dynamic Filters, Div. Michigan Dynamics, Inc., Garden City, MI.)

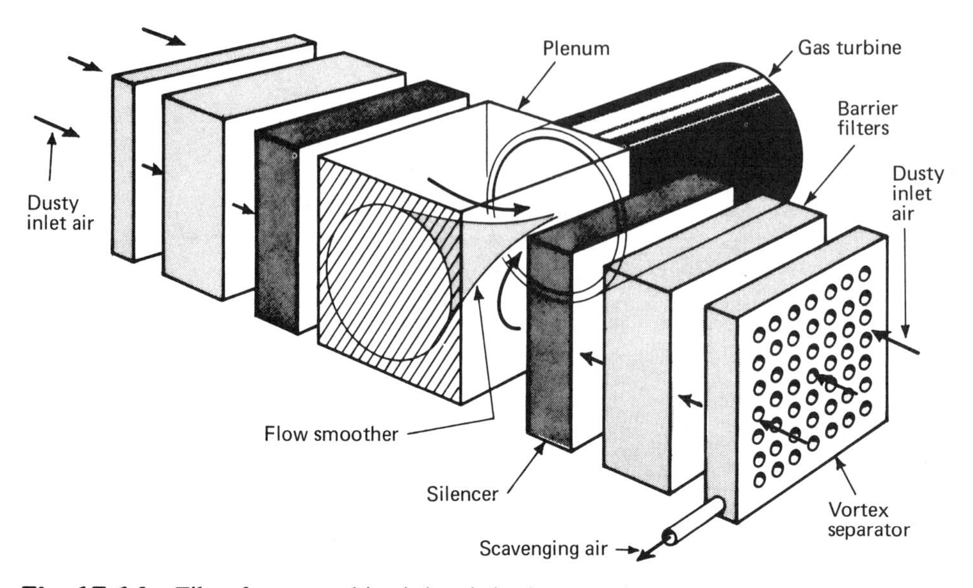

Fig. 17.14 Filter for gas turbine inlet air in dusty environment needs several stages. (Courtesy of Pall Corp., Glen Cove, NY.)

flowpipe to catch the dirt-free air which discharges from the center of the stream, and a means of sucking out the dirt-laden outer layer of air which hugs the tube inner wall (see drawings). The last is called scavenging airflow.

The effectiveness of particle removal is greatly influenced by the amount of suction available to pull out the contaminant-laden scavenging air. If a relative vacuum is available or can be readily created, there is no problem. Blowers can be used, but they are bulky. In most instances, the system has to generate its own vacuum by bleeding off pressurized air (10 psig or more) from someplace in the engine system and using it in an ejector.

An ejector creates a partial vacuum by venturi action while inducing up to 20 times the nozzle flow. In a typical gas turbine, the ejector flow plus the air scavenged out of the vortex tubes adds up to about 10% of the total airflow into the cycle. That is, 110% of the cycle airflow is drawn through the vortex tubes, and 10% is discharged.

Almost all heavy particles over 100 μm in size are removed. The finer the particle, the more likely it is to remain in the outlet airstream. Good efficiency is considered anything above 92% removal, based on standard test dust samples MIL-E-5007 test sand and AC coarse and fine test dust. Higher removal rates occur when two of the vortex separators are placed in series. Then 98% particle removal is attainable.

Conventional barrier filters often are applied in addition to the vortex types when 99.9% and higher efficiency is desired. The vortex is the first stage and removes most of the dirt that would clog the filter. It has been shown by test that barrier filter life can be increased up to 50 times with this combination. Total clean pressure loss through both stages is measured in inches of water—generally 2 to 4 in.

When liquid spray is a problem, a 2-in.-deep mist eliminator can be used downstream of the vortex separator, making approximately a 5-in.-deep package. The vortex section removes most of the liquid droplets down to 20 μm. The mist eliminator removes smaller droplets with an efficiency of 99% for 4- to 30-μm droplets and 97% for 1.7- to 4-μm droplets.

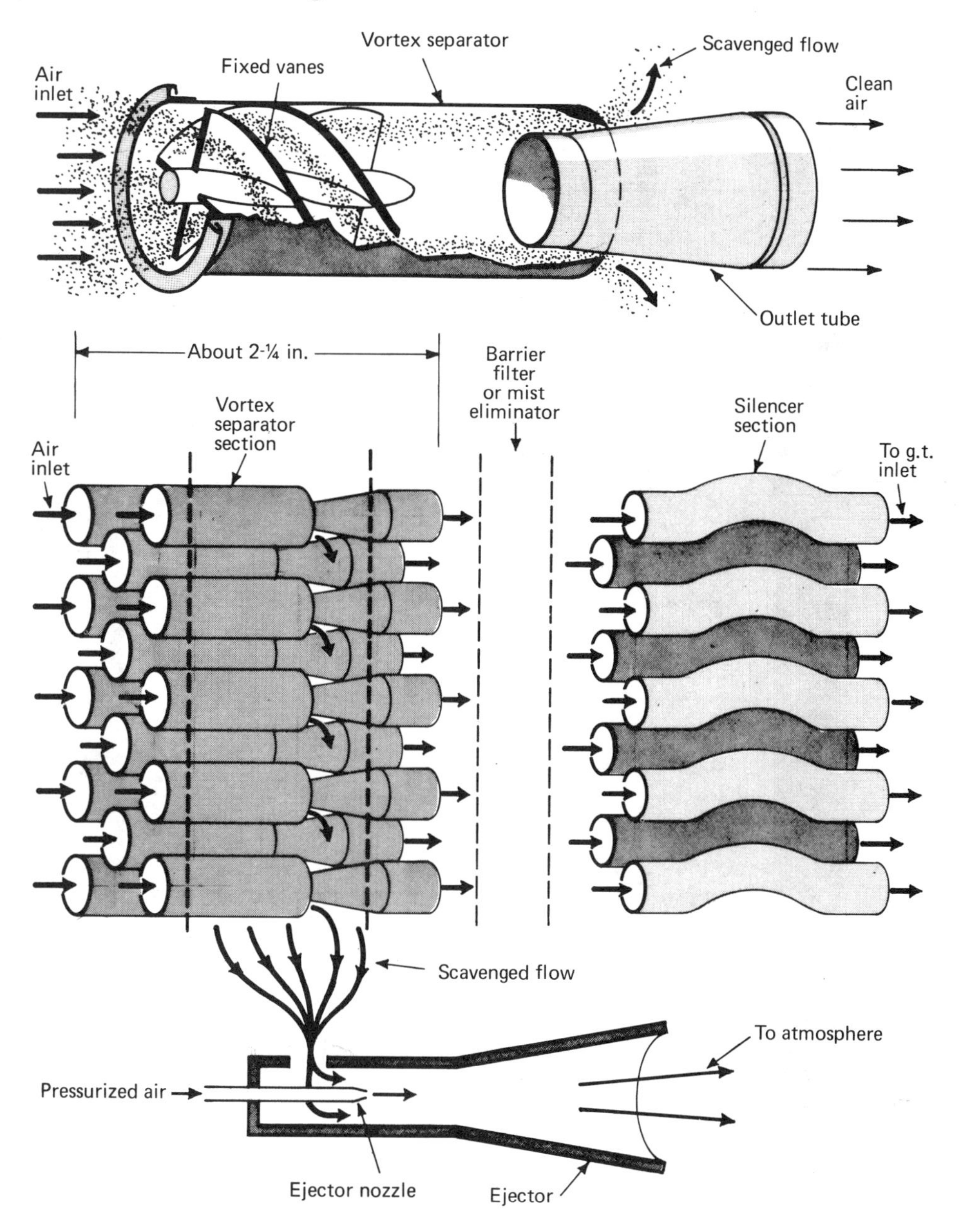

Fig. 17.15 Details of design of vortex separator in Fig. 17.14.

Quiet the whine: Silencers for high-flow inlets are not just padded pipes but are sophisticated acoustical chambers. The sound wavelengths to be attenuated must be known, and resonator principles must be applied. The Pall designs feature airflow tubes in series with the outlet tubes of the vortex separators. The silencer tubes have a length and bend radius that precisely match the wavelength of the generated sound and dissipate it.

Experience is critical when designing silencers. Sound generation is one of the most complex of the engineering phenomena and is not fully understood. The ideal

	METALS							
	Sintered Particles		Edge Type			Mesh and Cloth		
						Sintered		Nonsintered
	Fiber	Powder	Disk	Ribbon	Wire	Wound on mandrel	Woven	Woven
General appearance of medium								
Filtering action	Primarily surface		Surface					
Removal ratings, microns (based on smallest particle size removed). 98% removal (nominal)	No data	2-60	40-125	25-500	75-700	2 and over		
100% removal (estimated absolute rating)	No data	4-60	No data	50-500	125-700	12 and over		
Applicability—assuming reasonable flow conditions, is medium suitable for these dirt concentrations? >5% of liquid volume	Limited application only					Recommended		
0.1 to 5%	Recommended							
<0.1% (clarification)	Recommended			Not recommended		Recommended		
Cleanability	Cleanable (re-usable) based on nondestructive methods such as: chemical, ultrasonic, backwash, blowback							

*Estimated data based on media only, not on manufactured cartridges or systems; removal ratings based on liquid — with gases, filter performance is usually better, particularly in low-micron ratings.

NONMETALS												
Fibers (natural and synthetic)				Granules and Powder						Edge Type		Mono-lithic
Compressed or packed												
Felted	Nonfelted	Woven Cloth	Wound Yarn	Sand	Sintered Plastic	Fired Porcelain	Fritted Glass	Bonded Carbon	Bonded Stone	Ribbon	Disk	Membrane
Depth and surface	Depth	Primarily surface	Depth	Depth	Surface	Primarily surface				Surface		
5-100	10 and over	25 and over	1-100	30 and over	3-15	C.2-25	0.1-125	10-70	5-100	Available: 40 only	0.5-10	0.1-12
No data			10-150	No data	13-20	0.2-25	0.1-125	No data			5-10	0.1-12
Not recommended		Recom-mended	Limited applic.	Not recom-mended	Limited application only							Not recom-mended
Recommended		Not Recom-mended	Recom-mended		Recommended							Limited applic.
Throw-away type		Cleanable (re-usable) by nondestructive methods								Throw-away type		

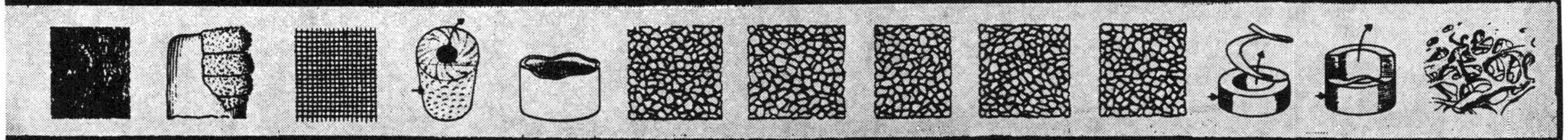

Fig. 17.16 Performance comparisons of typical filter-media constructions.

silencer will match the operating conditions of the engine in a very limited range of speed and load. Outside that range it not only will be ineffective as a silencer but might even amplify the sound.

Gas turbine inlets, for example, generate sound most strongly in the 4000- to 16,000-Hz range, with 8000 Hz the most typical.

MEDIA FOR ALL FILTERS

The porous and woven materials (called media) in filters have been continuously developed over the years to solve contamination-removal problems. This summary covers most of them (see Fig. 17.16).

Broadly, there are two types of media: surface and depth. In surface media (wire mesh, membrane, and simple edge-type are three examples) dirt is removed by a plane surface with fairly uniform orifices. Depth media (paper, felt, glass fiber, sintered powders, matted wool, and wound spools are examples) depend on tortuous paths through the material to remove particles.

With surface-type media, the number of openings available for clogging determines the time between cleanings for a given, allowable pressure drop. This cleaning is especially important where dirt particles are small and compressible: Relatively small accumulation of cake gives a disproportionate increase in pressure drop. It is possible to convolute the surface (as done with wire mesh) to give up to nine times the surface area within a given envelope volume compared with a simple cylinder.

Depth-type media have a high dirt capacity for a given filter volume, particularly if concentration of solids is small; the particles work themselves into interstices between fibers. However, for large concentrations of solids, cake buildup may be quite rapid, and the filter becomes essentially surface type.

Each type—surface and depth—has some characteristics of the other, and simple categorizing is not always possible. The following examples do not attempt to show direct distinction. Also, most comments apply to both metals and nonmetals.

Granular (porous) media have adjacent grains and these particles usually are joined to form a porous, rigid structure. Sintered powder is a good example of a joined-particle medium; sand is not, it is often used in loose form. The grains can be joined with a compatible adhesive or by sintering at temperatures slightly below melting point of the particle material. Sintering may be done with or without the simultaneous application of pressure. The degree of filtration depends on compression and on smallness of the grain chosen. Dirt particles are trapped in the interstices.

Media made of bonded granular materials are normally thicker in cross section (for manufacturing reasons) than woven-wire media, for the same fineness of filtration. Pore sizes are more uniform in granular media than in ordinary unsintered square-weave mesh (Fig. 17.9), but special interlocking weaves (dutch twill, for example) overcome this disadvantage.

Smaller contaminant particles can be filtered out by granular media than by woven types, but there is wider variation in particle size stopped and a greater tendency toward contaminant migration. Extreme fineness is possible: A leached-glass porous medium has a removal rating better than 0.01 micron, but it is not an industrial-type filter. Granular media are most suitable where solids concentrations are very low.

Porous depth-type filters can be cleaned to a limited extent by backwashing. Progressive fouling will occur after many backwash cycles, but the element can be restored to its original state with a suitable cleaning solution.

Edge-type filter media may be either metal or nonmetal. The fluid filters between edges of flat discs or wires of circular or other cross sections. Spacing determines the degree of filtration. In the disc type, spacing is with alternate discs that are smaller in dimension or with a central ring structure that has radial projections. The wire type is spaced by winding the wire around a mandrel or by using projections on the wire itself.

Long particles can get through an edge filter if one dimension (diameter or thickness) of the particle is smaller than the edge opening. (Round or cubical particles are stopped.) If downstream devices have small orifices or small-clearance moving parts, just a few fibers passing downstream can readily bridge the orifice or clearance, restricting flow or movement. Composite layer filters will solve this problem.

Woven media are cloths of wire or yarn (natural or synthetic). Both are produced on looms and use loom terminology: *Warp* is the wire or yarn that runs the length of the cloth; *shute* (sometimes called *woof*) runs perpendicular to and is woven into the warp by means of a shuttle.

Many types of weaves are available. The most common are square (plain), dutch, twill, and twilled-dutch. Plain and twilled weaves present straight-through openings to fluid flow; twilled-dutch and dutch weaves present a somewhat tortuous path and hence finer filtration.

Wire cloths may be sintered or unsintered. Sintered wire cloth has warp and shute wires sintered together where they cross, fixing the pore size permanently. In addition, sintering allows integral bonding of coarser meshes to finer facings to produce a stronger composite.

Woven cloths are most often used where the concentration of solid is high enough to permit rapid cake buildup for finer filtration. They also work without a cake—typical are those wire-cloth media having nominal (98%) removal ratings about 70 microns and smaller. If cake is used to improve filtration, recirculation (until cake builds up) will be necessary before the best rating is achieved.

Woven yarn media are basically flexible, and changes in pressure (including instantaneous reversal) can temporarily displace filter cake away from the filtering surface. Fine solids will then pass through the filter—bypassing the cake—until the displaced cake has settled again.

Wound wire mesh is a form of cloth made by helically cross-winding a fine flattened wire on a mandrel. Wires are then sintered where they cross, and finished mesh can be pulled from the mandrel to be slit into sheet stock or kept as a cylinder.

Felted media are masses of fibers matted together to produce a porous structure. Good felts are carefully designed to have interlocking fibers. They may or may not be resin-bonded. The degree of filtration obtained, regardless of the fiber type, depends on fiber diameter and length. Smaller fibers give finer filtration. Paper filters are similar in structure to felted media.

Felted media are usually of the depth type, are thrown away after use, and are limited to relatively low pressure differentials. Media migration will occur under some circumstances, particularly if fibers are loosely interlocked.

Unfelted fibers are initially loose but are packed into a woven or knitted bag. Cotton or other cellulosic fibers are the types usually chosen, and the major application is for in-plant oil filtering. Application is limited by temperature and pressure, and media migration is likely to occur. Filtering ability is improved with fillers such as diatomaceous earth.

Wound yarn media are commonly made in cylindrical form, helically wound around a support core that is part of the filter. The weave gives a diamond-shaped pattern externally, without pores in the strict sense of the term. Filtration here is the depth type; it takes place through (and in) tight channels formed by nap (furry surface) of the yarn.

Membrane media essentially are thin porous sheets produced by the evaporation of solvents from a solution of cellulose acetate or cellulose nitrate. The resulting structure is monolithic (a single piece). Because it is not a fused or bonded mass of fibers or granules, it is not as subject to media migration.

Composites of various media improve the overall characteristics for some applications. Without going into details of manufacturers' special products, here are several useful combinations: edge-type media with metal powder (powder is bonded to itself and to the edge-type medium by sintering), sintered stainless steel mesh with metal powder, and glass fiber with membrane. In the first two the coarser medium is downstream, and the finer is upstream. The two media are bonded together to prevent bypass leakage around one or the other.

For glass fiber and membrane, the fiber is upstream, and is not bonded to the membrane.

There are also magnetic composite filters which combine mechanical separation of nonmagnetic solids with magnetic removal of iron particles. Two common constructions are Alnico magnets with ordinary mesh and mesh with magnetic wires. Extremely fine filtration is possible. The efficiency of the magnetic portion is inversely proportional to the velocity of fluid past the magnet. One drawback is that as the filter cake increases in thickness during onstream time, particles are more likely to break loose.

Using the Performance Table

Just the filtering media, not the finished filter assemblies, are put under scrutiny here. Values given in Fig. 17.16 are practical for the media but might look ridiculous if applied indiscriminately to any given product. For instance, the 5% solids-concentration level will pile up 320 ft^3 of filter cake in one workday (8 hr) if flow is 100 gpm and the filter is 100% effective. Such an application is not normally recommended. However, the low-flow-through cloth media used in rotary vacuum-operated filters or in plate-and-frame units are certainly suitable for high concentrations of solids. And for many filter assemblies there are ways (such as mechanical scrapers) to automatically and continuously clean off the filter cake.

Another example: Take the 0.1%-and-below solids-concentration level. Here it might seem that individual size and not concentration of particles is the only logical basis for comparing filters. This is not necessarily so. Some coarse filtering elements such as edge type and cloth depend on cake buildup to achieve the manufacturer's rating. If concentrations are too low, bridging of the pores by the particles might not occur. In finer media, dirt-particle size is critical.

In-between concentrations such as 1.0% can be handled by almost any filter, but the table points out certain practical limitations. Sand filters, for example, are not recommended here. They have a relatively small apparent filter surface, which cannot be conveniently increased to any extent. Therefore sand filters are used primarily for filtering water, where solids concentrations are low—measured in parts per million.

18

Tubing, Piping, Hose, Fittings

Tubing, piping, hose, and fittings often are the most exposed parts of a pneumatic or hydraulic system, vehicle, or machine design. The tubes and pipes themselves are not likely to cause problems (other than bad cleaning) because industry standards are well established. It's the interconnections and flexible hose that provide the challenge. If there are to be any leaks or flaws, they probably will develop there.

Most of this chapter will concentrate on trouble spots, giving guidance in two areas: how to avoid unnecessary interconnections; and how to keep them leak-free and safe when they can't be avoided.

TUBING

Tubing has an advantage over threaded pipe in that it can be easily bent, comes in more sizes and materials, and requires fewer pieces and fewer fittings. Also, it can be cut, flared, and fitted in the field and makes a neater system with smoother flow and less chance of leakage (see the sections on fittings and on tube bending at end of this chapter).

Normal applications of tubing in hydraulic systems include pressures to 6000 psi in sizes up to 2-in. outside diameter. Nominal dimensions are given as fractions of an inch, available in one-sixteenths from the ⅛-in. to the ⅜-in. size and in one-eighths up to the 1-in. size. Metric sizes are available as well.

The basic material is steel, and it normally is electric-welded out of cold-rolled strip steel or made by the cold-drawing of pierced or hot-extruded billets from dead-soft steel.

Tubing wall thickness can be almost any dimension from thin to thick, but the most common dimensions are indicated in Table 18.1. The choice of wall thickness must be based on a combination of strength and fluid flow capacity. Two basic equations are

$$P = \frac{2ST}{(OD)} \qquad (18.1)$$

Table 18.1 Steel Tubing Dimensions

Nominal diameter, in. (OD)	ID = OD − 2W, where W = wall thicknesses, in.
Available in wide variety, typically, 1/8, 3/16, 1/4, 5/16, 3/8, 1/2, 5/8, 3/4, 7/8, 1, 1 1/4, 1 1/2, and 2 in. in OD (also in metric, typically from 4 mm to 50 mm)	Dozens of ID dimensions are available, from 0.055 in. (1/8-in. tube, 0.035-in. wall) to 1.870 in. (2-in. tube, 0.065-in. wall). Typical values for W: 0.020, 0.028, 0.035, 0.049, 0.058, 0.065, 0.072, 0.083, 0.095, 0.109, 0.120, 0.134, 0.148, 0.165, 0.180, 0.188, 0.203, 0.220, 0.238, 0.250, and 0.259 in. See the text for the method to select W.

where P is maximum working pressure, psi; S is allowable metal stress, psi; T is tube wall thickness, in.; and OD is tube outside diameter, in.; and

$$Q = 3.12 \times VA \tag{18.2}$$

where Q is flow, gpm; V is allowable flow velocity, ft/sec; and A is internal flow area, in.^{2} Area $A = \pi(ID)^2/4$. The constant 3.12 converts Q to gpm.

Other considerations such as the safety factor, pressure loss, temperature, and external mechanical stress must not be ignored.

A typical example would be to design for one-sixth of the actual tensile strength with a flow velocity of no more than 15 ft/sec.

PIPING

As with tubing, the outside diameter of pipe is fixed; the inside diameter varies with the wall thickness (Table 18.2). The difference, however, is that the nominal dimension of a pipe (⅛ in., ¼ in., etc.) is not related consistently to either OD or ID. It

Table 18.2 Steel Pipe Dimensions (Schedules 40 and 80)

Nominal size, in.	OD	ID		Wall thickness	
		Sched. 40	Sched. 80	Sched. 40	Sched. 80
1/8	0.405	0.269	0.215	0.068	0.095
1/4	0.540	0.364	0.302	0.088	0.119
3/8	0.675	0.493	0.423	0.091	0.126
1/2	0.840	0.622	0.546	0.109	0.147
3/4	1.050	0.824	0.742	0.113	0.154
1	1.315	1.049	0.957	0.133	0.179
1 1/4	1.660	1.380	1.278	0.140	0.191
1 1/2	1.900	1.610	1.500	0.145	0.200
2	2.375	2.067	1.939	0.154	0.218

seems to match the ID of Schedule 80 (heavy-duty) pipe best, at least in sizes from ¾-in. nominal diameter and up.

Pipe generally is selected over tubing for the larger hydraulic systems with long, permanent straight runs or where the components are threaded for pipe. Pipe is taper-threaded on its outside diameter for screwing into a tapped hole or fitting.

Extensive tables of pipe performance and dimensions for standard (Schedule 40) through heavy and extra-heavy pipe (Schedules 80 to 160) are readily available in handbooks.

Whatever problems occur probably will be caused by wrong selection, wrong quality of internal surface, plus bad cleaning and preparation. Air piping gives the most trouble.

For example, Automatic Valve Co. (Novi, MI) made a study of shop air systems in the United States, Great Britain, Europe, and South America and observed great waste of money and manpower in maintaining pneumatic valves, cylinders, and actuators against built-in dirt and mill scale in the piping itself.

Frequently the piping specs are for Schedule 40 black iron pipe per ASTM A-53 Grade B, A-106 or A-120. Nowhere in the spec does it specify level of cleanliness, such as freedom from mill scale, rust, pipethread chips, Teflon tape, and other debris. Studies reveal that 65% to 85% of pneumatic (and hydraulic) problems are directly traceable to pipe sediment.

What good does it do to stick a dozen filters in the main air line if the short run from valve to cylinder is rusty? One engineer said that it's not really too much of a problem because lubricated shop air eventually coats the internal pipe wall with a coagulating layer of oil; it stops the rust and traps the loose scale. That's using one problem to solve another. Maybe we should try to get rid of the excess oil as well as the scale.

Anyway, the company suggests that system designers at least demand pickling of the pipe, cleaning it thoroughly, and capping it during shipment. It only adds 15% to 30% to the cost. Also, if cold mandrel drawn welded steel tube ANSI B93.4 is chosen, it's delivered free of scale.

Circle Seal Controls (Anaheim, CA) offers the ultimate solution: stainless steel air lines throughout the plant. For that company, savings in maintenance have balanced the extra cost of the piping. The food, chemicals, and hospital industries came to the same conclusion decades ago.

RUBBER AND PLASTIC HOSE

Most of the hydraulic hose sold today is of reinforced synthetic rubber, which in this context means any of a variety of rubber-like elastomers such as neoprene and nitrile rubber. SAE standards define the types of construction, including a selection of wire braids, textile braids, spiral wraps, inner cores, and protective covers. The standards also qualify the performance. The accompanying drawings (Figs. 18.1 through 18.3) illustrate basic hose styles.

Plastic hose differs from rubber hose mainly in the hose materials and not so much in construction. So-called plastic hose is made from thermoplastics rather than from thermosets. The thermosets require a final vulcanizing step. Thermoplastics, including nylon, urethane, and Hytrel, on the other hand, do not have to be vulcanized.

The published rated pressure performance of synthetic rubber and thermoplastic hose is the same, and the multiple-ply construction is similar. The final products—

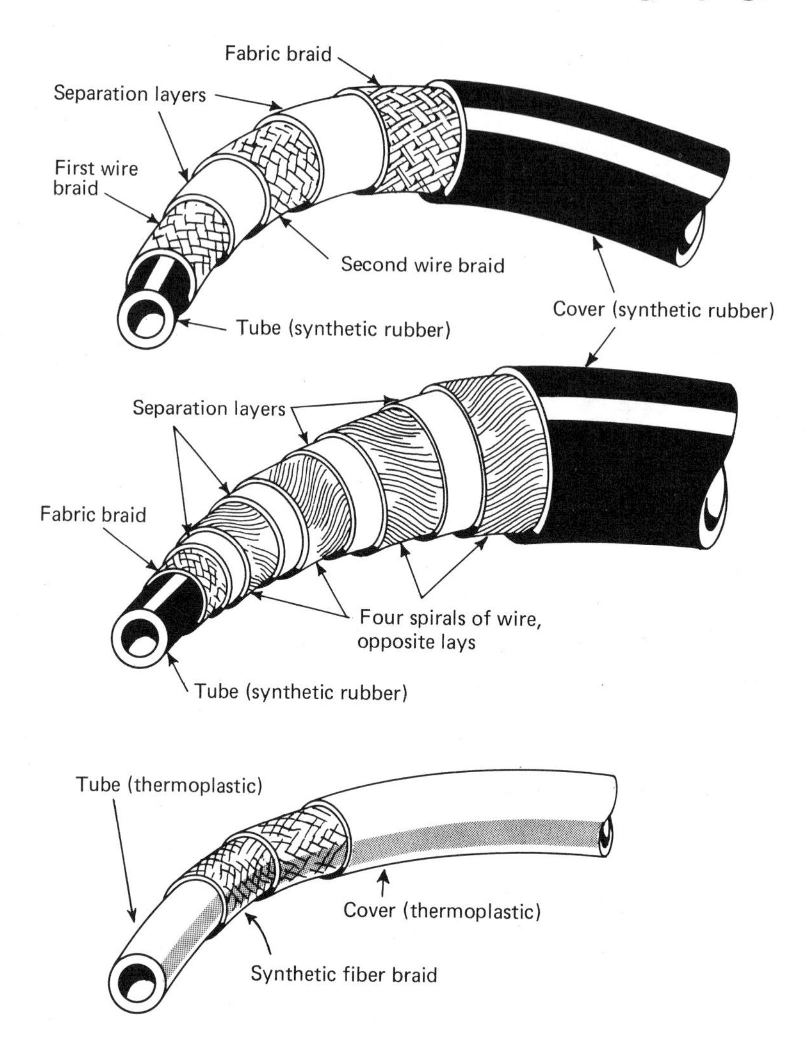

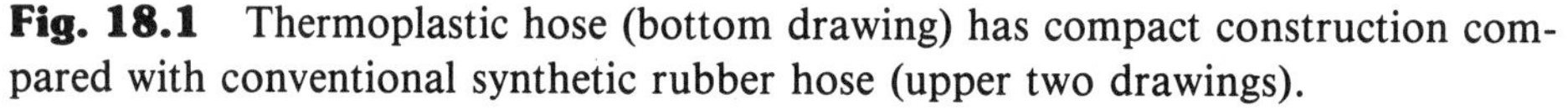
Fig. 18.1 Thermoplastic hose (bottom drawing) has compact construction compared with conventional synthetic rubber hose (upper two drawings).

hose-and-fitting assemblies—are intended to be directly competitive, depending on certain advantages or disadvantages of plastic vs. rubber relating to temperature, abrasion, resilience, chemical attack, and so forth.

The SAE standards include plastic designations as well as rubber, although not as many. Both types of hose (rubber and plastic) are widely available. The rubber hose has been around half a century and the plastic hose a decade or so. But only within the past half dozen years has the plastic challenged the rubber hose seriously for high-pressure hydraulic applications.

The plastics now available are better than those of a decade ago. Coupling design is improved. And manufacturing techniques are more reliable. But be careful: The inherent qualities of rubber vs. plastic still make the difference. Here they are:

Rubber hose is more tolerant of intermittent high temperatures, because it is a thermoset and therefore doesn't melt—it just degrades gradually. Any size from very small to very large can be made, and there is a wealth of experience behind each size. Also, rubber is more resilient, doesn't cold-flow, is not very notch-sensitive, and therefore is easier to couple fittings to.

Disadvantages are that rubber hose is much heavier than plastic hose for the same capacity, has less tolerance to certain chemicals and fluids, swells in oil, contains carbon black, and has higher electrical conductivity—a hazard in some applications. Also, the interlocking wire braid in reinforced hose somtimes breaks in fatigue if the hose is flexed too far and too frequently. One cure is spiral wrapping (Fig. 18.1) where the wires don't cross each other.

Plastic hose is much lighter and smaller than rubber hose for the same size and rating (Fig. 18.2). The plastics are inert to most fluids and chemicals, are not affected by ozone, do not age, are less permeable to gases, don't swell as much in oil, and don't conduct electricity. Plastic braid flexes more readily than wire braid does, and fatigue is no problem.

Plastic hose can be extruded in continuous lengths because no mandrel is needed (rubber hose length is limited by the mandrel that is required at the center to add stiffness to the relatively soft, unvulcanized rubber; the vulcanization is a later step). Plastic is more readily colored, because it doesn't have carbon black in it, like rubber does. Plastic hose is inherently cleaner, both in manufacture and in subsequent final assembly, where the harder, smoother surface is an advantage. And there are various plastics for extreme temperatures: Teflon at 450 F, for example.

Disadvantages of plastic hose are that it is less resilient than rubber, it cold-flows,

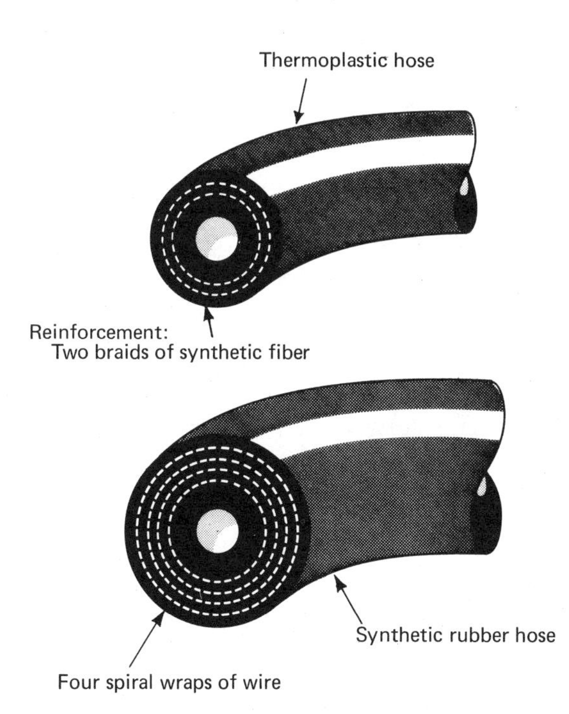

Fig. 18.2 Thermoplastic hose (top drawing) can be smaller for the same pressure rating compared with synthetic rubber hose. Shown: 1/4-in. hose for 10,000 psi.

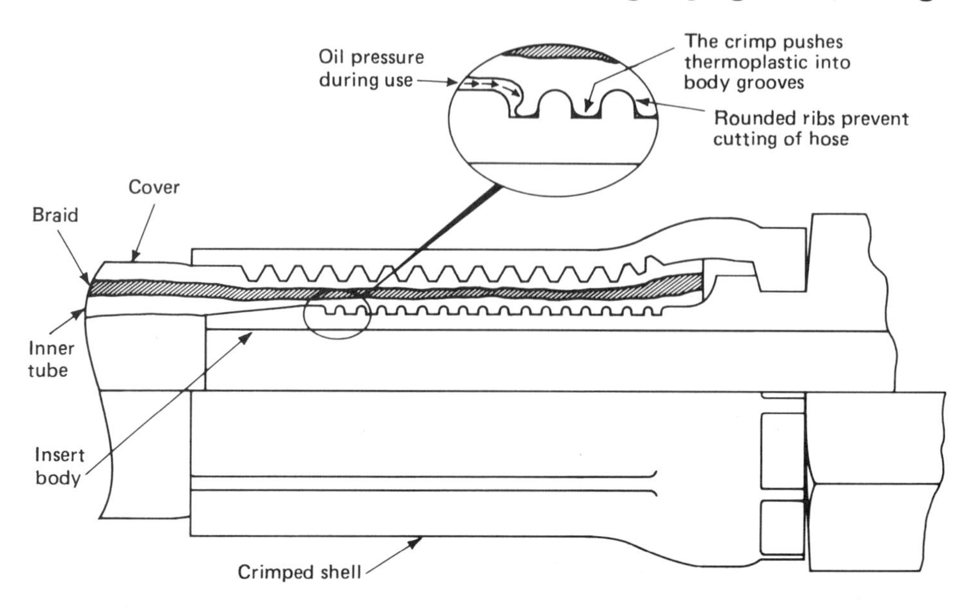

Fig. 18.3 Crimped fitting for plastic hose has close clearances, rounded edges, special grooves, and longer length to prevent cold flow of plastic and cutting of surfaces.

and it is more notch-sensitive. Fittings must be designed to fit accurately so that there is not undue stretching of the hose during insertion. The fitting must securely clamp and confine the plastic when crimped or swaged yet not cut the plastic surface. Also, plastic has a softening point with temperature.

One problem in the past with some nylon hose was that low humidity dried it out and made it somewhat brittle, and high humidity made it somewhat soft. It isn't a problem in working hydraulic systems, however. Anyway, Nylon 11, made in France and the United States from castor oil, eliminated those humidity problems. DuPont's Hytrel polyester (described below) was developed to compete with Nylon 11. Additionally Hytrel is petroleum-based and has the potential of mass production.

Tough Plastic Hose

The combination of Nylon 11 or Hytrel in the inner tube and Kevlar in the reinforcing braid created excitement in the hydraulic hose field. The first product offered was a 3/16-in. ID 10,000-psi rated (40,000-psi burst) hose, by Imperial Eastman and Samuel Moore.

Burndy Corp. (Norwalk, CT) specified the I.E. version on its line of hand-held hydraulic tools for in-the-field crimping of electrical connections. The 40,000-psi burst rating gives the hose a 4 : 1 safety margin, first in industry for plastic hose.

Kevlar fiber (formerly known as Fiber B) is an aramid (aromatic polyamid) with a tensile strength of as much as 525,000 psi and temperature resistance up to 500 F for short-time exposures. Pound for pound it is far stronger than steel. DuPont says that Kevlar is the first synthetic fiber they ever developed specifically for the industrial market.

Hytrel is a polyester elastomer that is thermoplastic above 400 F. It retains considerable strength and resilience from −65 to 300 F, does not require curing, needs no plasticizers, and is resistant to swelling in oils, solvents, and hydraulic fluids. It is a non-cross-linked polymer but exhibits cross-linked characteristics.

Gripping the Plastic

The coupling design for plastic hose is a different technology than that for rubber hose. In past years some difficulty was experienced in keeping the end couplings for plastic hose from leaking, but now there are suitable fittings and techniques.

What's the problem? Mostly it is the cold flow, combined with notch sensitivity, limited resilience, and higher coefficient of thermal expansion for plastic. An end fitting for plastic hose must trap the hose material and keep it from relaxing during the life of the hose and must have rounded edges to avoid cutting the surface, close tolerances to ensure tight initial manual fit without undue stretch of the hose, and built-in compression margin to allow for the expansion and shrinking of the plastic with temperature.

Plastic, once compressed or distorted for a time, will not return to its original shape as readily as rubber will. The fitting must squeeze the plastic into grooves that retain the material and prevent its further movement. The pressure itself helps by enhancing the crystalline characteristics of the plastic and reducing the tendency to cold flow. Figure 18.3 is based on an actual fitting and shows design features.

Swaging or crimping of fittings under tightly controlled conditions is recommended for plastic hose. Screw or clamp-type reuseable fittings are available, but plastic is not as forgiving of careless workmanship as rubber. But plastic hose can be made with tighter tolerances than rubber hose, so assembly is more predictable. There are many portable crimping or swaging tools available for making up factory-quality hose-and-fitting assemblies.

Special Hose Types

Almost any combination of materials can be made up into a hose if a customer wants it. Available are silicone hose for hot engine compartments, Teflon hose for aircraft, and lined hose for chemical plants. One unique idea is a coaxial hose, with the high-pressure supply hose inside the return hose.

Another is heated hose. The hoses are constructed with electrical heating wires vulcanized within the wall structure of the hose. Uniform temperature can be maintained in any length.

One example: The Technical Heaters Inc. (San Fernando, CA) model 212 hi-temp hose (Fig. 18.4) has a Teflon core, stainless steel braid, fiberglass-reinforced silicone rubber bonding layer for the heating element, thick thermal insulation, and a scuff-resistant jacket. It maintains 450 F when controlled voltage is applied.

The ¼-in.-ID size can sustain 1000-psi working pressure at 400 F. The core may be stainless steel tubing or metal hose if desired; then the maximum pressure of the ¼-in.-diameter size is 1500 psi, and the maximum temperature is 600 F.

Another unusual hose is a fuel line with a heater cable inside, for cold weather starts (Fig. 18.5).

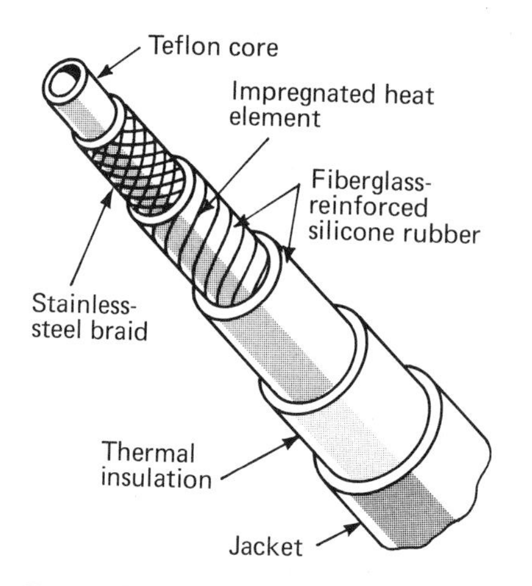

Fig. 18.4 Heated hose helps move viscous fluids. This one is rated at 1000 psi. (Courtesy of Technical Heaters, Inc., San Fernando, CA.)

Hose Application Problems

Table 18.3 itemizes what can go wrong with rubber or plastic hose. In addition to those tips, here are some case histories:

A mechanic in a typical mine hates to remove 12 hoses (Fig. 18.6) on a machine to get at a leaking fitting—it's easier to keep filling the reservoir.

It's not uncommon to cluster a dozen hydraulic hoses into one manifolded assembly. The bottom hose can't be repaired without removing many of the top 11. The top hoses make great footholds for mechanics and are abused by showering rock and dropped tools.

Jeffrey Mining Co. makes several suggestions. First, if hoses are mandatory, use fittings that don't require wrenching. For example, specify a Dayco/Stecko-type

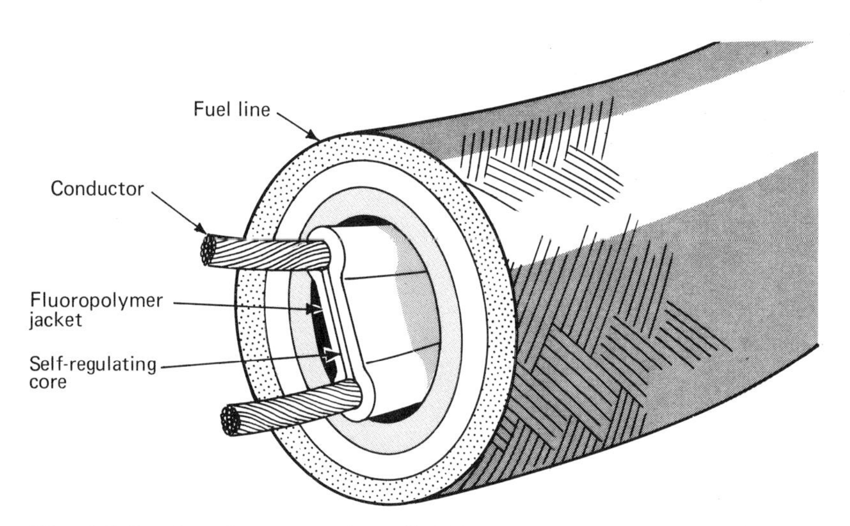

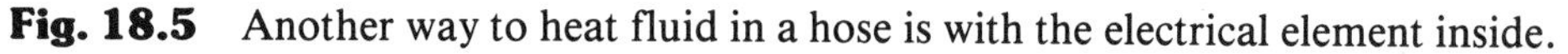
Fig. 18.5 Another way to heat fluid in a hose is with the electrical element inside.

Table 18.3 Misapplication of Hydraulic Hose is Avoided with This Design Check-off List

The hazards	The consequences, with symptoms & analysis
Wrong fluid	Inner liner swells, deteriorates, or partially washes out. **Analysis:** Never take compatibility for granted. Heat can be the catalyst that makes an otherwise safe fluid incompatible. Test the hose ahead of time with the fluid at max operating temp.
Wrong fitting or crimp	Fitting blows off end of hose. **Analysis:** Wrong fitting, or wrong setting on crimping machine, can cause it. Also, worn out screw-together fitting, or worn out swaging dies can be the culprits. Also see *too-short hose.*
Over pressure	Hose bursts in one or several places, but not with multiple breaks all along the hose. **Analysis:** Pressure exceeded minimum burst strength. See next item.
High frequency pulses	Hose bursts and leaves random broken wires the entire length of hose. **Analysis:** A high-frequency pressure impulse condition can cause it. SAE requirements for double wire braid are 200,000 cycles at 133% of working pressure. A better bet would be a spiral-wrapped hose: four layers endure 400,000 cycles up to 200 F.
Over heating	Inner liners soften and fail in one of two ways: internal collapse; or loss of fitting. **Analysis:** Teflon tubing, being thermoplastic, is not rubber-like; excessive heat combined with suction can fold it longitudinally. Radial tension of the braid, and hot-to-cold cycling, add to the problem. Nylon tubing under compression softens and flows out of compression area at over 200 F, loosening fitting.
Aerated oil (and heat)	Inner liner is very hard, and has cracked. **Analysis:** Heat can leach out the tube's plasticizer. Aerated oil causes oxidation of rubber, accelerated by heat.
Over chilling	Hose shows internal and external cracks, yet remains soft at room temperature. **Analysis:** Probable cause is flexing during a period of extreme cold. Choose a material that remains flexible at the lowest temperature you expect. Teflon is rated to −100 F.
Over suction	Hose liner is sucked inward in one of several ways: broken completely loose, broken or punctured; and collapsed and folded. **Analysis:** High vacuum on a hose not designed for it can break loose the liner and pile it up at one end of the hose. Even a vacuum hose can fail if kinked or bent too sharply. Another mode is entrapped gas that has effused into the liner pores during high pressure, then expands explosively when hose pressure is suddenly dropped to zero. A non-porous inner liner will solve it, but make sure there are no pinhole leaks. Still another mode is the failure of thermoplastic (e.g.: Teflon) caused by sharp bending that collapses the tubing.
Too short hose	Hose fails when pressurized. **Analysis:** It can fail because it inherently tries to shorten and cannot. A spiral reinforced hose, for example, can burst and split open with the wire exploded out and badly entangled.
Bending	Hose bursts on outside bend, and appears elliptical. **Analysis:** Tight bends not only can damage hose, but cut down on flow capacity, perhaps overheating pump.
Twisting & kinking	Hose flattens out in one or two areas, looks kinked, becomes very hard downstream of kink, or can burst. **Analysis:** Torquing of hose can tear loose reinforcing layers, allowing lining to burst through the enlarged gaps. Kinking or squeezing can create flow restrictions that throttle liquid to point of cavitation, and heat up and oxidize the liner downstream of kink.
Poor support	Hose stretches because of internal weight of fluid, and pulls off fitting. **Analysis:** Very long lengths of hose can weigh a lot when filled, and should be suspended carefully.
Poor external protection	Hose bursts, exposing rusty reinforcement. **Analysis:** Many external forces or environments can break through the protective outer cover of poorly applied hose: abrasion, cutting, acid, steam, brine, or heat & cold. Or moisture can penetrate an improperly attached fitting, work its way along the hose, and cause rusting many inches away from the fitting.
Inner pinholes & porosity	Blisters form in outer cover. **Analysis:** Pinhole leaks or porosity in the inner liner can allow high pressure liquid to leak, or gas to diffuse, under the cover That's another good reason to talk over any new application with your hose supplier.
High velocity erosion	Profuse leakage occurs, without actual bursting. **Analysis:** Inner liner can be gouged through to the wire braid for about 2 in. along the surface, caused by a high velocity needle-like fluid stream being emitted from an orifice and impinging on one section of the liner. Contaminant particles enhance the effect. Make sure the hose comes off straight from any port that is orificed.
High-voltage static	Teflon hose develops pin-hole leaks. **Analysis:** A petroleum-base fluid with low viscosity flowing at high velocity can generate high static voltage. The voltage seeks ground through the Teflon to the braid, causing arcing. One answer is to specify carbon-filled Teflon that can drain off the static electricity.
Old age	Outer surface is crazed and deteriorated, and hose might burst. **Analysis:** Old age, weathering, and ozone can take their toll of many common materials.

(Courtesy of Aeroquip Corp., Jackson, MS.)

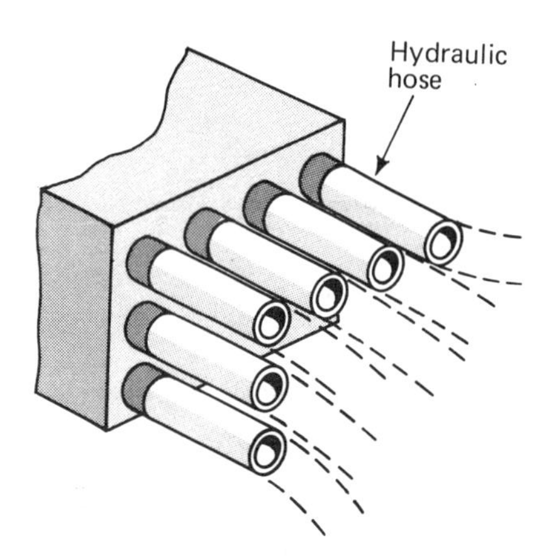

Fig. 18.6 Multiple hoses solve space problems but are hard to get a wrench on.

fitting which is a quick-disconnect that plugs in and then is held by a staple-shaped detent that's pushed into mating holes and grooves (see the section on quick connects following). The fitting is sealed with an O-ring. The O-ring is not clamped as in some SAE fittings but rides free in the internal groove and relies on pressure to seal. The big advantage is quick disassembly.

Another answer (not usually possible or practical) is to protect the hose by designing a ledge over it.

What about underrating? A backhoe application at 2000-psi rating is much more damaging to a hose and fitting than is a cultivator application at the same pressure, because the backhoe operates there most of the time and gets gratuitous pulses at much higher pressures. For the same life, the backhoe needs a 3000-psi hose. Not too many system designers want to spend the extra dollars. For best choices, specify acceptance tests to match the applications.

Then there is the problem of generated heat. Pressure drop is continuously converted into heat in any system, so the problem is not only one of economy but also of temperature limits.

As an example, trailing lines on a farm tractor are activated only occasionally, to raise or lower the implement being pulled. Also, the leveling pads on a crane are activated infrequently. Hydraulic lines in both instances can be small, with high pressure drop, without overheating.

Other lines such as the pressure line from the pump to the directional valves or the pressure hoses on the swing circuit of a back hoe will have fluid flowing through them constantly and need large lines. The trick is to strike a balance.

Also, remember the liability angle. Burst hose on a farm tractor might be the fault of a careless installer, and not the responsibility of the hose supplier or system designer—but will a jury understand this?

Tips: Make sure that the hose manufacturer and the fitting manufacturer know each other and mutually recommend the marriage; remember that hose pulsations can stress adjacent tubing, particularly bends; and pick a hose with long life under the worst conditions of pulsation and temperature.

METAL HOSE

When temperatures or the environment are too harsh for rubber hose, specify flexible metal hose. There are two main types: corrugated (Fig. 18.7) and interlocked (Fig. 18.8).

Corrugated Metal Hose (Braided)

The maximum working pressure rating of the hose assembly under given service conditions depends on the tensile strength of the braided-wire jacket, the ability of the corrugated tube (internal-pressure carrier) to resist permanent deformation under pressure while restrained by the braid, the operating temperature, the pulse or shock loading of the hose, and the method used to attach the end fittings. How these factors influence the final pressure rating of hose assembly can be expressed in this empirical relationship:

$$P_{rated} = F_T \times F_S \times F_E \times P_{nominal}$$

where F_T, F_S, and F_E are factors for temperature, shock or pulse loading, and end-fitting effect, respectively.

Nominal hose pressure $P_{nominal}$ in the equation is, by definition, the lesser of these two quantities: one-third of fluid pressure P_B (calculated) that will burst the external braid in an axial direction or four-fifths of pressure P_D (determined by test) that will start permanent deformation of the inner corrugated tube.

For a single-layer braided hose, the axial burst pressure obeys this relationship:

$$P_B = \frac{1.57Nd^2T \sin\Delta}{A}$$

where P_B = fluid pressure, psi, that will burst the braid in an axial direction; N = number of braid wires spiraling about hose in one direction; d = diameter of wire, in.; T = effective tensile strength of braid wire, psi; A = total (not annular) cross-sectional area of hose, in^2, based on mean diameter; and Δ = braid angle, deg, measured from a perpendicular to the hose centerline.

This axial-burst-pressure equation is mathematically exact and makes two major assumptions: (1) All of the axial pressure force is sustained by braid; (2) all the radial pressure force is sustained by the corrugated tube. Typical values for wire ten-

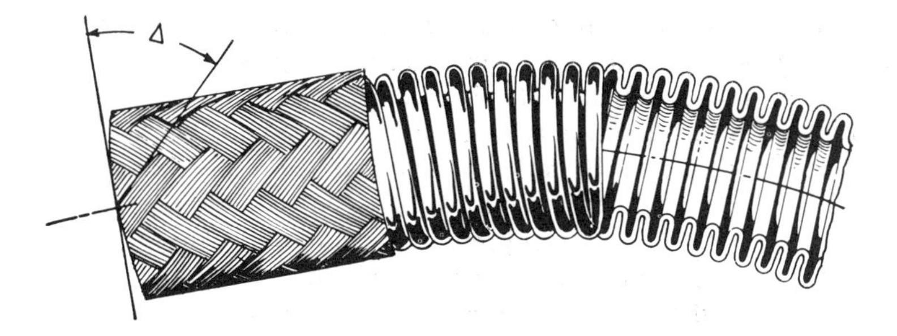

Fig. 18.7 Braided corrugated metal hose gets its axial burst strength from the braid. See the accompanying method for calculating this strength.

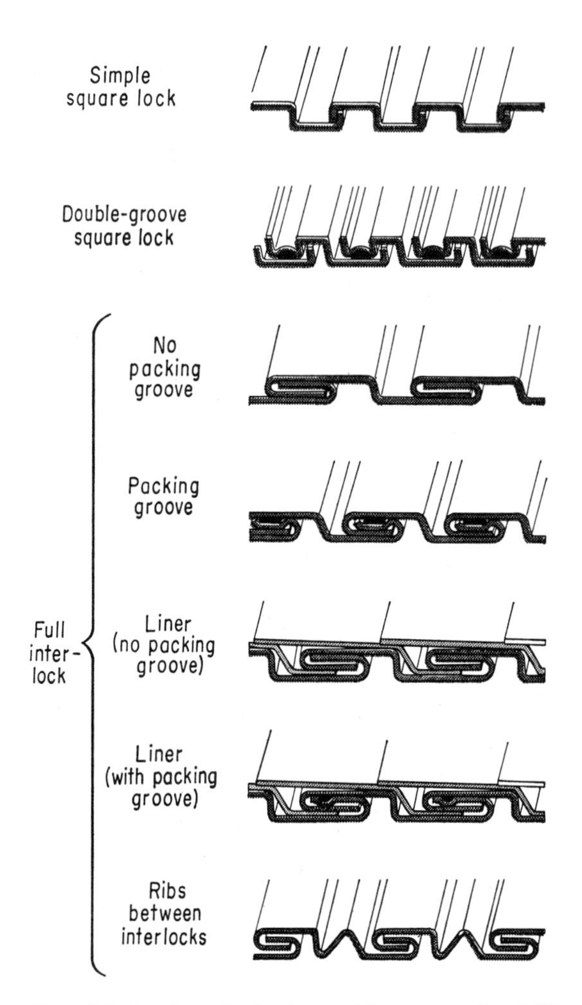

Fig. 18.8 Interlocked metal hose gets flexibility from the sliding between elements of the interlock. (Courtesy of Flexonics Div., UOP, Bartlett, IL.)

sile strength T are 40,000 psi for commercial and phosphor bronze, 50,000 for mild steel, and 115,000 for AISI (American Iron and Steel Institute) 300-series stainless steels. Published maximum hose pressure ratings must show values of T assumed.

For a double-braided hose, the same axial-burst equation applies, but the equivalent axial tensile strength of the combined jacket is assumed to be 1.8 times the value for a single-braided hose. This assumption is valid (and conservative) if braid wire can elongate at least 15% (based on a 10-in. length) when tested prior to braiding.

Summarizing: To find $P_{nominal}$ in the hose-assembly rating equation, compare one-third of the final value of hose axial-burst pressure P_B with four-fifths of the radial-deformation pressure P_D determined by test. The lesser of the two will be $P_{nominal}$.

Temperature factor F_T is defined as the ratio of the 0.2% offset yield strength for the hose metal at the elevated (or reduced) temperature to its yield strength at 70 F. See Fig. 18.9 for typical values at elevated temperatures. The factor will increase above 1.0 at lowered temperatures for some metals, but to be conservative, use values no greater than 1.0.

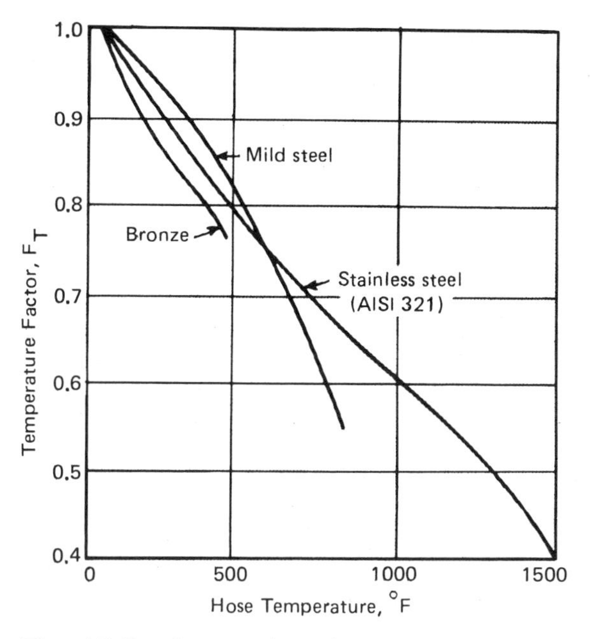

Fig. 18.9 Corrected maximum rated pressure equals maximum rated pressure at 70 F times temperature Factor F_T.

Cryogenic hose (Fig. 18.10) for liquid nitrogen gas (LNG) and liquid oxygen (LOX) service is a special case not covered by Fig. 18.9.

Pulse or shock factor F_S cannot be determined by simple calculation but is an empirical value. Typical values for F_S range from 0.5 (moderate pulses) to 0.2 (severe shock waves). For a manufacturer to estimate the probable derating required,

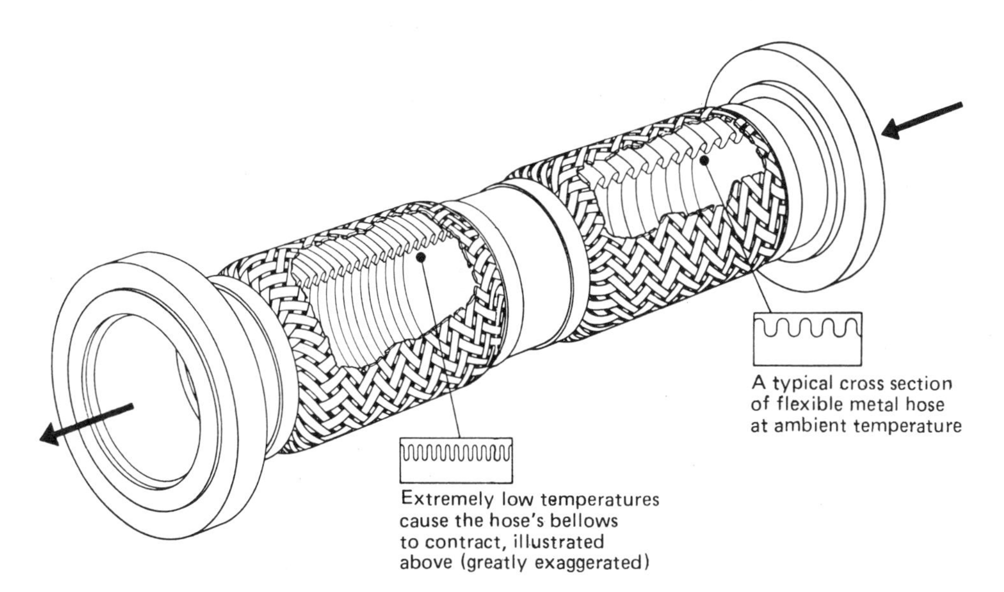

Fig. 18.10 Metal hose for cryogenic liquids such as LNG is built to absorb extreme contraction. (Courtesy of Dayco Corp., Dayton, OH.)

he or she must know the peak pressure, rate of pressure rise, number of pulses per second, and expected total operating cycles per year. He or she will probably recommend a metal hose specially designed for shock or impulse service.

End-fitting factor F_E is 1.0 for a standard brazed or silver-brazed connection and varies from 0.6 to 2.0 for other attaching methods. For heliarc-welding, the factor is 0.8 to 1.0; atomic-hydrogen welding, 0.6 to 0.8; soft solder, 1.0 to 2.0; and a mechanical coupling, 1.0 to 1.8. The design of the end fitting itself can influence the factor's value, but to a lesser extent.

Unbraided, Unrestrained Type

The maximum pressure rating for unbraided corrugated hose when free to elongate depends only on permanent axial elongation and temperature. Hose-deformation pressure P_D and end-fitting design are unimportant, because the pressure levels permissable in unbraided corrugated tubing are relatively low—axial deformation will occur before the other factors come into play. Shock factor F_S is left out because pulsing flows are not normally recommended for unbraided hose. The relationship between rated and nominal pressure in a hose assembly is

$$P_{rated} = F_T \times P_{nominal}$$

The temperature factor F_T is the same as for braided hose. $P_{nominal}$ is that pressure at which permanent elongation (set) equals 0.25% at 70 F, based on this equation:

$$\text{permanent elongation, } \% = 100 \times (L_F - L_O)/L_O$$

where L_F is final free length of a hose subjected to, and released from, the test pressure and L_O is the original free length. The test is run at room temperature. The hose rests on a smooth flat surface with one end free to move. Both hose ends are blanked, but the fixed end has a pressure connection attached. Free length includes active corrugations only—soldered, brazed, or clamped corrugations are not included.

Unbraided, Restrained Type

With this hose, ends are fixed axially to prevent overall elongation. It can often be rated considerably higher than unrestrained hose for special applications. However, the free length, offset, bend radius, stiffness, and other application requirements are so varied and interrelated that no simple standard can be written.

INTERLOCKED HOSE

Interlocked hose is made both for pressurized and nonpressurized service. Both kinds have temperature limits that depend on the packing materials and metals used (Table 18.4). But only the pressurized type has pressure-temperature standards similar to those for corrugated hose.

Pressure Type

The maximum pressure rating for an assembly depends on hose burst strength, temperature, and leakage. Axial deformation and end-fitting design are not considered in the maximum pressure rating, but some end-fitting leakage may occur if packing is

Table 18.4 Temperature Limits of Metal Hose Materials

Material	Max temp, °F
Aluminum	600
Asbestos packing, grade:	
commercial	400
Underwriters	450
A	550
AA	600
AAA	750
AAAA	900
Brazing alloys:	
copper base	850
silver (AMS 4770)	750
silver (AMS 4772)	850
Brass	450
Bronze	450
Cotton-cord packing	200
Galvanizing	450
Monel	800
Neoprene packing	250
Solder, soft	
60-40% lead-tin	250
95-5% lead-tin	350
Steel, mild	850
Steel, stainless AISI type:	
302, 304	850
316	850
316 ELC	1500
321, 347	1500

damaged by welding or brazing. This rating relationship for hose assemblies applies:

$$P_{rated} = F_T \times P_{nominal}$$

where temperature factor F_T is the same as for corrugated hose, but $P_{nominal}$ is defined as the lesser of these two values: one-third the burst pressure at 70F or one-third the minimum weep (leakage) pressure.

Burst pressure will vary from test to test; the value selected can be an average value but must be at least 15–20% higher than the minimum burst pressure. Minimum weep pressure is the lowest at which there is no continuous leakage of water (or SAE 10 oil) at 70 F through the hose convolutions. (Continuous leakage means that drops of fluid form on the outside of a hose bent to its minimum radius within 1 min.) However, most interlocking hose will burst before leaking.

Extra Performance

Higher-than-standard pressures and temperatures are allowed in special circumstances if the manufacturer knows that life and reliability will be adequate. Here are typical circumstances.

Intermittent temperatures: If the hose is at peak temperature only part of the time, then the hose-deterioration rate will be slower. This is important when failure is related to excessive oxidation, carburization, graphitization, diffusion of injurious gas into the hose metal, metallurgical phase changes, or other time-dependent changes.

A somewhat higher peak temperature can therefore be recommended by the hose manufacturer.

Favorable environment: If the atmosphere (or fluid) is essentially noncorrosive, even at high temperatures, then temperature-induced metallurgical phase changes in the hose metal are sometimes tolerable. For example, an AISI type-304 stainless steel hose suffers from carbide precipitation at grain boundaries at 1200 F—usually undesirable because the metal loses its resistance to corrosion. But in a noncorrosive environment, the hose will not deteriorate and can be rated at the high temperature.

Temperature gradient: The limiting temperature is not always the highest measured temperature, because cooler portions of the hose cross section can add strength to the whole. An uninsulated hose will have a slight temperature gradient from inside to outside, and the fluid temperature can safely be somewhat above the maximum service temperature of the hose. An insulated hose, however, has negligible temperature gradient, and the fluid must be at rated temperature.

Tube-wall thickness: A thick-walled hose exposed to a corrosive environment will corrode at the same rate as one with a thin wall. But for the same depth of penetration, the thick wall retains more of its original mechanical strength. Therefore, in corrosive environments, thick-walled hose can be rated for slightly higher temperatures or for longer service at the same temperatures.

LEAK-FREE CONNECTIONS

The basic problem is oil leakage, but the same reasoning applies to pneumatic leaks.

Ports and connections operating under hydraulic pressure show up any flaws in material or design by dripping, seeping (slow formation of drips), or weeping (wetted area with leak point uncertain). All the users know about this and blame vehicle and equipment designers, who in turn blame the fitting and component manufacturers, who blame the users.

It's not fair, each says. But there are no rewards or punishments in engineering: only consequences, as in nature.

The facts, which can be uncovered by anyone who takes the time to talk to dozens of makers, appliers, and users of fittings and connections, are clear. Every kind of fitting will work perfectly within the intended performance range if it is manufactured and applied properly. Even a metal-to-metal face seal can be made leak-free, but other ways are easier and cheaper.

The hue and cry in industry at the moment is to "get rid of pipethreads and go to SAE straight threads with O-rings." We mentioned this to a manufacturer of pipe fittings and had our heads handed to us. We mentioned it to a cylinder manufacturer, and he responded with detailed design drawings on techniques to enhance pipethread sealing. And a pipethread sealant manufacturer sent us a sample of a coated pipe fitting that is claimed to have solved all the problems of leaking fittings on trucks.

What's the Problem

Not many equipment or vehicle engineers want to make a career out of designing and redesigning all the connections in a system, because it means painstakingly seeking the best possible combination of types, sizes, materials, intermediate supports, bend radii, flexibility, fatigue life, resonant frequencies, tolerances, finishes, porosity, clearances, wrench torques, leak specs, temperature limits, pressure peak limits, and safety factors.

They'd rather stick with whatever they're using—which seem to be mostly pipethreads—and somehow make them work. Buyers would save money in the long run in many instances by plunging into the challenges of the previous paragraph. Certain industries have succeeded in solving the problems.

The aircraft industry has just about eliminated line leakage by making all possible connections permanent by welding, brazing, swaging, or shrink-fitting (described later). None are cheap, but the air force and commercial airlines cannot afford leaks at 50,000 ft.

The tractor and heavy equipment industries tend toward O-ring face-type seals such as the SAE straight-thread O-ring port (Fig. 18.11) and the SAE split-flange type (Fig. 18.12). Caterpillar, for example, got rid of pipethreads decades ago and concentrates on split-flange fittings except for small-diameter tubing.

The trucking industry deals with much lower pressures (except for diesel fuel injectors (at extreme pressures) and won't spend the money for fancy fittings. Pipethreads are common, and the goal is to make them leak-free and reliable with pastes, dopes, coatings, or inserts. Some examples are described in the section on separable connections.

Industrial cylinders, pumps, and valves are in a special category because the manufacturers have total control of the ports. Miller Fluid Power, for example, offers cylinder pipethread ports with a taper slightly steeper than the fitting (⅞ in./ft for the port, against ¾ in./ft for the standard fitting that mates with it). It creates a joint with less radial play.

SAE straight-thread O-ring ports also are widely available from manufacturers of industrial cylinders, pumps, hydraulic motors, and valves.

Automotive manufacturing plants now have a special problem: Not only are they faced with buying and using equipment of every description but have to seal against high water base fluids (HWBF). Water is far less viscous than oil, so all their leakage problems are multiplied by 10 (see Chap. 1).

Nuclear power plants, semiconductor processing plants, warships, and other high-technology systems where leakage of any sort is not allowed have the most reliable connections regardless of cost. Hermetic fluid circuits are common, and many are tested with light gas molecules such as helium. Permanent welded, brazed, swaged, shrunk, and (for the final separable connection) highly engineered screw-type compression fittings often are chosen.

Permanent Connections

The best way to get rid of leaks is to get rid of separable connections wherever practical, and that's what a lot of industries are doing. Welding and brazing are skill-intensive and can leave a trail of residue and fumes. However, they work, and there are volumes written on the subject, so we'll move on to the swaging-and-shrinking department.

Swaging means deforming the tube or pipe by uniformly compressing the fitting into nearly perfect contact by actually flowing the metal. The operation can be mechanical, hydraulic, magnetic, or thermal. All have proved to be reliable, and in many cases the joints have greater strength than the tubing.

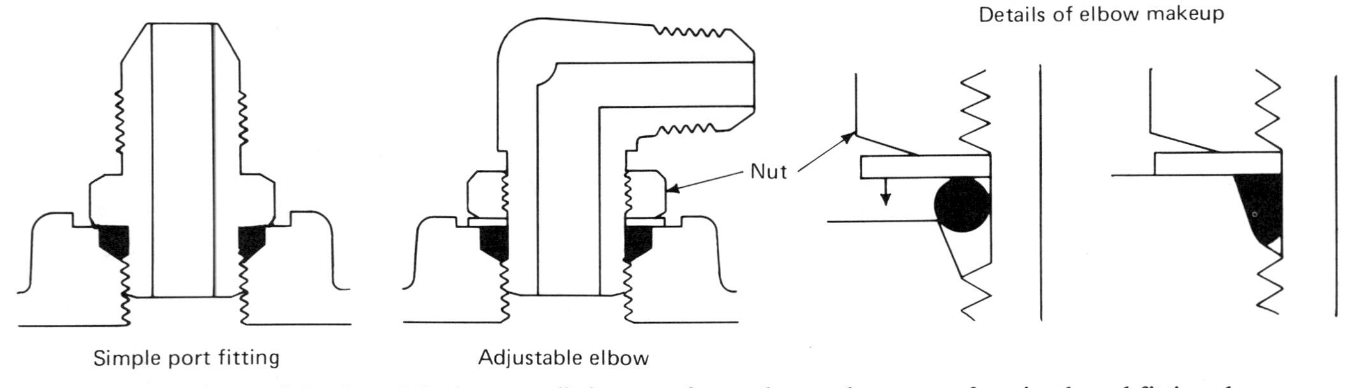

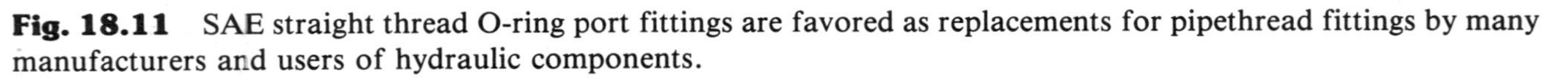

Fig. 18.11 SAE straight thread O-ring port fittings are favored as replacements for pipethread fittings by many manufacturers and users of hydraulic components.

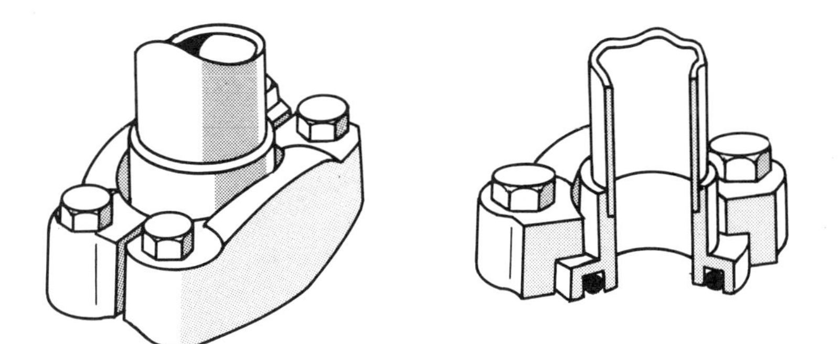

Fig. 18.12 SAE split-flange O-ring port fitting is widely used on heavy equipment such as tractors and bulldozers.

In one form of mechanical swaging, the tube is inserted in the fitting and then expanded by internal rollers until it is swaged completely into the inner diameter of the fitting. The fitting surface is serrated or grooved to receive the flowing metal and to help lock and seal the surface. The tube-expanding can be done with internal hydraulic pressure just as well as mechanically.

External hydraulic swaging is available, too. The fitting can be compressed inward against the tubing instead of vice versa, using a hydraulic hand tool. Deutsch offers a system called Permaswage™.

Cryogenic swaging or shrink-forming (Fig. 18.13) is the most interesting type. Raychem Corp. (Menlo Park, CA) developed it in the late 1960s and calls it Cryofit™.

It's been recently improved to the point where evaluations are being made for 800-psi hydraulic systems in LWH (lightweight hydraulic) aircraft systems. At the moment, it is used in 6000-psi systems and apparently works well. Process industries like it because it's a clean method and leaktight—even against low-molecular-weight gases.

Briefly, the technique is to machine the fitting or sleeve undersize out of a special nickel-titanium alloy and drop it into liquid nitrogen. While the sleeve is in the liquid

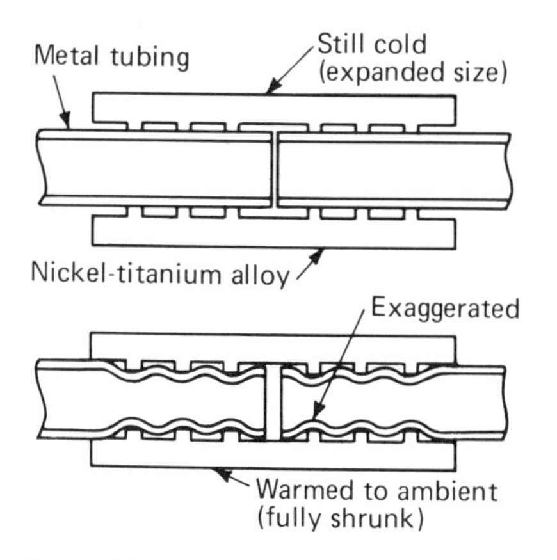

Fig. 18.13 Cryogenic sleeve fitting stays expanded until warmed. (Courtesy of Raychem Corp., Menlo Park, CA.)

nitrogen, it is mandrel-expanded to a larger inside diameter. The uniqueness is that when you remove the sleeve from the nitrogen, instead of it remaining in a deformed state as it warms up, it forgets the deformation and returns to its original precise dimensions—just as if you had never deformed it.

The benefit is that you can store a lot of sleeves (hundreds or more) with enlarged bores in liquid nitrogen cannisters until you need them. Then you remove one and quickly slip it over the end of the tubes or pipes you want to join. The sleeve fits loosely only so long as it remains cold and quickly shrinks back to original size as it passes through a critical transition temperature. It locks itself with tremendous force onto the tubing.

If galvanic problems are anticipated, an inner sleeve is added by Raychem. The sleeve can be machined out of any compatible material, and because it is not part of the shrinkable main sleeve, it doesn't have to be mandrel-expanded in the liquid nitrogen. Thus sharper teeth for better gripping are possible—helpful for tubing with special problems such as thinner walls or harder surfaces.

Contamination is a minor problem. Sand and other particles must be cleaned out; otherwise the surfaces cannot be forced into complete contact. But oils and greases don't seem to hinder the joint makeup or reliability.

Another development by Raychem is heat-shrink swaging (Fig. 18.14). It uses a special sleeve of copper alloy instead of nickel-titanium. Application of blowtorch heat for a brief period is sufficient to shrink the sleeve over the tube for a tight hermetic seal. The greatest use at this time is in refrigeration systems. Raychem calls its product PermaCouple™ and the copper alloy Betalloy™.

Swaging also can be done electromagnetically. The fitting is placed over the tubing; both are preheated to a temperature below melting and are drawn together by inducing an extremely high magnetic force between them. The joint is made instantly, and the forces and other factors are sufficient to cause diffusion or coalescence (coming together into one) across the joint. One such process is called Magneform™, offered by Maxwell Labs (San Diego, CA).

The short heating cycle (conventional induction heating) retards oxide buildup at the joint's interface. The magnetic attraction is created by magnetically inducing

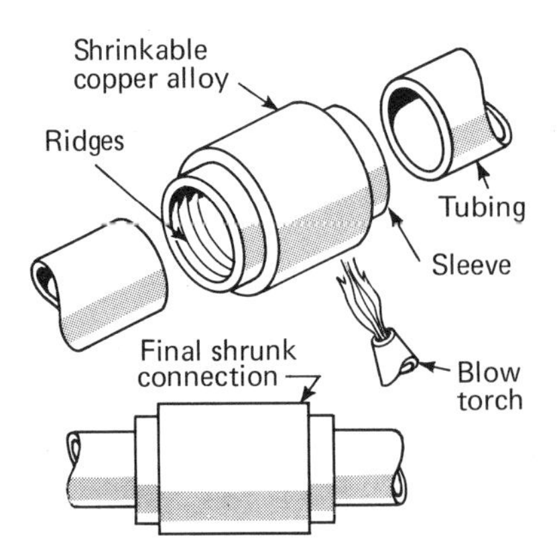

Fig. 18.14 Flame-shrunk fitting is similar in purpose to the cryogenic type. (Courtesy of Raychem Corp., Menlo Park, CA.)

extremely high dc current pulses through the tube and sleeve simultaneously, close to the surface, in parallel, and in the same direction. Surface-to-surface bonding forces of up to 50,000 psi are developed. The equipment is somewhat elaborate for a small shop and is not widely used for fittings.

Plugs expanded with taper pins are another form of swaging, used to seal drilled passages in castings. The Lee Plug, for example, is a grooved hollow cylinder (Fig. 18.15) that locks into the wall of an accurately reamed hole and seals at each rib. The 45-deg bevel at the bottom of the hole positions the grooved plug and must be precise. The ribs displace metal and can withstand up to 28,000 psi if assembled exactly according to manufacturer's instructions.

Separable Fittings

There need not be any leaks if each separable fitting is built, assembled, applied, and operated in strict accordance with good design practice. The problem is they often aren't, so pick the best odds.

Start with life-cycle costing. Consider the initial cost; assembly time; the time to fix leaky joints on performance test; the cost to clean up leaks, replace oil, and retighten fittings for life of system; and the cost of future lost business if you make the wrong decisions. Keep those in mind as we recount specific comments made by fittings experts.

Design the runs of tubing or piping with the fittings in mind. Resonant vibration can loosen any fitting (two-bite flareless compression fittings help in damping), unsupported lengths can create bending stresses, hidden fittings can't be checked or tightened, exposed runs make great footrests, and misaligned tube ends or side loads can highly stress fittings.

Bracket support design is an art. For example, a bracket located at a vibration node won't help much, so put the brackets at the excursion points even if it leaves a fitting hanging free.

Avoid fittings that turn the tubing (by friction against the flare, for example) during makeup, because it might loosen the next fitting down the line.

Watch out for effects from one component of a system on another: For example, a flexible hose designed to absorb vibrations might protect itself well enough, but the transmitted pulses can impose a rapid bending moment cycle on an attached tubing bend, fracturing it.

Double check system pressure levels, including peaks, to make sure the fittings aren't underdesigned. Treat fittings carefully in storage and in transit, or surface scratches might make them impossible to seal. Fittings are relatively soft and don't belong in a parts bucket.

Demand impeccable assembly practice or suggest drip pans under every system. Any imperfection in a joint (burrs, dents, scratches, dirt, rust, unsquare ends, out-of-roundness, nonsnug makeup) can cause leaks. Mismatched fitting halves are a constant hazard (there are dozens of look-alikes and many different flare angles that won't fit together). Bite-type compression fittings must be inspected 360 deg before final makeup to make sure surface hard spots are fully penetrated.

The tube wall thickness must be specified, or it will buckle (if too thin) or won't flare (if too thick). Surface finish of tube and fitting must be smooth, not pebbled or pockmarked. Degrees of wrench turn must be specified for compression fittings.

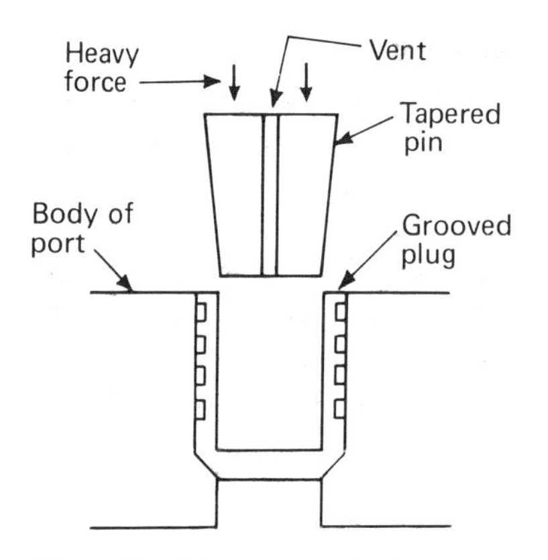

Fig. 18.15 Swaged plug expands into body material for a permanent tight seal. (Courtesy of Lee Co., Westbrook, CT.)

The compression fitting ferrules should be preset onto the tubing where possible (tools are available) to avoid possible mistorquing of the nut in assembly. Apply lubricant to reduce friction during flaring or compression, particularly with stainless steel tubing and fittings.

That's enough of the general warnings given by experienced makers and users of fittings. Let's examine specific types in more detail. Only the following types induced heated discussions (pro and con): O-ring face seals and O-ring chamfer seals, pipethreads, flared tubing, flareless tubing (compression fittings), and hose fittings.

There are other kinds of separable connections, such as flange-and-gasket, metal-to-metal diaphragm lip (Resistoflex Dynatube™), push-grip (plastic tubing), quick-connect, subplate interface, radial groove O-ring, metal O-ring and V-ring, European metal-washer seals (similar to SAE straight thread O-ring port fitting except that they have a metal seal and a machined washer), and hybrids of all sorts, but they weren't the subject of heated comments.

O-Ring Port Campaign

The National Fluid Power Association (NFPA) formed a national committee of fitting and joint experts to achieve two goals:

- Reduce hydraulic port leakage at pressures up to 6000 psi through an industry-wide coordinated conversion from ports where sealing depends on pipethreads to ports where sealing is done by means other than thread interference.
- Adopt a common hydraulic port connection configuration in the United States and in the world (see Fig. 18.11 for the prime contender). Also see Figs. 18.16 (an international metric port, ISO 6149) and 18.17 (a hydraulic motor with ISO ports).

And almost in the same breath, Parker Hannifin Corp. (a leading supplier of fluid-power components and fittings) issued a press release offering its pumps, valves, subplates, filters, cylinders, and fittings " . . . with SAE straight-thread ports at the same price and delivery time as those with tapered pipethread ports." The last time any manufacturer tried that bold approach was years back when Snap-Tite made a

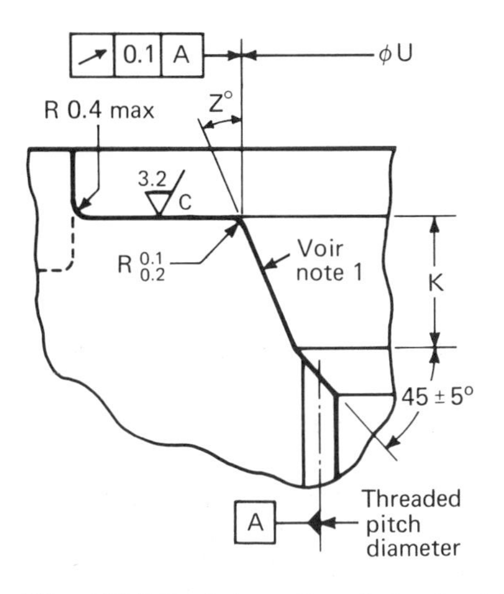

Fig. 18.16 International standard ISO 6149 for a metric threaded port.

similar offer—but the buyers then were still happy with pipethreads. Today the climate is more favorable for O-rings.

Two other old performers have been on-stage for years: the SAE split-flange O-ring face seal (Fig. 18.12) and the CPV (combination pump valve) fitting (Fig. 18.18). Each is a static O-ring face seal and relies on the O-ring for sealing. There

Fig. 18.17 ISO port on a typical hydraulic motor (Courtesy of Ross/TRW, Lafayette, IN) has spot faces for O-ring fittings (see Fig. 18.11).

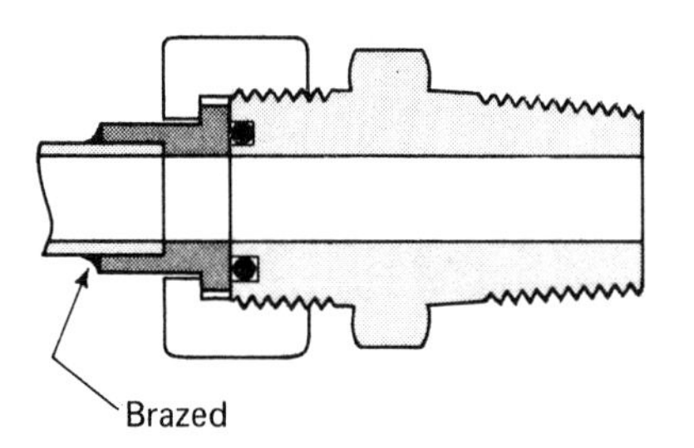

Fig. 18.18 O-ring face seal (CPV) for fittings has the same principle as Fig. 18.12.

are other face-type seals, but those two were mentioned most. For example, the O-rings can be replaced with lip-type seals.

The O-ring has an interference fit to seal against low pressures and is pressure-augmented at high pressures—good up to 6000 psi and more for certain applications and to 3000 psi for most. The bolts and threads are only for holding the seal in position.

The split-flange type (Anchor Coupling is one example) is widely used in heavy-duty off-road equipment such as tractors. The CPV type—now also offered by Parker Hannifin and by Cajon—is found more often in process plants and naval warships. The fundamental principles are the same.

The O-ring will seal surface imperfections such as porosity very well and is reliable in most applications. The O-ring behaves almost as a viscous liquid. Metal-to-metal seals are not as good against poor surfaces.

The chief problem is inattention to detail in assembly and application. The CPV is less of a problem because the fitting has a single thread, rather than four hold-down bolts, and is less likely to be cocked or wrongly torqued.

High shock pressures can stretch split-flange hold-down bolts enough to cause O-ring extrusion and subsequent nibbling. Undertorqued bolts can vibrate loose, and there doesn't seem to be a general awareness of the need for proper joint makeup. The flange can tilt if unevenly bolted.

It's not possible to check an O-ring joint once it is pulled together, which is a disadvantage. However, a careful check of all dimensions and surfaces is reasonable assurance that the joint is right. One problem, encountered chiefly in suction lines, is the possibility of sucking the O-ring out of the groove. Proper groove design will prevent it.

For chamfered port seals, follow very cautious procedures when inserting the O-ring, in bringing down the nut and backup washer, and in capturing the O-ring into the chamfer. Pinching destroys O-rings. Lubricant is advised.

Specify reasonable temperature limits in operation, because overtemperature can vulcanize elastomers. Also, remember that O-ring ports have spot-facing and take up extra diameter compared with pipefittings, so allow for that in sizing the pumps, valves, and cylinders.

Pipefitters, Beware

Nobody claimed that pipethreads were perfect, and most critics said it's impractical to cut precision pipethreads routinely. The ideal pipethread makes contact on both tip and flank and is designated NPTF, or "dryseal." Some industry spokesmen recommend that only the NPTF type be used for hydraulic systems and that the working pressure be limited to 1500 psi. Even then, pipe compound (dope) should be applied to the male threads.

Users of NPTF and NPT have discovered that although the fittings can be reused several times, they become harder to torque down after each application. Sometimes a 2-ft length of pipe is attached to the wrench handle to increase the torque, endangering the port housing while failing to stop the leak. Each makeup forces the threads to yield a little more, and a new yield area must be created. Port design must allow for possible distortion if the fitting is overwrenched.

There are other ways to stop leaks in pipefittings, but they all involve extra parts, coatings, pastes, or design modifications. Dedicated users believe the extra effort is worthwhile because pipethread fittings are cheaper than most other types, and there are an awful lot of fittings on a machine or vehicle.

Several engineers told of Loctite coatings and similar coatings from other manufacturers. Apparently they seal well, and the major complaint was that if the two halves of the joint are of materials with different coefficients of thermal expansion, the coatings might crack. Steel-on-steel works all right.

One designer recommended an "insoluble plastic lead seal" pipe dope made by John Crane. Another said he had success with molten solder applied to the threads—it worked better than plastics at extreme pressures.

Teflon tape around the threads seals well but is outlawed in some hydraulic applications because pieces of the tape find their way into the system and clog it. Teflon pastes also are available.

Precoatings for hydraulic fittings on Ford trucks have been developed by Loctite. One is called Vibra-Seal 503. It's a polymeric material with a shelf life of several years, enough lubricity to ease fitting makeup, and sufficient strength to survive five reuses. It works with pipethreads or straight threads and allows the fitting to be loosened slightly for positioning of elbows and tees.

Separate sealing nuts called Tru-Seal (Fig. 18.19) are offered by Miller Fluid Power. The nuts have a special bonded Teflon insert and function similarly to the SAE O-ring port fitting. That is, the insert does the sealing, and the threads do the

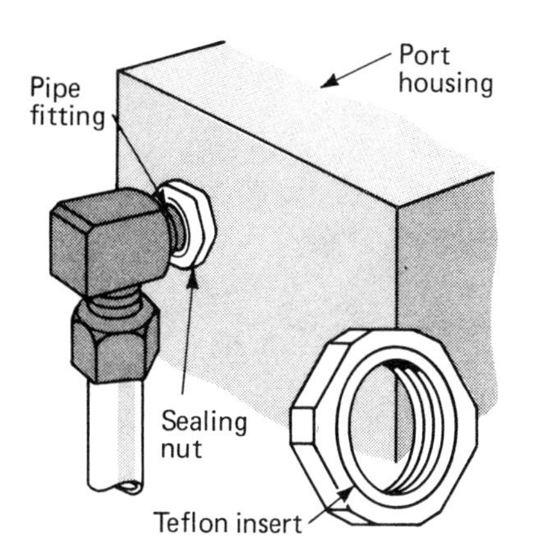

Fig. 18.19 Sealing nut with a Teflon insert for a pipethread does not rely on thread to seal. (Courtesy of Miller Fluid Power Corp., Bensenville, IL.)

holding. Elbows and tees can be positioned in any direction before the seal nut is tightened. For most efficient sealing, the port should be spot-faced.

Flared Tube Fittings

The principle is ingenious: The tubing (usually steel for hydraulic systems) is preflared in a special fixture and becomes its own deformable sealing flange when the fitting is assembled and tightened (Fig. 18.20). It works and is widely used in hydraulic systems up to about 3000 psi or more, depending on the tubing diameter and wall thickness. But the system is misleadingly simple and prone to errors in handling and assembly.

To start with, there are too many kinds available, and mismatching is always a danger. For example, the flare angle can be 37 deg (the most popular for hydraulic systems) or 45 deg, and a fitting won't seal if a 45-deg nut is mistakenly screwed onto a 37-deg male body. If you include European designs, there are 30-deg and 60-deg flares as well.

The tubing must be cut with a tube cutter and not a hacksaw, or there's the danger of torn edges, poor squareness, out-of-roundness, and embedded cuttings in the sealing surface. A hacksaw also work-hardens the steel and makes it more difficult to flare.

Assuming that the flare is correct, the joint is clean, the seats aren't rough or cracked, the parts are assembled snugly, there are no nicks in the seats, and all the right parts and materials are used, there is still the final problem of applying the right torque to make up the fitting. The joint is hidden, so how do you know it's tight?

Specify this: Mark a line lengthwise on the nut and adaptor after the fitting is fingertight. Then it will be easy to see how far the nut turns when it's wrenched, and there are guidelines for each size and type of fitting. For example, a 1-in. fitting should be turned the equivalent of about one hex flat (60 deg) to be sure the joint is tight.

Flareless Fittings

Some steel tubing is too heavy to flare, so several industrial and military standards define a flareless version (Fig. 18.21). A special ferrule with tapered sealing surfaces

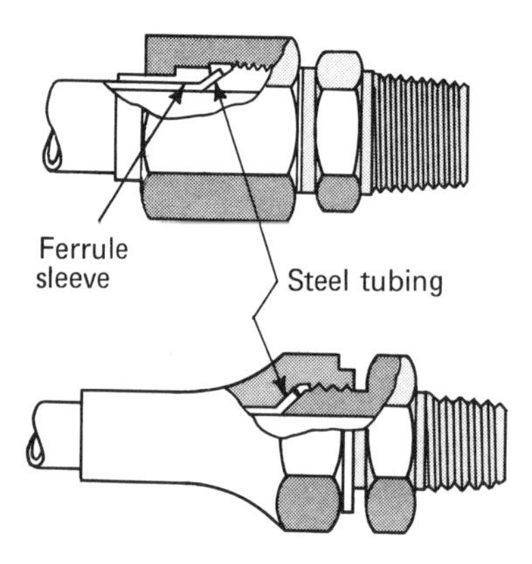

Fig. 18.20 Flared tubing fittings are widely used. Shown are three-piece and two-piece types.

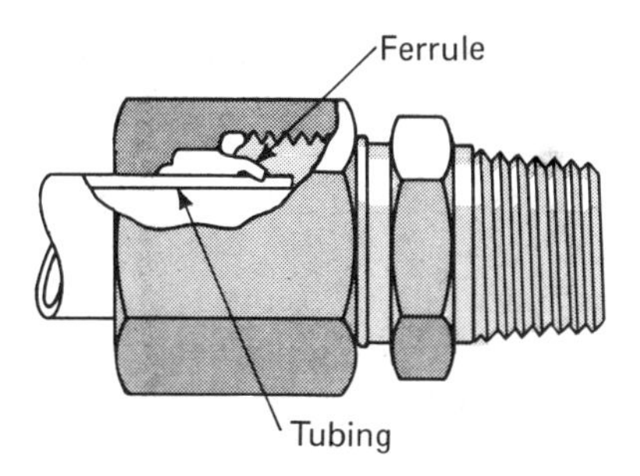

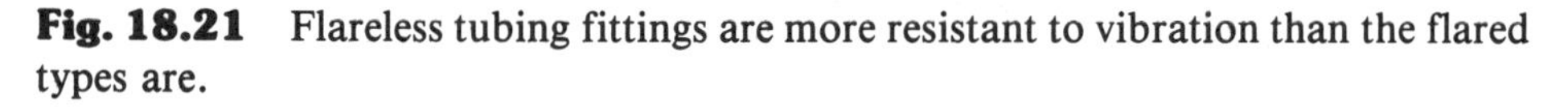
Fig. 18.21 Flareless tubing fittings are more resistant to vibration than the flared types are.

is swaged onto the end of the tube, and this ferrule does the sealing against the tapers of the nut and body of the fitting.

The ferrule can be preset in a separate fixture before the fitting is assembled, or it can be assembled loose within the fitting to be self-swaged onto the tubing as the nut is drawn up. In either case, the ferrule's hardened inner edge penetrates the surface of the tubing for a positive lock. The tubing will not move axially but may still rotate within the ferrule without loss of sealing.

Assembly requires great care, just as with the flare-type fitting. Overtorque can cause damage, and both the tube and sleeve must be replaced. A compensating advantage is that at least the fitting can be disassembled and examined visually to see if everything is all right.

There are many proprietary versions that offer multiple biting edges (Legris, for example), extra flexing of the ferrule to absorb vibration and extra segments in the ferrule to increase the wedging and holding (Swagelok), lips on the ferrule to prevent severing of the tubing during extreme torquing, and special fittings that are machined with a spherical or tapered surface to substitute for the ferrule-and-tubing half of the connector (Hoke). Many of these have been successful for decades – where assembled and applied in accordance with detailed instructions. The chemical processing industry has been a great proving ground, but industrial users don't always want to pay the price.

In Europe, the flareless fitting outsells the flared type by a wide margin. In the United States, the 37-deg flared fitting predominates over the flareless type. It is expected by some experts that the flareless types will gain prominence eventually because they withstand pressure and vibration better than the flared types.

Quick-Connect Fittings

The double-valve arrangement in many quick-connect couplings lets you join together two parts of a hydraulic (or pneumatic) system about as easily as plugging an appliance into an electric outlet. When the two parts mate, a valve in each automatically opens, and the liquid or gas flows through the union immediately. Disconnecting is just as simple – when the halves of the couplings are disengaged, the two valves close, stopping the flow.

Two other kinds of quick-connect couplings are available. One is plain (straight-though); it has no valves. This type has been used for years at your local gas station to join the rubber air hose to the pipe from the compressor. The other has a valve in its coupler; its simpler half, the nipple, is plain. This combination can be economical where many units containing nipples, such as a set of pneumatic power tools, connect to a single coupler.

You can get quick-connect couplings for various temperatures from −300 to 500 F. Some are designed to withstand vibration from 100 to 2400 cps over extended periods of time, pressures to 3000 to 5000 psi, and as many as 1000 connects and disconnects. Most commercial applications have far less stringent requirements.

Sizes generally range from ¼ to 1½ in. (pipethread size of the fittings); for special applications they go up to 3 or even 4 in.

There are couplings for practically every gas or liquid—corrosive and noncorrosive, organic (fuels) and nonorganic (acids, alkalis).

How They Lock. Different manufacturers use different methods to lock the coupling halves. The following four types are the most popular:

Balls in races: The coupler is designed to hold a number of circumferentially spaced balls (Fig. 18.22). An inner race retains the balls when the assembly is coupled.

Pins and grooves: A radial cam drives locking pins into a retaining groove in the nipple; spring-loaded detent secures the assembly. Sometimes cams are mounted around the internal circumference of the nipple. A rotation of about 90° draws the two parts together; they are then locked by a spring detent.

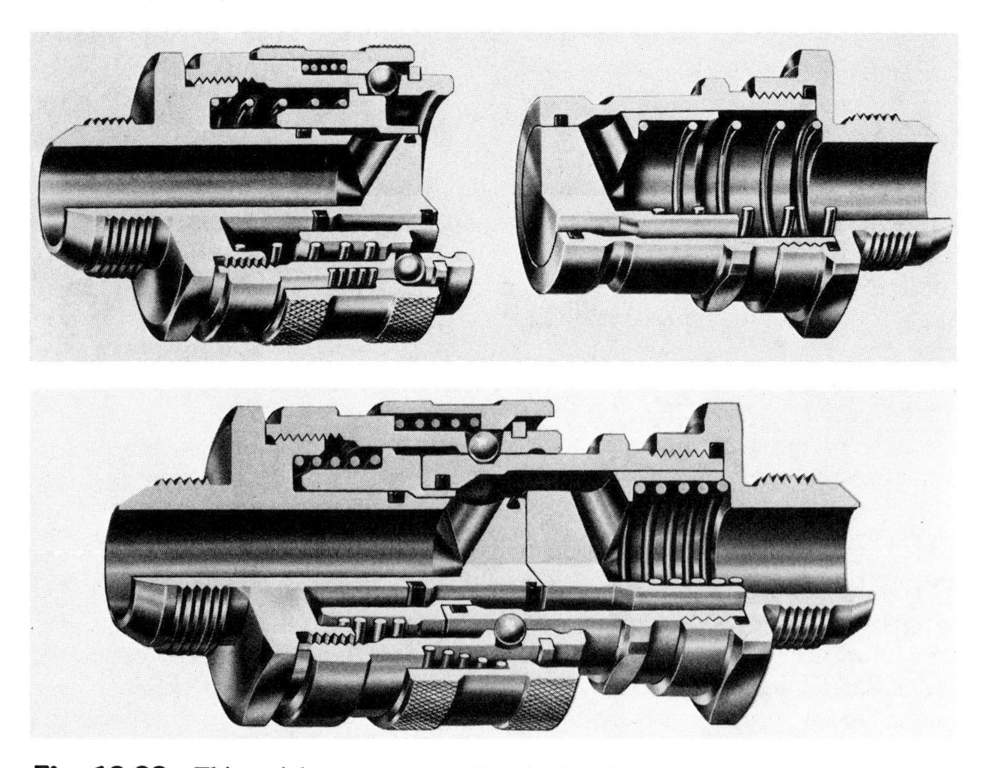

Fig. 18.22 This quick-connect coupling is double-valved, lap-sealed, and ball-detented. Drawings show coupling before and after insertion. (Courtesy of Snap-Tite, Inc., Union City, PA.)

Dogs in the coupler body: The dogs pivot into a locking groove in the nipple as the locking sleeve is moved into position.

Simple threads: The nipple contains the male and the coupler has the female threads (usually they are multiple-lead to reduce the number of turns required). The coupler sleeve is screwed onto the nipple and then locked by a spring-loaded detent or ratchet.

A fifth type is the staple-lock fitting (Dayco-Stecko), shown in Fig. 18.23. The locking means is self-explanatory.

How They Are Sealed. *Lap seal* uses a radial packing seal. The coupler and the nipple overlap to form the seal. This design provides tighter sealing as pressure increases. The lap seal is not suited to low-pressure or vacuum applications.

Butt seal: Here the packing seal is at the end of the nipple, forming a butt with the coupler. This method is suitable for all pressures, including vacuum.

Materials and Finishes. Couplings are usually made of brass plated with nickel and chrome; anodized aluminum; steel plated with cadmium, cadmium chromate, or nickel; or stainless steel, Types 303 and 316.

Seal materials are generally natural or synthetic rubber, depending on type, temperature, and pressure of the fluid.

How They Perform. The speed of disconnect is generally a fraction of a second and is accelerated by the internal line pressure. The check-valves are usually spring-loaded to ensure rapid action.

Couplings can be joined or separated by hand, but for safety, larger sizes are limited to lower internal pressures. For instance, a person can safely connect a ¼-in. coupling at 200-psi internal pressure or a 1-in. unit at about 65 psi. Systems handling higher pressures are usually manually connected and disconnected while internal pressure is reduced. They can be remotely disconnected under high pressure by using solenoid valves, lanyards, or other means.

How Much Liquid Spills? When an ordinary coupling is disconnected under pressure, some of the liquid will spill. But not much. For instance, ¼-in. couplings, depending on design, will spill anywhere from 6 drops to about 1.5 cc during the disconnect cycle, with an internal pressure of 60 psi. Stepping up to 1½ in., with an internal pressure of about 10 psi, spillage will be about 10 times as great.

Trends in quick-disconnects, according to Snap-Tite (Union City, PA), are as follows: demand for virtually zero leakage at any step from insertion to removal, more requests for "push-to-connect" types for ease of installation, and increased interest in flush-valved couplings that have no unnecessary crevices to catch dirt while disconnected and exposed to atmosphere.

The greatest abuse of quick-disconnects (as with other fittings and conduits) is in not paying attention to rated surge-shock cycles. Users seem aware of cycling as it relates to pumps and cylinders, but they forget about couplings.

The NFPA has a good test standard on the subject: T3.8.70.2, *test conditions and procedures for hydraulic fluid power quick disconnect couplings.* It's similar to the one for tube fittings (T3.8.70.3) in that it includes a pressure impulse graph to define just what constitutes a pulse.

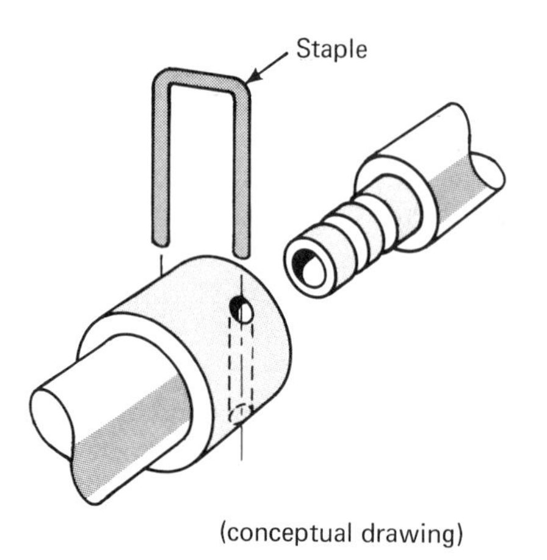

Fig. 18.23 This quick-connect coupling is unvalved and staple-locked.

Working pressures to 5000 psi are not unusual, and the connectors are expected to be leaktight and safe. Test benches and jacks have even higher pressures—to 10,000 psi.

Contamination is another problem area with users, especially in double-valved (double-shutoff) applications. A particle of dirt can prevent a detent from locking and cause a later accidental opening of the disconnect.

Quick-Insert Flexible Tubing

Low- to moderate-pressure pneumatic (or hydraulic) control systems can be quickly assembled with various push-in tubing connections. One—the Legris method—is pictured in Fig. 18.24.

The initial design was intended for pneumatic and hydraulic systems with ⅛- to ½-in. tubing at pressures to 280 psi and temperatures from 5 to 160 F.

Chinese finger puzzle: Each fitting in the line of a dozen or so types is completely self-contained, and the tubing needs only to be pushed into the hole for complete assembly. An O-ring fits around the OD of the tubing, and a slotted brass collet, lightly preloaded against the tubing OD, serves the dual function of supporting the tubing and gripping it against pull-out force.

The tube butts against the large fitting orifice. Flow is unrestricted because sealing and clamping are on the tubing OD.

The locking slope in the fitting body forces the collet hooks against the OD of the tubing when either internal pressure or external pull force acts in the direction to eject the tubing. The tubing will not pull loose under any operating conditions. In fact, like the Chinese finger puzzle, the harder the pull, the tighter the holding force. Fingernail pressure inward against the collet releases the collet hooks and frees the tubing for easy manual withdrawal at zero pressure.

The O-ring ID is slightly less than the tubing OD, so insertion of the tubing automatically places light sealing force at the interface. Internal pressurization of the system forces the O-ring against the end of the collet and increases interface contact area to include both the tubing and the fitting body. The sealing effect increases with pressure, as in conventional O-ring seals.

Ninety percent of the European applications for the fitting systems have Nylon 11 tubing, invented in France. It has a strength and resiliency that works well with the clamping-sealing scheme of the system. Polyethylene tubing also will work if the proper grades and sizes are selected, but the company sometimes recommends that a thin-walled metal "top hat" be inserted into the tubing for reinforcement.

A unique feature of the majority of the line of fittings is an internal hex in the flow path (Fig. 18.25). An Allen wrench fits the hex, so an open-end or socket wrench is not needed. Such fittings can be located on much closer centers than are allowed with external wrenching.

One of the fittings offered is a plug-in Y (Fig. 18.26). It's unlike any other Y because the base end is made in the form of tubing and can be inserted into any other of the fittings just as if it were tubing. Without using any tubing, three of the plug-in Ys can be cascaded to give four new tubing ports.

Another fitting is a press-in cartridge (Fig. 18.27) for making your own manifolds or special parts. The cartridge innards have the same elements as the regular fittings, including the O-ring and collet. However, the cartridge has a barbed outer sleeve that can be pressed into unthreaded stepped holes in a manifold, valve body, pipe, or any control device. The manifold bodies can be made of any of many types of metal or plastic. The hole can be machined to the necessary tolerance in one operation with

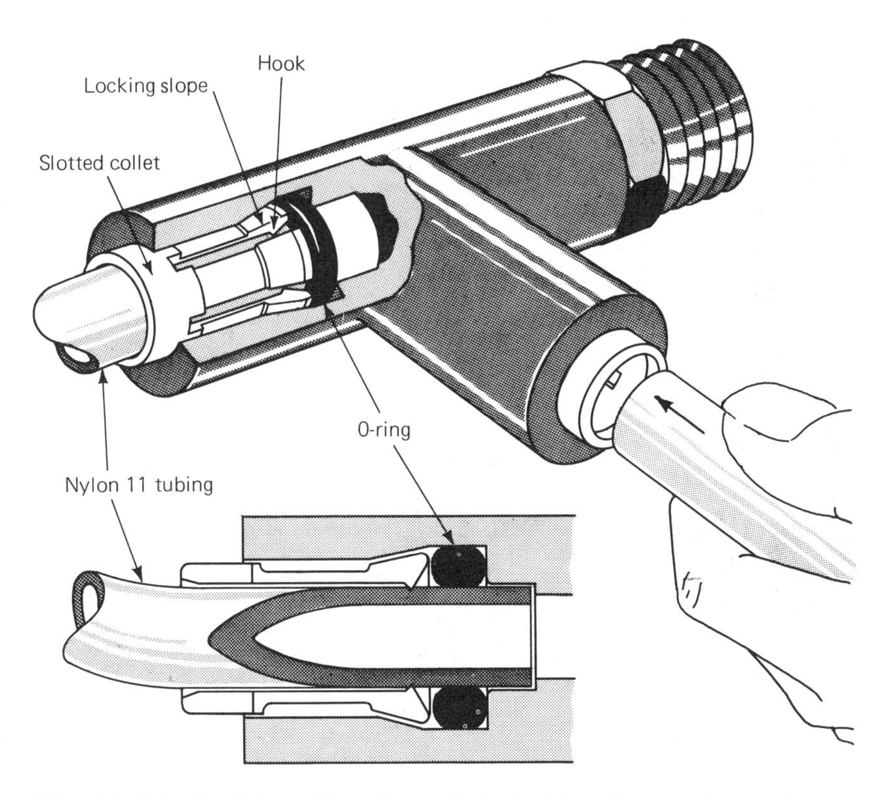

Fig. 18.24 Push-in tubing fitting is locked in with a collet and sealed with an O-ring. Tubing will not pull out unless the collet is held in with a fingernail or tool. Artist's conception only. (Courtesy of Legris Inc., Rochester, NY.)

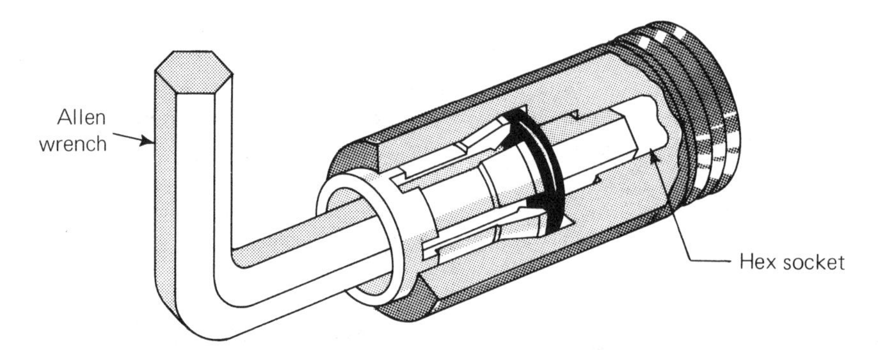

Fig. 18.25 Wrench fits inside the collet for the screw-type version (see Fig. 18.24).

a formed tool or drill or can be high-pressure-cast when the part is made. A carrier is used to hold the pieces of the cartridge together during insertion.

Swivel Joints

A rival to flexible hose is piping with joints that allow rotation or other relative movement yet seal the contained fluid. They are constructed of steel, cast iron, bronze,

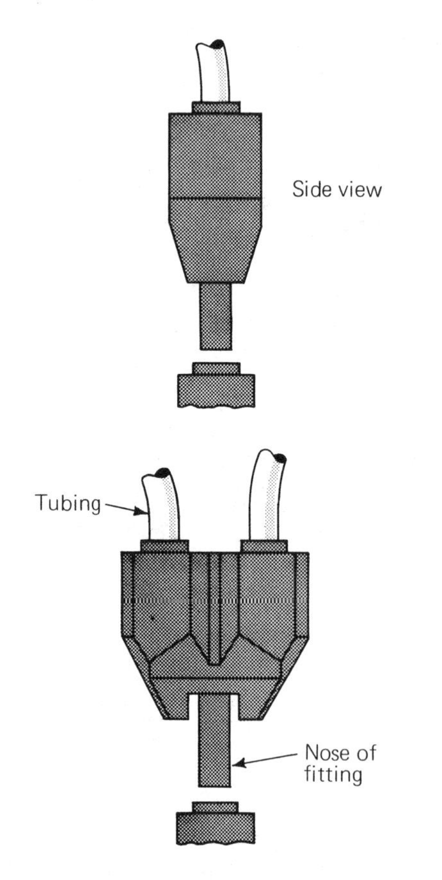

Fig. 18.26 Adaptors have push-in feature for the nose of the fitting as well as for the tubing (see Fig. 18.24).

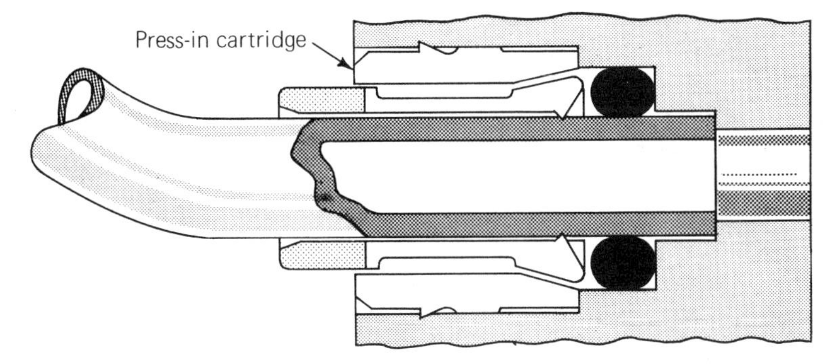

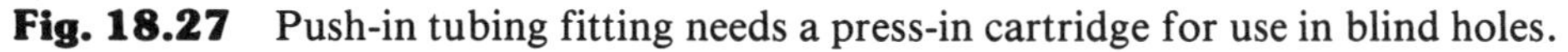

Fig. 18.27 Push-in tubing fitting needs a press-in cartridge for use in blind holes.

aluminum, or other rigid materials and can usually withstand the same environment, pressures, and types of service as the rest of the piping system. Low-pressure joints operate from 28-in. vacuum to 1000 psi and high-pressure joints to over 15,000 psi. Most joints will work at 250 F, and special designs permit as low as -300 F or high as 1000 F. Corrosion is not too great a problem because there is a lot of material to corrode away. Small-radius bends are easily accommodated because the rotating joint is not part of the bend and therefore does not limit the bend radius significantly. A secondary advantage of a single-plane swivel is its movement in a predetermined arc and plane.

Almost any degree of freedom of movement can be obtained by combining two or more joints (Fig. 18.28). Single-plane swivels are common, but multiplane, spher-

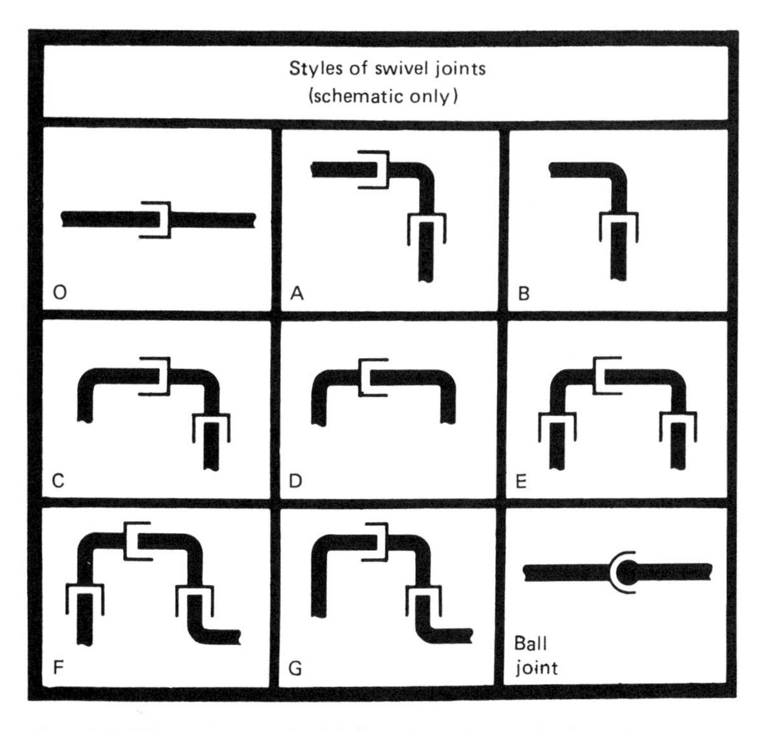

Fig. 18.28 These swivel joints have letter designations referred to in Fig. 18.30.

ical, or combinations of swivels and hose are available. Seven swivel joints will give unlimited flexibility. Sizes from ¼ to over 36 in. in diameter can be built.

Two limitations to swivel joints are difficulty in maintaining a leaktight rotating or sliding seal, particularly when installation and maintenance personnel forget that these are precision devices and require special care, and relatively few planes of rotation available to any type except the spherical.

The joint itself (Fig. 18.29) may swivel on ball bearings, a spherical joint, thrust plates, or a bushing. The seal packing may also serve as the bearing in some applications. Full 360° rotation in the plane of the swivel is usually available, and the spherical (ball-joint) type may also pivot a total of 40° about its ball center.

Ball bearings are used to reduce torque without hurting smoothness. They have less wear than sliding bearings and can carry thrust and radial loads simultaneously. Thrust loads come from thermal expansion and other forces exerted by the piping on the coupling and from the internal fluid pressure. Radial loads are caused by weight of the piping, which also contributes a bending moment at the joint. With proper spacing between bearing races and careful choice of heavy-duty rolling elements, up to 90,000 in.-lb can be handled by a nominal 4-in.-diameter swivel joint.

Thrust bearings (shoulders and washers) cannot take radial loads and will not rotate freely if the axial load is heavy or cantilevered. Applications are for small-size hydraulic lines.

Sleeve bearings are satisfactory for light radial loads. Heavier loads increase torque considerably and accelerate wear. Axial loads cannot be taken.

Spherical-type fittings have circular gaskets that seal the joint and also act as spherical bearings. Each joint is self-aligning, but a single one should not be depended on to solve misalignment in other parts of the piping system. However, two will often correct misalignment better than three or more single-plane swivels.

Continuous rotation swivel joints are needed when fluid from a stationary source is fed into a revolving drum. Special swivel fittings may be required for fast, continuous rotation, and knowledge of peripheral speed is important. The larger the joint diameter, the lower the allowed maximum rpm—this is to limit surface speed and help packing life. There are small joints designed for speeds to 10,000 rpm, pressures to 250 psi, and temperatures to 400 F.

Fluid type influences material selection for swivel joints. Any metal is satisfactory for ordinary oils, alcohols, aromatics (benzene, napthalene, toluene), esters (acetates), natural gas, kerosene, and so forth. Stainless steel, monel metal, or nickel-bronze should be specified for acids, mercury, molasses, and sewage. Where allowed, asbestos chevron-type packing is suitable for the aromatics and esters, but Teflon or synthetic rubbers will be required for most of the other fluids mentioned above. The subtle effect of temperature, plus compatability problems, suggests that experts be consulted to assure proper material selection.

3-D Applications. A scale model of the proposed installation will probably save time, but there are economical alternatives to figuring the best layout and design. The accompanying example (Fig. 18.30) is one of them.

TUBE BENDING

Rigid bent tubing for flowing liquid or for the protection of wiring is needed in cars, aircraft, and equipment of all sorts. The problem is how to figure the exact bends

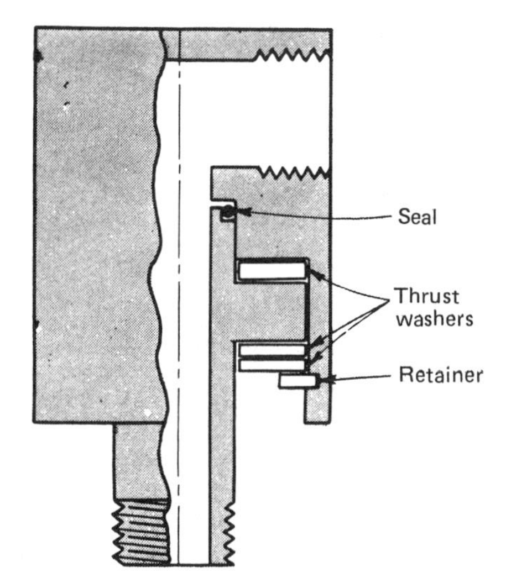

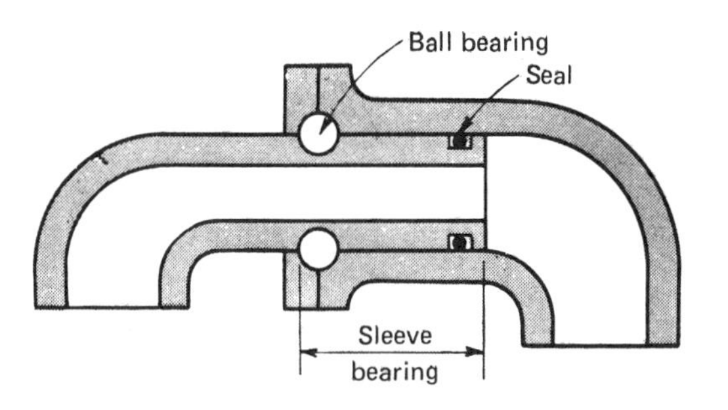

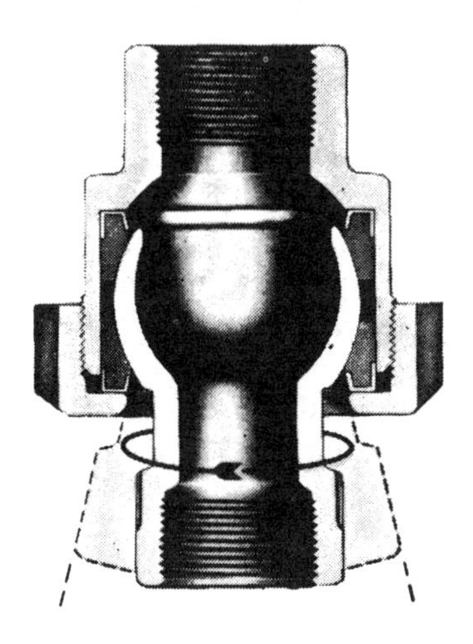

Fig. 18.29 Many swivel joints work on one of these three principles.

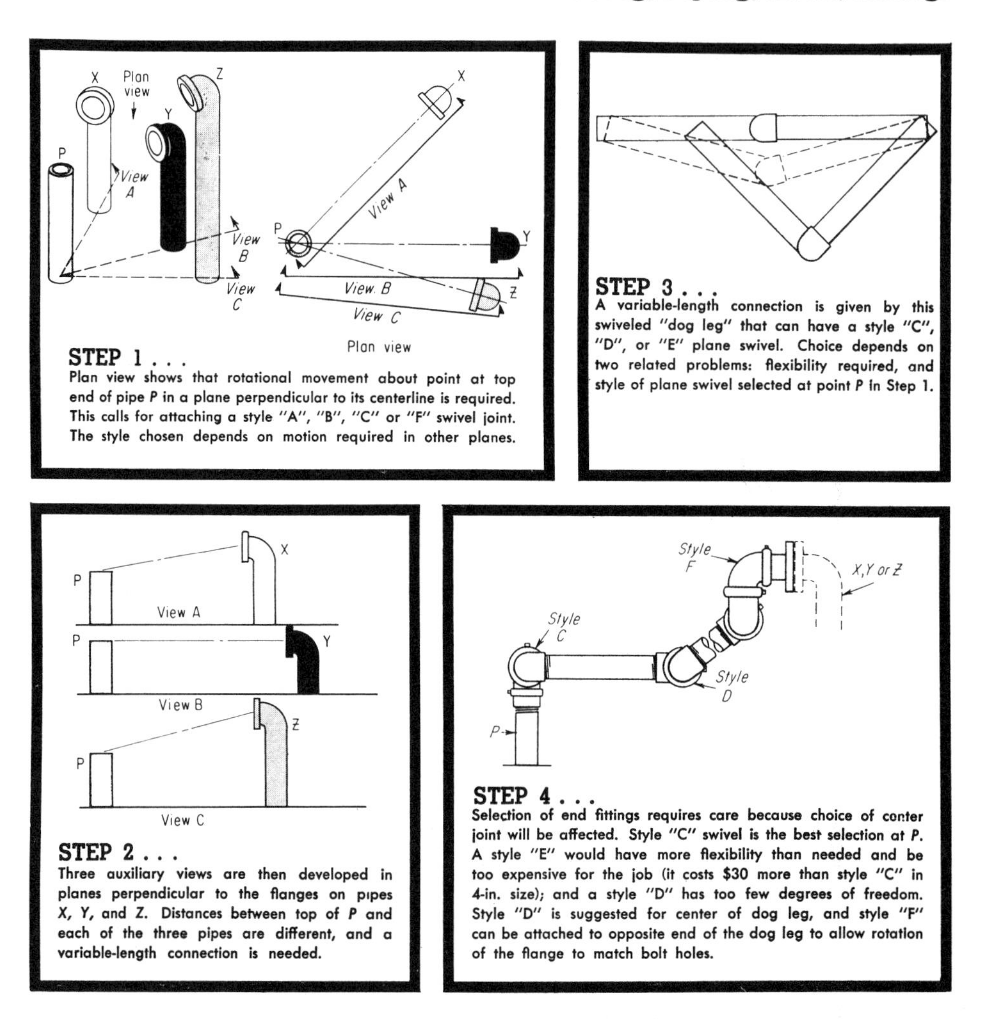

Fig. 18.30 This simple three-dimensional design problem is typical: Pipe P is a feed line that must pass fluid to any one of three receiving pipes: X, Y, or Z. Choose a single, flexible-line connection to serve the three pipes. See Fig. 18.28 for an explanation of styles.

and write an instruction that the bending machine operators can follow. This would apply to rods also. The accompanying method was developed by David E. Brown, Geosource Inc. (Erie, PA).

The knowns are fairly straightforward: Start with a system layout or assembly drawing showing fitting points that must be connected. Analyze the available space and choose the optimum path, making sure it is a series of straight-line segments with as few bends as possible.

Then it gets harder. Orthographic views on a detail drawing (Fig. 18.31) will show the desired tubing run, but making up the segments accurately on a tube bender is an art. At least six factors must be considered: distance between bends, angle of each bend, tubing rotation between bends, stretch, springback, and required blank lengths.

Bending machines work in this sequence: The proper length blank is placed in the machine; first straight length (an end) is extended (a in Fig. 18.32); first bend is made (b); next straight length is extended (c); tube is rotated to put next bend into plane of motion of bending head (d); second bend is made (e); and same steps are repeated for rest of bends (Fig. 18.33). The parameters are the same for a tube or solid rod.

Calculation Procedure

The accompanying equations (Table 18.5) are self-explanatory, but the reasons for the sequences and an explanation of the symbols and coordinates will be helpful.

Points on the tube, including bends and end points, are designated P, with the coordinates given as subscripts or in parentheses. The starting point is P_1, the intermediate points are P_2, P_3, etc., and the final point is P_F. The starting point is fully described as $P_1(X_1, Y_1, Z_1)$, and the final point is $P_F(X_F, Y_F, Z_F)$.

The projected length of a segment along a given axis is designated L, with the chosen point-to-point designations and axes given as subscripts. For example, the projected length of P_1 to P_2 along the Z axis is L_{12Z} or Z_2-Z_1. Another way of writing the expression is Z_{02}-Z_{01}, which means the same thing but is a little easier to program on

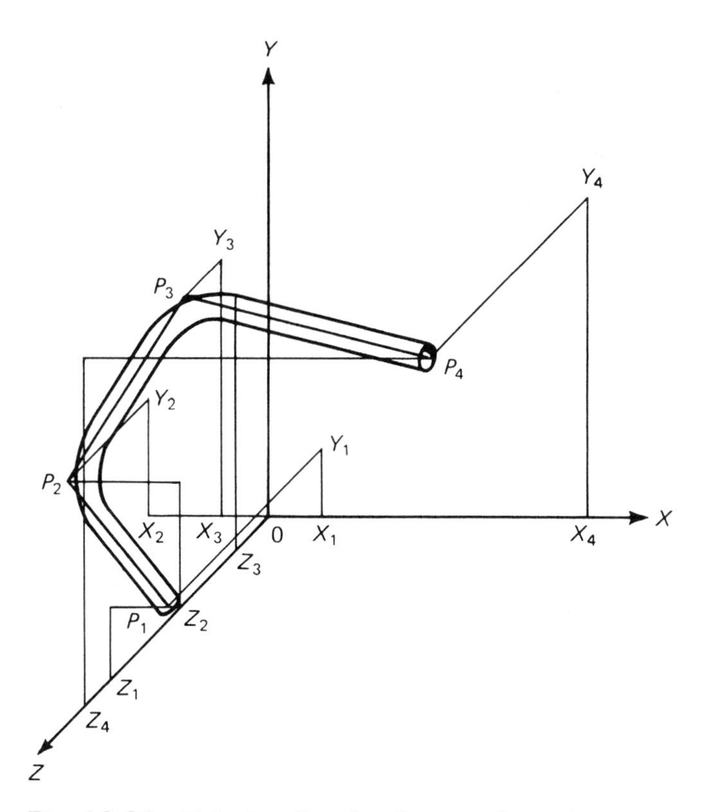

Fig. 18.31 Tube-bending drawings are shown in a conventional right-hand orthographic projection coordinate system.

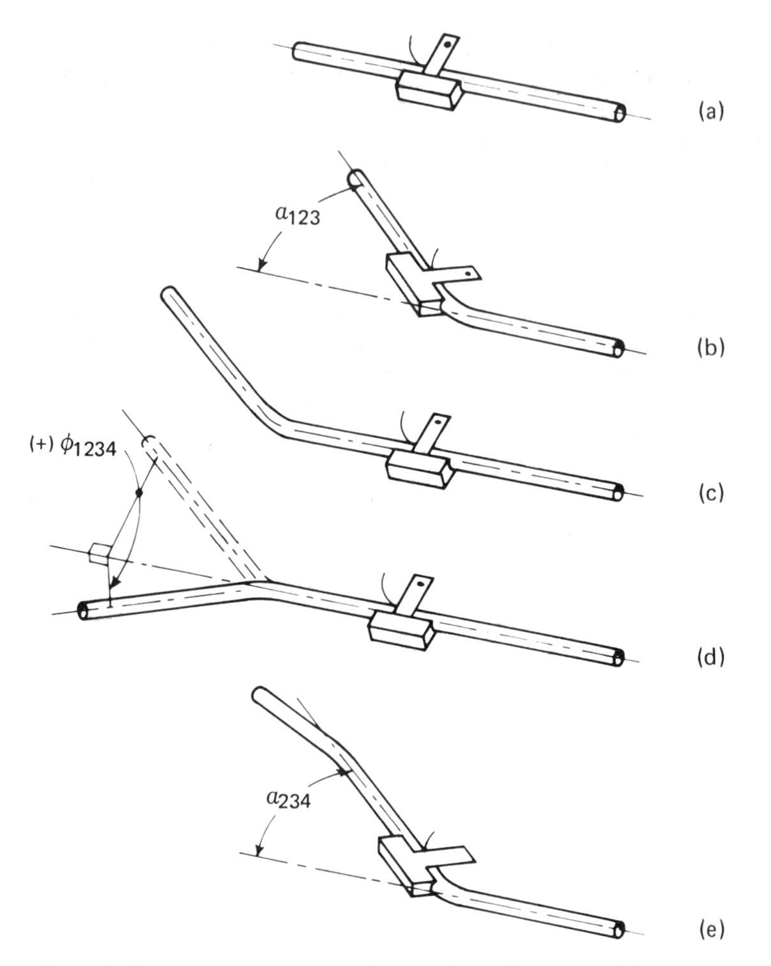

Fig. 18.32 Bends are made in a series of precalculated steps involving linear and rotary motion. (Courtesy of Geosource, Inc., Erie, PA.)

the computer. We've used these and similar expressions both ways interchangeably in the text and drawings.

The actual length of a segment between bends, ignoring bend radii, is the square root of the sum of the squares of the projected lengths. For example, $L_{12} = (L_{12X}^2 + L_{12Y}^2 + L_{12Z}^2)^{0.5}$.

Bends in a single plane involve three points: say P_1, P_2, and P_3 for the first bend, and P_2, P_3, and P_4 for the second bend. The bend angle is designated α. For example, the bend angle of points 1, 2, and 3 is α_{123}, and for the bend 2, 3, and 4 it's α_{234}. See Table 18.5 for the math.

Rotation (twist) of the tube between bends (Fig. 18.34) involves four points: say P_1, P_2, P_3, and P_4. The math is a little more involved but can be followed step by step in the equations box and the example (Table 18.6). The symbol is ϕ. For example, ϕ_{1234} is the tube twist between bend 123 and bend 234 (Fig. 18.33).

Two methods of determining rotation between bends are included in the equations. The first is the coordinate rotation method; the second is the plane rotation method. Each has its advantages and will be separately described.

The coordinate rotation method (Fig. 18.33) is adaptable to computerization because the direction of rotation can be computed directly without use of reference drawings. Angle β is the defined rotation of a given pictorial view to enable a draftsman to draw the true length of a desired segment. An example is view 1. The dimensions for drawing or checking conventional auxiliary views are a byproduct of this work.

The plane rotation method, which is simpler, is good for designing tube configurations that lie in prescribed planes or are coplanar with other bent tubes, as in heat exchangers.

A special case for the coordinates in the tube, bending equations, section D, is when $X_3 = Y_3 = Z_3$ (or $X_{03} = Y_{03} = Z_{03}$, which means the same). For that condition, use these values for A, B, and C:

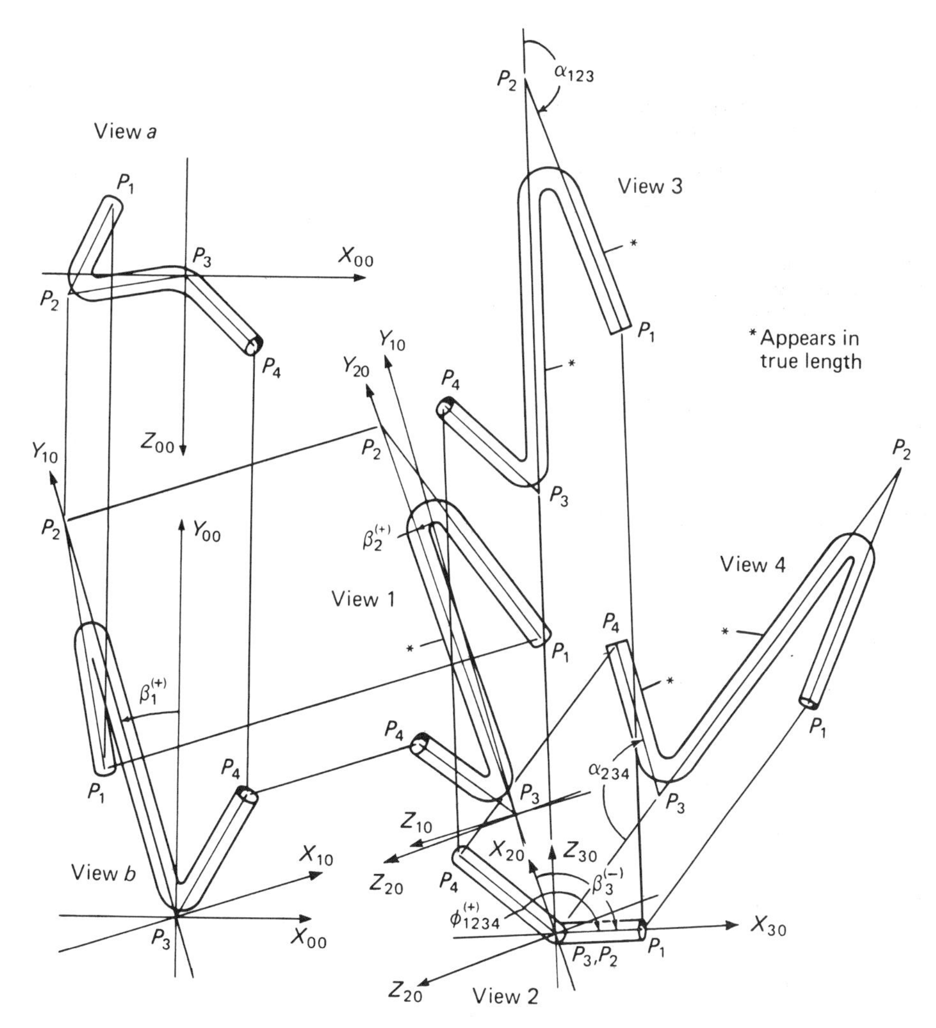

Fig. 18.33 Coordinate rotation method for tube bending is designed for computer calculation. Views a and b are conventional orthographic views, using the original reference coordinate system. The plus and minus signs over the angles indicate positive or negative. The symbols are explained in the text.

Table 18.5 Tube-Bending Equations

A. Lengths between bend intersection points (ignoring bend radii)

Let $L_{12X} = X_2 - X_1$

$L_{12Y} = Y_2 - Y_1$

$L_{12Z} = Z_2 - Z_1$ etc. for other points

The lengths between points of intersection are

$$L_{12} = (L_{12X}^2 + L_{12Y}^2 + L_{12Z}^2)^{0.5} \text{ etc.}$$

B. Bend magnitudes (see Fig. 18.32 (b)

In calculating bend magnitudes, three successive points are evaluated for each bend, viz:

P_1, P_2 and P_3 for the first bend

P_2, P_3 and P_4 for the second bend, etc.

$$\alpha_{123} = \cos^{-1}\left[\frac{L_{12X}L_{23X} + L_{12Y}L_{23Y} + L_{12Z}L_{23Z}}{(L_{12})(L_{23})}\right]$$

This gives the bend magnitude of the first point. Repeat for each following required bend, as follows:

$$\alpha_{234} = \cos^{-1}\left[\frac{L_{23X}L_{34X} + L_{23Y}L_{34Y} + L_{23Z}L_{34Z}}{(L_{23})(L_{34})}\right], \text{ etc.}$$

C. Rotation between bends: coordinate rotation method

Position the origin of the reference system at the third of the four points being considered, viz: $P_3\ (X_{03} = Y_{03} = Z_{03} = 0)$
The following equations should be used in the sequence shown:

1. If $X_{02} = 0$, skip to step 2

$X_{11} = X_{01}B + Y_{01}A \quad X_{12} = X_{02}B + Y_{02}A \quad X_{14} = X_{04}B + Y_{04}A$

$Y_{11} = -X_{01}A + Y_{01}B \quad Y_{12} = -X_{02}A + Y_{02}B \quad Y_{14} = -X_{04}A + Y_{04}B$

$Z_{11} = Z_{01} \quad Z_{12} = Z_{02} \quad Z_{14} = Z_{04}$

$$\text{where } \beta_1 = \frac{-X_{02}}{|X_{02}|}\left[\cos^{-1}\left[\frac{Y_{02}}{(Y_{02}^2 + X_{02}^2)^{0.5}}\right]\right]$$

and $A = \sin\beta_1$; $B = \cos\beta_1$

2. If $Z_{12} = 0$, skip to step 3

$X_{21} = X_{11} \quad X_{22} = X_{12} \quad X_{24} = X_{14}$

$Y_{21} = Y_{11}D + Z_{11}C \quad Y_{22} = Y_{12}D + Z_{12}C \quad Y_{24} = Y_{14}D + Z_{14}C$

$Z_{21} = -Y_{11}C + Z_{11}D \quad Z_{22} = -Y_{12}C + Z_{12}D \quad Z_{24} = -Y_{14}C + Z_{14}D$

$$\text{where } \beta_2 = \frac{Z_{12}}{|Z_{12}|}\left[\cos^{-1}\left[\frac{Y_{12}}{(Y_{12}^2 + Z_{12}^2)^{0.5}}\right]\right]$$

and $C = \sin\beta_2$; $D = \cos\beta_2$

3. If $Z_{21} = 0$, skip to step 4

$X_{31} = X_{21}F + Z_{21}E \quad X_{32} = X_{22}F + Z_{22}E \quad X_{34} = X_{24}F + Z_{24}E$

$Y_{31} = Y_{21} \quad Y_{32} = Y_{22} \quad Y_{34} = Y_{24}$

$Z_{31} = -X_{21}E + Z_{21}F \quad Z_{32} = -X_{22}E + Z_{22}F \quad Z_{34} = -X_{24}E + Z_{24}F$

$$\text{where } \beta_3 = \frac{Z_{21}}{|Z_{21}|}\left[\cos^{-1}\left[\frac{X_{21}}{(X_{21}^2 + Z_{21}^2)^{0.5}}\right]\right]$$

and $E = \sin\beta_3$; $F = \cos\beta_3$

4. Rotation angle between bends (18.34)

$$\phi_{1234} = \frac{Z_{34}}{|Z_{34}|}\left[\cos^{-1}\left[\frac{X_{34}}{(X_{34}^2 + Z_{34}^2)^{0.5}}\right]\right]$$

Where ϕ is the $\leqslant 180°$ angle between the plane containing (P_1, P_2) and the plane containing (P_3, P_4). When viewed down the line L_{23} in the direction from P_3 toward P_4:

Positive ϕ indicates a clockwise rotation from P_4 to P_1
Negative ϕ indicates a counter-clockwise rotation from P_4 to P_1

D. Plane rotation method (see text)

In calculating rotations between bends, four successive points are evaluated for each rotation. The first rotation (between the first and second bends) evaluates points P_1, P_2, P_3 and P_4. The second rotation (between the second and third bends) evaluates points P_2, P_3, P_4 and P_5 (see Fig. 18.31). The rotation required between bends α_{123} and α_{234} is:

$$\phi_{1234} = \cos^{-1}\left[\frac{A_1A_2 + B_1B_2 + C_1C_2}{(A_1^2 + B_1^2 + C_1^2)^{0.5}\,(A_2^2 + B_2^2 + C_2^2)^{0.5}}\right]$$

where:
$$A_1 = (Y_1Z_2 + Y_2Z_3 + Y_3Z_1 - Y_1Z_3 - Y_2Z_1 - Y_3Z_2)$$
$$B_1 = (X_1Z_3 + X_2Z_1 + X_3Z_2 - X_1Z_2 - X_2Z_3 - X_3Z_1)$$
$$C_1 = (X_1Y_2 + X_2Y_3 + X_3Y_1 - X_1Y_3 - X_2Y_1 - X_3Y_2)$$

and:
$$A_2 = (Y_2Z_3 + Y_3Z_4 + Y_4Z_2 - Y_2Z_4 - Y_3Z_2 - Y_4Z_3)$$
$$B_2 = (X_2Z_4 + X_3Z_2 + X_4Z_3 - X_2Z_3 - X_3Z_4 - X_4Z_2)$$
$$C_2 = (X_2Y_3 + X_3Y_4 + X_4Y_2 - X_2Y_4 - X_3Y_2 - X_4Y_3)$$

Refer to conventional orthographic and auxiliary views to determine the direction, clockwise or counter-clockwise, of this value. Also see text for simplified special case

The coefficients A_1, B_1 and C_1 above, in combination with:
$$D_1 = (X_1Y_3Z_2 + X_2Y_1Z_3 + X_3Y_2Z_1 - X_1Y_2Z_3 - X_2Y_3Z_1 - X_3Y_1Z_2)$$
form an equation for the plane containing P_1, P_2 and P_3 of the form:
$$A_1X + B_1Y + C_1Z + D_1 = 0$$
Likewise, the coefficients A_2, B_2 and C_2 above, in combination with:
$$D_2 = (X_2Y_4Z_3 + X_3Y_2Z_4 + X_4Y_3Z_2 - X_2Y_3Z_4 - X_3Y_4Z_2 - X_4Y_2Z_3)$$
form an equation for the plane containing P_2, P_3 and P_4 of the form:
$$A_2X + B_2Y + C_2Z + D_2 = 0$$
and the angle at which the half-planes containing L_{12} and L_{34} intersect is ϕ_{1234}

E. Required blank length and length between bends (Fig. 18.35)

For bent rods where all bend radii (R) are equal:

1. Blank length, L_B

$$L_B = (L_{12} + L_{23} + \ldots) + [R(\alpha_{123} + \alpha_{234} + \ldots) - 2R(\tan(\alpha_{123}/2) + \tan(\alpha_{234}/2) + \ldots) - \sum_{F=1}^{N} e_F]$$

2. Straight length between bends, L'

$$L'_{12} = L_{12} - R[\tan(\alpha_{123}/2)]$$

$$L'_{23} = L_{23} - R[\tan(\alpha_{123}/2) + \tan(\alpha_{234}/2)]$$

where N = number of bends; α = radians

Table 18.6 Example of Tube Bending

1. The coordinates of the points on the rod illustration in Fig. **18.33** are:

	P_1	P_2	P_3	P_4
X	−7.5	−12.5	0	7.5
Y	15.0	40.0	0	12.5
Z	−7.5	2.5	0	7.5

2. Using the equations above, the lengths are $L_{12} = 27.386$, $L_{23} = 41.982$, $L_{34} = 16.394$

3. Using the equations above, the bend magnitude for the two bends are: $\alpha_{123} = 161° 03' 47''$; $\alpha_{234} = 128° 08' 07''$

4. Using the equations in C, the plane coefficients are:

$A_1 = 337.5$ $\quad A_2 = -268.75$

$B_1 = 112.5$ $\quad B_2 = -112.5$

$C_1 = -112.5$ $\quad C_2 = 456.25$

$D_1 = 0$ $\quad D_2 = 0$

yielding:

$337.5X + 112.5Y - 112.5Z = 0;$

$-268.75X - 112.5Y + 456.25Z = 0;$

and a rotation of $\phi_{1234} = 139° 58' 54''$

5. Using the equations in D, it is found that:

$\phi_{1234} = +139° 58' 54''$ (clockwise)

and:

		P_1	P_2	P_3	P_4
$\beta_1 = 17° 21'$					
	X	−2.68	0	0	10.89
	Y	16.55	41.91	0	9.69
	Z	−7.50	2.50	0	7.50
$\beta_2 = 3° 24'$					
	X	−2.68	0	0	10.89
	Y	16.07	41.98	0	10.12
	Z	−8.47	0	0	6.91
$\beta_3 = -107° 35'$					
	X	8.88	0	0	−9.88
	Y	16.07	41.98	0	10.12
	Z	0	0	0	8.29

6. Using the equations above, given $R = 2{,}0$

$L'_{12} = 15.394$ $\quad L'_{23} = 25.877$ $\quad L'_{34} = 12.281$

$L_B = 63.644$ (ignoring rod stretch)

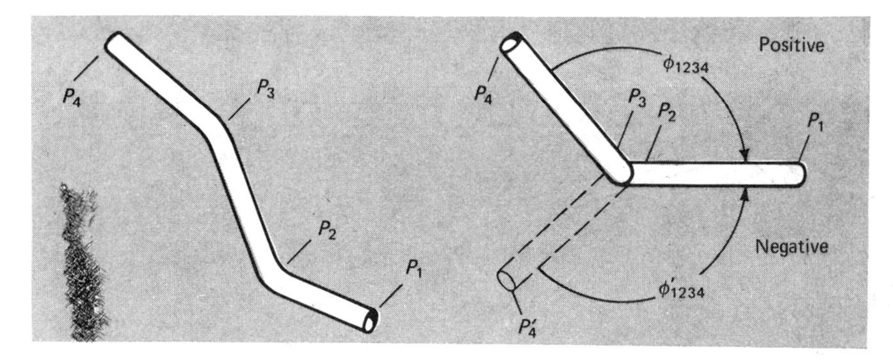

Fig. 18.34 Rotation angle between bends involves four points and needs somewhat complex math to figure the twist. See the text.

$A_1 = (Y_1Z_2\text{-}Y_2Z_1)$
$B_1 = (X_2Z_1\text{-}X_1Z_2)$
$C_1 = (X_1Y_2\text{-}X_2Y_1)$
$A_2 = (Y_4Z_2\text{-}Y_2Z_4)$
$B_2 = (X_2Z_4\text{-}X_4Z_2)$
$C_2 = (X_4Y_2\text{-}X_2Y_4)$

Most of the computations can be done on a hand calculator, but here are some tips. If zero divided by zero gives an error signal, add an arbitrary small number (1×10^{-6}) to the numerator and denominator. Also note we have designated the absolute values in the denominators for equations of coordinate rotation. The fraction is always 1.0, plus or minus, but the sign is changed to match the quadrant. This makes computer programming easier.

Rod stretch and springback: When material is plastically formed by the wiping motion of machine elements in Fig. 18.32, it undergoes some elongation, as measured along the tube or rod centerline in the region of the curved bend. Generally this stretch will be proportional to the magnitude of the bend: $e_F = K\alpha_F R$, where e_F = stretch in bending, in.; K = empirically derived constant ($K < 1$); α_F = bend magnitude, radius and deg; and R = bend radius at rod centerline.

The result of stretch is to require a shorter rod blank length (Fig. 18.35) by the amount of the sum of bend stretch values along its lengths.

Some springback will occur. The angular springback value will be a function of material modulus, bend magnitude, material yield strength, and forming method. The relationship is $\alpha_{mi} = K\alpha_{mi}$, where α_{mi} = machine bend magnitude, radius or deg; K = an empirically derived constant; and α_{mi} = bend magnitude required, radians or deg. This will require a machine angular bend setting slightly larger than the required bend.

Auxiliary views: Apart from the conventional orthographic views, four auxiliary views (Fig. 18.33) are required to show in true length for dimensioning purposes the elements of a four-point bent tube (two bends and one rotation).

The first auxiliary view (view 1) is projected perpendicular to line (P_2, P_3) in view b to give true length. The coordinates are given above.

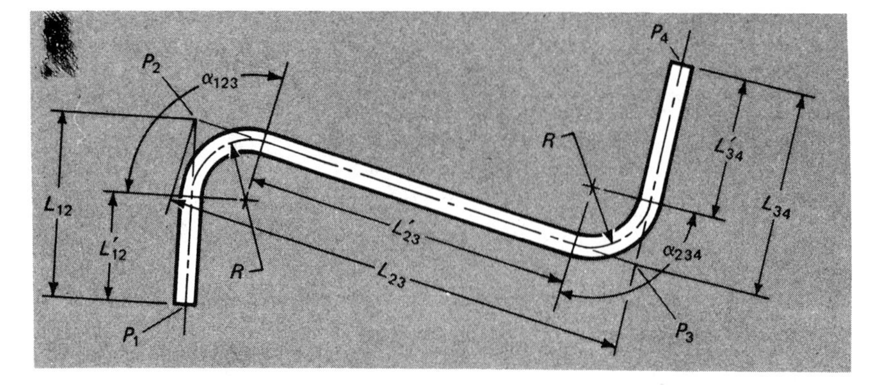

Fig. 18.35 Length between bends and the required blank length are two of the goals of the computation. Don't forget to allow for normal tubing stretch.

The second auxiliary view (view 2) is projected parallel to line (P_2, P_3) in view 1. This view will show line (P_2, P_3) as a point and the true magnitude of ϕ_{1234}.

The third and fourth auxiliary views (views 3 and 4) are projected perpendicular to lines (P_1, P_2) and (P_3, P_4) in view 2. These views will show, respectively, the true magnitudes L_{12}, α_{13} and L_{34}, α_{234}.

19

Dynamic and Static Sealing

Sealing against oil or air leakage is too complex a subject to be covered in a single chapter, so we've broken it into three:

- Chapter 19 (this chapter): Dynamic and Static Sealing (typical elastomer and plastic seal materials, special-purpose materials, friction and wear, and static seals and gaskets)
- Chapter 20: Solving Cylinder Rod Leakage
- Chapter 21: Special rotary seals (mechanical face seals, labyrinth seals, and magnetic fluid seals)

The performance requirements faced by all seals are typified by the hypothetical lip seals shown in Figs. 19.1 and 19.2

There may be design flaws worse than leakage, but none grip the attention more than a puddle does, unless it's the hiss of escaping air, gas, steam, or chemical vapor. The aircraft industry solved the drip problem decades ago by superior design because it had to: An empty hydraulic system won't land a plane.

It's about time we solve the problem for conventional vehicles, systems, machines, and equipment. Mobil Oil Corp. (New York, NY) made a good start by creating the Hydraulic Fluid Index (HFI). They calculated that a typical manufacturing plant leaks or discards annually four times more oil than its machines actually hold (HFI equals 4.0). Oil doesn't usually wear out, so the goal should be HFI equals zero. Similar figures can be cited for plant air.

Good design, combined with good seal materials, can help prevent leaks. Let's start with the materials.

THE SEALING MATERIALS ART

Selection is an art because most applications are compromises, such as resilience vs. stiffness, softness vs. hardness, low temperature vs. high (Fig. 19.3), and performance vs. cost.

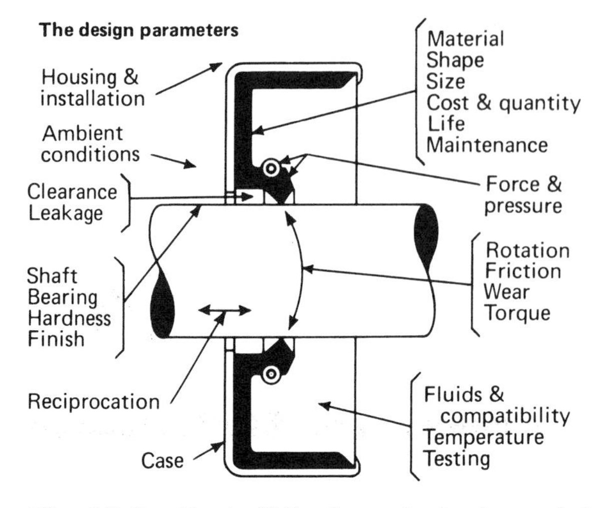

Fig. 19.1 Checkoff list for seal selection and design.

Some of the best-known materials are nitriles, urethanes, and fluorocarbons (including elastomers and plastics). There also are dozens of special materials and proprietary blends to solve specific sealing problems.

The term elastomer is loosely defined as a material that can stretch up to twice its length and still return to approximately its original length. Plastics tend to cold-flow, but the exceptions are sufficient to leave the definitions in doubt.

And don't forget leather. According to Auburn Manuf. Co. (Middletown, CT), this natural material is pliable, tough, and highly resistant to abrasion and wear. Tensile strength is up to 3500 psi. Leather seals impregnated with Teflon can operate at temperatures to 500 F and higher and seal readily against hydraulic fluids at over 15,000 psi; resistance to oil, water, and synthetic fluids is excellent. Leather is not widely used today but mostly because synthetic materials cost less.

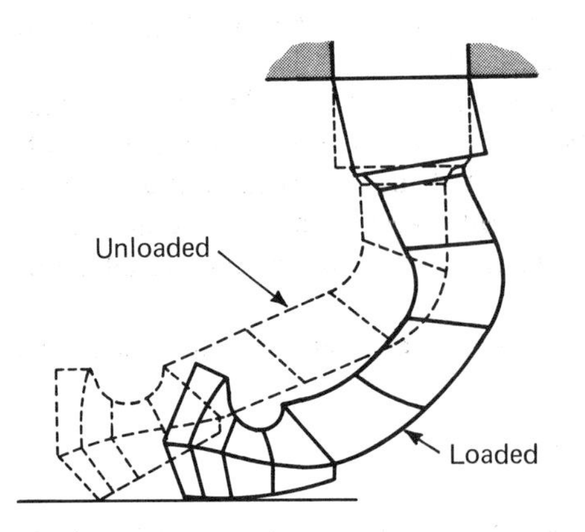

Fig. 19.2 Flexing of lip seal is examined with computer help. (Courtesy of MTI, Latham, NY.)

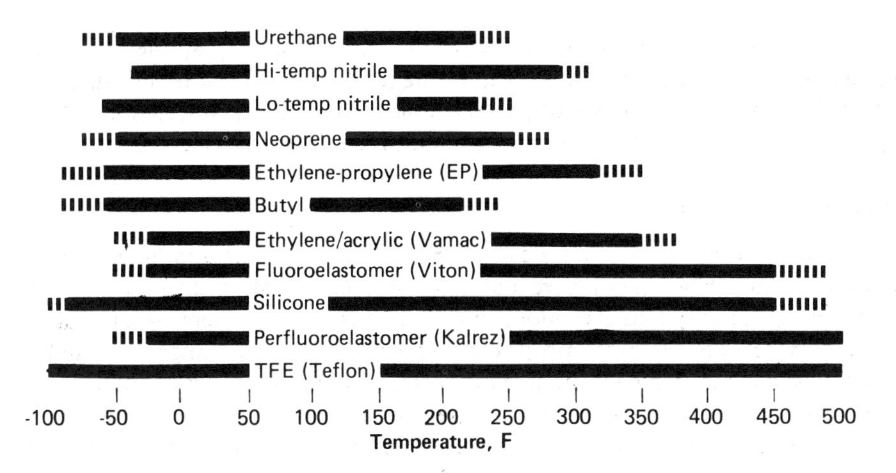

Fig. 19.3 Temperature limits of typical seal materials. Broken lines are for intermittent service.

Nitrile Rubber Seals

Nitrile (Buna N) seals are widely used in industry because of excellent resistance to oil, a good practical range of working temperatures (−40 F to as high as 275 F), and low cost. Oil on the surface of nitrile rubber even helps prevent oxidation and aging, which is a problem when nitriles are exposed to air and sunlight.

Performance in fluid power systems is generally excellent. For example, Parker Seal Co. ran a series of brutal tests to determine how much fluid pressure a common nitrile O-ring can withstand without extruding into the gap. To make the tests more meaningful, they held the temperature at 160 F and cycled the pressure 100,000 times at each value. They also related the data at 160 F to previous static tests at other temperatures and came up with correction factors. The results of tests and studies are summarized in Fig. 19.4.

The temperature correction applies in both directions: up or down. For temperatures higher than 160 F, reduce the value of pressure by the percentage indicated; for

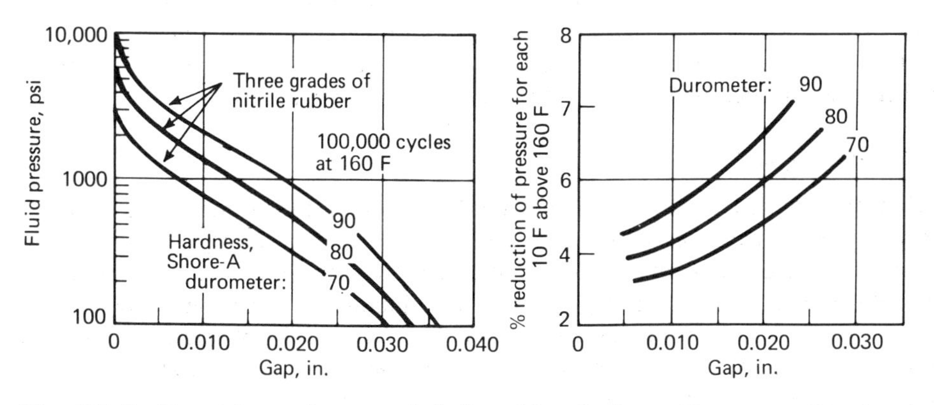

Fig. 19.4 Extrusion resistance of nitrile rubber O-rings. (Courtesy of Parker Seal Group, Lexington, KY.)

lower temperatures, raise it. Also, the tests may be qualitatively applied to other elastomers besides nitrile so long as they have the same range of Shore durometer A hardness (70, 80, and 90) and have roughly the same temperature characteristics. However, great care must be taken if you seek quantitative information.

Urethane Resists Abrasion

Polyurethane elastomers, commonly called urethanes, bridge the gap between synthetic rubbers and plastics. They have a unique combination of hardness, resilience, and load-bearing capacity plus exceptional abrasion resistance. The temperature rating is moderate: 200 F. The mixes are proprietary and result in a wide range of available properties. The material can be formulated for extrusion, molding, casting, or machining. Hardness can be as soft as a pencil eraser or as hard as a bowling ball.

Urethane can significantly outwear most rubbers and plastics and even some metals. Also, the coefficient of friction is fairly low and gets even lower as hardness is increased. Urethane seals enjoy wide application as rod scrapers and other dynamic seals in harsh environments. Some grades of urethane are used in high-pressure hydraulic hose, which further attests to its strength and its ability to take abuse.

Viton Fluoroelastomer

DuPont Co. (Wilmington, DE) supplies this particular material to the seal manufacturers. It's fairly expensive but is being recognized for certain performance advantages that make it economical in numerous applications. It even replaces low-cost seals where reliability outweighs initial cost. A similar material is 3M's Fluerel.

The graphs in Fig. 19.5 depict important properties of Viton. The normal maximum operating temperature is 400 to 600 F for intermittent service, and some hardness and resilience are retained at that high level. Even if the high operating temperature isn't needed, the hot tear strength of Viton makes it easy to strip from a mold without damage, enabling intricate shapes to be formed. Dynamic properties and low-temperature performance also are good. Thick sections retain enough flexibility for sealing down to −25 F and thin sections (say 0.075 in.) down to −40 F. These values are only guidelines, of course, and must be qualified for each design.

Lower-cost substitutes are sought. A relatively recent example is ethylene/acrylic (Vamac by DuPont is an example). Basically, it is a copolymer of ethylene and methyl acrylate and can seal effectively from −65 F to +350 F. It can be molded by conventional compression, transfer, and injection methods. It is expected to replace higher-cost fluorocarbon seals in some automotive applications.

Teflon Fluoroplastic

Teflon (DuPont), Halon (Allied), and similar fluoroplastics from other companies have two key properties that make them interesting for sealing: They work at temperatures from cryogenic to 600 F, and they are self-lubricating. The prime disadvantages are creep (cold flow) and limited techniques for forming the parts. The creep is taken care of by designing the seal to either trap the plastic flow and thus contain it at the seal area (see the section on static seals) or to add a metal or elastomer spring to maintain sealing force despite the creep. Teflon seal blending and manufacture is an art akin to powder metallurgy and has been perfected by the specialists.

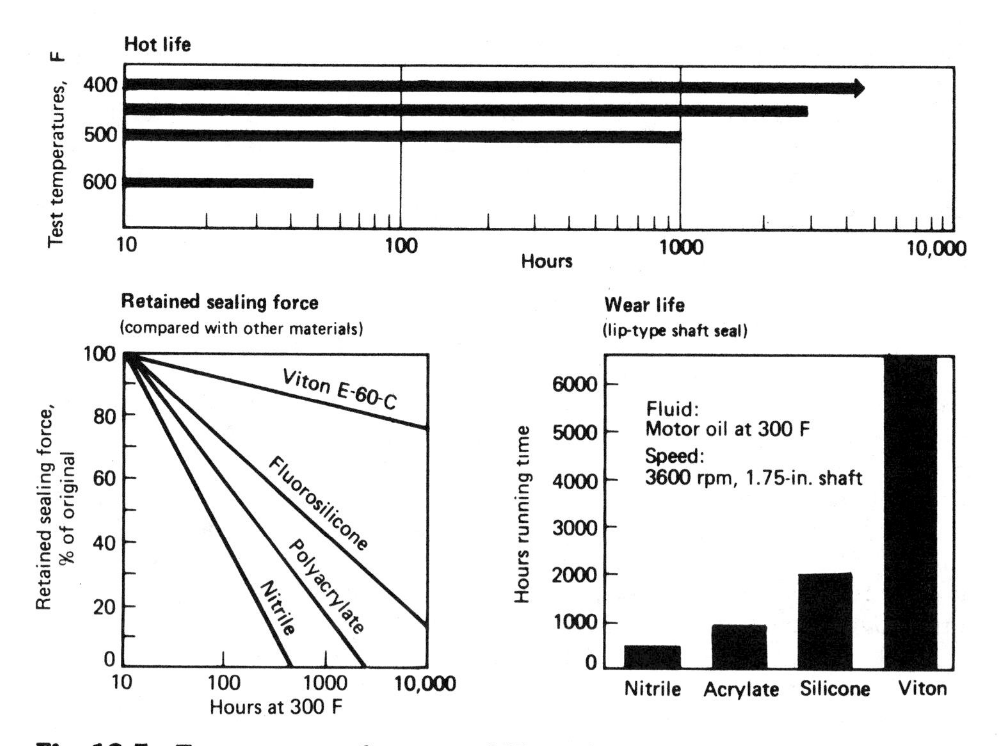

Fig. 19.5 Temperature performance of Viton Fluoroelastomers. (Courtesy of Du-Pont Co., Wilmington, DE.)

One specialist in manufacturing Teflon seals is Fluorocarbon Co.'s Mechanical Seal Div. (Carson, CA). Company engineers point out that with proper combinations of strengtheners such as graphite, glass fiber, metal oxide, bronze, and molydenum disulfide the basic Teflon can be given a wide range of strengths and other properties.

Special-Purpose Materials

Many applications require special materials, blends, and custom designs to satisfy unusual operating conditions. The following is only a sampling but shows the problems faced.

Temperature Limits. Elastomers and plastics lose resilience and hardness as temperature rises, and most seal materials expand with temperature at a far greater rate than metals do. Zero to 200 F is a fairly safe range for almost any application. Beyond those limits, be sure to consult experts.

The temperature chart (Fig. 19.3) is a guide for extending the limits somewhat but is based on ordinary hydraulic fluids. For example, EP rubber (ethylene-propylene) performs well in steam up to 400 F but is limited to 250 F in phosphate ester hydraulic systems and is not usually recommended for petroleum-based fluids.

Temperature performance also is affected by positioning of the sealing surface. As an extreme example, a particular fluorocarbon O-ring will seal to − 80 F as a rod seal (sealing surface on the ID of the O-ring) but only to − 10 F as a piston seal (sealing on the OD). All these numbers are affected of course by pressure level, seal composition, and operating conditions.

For extremes in temperature, there are *silicone* rubbers, *polyimide* plastics, and *perfluoroelastomers.*

Some silicone compounds remain flexible down to – 150 F; others remain strong up to + 700 F. However, applications are limited because most silicones do not resist petroleum fluids well, compared with the much lower-cost nitrile rubbers. Most applications are for special static seals where specific properties of silicone are needed.

Silicone requires special manufacturing techniques. In the process of molding, it is sensitive to contaminants such as carbon blacks in the air and must have its own manufacturing area. Freezing to remove flash from molded silicone seals in liquid nitrogen is not too effective because silicone doesn't freeze very brittle. Silicone can be molded with wall thicknesses as thin as 0.009 in. if that's the need.

Polyimide is a special plastic. It is stiffer than Teflon and has much higher temperature resistance and much less creep. It is not a fluoroplastic yet has self-lube properties like those of Teflon. DuPont markets a version called Vespel. The company usually does not sell this polyimide directly but sells precision parts, compression-formed and sintered using methods similar to powder metallurgy techniques. In working with customer specifications, such seals can be made to withstand 600 F continuously and 900 F intermittently without degradation.

Vespel is too stiff to be considered truly resilient but is flexible enough for sealing if properly dimensioned. It is popular for high-pressure valve seals in nuclear applications because it can withstand creep and radiation better than Teflon can. Manufacturers of similar materials are Rhodia, Inc. (Kinel), Ciba-Geigy (P-13-N), Upjohn (Polyimide 2080), and Amoco (Torlon).

Industrial valves also take advantage of the physical properties of the stiff polyimides. For instance, a line of check valves by Circle Seal Corp. (Anaheim, CA) has polyimide seals to increase reliability and life. Engineers at Circle Seal note that the material is more expensive than other plastics but has high resistance to compressive stress—a critical parameter.

Another high-temperature material that is not sold in bulk but as finished parts is perfluoroelastomer (Kalrez by DuPont is an example). Costs are high, but the material is selected where no other elastomer will work at such high temperatures (to 600 F).

Dual-Material Seals. Grover/Universal Seal (Milwaukee, WI) solves conflicting requirements of resilience on the one hand and hardness on the other with Uniring, a dual-durometer one-piece sealing ring that does both. The sealing surface of the seal is hard (97 durometer) urethane; the backup is softer (45 durometer) urethane. Auburn (Middletown, CT) offers its Uni-Seal U-cups of dual-durometer urethane and provides impregnates such as TFE and molybdenum disulfide to lubricate the surface.

Greene, Tweed (North Wales, PA) inserts a nitrile rubber O-ring within a urethane U-cup to achieve a spring-loaded rod and piston seal (Fig. 19.6). The O-ring assures sealing at low fluid pressure, and the sealing force increases as pressure rises. Microdot Inc. (Broadview, IL) offers a similar seal with a multilobe ring instead of the O-ring. Parker energizes its version with an O-ring and molds the seal from a self-lube urethane compound.

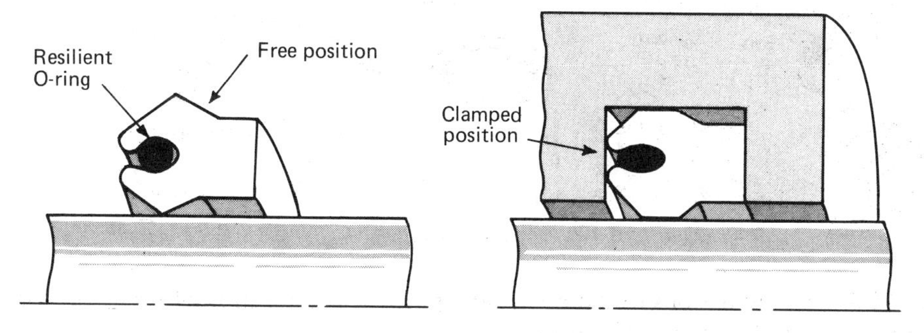

Fig. 19.6 O-ring as spring inside U-seal exploits the benefits of two materials. (Courtesy of Greene, Tweed, North Wales, PA.)

Minnesota Rubber Co. bonds Teflon to rubber to combine lubricity with resilience, eliminating the need for a separate spring or elastomer ring. Garlock Co. bonds Teflon to silicone (Fig. 19.7). A more conventional way to combine the qualities of two materials is to make the seals in two parts (Figs. 19.8 and 19.9). Several manufacturers do this.

Bal Seal developed a high-pressure (to 25,000 psi) reciprocating seal made of graphite-carbon-Teflon energized with a canted loading spring. Supporting rings made of graphite-impregnated reinforced polymer prevent extrusion of the seal into the clearance.

Some manufacturers such as RCF Seals & Couplings (South Gate, CA) specialize in rubber-coated fabric piston rings and rotary shaft seals. The fabric can be fiberglass and the rubber silicone, giving the composite tolerance to temperatures from −170 to 800 F. Fiber alignment adds stiffness and prevents twisting.

Translucent urethane is featured by Hercules (Alden, NY) in rod or piston seals. The advantage is that inspection for defects can be made visually. The company also specializes in a U-lip seal design wherein the lips have compression lobes molded at the inside edges of each tip, serving as an elastomer spring during installation.

Ever hear of the shrink-stretch syndrome suffered by elastomers? Called the Gow-Joule effect, it's the tendency of an elastomer to shrink a little when heated, if

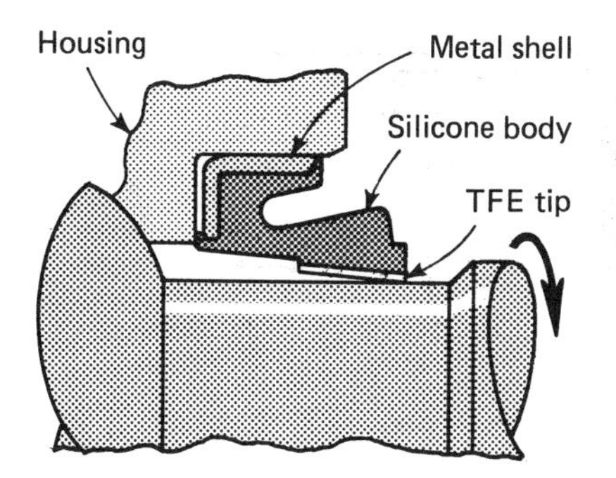

Fig. 19.7 Lubricity of TFE tip combined with the resilience of silicone creates a seal better than either alone. (Courtesy of Garlock Inc., Palmyra, NY.)

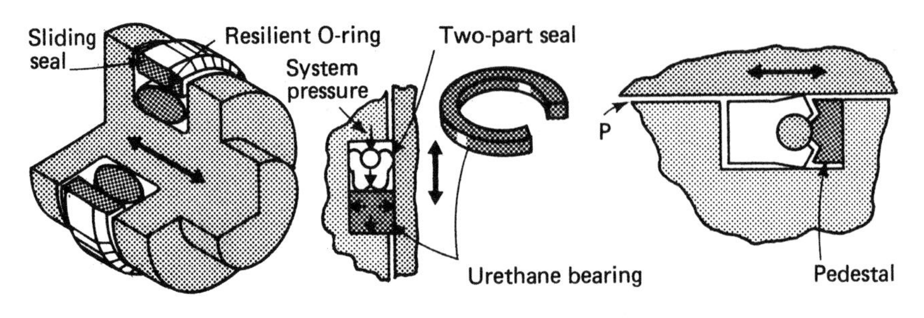

Fig. 19.8 Combinations of materials, and proper concentration of pressures and forces, create seals for special purposes.

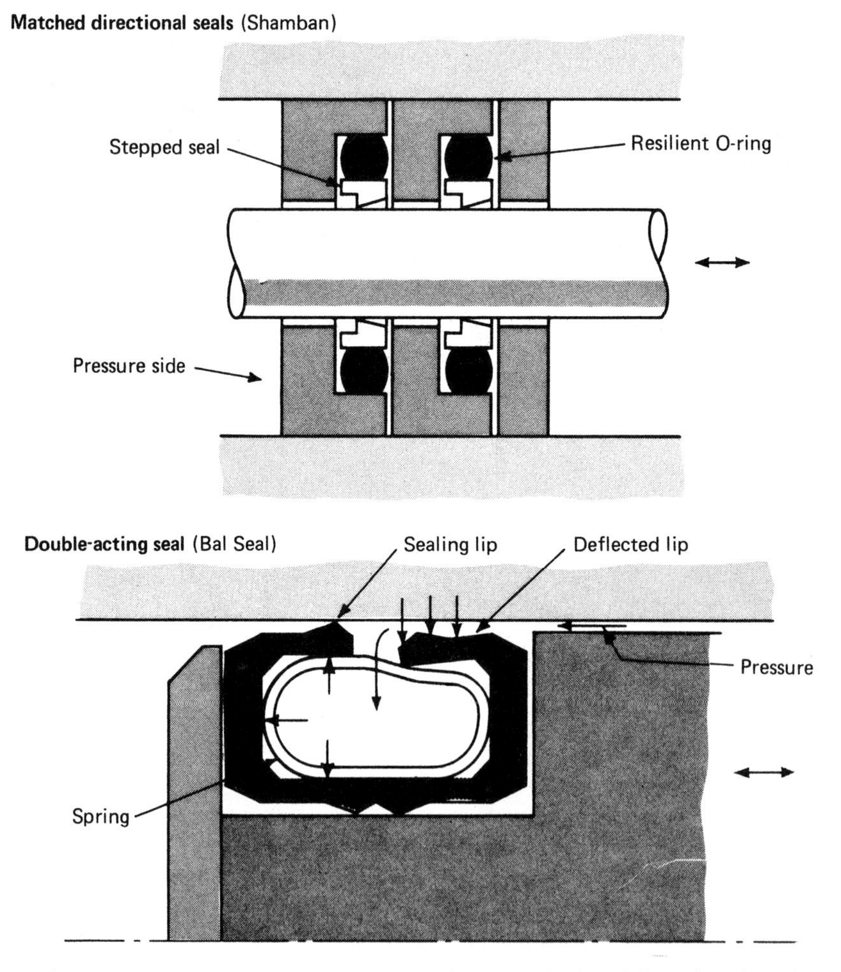

Fig. 19.9 Pressure trap is avoided with matched unidirectional seals or double-acting seals.

it had been stretched during installation! It's a clever way to get a little squeeze on the ID of an O-ring as temperature goes up in actual service, but the effect diminishes with time.

Customizing the Seal

A hundred dollar custom seal that saves thousands of dollars in maintenance, and tens of thousands in downtime, is a good investment. It puzzles seal manufacturers why more machine designers don't exploit the wealth of experience that suppliers have. At least send them a sketch and description of the application before finalizing the design. A common mistake is to leave too little room for the seal.

Think about maintenance and repair before approving those final detail drawings. Ask yourself how long the seal will last and what a mechanic has to do to replace it. If your competitors' products are easier to maintain, your company is in big trouble.

A semicustom design is the Aeroquip floating seal (Fig. 19.10), an adaptation of the company's Omniseal. It has the dual function of sealing an axial tube (or shaft) yet allowing radial movement to compensate for eccentricity of the tube.

The pressure-expanding carrier ring (Fig. 19.11) is a concept by Lehigh University. It is intended to seal against superpressures of 200,000 psi and up. The O-rings do the actual sealing. The carrier ring is forced to expand radially by the extreme pressures within the cylinder, and this helps keep the O-rings in contact with the expanding cylinder.

Some custom designs are highly specialized and costly. One example, highly developmental, is the triangular-asperities seal (Fig. 19.12) proposed by Dennis Lee Otto of Timken Co. in an SAE paper. His idea is to etch microscopic paths onto the surface of the metal shaft or wear ring that the elastomer seal rubs against, leaving thousands of tiny flat-topped islands (asperities). The paths aid in building and maintaining a thin hydrodynamic film between the asperities and the seal. The theory is based on the fact that microirregularities in a smooth surface are what promote a lubricating film. Otto believes that proper design of these irregularities will give ex-

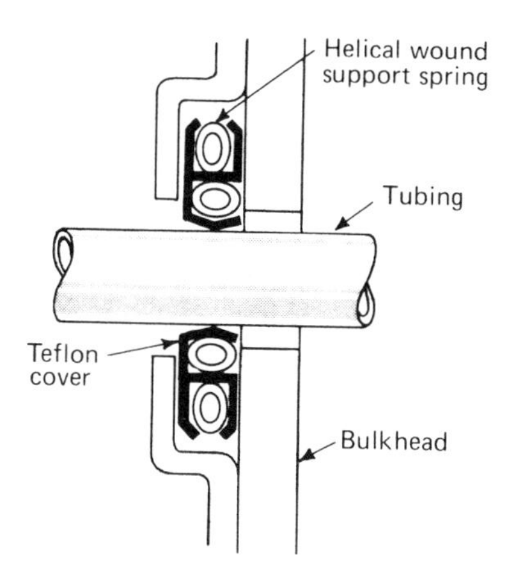

Fig. 19.10 Special seal for tubing or shaft that is eccentric or floating. (Courtesy of Aeroquip Corp., Jackson, MS.)

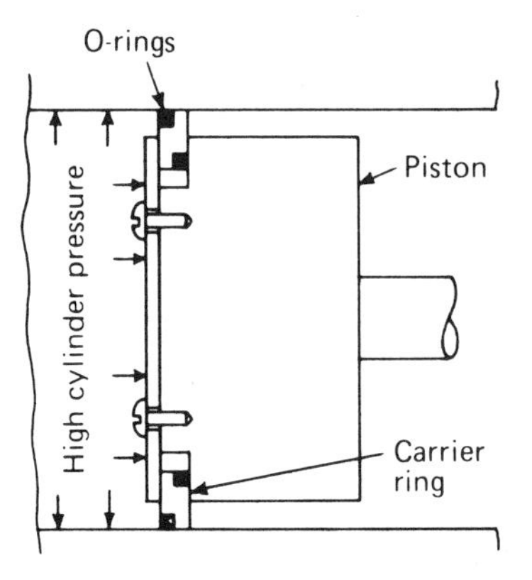

Fig. 19.11 Carrier ring expands under extreme pressure to hold O-rings against sealing surfaces. (Courtesy of Lehigh University, Bethlehem, PA.)

actly the desired film thickness. The depth of the etch he proposes is less than 0.0001 in. He has experimented with a wide variety of island shapes.

Another unusual concept is to peen the surface against which the seal will rub. General Motors Research Laboratories (Warren, MI) determined from extensive tests that if the surface of a rotating shaft were peened (Fig. 19.13) rather than plunge-ground it would substantially reduce frictional losses and wear of a conventional lip-type radial oil seal in contact with the shaft.

A major reason for the reduced friction of the peened surface is that it has gen-

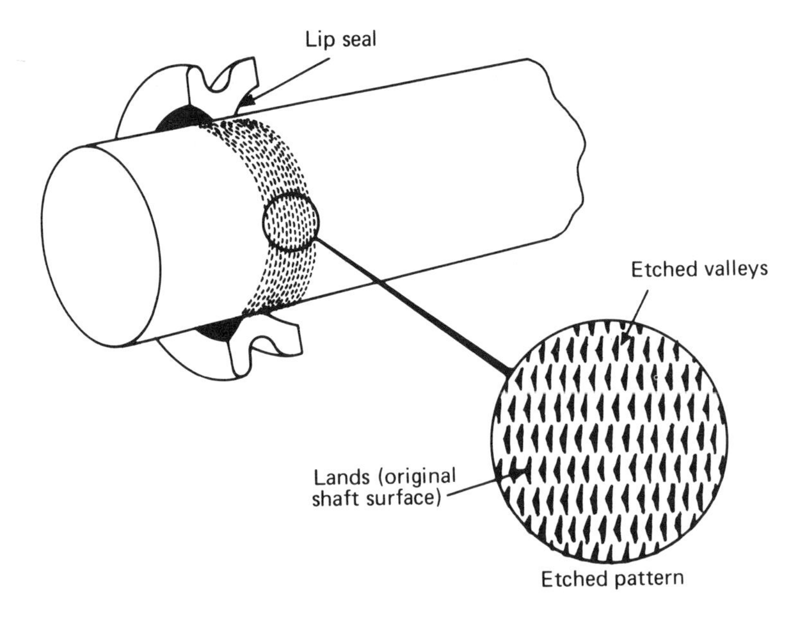

Fig. 19.12 Microscopic channels etched into the surface of the shaft create triangular asperities (islands) that serve as bearing surfaces for the lip seal. The concept is experimental. (Courtesy of Timken Co., Inc., Canton, OH.)

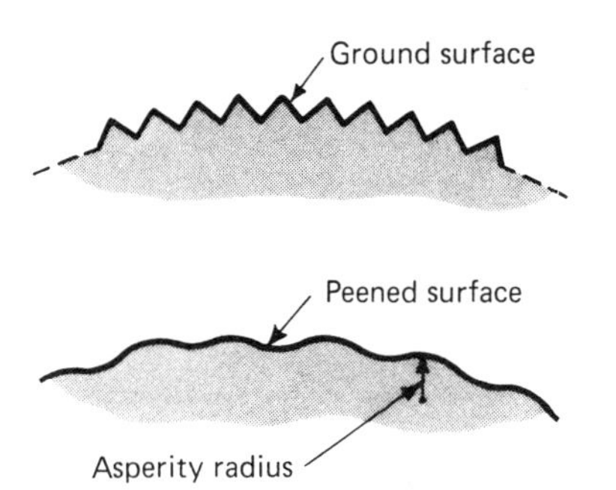

Fig. 19.13 Peened surface on a shaft is easier on a seal than a ground surface is.

tle slopes, that is, a high radius of curvature. Enough pockets of microscopic size still remain in the shaft surface to help trap oil, which also reduces friction.

After the usual engineering trade-offs, the following design and manufacturing procedure was tested at GM and found adequate. The shaft surface is shot-peened with a flap-brush wheel, which is an economical substitute for the normal expensive blast peening.

The flap brush has nonwoven radially extending fiber flaps impregnated with cast steel shot and bonded to a rigid fiber or steel core. Shot sizes chosen were from 0.3 to 0.8 mm, ensuring the correct amount of indentation in the shaft surface.

The spacing and size of the asperities on the peened surface must be carefully designed to match the seal material chosen. The direction of slope of the depressions is critical too (Fig. 19.14).

Several other examples of custom seals are shown in Figs. 19.15 through 19.17.

FRICTION IN HYDRAULIC SYSTEMS

Energy-wise, friction is worse than leakage. To reduce leakage of lube oil by tightening down sealing glands will greatly increase the friction. The amount of fuel oil needed to generate the electrical energy to drive the motors to overcome the friction is greater than the oil leakage. The answer is to redesign the seals.

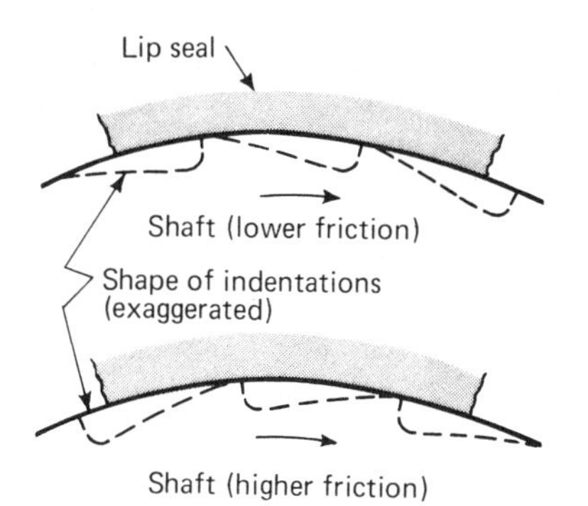

Fig. 19.14 Shape and direction of peened indentations are factors in shaft seal friction.

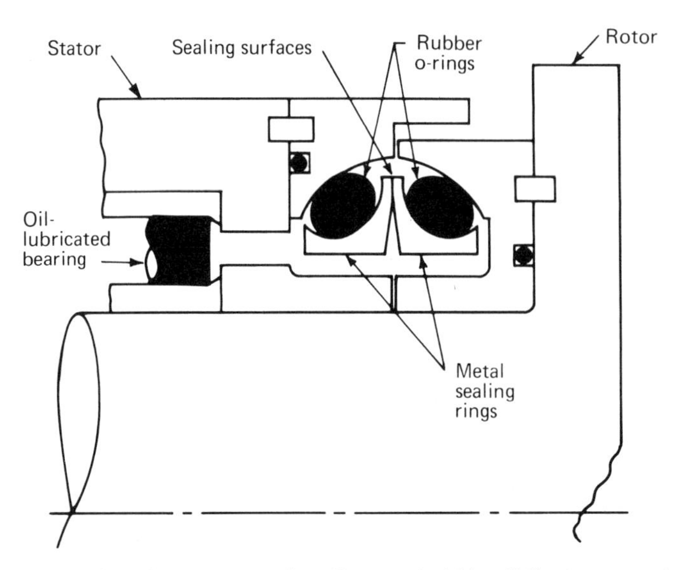

Fig. 19.15 Metal sealing lips are held in sliding contact by rubber O-rings and will seal despite flexing. (Courtesy of Caterpillar Industrial Products, Peoria, IL.)

The ultimate—a seal that rides on a film of oil that is not allowed to escape—is possible only in special instances such as face seals and magnetic-centrifugal hybrid seals. The others only approach the ideal.

For example, if a rod seal floats efficiently on a film of oil, it's called hydroplaning, and the design engineer better correct it or the oil will leak out on every stroke. The biggest problem is during the extend stroke when rod seal pressure is lowest and the seal contact force is least.

In contrast, if the rod is wiped too dry, metal-to-metal wear or abrasive wear from hard particles can result. Anyway, most rods move slowly and intermittently and cannot develop hydrodynamic lubrication. See Chap. 20 for details.

The answer generally accepted is to specify a rod surface finish that is smooth enough to prevent seal wear but rough enough to hold lube. The recommended finish is about 15 μin. rms. Then the surface won't be wiped completely dry. As explained

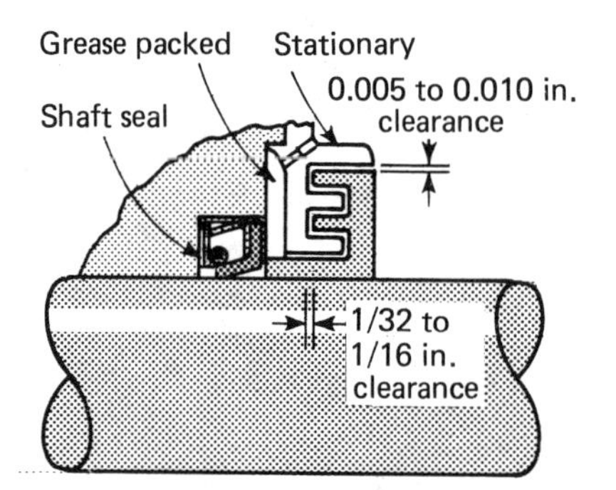

Fig. 19.16 Grease-packed labyrinth protects the radial lip in hybrid seal. (Courtesy of Albert Trostel Packings, Ltd., Lake Geneva, WI.)

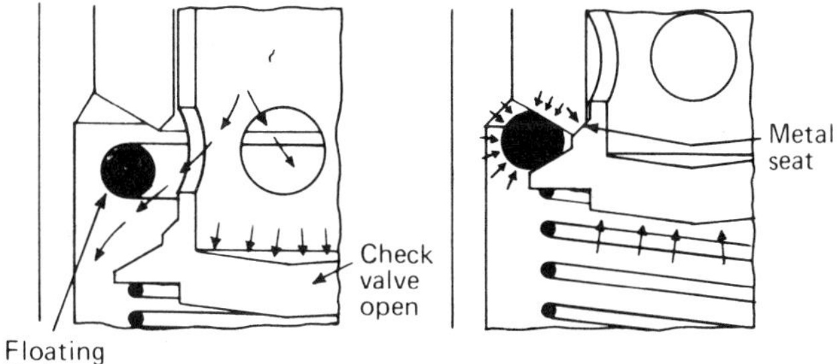

Fig. 19.17 In a combination seal for a check valve, the metal seat carries the load, and the O-ring does the final sealing. (Courtesy of Circle Seal Corp., Anaheim, CA.)

before, some experimenters actually have etched the surface of a rotating shaft to leave shallow triangular valleys to hold oil (Fig. 19.12) and to peen the surface with shot or glass beads, yielding an orange-peel finish of 10 to 20μin. in depth (Fig. 19.13).

Still another idea is to cover or coat an ordinary resilient seal with Teflon or other slippery fluorinated plastic. The inner core provides the sealing force or squeeze, while the coating reduces the need for lubricant. Several companies offer Teflon-coated O-rings and Teflon-tipped lip seals.

Lip seals in general have lower friction than squeeze seals but need fluid pressure for full energization. Squeeze seals (O-rings are an example) seal effectively at low and high pressures but exert high frictional force.

Slippery seals (Figs. 19.5 through 19.9) also can be spring-loaded; numerous designs have been created over the years. The advantage of a spring is that it can withstand higher temperatures and can be tailored to give exactly the loading needed. Cost usually is higher than that for a slippery seal with an elastomer core.

Leather packings have lower friction than most elastomers and sometimes are recommended. Leather is fibrous and tough yet pliable enough to avoid scoring or abrading the rod. Leather actually tends to polish the surface of the metal it seals against. It also can be heavily impregnated for special service.

Many other factors influence friction: temperature, pressure, and so forth. For example, the oil sump temperature indirectly affects the friction of elastomer seals by thermally expanding the seal and changing its contact pressure against the rod or shaft. Also, hotter oil has lower viscosity, and this reduces drag. Friction, in turn, affects wear, discussed in the following section.

Where friction is unavoidable, make sure there is a good thermal path to carry away the frictional heat. Don't place the seal too near any source of heat, such as a hot bearing.

Rotary Shaft Seal Friction

CR Industries (Elgin, IL) pointed out that coefficients of running friction in lip-type rotating shaft seals, such as those in engine transmissions, are fairly high, in the range of 0.1 to 1.0 and typically at 0.5. The frictional drag of a typical lip seal dissipates about 0.06 brake hp (40 W). If you multiply this by the hundreds of millions of seals on cars, trucks, and off-the-road vehicles, it is equivalent to 5 billion gal of fuel oil annually, just to overcome frictional drag. That is roughly the oil output of Venezuela and Kuwait for 2 weeks.

A special study on frictional power loss was made by Simrit (Des Plaines, IL); the results are shown in Fig. 19.18, and it's worth doing something about. Other specifics of the problem are summarized in the accompanying graphs for temperature rise in rotary shaft lip seals, developed by CR Industries (Fig. 19.19). The frictional drag produces a torsional moment, and the power dissipated is proportional to torque times speed. A good measure of this power is the temperature under the lip of the seal, assuming that the shaft size is the same. A larger shaft dissipates heat faster than a smaller shaft and gives a different but related set of temperature readings. The graphs are based on a 3-in.-diameter shaft.

To guide the lubricant, and to increase the effective area of the shaft for dissipating heat, CR Industries has developed a lip seal with a wavy pattern on the surface (Fig. 19.20). This design, trademarked Waveseal, is a smooth-lip, birotational, hydrodynamic radial-lip seal, designed to pump lubricant back against pressure regardless of the direction of shaft rotation.

The seal's contact with the shaft surface is not as an intersection with a plane, but it projects a wavy pattern—a shallow sine wave—on the shaft. Thus the swept width of the contact area is greater than with a conventional straight lip seal, yet the actual contact width at any point is still narrow. So the friction is low, yet the shaft surface available for cooling is high. The underlip temperature rise is at least 25% less than for standard seals.

The wave pattern also has a snowplow effect, pushing the lubricant back toward

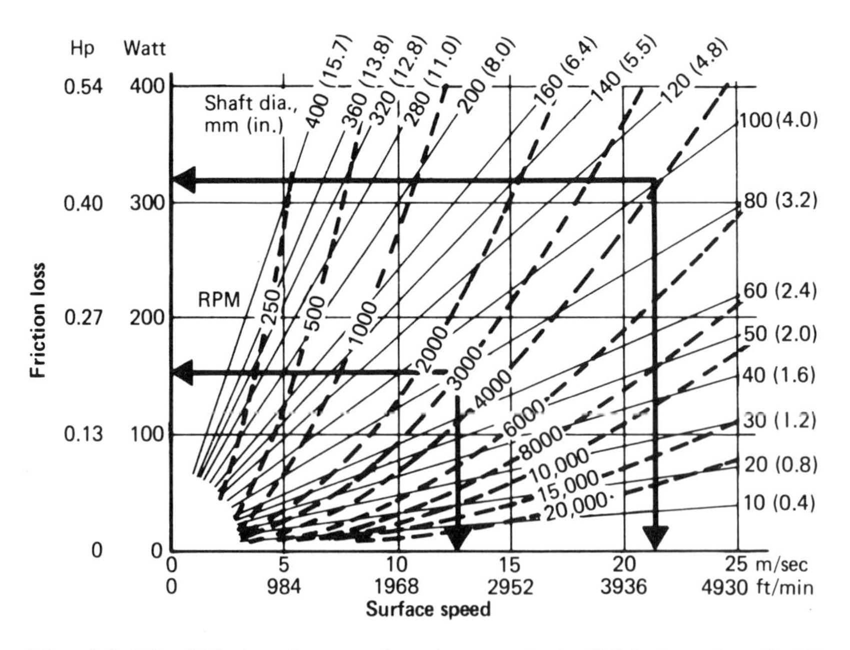

Fig. 19.18 Frictional power loss in a typical oil-lubricated radial lip seal for a rotating shaft. Data are based on SAE-20 engine oil at 100 C. (Courtesy of Simrit, Des Plaines, IL.)

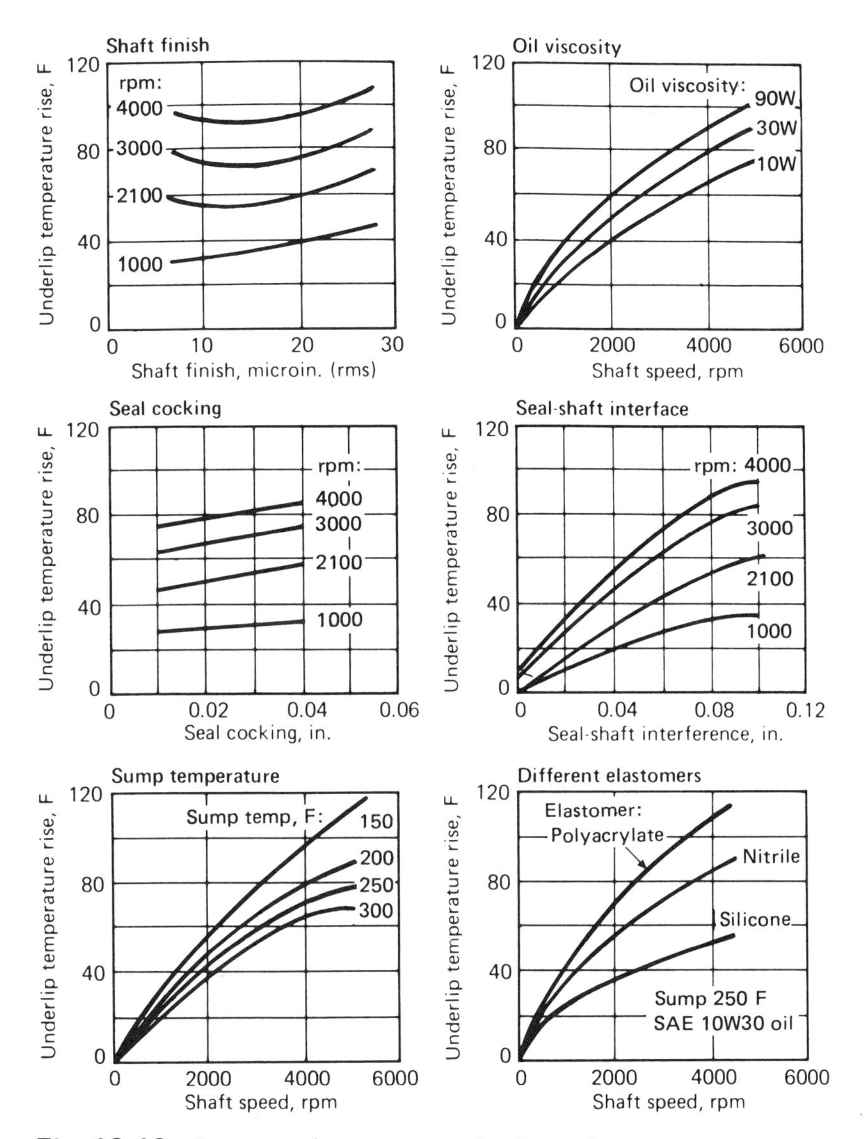

Fig. 19.19 Rotary seal temperature rise for various operating conditions. Each is based on a 3-in.-diameter shaft, nitrile seal, 200 F sump, and 30W oil, except as noted. (Courtesy of CR Industries, Elgin, IL.)

the pressure side. The company says the pumping effect continues despite seal wear and is three times greater than that for conventional seals.

Oil sump temperature has a complex effect on seal lip friction, lowering it. Hot oil thermally expands the seal and reduces interference. The heat also lowers the modulus of elasticity of the seal material and reduces load pressure on the shaft. Finally, the hotter oil has lower viscosity and further reduces frictional drag. The proper choice of seal material will balance out some of the conditions encountered, helping achieve reasonable leakage, good life, and minimum drag.

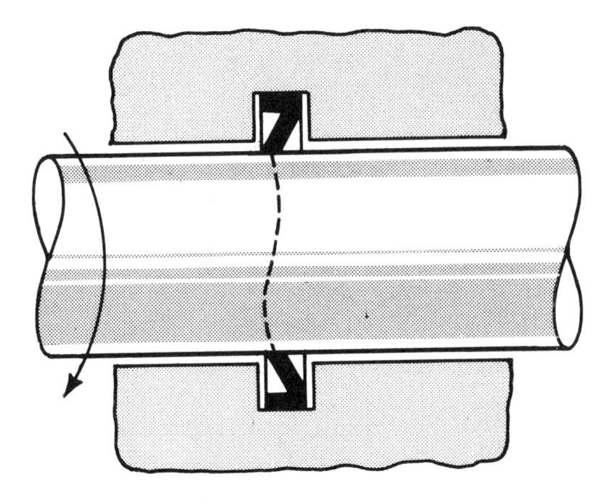

Fig. 19.20 Wave-shaped lip of seal increases the swept area of contact. (Courtesy of CR Industries, Elgin, IL.)

Cylinder Rod Friction

Axial motion has its own stable of problems, including reciprocation and slow speed. Disogrin Industries (Manchester, NH) notes that engineers are too ready to accept the rule-of-thumb that "frictional losses in hydraulic piston and rod seals represent an energy loss of less than 5% and can be neglected." The company took a close look at what that 5% meant and sounded an alarm.

There are about 20 million hydraulic cylinders at work in the United States today, and the average frictional power dissipation (based on a 1-in. rod, 2-in. bore, 12-in. stroke, 1500-psi operating pressure, 20-ft/min speed, and about a 4000-lb operating force) is about 0.2 hp. Working backwards to the power plant, and assuming normal load cycles and efficiencies, Disogren estimates that it takes about 7 million barrels of fuel oil a year to make up that kind of power loss.

The leakage loss is another matter. Computations show that the 20 million cylinders leak about 0.7 million barrels per year, which is only one-tenth of the crude oil demand for overcoming friction. Leakage is less of an energy waster than friction. Seal designers must reduce friction by a combination of clever configurations and techniques, including some allowable leakage to assure seal lubrication.

The company suggests special molded or machined grooves in the lip to increase flexibility and hold lube and smaller cylinders at higher pressures to reduce the percentage of the friction.

Another example: O-rings have a very small contact area and should be used where loads are light enough to allow them. However, for many industrial applications where pressures are high and life must be long, U-cups are chosen. Fabric reinforcement in U-cups helps reduce friction by holding more lubricant, up to about 4500-psi pressure. At higher pressures, the lubricant is squeezed out, and the entire U-cup makes intimate contact with the metal surface. The accompanying graphs (Fig. 19.21) show how seal design, additives, pressure, speed, and temperature affect seal friction in pistons and rods.

Friction can be helpful in certain circumstances. For example, a rod seal is less likely to extrude into the gap if the friction force is designed to oppose the pressure force (Fig. 19.22).

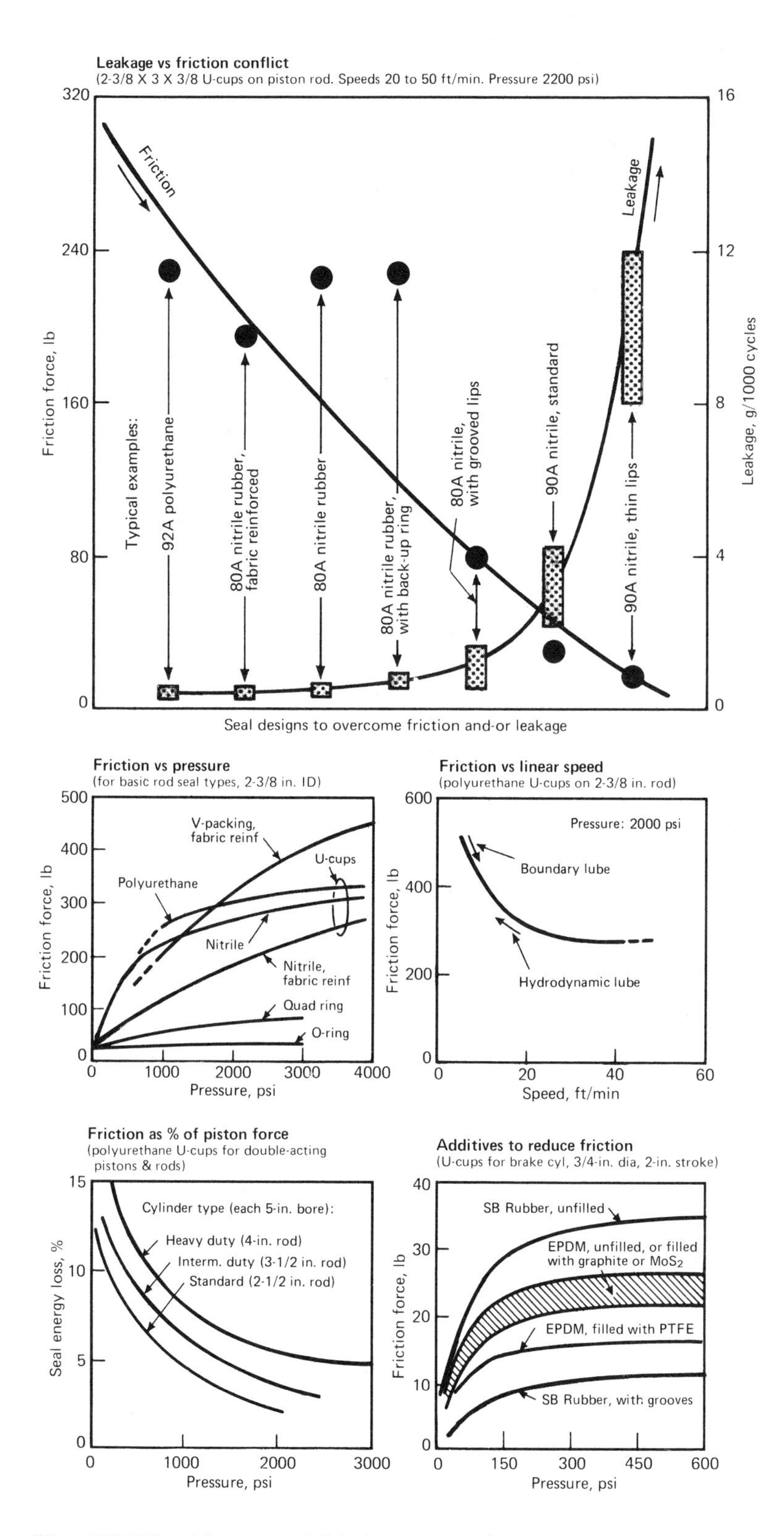

Fig. 19.21 Linear seal friction for various operating conditions. (Courtesy of Disogren Industries, Manchester, NH.)

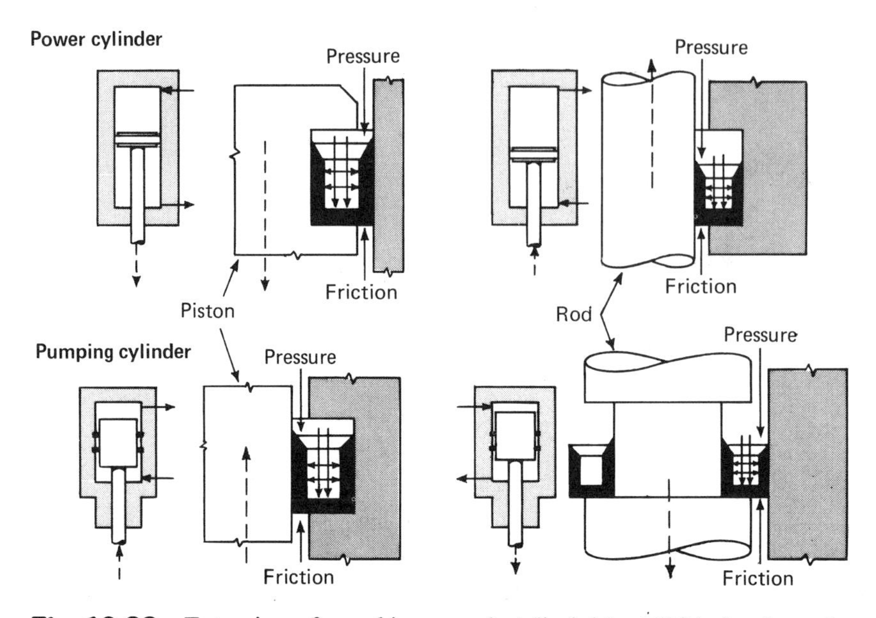

Fig. 19.22 Extrusion of a seal is somewhat diminished if friction is made to oppose pressure force. (Courtesy of Parker Seal Group, Lexington, KY.)

Calculating O-Ring Friction

A nomographic method quickly estimates dynamic linear friction of an elastomeric O-ring piston or rod seal (Fig. 19.23). Parameters include differential pressure and O-ring squeeze. The method was developed by L. J. Martini, Mechanical Engineer, Naval Ocean Systems Ctr (San Diego, CA).

To explain the method, we'll calculate a before-and-after O-ring selection based on a nominal 3¼-in. piston. Two nomographs (Fig. 19.24) are needed. Nomograph A computes friction caused by differential pressure across the O-ring sealing surface, while nomograph B gives friction caused by squeeze.

Total dynamic friction is the sum of the two. Static friction usually is 1.5 to 3 times the total dynamic friction. Dynamic hysteresis or system hysteresis is the dynamic friction force divided by the working force acting on the piston.

Two examples are worked out (Tables 19.1, 19.2, and 19.3 and Fig. 19.25). The first example is of a conventionally sized O-ring, based on standard catalog recommendations. It turns out to have considerable dynamic friction. The second example is of an O-ring with smaller cross section but with less friction. The trade-off is that greater care must be taken in designing with the smaller cross section to make sure the squeeze is sufficient at all tolerances for sealing at low differential pressure. The advantage of the nomographs is that effects of design changes can be quickly recognized.

The nomographs and examples are derived from theoretical models and empirical data. The data are based on tests of standard elastomeric O-rings (except Teflon) reciprocating against 15 rms finish chrome-plated surfaces at speeds greater than 1 fpm, lubricated with hydraulic oil MIL-H-5606, at room temperature. The original data are documented in *O-Ring Handbook OR5700*, Jan. 1977, Parker Seal Co. (Lexington, KY). This handbook also covers O-ring theory, design, and application. Also see the *Handbook of Plastic Products Design* (Marcel Dekker, New York).

The examples: The calculating steps are tabulated in the tables and are for the most part self-explanatory. However, certain interesting observations should be made. The O-ring stretch is only about 1% in the first example yet is 3% or 4% in the second. This was purposely done to make the thinner O-ring hug the piston tightly to help prevent spiral failure.

Squeeze is another problem area. The nominal value is 8% to 13% in the first example, which is the minimum recommended. But the second example is 11% for a tight tolerance stackup and only 1.5% for a loose tolerance stackup. Just be careful there is at least some squeeze, even if it is only 1% or so. Once the squeeze is known, recompute all steps.

And keep aware that friction between an O-ring and any bearing surface is caused by asperities, or microscopic hills and valleys, at the interface. Even though an O-ring is much softer than the metal surface it slides against, it will wear away the asperities of the metal. One effect is polishing. Another is wear and leakage if friction is high and the metal is soft.

To minimize friction and wear, try these: smoother finishes (to 15 rms) on moving parts, adequate speed (over 1 fpm), smaller cross section of O-ring, less squeeze, softer O-ring surface, and better lubrication.

Watch out for spiral failure, sometimes aggravated by reduced cross section, softer O-ring surface, and temperature changes as recommended above.

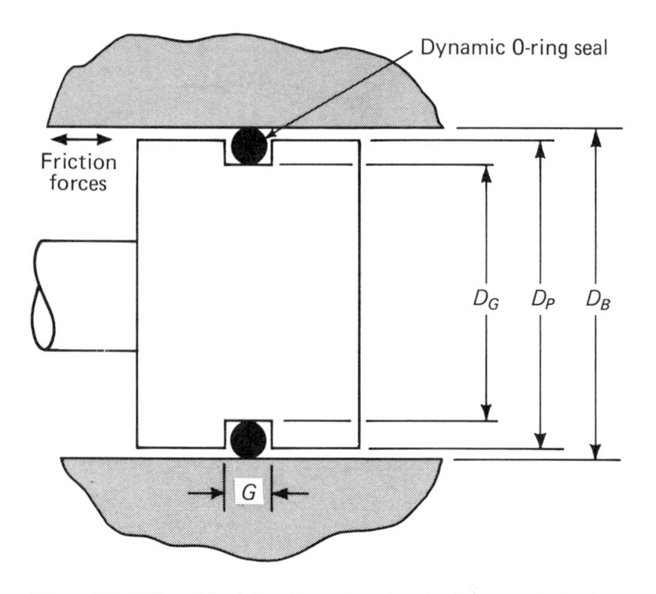

Fig. 19.23 Model of a simple O-ring-sealed piston used in the accompanying nomographs and calculation procedure. (Courtesy of Naval Ocean Systems Center, San Diego, CA.)

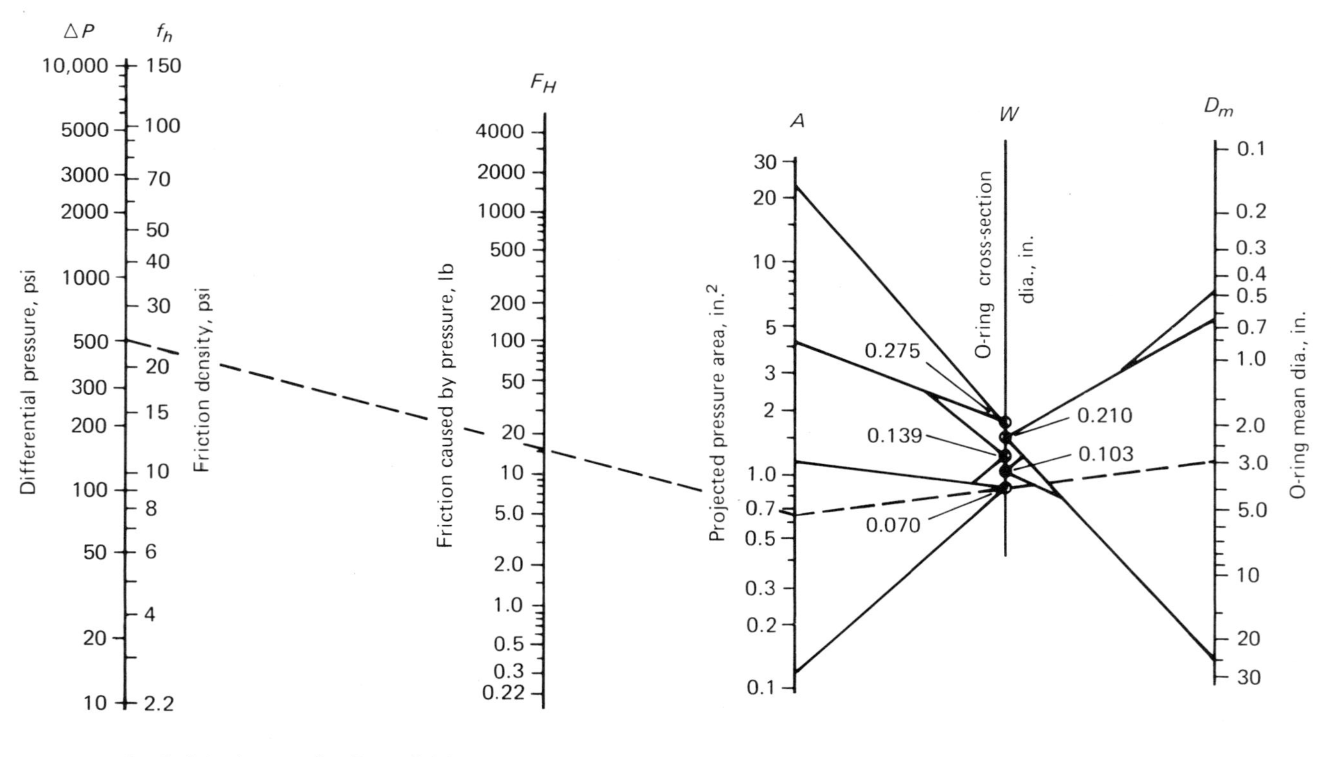

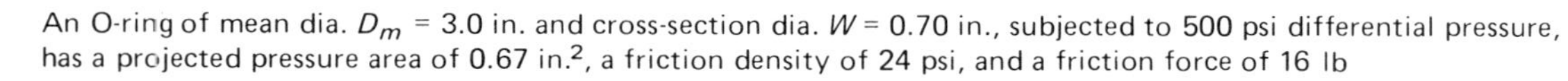
An O-ring of mean dia. D_m = 3.0 in. and cross-section dia. W = 0.70 in., subjected to 500 psi differential pressure, has a projected pressure area of 0.67 in.^{2}, a friction density of 24 psi, and a friction force of 16 lb

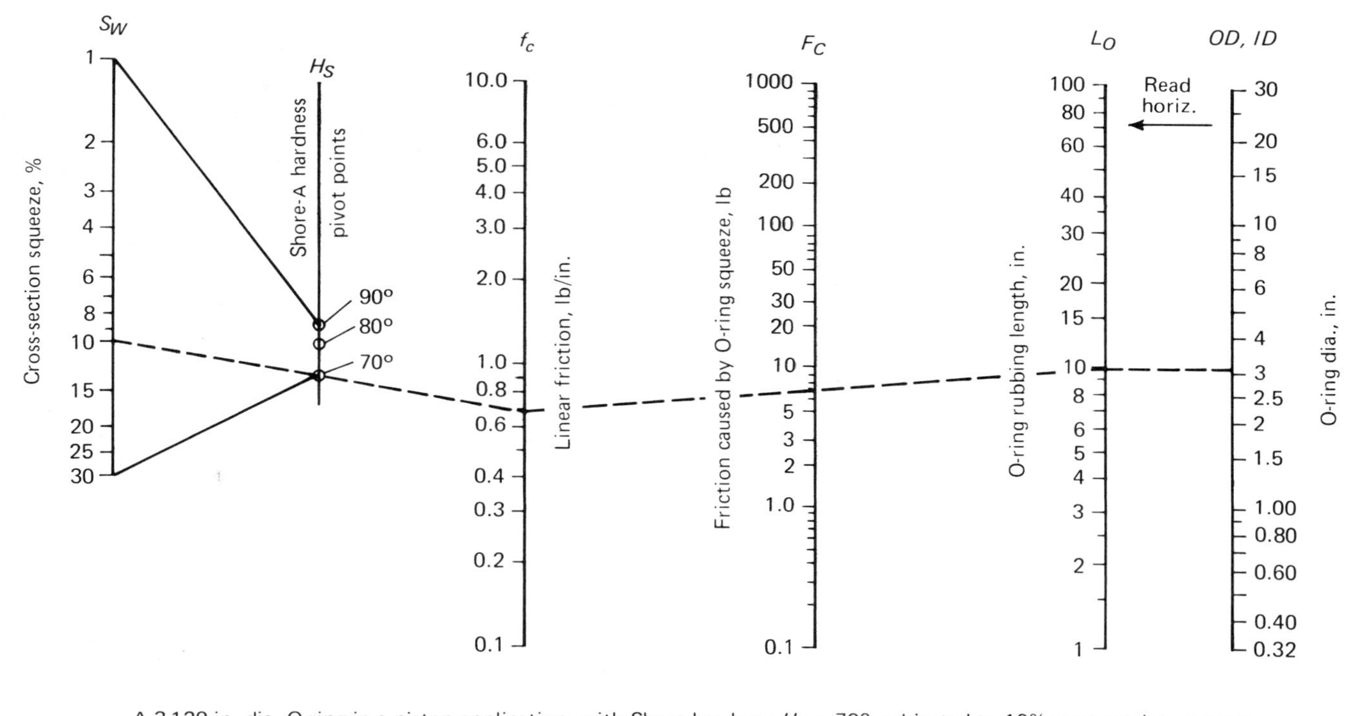

A 3.129 in. dia. O-ring in a piston application, with Shore hardness H_S = 70°, subjected to 10% cross-section squeeze, has a rubbing length L_O of 9.8 in., and will result in a linear friction of 0.68 lb/in., and a friction force of 6.7 lb

Fig. 19.24 Two nomographs for predicting the friction of an O-ring (See Fig. 19.23 for the model).

Table 19.1 Basic Dimensions of the Piston and O-Ring Used in Friction Calculations (See Fig. 19.23)

	Example 1	Example 2
Groove dia, D_G, in.	2.880 +.000, −.002	3.137 +.000, −.002
Piston dia, D_P, in.	3.247 +.000, −.001	3.258 +.000, −.001
Bore dia, D_B, in.	3.250 +.002, −.000	3.261 +.002, −.000
Groove width, G, in.	0.281 +.005, −.000	0.093 +.005, −.000
O-ring ID, in.	2.850 ±.015	2.989 ±.015
Cross section, W, in.	0.210 ±.005	0.070 ±.003

FRICTION IN PNEUMATIC SYSTEMS

The easiest way to reduce seal friction is with lubricant introduced into the airlines supplying the system. The toughest solution is to run totally dry, and that is what OSHA seems to have in mind. A practical solution lies in between.

Let's take the worse case: no seal lubrication permitted. That's really a matter of definition, because there almost always is some lubricant even if it is a film of air, an almost imperceptible coating of grease, an oil-soaked wick, a hidden pocket of nonvolatile oil in the piston or seal, a solid lubricant such as molydisulfide impregnated into the sealing surfaces, or just a microscopically small coating of Teflon retained by a rod surface rubbed by a dry seal.

Table 19.2 Calculation Examples of O-Ring Friction Using the Nomographs in Fig. 19.24

Dimensions & forces, nominal	Example 1 (standard)	Example 2 (special)
O-ring type (Parker)	AS-568-336	AS-568-041
Piston dia., in.	3-¼	3-¼
Differential P, ΔP, psi	500	500
O-ring dimensions, in.:		
ID (free dia.)	2.850	2.989
W (see sketch)	0.210	0.070
Mean dia. = ID + W	3.06	3.06
Nomograph A (ΔP):		
Annulus area, A, in.^{2}	2.0	0.65
Friction density, f, psi	24	24
Friction force, F, lb	45 (from ΔP)	16 (from ΔP)
Nomograph B (squeeze):		
O-ring OD = ID + $2W$	3.270	3.129
Cross-section squeeze %:		
$S_W = \dfrac{W\text{-}.5\ (D_B - D_G)}{W}$	$\dfrac{.210-.5(3.250-2.880)}{.210}$	$\dfrac{.070-.5(3.261-3.137)}{.070}$
(nominal value)	= 11.9%	= 10%
O-ring Shore hardness H_s	70	70
Rubbing length, L_o, in.	10	9.8
Linear friction, f_c, lb/in.	0.72	0.68
Friction force, F_c, lb	7 (from squeeze)	6.7 (from squeeze)
Total dynamic friction, lb	45 + 7 = 52 (ans)	16 + 6.7 = 23 (ans)
Static friction (×3)	156	69
Hyd force on piston, lb	4148 = area x ΔP	4148
Dynamic hysteresis:		
Friction/Hyd force	52/4148 = 1.25%	23/4148 = 0.55%

Table 19.3 Refinement of the O-Ring Friction Calculation using the Stretch Graph in Fig. 19.25

Double check on squeeze	Example 1 (standard)	Example 2 (special)
Diametral stretch, %:		
a) Max piston, min O-ring	(2.880 − 2.835) ÷ 2.835 = 1.59	(3.137 − 2.974) ÷ 2.974 = 5.5
b) Min piston, max O-ring	(2.878 − 2.865) ÷ 2.865 = 0.45	(3.135 − 3.004) ÷ 3.004 = 4.4
Percent squeeze	1.5% (a); 0.5% (b)	4% (a); 3% (b)
Actual W, in. (max)	.215 (1 − .015) = .2118 (a)	.073 (1 − .04) = .0701 (a)
(max)	.215 (1 − .005) = .2139 (b)	.073 (1 − .03) = .0708 (b)
Actual W, in. (min)	.205 (1 − .015) = .2019 (a)	.067 (1 − .04) = .0643 (a)
(min)	.205 (1 − .005) = .2040 (b)	.067 (1 − .03) = .0650 (b)
Corrected squeeze (max) (large W, small bore, and small piston)	$\frac{.2139 - .5\,(3.250 - 2.878)}{.2139}$ = 13 % (close to estim.)	$\frac{.0708 - .5\,(3.261 - 3.135)}{.0708}$ = 11% (close to estim.)
Corrected squeeze (min) (small W, large bore, and small piston)	$\frac{.2040 - .5\,(3.252 - 2.878)}{.2040}$ = 8.3%	$\frac{.0650 - .5\,(3.263 - 3.135)}{.0650}$ = 1.54%

Some mechanical face seals for rotary compressors rely on the hydrodynamic pressure of the gas to maintain seal clearance. In effect they are marginally lubricated gas thrust bearings and require great care in design.

Dry-seal pistons are simpler. TFE-composite piston rings and seals have self-lubricating characteristics and need only to be protected from excessive heat and wear by proper choice of linear speeds (slow as possible), avoidance of side loads, and precise dimensioning and finishing of all parts.

Just remember that dry seals run hotter than lubricated seals, expand more, wear

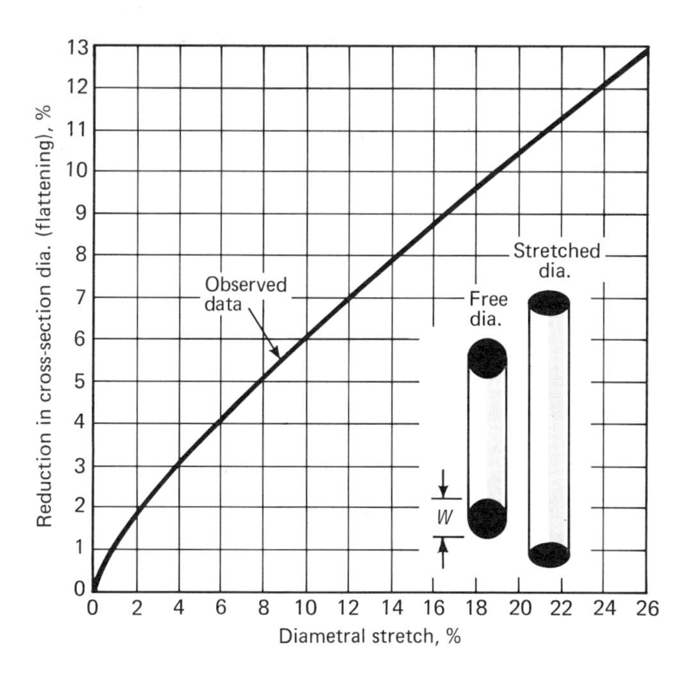

Fig. 19.25 Refinements to the O-ring friction calculation can be made by considering the effects of diametral stretch.

faster, and are less tolerant of careless installation. Just the same, some manufacturers offer them on products. See Chap. 15.

WEAR AND FAILURE (ALL SYSTEMS)

Seals should be limited to sealing and should not double as bearings if long life is the goal. Wear rings are better suited to the job of supporting side loads and centering the rod or shaft.

O-rings and other squeeze seals sustain a sealing load which is similar to a bearing load, but even O-rings should be protected from eccentricities in rods and shafts and from other distorting influences. The problem is pronounced in rotary seals where the line of sealing contact is circumferential and can wear a groove in the shaft. O-rings for rotary sealing often are narrow gage to reduce friction, but this intensifies the problem of wear.

A partial answer is to slant the O-ring so that the line of sealing contact is not in a plane perpendicular to the shaft (Fig. 19.26). The result is a wider swept area. Lubrication is drawn in more readily because of the intermittent (sinusoidal) contact at any given point, and wear can be less.

The swept area should be at least twice the contact width of the O-ring, and leakage is expected to be by shear action rather than from fluid pressure. Lip seals have less leakage than slanted O-rings but cannot sustain as high a fluid pressure.

A similar and enhanced effect is obtained with elastomer seals that are mounted straight but have a wavy line of contact. CR industries (Elgin, IL) offers a radial lip seal called Waveseal, and it features a narrow-width one-cycle sinusoid (Fig. 19.20). The waviness tends to pump oil back against pressure and keep it from leaking out. A secondary advantage is that the increased swept area gives more area for cooling.

Grit-induced wear is more of a problem than clean frictional wear. Where grit next to the seal is unavoidable, accommodate it by making one seal surface soft enough to catch and hold the particles or make both seal surfaces hard enough to grind up and pass the particles. Babbitt metal is a good absorber, and silicon carbide is hard to beat as a particle chewer in mechanical face seals.

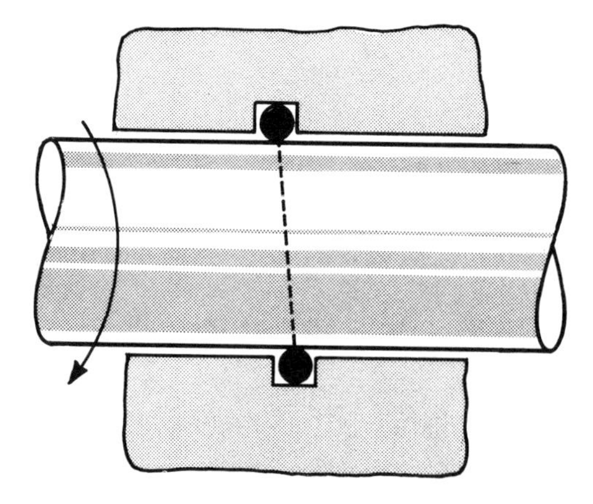

Fig. 19.26 Canted seal has a wider swept area and is less likely to wear out the shaft.

Some slurry pumps couldn't work long without face seals of silicon carbide, tungsten carbide, aluminum oxide, or some other extremely hard material. If all else fails, inject clean fluid continuously into the seal cavity.

Other modes of failure include elastomer seal extrusion, spiral failure (where the sealing ring twists within the groove), pressure-trap blowout, face-seal wobble (plus fretting of secondary seals), fatigue, fracture, and thermal runaway.

Extrusion of elastomer seals into the shaft-to-bore clearance can be prevented by paying attention to known facts. The extrusion results from a combination of these adverse conditions: a large clearance, heat, high pressure differential, frictional drag in the same direction as the pressure differential, and seal materials that are soft and have low strength, low resilience, and high stress relaxation.

Remember that a centered shaft has half the clearance gap of a shaft pushed to one side. Wear rings can keep pistons centered. Anti-extrusion rings can back up weak seals. A seal cavity for a cylinder can be placed so that the frictional drag force is in a direction that opposes the pressure force (Fig. 19.22).

Very large gaps can be accommodated with ring-shaped seals of special compositions that give them resilience, strength, and lubricity. In extreme cases, radial pressure may be applied to force the seal outward against the cylinder bore or inward against the rod.

Spiral failure of an O-ring happens in cylinders if frictional force prevents the ring from sliding. Also, friction and pressure can buckle lip seals.

Stacked V-ring packings sometimes fail because the first ring in the stack gets most of the lubricant and the succeeding seals in the stack run dry.

The best way to avoid spiraling and buckling of seals is to make sure that the seal fills as much of the cavity as possible. In that way each seal braces itself in every direction. Then be sure to involve the seal supplier in the problem so that he or she can recommend the best combination of resilience and surface hardness for the dynamics of your application.

Pressure trap can be an insidious flaw in an otherwise carefully designed sealing system. It happens when seals are arranged in series on a reciprocating rod or piston but without enough thought about the pumping action that occurs when the rod oscillates.

If each seal acts as a pumping check valve in the direction that builds up pressure between the seals (it occasionally happens), the between-seal pressure can rise to several times maximum system pressure. If the pressure isn't vented or the seals redesigned, blowout can occur. One answer was illustrated in Fig. 19.9.

Thermal runaway is a problem where initial thermal distortion causes friction, friction causes heat, heat increases the distortion, and so forth until failure. Remember that the thermal coefficient of expansion of elastomers is very high compared with metals. Good sealing manuals recommend materials, finishes, dimensions, tolerances, and designs to avoid the problems entirely. It still might be necessary to run your own tests, just to be sure.

STATIC SEALS

A seal that is locked in place and doesn't spend its life sliding on an oil film—or worse yet, on bare metal—is not nearly as hard to pick or design as a dynamic seal.

Much has been written on conventional gaskets, static O-rings, and sealants, so we'll limit this chapter to the unusual types. Examples are hollow metal O-rings for high-temperature flange service, penetration-type metal-to-metal seals, captive-plastic flange seals, liquid fluoroelastomers, and do-it-yourself interlocking gaskets. None are the typical shop-maintenance item. See Fig. 19.27 for examples.

Hollow Metal O-Rings (Fig. 19.28)

Like solid metal gaskets, hollow metal O-rings seal at high temperatures—easily over 500 F and sometimes over 2000 F when made of special materials. More resilient than solid gaskets, they seal along a narrower contact area and require only a fraction of the bolting force.

Like elastomer O-rings, hollow metal O-rings deform under fluid pressure (it takes higher pressure, of course) and are forced snugly against the sealing-groove surfaces when system pressure increases. Where an elastomer O-ring can be used, a metal O-ring can be too.

O-Ring Design. There are three types: plain hollow, pressurized hermetic, and self-energized. The *plain hollow* metal O-ring is a tube formed into a circle, with ends welded together. It can withstand system pressures of 100 to 400 psi without collapsing, depending on tube wall thickness.

The *pressurized hermetic* O-ring contains gas to keep the tube from collapsing under high operating pressures. A typical charge of 600 psi will support a system of the same pressure.

The *self-energized* or *vented* O-ring is perforated with holes on the fluid-pressure side to equalize operating pressure across the tube wall. Most adaptable of the three types, it best supports high system pressures.

Most O-rings are made of circular tubing, though even square tubing can be used. Normally the tubing is formed into a ring, but it can also be formed into oblongs or squares or into nonplanar shapes to fit irregular flanges. Round shapes are cheapest and should be chosen where possible. The larger the diameter of the tubing (over ⅛ in.), the easier it is to make.

Platings and coatings on O-rings improve sealing and are usually recommended. Tips for choosing coatings and surface finish for many typical applications are given in Table 19.4.

Making any metal O-ring is a custom job. The final dimensions after manufacture must be accurate to mils, despite the numerous steps involved. The tube must be formed into a circle, cut, sized, deburred, cleaned, resistance-butt-welded, ground, polished, and sized again. No filler is used during butt-welding—this helps to keep the joint as strong as the rest of the O-ring.

Applying the O-Ring. Do it in three steps: Determine the worst operating condition in terms of temperature, pressure, type of fluid being sealed, and general environment; select O-ring material and dimensions to meet the operating requirements; then design the joint.

Experience has shown that tubing with a minimum ultimate tensile strength of 85,000 psi gives the best performance, particularly for sealing gas. Whenever possible, choose a metal with at least that strength throughout the operating temperature range. When no such metal is available, the O-rings can still be made to work, but you must experiment to find whether the seal will match the requirements. The main

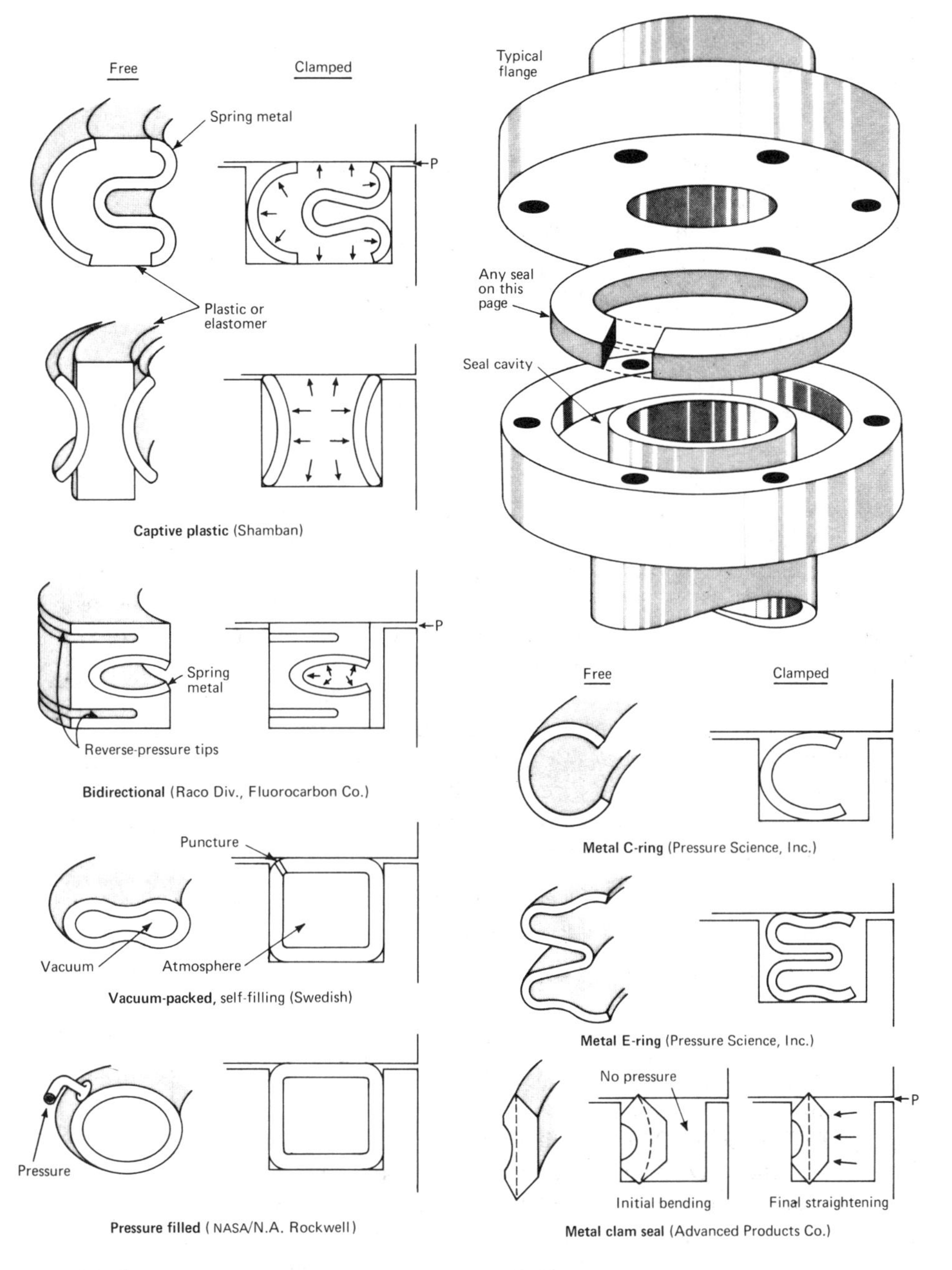

Fig. 19.27 Flange seals for special applications.

problem at high temperature is loss of resiliency (spring-back), which reduces the sealing force of the O-ring against the flange surface.

Another rule: Don't choose an O-ring metal with a coefficient of thermal expansion that differs considerably from that of the flange. The sealing faces might scuff, causing leakage.

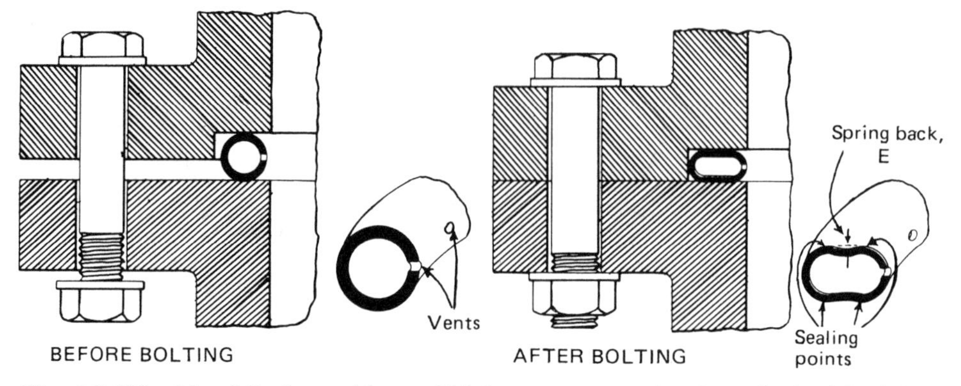

Fig. 19.28 Metal O-rings withstand high temperatures but have limited flexibility and require high clamping forces.

Pressure: System pressure can be as high as you want for most tubing sizes if the seal is self-energizing. A pressure of 50,000 psi is not unusual.

The initial sealing is mechanical, but as system pressure increases, the fluid forces become greater; the tubing deforms into the groove cavity and seals even more positively. The O-ring can withstand any pressure the flange can, and the higher the pressure, the better the seal. Make sure there is at least 25% margin in bolt strength to prevent flange separation during unexpected pressure surges.

Gases vs. liquids: Gases are harder to seal than liquids. Liquid molecules are larger; this and capillary attraction help block minute leakage paths.

Surface finish: Hardness, O-ring wall thickness, the character of the machined surfaces, and the type of fluid determine the choice of finish. Coatings or platings on the O-ring surface will help substantially and are essential on gas seals. Here are some general rules for all O-ring seals:

- Specify tool marks to be circumferential, because spiral, radial, and transverse marks are potential passageways across the seal.
- Clean surfaces thoroughly. For gas seals, use solvents for cleaning O-ring and flange surfaces.
- Flange surface finish should be in accordance with the following:
  Plated or coated O-ring, against a turned-finish surface, sealing heavy lube oil or tar—63 μin.
  Teflon coated O-ring sealing ordinary hydraulic oil—63 μin.
  Nonplated O-ring sealing heavy lube or tar—32 μin.
  Plated or coated O-ring sealing low-viscosity liquids such as water—32 μin.
  Teflon-coated O-ring sealing gases—32 μin.
  Precious-metal-plated O-ring sealing gases—16 μin.

If in doubt, specify a 32 μin. finish and use coated or plated O-rings.

O-ring squeeze: Tube dimensions, compression forces, spring-back data, and groove dimensions are not arbitrary, nor can they be easily calculated. They must be based on tests.

Note: The O-ring is squeezed beyond the elastic limit during installation, and the spring-back is only a small fraction of the initial deformation. The groove dimensions must be accurate; the force margin is not very great, and all of the elastic strength must be exploited. If the O-ring is compressed too far, excessive deformation reduces the spring-back. If the compression is too slight, the full elastic force is not achieved.

The problem is solved by using the self-energized (vented) seal—it supplies the sealing force hydraulically, provided the initial mechanical seal has been made. (At the other extreme are the solid gaskets: The fluid normally does not help seal the gasket, and spring-back is negligible). The ratio of mechanical seating force to fluid-pressure seating force for the initial stage of sealing can be controlled to some extent in an O-ring seal by selection of the wall thickness, outside diameter, and metal of the tubing from which the O-ring is made.

Metal Alphabet Rings

Besides metal O-rings, there are C-rings, E-rings, K-rings, U-rings, and V-rings, each named for the shape of its cross section.

The C-rings and E-rings in particular, are making headway today (Fig. 19.27).

Table 19.4 Matching Metal O-Rings to Environment

APPLICATION	O-RING MATERIAL
Cryogenic temperature only	Tubing — 304 or 304L stainless steel Coating — Teflon, Kel-F, or silver plating
Ambient (moderate temperature)	Tubing and coating — unlimited selection of materials, depending on media to be sealed
Cryogenic to high temperature	Tubing — Inconel or Inconel X Coating — silver, gold, or platinum
High temperature only	Tubing — Inconel, Inconel X, or refractory metals Coating — silver, gold, or platinum plating
Nuclear environments	Tubing — 347 stainless steel Coating — silver, gold, or platinum plating
Corrosive environments	Tubing — 316 or C20 stainless steel Coating — Teflon or Kel-F
Liquid metals (NaK, Na, K)	Tubing — 18-8 stainless steel, Inconel, or Inconel X Coating — silver or copper plating
Vacuum (ambient temperature)	Tubing — Unlimited material selection Coating — Teflon, Kel-F, or silver plating
Vacuum (high temperature)	Tubing — Inconel or Inconel X Coating — silver, gold, or platinum plating

A C-ring in essence is an O-ring with a section left out. It becomes much more flexible that way yet retains the advantages of the rounded O-ring surface contact, which does not mar the flange surface. Spring force provides initial sealing; system pressure augments the sealing force. The E-ring and some of the other rounded shapes have similar qualities.

The flange surfaces are not penetrated, so the sealing action is much like that of a conventional elastomer O-ring or lip seal. Sealing surfaces must be much smoother and flatter, however. Hydrodyne Co. engineers suggest that cavity surfaces be flat within about 0.00005 in./in. of circumference for optimum performance and within about 0.00002 in./in. for general service, based on metal-plated rings. Teflon-coated rings will seal against 0.0005 in./in. flatness.

Penetrating Metal Seals

Some metal seals must penetrate the flange surface to seal. An interesting example is the Advanced Products Co. Clam Seal (Fig. 19.27). It bites into the surface initially when the cover is bolted down. Then the internal pressure from the system straightens and elongates the clam and forces it to bite even deeper. The clam metal must be at least 30% harder than the flange surface to penetrate properly. The seals are reusable if the parts are placed exactly in the same positions each time, which is accomplished automatically if the grooves are machined in accordance with the manufacturers recommendations. Silver or cadmium platings are standard, and others are available. Optimum sealing requires a loading of 1300 lb/in. of circumference. A similar seal, for pipe, is shown in Fig. 19.29.

The bimetal cover gasket (Fig. 19.30) is made of a balanced combination of metals so that extreme changes in temperature are compensated for by the different coefficients of thermal expansion, and the high-pressure cone seal (Fig. 19.31) is intended for superpressure applications of 200,000 psi and more.

Captive Plastic Seals

Cold flow and lack of resilience need not be a disadvantage in a seal if you can hold the plastic in place and keep it under compression. The first sketches in Fig. 19.27 show how some designers have accomplished it.

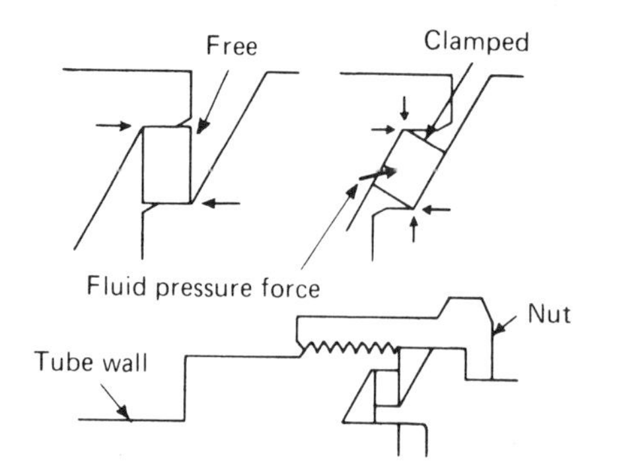

Fig. 19.29 Cocking of the metal ring seal when clamped ensures high contact pressure. Internal fluid pressure augments sealing. (Courtesy of Gamah, Div. Stanley Aviation, Denver, CO.)

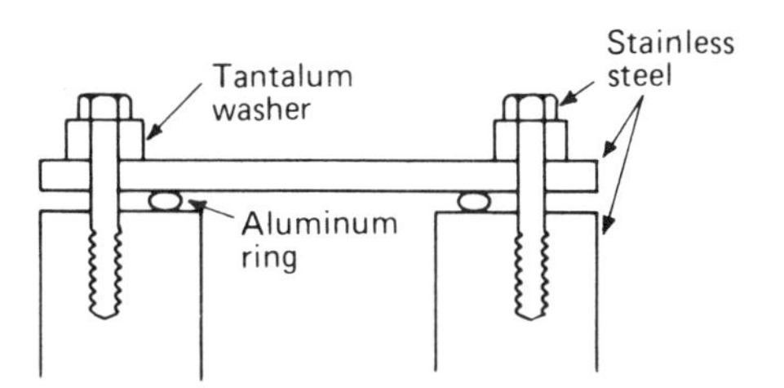

Fig. 19.30 Bimetal cover gasket has beneficial differential expansion to maintain seal contact forces. (Courtesy of NASA.)

In the captive seals sketched (W. S. Shamban & Co.), fluoroplastics such as Teflon (DuPont), Kel-F (3M), and Halon (Allied) are mechanically compressed far beyond the yield point and are forced against the sealing surfaces. Spring metal retainers hold the plastic and provide enough resilience to maintain the sealing force if the plastic shrinks or the sealing surfaces separate slightly. According to Shamban engineers, a 4-in.-diameter captive seal can hold gas pressures up to 12,500 psi. Liquid pressures can be much higher.

The properties of fluoroplastics, except for the tendency to cold-flow, are remarkable. They retain some ductility down to −420 F and will operate up to 500 F. Furthermore, they are unaffected by most fluids. For higher temperatures, it's better to use soft metals such as annealed nickel or copper.

Even the high shrinkage of fluoroplastics with temperature can be compensated for. Teflon can be compressed to 45,000 psi at room temperature, held captive as shown in the sketches, and cooled down to cryogenic temperature. It will shrink considerably, of course, but not as much as the original precompression at 45,000 psi. So there is still sealing force left at 75 K (135 R).

Liquid Gaskets

A latex form of DuPont's Viton fluoroelastomer is used as a binder for fibers for producing high-performance gaskets and packings. Gaskets made this way remain flexible and useful to 500 F, whereas gaskets made with conventional latex materials are limited to about 300 F. Armstrong Cork Co. (Lancaster, PA) has made engine head gaskets that need no metal cladding.

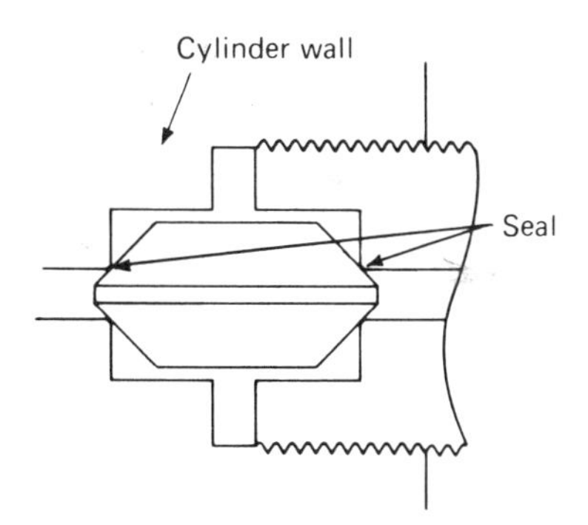

Fig. 19.31 Metal cone can seal against extreme pressures. (Courtesy of Harwood Engineering, Walpole, MA.)

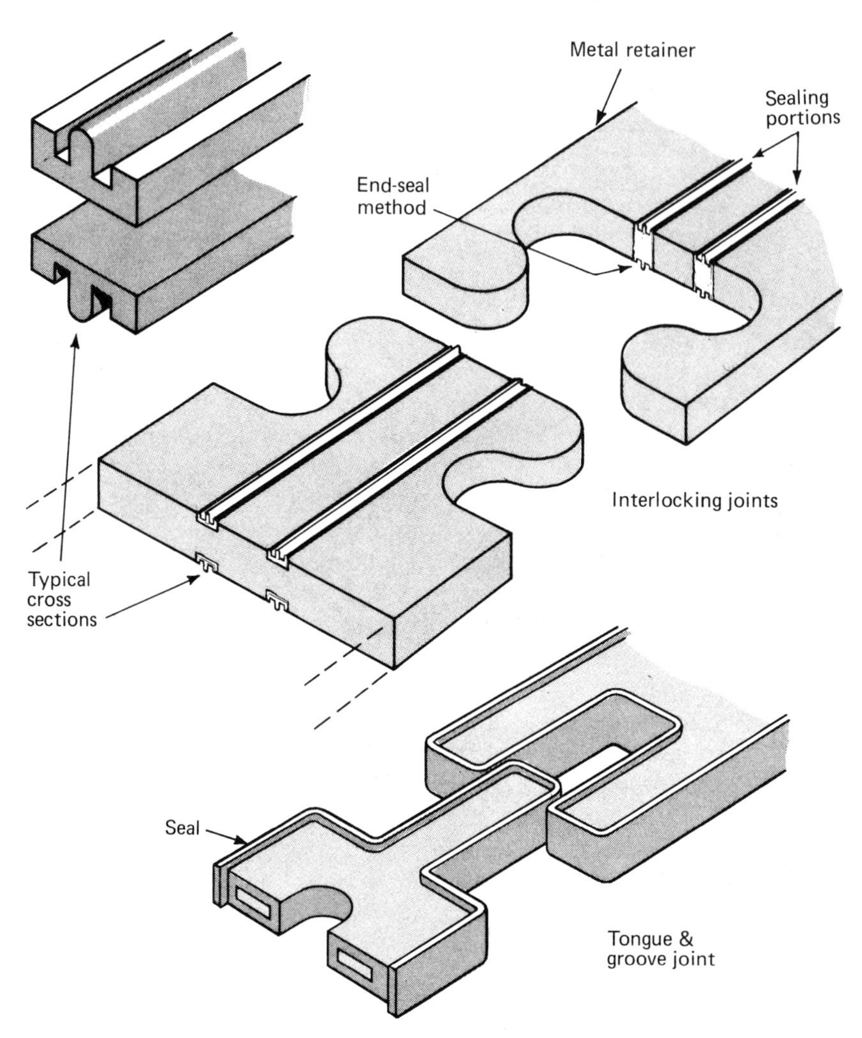

Fig. 19.32 Interlocking gaskets can be assembled into extremely large sizes. (Courtesy of Parker Seal Group, Lexington, KY.)

Dow Corning (Midland, MI) and General Electric Silicone Dept. (Waterford, NY) have a great answer for static joints that are too large or too complex for manufactured seals. They dispense a silicone paste as a bead along the joint and vulcanize it in place either at room temperature or at some designated temperature and humidity. A big advantage is that the silicone is forced into every crevice when the joint is closed, usually before the silicone has been allowed to fully cure. A computer program can be written to produce any bead path for mass production of machines and parts with gaskets formed in place.

Large Spliced Gaskets

Seals larger than 8-in. diameter are difficult to mold because of the huge presses involved. However, it's not impossible; Minnesota Rubber says they have molds up to 22 in. in diameter and that for even larger sizes splicing is used.

Parker Seal disclosed several splicing methods to us, including one for making large-diameter O-rings out of continuous lengths by scarf-cutting the mating ends and bonding them. Another method (for gaskets) is to create interlocking joints (Fig. 19.32).

20

Solving Cylinder Rod Leakage

Intensive discussions with dozens of designers, manufacturers, distributors, application engineers, and ultimate users of hydraulic cylinders and seals, over a period of several years, disclosed several interesting facts:

- Leak-free rods are possible.
- Most users won't pay the price.
- The price is custom design and eternal vigilance.

The designs pictured in Figs. 20.1 through 20.14 are industry's answers to some of the problems.

Seal manufacturers' catalogs generally avoid the subject of custom design and instead extol the virtues of standard seals—thousands of styles and materials, including modular packages bonded to metal.

They in fact have the answers if the cylinder designer stays within the guidelines of performance, including dimensional tolerances, and considers every type of seal before deciding. The danger is to believe that every application is the same, or even similar, and that any rugged seal that fits the gland will suffice.

It's difficult to carefully examine all the alternatives, and no single supplier covers every type. Even major distributors can't carry them all.

Then, assuming that the final shipped product—machine, vehicle, or whatever—is leak-free, will it remain so? Will the user keep the faith and maintain it as-new?

Idea: It would be great if the final product manufacturer would attach a label to each cylinder denoting the limiting conditions to which the seal was designed or specified. Crucial seal information is material, manufacturer, part number, fluids allowed, pressures, temperatures, stroke, speed, and recommended maintenance. At the least, it would make the user think twice before arbitrarily intensifying or speeding up the duty cycle.

A case in point: It apparently is common practice for agricultural users to add implements or increase attachment sizes without replacing or modifying operating

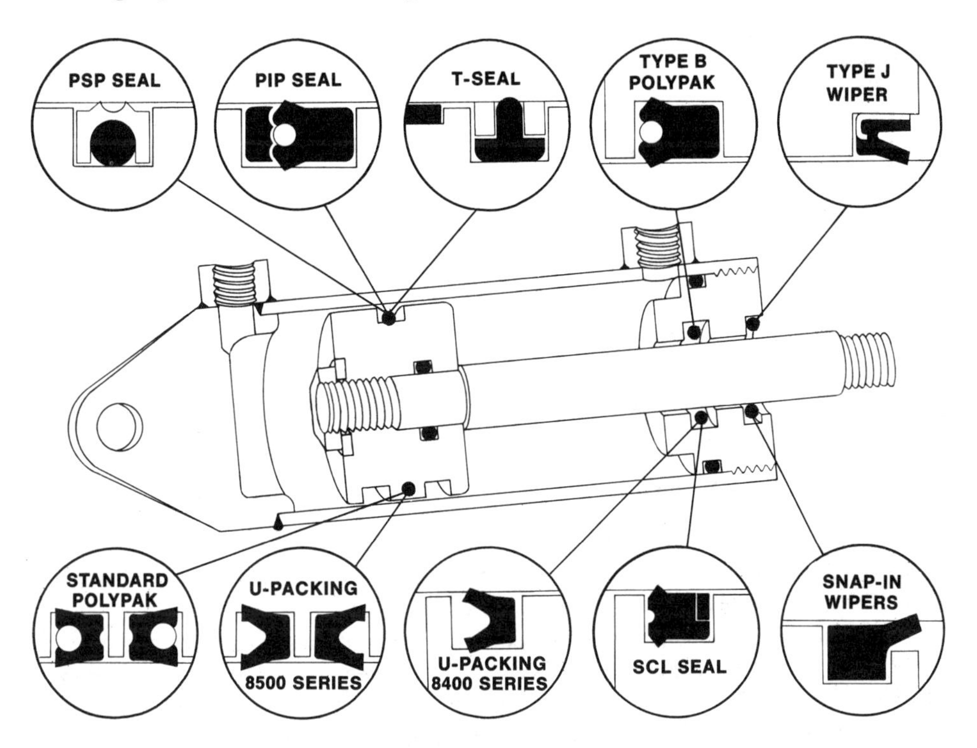

Fig. 20.1 Typical choices for rod and piston seals. (Courtesy of Parker Seal Group, Lexington, KY.)

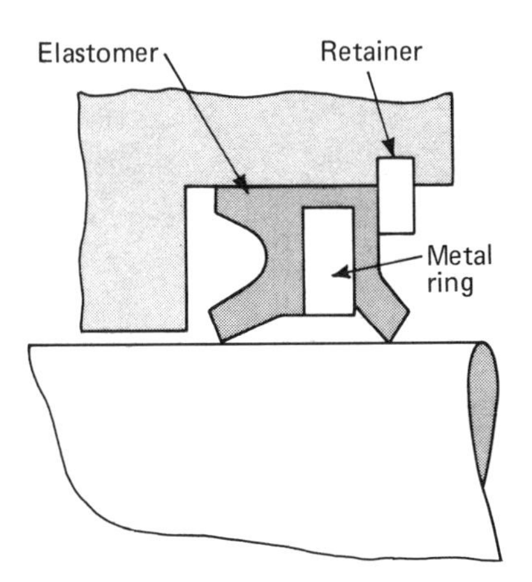

Fig. 20.2 Wiper seal assembly is outboard of the rod pressure seal (not shown) to back up the pressure seal and to keep out atmospheric dirt. (Courtesy of Crane Packing Co., Inc., Morton Grove, IL.)

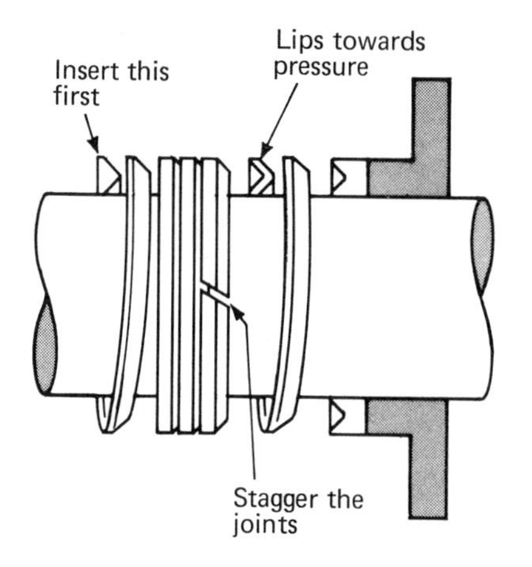

Fig. 20.3 Stacked V-rings can be cut and wrapped around a rod. (Courtesy of Garlock, Inc., Palmyra, NY.)

cylinders. But a seal that lasts 2 years against 2500-psi peaks might last 2 months against 3500-psi peaks or 2 days against 5000-psi peaks.

In many industries that we explored, typical maintenance consists of throwing away the seal when the leakage becomes excessive and replacing it in one of these ways: with the same brand and model number if it's available, with one that looks the same, or with another size or type of seal if some way can be found to make it fit.

Horror stories abound. Some users discard an entire cylinder rather than struggle to find the right seal replacement. Sometimes an acetylene burning torch is applied to the rod to free the cylinder, and a new rod-and-cylinder is welded on in its place. Even repairable cylinders can be dismal wrecks, with the rods worn dumb-bell shape by dry seals or scored like a rasp file by foreign objects.

Agricultural cylinders lie exposed to weather for long periods and often rust.

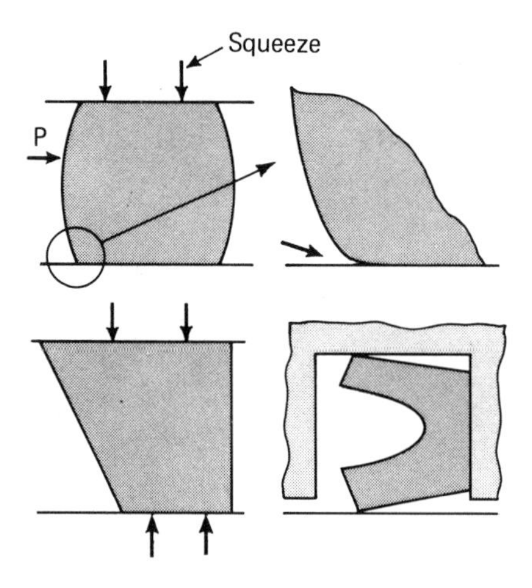

Fig. 20.4 Stretched nose of square seal (upper sketches) lets oil intrude under the slightly rounded edge. Seals with special precut angles and sharper edges block the oil better.

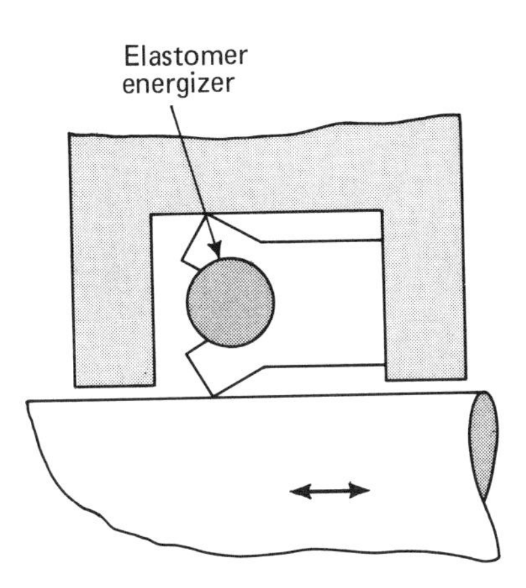

Fig. 20.5 Elastomer O-ring provides heavy preload for seal lips. (Courtesy of Parker Seal Co., Lexington, KY.)

Construction cylinders are shock-loaded or otherwise abused whenever the operator thinks he can break a load free that way.

Very few seal or cylinder manufacturers deal directly with the ultimate user, so most of the damage from these atrocities is repaired in obscure shops.

On the bright side, the few manufacturers and users who exert strict control over design, application, and maintenance claim relatively leak-free performance. We believe them.

IT CAN BE DONE

It's been proved time and again that a snug reciprocating seal of good design will not leak. Thousands of shock absorbers, liquid springs, hydraulic jacks, and premium dedicated cylinders are ample evidence—some stay leak-free for years. Those devices are specially designed and highly tested, and the operation is predictable.

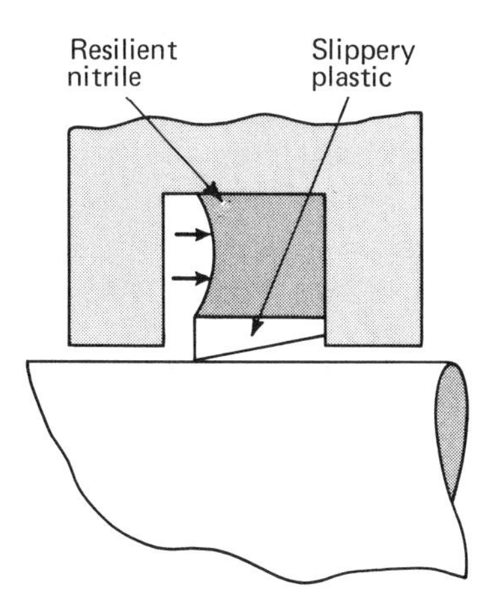

Fig. 20.6 Fluoroplastic seal lip is not very resilient and needs a resilient energizer. (Courtesy of Simrit Des Plaines, IL.)

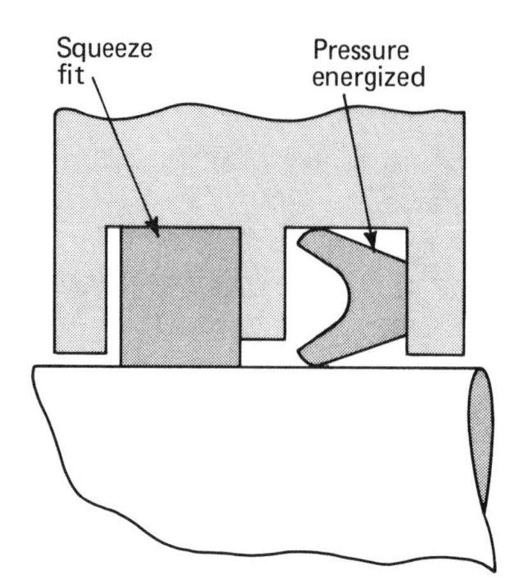

Fig. 20.7 Square buffer seal at the left takes care of low pressures and then lifts to pass high pressure to the conventional lip seal.

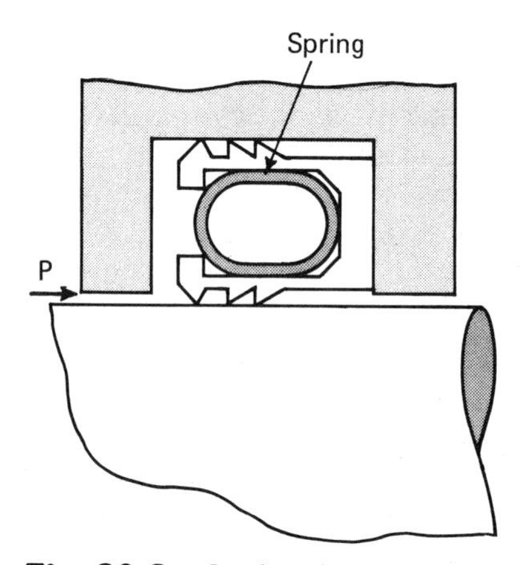

Fig. 20.8 Oval spring energizes the lips during low-pressure periods of the cycle. (Courtesy of Bal Seal Co., Tustin, CA.)

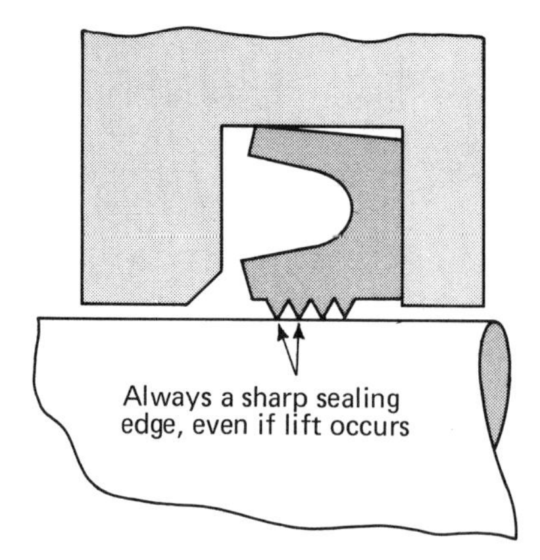

Fig. 20.9 Serrated flexible lip has extra edges to stop oil in case one or two lift off the rod. (Courtesy of Parker Seal Co., Lexington, KY.)

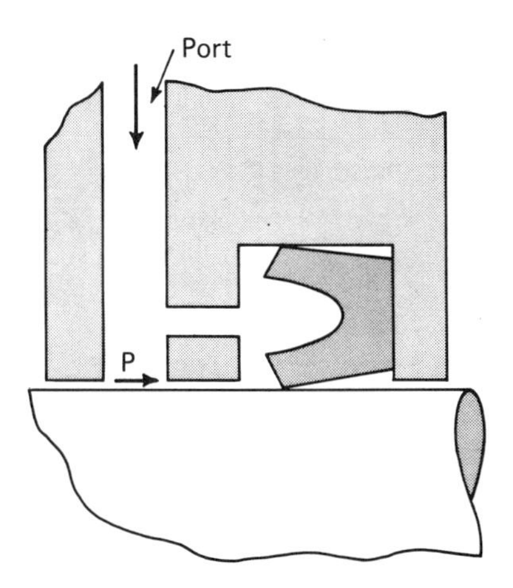

Fig. 20.10 Bypass hole from the port to the seal ensures quick energization of lips. (Courtesy of Hydro-Line Manufacturing Co., Rockford, IL.)

But ordinary hydraulic cylinders are beset with a broader variety of tough conditions, such as temperature extremes, rapid cycling, long strokes, great pressure variations from zero to thousands of psi, high side loads, rust, deteriorating surfaces, atmospheric dirt and chemicals, unusual hydraulic fluids, vibration, abuse, and bad application or maintenance.

A well-engineered cylinder-and-seal can eliminate or solve those problems one by one. Certain requirements are obvious: compatibility of the seal with all the expected fluids, pressures, temperatures, environments, and general operating cycles. The catalogs give good advice in those areas.

The harder requirements or decisions are to determine proper values for seal preload for nonpressurized parts of the cycle, rod surface materials and finishes, seal surface hardness, seal resilience, lubricity, rod force against seal lip during extend,

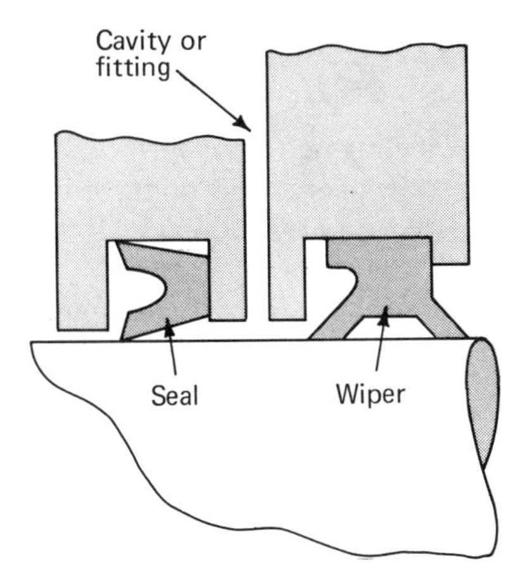

Fig. 20.11 Vertical hole can be a standpipe for oil trapped between seals or a cavity for grease.

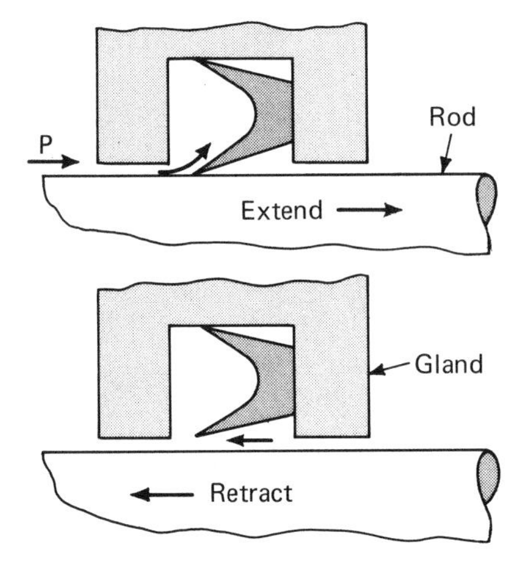

Fig. 20.12 Flexible lip seal blocks oil in the extend direction (upper sketch) but lets the rod carry oil film back, even against pressure, during retraction.

lip angle, effects of hoarfrost or ice on rod during retract, level of contamination permitted in oil, length and speed of individual strokes, expected shock loads and peak pressures, total linear travel per hour, friction coefficient and heat buildup, effects of alternate hydroplaning and boundary-level lubrication, hours (or years) between maintenance periods, allowable leakage, and eccentricity.

All those design conditions must be discussed with experts at each corner of the design triangle: the seal designer, cylinder or equipment designer, and ultimate user. Follow-ups must be scheduled after delivery.

Note that we did not include *allowable cost*. The reason is that a stock seal of good manufacture costs as little as 50¢ and will fit into the gland of most standard cylinders with ease. A custom seal—or even a special seal—can cost 5 to 100 times as much, including installation. If low initial cost is your chief objective, read no further. If top performance and best life cycle cost are your objectives, study this chapter carefully.

THE BASIC DESIGN PROBLEM

A conventional cylinder rod extends with a film of oil clinging to the surface, and some of this is wiped off as the rod retracts. The best seals theoretically limit the film to an almost imperceptible thickness on the extend stroke and let the film be drawn back into the cylinder on thc rctract stroke.

The term "dry rod" is usually a misnomer. There probably is no such thing as a totally dry rod—only varying degrees of dampness. A truly dry rod would have no lubricant and would wear out the seals. There are applications that approach this, but it's not the intended thrust of the discussion.

A rod that is merely damp is considered "dry" in many industrial applications. In less critical applications—say agriculture—a rod is considered dry if it doesn't drip noticeably. Even aircraft cylinders weep, but they don't reciprocate constantly as industrial cylinders do and normally don't accumulate enough oil on the rod to cause trouble.

Sealing against pressure is not the problem. Most manufacturers can do it easily. It's the continual rod extensions and retractions along with periods of low pressure that bring on the leaks.

The oil pressure against the seal usually is lowest on the extend stroke, thus unloading the seal to the point where it might float too lightly on the rod. The pressure usually is highest on the retract stroke, causing the seal to grip the rod more tightly and scrape off the oil.

The Rod Surface

Surface chemistry and physics are not exploited as well as they could be. Some fluids protect the surface; others deteriorate it. An alkaline fluid protects iron but etches aluminum.

Extreme rod smoothness (a few micrometers) is good for reducing seal friction and wear if a film of lubricant is maintained, but it's bad for seals if the film is scraped

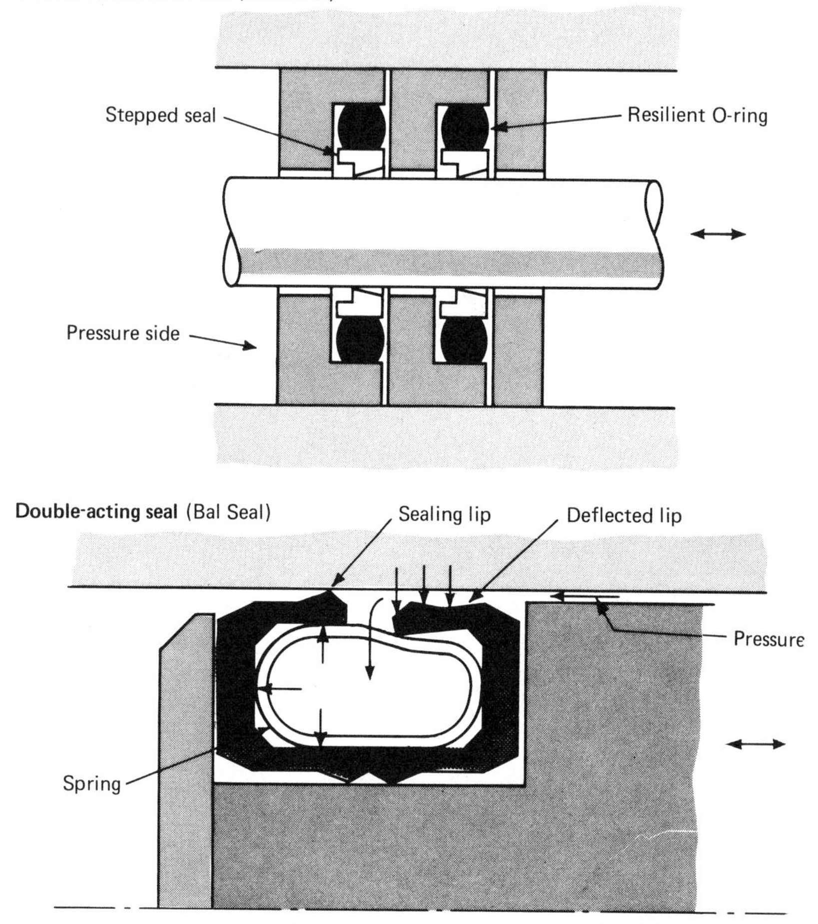

Fig. 20.13 Pressure trap is avoided with matched unidirectional seals or double-acting seals.

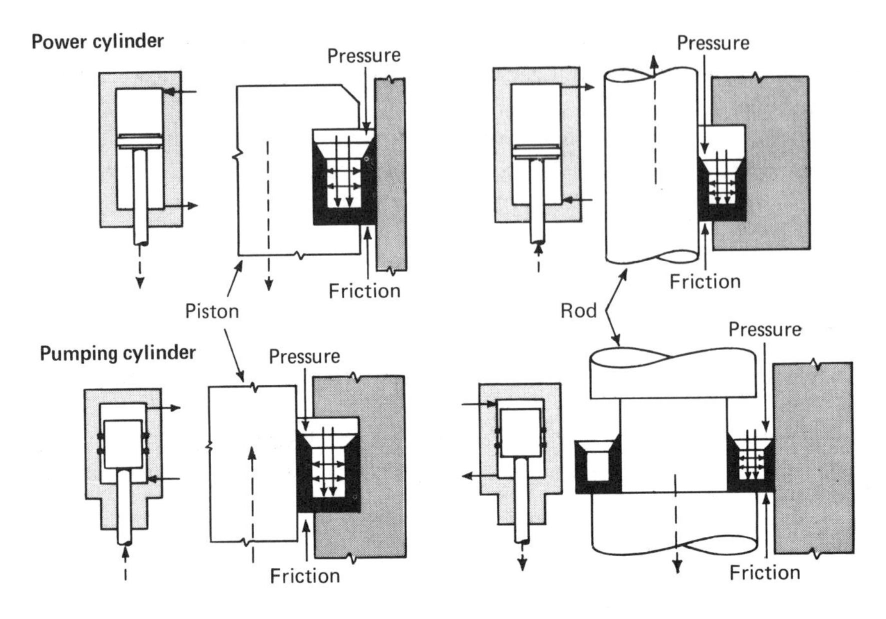

Fig. 20.14 Extrusion of the seal is somewhat diminished if friction is made to oppose the pressure force. (Courtesy of Parker Seal Group, Lexington, KY.)

off during the retract cycle. A compromise is an orange-peel finish of about 15 to 20 μm that retains at least some lubricant.

Some rod metals and platings are polarized (alkaline vs. acid): They will repel certain hydraulic fluids and tenaciously retain others (wetting). Petroleum oil generally is neutral (nonpolarized) and will wet most metals, but phosphate esters are not neutral.

If wetting is required, do your chemistry homework before specifying the seal design. Additives are available to increase the wetting of fluids. Remember that a tenacious lubricant film will return into the cylinder during retract, even against high cylinder pressure. That phenomenon can be used to reduce rod leakage.

Thin chrome plating often is porous and cracked, even though apparently smooth. The steel beneath will rust, and the cracks will wear the seal. If a hard, smooth, and unbroken surface is needed, be sure to specify it. It won't be cheap.

Seal materials are very important, and most of them are proprietary. The manufacturers keep computer records of the ingredients for each improved mix and can duplicate them exactly at any time in the future. Similar appearance does not mean similar performance, even if the basic ingredients seem about the same.

The Seal Surface

Seal surface hardness, body resilience, tear strength, porosity, temperature limits, and so forth are properties that can be altered with cookbook techniques. Just don't lose the cookbook. Surface hardness helps increase the allowed pressure level. Up to 90 durometer (Shore-A hardness) is usual.

Cast iron rings are sometimes used as secondaries for worn seals. The cast iron has the quality of polishing itself in service, and by the time the elastomer seal is worn out, the cast iron has a pretty good surface and fit and will limit piston or rod leakage.

Fluoroplastics have inherent lubricity but tend to cold-flow. So a combination of a fluoroplastic coating or shell over a resilient rubber or spring energizer solves both problems. Add to these a backup anti-extrusion ring of glass-filled nylon, and exceptional performance at high speeds and pressures is possible.

Neoprene rubber is resilient enough to seal even where fluid pressures drop to low values but tends to wear more rapidly than some other seal materials. Polyurethane wears well but isn't as good a seal as neoprene at low pressures.

Polyurethane impregnated with molybdenum disulfide solid lubricant is a popular seal material for starved-lubricant service. Nitriles and neoprenes depend more on rod lubricant.

Fabric-backed V-rings have the inherent advantage of wicking the oil into the cloth. Molded urethane seals don't need fabric for strength, but the rod must have at least a trace of lubricant or the friction will cause seal crazing and wear. Additives to the urethane help lubricate it somewhat.

Seal squeeze requirements vary with the seal material. Some companies aim for about 8% squeeze for nitriles but more than that with urethanes. The art of calculating squeeze and frictional force was described in Chap. 19; it applies only to O-rings, however.

Rod speed greatly affects hydrodynamics and wear at the seal lip. As a rule, 1 in./s is considered slow, 1 ft/s fast, and anything beyond a few ft/s very fast. Rods moving at 1 ft/s can get too hot to be touched.

Fast rods riding on a film of oil have low friction and can save a lot of energy—far more energy than is lost in the oil dripping off the end of the rod. However, if the goal is zero drips, you might have to live with the friction.

Contamination wreaks havoc on some seals. Yet fluid cleanliness is hard to maintain in a cylinder that reciprocates in dirty environments. Remember that a rod wiper seal (Fig. 20.2) that faithfully keeps out all contamination also scrapes off the oil film. It puts the cylinder designer on the horns of a dilemma. The best answer is to prevent any oil from traveling with the extending rod, so that efficient scraper-wiper seals are allowed.

Some cylinder users take the opposite view: They purposely specify inefficient wipers so that the oil film can travel in-and-out with the rod and not be scraped onto the ground. Unfortunately, the dirt will deposit on the oil film and get carried past the wiper into the cylinder gland. Short seal life and/or frequent maintenance is the inevitable result. Good filtering of the hydraulic system will help.

Test, Test, Test

There is no substitute for building and trying the total sealing system, despite what some seals catalogs promise.

Users such as automotive companies add a few extra twists, such as special air tests on hydraulic seals to see if the lips lift and performance tests spanning several years before approving new concepts. Some companies run wear tests on all seal types. 200,000 feet total travel is one figure given.

A typical rod-drip test is to initially clean off the rod, run the cylinder through 1000 cycles, and then wipe the rod with a clean cloth. Oil film will show on the cloth, even though it might not be visible on the rod. The test calibration is up to the user. One engineer wipes the accumulated film onto a face tissue and weighs it.

The crucial test is at low cylinder pressures. Any robust seal can stop leaks if there's high enough fluid pressure to force the lips against the rod. But careful preload is essential if the extend stroke is at low pressure.

How much leakage is acceptable? Here are opinions: A drop of oil on the ground every 1000 cycles or so is not unreasonable—unless the cylinder cycles 60 times a minute, day and night . . . or is located in a food factory. Farmers accept 1 drop in less than 1000 cycles, according to cylinder manufacturers.

Compression-set tests show whether a seal has resilience but might be misleading if the seal has additional means of energization, such as spring loading or elastomer-ring loading.

Footprints of seals taken wtih laser light during reciprocation in a test fixture show where an oil film is supporting the seal and where contact is dry. A long stroke cylinder develops a varying footprint as the stroke progresses and can't be compared directly to a short stroke cylinder.

Tolerance stackups in cylinder-seal assemblies can be very large. The molded seal itself can have 0.030 in. or more variation; the assembled gland and rod add more. Reliance on theoretical squeeze will be illusory unless selective assembly or special re-working is exploited.

Sometimes the cylinder manufacturer machines and grinds the seal tips for best fit. More often he or she simply asks the supplier to hold special tolerances or even designs the initial seal for the supplier.

Cylinder manufacturers can have elaborate in-house inspection of all seals, including GO, NO-GO gaging of seal ID. A seal too tight or too loose on the rod will not perform properly, and the margin for error is small. That's why certain cylinder manufacturers mark the mold number on the seal, along with the part number and company of origin.

Despite best efforts, the only place where fine-tuning of the design can occur is on the final machine as it is manufactured and tested. Once in the user's hands, it will be worked rather than played with, and delicate balances are lost. There better be margin in the design, or it's puddle city.

Life figures are rarely offered by seals manufacturers because the eventual application is a great unknown. Only the cylinder manufacturer can offer such a warranty, and then only after extensive testing in cooperation with the user and the seal supplier.

The average seal supplier does not do extensive in-house testing because there are too many potential variables.

Universal standards of design and performance are not available, although there are dimensional standards for sealing glands and for certain O-rings and packing. The International Standards Organization (ISO) has some general standards, but none of the U.S. companies we talked to were following them.

Design Concepts

A lot of seal technology is in the manufacture of the seals and the blending of the seal materials. The resulting choices are amply described in company catalogs. However, we were able to identify three different design approaches:

- Seal the rod so tightly that nothing passes the lips in either direction, at high pressure or low pressure.
- Seal more gently, and drain off the slight leakage.

- Seal gently, and exploit the clinging quality of the oil film to carry it back on retract stroke.

Tight seals include stacked V-rings (Fig. 20.3) with tightened glands and heavy-duty lip seals that are spring or elastomer energized. A good interference fit is needed, and lubrication is minimal in either case because oil is prevented from passing under the lips. Seal materials with good lubricity are essential.

V-packing has great theoretical sealing ability at low to high pressures but is bulky, costs more, and has higher friction than one-piece lip seals and is sensitive to poor assembly techniques—such as overtorquing the gland bolts. One answer is to specify precision shims and fixed gland length, so that the bolts can be drawn all the way down.

An advantage of V-ring packing is that the rings can be cut and wrapped around the rod without disassembling the cylinder.

Friction can be reduced by specifying that one or more of the rings be of special bearing material with good lubricity. Low-pressure sealing is enhanced if one of the rings is elastic.

If the stack is designed wrong, the pressure load will fall to the first seal, and the remaining seals will see very little pressure and might even run dry. Consult the experts.

Buckling of stacked lip seals (and twisting of single seals) can cause problems too. A combination of friction, pressure unbalance across the seal cavity, eccentricity, excessive shaft-to-bore clearance, and wrong seal design can trigger it.

Tight single-element lip seals (Figs. 20.4, 20.5, 20.6) are often found on heavy-duty off-road equipment. The design trick is to achieve the needed interference fit over the rod without excess friction or stick-slip. Lubricity additives to the seal materials help. A rod bearing inboard of the seal is recommended to keep the seal lips concentric and evenly loaded.

A variation on a single tight seal is a more conventional lip seal with a rubber ring or buffer seal ahead of it, in its own groove. The inboard ring is assembled with some interference so that it seals against low pressure yet lifts slightly away from the rod when cylinder pressure increases (Fig. 20.7).

The lip seal does its job when the pressurized fluid leaks past the buffer ring and energizes the lip. The ample cavity between the two seals remains pressurized during the extend and retract strokes and keeps the lip loaded.

Gently loaded seals mean those with flexible lips (Figs. 20.8, 20.9, and 20.10). Most cylinder seals sold fall into this category because installation is simple and performance normally is reliable.

The problem is to eliminate that pesky film of oil that sometimes slips under the seal and past the wiper and drips onto the floor. Installing a small drain in the cavity between the seal and the wiper will solve it, but there's not always room for the tubing and fittings.

A more interesting answer is to exploit the clinging tendency of an oil film and let the rod pump it back against high cylinder pressure. It really works.

First, create a small reservoir cavity (Fig. 20.11) between the seal and the wiper. A simple vertical hole covered with a cap or an empty grease cup will be sufficient for rods of short to medium stroke. In extreme cases, a small reservoir can be mounted above the cavity.

As the rod extends, assuming that there is some back pressure in the cylinder (easy to achieve), the seal lip will be energized enough to retain the oil within the cylinder. A small film might slip through, but it is momentarily trapped within the cavity by the wiper.

On the return stroke, the film clings to the rod and is carried under the seal lip, which is easily lifted because it is facing the reverse direction (Fig. 20.12). Tests have shown that this pumping action can force the oil into a cylinder at pressures over 5000 psi. The action is present even if the cavity is continuously drained, but then only part of the oil returns to the cylinder.

Another way to eliminate external drips is to pack the cavity (Fig. 20.11) with grease. The seals should be double-acting to keep the grease in the cavity. It helps load the seal lips and provides lubrication to make up for the lack of leakage oil through the seal. This technique is used in some of the cylinders in the Boeing 767 and 757 aircraft.

Avoid Traps

Don't get yourself into a pressure trap. Two facing seals on a piston or rod will trap fluid between them if not designed properly and in some instances will generate extreme pressures by a pumping action.

How do you avoid it?: Prevent backward insertion of unidirectional lip seals; don't use two bidirectional seals together; or vent the space between the seals if pressure buildup is unavoidable.

Shamban (Los Angeles, CA), for example, has developed a sealing system called StepSeal that combines an O-ring (for its resilience) and a proprietary self-lube PTFE compound for the seal (Fig. 20.13). Breakout friction can be less than running friction, preventing stick-slip.

The sealing element is a lip seal that blocks leakage under pressure in one direction but lets the oil film pass under the lip in the other. Two of them in tandem on a reciprocating rod will prevent loss of oil yet will not build up a pressure trap between the seals.

Another way to prevent pressure trap is to install one double-acting seal in place of two opposite seals. One such seal (Fig. 20.13) is made by Bal Seal (Tustin, CA). The lip of the seal nearest the pressure buckles to release, and the opposite lip is pressure energized to seal. An integral spring keeps both tips in contact with the sealing surface at low pressures.

Another problem is seal extrusion into the clearance between a moving rod (or shaft) and the bore. Seal designers have many options to solve it: higher hardness (durometer) of the seal surface; backup rings to support the seal against the clearance; closer tolerance on roundness and clearance of the shaft and bore; good concentricity by use of bearing rings on the piston or rod, so that one side of the shaft doesn't get the whole clearance; vertical mounting of large cylinders so that the weights of the rod and piston don't cause an eccentric clearance; limited side loading on the extended rod for the same reason; and sturdy cylinders that don't stretch excessively under pressure. (Also see Chaps. 5 and 19.)

Last but not least, design the shaft or rod motion so that the friction force tends to oppose extrusion of the seal into the clearance. Sometimes by putting the seal groove in the rod instead of the bore, a more favorable force balance will result (Fig. 20.14).

21

Rotary Seals: Mechanical, Labyrinth, Magnetic, Self-Pumping

Mechanical seals are rotary face seals, often used on turbomachinery; labyrinth seals are noncontracting seals for turbomachinery (where rubbing contact is not allowed); the magnetic seals covered here are those that seal with ferrohydrodynamic fluids; and the self-pumping seals have helices or spirals that generate oil pressure at running speed.

MECHANICAL FACE SEALS

Two types of mechanical face seals are discussed: spring-loaded carbon-to-metal or ceramic seals (Fig. 21.1) used in such applications as centrifugal pumps for the chemical process industry and elastomer-loaded metal-to-metal seals (Fig. 21.2) for sealing wheel bearings in off-the-road equipment.

Both types have a stationary face and a rotating face, and the sealing takes place at the interface. The seal in Fig. 21.1 is designed to maintain the faces strictly perpendicular to the shaft axis; the seal in Fig. 21.2 is flexible enough to follow wheel wobble.

We'll describe chiefly the seal in Fig. 21.1; however, the theory applies in many ways to all dynamic rotary seals.

Face seals of the design in Fig. 21.1 have two chief parts: a seal face that rotates with the shaft and a mating seal face that is stationary with the housing. The object is to keep each face exactly perpendicular to the shaft axis and to keep the clearance between the faces extremely small. For some of the difficulties, see *face-seal wobble,* Fig. 21.3.

Carbon-graphite is one material for the replaceable face, and it works well against hard steel, tungsten-carbide, silicon-carbide, or a ceramic such as aluminum oxide. In most seals the surfaces are ground perfectly flat (within 0.000023-in. tolerance) so that a lubricating film of oil or of the fluid being pumped covers all the mating surfaces.

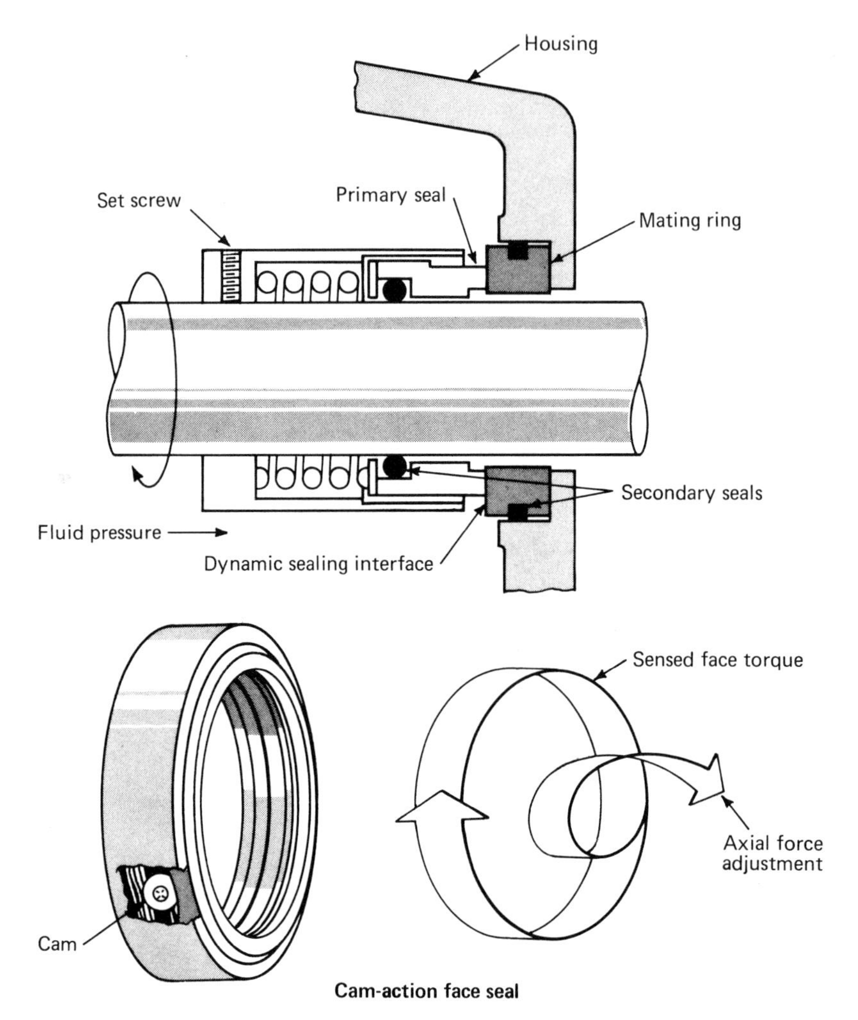

Fig. 21.1 Typical mechanical rotary face seal is shown in the upper drawing. A novel variation is cam-action built into the cartridge to automatically vary the face contact pressure in response to sensed face torque. (Courtesy of Rexnord-Cartiseal, Wheeling, IL.)

Ideally the faces never touch yet are close enough so that the surface tension of the lubricant film is sufficient to prevent flow-through. In special designs the shapes and surfaces of the faces are slightly altered to create a pumping action. For example, some have an almost imperceptible helix etched into the rotating face (Fig. 21.4), and one of the surfaces is tapered to enhance the pumping action.

Such precision seals are valuable where leakage is intolerable, as in certain chemical processes or nuclear power plants. They work best at the designed speeds and pressures, and in some instances the hydrodynamic pressure of the lubricant film will balance the internal pressure working against the seal assembly.

In the real world of hydraulic piston and rod sealing and in many other intermittent linear and rotary sealing applications, it is not practical to apply hydrodynamic sealing to every surface. For better ways, see Chap. 20.

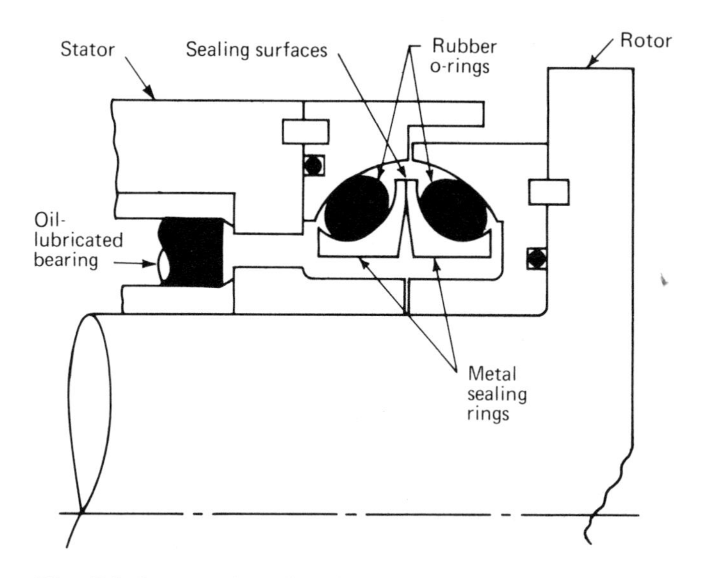

Fig. 21.2 Metal sealing lips are held in sliding contact by rubber O-rings and will seal despite flexing. (Courtesy of Caterpillar Industrial Products, Peoria, IL.)

Special Problems

Actual designs of face seals are not as simple as the upper drawing in Fig. 21.1 and involve special sealing materials for high velocity, high pressure, high temperature, and corrosive fluid media. Books have been written on the subject.

Improvements come gradually. Typical are better materials such as different carbides for long wear life of the stationary ring, better carbon for the wearing faces, tough elastomers used as loading springs, and cartridge-type designs that press right onto the shaft.

Rexnord/Cartriseal (Wheeling, IL) came up with a novel design: a high-pressure mechanical face seal with a camming action built into the cartridge so that when frictional torque increases, the seal-face contact force is automatically reduced to limit this torque (lower drawing in Fig. 21.1).

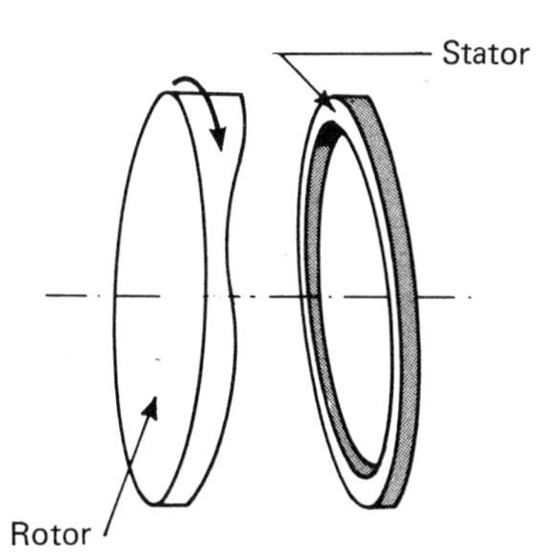

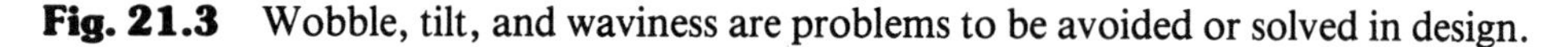
Fig. 21.3 Wobble, tilt, and waviness are problems to be avoided or solved in design.

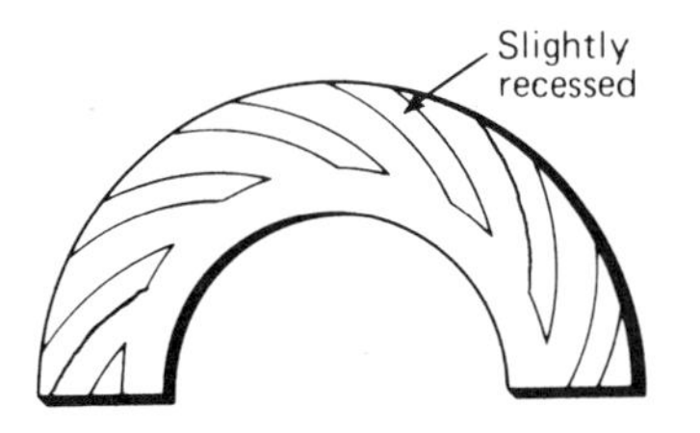

Fig. 21.4 Self-pumping face seal has spiral grooves to force leakage oil back against pressure. (Courtesy of Crane Packing Co., Inc., Morton Grove, IL.)

Where abrasive friction is unavoidable in any face seal, one answer is to make the nonreplaceable sealing surface out of extremely hard material. For example, Sealol/Rotary Seal Corp. (Chicago, IL) offers silicon carbide face seals that are 50% harder than tungsten carbide and last under severe conditions of temperature and pressure.

Chemistry is important too: Chesterton has a silica-free face seal that avoids reactions with certain chemical media.

Wear of the secondary elastomer seals in a mechanical face seal is a problem when there is flexing and sliding. Gits Brothers (Bedford Park, IL) solved this by substituting an elastomer C-ring for the conventional O-ring. The C-ring rolls as it flexes, instead of sliding. Wear life has increased considerably. Also, the force to compress the C-ring is less, resulting in lighter auxiliary springs and higher maximum speed for the total seal.

Thermal shock is damaging when too sudden to allow close-clearance sealing assemblies to respond uniformly. It can be a very serious problem in high-speed mechanical face seals. Certain face seal materials are better than others in this service. Silicon carbide has good thermal conductivity and seems to withstand thermal shock better than other materials. Carbon-graphite also is thermally stable and according to Pure Carbon Co. (St Marys, PA) can be hardened by surface conversion to silicon carbide.

A mechanical face seal wobbles if it's not square (Fig. 21.3). If the self-aligning half can't track fast enough, the seal can destroy itself by the rubbing of the sealing faces. Even if the tracking is fast enough, the static secondary seal can wear out quickly because of the short but rapid axial oscillation of the self-aligning seal. Short strokes can be worse than long strokes because there's less swept area for cooling and for picking up lubricant.

A small wobble, and even some waviness of the surface, is said to be an advantage if designed into the seal. They produce a pumping action to keep the surfaces flushed with lubricant.A phenomenon known as half-frequency whirl also enters into the picture. It's a form of hydrodynamic instability but can be damped by proper choice of elastomers in the secondary seals.

The experts are not 100% in agreement over the importance of all of this, and the conclusion is that trying to exploit waviness is risky: Square, flat seals are simpler and more reliable.

It's been found easier to hold squareness in a face that rotates with the shaft than in a face attached to the housing. One reason is that when the seal is assembled into the housing there are uncertainties such as flange surface contamination and uneven tightening of flange bolts.

LABYRINTH SHAFT SEALS

Holding fluid behind a labyrinth shaft seal (Fig. 21.5) is a little like stopping the wind with a door ajar. You won't stop all of it. But a labyrinth or some other clearance seal may be your only choice when the rubbing friction and wear of a conventional mechanical seal can't be tolerated.

Shaft seals are primarily to reduce leakage of fluid from the system, and the greater the obstruction to flow, the more efficient the seal. Although only basic designs are sketched (Figs. 21.6 through 21.13), there are hundreds of variants. Also see the symbols for equations (Table 21.1), developed by Louis Dodge.

Some companies, including Kentucky Metals Inc., and Clark Bros. Co. have built special labyrinth shaft seals made of honeycomb metal to seal low-pressure gases. The advantage is that the rotating shaft can easily machine away any interference, assuring minimum operating clearance.

The majority of labyrinth seals are machined of bronze or a similar alloy. They can be made accurately, they are mechanically strong, and they withstand high temperature. But the metal selected must not gall or melt during wear-in. There must be no residue or distortion to lessen the effectiveness of the seal. For high-speed turbine shaft seals, the stationary labyrinth teeth often are machined of an alloy that effectively sublimates (evaporates without melting) when rubbed by the shaft, which is usually of a different material. After the wear-in period, the seals are noncontacting, and the final operating clearance depends on thermal expansion and eccentricity of the shaft relative to the seal teeth.

There is a limit to the complexity of the seal. Tests indicate a diminishing advantage beyond a certain number of stages, particularly where the shaft doesn't stay

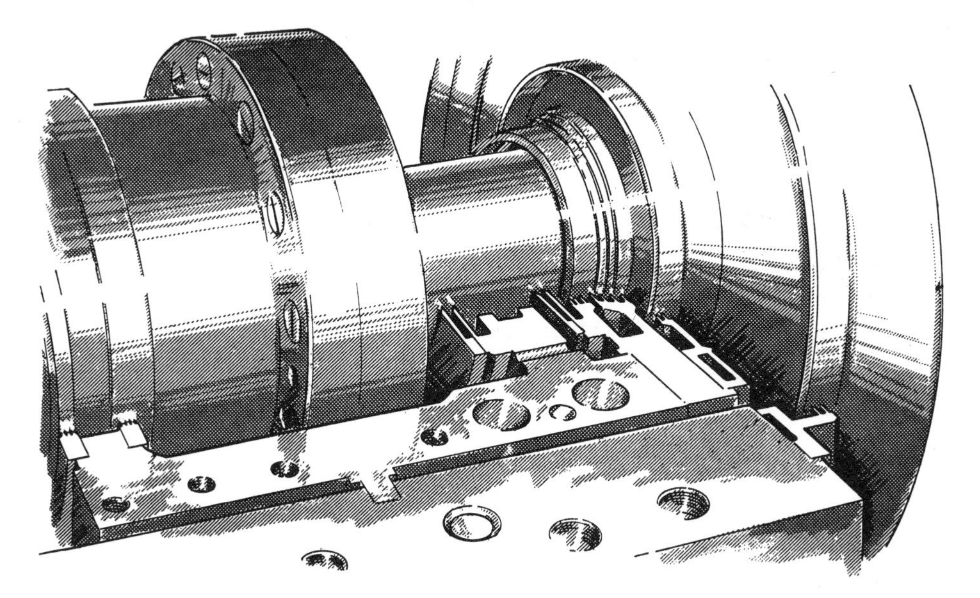

Fig. 21.5 Labyrinth seals are necessary where contact is not allowed (GE gas turbine).

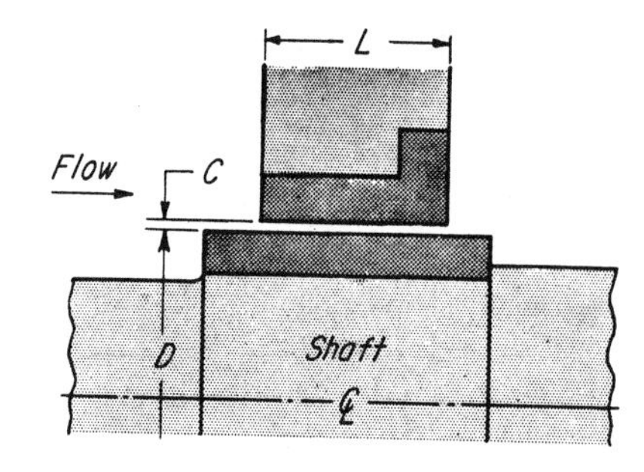

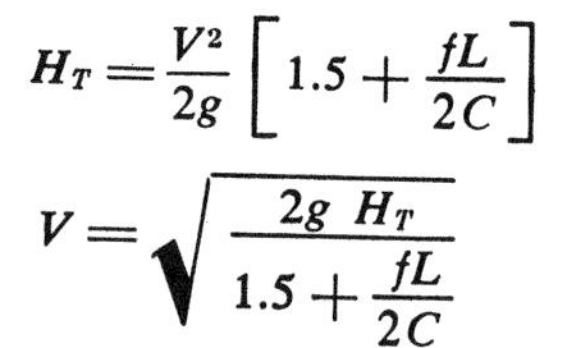

$$H_T = \frac{V^2}{2g}\left[1.5 + \frac{fL}{2C}\right]$$

$$V = \sqrt{\frac{2g\ H_T}{1.5 + \frac{fL}{2C}}}$$

For centrifugal pumps, and other liquid-handling applications. Has lowest cost.

Fig. 21.6 Sleeve-type positive clearance seal (see Table 21.1 for nomenclature).

centered according to calculations. In such cases, the simpler labyrinths are as good as the more serpentine and much less expensive.

The tapered teeth in Figs. 21.9 and 21.12 are considered the most efficient for sealing gas. Usually the teeth are rings inserted into the seal sleeve and staked. They are tapered to an edge of about 0.010 in. at the shaft diameter, and no clearance is allowed. The tips are quickly ground off when the machine begins to operate. Water lubrication can be used to prevent excessive heating during wear-in.

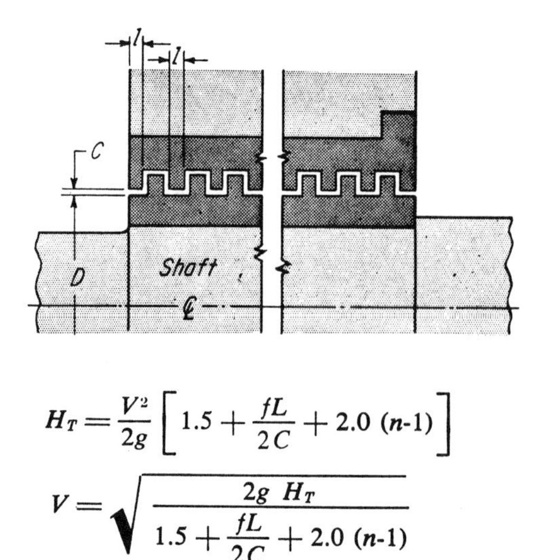

$$H_T = \frac{V^2}{2g}\left[1.5 + \frac{fL}{2C} + 2.0\ (n\text{-}1)\right]$$

$$V = \sqrt{\frac{2g\ H_T}{1.5 + \frac{fL}{2C} + 2.0\ (n\text{-}1)}}$$

For sealing liquid and gas. Requires split assembly, but has lower leakage.

Fig. 21.7 Interlocking labyrinth seal.

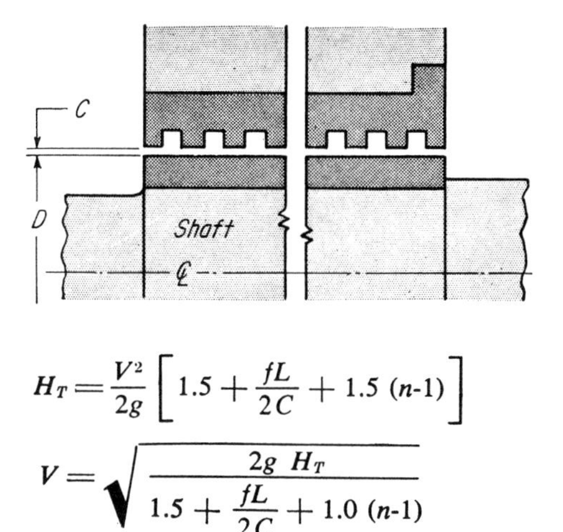

$$H_T = \frac{V^2}{2g}\left[1.5 + \frac{fL}{2C} + 1.5\ (n\text{-}1)\right]$$

$$V = \sqrt{\frac{2g\ H_T}{1.5 + \frac{fL}{2C} + 1.0\ (n\text{-}1)}}$$

Compromise between Fig 1 and Fig 2. Seals liquid and gas. No split necessary.

Fig. 21.8 Noninterlocking labyrinth seal.

Seal Calculations

In the hydraulic field, it is conventional practice to set minimum and maximum limits for the clearance C between the shaft and seal. The values depend on the strength of the rotating shaft and its deflection. As proved by tests, speed of the shaft has a negligible effect on performance of the seal. Typical recommended values of clearance are plotted in Fig. 21.10.

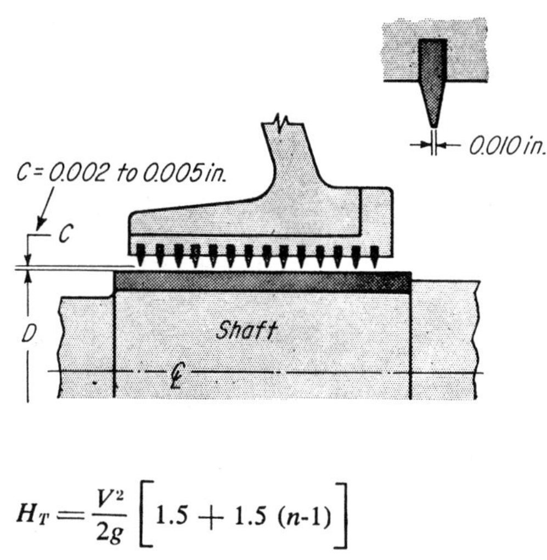

$$H_T = \frac{V^2}{2g}\left[1.5 + 1.5\ (n\text{-}1)\right]$$

$$V = \sqrt{\frac{2g\ H_T}{1.5 + 1.5\ (n\text{-}1)}}$$

For low-leakage seals. Commonly used on steam turbines and gas turbines.

Fig. 21.9 Tapered-teeth labyrinth seal.

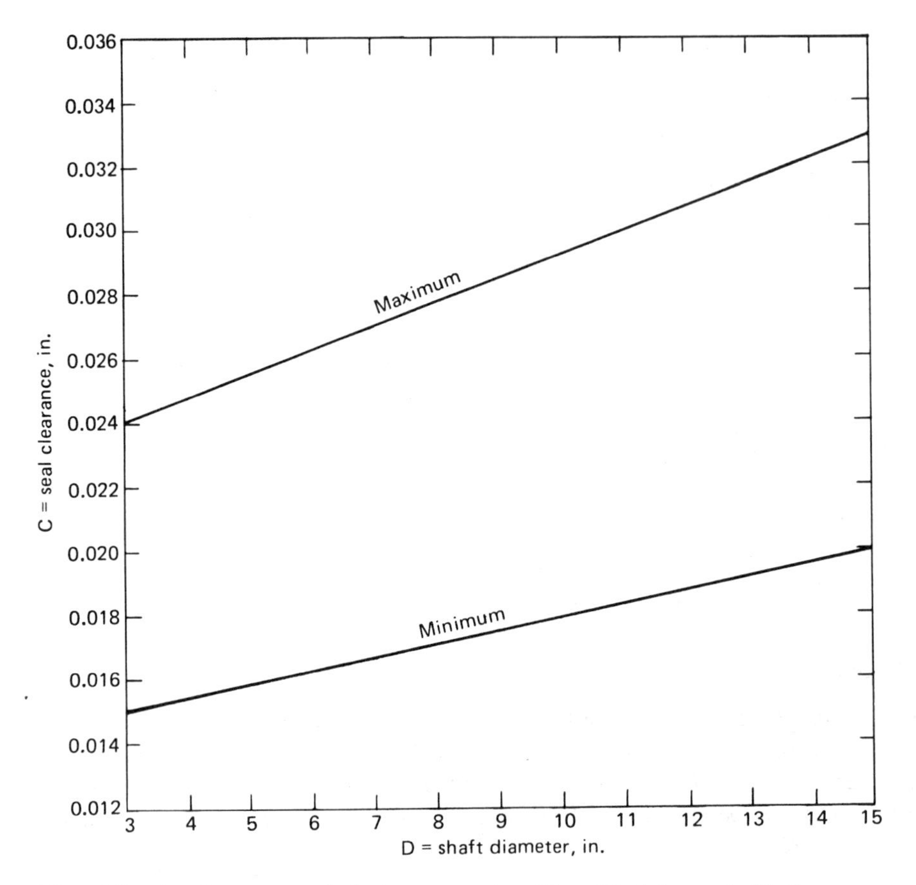

Fig. 21.10 Recommended clearances for Figs. 21.6 through 21.8.

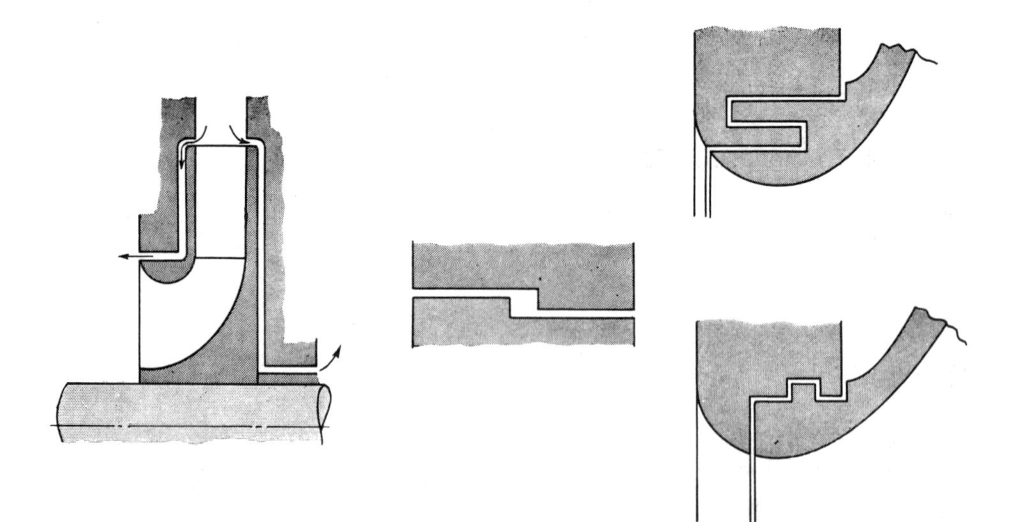

Fig. 21.11 Typical variants of sleeve seals.

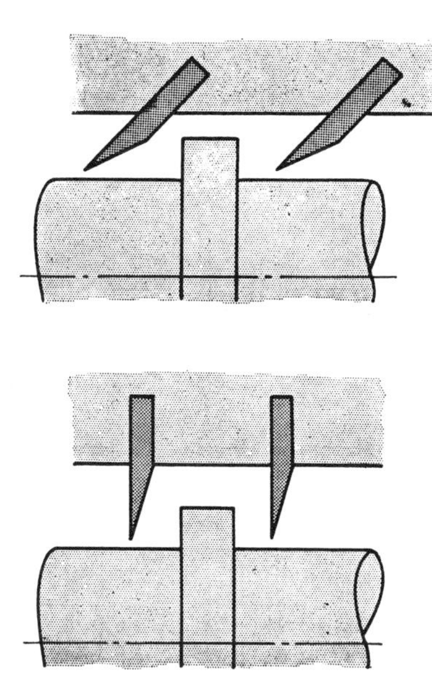

Fig. 21.12 Knife-edge labyrinths.

The head loss and flow equations in each figure are approximate and cannot take the place of tests. They show trends accurately, however, and are workable once you've applied correction factors for your own designs.

Briefly, the effective total resistance to flow, measured in feet of lost head, is the sum of three types of unit resistance: turbulent conversion of initial static pressure to velocity, wall friction, and turbulence caused by abrupt changes of section in the flow path.

The first two, H_V and H_F, are explained sufficiently by the equations and symbols in Table 21.1. The third type of resistance includes both H_E and H_S and is explained as follows: The first abrupt change of section is at the entrance, and head loss according to the conventional entrance equation is $H_E = 0.5V^2/2g$. The succeeding changes in section depend on the number of edges in each labyrinth. For Fig. 21.7, there are four edges per stage and a rough estimate of head loss $H_S = 4 \times 0.5V^2/2g = 2V^2/2g$. For Fig. 21.8 there are two edges per stage, and the relationship is $H_S = 1.0V^2/2g$. The summation is

$$H_T = H_V + H_F + H_E + \Sigma H_S \tag{21.1}$$

where the value of ΣH_S is explained in the equations and symbols table.

To express Eq. (21.1) in terms of velocity, insert the known expressions for head loss given in the table. This useful relationship results:

$$H_T = (V^2/2g)[1 + fL/2C + 0.5 + a(n - 1)] \tag{21.2}$$

where $a = 2$ for the seal of Fig. 21.7, and $a = 1$ for the seals of Figs. 21.8 and 21.9.

Equation (21.2) applies to gases and liquids. For gases the accuracy is problematical because velocity and density change from stage to stage, and you have to assume average values.

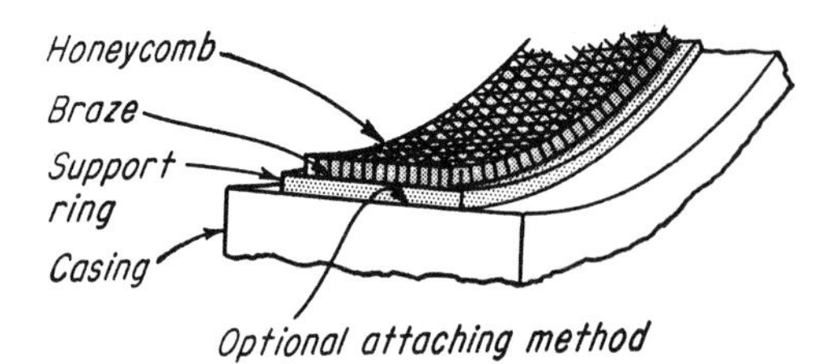

The approximate equation is:

$$H_T = \frac{V^2}{2g}[1.5 + 1.5(n - 1)]$$

where H_T = total head loss, ft, V = fluid velocity, ft/sec, and n = number of cell walls intercepted by the camber of the seal edge (length in direction of flow).

Fig. 21.13 Honeycomb labyrinth seal.

Leakage flow at any given point, for gas or liquid, is simply the average velocity V at that point times the cross-sectional flow area A: Q = VA.

Honeycomb seals (Fig. 21.13) are a special case. The approximate equation is

$$H_T = \frac{V^2}{2g}[1.5 + 1.5(n - 1)]$$

where H_T = total head loss, ft; V = fluid velocity, ft/sec; and n = number of cell walls intercepted by the camber of the seal edge (length in direction of flow).

One way to improve the accuracy of calculation is to use trial and error. Assume a Reynolds number, calculate the friction factor, and compute leakage flow. Using the computed flow, check out the assumed Reynolds number, correct it if necessary, and try again. The graph (Fig. 21.14) gives viscosity values for calculating Reynolds number.

Examples (Water)

1—Sleeve seal (Fig. 21.6)

Known pressure drop H_T = 300 psi = 693-ft head. D = 10 in., L = 2 in., C = 0.020 in., and Reynolds number $N_R = 2 \times 10^5 CV$. Velocity is assumed to be 180 ft/sec through the clearance, which is 85% of the free discharge velocity $\sqrt{2g \times 693}$.

Then $N_R = 2 \times 10^5 \times 0.020/12 \times 180 = 6 \times 10^4$, $f = 0.316/(6 \times 10^4)^{1/4}$, $V = \sqrt{62.4 \times 693/(1.5 + 0.020 \times 2/2 \times 0.02)}$ = 134 ft/sec, $A = 10\pi \times 0.020/144 = 0.00436$ ft^2, and Q = 134 × 0.00436 = 0.58 ft^3/sec.

2—Labyrinth seal (Fig. 21.8)

Known pressure drop H_T = 300 psi = 693 ft. D = 10 in., L = 0.125 in., C = 0.02 in., n = 8 stages. Assume the velocity is 35% of free discharge.

Velocity $V=0.35\sqrt{2g\times963}=74$ ft/sec. $N_R=2\times10^5\times0.02/12\times74=2.48\times10^4$. $f=0.316/(248\times10^4)^{1/4}=0.025$. $V=\sqrt{62.4\times693/(1.5+0.025\times0.125/2\times0.02+1.0\times7)}$ $=72$ ft/sec. $Q=72\times0.00436=0.314$ ft³/sec.

LIQUID MAGNETIC SEALS

In working with vacuum or pressure chambers, one problem occurs in providing a rotating shaft through the wall to operate equipment inside. Most dynamic seals, including O-rings, are bound to leak eventually. Nutating bellows-sealed shafting is vacuum-tight, but that's a complicated way to transmit rotary motion, and the torque and speed levels are limited. Magnetic couplings working through a solid wall also are vacuum-tight, but it's an expensive way to transmit torque.

A better way, according to Ferrofluidics Corp. (Burlington, MA) is to fill the gap between shaft and hole with magnetic fluid (Fig. 21.15) and use permanent magnets to hold the fluid in position.

Encyclopedias call the phenomenon ferrohydrodynamics. Extremely small magnetic particles (about 100 Angstroms) are colloidally dispersed in a carrier fluid. Quadrillions of them are crowded into one cubic centimeter, yet they are kept from clustering together by a monomolecular surface coating. The colloidal suspension retains its liquid properties even when highly magnetized, because the particles are almost molecular-size. Earlier types of magnetic clutch fluids solidify when magnetized.

A permanent magnetic field, applied as shown in the diagram, will hold the magnetic fluid in contact with the pole tips on the rotating shaft and the stationary surface of the magnetic pole blocks. The fluid is a complete barrier to gases attempting to pass in either direction yet offers only small viscous resistance to shaft rotation.

The company's product, called the Ferrometic rotary feed-through, has been extensively tested at many speeds, pressures, and temperatures. Claims are as follows: negligible leakage (less than 10^{-11} standard cc/sec of helium against a vacuum), speeds to 10,000 rpm (continuous), torques to 150 in.-lb, power to 30 hp, axial loading to 400 lb, and radial loading to 50 lb, cantilevered at the farthest point.

Tests show that essentially no leakage occurs at pressure differentials to 60 psi,

Table 21.1 Nomenclature for Positive-Clearance Seal Equations (See Figs. 21.6 through 21.13)

Symbol	Definition	Symbol	Definition
A	= area of annular clearance, ft² = $D\pi C$	H_F	= friction head loss, ft = $f\frac{L}{2C}\frac{V^2}{2g}$
C	= radial clearance, in.	H_T	= total head loss of all stages, ft $H_T = H_E + \Sigma H_S + H_V + H_F$
D	= diameter of shaft, in.	L	= total axial length of shaft seal, in.
f	= friction factor (dimensionless) $f=64/N_R$ for laminar flow ($N_R<2320$) $f=0.316/N_R{}^{0.25}$ for turbulent flow	n	= number of seals or teeth $n-1$ = number of stages or spaces
H_E	= entrance head loss, ft = $0.5\ V^2/2g$	N_R	= Reynolds number, dimensionless $N_R = 2CV/12\,\upsilon = 2\times10^5\,CV$ for 80 F water
H_S	= head loss for one stage, ft $H_S=2V^2/2g$ for Fig 2 (4 edges) $H_S=V^2/2g$ for Figs 3 and 4 (2 edges) $\Sigma H_S=H_S\,(n-1)$	Q	= flow, ft³/sec = VA
H_V	= velocity head, ft = $V^2/2g$	V	= fluid velocity, ft/sec
		υ	= kinematic viscosity, ft²/sec

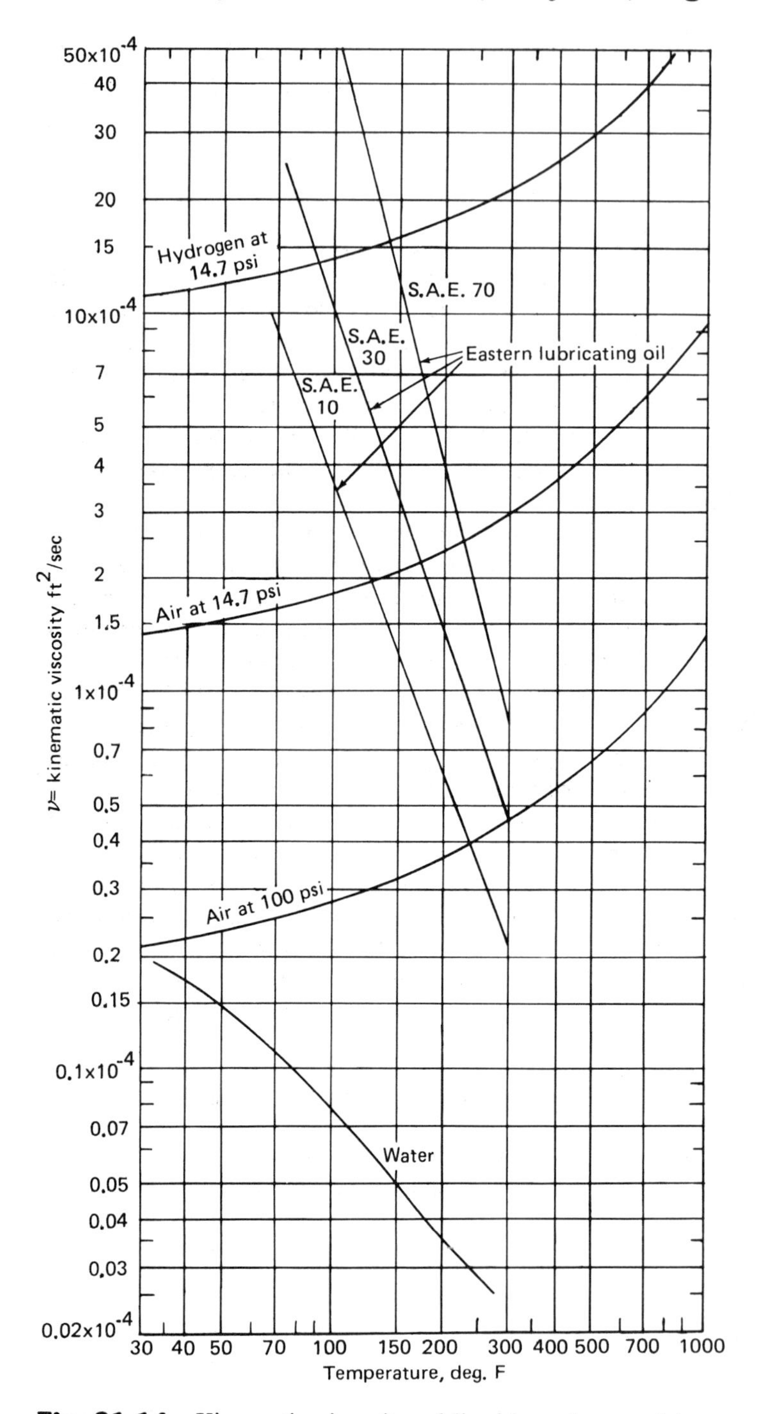

Fig. 21.14 Kinematic viscosity of liquids and gases (also see Chaps. 1 and 11).

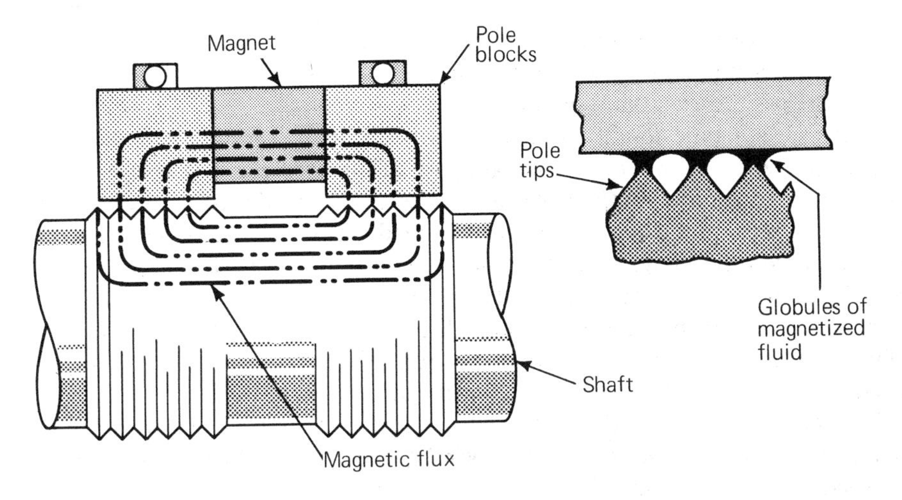

Fig. 21.15 Magnetic-liquid seal works by bridging the gap between the permanent-magnet poles of the stator and the pointed ridges of the rotating steel shaft.

and rotating life without leakage against a vacuum is better than 600 million revolutions. Some units have been tested at 3450 rpm for 3000 hr without leakage. Certain models have been run successfully at 60,000 rpm for long periods.

A variation on the magnetic seal is an experimental dual-mode seal that is partly magnetic and partly centrifugal (Fig. 21.16).

The magnetic fluid is a diester carrier liquid with a ferric oxide suspension, retained by a number of electrically magnetized poles (shoulders) in the rotor face. Each

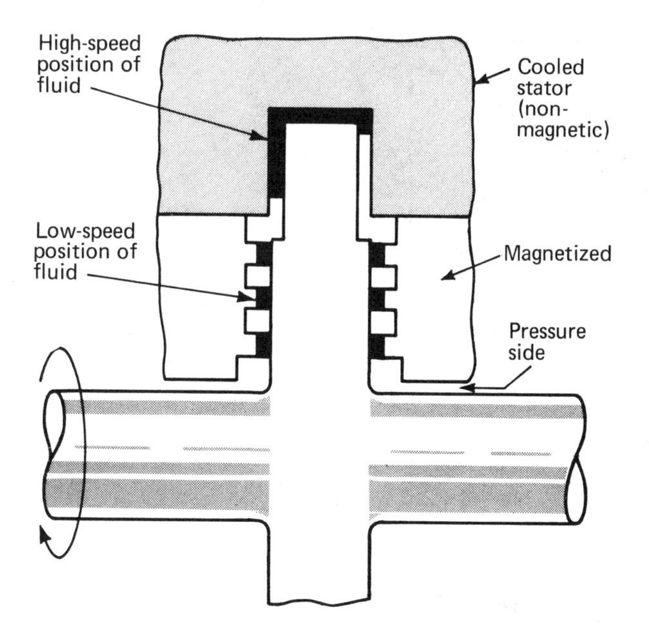

Fig. 21.16 Experimental combination magnetic-liquid static seal and centrifugal seal works at high speed and low speed. (Courtesy of MTI, Latham, NY.)

pole withstands about 4-psi differential at the magnetic intensity chosen: 7300 gauss to 8500 gauss at the pole tips, which approximates the residual flux density of samarium cobalt, the desired eventual pole material.

At zero speed (static condition) and at low to moderate speeds, the sealing occurs only at the pole tips, where beads of magnetic fluid are held. The level of pressure is determined by how many poles are in series (six in the rotor illustrated).

The centrifugal mode begins when rotor speed is sufficient to break loose the beads of fluid from the magnetized pole tips—about 11,000 rpm in the tests—and to throw the fluid into the outermost cavity. Above 11,000 rpm the sealing effect increases as the square of rpm, reaching 30-psi differential at 13,000 rpm (Fig. 21.17).

The effect is reversible, but some of the fluid fails to return to the magnetized poles. The static sealing effect drops to 12.5 psi and stabilizes.

The fluid is lost because the magnetizing forces aren't strong enough to draw against centrifugal force. It was found that several of the poles would lose beads, usually on the same side of the rotor.

To test the concept, a 75-hp adjustable-speed motor supplied the motive power, and air pressure supplied the seal test load. A system of sensors detected the transfer of fluid from the static to the centrifugal mode and the advent of seal breakdown.

No airflow through the seal was detected until seal breakdown actually occurred. Fluid transfer was readily detected by sensing temperatures in the centrifugal chamber: The liquid always got hot because of the shearing action.

Cooling water inside the housing prevented the liquid from getting too hot. This will be a design problem at the higher speeds.

Other design problems to be solved are the clearance dimensions and the centrifugal-to-static transfer. Clearances must be close enough to develop strong magnetic

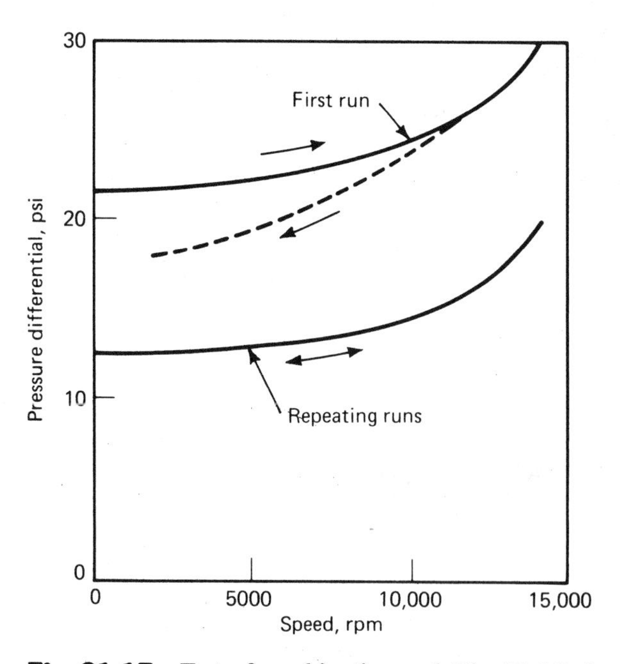

Fig. 21.17 Test of combination seal (Fig. 21.16) shows that performance diminishes somewhat in operation.

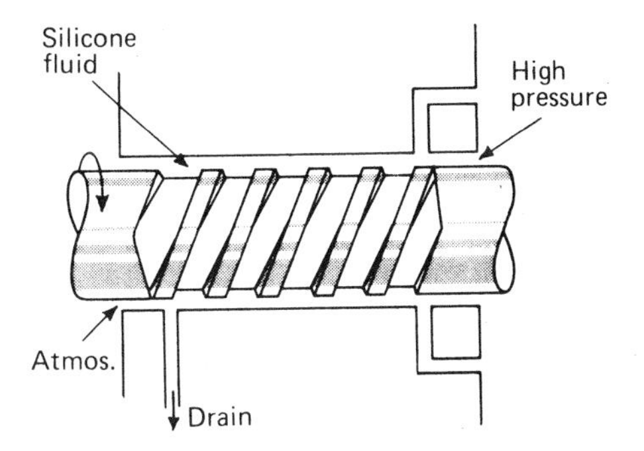

Fig. 21.18 Self-pumping shaft seal uses helix to force fluid back against pressure. (Courtesy of General Motors Research Laboratories, Warren, MI.)

forces to hold the sealing beads yet must be wide enough to prevent rubbing under a wide range of temperature, pressure, and speed.

Clearance volumes at the static and centrifugal areas must match so that each can hold all of the fluid.

The biggest problem is to force the magnetic fluid to return to the static position while the rotor is still at moderate speed. The fluid will remain at the outer diameter until a very low speed.

SELF-PUMPING SEALS

Shaft rotation can be exploited to keep fluids from leaking past a seal. A spiral or helix, machined or etched into the surface of a seal (Figs. 21.4 and 21.18), works as a pump and develops a pressure head to force the fluid back against system pressure. Versions may be found in centrifugal pump applications and in automotive transmissions, and they work.

The chief problem is that a minimum shaft speed is required. Below that speed, the pumping effect is not sufficient to oppose the leakage pressure.

22

Performance and Testing

This chapter summarizes the levels of expected performance in all fluid power systems, explains how to test for them, and adds details in certain critical areas: flow measurement, remote testing, pressure-fatigue life, noise reduction, and speed control.

BASIC PERFORMANCE LEVELS

Key parameters include pressure, temperature, contamination tolerance, and dynamics of flow control. Most of the discussions will deal with mobile equipment (the proving ground for many engineered systems) but will apply in degree to machine tools, aircraft, and processes.

Pressures for typical mobile hydraulic systems (transportation, mining, agricultural, construction, road-building, and materials handling, for example) range from less than 1000 psig to over 6000 psig in normal steady operation, with somewhat higher values allowed intermittently.

Predictions

A decade or so hence, according to studies (Fig. 22.1) made during the 1970s at MSOE (Milwaukee School of Engineering), many engineers predict a considerable rise in operating pressures. Hydrostatic transmissions probably will run at 7000 psi to improve efficiency and performance, and some predict 8000 to 10,000 psi. The problem is that a crossover point in total system weight vs. operating pressure occurs not much over 8000 psi because of thicker walls in piping and components. Also, leakage, heat generation, component fatigue, cylinder column loading, and reliability become design problems. Liability suits might crop up because the advanced technology is beyond the ken of the average user and service support.

New higher-strength materials could change the whole picture. A number of successful applications of 8000- to 10,000-psi systems in workhorse mobile equipment

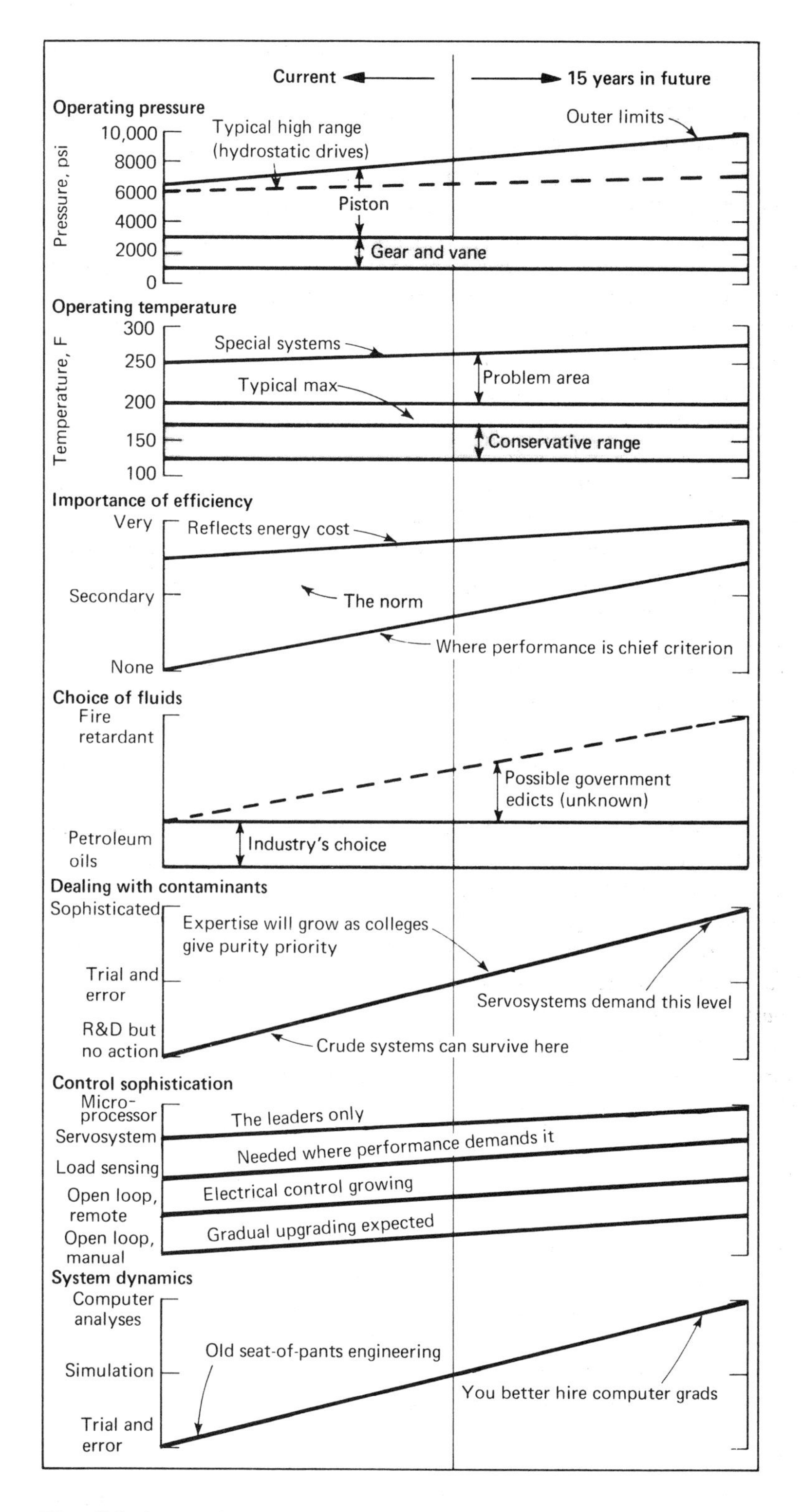

Fig. 22.1 Performance survey (1978): mobile equipment users reported performance and made predictions.

would serve as technological proofs. Innovation then would be under less threat, and higher efficiency designs would be encouraged.

Europeans used to be somewhat ahead of U.S. engineers in high-pressure systems but have backed down closer to the U.S. levels because of the need for better reliability and the danger of confronting liability suits in the U.S. markets.

Temperature in any hydraulic system becomes a concern above 200 F, and even with improved fluids, hoses, and seals, few engineers predict more than 250 F operating temperatures. Problems include deterioration of seals and hoses, decreased lubricity, fluid breakdown, operator hazards, increased leakage, wasted heat energy, thermal expansion, and varnishing of critical surfaces. Oil coolers can keep the problem somewhat under control.

Efficiency becomes important as fuel costs rise or where fuel storage is limited—as on aircraft. Many designers ignore efficiency in the quest for performance, reliability, and price but admit that pump and system efficiency will have far more importance during the coming years. Gear pumps in the 83 to 87% efficiency range are expected. Some designers even see efficiencies up to 92%. Vane pumps will probably hover closer to the 80% efficiency range. Piston types might climb up into the 88 to 92% range or higher.

Fire-retardant fluids are a necessary expense (and problem) in many applications, but the consensus in the mobile hydraulics industry is that they wish it were all a bad dream and would go away. A quote from the mining segment probably epitomizes the general feeling: "The very real apprehension is that we may suddenly face a government edict to use fire-retardant fluids across the board with an impossibly short lead time." Even water-based fluids have a myriad of problems (Chap. 1).

Contaminants are a special problem. Everybody knows something about contamination, and there is sufficient technology available to pinpoint the sources of contamination, the kinds of damage that will result, and the means of reducing them. However, the cure is rigorous attention to overwhelming detail during every step of design, manufacture, and operation of the equipment (Chap. 17), and the world doesn't seem ready for this.

The answer is a massive education program in every area from the college classroom to the shop floor, so that every designer and user of hydraulic and pneumatic equipment is "Mr. Clean" from the start.

System Dynamics

Designers are moving gradually away from seat-of-the-pants designs where the chief goal has been to get everything working and shipped. Figure 22.1 shows the logical progression from manual open-loop control to the present ultimate of microprocessor control of all machine functions from a central panel. Pressure-compensated variable-displacement pumps (piston type) are accepted as a present and future component.

Load-sensing controls are popular now and will become more so. They modulate pump flow in response to pressure, flow, or power at the load and will save a lot of energy.

Electrically modulated control valves (open loop) are not universally used, but there is a variety of lower-cost linear solenoids heading onto the market. Full-fledged electrohydraulic servosystems are the next step after load-sensing systems and electrically modulated valves. Cost is an obstacle, but wider demand will lower prices.

The most profound change coming about is the integration of the microprocessor into fluid power control. The art of applying microprocessors is difficult for engineers who are used to simple relay logic, but learn it they must or hire someone to help. The alternative is obsolescence.

The computer has still another role to play in fluid power: original system design and simulation. Universities and industry have developed math models for many hydraulic circuit functions; performance prediction will become a reality when enough computer-oriented engineers drift into the fluid power field and exert their influence.

INSTRUMENTS AND MEASUREMENTS

This section takes an overview of common test instruments—flowmeters, manometers, pressure gages, thermometers, tachometers, and strain gages—and evaluates accuracy and performance of fluid power instrumentation in general.

Perceived vs. Real Accuracy

The Milwaukee School of Engineering, in 1974, completed a study to ascertain if typical engineers really know how accurate their instruments are. Conclusion from the questionnaires: They don't (Fig. 22.2). Almost all the bars on that graph are too far left and are not based on real-life calibrations.

The results were an eye-opener, putting the majority of the companies into amateur status when it comes to measuring true flow, pressure, temperature, power, torque, rpm, force, viscosity, sound, and all the related parameters. (An exceptional few companies are doing an outstanding job of upgrading testing in general: We are not addressing this section to them.)

It's not that the final products won't do the job—they probably will. It's just that the test data are not as accurate as they could be, and too much reliance is placed in the skilled test technicians who "sense" when the test data are OK.

Electronic instruments have made inroads, but the majority of the measurements still are made with mechanical devices such as bourdon-tube gages, manometers, and thermometers. When engineers were asked for estimates of accuracy, they stated very optimistic levels but had relatively few programs for verifying the assumptions. These additional facts came to light: Piezometer pressure taps are not widely used, and only for low-pressure measurements. Mechanical gages are popular, and some labs insist that a 1% accurate gage retains that relative accuracy at the lower scale as well as at the upper—a false assumption but temporarily comforting when you are trying to meet test specs.

Complex instruments such as flow, torque, and speed detectors are almost always exploited throughout the whole range, even though better accuracy would result if the instruments were changed each time a lower scale was needed. Flowmeter calibration, because of cost, is almost nonexistent.

Traceability (verification of instrument accuracy back to an NBS standard) is not widely understood or applied, yet top accuracy often is assumed.

Another culprit is faulty measurement of electric motor torque. Many test engineers rely on calibrated induction motor performance graphs for computation, but such data are based on two test points: locked-rotor and no-load, with an assumed characteristic shape. Good accuracy is not possible.

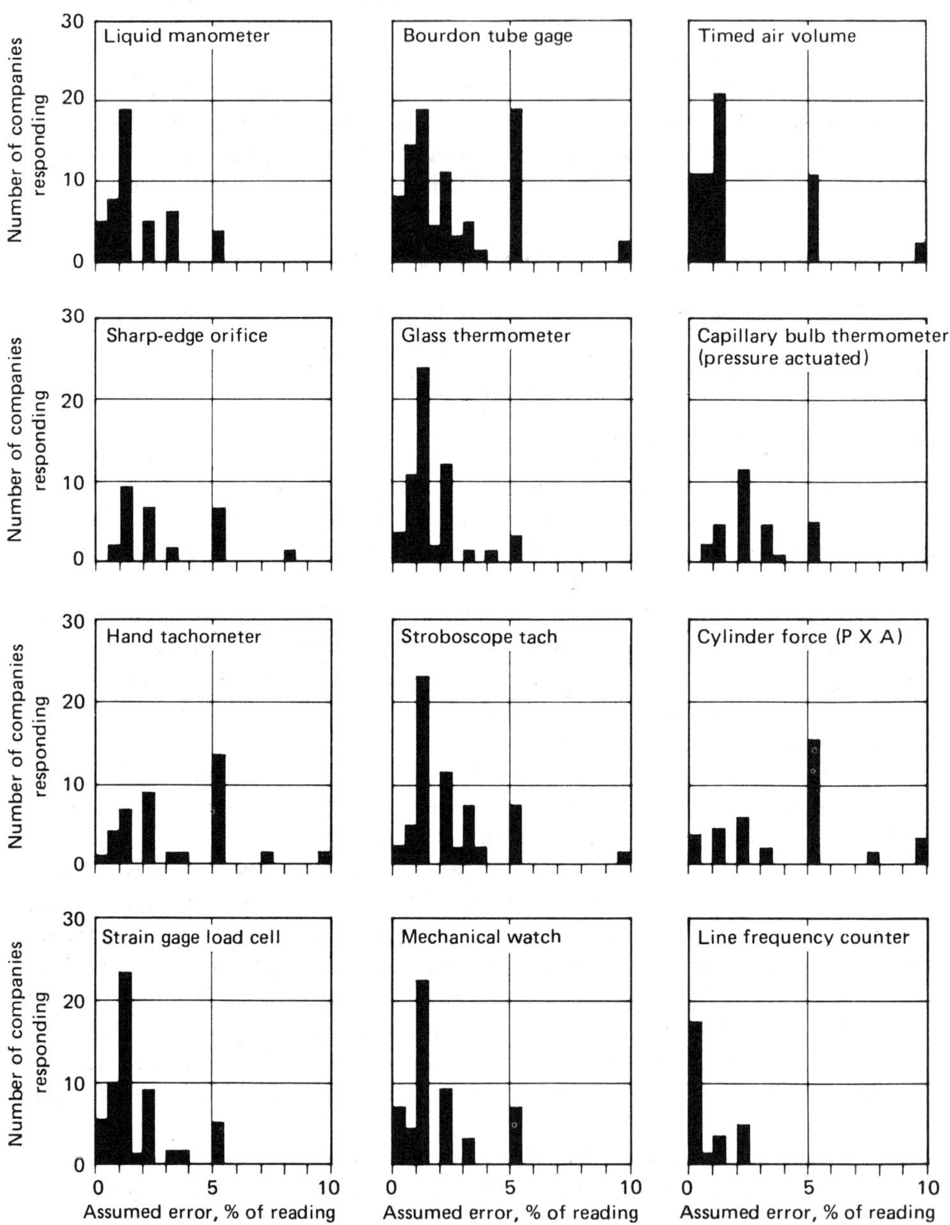

Fig. 22.2 Accuracy survey (1974): test engineers were too optimistic in judging calibration.

Pulsating flows also are a problem. Most hydraulic systems have ripples that sometimes reach 20% of the mean value, and it is difficult to select a steady-state value to use in calculations of power and efficiency. You really need high-response instrumentation that can follow the wave shape of the pulsation and continuous digitizing techniques to yield data that can be analyzed by a computer.

Shortcut methods are available if moderate accuracy will suffice. The most common method is to "smooth out" the readings by selecting low-response instruments. But there's a question: Are they "average" or "rms" or something else? Experience has shown that the final result often is random and not necessarily predictable. A better technique is to use high-response instruments such as pressure transducers in conjunction with a shaft-operated pulse counter set for discrete intervals of time. Each reading is converted to a frequency-modulated signal, and the data are combined to give a good average value for the selected time interval.

One way to achieve reliable test procedures is to send a test component to several labs (round-robin testing) and require that the results be consistent. Figure 22.3, for an hydraulic pump, shows how wide the deviations can become if the procedures are wrong.

Another way is repeated testing to achieve a reliable average. The low-speed shaft torque example (Fig. 22.4) proved consistent when 40 random positions were recorded.

Computers can help by analyzing data in real time, automatically correcting for known instrument errors. Some of the most successful test programs have been in the measurement of contamination in oil (Chap. 17) and pressure-induced fatigue in valves and other components (see a following section for details).

Flowmeters

Figures 22.5 through 22.10 show examples of the flowmetering art. Each is self-explanatory; choice depends on the desired flow rate and accuracy.

Conventional rotary meters, orifices, and tapered-tube rotameters are not pictured; but the principles were touched upon in Chapters 9 and 16 (pressure losses), and Chapter 6 (hydraulic motors). Positive-displacement meters such as lobed rotors can be the most accurate because the actual flow is directly measured. Area-differential meters such as venturis, tapered-tube rotameters, and orifices are simplest but require careful calibration and are limited to a narrow flow range for each size of meter. In between are turbine meters (velocity sensing), charged-particle meters (the fluid

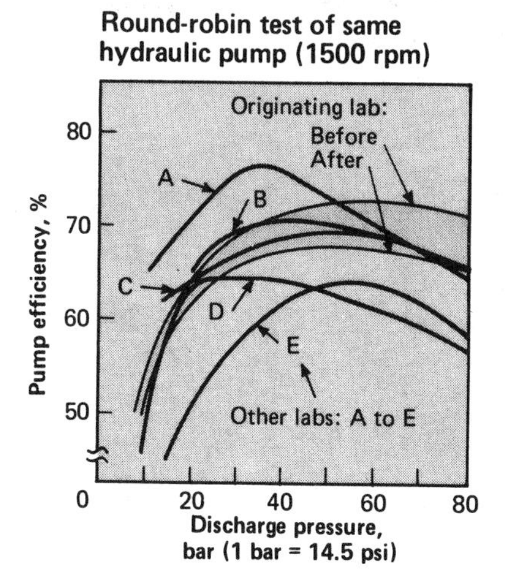

Fig. 22.3 Repeatability round robin: same pump sent to six labs, with inconsistent results.

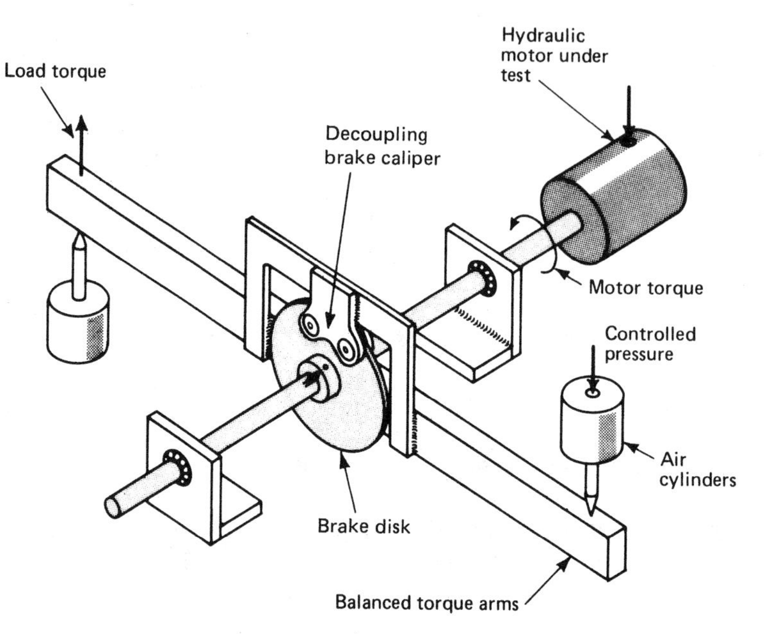

Fig. 22.4 True starting torque and not locked-shaft torque is measured with rig that allows some shaft rotation without altering effective moment significantly (Milwaukee School of Engineering).

must contain the particles), sonic meters (round-trip travel time of an oblique sound wave is influenced by stream velocity), and thermal meters (mass flow of fluid measured with calibrated temperature sensors).

Density, Viscosity, Mass

Closely related to flowmeters are special meters for sensing density, viscosity, and mass flow rate. Figures 22.11 through 22.13 are some examples.

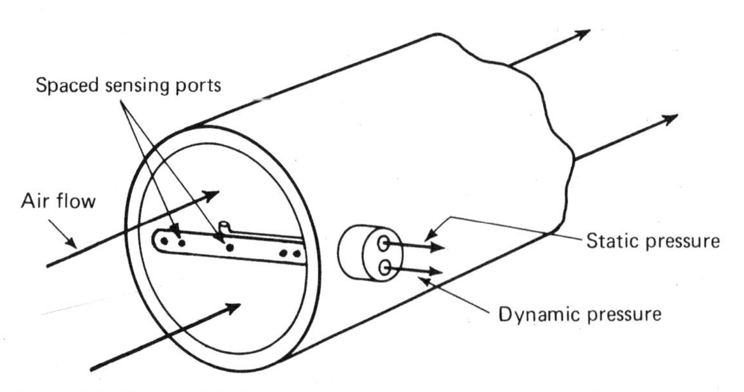

Fig. 22.5 Multiple pitot tube measures average flow in any fluid if dynamic and static ports are spaced to take into account parabolic profile of stream.

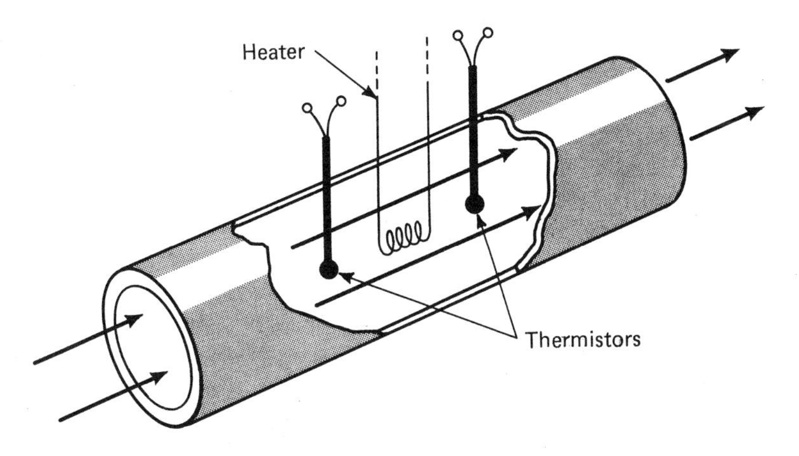

Fig. 22.6 Thermal-type flow sensor measures change in temperature between two thermistors caused by change in fluid flow against heater.

SPECIAL TEST TECHNIQUES

Special problems occur in any area of testing. We picked three typical categories for comment: remote measurement, digital analysis, and tests of flexible hose.

Remote Testing

Electronic engineers have provided tools for any desired level of remote and automatic testing. But sensitive readings are greatly influenced by external disturbances. For example, at 125 C a typical thermocouple generates barely 5 or 6 mV, which can easily be distorted or lost if transmitted over any distance. Similar problems are presented by most other low-energy sensors, including pressure transducers, angle indicators, strain gages, liquid level detectors, thermal bridges, and proximity devices.

Extreme system-design measures have been customary, involving highly shielded wiring, complex signal processing, and high-gain amplification of analog signals to overcome effects of ground loops and local electrical interference. The problem for years was handled on a custom-design basis, with standard sensors at one end, elaborate readout instruments at the other, and do-it-yourself amplifying, protecting, and

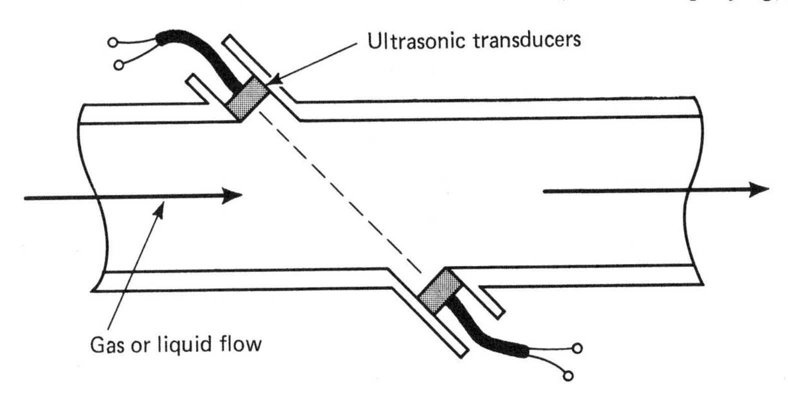

Fig. 22.7 Ultrasonic flowmeter sends and receives signals across stream: difference in time lapse is measure of volume flow.

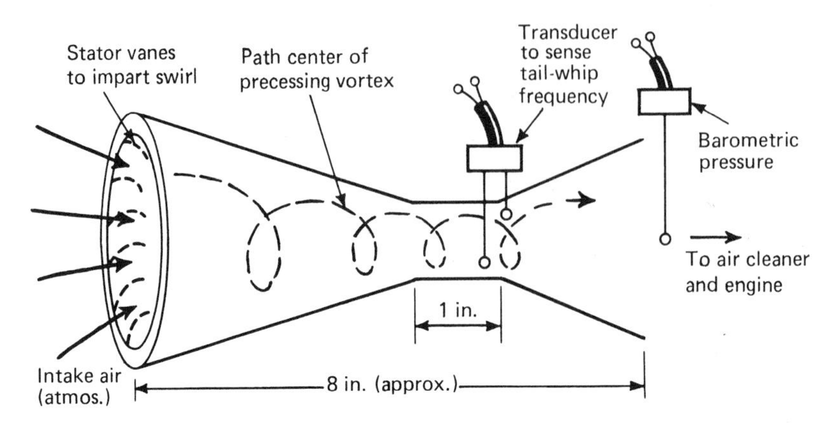

Fig. 22.8 Precessing vortex flowmeter has tailwhip frequency proportional to flow.

transmitting circuitry in between. Only highly skilled systems engineers and technicians could do it.

Now, solid-state designers have created miniature modules (Fig. 22.14) that conver the sensor signals at their source to strong, standard digital codes that are not affected by local electrical interference and can be read directly by teletypewriter instruments, or computers.

The modules work in two directions. They send sensor signal codes to the teletypewriter and return instructions in the same code to the sensors or to associated control devices such as digital or analog operators for valves, motors, and pumps.

The teletypewriter (or a computer, if you prefer) automatically interrogates the right sensor when you push the appropriate keys and types out the response. There's already an established code for this, called ASCII for American Standard Code for Information Interchange (Fig. 22.15 and Tables 22.1 and 22.2).

Simple twisted-wire pairs can transmit the coded signals without interference because the pulses are 20-mA current bursts rather than voltage signals. Years of experience worldwide have shown this to be so. Only lightning and massive power surges (megawatts) will distort coded current pulses of this magnitude. Furthermore, opti-

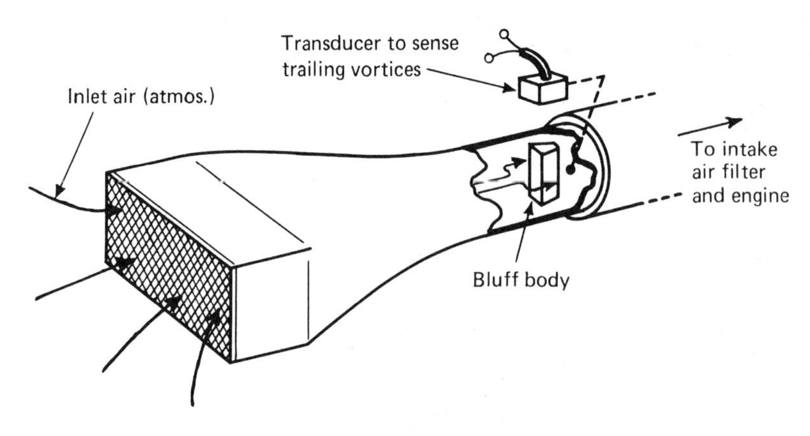

Fig. 22.9 Trailing vortices flowmeter has bluff body that sheds vortices at frequency proportional to air flow.

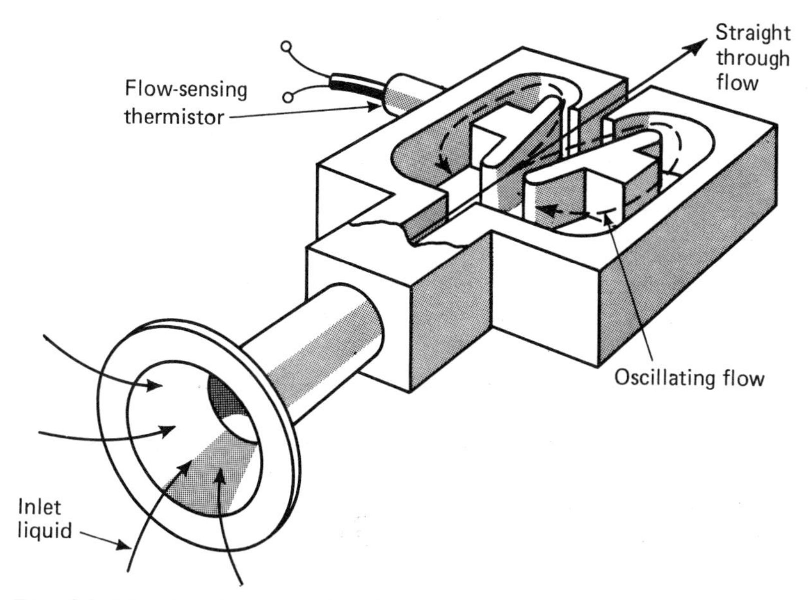

Fig. 22.10 Fluidic wall-effect flowmeter generates branch flow that oscillates with frequency proportional to volume flow.

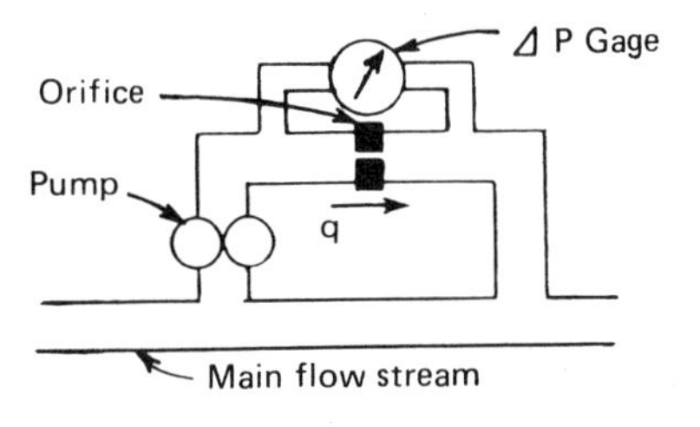

Fig. 22.11 Density sensor: When pump flow is held constant, density of fluid is proportional to orifice pressure drop.

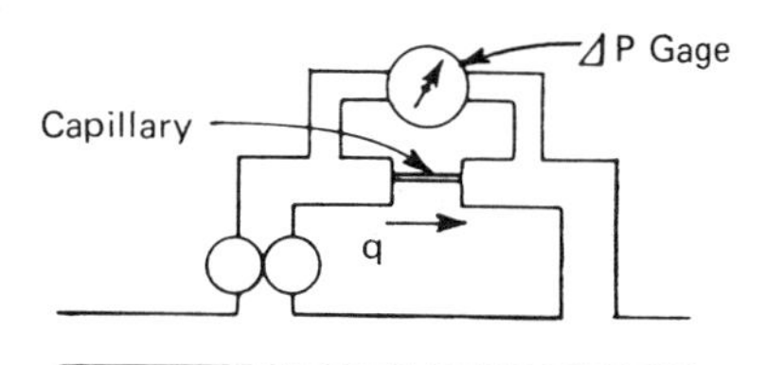

Fig. 22.12 Viscosity sensor: When pump flow is held constant, viscosity of fluid is proportional to capillary pressure drop.

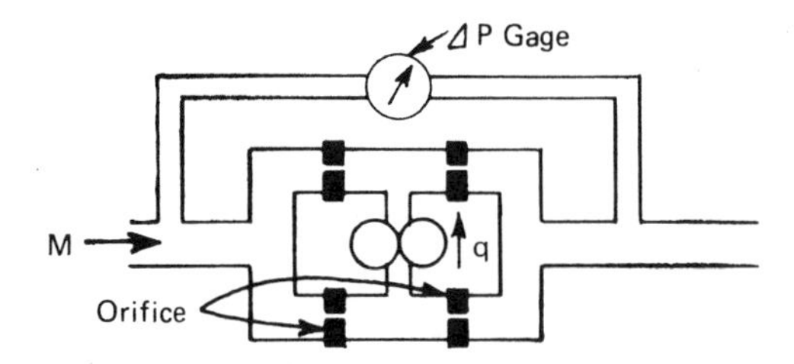

Fig. 22.13 Mass flow sensor: When pump flow is held constant, net mass flow through orifices tends to vary directly with overall pressure drop.

cal isolation with LEDs (light-emitting diodes) separates the 20-mA pulse circuits from the module circuits (Fig. 22.16).

Three modules comprise a typical basic system. The first module, more or less standard with every electronic manufacturer, is either an A/D (analog-to-digital) converter or a digital panel meter. It is located next to each sensor or control (at the right in Fig. 22.14) and converts analog voltage to digital form (binary code).

The second module is either a transmitter or a receiver. The transmitter transmits to the teletypewriter or computer, and the receiver receives from the teletypewriter or computer, depending on which you want. The transmitter converts the binary code from the A/D converter into a string of coded pulses suitable for the teletype-

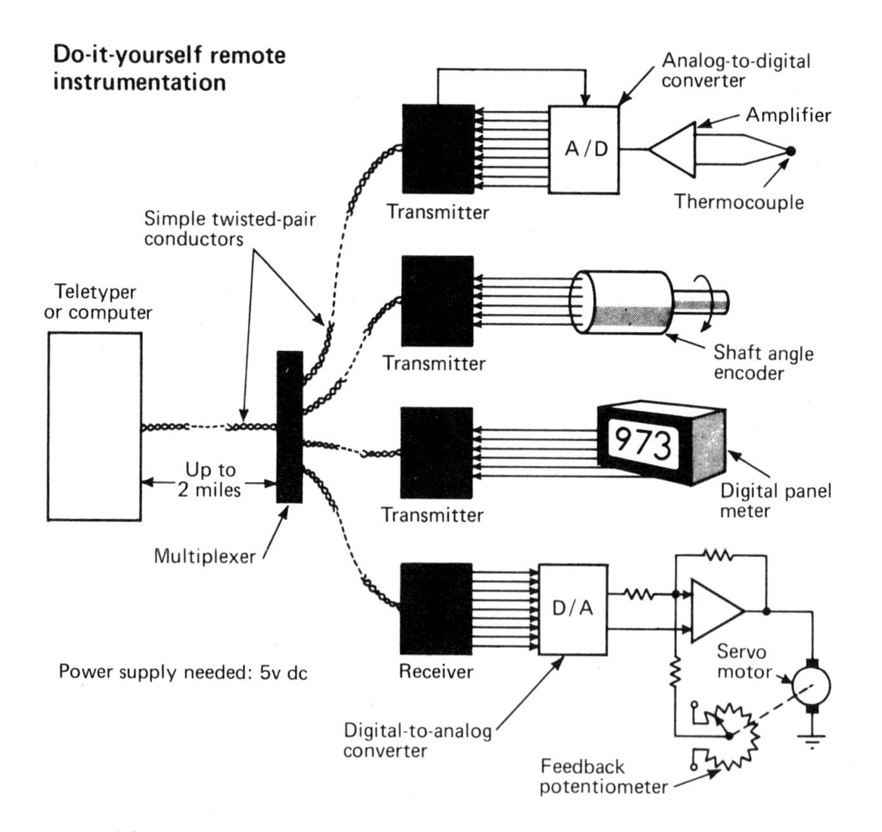

Fig. 22.14 Pulsed digital code is easier than analog to transmit long distances.

Pulses that comprise teletype code

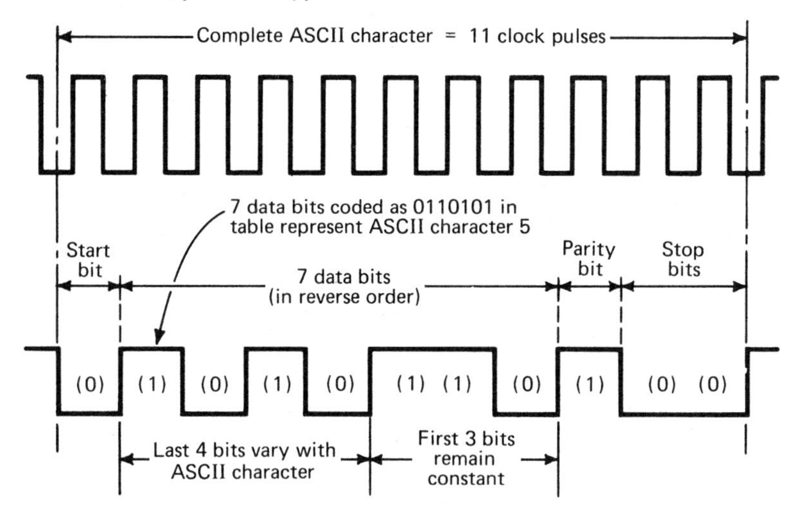

Fig. 22.15 Teletypewriter code has well-defined, hard-to-distort pulses.

writer. The receiver does the opposite: It receives teletypewriter pulses and then sends them on as binary code to the sensors or controls.

The third module is a multiplexer and is used only if more than a single sensor or device is being monitored. Multiplexing is the sharing of many signals in a single device by accepting them sequentially. Only one multiplexer is needed for a large number of transmitters and receivers.

Table 22.1 The ASCII Code is a Series of Binary Numbers

Numeral	ASCII code
0	1 0 1 1 0 0 0 0
1	1 0 1 1 0 0 0 1
2	1 0 1 1 0 0 1 0
3	1 0 1 1 0 0 1 1
4	1 0 1 1 0 1 0 0
5	1 0 1 1 0 1 0 1
6	1 0 1 1 0 1 1 0
7	1 0 1 1 0 1 1 1
8	1 0 1 1 1 0 0 0
9	1 0 1 1 1 0 0 1

These do not vary for numerals

Table 22.2 Typewriter Characters Can Designate Additional Information

Character	ASCII code	Transmitter response	Receiver response	Multiplexer response
?	0 1 1 1 1 1 1	Update A/D and transmit result	Control pulse	–
/	0 1 0 1 1 1 1	Transmit register content	Control pulse	–
=	0 1 1 1 1 0 1	Control pulse	Register cleared	–
#	0 1 0 0 0 1 1	–	–	Address alert

None of the modules is expensive, at least compared with older methods of monitoring multiple remote data sources.

Digital Analysis

The microprocessor revolution has given impetus to versatile instruments, such as digital oscilloscopes that store pump dynamic pressure waveforms in memory, to be analyzed later at length. An operator's key can call for the measurement of peak-to-peak pressure differential between any two points or an average of accumulated data. Other keystrokes will order the computer to make calculations of power and plot the results—all while the test proceeds.

Hose Tests

Hose has to be flexible yet tough enough to stand more abuse than almost any other component.

One school of thought (the hose user) wants hose testing to match the expected operating conditions—including flexing combined with pulsing pressure—as closely as possible. Another school (the hose maker) prefers universal tests not tied down to specific applications but designed to ensure uniform hose quality.

Whether or not agreement is ever reached, all the operating conditions of a hose

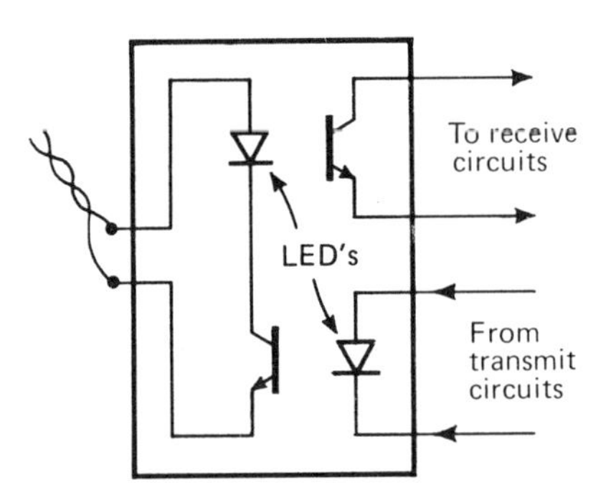

Fig. 22.16 Optical isolation keeps circuits electrically separated.

ought to be considered when developing a test. The key ones are temperature, pressure, flexing, abrasion, compatibility with oil, and durability.

SAE hose-test standards run most of the gamut, with tests for pressure impulse (Fig. 22.17), static burst, cold flex, shortening under pressure, oil and ozone resistance, vacuum, and electrical conductivity. Most of the tests, however, are meant to determine a single quality of a single hose, or in some cases a few combinations.

But what's really needed is a selective combination of tests, one after the other or simultaneously, on the same hose sample. For example, why not age the hose before running impulse tests? Hose manufacturers have begun to do this on a systematic basis.

Needed is a series of tests, tailored to the application, on each hose. One example of a sequence is (1) change in length, (2) proof pressure, (3) aging in oil or in air, (4) another proof test, (5) cold flex, (6) impulse pressure, and (7) impulse and flex simultaneously. Some of the samples ought to be tested to destruction, particularly during the impulse and impulse-flex steps, so that ultimate capability is known.

The accompanying Figs. 22.18 through 22.20 are self-explained examples of needed tests.

PRESSURE FATIGUE STANDARDS

Two kinds of pressure ratings are required: static and cyclic. The first is a simple gradual test to determine if the rated static pressure can cause structural failure, surface cracking, or sufficient deformation of the pressure envelope to induce malfunction or allow excessive leakage. It's a straightforward problem.

Cyclic testing is another matter, and not until the early 1970s was it addressed in a coordinated manner. The work is still going on, and a standard technique has been developed under the auspices of the NFPA (National Fluid Power Assoc.) in Milwaukee, Wisconsin.

A task force of about three dozen engineers from two dozen major producers of fluid power components in the NFPA completed the technical analysis and sug-

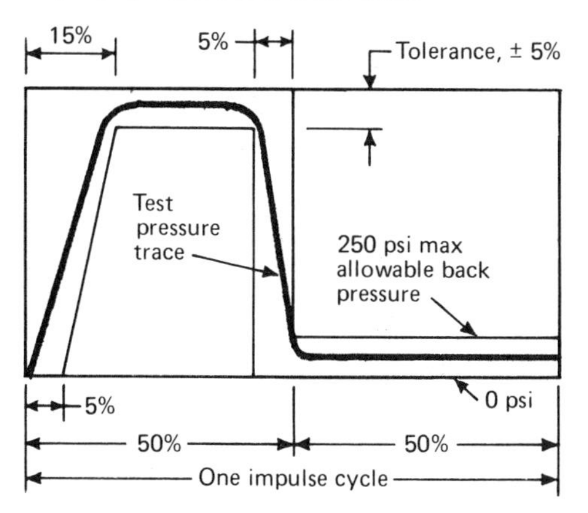

Fig. 22.17 Test pressure pulses are defined by industry standards.

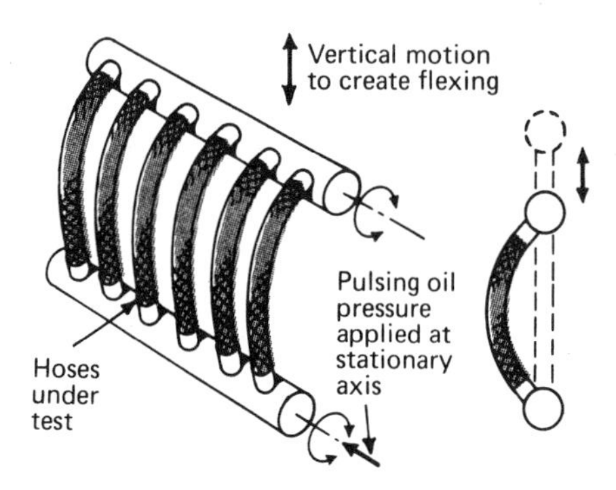

Fig. 22.18 Flexing apparatus tests pressure and flexing simultaneously.

gested a test procedure (Fig. 22.21); and the NFPA issued Recommended Standard NFPA/T2.6.1-1974 (R1982). The force was cochaired by Reed Schroeder, president of Schroeder Brothers, McKees Rocks, Pennsylvania, and Stan Skaistis, director of R&D, Sperry Vickers, Troy, Michigan.

Burst, Crack, or Leak

The purpose of the document was to provide a standard method for verifying the pressure rating of any fluid power component's pressurized envelope. The proposed pressure ratings will supplement but not replace those now in use that are based on per-

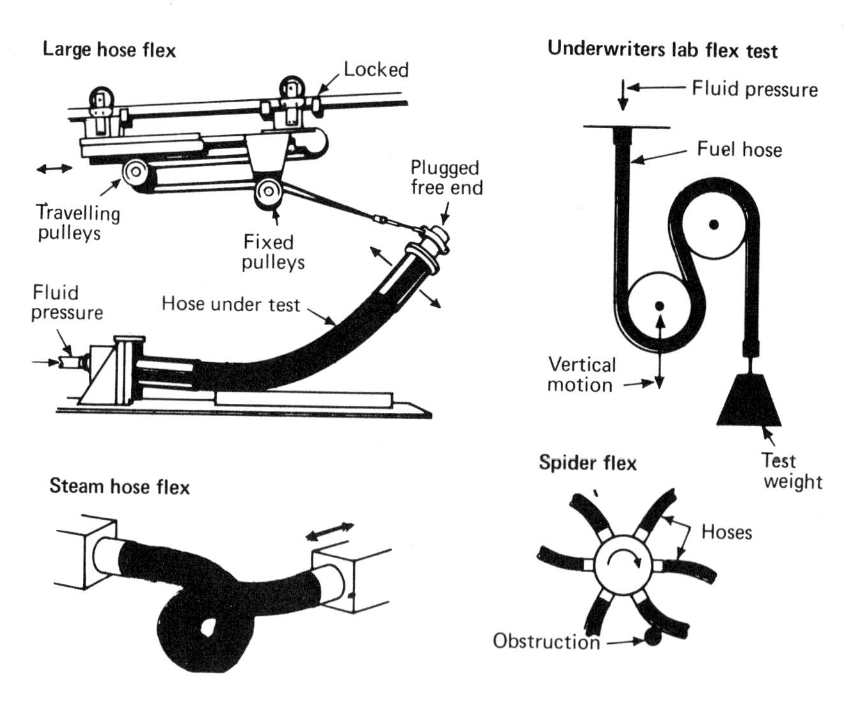

Fig. 22.19 Special tests examine specific flexing performance criteria (Gates Rubber, Denver, CO).

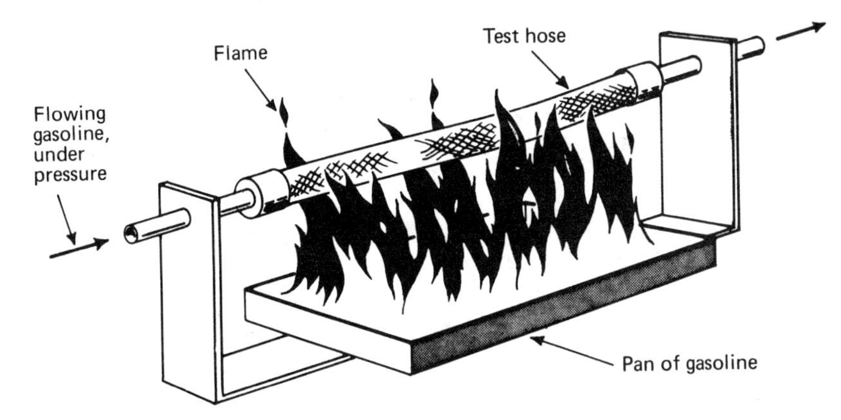

Fig. 22.20 Severe flame test is designed to qualify fuel hose.

formance. Only metal devices were included in the initial work because there was a large body of static and dynamic stress data available on metals.

Regardless of the material, any verification method has to be partly statistical because it is impossible to test every product. The committee recognized that strengths in a batch of supposedly identical parts fluctuate over a wide range because of dimensional and material variations. If the test engineer happens to pick one of the strongest and test it, he or she could be misled into concluding that all are equally strong. By using the large volume of material strength data found in government reports, the committee found a way to keep risks at an acceptable level in verifying the strength of parts without being too conservative.

Help was sought among experts in hydraulics, pneumatics, stress analysis, metallurgy, mathematics, and statistics. The final method recognizes the importance of fatigue by incorporating a zero-to-maximum pressurization, repeated 10^7 times. The number 10^7 was selected because experience has shown that if the test specimen will last that long, it will go on to have extremely long life (see Fig. 22.22: S-N diagram).

One alternative is allowed: The test may be continued to only 10^6 cycles but at an appropriately higher pressure to compensate for the shorter number of cycles. One of the problems that the task force encountered was determining what that higher pressure must be. The results are in the accompanying text and in Table 22.3.

Fewer cycles than 10^6 are not allowed, even with correction factors, because of several metallurgical phenomena extensively documented during the past several decades and applied to fluid power pressure containers more recently. Here's one of the most important of the phenomena.

False Strength

Cycling above certain critical stress levels for each material will cause localized yielding and subsequent compressive residual stresses that strengthen it. Failure is delayed—dramatically in many instances. This in itself is not bad, but it gives misleading life predictions for some future component in service where the cycling stresses are lower but occur more often.

The dramatic effect of yielding on metal surfaces was disclosed in the article *Shot Peening* (*Product Engineering,* Aug. 1972, p. 31). Residual compressive stresses

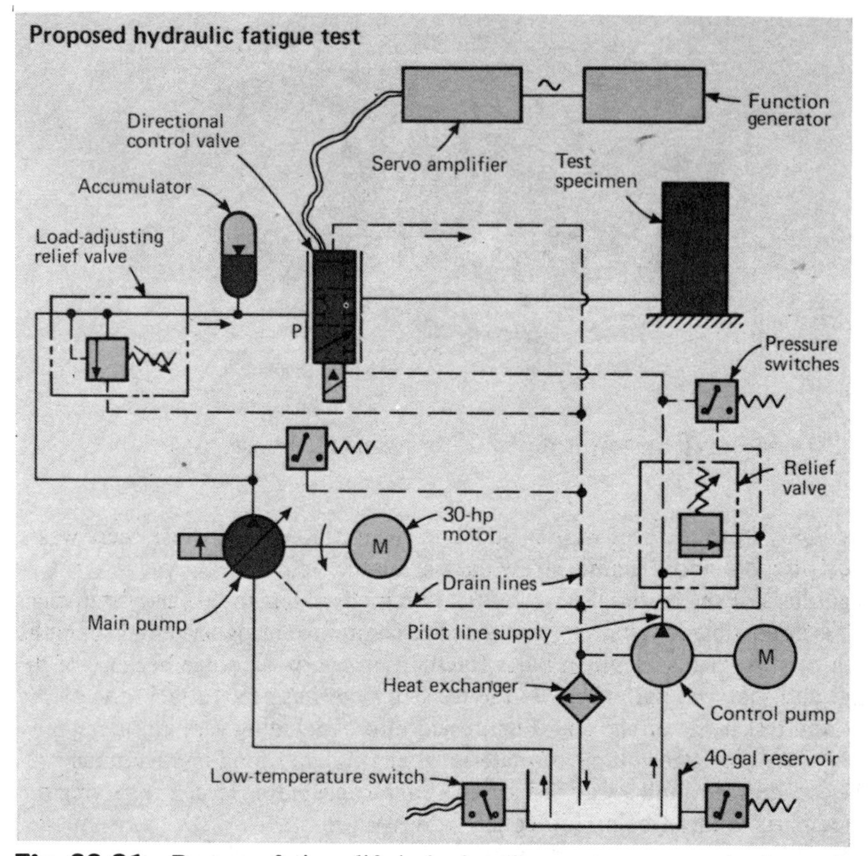

Fig. 22.21 Pressure fatigue life in hydraulic components can be determined with this test setup.

are induced below the surface of the metal to 0.010- to 0.040-in. depth, boosting fatigue strength greatly. Cycling of pressure isn't exactly the same, but the results are alike.

As an example, say a pump is rated to withstand 10^7 cycles of 10,000-psi internal pressure. The manufacturer decides, on the basis of limited knowledge, to shorten the test time by increasing the cycling pressure to 15,000 psi and ending the test at 10^5 cycles. He could have based this on some previous data showing that if simple specimens will survive 10^5 cycle at 150% of a given load, they will survive 10^7 cycles at the given load. So, the manufacturer sees a chance to save weeks of test time.

The premise could be false, because yielding might occur in certain highly stressed localized portions of the product as fluid pressure approaches 15,000 psi. This phenomenon is more likely to occur if the surfaces are rough or notched. Since the high-stressed regions are restrained by the nearby lower-stressed material, this yielding produces favorable, compressive residual stresses.

Subsequent alternating from 0- to 15,000-psi fluid pressure then produces a less severe stressing than would occur if the yielding had not occurred previously. As a result, the fatigue life of the part with this yielding will be extended and might exceed 10^5 cycles.

In an alternate test at 10,000 psi on an identical pump, the beneficial yielding

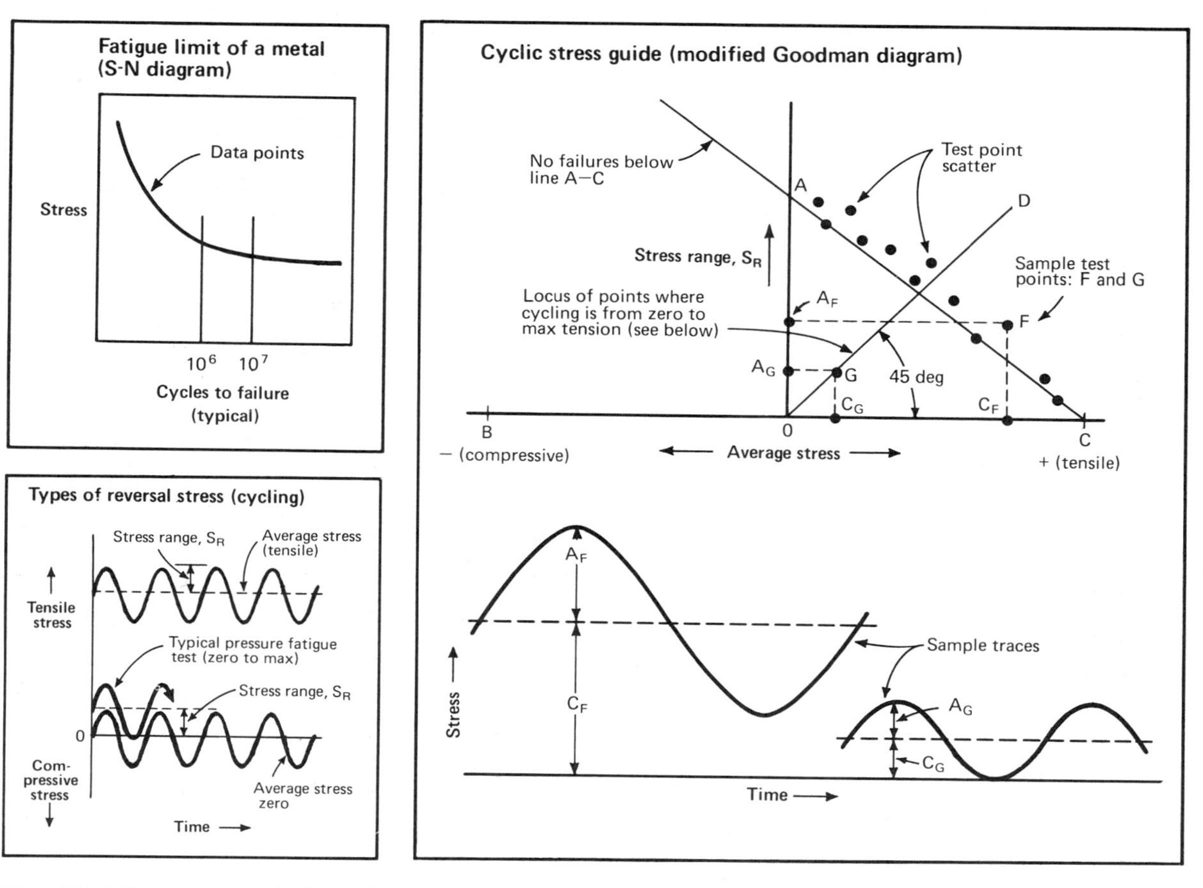

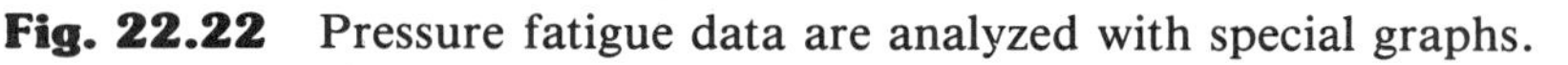

Fig. 22.22 Pressure fatigue data are analyzed with special graphs.

Table 22.3 Fatigue Correction Factors Improve Test Accuracy. All Values are Approximate

---

Rated fatigue pressure—Pressure a component can sustain for 10^7 zero-to-max cycles without failure. It generally will be much lower than the conventional rated static pressure.

Test duration factor, K_N—Multiplier applied to rated fatigue pressure to determine what the cyclic test pressure would have to be to shorten the test life to 10^6 cycles from 10^7 cycles.

Variability factor, K_V—Multiplier applied to rated fatigue pressure to compensate for variability in fatigue strength of metals. It also is applied to rated static pressure (see equations).

Assurance level—Minimum percentage of components that will meet the rated fatigue pressure conditions.

Confidence level—The percentage chance the manufacturer has of being certain that his components actually meet the assurance level he has established by his tests.

Goodman diagram, S-N diagram, reversal stress diagram—see graphs and text.

Equations—

$P_{CT} = K_N K_V P_{RF}$

$P_{ST} = K_V P_{RS}$

Where: P_{CT} = cyclic test pressure
P_{ST} = static test pressure
P_{RF} = rated fatigue pressure
P_{RS} = rated static pressure
K_N = test duration factor
K_V = variability factor

---

would not occur, and the pump might fall in fatigue before it reached 10^7 cycles. This is a fact that has been well documented over the years but never brought to bear directly on fluid-power testing standards before. The dividing line is at a few million cycles: The further below this value tests are run, the higher is the probability that beneficial yielding will give an unconservative estimate of fatigue strength.

Much of the supporting background data for the proposed standard are from years of tests on rotating cylindrical beams, where complete reversal of stress (tension-to-compression) occurs every cycle. There is justification for applying these data to half-reversal (from zero to tensile) if the modified Goodman diagram (Fig. 22.22) is considered.

Modified Goodman Diagram. Decades ago, many experimenters observed that there was a simple correlation among stress failure data points throughout the whole range of tensile stress. There always must be some tension to cause fatigue, so pure compression doesn't apply.

In simple terms, if the ultimate tensile stress is known and you know the failure stress for cyclic full reversal, you can plot those points on a special diagram that gives you a minimum failure line for any combination of tensile, compressive, and cyclic stresses.

It works like this (Fig. 22.22): point C is the ultimate strength in tension, and

Table 22.3 (Continued)

Test duration factor, K_N

Test cycles	Ferrous	Non-ferrous
10 million	1.0	1.0
1 million	1.15	1.25

Variability factor, K_V

Assurance levels (text)	Number of units tested		
	1	2	5
Iron			
99.9%	1.90	1.42	1.22
99	1.55	1.30	1.15
90	1.30	1.15	1.07
Aluminum, magnesium, and steel			
99.9%	1.55	1.30	1.15
99	1.45	1.25	1.12
90	1.30	1.15	1.06
Copper-based alloys			
99.9%	1.50	1.20	1.09
99	1.35	1.15	1.07
90	1.20	1.10	1.02
Stainless steel			
99.9%	1.30	1.17	1.09
99	1.25	1.15	1.07
90	1.17	1.10	1.02

point A is the failure point for complete stress reversal from compression to tension. Points on the abscissa are average tensile and average compressive stresses. Goodman observed that no failures occurred below the line A-C.

The sample traces accompanying the Goodman diagram show two typical examples of stress cycling. The first shows a part held always in tension, with the stress reversal occurring around an average tensile stress. The second shows a part held in tension at the top of cycle but dropped to zero stress at the bottom. The test points are plotted as F and G on the diagram.

The first example (always in tension) is not a typical proposed test but points out a principle. If point F happens to fall above line A-C, a failure might occur. The average tensile stress is C_F, and the stress range is A_F.

The second example (from zero to full tension) is more typical of a fluid-power test. Note that point G falls on line O-D. That line is of particular importance to the writers of the proposed standard. It is the locus of all points where the stresses cycle between zero and a maximum, so that the average tensile stress C_G and stress range A_G are equal. Failure is not expected to occur at point G but would possibly occur farther up on line O-D if the test pressures were increased to intersect line A-C.

Stress Correction Factors. Not every material behaves the same under stress, nor does any given material behave very simply when stress conditions are changed. The most significant part of the NFPA study is the creation of two correction factors (Table 22.3) to compensate for different materials of construction and for variations in stress cycling life.

K_V, the variability correction factor, is a pressure multiplier to compensate for the fact that fatigue strength varies from part to part. Different values for the factors account for one material being less predictable than another. A higher factor means that a higher test pressure is needed to be certain that the rated pressure is conservative enough. The broad spread in K_V values comes from analysis of fatigue test data found in the literature.

K_N, the test duration factor, is a pressure multiplier to compensate for fewer cycles being run, to save test time. Only two cycling periods are allowed: 10^7 and 10^6. The base period is 10^7 cycles. The alternate period is 10^6 cycles, but here the test pressure must be multiplied by K_N to allow for the fatigue strength at this life being higher. For static tests, K_V is selected. For cycling tests, both multipliers (K_V and K_N) must be used. The equations are in Table 22.3.

There are several ways to make your estimates more reliable. The more samples you test, the lower is the chance of being misled by the stronger samples, so a lower allowance (K_V) is allowed. Irons are less predictable than stainless steels, so to have equal assurance levels you assign them to a higher value of K_V.

So-called "confidence" levels are a statistical addendum to the "assurance" table. Whereas assurance level means a minimum percentage of operating units that will last 10^7 cycles at rated fatigue pressure, the confidence level means the percentage chance a manufacturer has of *really* meeting the assurance level in the field. The entire K_V table is based on a 90% confidence level. If you want to increase your batting average to a 99% confidence level, double the called-for number of test samples.

Skaistis of Vickers points out that a 90% assurance level doesn't mean that 10% of your products will fail in the field. It only means that 10% would theoretically fail if the user cycled them at rated fatigue pressure 10^7 times. This is unlikely, so the field failure rate of your product will be far less than the worst prediction. In cases where stresses might actually cycle in this manner, you have to consider using assurance levels of 99% or even 99.9%.

Temperature effects are a possible addition to the method. Right now it is based primarily on room temperatures, with an allowable range from −20 to 200 F using the given materials.

Faster Tests. Most test people when they hear of a 10 million cycle or even a million cycle test think of months of testing. It doesn't take that long. You can apply up to 30 pressure pulses per second, automatically and unattended 24 hr a day. A typical test setup is shown in Fig. 22.21. Under these conditions, it can take as little as 4 days to generate 10 million cycles (Fig. 22.23).

In some versions of a proposed standard, the cycling rate limit is adjusted to the highest rate that can be counted on to give repeatable pulses under ideal conditions. Stable operation is important, and this will limit cycling rates in most tests to from 8 to 15 pulses per second. Most tests will be completed in less than 2 weeks.

The tests are easy to monitor with ordinary oscilloscopes. Oscillations superimposed on the test pulses will not contribute to fatigue failure unless the range of the oscillations (Fig. 22.24) exceeds 60% of the test pulse height, according to Skaistis. Keep in mind that the test pressure is the highest peak felt by the test piece.

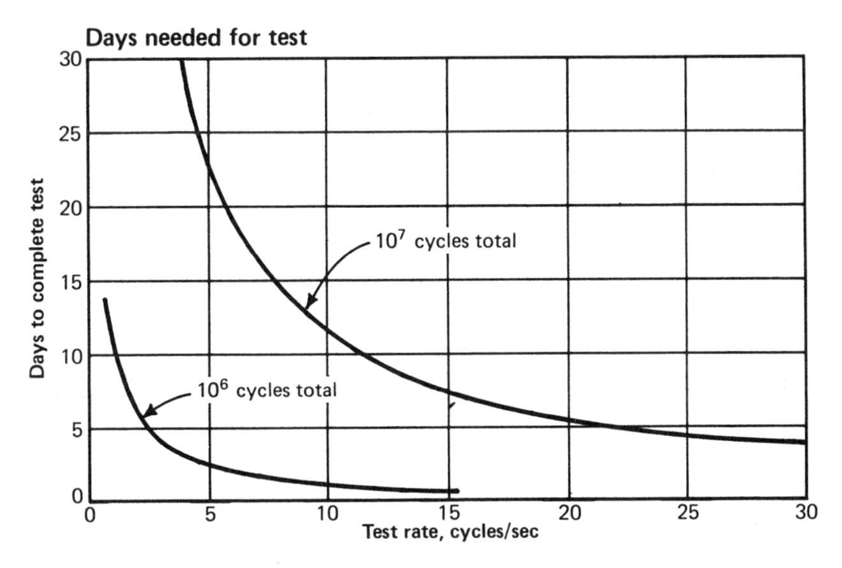

Fig. 22.23 Millions of test cycles are quickly applied.

A trick for raising the cycling rate is to displace oil in the pressurized cavities with metal shot or loose-fitting metal slugs. This reduces the input flow needed to compress the remaining oil in building up pressure. Another trick is to restrain all operational motions, such as that of a valve or cylinder.

Pressure travels through oil at the speed of sound (about 50,000 in./sec); stresses travel through metal at about three times this speed. Even with cycling rates as fast as 30/sec, the full fatigue effect of the pressure pulses still will be felt. There can be rare exceptions: One is when pressure is monitored in the pressure supply line instead of at the test piece; another is when very large units are tested; a third occurs with some units having bolted joints.

Verification of the accuracy is easily accomplished by comparing the signal from a strain gage on the test part, not necessarily at the point of highest stress, with the pressure trace. If the ratio between these signals is the same at the test conditions as it is statically, the measurements are valid.

NOISE CONTROL

The noise in fluid power machinery—both hydraulic and pneumatic—can be substantially reduced by application of known design principles.

In hydraulic equipment (a harvesting machine is an example to be covered), the noise originates primarily in the pulsations of reciprocating power elements, such as pistons, which rapidly pressurize and exhaust the hydraulic fluid. Secondary sources of noise include throttling, turbulence, cavitation, and chattering in pumps, motors, and valves. Bubbles of entrained air and even vaporized water create noise when they collapse violently.

Similarly, noise in pneumatic systems stems from the pulsations of reciprocating compressor elements plus the throttling, turbulence, and chattering of air motors and valves. The greatest problems of turbulence are in high-pressure gas valves, which throttle thousands of psi down to greatly reduced pressures, creating loud whining

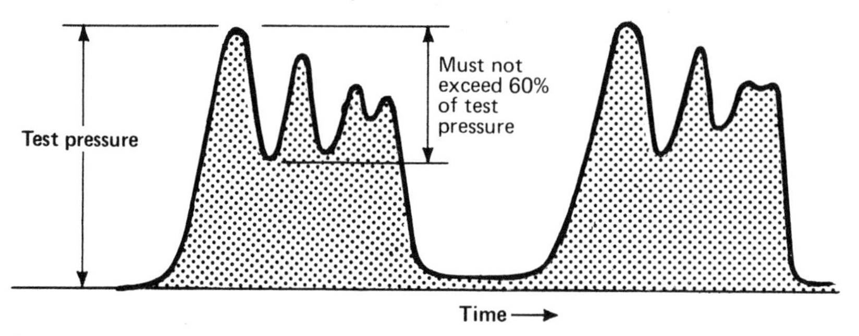

Fig. 22.24 Fatigue test pulse pressures don't have to be perfectly formed, but must not oscillate more than 60% of peak value.

and roaring. For example, a 2-in.-diameter angle valve, when reducing gas pressure from 4000 to 2275 psi, will dissipate more than 5 kW of energy, converting it to noise. This is about the equivalent of the level of sound next to a jet engine operating at full power.

Mechanical noise from gearing and bearings adds to the din, and many of the sounds tend to be amplified and transmitted by the machine structure, the fluid itself, and the surrounding air.

Theory of Noise

A heard sound is the result of pressure variation superimposed on the ambient atmospheric pressure. Its amplitude is sound pressure in millionths of a bar (microbar), where a bar is 14.5 psi—essentially atmospheric pressure.

Sound power is a measurement of the energy (in watts) that sound dissipates in the elastic medium—gas, liquid, solid, or plasma.

In practice, neither of these is used directly. Instead, sound power *level* and sound pressure *level* are used. These are ratios, relating the sound pressure and power to reference quantities. The ratios are expressed on a log scale because of their range. Typical levels, in decibels, are noted in Table 22.4 and Fig. 22.25.

While the reference quantity for sound pressure level is usually 0.0002 μbar (a microbar is 1 dyne/cm^2) and that for sound power level 10^{-12}W, it is very important to know—and state—the reference quantity in every case; otherwise, the decibel figures cannot be compared. (Some people, for example, use 10^{-13} W instead of 10^{-12}.)

How We Hear. The human ear is a transducer that responds to frequencies between roughly 20 and 15,000 Hz. Its highest sensitivity has evolved in the 3000-Hz range of human speech. In terms of sound pressures, it responds from 0.0002 dyne/cm^2 (microbars) as the low threshold to more than a million times that pressure. The logarithmic decibel scale usually is referred to 0.0002 dyne/cm^2 as the base for sound measurement in sound pressure level:

$$\text{dB} = 20 \log \frac{\text{sound pressure}}{\text{reference pressure}}$$

You could make a linear broadband measurement of sound pressure level with a condenser-type microphone and a meter, but it would be a purely physical meas-

Table 22.4 Decibel Levels are Guidelines for Comparing Noise Sources

Some quick guidelines: Threshold of audibility, 0 dB. Annoyance threshold for intermittent sounds in community, 50–90 dB. Intolerable for phone use, 80 dB. Discomfort threshold, 110 dB. Pain threshold, 120 dB. Short exposure that can cause permanent hearing loss, 150 dB.

Some examples: Whisper, 20 dB. Quiet room, 40 dB. Quiet street, 50 dB. Normal conversation, 50–60 dB. Sports car, truck, shouted conversation, 90 dB. Electric blender, 93 dB. Pneumatic jackhammer, 94 dB. Loud outboard motor, 102 dB. Loud power mower, 107 dB. Jet plane (at passenger ramp), 117 dB. Thunderclap, 120 dB. Diesel engine room, nearby air-raid siren, 125 dB. Machine gun, riveting machine, 130 dB. Jet takeoff (at close range on the ground), 150 dB.

urement, not directly related to human hearing of the sound. The ear hears certain frequencies more acutely than others, so the meters must be modified.

Several broadband sound-level meters employ weighting networks to simulate the effect on the human ear: the dB(A), dB(B), dB(C), dB(N), and—recommended by the International Electrotechnical Commission—dB(D). The dB(A) is usually considered the one simple meter most useful for law enforcement, responding most like the human ear.

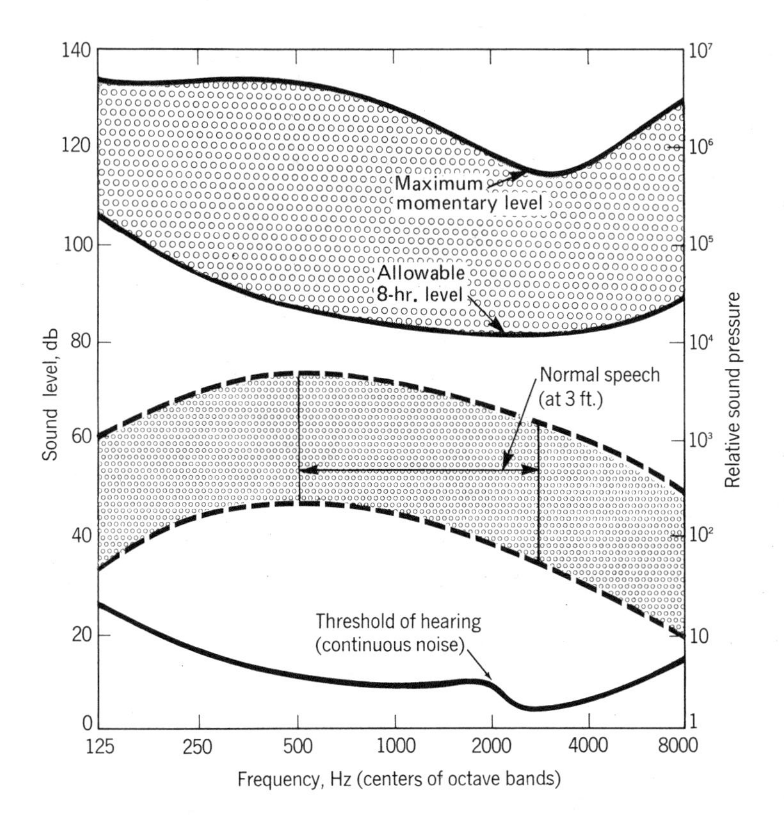

Fig. 22.25 Noise range in this approximate graph is based on OSHA limits.

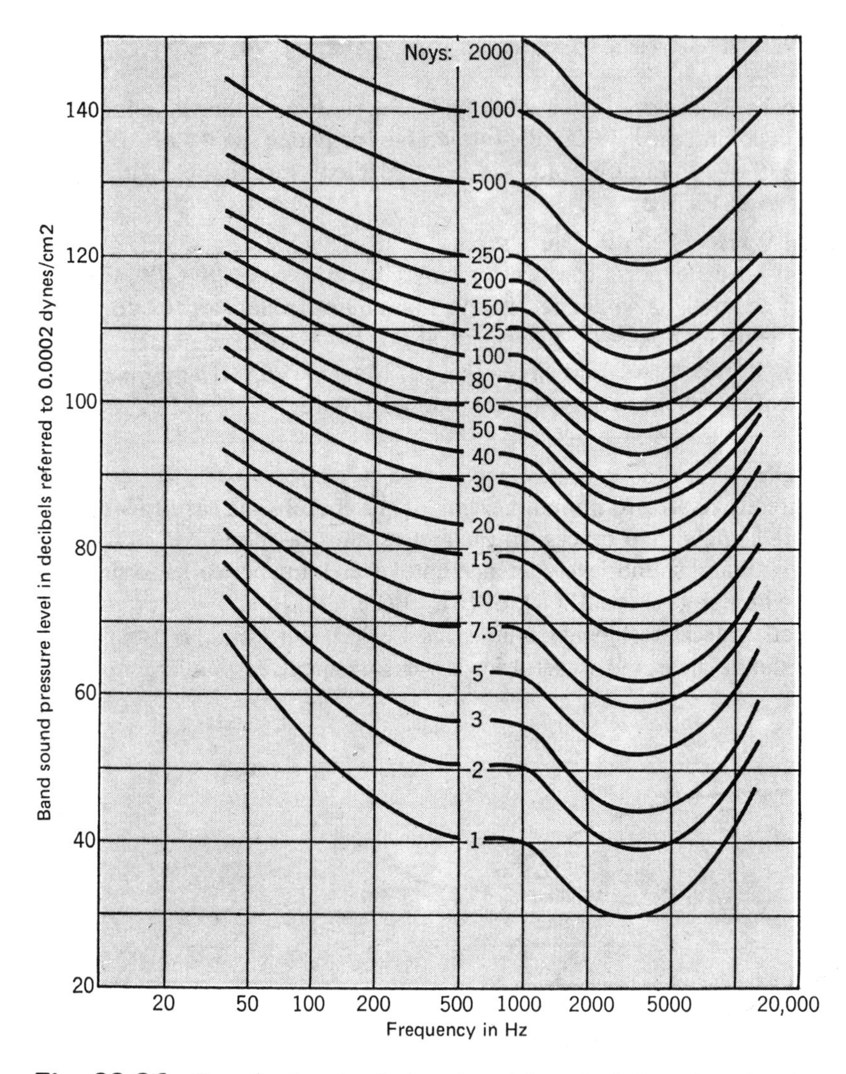

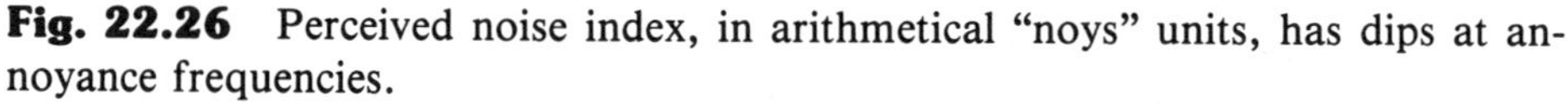

Fig. 22.26 Perceived noise index, in arithmetical "noys" units, has dips at annoyance frequencies.

More complex measurements involve breaking the sound frequency spectrum into octave or ⅓-octave bands and measuring the band pressure levels through filter networks with elaborate equipment. These data are the basis for calculating a loudness index (in phons or sones) or a perceived noise index.

Phons are obtained by measuring the loudness of any sound against a pure 1-kHz tone; at equal loudness, phons equal the measured dB level of the reference tone. Sones are directly proportional to loudness of sound as subjectively heard by the human ear, based on a scale of 1 sone = 40 phons and 2 sones = about 50 phons. Some experts consider sones the most understandable scales.

Perceived noise (PN) scales are based on comparison of noise with a "white" or random noise centered at 1 kHz. They are an attempt to measure "noisiness" rather than loudness, seeking to rate the degree of human annoyance. Noise scales are similar to those of phons and sones but are called PNdB (logarithmic) and noys (arithmetical). PNdB is used to measure aircraft noise, though the noys scale (Fig. 22.26) would be more understandable.

For single-number noise ratings, there are noise criterion (NC) curves for noise levels in offices; the International Standardization Organization (ISO) has noise rating (NR) curves (Fig. 22.27) that relate the spectrum to the dB level at 1 kHz.

Sound Power. There is no direct way to measure the power of sound once it is dissipated from the source. Many experimenters estimate sound power by comparing noise levels of the component under test with the noise levels of a calibrated sound generator, whose input power is known, placed in the same location. Also, there are empirical equations for converting sound pressure to sound power, but none seem to be reliable except as a guide.

Radiation Surfaces. One dramatic fact stands out: A sound cannot be radiated easily from the source unless the radiating surface has a side dimension at least roughly equivalent to one wavelength (Fig. 22.28).

You can prove this point by striking a tuning fork: Observe that its loudness is negligible even though its vibration amplitude is high. The reason is that the length

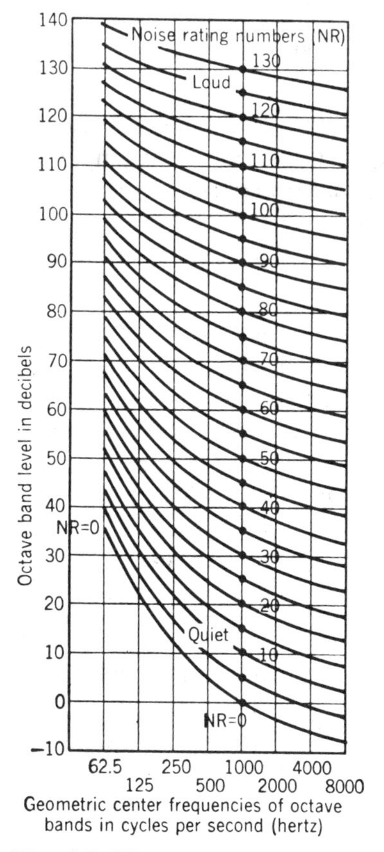

Fig. 22.27 Another rating method is the International Noise Rating Number (NR).

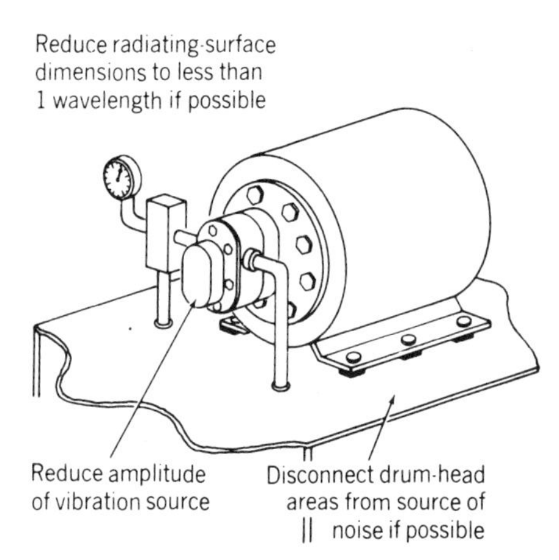

Fig. 22.28 Hydraulic motor can be quieted somewhat by isolating the noise source from large radiating surfaces.

of the tuning fork is much less than the wavelength of the musical note it is creating.

A piano, however, has a large sounding board and will much more readily transmit the same note. Take middle C on the piano. The frequency is 256 Hz and the wavelength about 4 ft; a piano has much more than a 4-ft dimension.

As a rule of thumb, fluid power engineers should memorize the wavelength of sound at 1000 Hz: It is approximately 1 ft.

When all this is related to pumps, compressors, and valves, it becomes evident that noise can be stifled at its source if all radiating surfaces approaching in dimension the wavelength of an offending noise frequency can be eliminated. If these radiating surfaces must remain, the alternatives are the following: reduce the amplitude of the vibrations; isolate small, severely vibrating portions of an otherwise stable assembly; or shield the hearer from the noise.

Noise Barriers. The wavelength rule of thumb also applies to sound barriers. To be effective, barriers must have dimensions of several wavelengths.

Maximum noise reduction can be achieved with an airtight enclosure around the source. Any openings in the enclosure must be kept as small as possible. Where they are necessary for ventilation or materials handling, sound control is achieved with appropriate mufflers at each of the openings. Sound waves find it difficult to escape through openings that are smaller than one-quarter of a wavelength.

Where total enclosures are not practical, then barrier screens, called shadow screens, can be installed (Fig. 22.29). Heavy walls of sound-deadening material are best.

Legal Noise Limits. Despite the controversy of measurements, OSHA requires that products be made in safe, clean, and *quiet* plants. According to the regulations, any noise registering above 85 dB in sound pressure is suspect. The wording is broad enough to permit louder sound levels but only for limited periods of time (Fig. 22.30).

Quieting Pumps and Drives

The inherent pulsations of reciprocating hydrostatic pumps and drives can be reduced or damped effectively in several ways:

- Limit the shaft speed
- Increase the mass of supporting housings
- Time the valving to match the pump chamber pressure at the moment of discharge
- Allow some slippage in the pumping pistons (to decrease shock pressures)
- Isolate the pumps or shield them from hearers' ears

Engineers at Rexroth (Lohr/Main, Germany) tested over 100 positive-displacement pumps of every type, including axial piston (swashplate and bent-axis), vane, and gear (internal and external), trying to find the common denominator of noise generation. They didn't fully succeed because of the complex interrelationships of the pumps, the pumped fluid, piping, valve characteristics, and resonant frequencies of the test stand. However, some trends were established (Fig. 22.31) and some conclusions made.

Small-displacement pumps proved to be considerably quieter than large-displacement pumps. The noise level of two identical pumps combined is much less than twice that of either pump alone. Speed has more effect on noise than does pressure or displacement. A slow-speed pump (under 1500 rpm) would appear to be a good buy if there is room for it. More compact pumps (such as swashplate) are noisier than larger pumps of the same performance (such as bent-axis) because the extra mass helps reduce vibration. Variable-delivery pumps are noisier than constant-delivery designs. The quietest pumps seem to be vane-in-stator types.

Additional reductions in noise will come through system tuning; careful selection of size, type, and quality of pump; and arrangement of components and piping. For the quietest results, try to keep displacement to 1.5-in^3 swept volume and speed to 1500 rpm, even if it means putting small pumps in parallel to achieve the desired flow.

Engineers at Parker Hannifin's Manatrol Div. (Elyria, OH) observed that positive-displacement-type hydraulic pumps generate most of their sound at the outlet port,

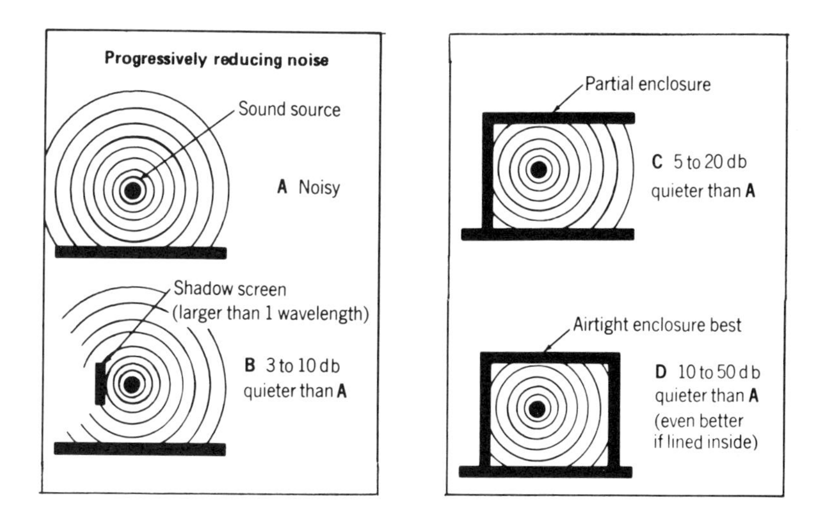

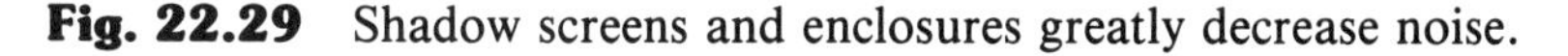

Fig. 22.29 Shadow screens and enclosures greatly decrease noise.

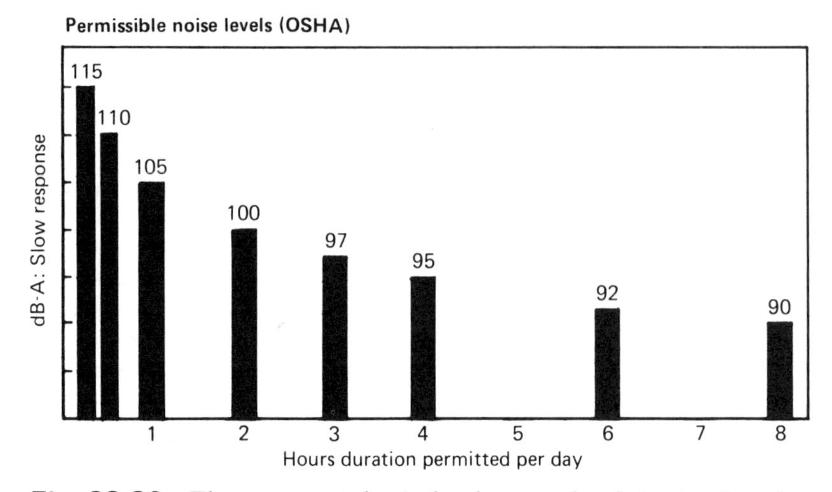

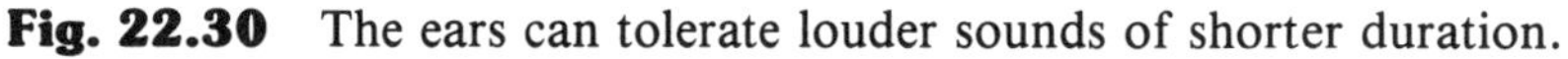
Fig. 22.30 The ears can tolerate louder sounds of shorter duration.

where fluids recombine and otherwise interact at the highest pressure. The noise is greatest when the line pressure is higher than the pump discharge pressure. The sudden compression as line pressure rushes into each oncoming pump chamber causes the problem. The ideal situation is when the discharge pressure is neither above nor below line pressure at the moment the pumping chamber is exposed to line pressure. Table 22.5 compares the relative quietness of pump types. Another example of hydraulic noise is the sound generated when trapped pockets of pressurized fluid within the pump are exposed to inlet pressure during the suction stroke.

Pressure ripple is another source of noise. It is the cyclic pressure change caused by compression and decompression at the pump outlet port. If ripple frequency coin-

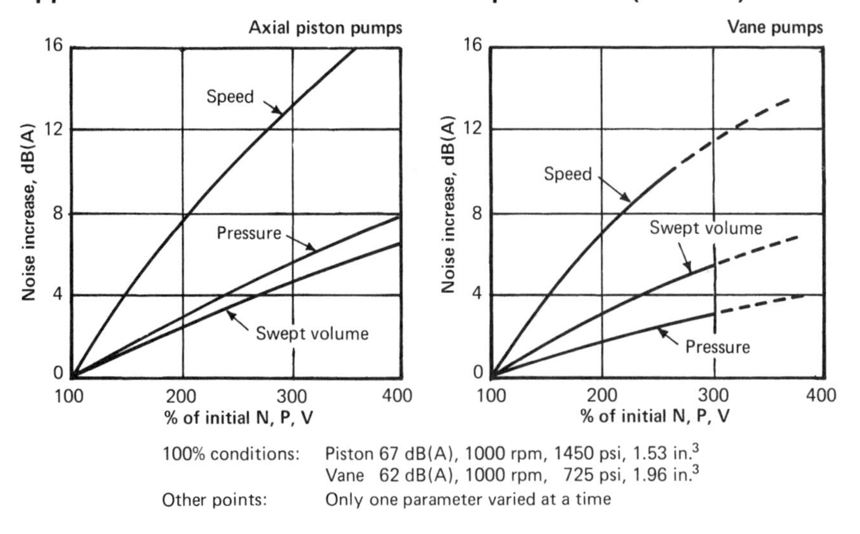

Fig. 22.31 Pump shaft speed has a greater effect than pressure or flow on noise.

Table 22.5 Each Pump Type has a Characteristic Range of Noise Level

Pump type	dB
Screw	72–78
Vane (industrial)	75–82
Axial piston	76–85
Gear (powdered metal)	78–88
Vane (mobile)	84–92
Gear (machined stock)	96–104

cides with mechanical resonant frequency elsewhere in the system, the amplitude of vibration can build up to excessive levels.

Cyclic pressure variations have another noise-making effect: torsional vibrations in the pump shaft. The vibrations are picked up by the piping and by all devices in the system.

Some experimenters have found that ripple can be damped somewhat by superimposing a higher frequency into the fluid, using a vibrator that is not in resonance with any fundamental frequency or its harmonic.

Turbulent flow also is noisy. If flow can be kept laminar, it will be much quieter.

Air dissolved in oil can create noise under certain circumstances. Mineral-based hydraulic oil contains about 8% dissolved air to begin with. High-velocity flow through a restriction will severely lower the static pressure in the oil and thus bring some of the air out of solution. Air bubbles form, usually in the pump suction region, and are carried along in the oil. The trouble occurs when the oil repressurizes, suddenly forcing the air bubbles back into solution. The sudden collapsing of the bubbles (cavitation) against the metal housing creates sound level increases of from 10 to 15 dB, and the sound is a shrill one.

Vibration, uncorrected, will loosen fittings and allow oil leaks, resulting in air being sucked in to replace the lost oil, thus aerating the oil with tiny bubbles. The bubbles cause more turbulence, which increases the noise and vibration. Control valve spools chatter and fail to respond smoothly, resulting in spongy or jumpy cylinder operation and more vibration. The leaking areas enlarge, more oil escapes, and the pump begins to ingest large air bubbles. The final result is excessive cavitation and noise.

The cavitation also can be caused by mechanical restrictions at the pump inlet. An example is a suction pipe that is cut off square where it faces the tank bottom, instead of being beveled.

Even if the cavitation is stopped by refilling the tank, removing the inlet restrictions, and realigning the couplings, the system might still be noisy because of the worn, loose parts. Clattering parts can boost sound levels 4 to 6 dB. If you don't correct the cavitation, the system can beat itself to death. Bearings can freeze up, pump housings can work-harden and crack, and pumping elements can become too battered to pump.

Quieting Hydraulic Systems

In hydraulic systems, you have many warnings before noise becomes excessive and system performance is destroyed (Table 22.6).

A low oil level indicates leaks. Filling the tank will reduce the noise level, sometimes as much as 2 to 4 dB. However, the leak warns you of something else: The lost

Table 22.6 Noise Problems in Hydraulic Systems can be Analyzed by Ear, Using Guidelines Developed by Parker Hannifin (Cleveland, OH)

Typical Noise and Source	Probable Cause of Noise	What to Check	How to Cure It	Potential Reduction (dB-A)
Noticeable increase in overall noise level at fittings, pipes, valves, and pump or motor (usually accompanied by spongy or jumping operation of hydraulic cylinders)	High flow turbulence caused by aeration. Air is being leaked into the system.	Look in oil reservoir. See if oil is "milky" or "frothy," but reservoir is full.	Find and repair source of air leak. Look especially for leaks in suction side at fittings, oil filter, fill-cap threads, fill-cap o-ring and pump seal. Remove air in system by bleeding at bleed points.	2-4
Same overall noise problem as above. (Same erratic cylinder operation	Turbulence caused by aeration brought on by oil leaking out of system while air is leaking in.	Check to see if oil in reservoir is frothy, but oil level is down.	Find and repair source of oil leak, refill oil reservoir, bleed lines. Operate system long enough to see if noise has stopped and operation is smooth. If not, find and repair air leak as above and operate again to check results.	2-4
Same overall noise as above continues, but not quite as loud after both above remedies have been tried	Turbulence caused by aeration because oil is retaining too high a percentage of air.	Check to see if oil reservoir is full, but still frothy.	Change to hydraulic oil with anti-foam additive. (Consult with system manufacturer and oil company to make sure new oil type is compatible with the system.)	1-3
Same overall noises as above in system that has water cooling unit in it.	Turbulence caused by aeration from air carried in water leaking into oil from cooling coils.	See if oil is cloudy or milky in reservoir. (Boil small quantity of oil—if it clears, cause is water in oil.)	Let system stand overnight. If cloudiness disappears, cause is air leak (find and repair). If cloudiness remains, cause is water homogenized in the oil. Find leak, repair it, drain oil, replace with new oil, bleed lines during checkout under no load.	2-4
Pump, motor, or valves make loud rattling or clanking noises under load when first started up . . . but noise disappears shortly.	Cold oil is too viscous—causes cavitation by drawing air from reservoir into the system.	Check oil in reservoir. See if oil is frothy, thick, and much cooler than when running normally.	Warm up the system with a pre-heater or by running it under no load till side of oil reservoir is hot to palm of hand. Bleed lines when system starts to quiet. (Ideal oil temperature is 120° F.)	10-15
Pump, motor, or valves for no apparent reason start making loud rattling or clanking noises. (Accompanied by erratic operation of cylinders.)	Cavitation caused by pronounced restriction in system immediately ahead of noisy component, or very low oil level.	Check oil level. If full, then check for foreign object (cleaning rag, bit of teflon tape, plugged oil filter, crimped suction inlet, etc.)	Fill oil reservoir. Find obstruction and correct it. Clean oil filter. Tighten all connections on suction side. Bleed lines during checkout under no load.	10-15
Single loud "plop" or "clank" repeating at irregular intervals in pump or hydraulic motor.	Single cavitation sound caused by one large air bubble collapsing.	Check oil for froth. Check suction connections for air leak.	Fill reservoir if low. Tighten all connections on suction side. Look for any restricting foreign object. Clean oil filter. Bleed lines on startup.	Peaks of 4-8
Increased noises from pump or hydraulic motor. (Usually accompanied by sluggish performance by cylinders.)	Worn parts in pump or motor caused by abrasive action of wear particles on rotating and working parts.)	Check oil temperature. Probably too high because of friction.	Remove metal particles in suspension; drain entire system, flush piping, clean all components, refill with new oil and bleed lines during checkout under no load. Repeat in 30 days. (If noise level is not reduced, replace worn parts or entire pump)	4-6
Increased noises from valves, usually chattering sound, sometimes sticking or erratic performance.	Worn spools or orifices caused by wear particles.	Check dimensions and clearances of spool and orifices.	Replace worn parts with new if practical, or replace entire valve unit. Remove metal particles from oil as above.	2-4
Loud slam travels through hydraulic system during erratic performance after changing to different type of hydraulic oil in system.	Hydraulic shock waves caused when sticking part suddenly overcomes the constricting force.	Check filter and valve or cylinder parts for sludge, corrosion and varnish.	Clean filter and parts with lacquer thinner. (Have oil analyzed—make changes in contents or switch oil type recommended.) Flush system thoroughly.	Peaks of 10-20

oil is being replaced with air, and that can cause aeration, with its attendant cavitation problems.

A high oil temperature warns of excessive friction. Contributing causes include sludge, varnish (oxidized oil), turbulence, aeration, excessive flow velocity, trapped metal particles, and rubbing or contacting parts. A secondary effect of continuing oil temperature is oil breakdown. This reduces lubricity and viscosity, and the thinned oil transmits noise more readily. High temperature also expands operating parts and might enlarge (or close up) gaps at interfaces, compounding the problem.

To check out temperature, put the palm of your hand against the side of the reservoir—carefully. Most people can stand 120 to 130 F for a few seconds, and this happens to be a safe temperature for reservoir oil. Pump temperature, particularly that of a variable-displacement pump, normally will be much higher, so don't use that surface as your guide.

Frothy or milky oil is a sign of entrained water or air. The water can come from condensation within the reservoir or from leaks in the water-to-oil heat exchanger. The water can be detected by boiling a small sample of the oil; if it clears, the contaminant was water. Air is detected another way: Let the reservoir stand overnight; if it clears, the contaminant was air.

If the oil is too hot, for any of the reasons discussed before, drain and flush the whole system, clean out the reservoir, and put in new oil. Run the system under no load and bleed it at all overhead bleed ports. If you reuse the old oil, you'll put the same contaminants back into the system. Keep out contaminants with the right filters.

If clogging of filters is considered unavoidable, then install a bypass filter to prevent loss of pump suction. Sometimes the clogging is invisible, caused by transparent varnishing of the filter element with overheated oil. The varnish can be removed with lacquer thinner.

Pressure pulsation sounds often can be dramatically reduced by lowering system pressure at strategic places, so long as it doesn't interfere with system operation. For instance, are you exploiting relief valves as load controls for cylinders? OK, but try lowering the settings to the minimum possible.

Coupling slapping noise is probably caused by coupling misalignments as high as 0.010 in. FIR (full indicator reading), which can be readily corrected. Sometimes a popping or crackling sound indicates severe misalignment of a motor-to-pump coupling. If necessary, substitute a rubber-faced coupling for quietest operation.

Fan noise usually can be traced to the fact that the fan and cover are thin metal. Replace the metal fan with plastic and the thin cover with cast iron, and the noise can be reduced as much as 5 to 7 dB. Plastic absorbs sounds instead of transmitting them.

Pipe-transmitted sounds can be stifled by substituting hose wherever there are sharp bends. The hose expands radially to absorb pressure pulsations. Some users have reported 12-dB noise reduction. Resilient mounts for the piping help also.

The pump/motor mounting plate on the reservoir can be made less of a transmitter of noise if a resilient mount is placed between it and the tank top. Further silencing is possible with stiffeners. The quietest reservoir is still and heavy, and if all else fails, put in a new reservoir made of thick steel plate. There's the possibility of reducing noise 6 to 9 dB with those techniques.

Larger pump displacements will help maintain flow at the lower rpm and still keep velocities low. Pressure lines should be kept below 15 ft/sec and suction lines below 5 ft/sec.

Acoustical filters such as accumulators, pulse filters, single-frequency tuning devices, reactive mufflers, and torroidal dampers will help quiet particularly noisy lines, especially at known fixed frequencies.

Working devices such as hydraulic cylinders can be quieted with dashpots and other load-decelerating devices. Replace direct-acting relief valves with pilot-operated or differential-area types. Avoid needle valves or low-cracking-pressure check valves, which are noisy.

Quieting Pneumatic Systems

Air compressors and pneumatic tools constitute two of the greatest noise challenges in an industrial plant. The compressors usually can be isolated or heavily insulated. The air tools can be muffled or exhausted out of earshot.

Muffling includes absorbing the sound in some porous material (watch out for icing if the tool is run continuously); acoustical filtering, such as an expansion chamber or resonating chamber; diffusing the air as it exhausts to break up the sound wave front; or forcing the exhaust to stretch a rubber sleeve surrounding the exhaust pipe, thus damping the noise.

Big gas valves in process plants are almost impossible to quiet, but great efforts have been made. The energy converted into noise is proportional to about the seventh or eighth power of the velocity. When the velocity is cut in half, the noise energy is reduced by a factor of about 250. A logical way to reduce velocity is to introduce restrictions in series with the valve seat to do the throttling more gradually.

Several examples (Fig. 22.32) show the basic principles. In each, the gas is forced to follow a serpentine flow path. Thus, the drops in pressure are gradual, rather than in 1000-psi jumps. Also, the outlet ports are purposely enlarged to reduce outlet velocity.

Much of the noise in a valve comes from the vibration of valve components. A lot of this is from the lateral movement of the valve plug relative to the guide surfaces. The frequency is usually less than 1500 Hz and is described as metallic rattling. Rattling is damaging to the valve as well as being noisy.

Another source of valve noise is mechanical resonance. It is usually a single-pitched tone with a frequency between 3000 and 7000 Hz. Resonance produces fatigue and failure in addition to noise.

Quieting Motors, Bearings, Gears

You can tell when the hydraulic and pneumatic components are quieted because then you can hear the electric motor noises.

Cyclic deflections result from magnetic interaction of the rotor and stator windings, slots, and housing. One way to reduce the problem a dB or two is to fill the stator slots with an inert plastic.

Mechanical noises spring from dynamic unbalance, mechanical misalignment, worn bearings, wrong fans, and rough gearing. Dynamic unbalance is readily discovered by running the motor at full speed, no load, and then shutting off the power. If the motor vibrates as it coasts down, then the fault is likely to be mechanical dynamic unbalance. This can be corrected by conventional motor-balancing techniques.

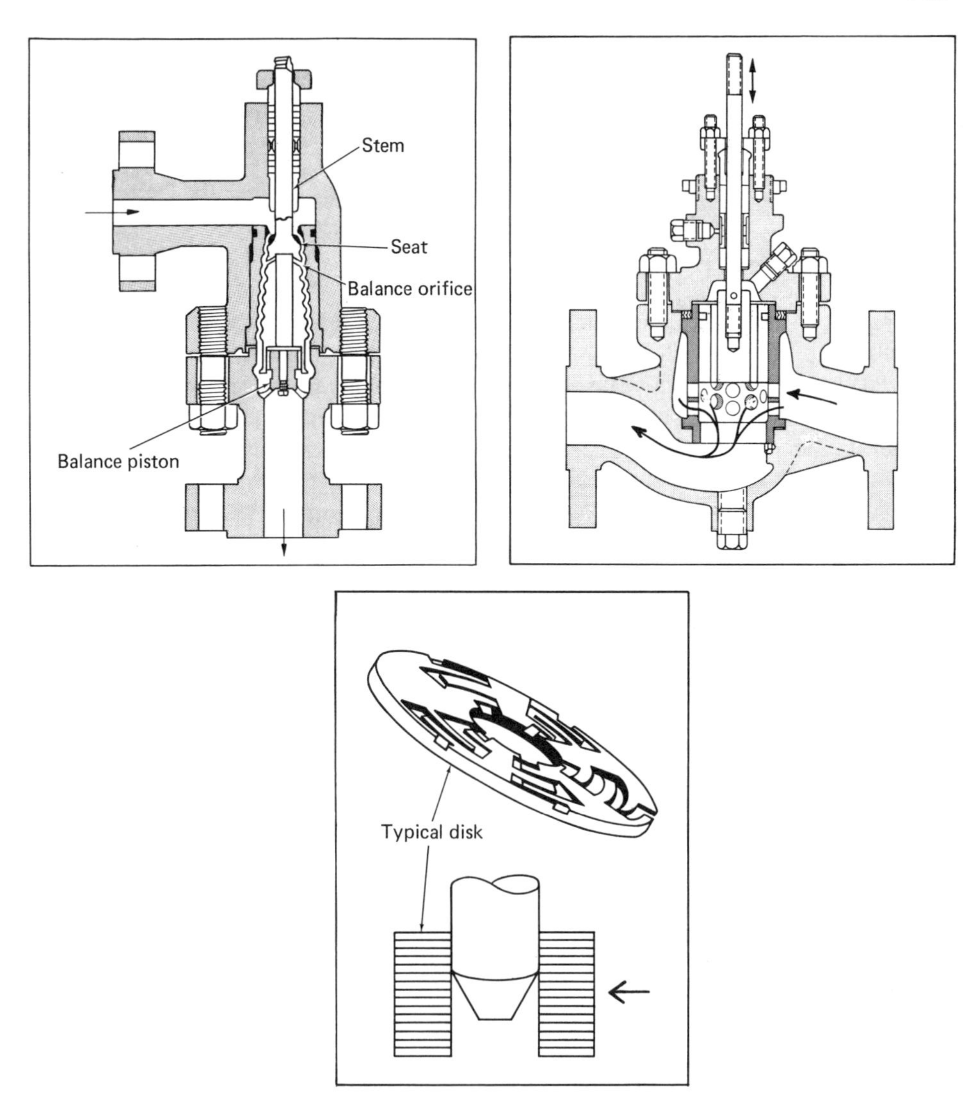

Fig. 22.32 Labyrinths and multiple orifices in gas valves help soften pressure drops and reduce noise.

Be sure the motor-to-pump couplings are aligned within 0.003 in. FIR and that the end caps and motor frame are aligned accurately.

Ball bearings are noisier than sleeve bearings, but each will be very noisy unless manufactured with precision. Ball bearings must be preloaded to eliminate rattling of loose balls. Preloads of 2 lb or less are sufficient for most fractional hp motors; higher preloads bring sound levels back up again. Ball bearings generate less than 60 dB of noise—well below OSHA limits.

Sleeve bearings are inherently quieter than ball bearings but must be made with close clearance to prevent knocking or pounding caused by radial vibration. The biggest problem is control of thrust washer noise, which is an intermittent scraping sound. Hot, thin oil increases both the radial and thrust problems.

Reduction of motor speed is very effective, if possible in your system. By replacing an 1800-rpm motor with a 1200-rpm motor, as much as 6 to 10 dB of noise can be dropped, because air turbulence is less. Also, vibration frequency is lower and thus more pleasant to the ear. For greatest silence, specify an oil-cooled motor, or at least a totally enclosed fan-cooled motor with filled stator slots, plastic fan, and cast iron shroud.

Harvesting Machine Example

Vickers (Troy, MI) completed a test study on an operating hydrostatically powered harvesting machine and dealt specifically with the noise sources.

The loudest noise is not during normal harvesting when all the equipment is operating at once but occurs during high-speed transport when only the propulsion drive is running. Furthermore, the hardest noise to silence is that generated within the cab and not the airborne external noise.

Sources for in-cab noise include mechanical vibration (structure-borne) and hydraulic pulsation (fluid-borne). Fluid-borne noise is taken care of best by not bringing the valves and piping into the cab in the first place. It is nearly impossible to predict how pump flow perturbations will react with the dynamic resonances of circuits. Flow-through gas-loaded accumulators help but are expensive. Flexible hose attenuates noise by absorbing the pulsations but also radiates noise if too lengthy (Fig. 22.33).

Stan Skaistis, who was manager of the project, suggests avoiding the fluid noise problem by remotely controlling the fluid power with electrical controllers within the cab. This technique is becoming more and more popular and neatly eliminates valve hiss and piping noise from the vicinity of the operator's ears.

Structure-borne noise is not so easily controlled. The pulsations left outside the cab by isolating the hydraulics retaliate by shaking the machine structure. It often is impractical to mount the entire cab on cushions, so these sounds enter through every solid joint and are radiated from the ample wall areas of the cab.

Tests show that noise inside the cab during the transport mode breaks down into readily recognizable frequencies (Fig. 22.34). For example, a 1080-Hz vibration is identified as the bevel gear in the main drive train because it has 26 teeth and turns at 2500 rpm (41.7 cps), thus meshing 1084 times/sec. A 450-Hz vibration would be the nine-piston pump operating at 3000 rpm (50 cps). Harmonics of these frequencies also appear: $2 \times 450 = 900$ Hz, for example.

The engine firing frequency (125 Hz) is missing from the spectrum because the sound is strong only outside the cab. However, mechanical vibrations induced by engine shaft unbalance (40 Hz) and angularity (80 Hz) are present and very strong inside the cab.

High frequencies (say above 1000 Hz) bother the ear more than low frequencies do, so when test data are taken the microphone noise signals purposely are filtered to reduce sharply the recorded strength as frequency drops from 1000 to 20 Hz.

For example, a 110-dB noise at 40Hz is no more annoying than a 75-dB noise at 1000 Hz, so the test acoustic filter automatically reduces the measured value by 35 Hz, in accordance with an international standard. The measurements are labeled dB(A), or just A-scale. The dB(A) values are the ones OSHA looks at.

Piston Pump Noise. Variable-displacement pumps generate noise in three ways:

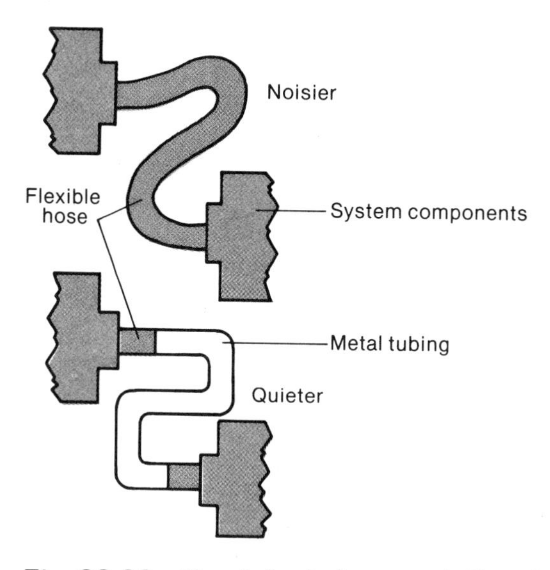

Fig. 22.33 Hose helps isolate sound. Short lengths of hose radiate less sound than long lengths, and are preferred.

pressure pulsation, flow pulsation (fluid-borne noise), and cyclic torque. The most troublesome is pressure pulsation internal to the pump, causing cyclic deflections of internal pump parts. These are mirrored by deflections in the pump housing and transmitted to the mounting. The greatest motion is along the piston axis. This noise can be reduced by mounting the pump resiliently (Figs. 22.35 and 22.36).

Fluid-borne noise is complex and needs various well-placed volumes such as accumulators, or even ordinary filters, to damp out the line noise. Sperry Vickers of-

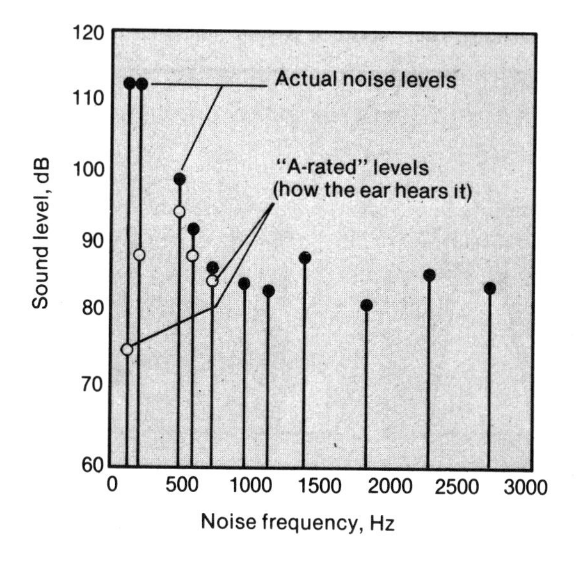

Fig. 22.34 Cab noise in harvester machine is less irritating at lower frequencies, so noise readings are corrected accordingly. These data taken at transport speed.

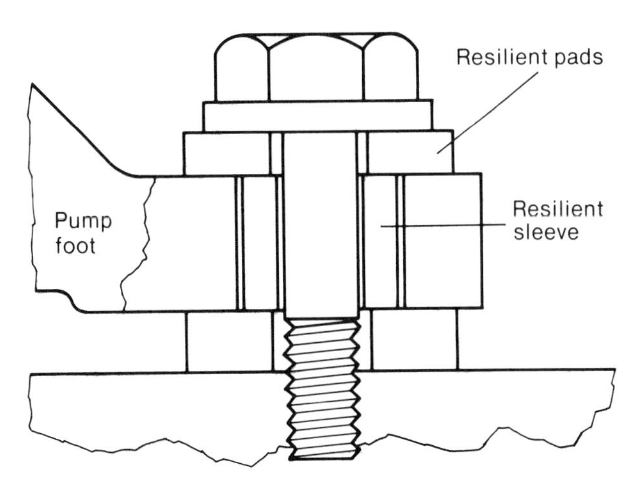

Fig. 22.35 Resilient pads and sleeves effectively isolate a hydrostatic pump assembly from the vehicle structure, and reduce noise.

fers a manual called "Sound Advice" that pretty much summarizes the range of fixes available. Also see Chapter 8 (Fluid Shock Absorbers and Accumulators).

Cyclic torque is characteristic of hydrostatic pumps and is well documented in textbooks and technical papers. Less understood is yoke moment noise.

The yoke (Fig. 22.37) is the mechanism that adjusts the angle of the swashplate in a variable-displacement pump. The pressurized piston forces are seldom arrayed symmetrically about the yoke axis, so when the pistons are pumping, they alternately add to and subtract from the forces holding the swashplate in the selected stroking position.

The cyclic forces should average out each revolution and not interfere with the operator's effort to adjust and hold the swashplate. However, the cycling has a fundamental frequency that is twice the pumping frequency, and this appears as vibration in the yoke linkage external to the pump. In the pump studied, a nine-piston model with a 2.5-in^3/revolution displacement, the cyclic forces on the yoke arm were 70 lb. The noise was diminished by inserting a rubber isolator.

Isolation Mounts. The pads should be soft enough to lower the natural frequency of each assembly to much less than the lowest vibration frequency to be attenuated. That's a rule of thumb for any application. A second statement of the rule is that natural frequency of the mounted assembly should give a wide margin to any vibration source frequency, say at least 1.7 times the nearest lower vibration frequency and less than 0.7 times the nearest higher vibration frequency. This will avoid amplification of those vibrations (resonance).

Figure 22.38 spots the key vibrations to be avoided in the harvester example. The shaded envelopes are "windows" where no important vibrations are expected and where natural frequencies of mounted assemblies should be placed if possible.

Remember that there are six vibrational modes: three linear and three rotational. Try to attach the pads symmetrically around and in line with the center of gravity of the assembly being cushioned (Fig. 22.36). This avoids coupled vibration modes, which are pitching moments created when a linear vibration force acting through the

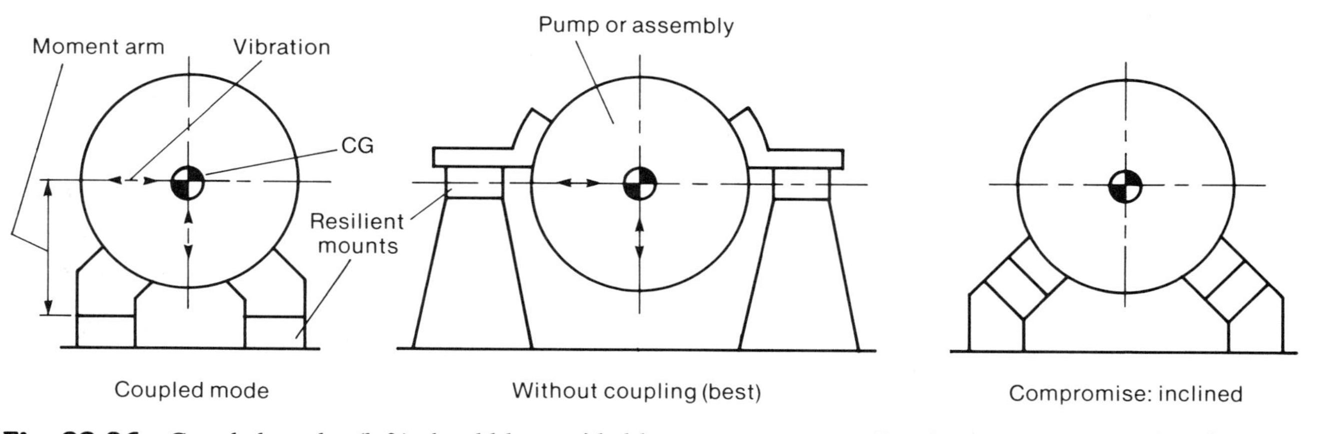

Fig. 22.36 Coupled modes (left) should be avoided because transverse vibration sets up rotary vibration as well, acting through moment arm.

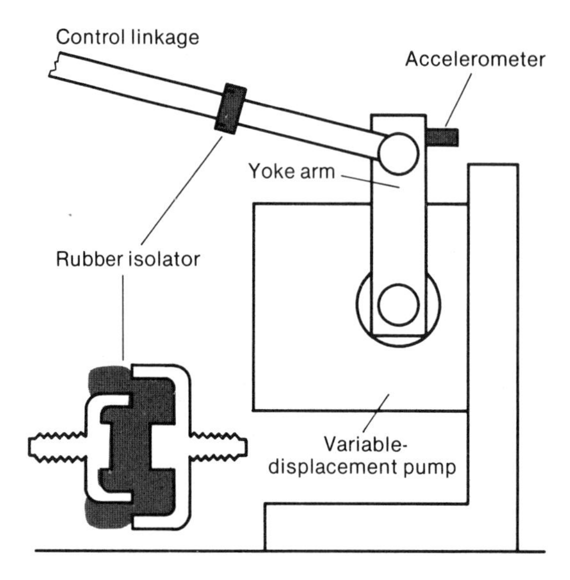

Fig. 22.37 Yoke moment noise is damped by isolating control linkage from yoke arm.

center of gravity is not in line with the mounting points. Usually it's not possible to eliminate coupled modes entirely, but the drawings show compromises.

SPEED CONTROL

The accompanying illustrations and captions (Figs. 22.39 through 22.46) are self-explanatory; they summarize the theory and describe simple mechanical and elec-

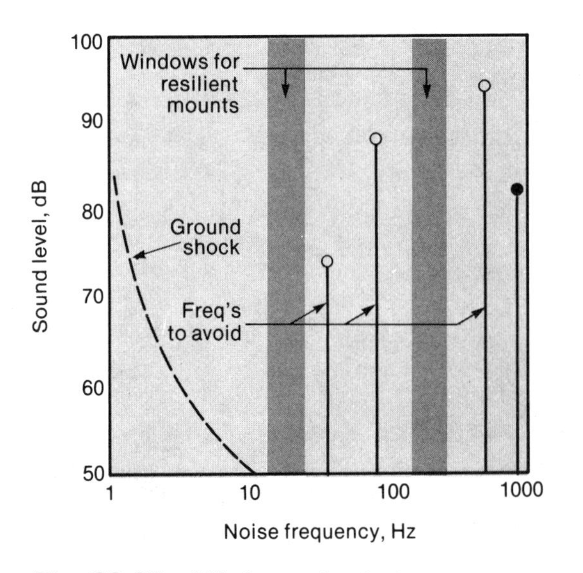

Fig. 22.38 Windows of safe frequencies that avoid all known noise frequencies are desirable for the mounts.

tromechanical speed-control components. The purpose is to help explain basic principles involved in controlling any drive system, including electrical, hydraulic, and pneumatic types. Initial study was by Beryl Boggs, Goodyear Atomic (Piketon, OH).

Prime Mover and Load

One affects the other in any system, and the mathematical relationships defining transient performance are complicated. A seemingly simple combination such as a diesel engine driving a generator requires rigorous mathematical analysis, as do the approximate expressions for time lag, signal amplification, throttle movement vs. load change, generator voltage transients, and torsional vibration.

The easiest loads to control are those with slow speed changes such as compressors and pumping engines, particularly when they have heavy flywheels or other inertia members that reduce transients.

You can determine the problem areas without math, however, and pick a suitable governor type. Here are the points to ponder.

Self-Regulation of Drive. A prime mover will help regulate itself if output torque tends to drop as speed increases (Fig. 22.40). Acceleration is lessened in proportion to the loss in torque, helping to stabilize the change. All turbines and gas engines have this characteristic and so do most gasoline engines except those with supercharging.

A supercharged gasoline engine has relatively poor self-regulation because an increase in speed improves the effectiveness of the supercharging compressor and actually can increase engine torque. This characteristic is useful in compensating for engine losses but makes the engine harder to govern.

A wound-rotor ac induction motor is self-regulating, but there are many special ac and dc motors that aren't as stable. Governed motors usually are designed and sold as complete drives.

Damping in Load. If the resisting torque of the driven load increases with speed, it helps limit speed transients that might otherwise cause instability. Compressors, blowers, pumps, and propellers have excellent damping characteristics because torque requirements increase as the square of the speed.

Response Lag. No governor can force an internal combustion engine to respond faster than the fuel can move from the throttle valve to the cylinder. True, some governing systems can *anticipate* a load requirement by sensing the rate at which the speed or the load is changing and can move fuel into the combustion chamber a little faster or slower to meet the anticipated load. But as a general rule, speed and load changes are limited by engine or motor characteristics.

Don't interpret this to mean that engines respond sluggishly; at full speed, even the slowest engines respond in a fraction of a second, and many go from no load to full load in 1 sec or less. The problem is that transients in the load being driven might occur much faster than that, causing overshoot and hunting. There is no problem where loads vary slowly, as in pipelines and conveyors.

Typical response bottlenecks in the prime movers are gas compressibility and intake manifold volume in a gas or gasoline engine, piston travel and ignition lag of any reciprocating IC engine, and time to complete the full firing sequence of any mul-

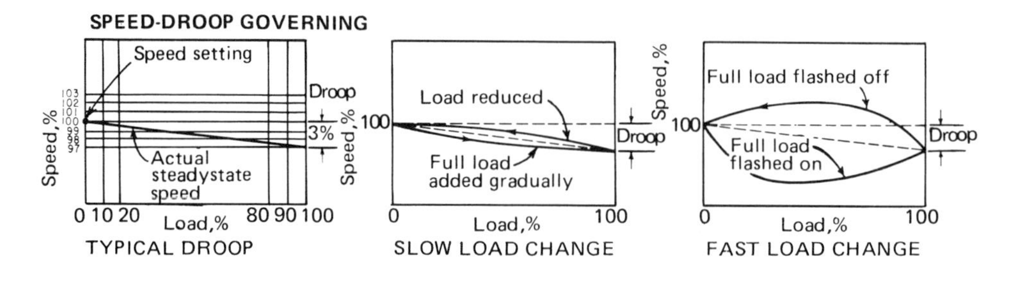

Speed-control language

CONTROL ELEMENTS

Speed governor = primary speed-sensitive element (NEMA definition). Its output signal can be force, motion, voltage, or even a binary number. Speed sensor is synonymous here with speed governor.

Relay = electric, pneumatic, hydraulic or other amplifier that responds to the speed governor output signal and raises its energy to a more useful level.

Speed-governing system = speed governor and all auxiliary control devices needed to regulate the source of energy and thus govern speed. Typical auxiliary elements are relays, servos, and control linkages.

Primary mover = the driving engine, turbine, motor, or other rotating power-producing machine whose speed is being controlled.

Load = the driven compressor, generator, pump, gearbox, conveyor drive shaft, mixer, propeller, hoisting winch, or other power-using machine.

SPEEDS AND ADJUSTMENTS

Constant speed = one pre-set speed. True constant speed is achieved only with servo-controlled systems, but for this article constant speed means the governor is *not* adjusted by remote control while the engine is running. Suitable for generators, compressors, mixers.

Variable speed = remotely controlled speed. Operator usually can adjust speed during running from idle to any speed up to maximum. Important for locomotives, tractors, and ships.

Overspeed = speed limit rather than speed control. Important where engine overspeed can destroy the machine, interfere with adjacent operations, or create a hazard. Some turbines have several overspeed governors, including centrifugal, electrical, and hydraulic. Where the governor shuts down the machine on overspeed, it is called an overspeed trip.

Two-speed = idle and normal. The idle speed setting can indicate that the start-up cycle is complete, and the normal speed setting can control normal operation.

Three-speed = idle, normal, and overspeed. One setting indicates completion of startup, one allows moderate control at or slightly below normal speed, and one prevents overspeed. Such a system can be formed entirely of speed-sensitive electric switches.

Load speed = tailshaft or driven-load speed. Load-speed control is often applied to buses and trucks to limit road speed. Also

Fig. 22.39 Speed governing has a language and theory of its own.

tipiston engine. Gasoline engines are slowest but can gain better response with a control valve at each cylinder. This eliminates the manifold time lag.

Diesel engines are fairly fast because fuel enters the cylinders immediately. The only appreciable lag—about one-third of a revolution—is the time required for the pistons to reach the firing position. Gas turbines are even faster—the fuel is injected and burned instantly, and there is always excess airflow to support combustion. Steam turbines are fast during transients because the stream is always at full pressure, ready to be throttled.

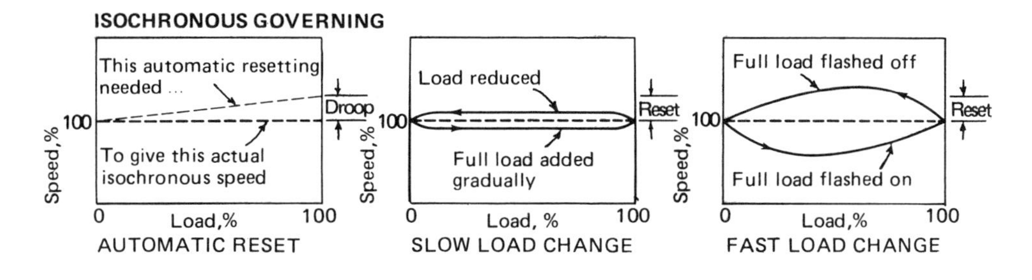

applied to the output shaft of a torque converter. The load-governor signal regulates energy input to the prime mover, which usually has a separate governor to prevent excessive speed.

PERFORMANCE

Regulation = amount of speed change during or caused by application of load. The graphs above and in Fig. 22.40 show the possible combinations.

Isochronous regulation holds speed constant at all loads. Speed-droop regulation entails a speed drop as load is added.

Most simple governing systems have speed droop—some as much as 1000 rpm. Also called *compensation*, speed droop is an easy and natural way to pick up load stably, because the speed change allows governor movement to actuate the engine throttle or control linkage directly.

More-elaborate governors have speed droop only during load changes with gradual automatic reset to the original speed after the load level is safely achieved.

A limited number of governors—mostly for turbine generators—have isochronous regulation, and some of these take up to 5 sec to return to the original speed after a load change.

Reset = automatic resetting of the called-for speed to compensate for droop. Makes isochronous speed control more practical. One way is to place a torque-sensing element between the engine and load with enough power output to increase the speed adjustment of the droop governor as the load is increased. This torque sensor anticipates a change in speed and acts toward correction before the speed change occurs. However, some change in speed is needed to actuate the throttle.

Stability = ability to maintain the set speed without undesirable speed oscillation despite load variations and other disturbances. This definition is unspecific because it hinges on the characteristics of the prime mover and the load.

Hunting = speed change without a change in load. Usually caused by worn governor parts or a design fault, such as too much end play.

Deadband = amount of speed change within which the governing system makes no measurable correction. Excessive deadband can result in hunting or instability.

Stalled work capacity = foot pounds of energy available from governor at normal speed to overcome resistance of the engine throttle and operate it through its full stroke. Stalled work capacity ought to exceed throttle work requirement by at least 50%.

Misfiring. Gas and gasoline engines are sensitive to fuel-air mixtures. The governor should have enough built-in time delay to permit occasional misfires without the control prematurely changing the fuel setting.

Flywheel Effect. Transient speed and torque changes are slowed by the combined inertia (WK^2) of the prime mover and load. High inertia makes it easier for a governor and its throttle linkage to control speed stably, because there is more time for corrective action.

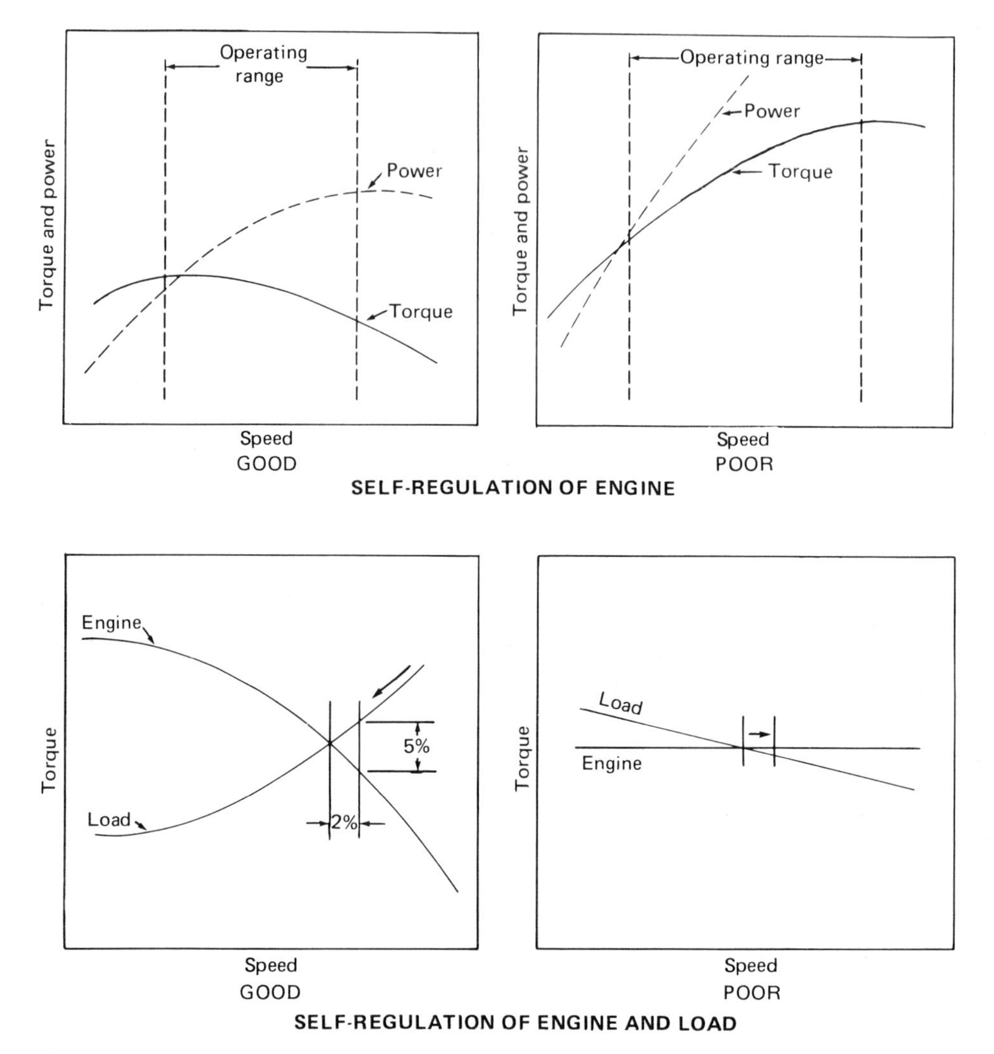

Fig. 22.40 The engine and load can be self-regulating if the speed-torque characteristics are right.

driving a constant-voltage dc generator is an example—the governor will have to provide the needed stability.

Part Load. Light loads will cause erratic action in gasoline engines and some other prime movers if they are designed primarily for full speed and full load. One answer is to include several intermediate speed positions of the governor, to be adjusted under actual operating conditions.

Drop to Idle Speed. It's difficult to adjust speed directly from full to idle in some engines because the engine and combustion characteristics vary greatly with speed. Take the diesel engine as an example. The fuel injection pump loses efficiency and displacement as speed goes down, and if you return the fuel metering shaft to idle position during full speed, the engine will slow down and die. To catch the engine on the way down to hold idle speed, you must advance the metering shaft to what normally would be the half to three-quarter wide open position. A governor can be designed to do this automatically.

Actuating Linkages. Because governor output force and movement don't necessarily match the desired engine throttle valve movement, the connecting linkages, particularly on IC engines, must correct the motion. A diesel is relatively easy to connect because injector-nozzle flow is proportional to stroke.

A gasoline engine with a butterfly throttle valve needs special linkages because the butterfly varies flow area more rapidly during the initial opening than at the wider positions. Another way to compensate is to design a special valve. Figure 22.41 shows the desired response in terms of engine torque.

Linkages must be without backlash and friction-free, or deadband and hunting can result.

Ambient Temperature. Remember that a spring-loaded governor will change about 1% in governed speed for a 100-deg change in ambient temperature.

The Speed Sensors

By definition a speed governor is the speed-sensing part of a speed governing system. Its output—force, pressure, or voltage—is subsequently modified or amplified to do the actual controlling. The accompanying sketches of speed governors mainly are of sensors, although some include relays, servos, and other auxiliary actuating elements.

The governor title—hydraulic, electrohydraulic, etc—does not always indicate the speed-sensing method.

Centrifugal sensors (Fig. 22.42) are the most common—they are simple and sensitive and have high output force. There is more published information on centrifugal sensors and governors than on all other types combined.

In operation, centrifugal flyweights develop a force proportional to the square of the speed, modified by linkages as required. Eight examples are shown in the figure.

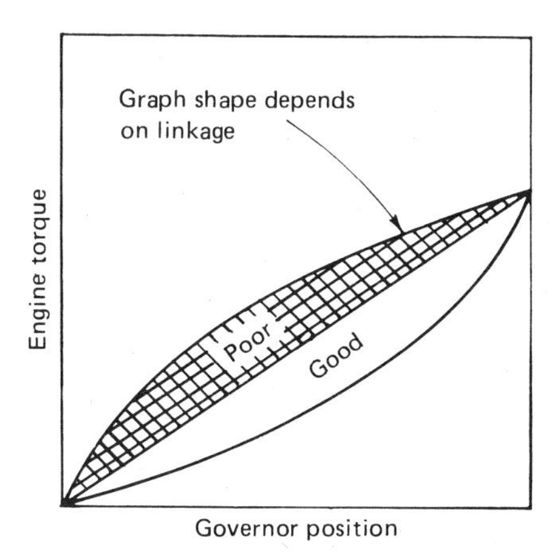

Fig. 22.41 Butterfly valve for gasoline engine should respond along lower curve.

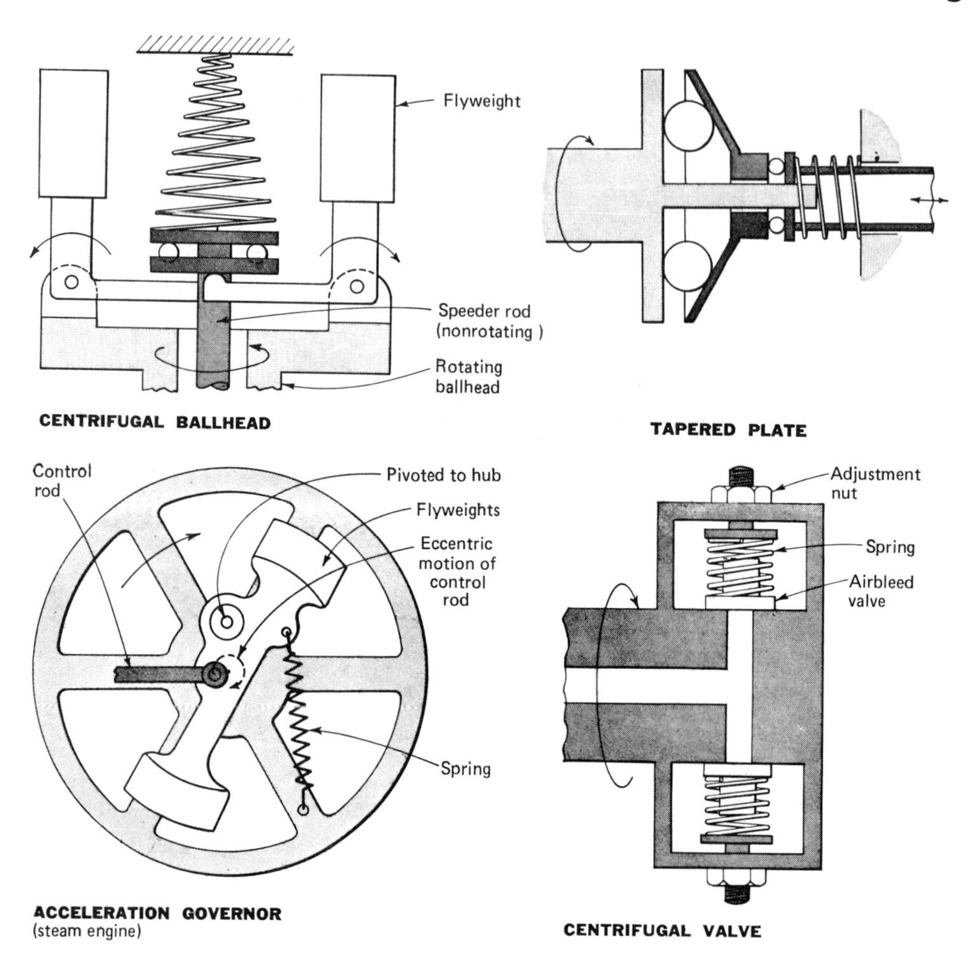

Fig. 22.42 Centrifugal sensors and governors may have any of these eight designs, and more.

Combined Stability. If either the prime mover or the load is self-regulating and stable, the combination will probably be stable. If neither is self-regulating—a diesel In small engines the flyweight movement can actuate the fuel throttle directly. Larger engines require amplifiers or relays, which gives rise to innumerable combinations of pilot pistons, linear actuators, dashpots, compensators, and gear boxes.

Centrifugal flyweight governors have inherent speed droop, stabilizing the movement of the throttle valve. And they are simple and inexpensive. Their disadvantages are: insufficient power for larger prime movers; no adjustment for droop, because speed droop is a fixed function of the regulating spring; limited variable-speed range; and friction in the linkage.

A typical flyweight vs. spring curve is shown in Fig. 22.43. Governing occurs at the balance between the spring force and centrifugal force of the flyweights as the weights move out. A nearly right-angle intersection is preferred for greatest accuracy.

Stalled work capacity at the control arm of a centrifugal governor is the differential between the energy of the weights spinning in closed position and the energy of the weights spinning in open position. The initial setting or compression of the

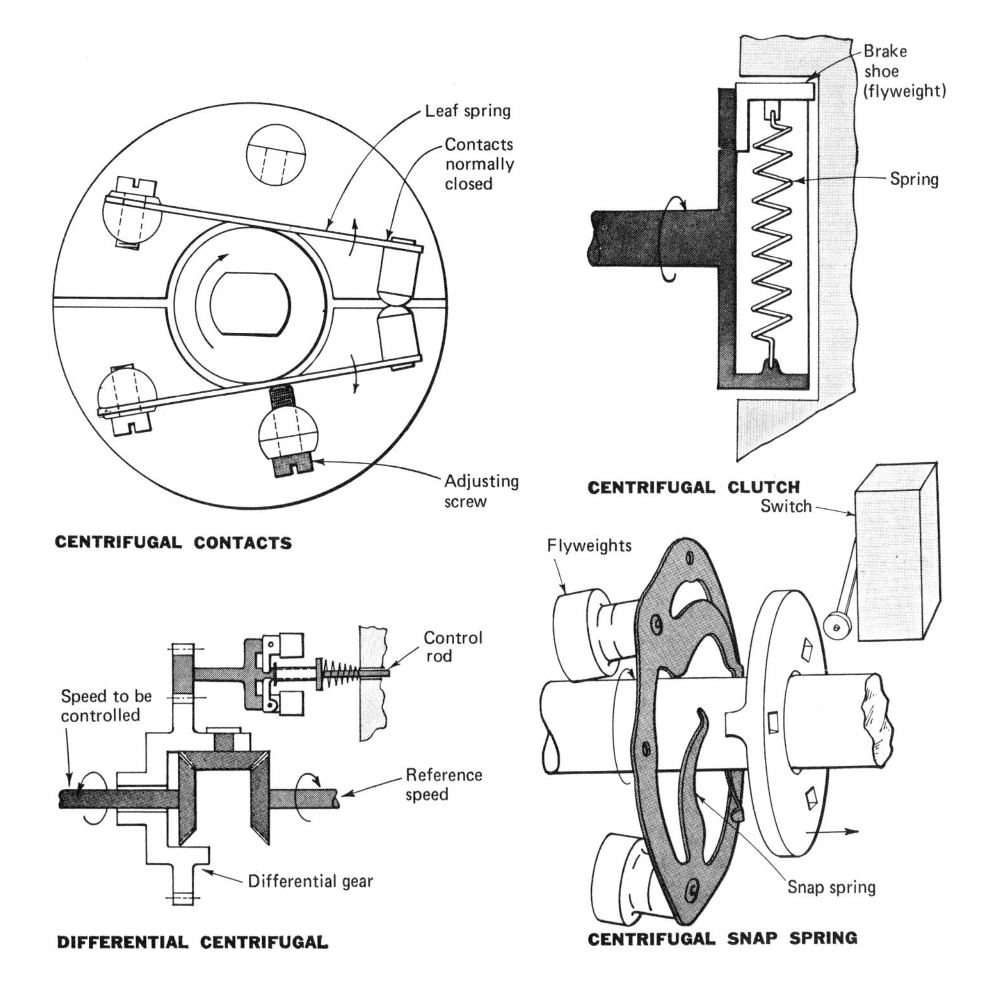

Fig. 22.42 (Continued)

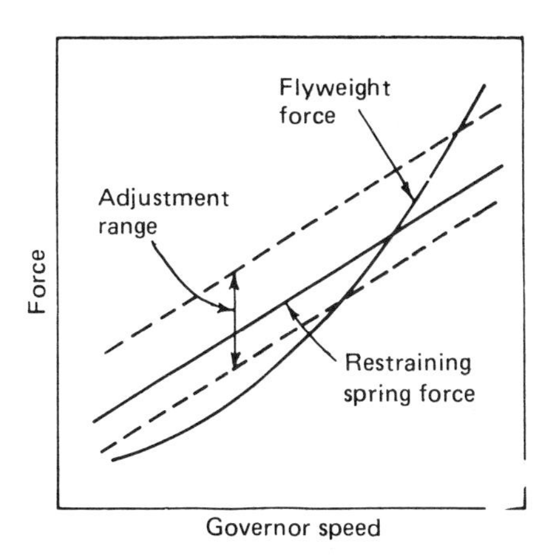

Fig. 22.43 Flyweight force must increase more rapidly than spring retaining force.

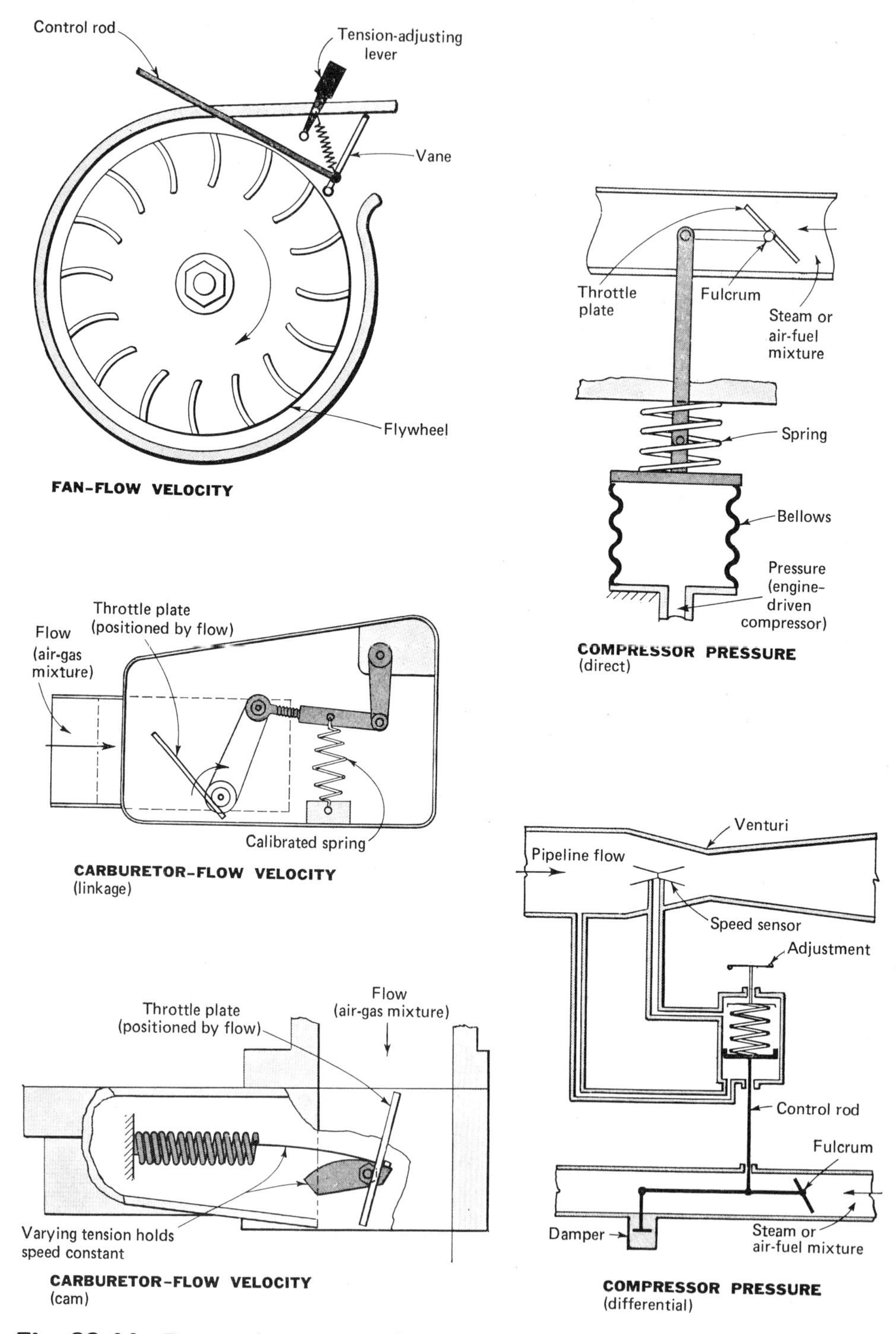

Fig. 22.44 Pneumatic sensors and governors exploit the pressure or velocity of cooling or combustion air to sense and control engine speed.

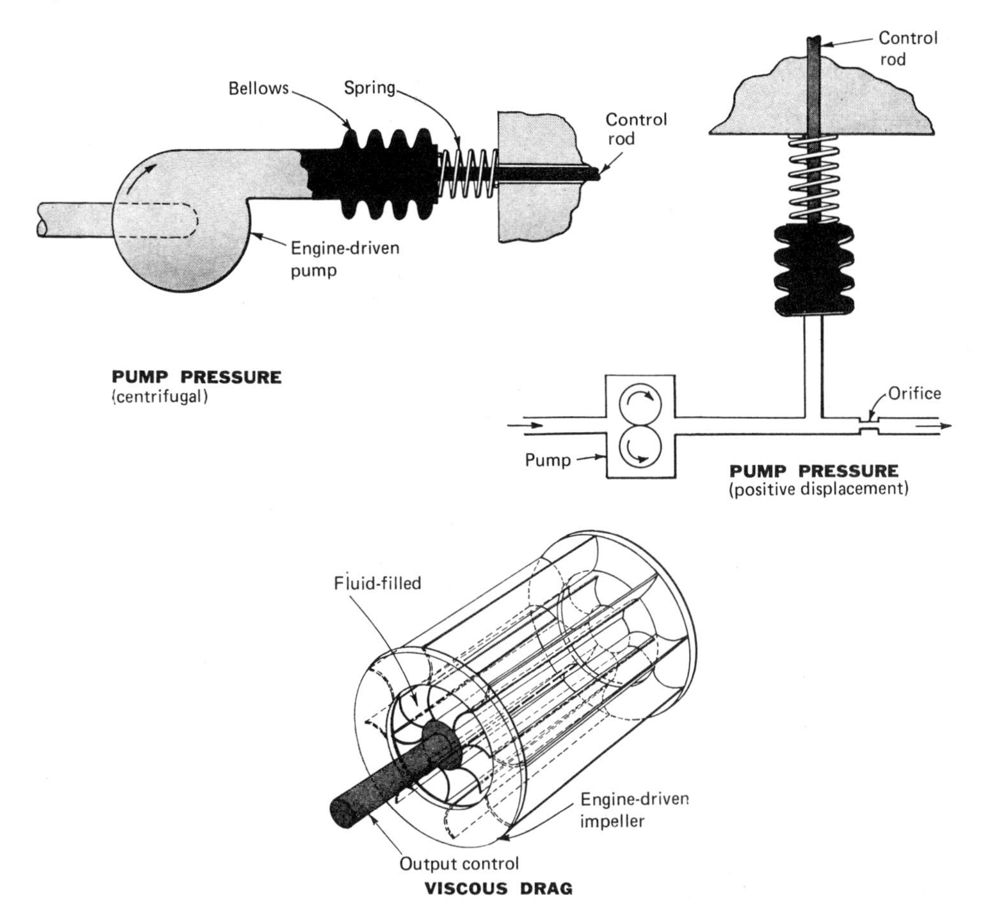

Fig. 22.45 Hydraulic sensors and governors exploit fluid pressure or viscosity to sense and control speed.

spring rate determines the speed at which the weights begin to move, and the spring rate determines the speed at which the weights are in full open position.

Where considerable power, sensitivity, and accuracy are required, a hydraulic servosystem usually is added. It has a flyball head, oil pump, hydraulic relay system with a control valve and one or more amplifying pistons, and a compensating mechanism to stabilize speed changes. These adjustments are provided: speed range, speed droop, steady-state speed regulation, and anti-hunting, and stability.

Centrifugal-hydraulic governors can use the oil to advantage in other ways, including as an accumulator to reduce pump size, an oil bath for flyweights to damp their motion, and a dashpot timer to reduce hunting by setting the proper distance for the throttle to move for a given change in load. The disadvantages of oil are possible leaks, contamination, and sluggish flow at low temperatures.

One variant of the ballhead is the tapered-plate or thrust centrifugal governor. As the balls move radially outward, one plate is moved axially to actuate the control. Usually there are grooves in the plates to maintain the balls in correct angular relation and dynamic balance. The grooves can be straight or spiral. Spiral grooves cause the balls to advance or retard in angular position as they move outward.

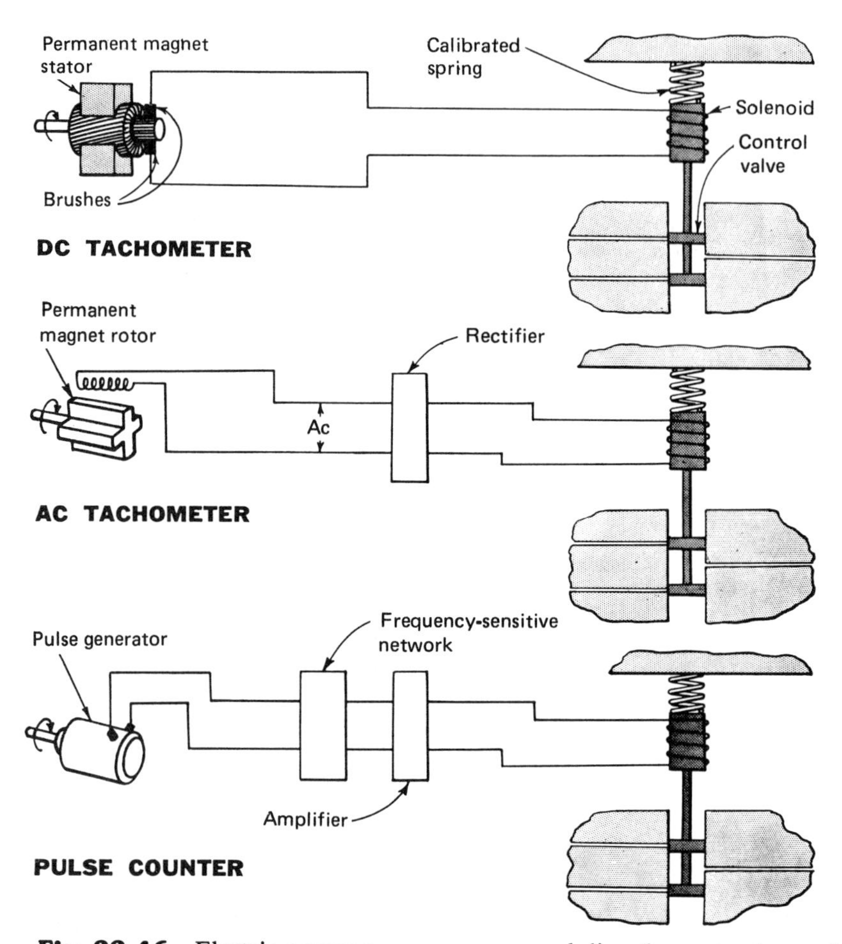

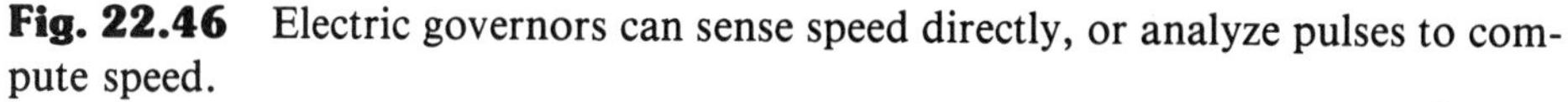

Fig. 22.46 Electric governors can sense speed directly, or analyze pulses to compute speed.

Another variant of the centrifugal governor is the inertial or acceleration governor. The flyweights sense the rate of change of speed in addition to sensing steady-state speed. For example, by reciprocating in a motion determined by the eccentricity of the flyweight arm, the control rod can regulate a steam valve. The eccentricity changes with speed and rate of change of speed, shortening the valve response time.

Step-up gearing of differentials makes it possible to control a slow-speed shaft of less than 100 rpm by a high-speed and therefore more effective centrifugal governor. An interesting example is the differential drive in Fig. 22.42. The governor senses the difference in speed between the controlled shaft and the higher-speed shaft.

A brake-type centrifugal governor is used on some small electric motors to control speed. The brake shoe is the flyweight and is restrained by spring force as shown in the figure. When speed is increased, the brake shoe touches the housing and slows the motor. Applications include spring-driven motion picture cameras.

In the centrifugal valve air-bleed governor the poppet valves are the flyweights. When the controlled speed is reached, the valves open and release air to reduce the pressure to an actuating piston, a diaphragm, or a servovalve. A centrifugal vacuum governor is similar except that the valve is held open by spring tension, and the centrifugal force action tends to close it.

Switch contacts are the flyweights in another form of centrifugal-speed sensor. The restraining force is a leaf spring containing the contact point, and motor speed is controlled by switching the current on and off or inserting resistance in the armature circuit as the speed varies around the set point. A special form has the centrifugal sensor separate from the contacts. At the set speed, the weights cause the snap-ring carrier to snap over, moving an actuator axially to operate the contacts.

Centrifugal-switch governors are used chiefly for the starting winding cutout on single-phase electric motors. Safety protection and speed control are other possible applications.

Pneumatic sensing devices (Fig. 22.44) are the most inexpensive, and also the most inaccurate, of all speed-measuring and governing methods, yet they are entirely adequate for many applications. Five types are shown in the figure.

Best known is the vane air-velocity governor used on small gasoline engines. The sensing vane is placed in the airstream of the engine cooling fan and is connected to the throttle so that an increase in speed, increasing airflow, will close the throttle valve.

Air-velocity governors also control the speed of larger gasoline engines. A vane or other obstruction is placed in the path of the gas-airflow between the carburetor and intake manifold, and as velocity increases, the vane is moved to close the opening. In the type shown in the figure, the force of the air-fuel mixture on the off-center throttle plate is balanced by spring tension holding the throttle in the open position. Increased engine speed, and the resulting increased air-fuel velocity, tends to close the throttle. Some designs have a variable-force spring; others have an auxiliary manifold-vacuum sensor to close the governor throttle at idle speeds. One variant achieves variable-spring force through a combination of cam and springs.

Pressure-sensing air governors utilize the static pressure of the air acting against a spring-loaded bellows, diaphragm, or valve. An ordinary pressure switch and a snap-operation valve are two of the relay techniques used. A venturi is another way: The differential pressure is sensed.

The pressure source is usually a compressor driven by the prime mover, and the pressure is approximately proportional to the square of compressor speed.

Typical applications of pressure-sensing governors are in process gas pumping, compressed air systems, and exhaust or inlet pressure control of steam turbines.

Pneumatic vacuum governors are common in reciprocating engines. They depend on a reduction in intake-manifold presssure as engine speed increases and an increase as engine load decreases. This vacuum is balanced against a spring-loaded diaphragm or other control element. It should be noted that if the engine has a vacuum-operated spark advance, it may be necessary to correct for the effect the governor has on the vacuum sensed by the spark control.

Applied to diesel engines, a vacuum governor has a venturi in the air-intake line; the pressure differential or partial vacuum operates the control.

Hydraulic sensors (Fig. 22.45) measure discharge pressure of an engine-driven pump. Pressure is proportional to the square of the speed of the pump in most designs, although there are special impellers with linear pressure-speed characteristics.

Straight vanes are better than curved vanes because the pressure is less affected by the volume flow. Low pressures are preferred over high because fluid friction is less.

Typical applications include farm tractors using diesel or gas engines, larger diesel engines, and small steam turbines. Hydraulic-governor sensing using other than pressure has had limited application and success.

One type of hydraulic-pressure sensor is a dead-ended bellows fed by a centrifugal pump. Another is a fixed orifice fed by a constant-displacement pump; the pressure drop across the orifice is proportional to the square of the speed.

Viscosity actuates one unusual speed-sensing device. The outer impeller is rotated by the engine; the inner impeller, connected to the throttle, is restrained by spring force.

Electric tachometers (Fig. 22.46) generate voltage proportional to speed; the output actuates a solenoid to move the control valve. Ac or dc designs are available. The ac version has no brushes on the rotating generator, but rectifiers must be added to the circuit. Dc tachometer generators have brushes, but no rectifiers are needed.

In some instances an ac generator and a frequency-sensing circuit measure the speed; the output is amplified to actuate the control. In one version the sensor compares generator frequency with a reference frequency; the difference or error is amplified for throttle control.

Vibrating-contact governors (not shown) are used on small dc motors to provide two regulated speeds in the 4000- to 15,000-rpm range. Each speed can be set by a separate adjusting screw. Two pairs of vibrating reed contacts control the adjusted speed.

Actually, any speed-measuring device can be used as the sensor of an electrical speed governor. This includes tachometers, speedometers, revolution counters, gear-tooth counters, pulse counters, vibrating reeds, and even stroboscopes.

Relays and Servos

Amplifiers and follow-up controls, called relays or servos, add muscle where needed in many speed-governor systems. The speed sensor is relieved of heavy work, and droop can be greatly reduced.

The centrifugal-vacuum governor is an example of a sensor combined with a relay. A centrifugal valve is the pilot, admitting air to a working diaphragm, with a partial vacuum on the opposite side.

Centrifugal-pneumatic governors, by the action of a centrifugal valve, release air from a balance chamber of a servovalve, moving the throttle toward the closed position.

Centrifugal-hydraulic governors have a centrifugal ballhead; a hydraulic servopiston admits or releases fluid from the throttle-actuating cylinder.

Other sensor-relay combinations include the hydraulic-relay governor, which is a hydraulic speed sensor and a hydraulic servomechanism control; electric-relay governors, with electric speed sensor and reversible electric-motor servo control; and electric-hydraulic governors, with electric speed-sensing and hydraulic servo control.

Hydraulic servopistons are much faster than electric servomotors and are most frequently selected. Electric-hydraulic and centrifugal-hydraulic governors are very common in steam turbine and gas turbine power applications.

Example 1: Gas Engine for Pumper

Requirements. Adjustable speed to control pipeline liquid flow. Speed changes must be slow to avoid engine stalling. No deadband is allowed because it would make control unstable, affecting pipeline and adjacent pumpers. Speed accuracy is not important.

Two important engine characteristics are the one-revolution dead time of a four-stroke cycle and possible misfiring from too rich or too lean a mixture. A flywheel is needed to provide kinetic energy during lag time or misfiring.

Choice of Governor. Centrifugal-hydraulic, hydraulic-hydraulic, and electric-hydraulic are desirable, in that order.

The electric-hydraulic system is the least desirable because electric power is not always available or dependable at the pumping station, and a special engine-driven generator may be necessary. The hydraulic-hydraulic system is less dependent on electricity, but speed sensing is less accurate, and flow variations will result. Also, force levels of most hydraulic sensors are somewhat less than for electric tachometers and much less than for centrifugal ballheads.

The centrifugal-hydraulic system is best. It will cost less, partly because the sensing ballhead has enough force and accuracy to actuate a relatively simple hydraulic servo. The ballhead alone, however, is not sufficient—the pumper-speed control demands more actuator power than that.

Delayed response is achieved with dashpots, intermittent pumping of working oil, or flow-control valves. These are governor design problems, however.

Fuel control vs. governor movement is a problem because the engine butterfly will have its own characteristics, and the governor linkages must be designed accordingly.

Torsional vibrations in the engine and gear trains must be analyzed to avoid later problems.

Special auxiliaries will be needed: solenoid shutdown for remote stopping, remote speed setting with a pneumatic or electric signal in conjunction with the pipeline control, and booster servomotors for setting fuel positions during the starting cycle.

Example 2: Gasoline Engine for Small Compressor

Requirements. Need is for: constant speed; fast governor response for rapidly applied load; and positive but small speed droop. The engine rating is 5 hp. The air compressor will have a control to release excess pressure.

Choice of Governor. Cost must be low to match the overall low cost of the small engine-compressor set. Consequently only the simple nonservo governors will do, and of these, only the ballhead and tapered plate centrifugals have enough power and accuracy to meet the low-droop requirement.

The ballhead is superior because it has smoother mechanical movement and higher output force.

Index

G

H